Lecture Notes in Civil Engineering

Volume 58

D1827306

Lecture Notes in Civil Engineering (LNCE) publishes the latest developments in Civil Engineering—quickly, informally and in top quality. Though original research reported in proceedings and post-proceedings represents the core of LNCE, edited volumes of exceptionally high quality and interest may also be considered for publication. Volumes published in LNCE embrace all aspects and subfields of, as well as new challenges in, Civil Engineering. Topics in the series include:

- Construction and Structural Mechanics
- Building Materials
- Concrete, Steel and Timber Structures
- Geotechnical Engineering
- Earthquake Engineering
- Coastal Engineering
- Ocean and Offshore Engineering; Ships and Floating Structures
- Hydraulics, Hydrology and Water Resources Engineering
- Environmental Engineering and Sustainability
- Structural Health and Monitoring
- Surveying and Geographical Information Systems
- Indoor Environments
- Transportation and Traffic
- Risk Analysis
- Safety and Security

To submit a proposal or request further information, please contact the appropriate Springer Editor:

- Mr. Pierpaolo Riva at pierpaolo.riva@springer.com (Europe and Americas);
- Ms. Swati Meherishi at swati.meherishi@springer.com (Asia—except China—and Australia/NZ);
- Ms. Li Shen at li.shen@springer.com (China).

Indexed by Scopus and Compendex

More information about this series at http://www.springer.com/series/15087

Sirajuddin Ahmed · S. M. Abbas · Hina Zia
Editors

Smart Cities—Opportunities and Challenges

Select Proceedings of ICSC 2019

 Springer

Editors
Sirajuddin Ahmed
Jamia Millia Islamia
New Delhi, India

S. M. Abbas
Jamia Millia Islamia
New Delhi, India

Hina Zia
Jamia Millia Islamia
New Delhi, India

ISSN 2366-2557 ISSN 2366-2565 (electronic)
Lecture Notes in Civil Engineering
ISBN 978-981-15-2547-6 ISBN 978-981-15-2545-2 (eBook)
https://doi.org/10.1007/978-981-15-2545-2

This Springer imprint is published by the registered company Springer Nature Singapore Pte Ltd.
The registered company address is: 152 Beach Road, #21-01/04 Gateway East, Singapore 189721, Singapore

Preface

"The 19th Century was a century of empires. The 20th century was a century of nation states. The 21st century will be a century of cities". This was said, around 30 years back, by Wellington E. Webb Mayor of Denver USA.

United Nations estimates that 70% of the people in the world would be living in urban areas by 2050 (88% in developed countries and 67% in developing countries). Cities are growing both in numbers and in size, thus creating unprecedented demand for resources such as energy and water along with services like education, health care, transport, communication and sanitation. Cities are also driving the economic growth.

In India, the level of urbanization is expected to increase to 40.76% in 2030 from 31.2% in 2011. It emerged as the world's fastest growing economy in 2018, with a GDP growth rate of 8.4%. The urban sector in India contributes around 70–75% to the GDP. This dependency of the national economy on the urban sector will get stronger with an increase in the rate of urbanization and will open new opportunities. Cities, today, enjoy more economical, political and technological power than ever before but are facing a number of challenges and threats to their sustainability. To address these challenges to sustainable growth, what we require are "Smart Cities".

The concept of Smart City varies from people to people, city to city and country to country, depending on the requirements of the city residents, level of development, willingness to change and reform. Interest in Smart Cities is driven by major challenges, including climate change, economic restructuring, ageing populations and pressures on public finances. A number of definitions of the term "Smart City" exist, but there is still no consensus on what a smart city is, since several synonyms of the word "smart" are often used interchangeably such as "intelligent" or "digital" or "innovating" or "knowledge".

To deal with the challenges and opportunities during the development of new Smart Cities and renovation of the old cities, authors were invited to write and present their researches/articles in the International Conference on *Smart Cities— Opportunities and Challenges* organised by Department of Civil Engineering, Jamia Millia Islamia (A Central University) New Delhi, India. The selected papers/articles are presented in this book.

This book contains chapters on urban planning and design, policies and financial management, environment, energy, transportation, smart material, sustainable development, information technologies, data management and urban sociology. Each chapter presents the research papers contributed by renowned researchers, professionals, policy makers of science, engineering, social, management and financial backgrounds. The research papers contribute towards improved governance and efficient management of infrastructure such as water, energy, transportation and housing for sustainable development, economic growth and better quality of life for its citizens, especially for developing nations.

This book will be useful for academicians, researchers, and policy makers interested in developing, planning, designing, managing and maintaining Smart Cities.

New Delhi, India

Sirajuddin Ahmed
S. M. Abbas
Hina Zia

Introduction

Cities for long are considered to be engines of growth. It is for this reason and the tremendous opportunities (along with a range of challenges) that cities offer that the population living in urban areas grew to 50% by 2008 and likely to grow to 70% by 2050 (UN Habitat 2009). Recognizing the ever-increasing relevance of cities in sustainable development and the urgent need to address the associated challenges, the United Nations in its post-2015 development agenda identified a new set of international development goals called the Sustainable Development Goals (SDG) with a vision of a more prosperous, sustainable and equitable world. For the first time, there is an exclusive goal on cities, Goal No. 11. The eleventh goal states "Make cities and human settlements inclusive, safe, resilient and sustainable". Out of 17 approved by the UN for 2016–2030, focuses on cities and calls for making cities and human settlements inclusive, safe, resilient and sustainable. Some of the specific targets under this goal include a focused approach to ensure the following by 2030[1]:

- Ensure access for all to adequate, safe and affordable housing and basic services, and upgrade slums.
- Provide access to safe, affordable, accessible and sustainable transport systems for all, improving road safety with special attention to the needs of vulnerable population (children, women, elderly, differently abled).
- Enhance inclusive and sustainable urbanization.
- Prevention of deaths and number of affected people and economic losses relative to GDP caused by disasters.
- Reduction of environmental impact of cities with special attention to air quality, municipal and other waste management.
- Provide universal access to safe, inclusive and accessible, green and public spaces, particularly for women and children, older persons and persons with disabilities.

---

[1]http://www.undp.org/.

- Support positive economic, social and environmental links between urban, peri-urban and rural areas by strengthening national and regional development planning.
- Integrated plans and policies towards inclusion, resource efficiency, mitigation and adaptation to climate change and resilience to disasters.

India is committed to achieve Sustainable Development Goals. Several of its national priorities are aligned with the SDG targets. "The SDG India Index: Baseline Report 2018" was released by NITI Aayog, Government of India, to understand and map the progress being made by States and Union Territories in moving towards SDG targets. For Goal No. 11, the report gives data on only two SDG targets out of ten. The overall performance of various States and UTs is not very good based on the limited data mapped for the targets.

India as a nation is on its path of fast pace of urbanization posing multiple stresses on the already overburdened infrastructure of the cities such as affordable housing, provision of clean water, waste management, transport-related services, access to affordable health services, lack of open spaces and poor air quality. The country has launched several programmes at the national and sub-national scale to tackle some of these burgeoning issues. One of the ambitious programmes in this direction is "Smart Cities Mission" which was launched in 2015 with the hope that the initiative itself will catalyze the creation of similar smart cities initiatives in various regions and parts of the country. Considering the different context of Indian cities, the Mission rightly defined the "Smartness" in Indian context and different from the one prevalent in other parts of the world like Europe and America.

"In the approach of the Smart Cities Mission, the objective is to promote cities that provide core infrastructure and give a decent quality of life to its citizens, a clean and sustainable environment and application of 'Smart' Solutions. The focus is on sustainable and inclusive development and the idea is to look at compact areas, create a replicable model which will act like a lighthouse to other aspiring cities".[2]

The core infrastructure covers everything from adequate water supply, electricity supply, sanitation, efficient mobility services, affordable housing, good governance, robust IT connectivity, sustainable environment, public safety and health and education. Adequate Smart Solutions will thereupon enable cities to use technology, information and data to improve infrastructure and services. Hundred cities from various States and Union Territories were selected as part of the Mission with 5151 projects proposed with a total approved budget of 48,000 crore[3] and currently under various stages of implementation.

The book address various themes including urban planning, disaster management and resilient cities, sustainable mobility systems, environmental quality, smart construction technologies, renewable energy technologies for smart city applications,

---

[2]http://www.smartcities.gov.in/.

[3]'What is the status of Smart city projects in India?, The Hindu, 17 July 2019.

sustainable water and waste management systems, Internet of Things, health and safety and community participation. Several of these papers present research ideas and laboratory-scale testing with potential to replicate and upscale to contribute to Smart Cities Mission and SDG of safe, inclusive, resilient and sustainable communities and cities.

Section related to urban planning, disaster management and resilient cities covers several contributions on a range of pertinent issues and possible solutions. Proposals for Smart cities like Jaipur and Tehri have been critically reviewed. Intelligent urbanism as an approach to urban planning and management specifically for bringing in smartness in cities has been explored in the context of Bhopal city. Paper on Low Carbon Smart cities in India emphasizes the need to integrate disaster and climate resilience framework into the development plans of cities. Relevance of Resilience Maturity Model, a five-stage model to incorporate resilience in smart cities and reduce disaster risks, has been discussed. Spatial-temporal analysis to promote mixed-use development and avoid the increase in urban heat island can be used to have more balanced growth. Urban flooding has been addressed in several papers. Use of weighted aggregated sum/product assessment to prioritize challenges and action plans can be very helpful in making proposals and implementation plans for smarter cities. Citing Bengaluru as a case, need for reorienting the material ecologies of cities has been discussed.

Urban mobility is increasingly becoming a huge challenge to deal with especially in the context of large cities and urban agglomerations. Application of Information and Communication Technologies in monitoring, operation and management of mobility services including parking systems and traffic forecasting has been discussed. Application of renewable solar energy with autonomous vehicles and highway gradient effects on hybrid electric vehicles performance dwell on pertinent issues pertaining to the future of mobility systems.

Section on materials and construction technologies is very extensive and caters to several wide-ranging issues. Corrosion of steel is a serious problem in infrastructure systems; experimental investigation of reinforced concrete corrugated beams strengthened with FRP sheets shows promising results. The effect of elevated temperature on the residual compressive strength of normal and high-strength concrete is helpful for developing appropriate solutions against fire effect. Use of waste coarse ceramic aggregates, incinerator ash and waste nanocarbon black in construction material has been explored. A comparative cost analysis of MMFX bars in Indian scenario explores the possibilities of using smarter materials. Soil blended with more than 5% fly ash is more durable and can increase the durability of roads and embankments. Potential benefits accrued from nanomaterials in various fields of applications hold immense hope. Use of smart construction technologies like "3-S" prefab technology for sustainable mass housing can help in meeting the "Housing for All" targets and can also result in cost savings. Performance assessment indexing of buildings through fuzzy-AHP methodology for predicting the survivability and performance of buildings in case of a likely earthquake has been studied. Use of thermal and optical investigation of lime mortar as a tool for retrofitting and conservation of architectural heritage is interesting given the large cultural and architectural heritage present in various parts of the country.

Many small and big cities of the country continue to face increasing air quality deterioration and pose serious concerns for the vulnerable population (elderly, children and diseased). The contributions by various authors on the above theme cover aspects like dynamic programming-based decision-making model for selecting optimal air pollution control technologies for an urban setting, spatio-temporal analysis of urban air quality using ARIMA model at a regional scale, role of particulate matter on air quality assessment of Delhi and ANN-based prediction of PM2.5.

Renewable energy technologies are already playing an increasingly important role in meeting the huge energy demands of cities and are likely to grow further. There are several contributions like control techniques to optimize PV system performance for smart energy applications, techno-economic feasibility analysis of hybrid RE system, review of dSPACE 1104 controller and its application in PV, effective grid-connected solar home-based system for Smart cities in India and MPP technique for solar PV module through modified PSO.

Provision of adequate quality and quantity of water to all is a huge challenge in cities, more so in the face of increasing climate change-related extreme events. WSN-based water channelization addresses the challenges of smart water grid and priority-based water supply for a smart water system. In spite of the addition of capacities for treatment of municipal and industrial wastewater, the gap between demand and supply is huge. There is a need to look for scaling up multiple options which are cost-effective and address the various contextual issues of Indian cities. Comparative study of treatment and performance in a membrane bioreactor and sequencing batch reactor for hospital wastewater in smart cities, removal of Pb from Industrial wastewater using CuO/Hg, forward osmosis exploring for waste/wastewater treatment has been discussed.

Feasibility of aquatic plants for nutrient removal from municipal sewage in Smart cities explores nature mimicking ways to treat municipal wastewater. India is also home to one of the largest informal waste recycling. India's lethal informal e-waste recycling: a case study of Delhi and NCR explores the possibilities of further recovery potential and need for EPR/appropriate policies to be adopted and implemented in cities of all types.

Cities need to be safe and healthy. Internet of Things (IoT) can play an important role in providing smart applications. The immense requirement of bandwidth serving the large data transfer in smart cities can be served by millimetre waves and emerge as a promising candidate for 5G networks. A paper examines the protocols, simulators and initial feasibility analysis of introducing mmwave communication. Another contribution discusses carbon nanotube-based input buffer for high-speed digital transmission. Challenges of IoT implementation and ways to address the same are discussed in another paper. IoT middleware platforms are further explored in a paper to help application developers in choosing a platform according to the application needs. In another paper, IR sensors have been used as test bed in the classroom environment for the analysis, design and implementation of a scalable automatic lighting control system. Such smart lighting control systems can be used in several other settings. In another paper, a STRIDE-based approach is proposed to help the system designer in framing security requirement and proposing possible solutions against specific threats related to smart cities and communities.

Community participation is fundamental to the success of smart cities. There are two contributions which specifically look into ways of empowering the community and ensuring their engagement in various ways.

This book thus attempts to look at a variety of issues currently faced by cities with special reference to Indian cities and offers immense possibilities of a whole range of solutions, not necessarily cost-intensive, to make smart cities a reality.

Contents

About the Editors

Dr. Sirajuddin Ahmed is working as a Professor in Department of Civil Engineering at Jamia Millia Islamia (A Central University). He obtained his Masters in Engineering from Delhi College of Engineering (DU). He was awarded Doctorate from University of Wales (U.K.) in 2008. His research interest includes water treatment, constructed wetlands and other natural wastewater treatment technologies, recycling & reuse of wastewater, urban utilities and services, sustainable development, environmental economics.

He has published 113 research papers in peer reviewed journals and conference proceedings. Dr. Ahmed has guided 10 PhD thesis, 54 M.Tech dissertations and also edited five books. He is a regular reviewer of international journals viz. Natural Resources and Conservation (USA), Atmospheric Pollution Research, Science of the Total Environment, Waste and Resources Management, African Journal of Environmental Science and Technology, Journal of Cleaner Production Elsevier, Current Science, Int. J. of Environment and Waste Management (IJEWM) and Indian Journal of Environmental Protection. Dr. Ahmed is Fellow member of Institution of Engineers (India) and Wessex Institute, UK.

Dr. S. M. Abbas is currently working as a Professor in the Department of Civil Engineering, Jamia Millia Islamia (A Central University), New Delhi, India. He obtained his B.Sc. Engineering (Civil) and M.Sc. Engineering (Building Engg.) from Zakir Husain College of Engineering & Technology, Aligarh Muslim University, Aligarh. Later he obtained PhD (Civil Engineering) from IIT Delhi. His major areas of research interest are Ground improvement techniques, rockfill modeling, Geotechnical Earthquake Engineering, Seismic response of foundations, Geotechnical Characterization of Pond Ash/ Flyash. He has published more than 30 research papers in peer reviewed journals and conferences. He got published two books in the area of Rockfill Materials. He got IJOG (an ASCE publication) best paper award. He also got excellent teaching award from the institution. He is a member of ASCE, ISET, IGS, and Fellow of ISRD.

Dr. Hina Zia is currently Professor and Dean at the Faculty of Architecture and Ekistics, Jamia Millia Islamia, (A Central University) at New Delhi, India. She is qualified as an architect and urban and rural planner. Her specific interest lies on business models on issues pertaining to growing urbanization. She has worked extensively on technical development and pilot implementation of LEED for cities- a performance based rating system by GBCI in order to mainstream cities and urban settlements of all scales towards livability, efficiency, sustainability and inclusiveness.

She has authored and contributed to several books and publications in the domain of sustainable built environment and has more than 25 papers in peer reviewed journals and conference proceedings. She was member of the Guidelines Development Group for Healthy Housing, WHO, member of the multi-stakeholder Advisory Committee, UN's 10YFP Programme on Sustainable Buildings & Construction, editorial board member of Renewable and Sustainable Energy: An International Journal (RSEJ) and is a registered reviewer with several Elsevier publications.

Application of ICT in Parking System

Abdul Ahad, Farhan Ahmad Kidwai, Yasir Khan and Wiqas Anwar

Abstract It is of utmost importance to modify all the material which is mostly used for personal and commercial purposes. At a certain instant of time, people are getting annoyed by the present parking system so a new smart technology is introduced which proves to be more economical and eco-friendly in all aspects. This report is presented to highlight the working performance of the smart parking system and the implementation of **Intelligent Transport System**. This report is also presented to analyze the reliability of the smart parking system.

Keywords Smart parking · Smart city · Pollution free · Prevention of car theft

1 Introduction

In transportation system, parking space plays a vital role. Each and every vehicle making a trip is in need of a parking space at the origin as well as at the destination, irrespective of how long the trip goes. There are numerous systems implemented for managing and controlling the parking scenario.

The problem in the parking system is the rapid increase in the motor field but not in the parking spaces. This problem is overcome by the use of smart technologies.

The smart parking system is the biggest change in the world of parking whose responsibility is to make parking system reliable and also helps in saving time and space. Also, this technology is the most convenient when considering the high population. Also, by using this technology, we can use all the different kinds of gadgets for safety, protection and scheduling of the vehicles.

The term smart city is broadly utilizing nowadays as it is turning into the fundamental requirements for the whole world. In this quick and enraged time, we

A. Ahad (✉) · F. A. Kidwai
Department of Civil Engineering, Jamia Millia Islamia, New Delhi 110025, India

Y. Khan · W. Anwar
Department of Civil Engineering, Mohammad Ali Jauhar University, Rampur, India

© Springer Nature Singapore Pte Ltd. 2020
S. Ahmed et al. (eds.), *Smart Cities—Opportunities and Challenges*,
Lecture Notes in Civil Engineering 58,
https://doi.org/10.1007/978-981-15-2545-2_1

require each and everything dependable with the efficient and better outcome. The solid transportation with no issue is the essential needs which can not be satisfied without making the smart urban communities.

One of the important things that are to be considered before designing any system is to make such systems eco-friendly and sustainable. The energy sources that we are using are not completely reliable and eco-friendly as all these sources are polluting the environment. The land and water pollution are due to various schemes like garbage management, pilgrims, deforestations, etc. The concept of smart city proves to be revolutionary for maintaining and managing each and everything rapidly. Smart city is used to discuss the implementation of modern technology in present urban life. This not only includes information and communication technologies (ICT) but also modern transport technologies. New transport systems as "smart" systems that improve the urban traffic and the inhabitants' mobility [1]. However, various other aspects referring to life in a city are mentioned in connection to the term smart city like security/safe, green, efficient, sustainable, energy, etc., as shown in Fig. 1. Mobility or transportation is one of the important needs of the smart city.

For a smart and efficient infrastructure, smart parking system is used. In this paper, a new system app, "**Park ON**" is based on the use of smart phones, sensors monitoring techniques with a sensor's camera to take photos to show the occupancy of car parks. By the image, a particular vacant space can be known and used to guide a driver to a car park. By implementing this system, the utilization of parking spaces will increase. This system uses the vacant parking space for parking purposes and renews the space when vacated by the user when the user leaves the parking area and transfers the billing data to the user with the help of communication module. It also plays an important role in finding the best possible path as per the present location of the user.

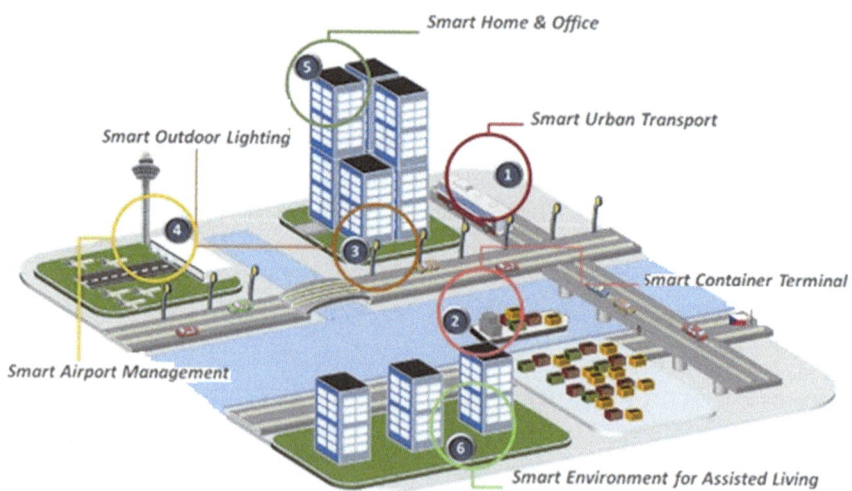

Fig. 1 Smart city requirements

Smart parking technology is one of the efficient ways to overcome the traffic problem and parking issues in city rods and ways. This technology proves to be more reliable and economical.

2 Merits of the Smart Parking

- This system is much safer and secure because of the faster response and digital system implementation.
- The reliability of this type of system is much higher than the traditional parking systems.
- The area requirement of smart parking system is lesser as compared to the traditional system.
- The utilization of GPS technology makes the smart parking system more accurate and also helps in allocating the vacant spaces.
- Due to the centralization of power, a connection is always established between the owner and the staff (Fig. 2).

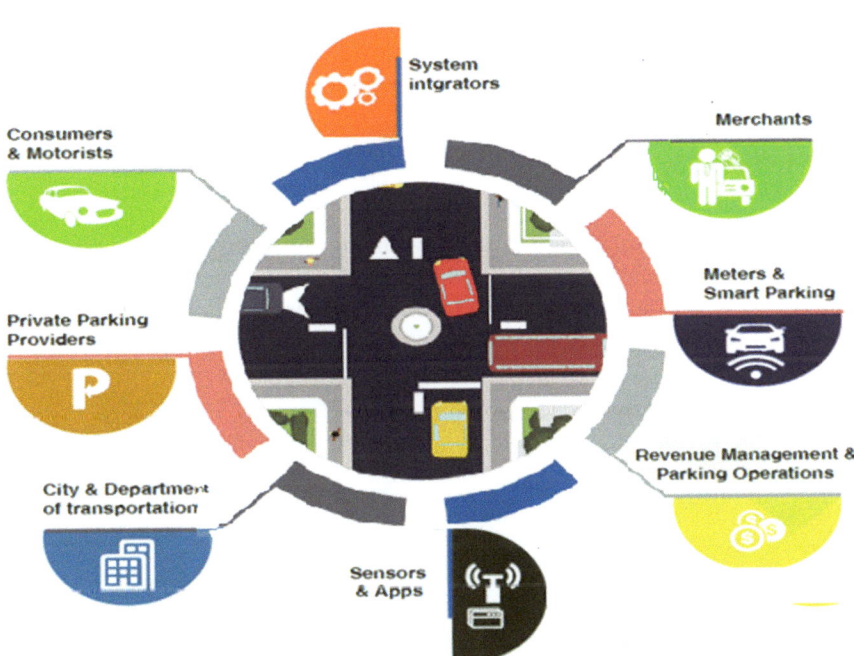

Fig. 2 Coordination of various stakeholders to perform a unique task

3 Module for the Parking Management App

Different modules are implemented in parking management app. The modules are as discussed below with the help of figures.

a. User Interface Module

This module is responsible for establishing a bond between the user and the parking authority which is the main reason this module is highly recommended.

b. Communication Module

This module is responsible for making all the SMS services that are required for conveying the information regarding the parking schedule between the user and the parking system (Fig. 3).

c. Function Module

This is the main module of the smart parking as it has the ability to cover all going things in the parking. The main responsibility of the function module is to look after the entire database system which is going on in the parking like communication, reserve parking, ongoing parking, etc.

d. Parking Space Controller Module

This module covers all the hardware communication and the sensors as it is necessary to know about the current status of parking machinery and system.

4 Methodology

The smart parking technology is the key solution to many problems regarding parking space, parking spots, long and irritable cues and much more. This is the only solution that handles cue system into a professional way and solves the

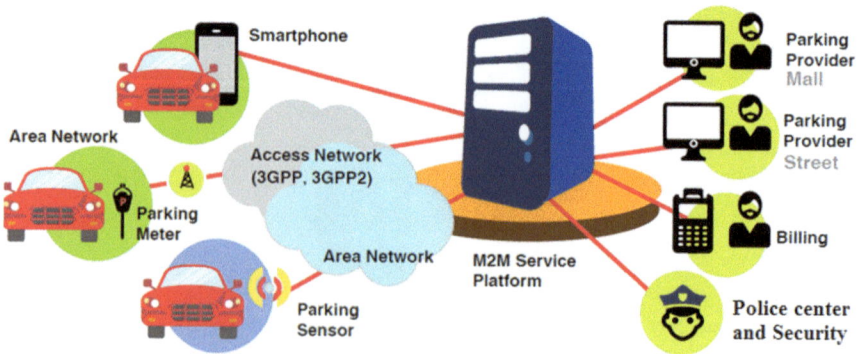

Fig. 3 Using procedure of "Park ON," android app

problem without any disturbance and interruption. A database system is used in the smart parking to increase the reliability of the system at all the time. Various apps like Park ON system are always connected with the parking system in the best possible way (Figs. 4, 5, 6 and 7).

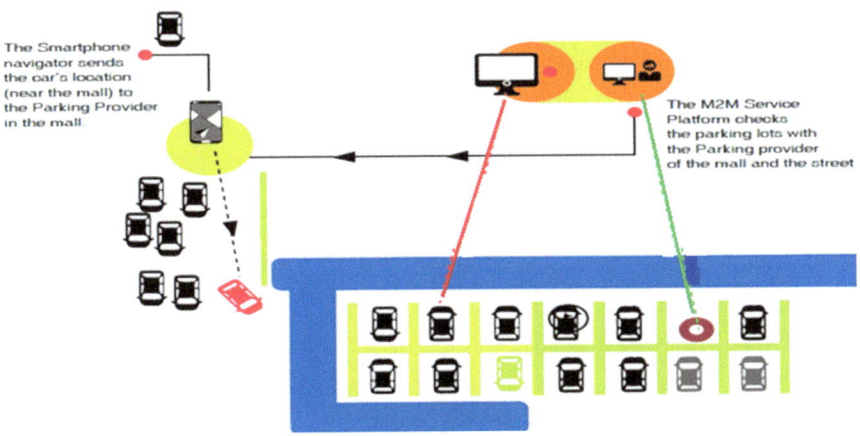

Fig. 4 Response of parking availability and reservation of parking

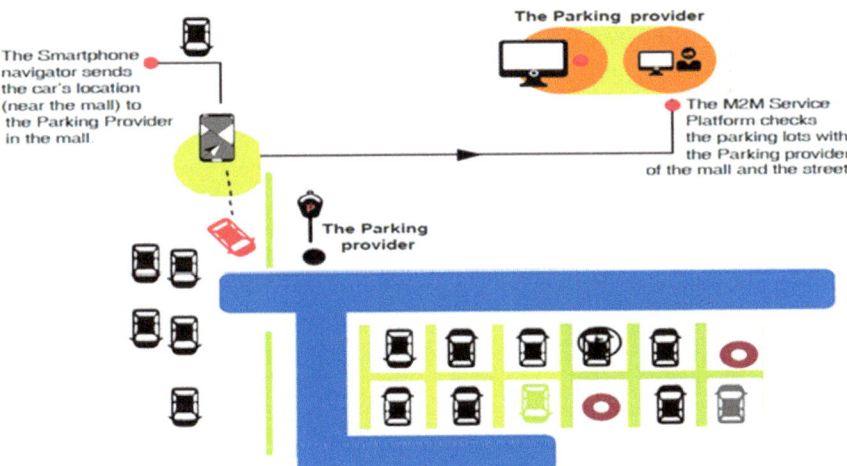

Fig. 5 Procedure of detecting of parking spaces

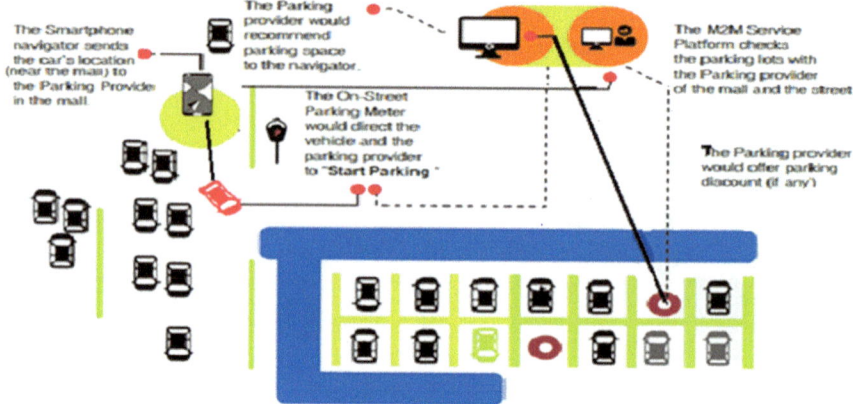

Fig. 6 Service procedure

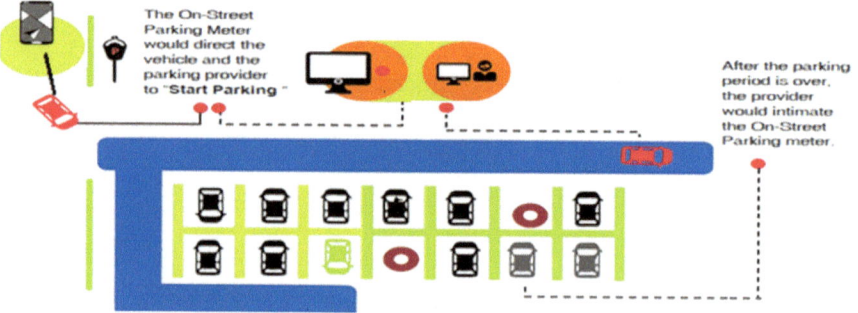

Fig. 7 Final stage of the parking service

5 Authorization for the "Park ON" App

- Only the authorized person has the right to get the reservation slot for parking with a confirmation number.
- A grace period is allotted for the reservation which has the time limit of 15 min. The reservation will be canceled if the customer is coming after the allotted grace period. There is no relaxation after the grace period even for that customer who demanded the reservation.
- One customer can avail more than one reservation.
- There will be an increase in rate if any customer increase the time slot reserved for him and the billing will be sent to that customer automatically with the help of communication module.

- Complete automation is done for car detecting and information is sent to the customer through messages.
- At the final level, the user will receive a final printout showing all the information regarding the timing and the user data.

6 Conclusion

The basic necessity of a smart city is smart parking as the rate of traffic is increasing day by day with a range of approx. One of the basic needs to make the smart city is smart parking as the traffic is increasing at all 20–40% globally. The key point considered in this report is the management of traffic by using Park ON app and smart parking. With the help of this report, it is concluded that both the technologies Park ON app and smart parking proves to be economical and eco-friendly by solving the issues related to traffic and parking.

Bibliography

1. Giuffrè T, Siniscalchi SM, Tesoriere G (2012) A novel architecture of parking management for smart cities. In: 5th international congress—sustainability of road infrastructures. Elsevier, pp 16–28
2. Wootton JR, Garcia-Ortiz A, Amin SM (1995) Intelligent transportation systems: a global perspective. Math Comput Model 22:259–268
3. Kafi MA, Challal Y, Djenouri D, Doudou M, Bouabdallah A, Badache N (2013) A study of wireless sensor networks for urban traffic monitoring: applications and architectures. In: 4th international conference on ambient systems, networks and technologies (ANT 2013). Elsevier
4. Faheem, Mahmud SA, Khan GM, Rahman M, Zafar H (2013) A survey of intelligent car parking system. J Appl Res Technol 11:714–726
5. Qian ZS, Rajagopal R (2013) Optimal parking pricing in general networks with provision of occupancy information. Procedia Soc Behav Sci 779–805
6. Ahad A et al (2016) Intelligent parking system. World J Eng Technol 4:160–167
7. Happiest Minds Technologies Pvt. Ltd. Smart parking
8. Ahad A et al (2017) Smart, sustainable infrastructure development. In: International conference on urbanization challenges in emerging economies. IIT Delhi, New Delhi, pp 769–774
9. Zhang X, Wan D (2010) Economic analysis of regional parking guidance system based on TIA. In: 2010 WASE international conference on information engineering (ICIE), 14–15 Aug 2010, pp 401–404
10. Al-Kharusi H, Al-Bahadly I (2014) Intelligent parking management system based on image processing. World J Eng Technol 2:55–67
11. Al-Kharusi H (2014) Intelligent car parking management system. Thesis, Master of Engineering

Enabling Technologies for Smart Energy Management in a Residential Sector: A Review

Mohini Yadav, Majid Jamil and M. Rizwan

Abstract Continuous rise in energy demand with exposure in the field of smart grid creates new opportunities for energy management in both residential and commercial sector to reduce energy demand. Smart energy management system incorporates the demand response tool to shift and reduce the energy requirement. Further, this system also schedules the energy usage effectively depending on environmental parameters, load consumption profile, user priority index and energy price. Deployment of smart meters creates several opportunities to control the load profile with demand response enabling appliances. Smart energy management has the potential to reduce the carbon emissions with cost-effective energy usage involving renewable energy sources and consumer perspectives. Due to this rising interest toward smart energy management technologies, a review article based on techniques involved in energy monitoring and controlling based on consumer behavior is presented. Further, the implementation of artificial intelligence techniques and optimization approaches involved in optimal load scheduling in a residential sector are also presented.

Keywords Demand response · Smart technologies · Energy management · Consumer's behavior

1 Introduction

With the continual rise in global warming and consumer energy demand, the smart home energy management has shown its existence in an energy sector from decades [1]. This system tries to cut down the electricity demand from peak load times and proves to be efficient for automatic electricity management in a residential sector [2].

M. Yadav (✉) · M. Jamil
Department of Electrical Engineering, Jamia Millia Islamia, Jamia Nagar,
New Delhi 110025, India

M. Rizwan
Electrical Engineering, Delhi Technological University, New Delhi, India
e-mail: rizwan@dce.ac.in

© Springer Nature Singapore Pte Ltd. 2020
S. Ahmed et al. (eds.), *Smart Cities—Opportunities and Challenges*,
Lecture Notes in Civil Engineering 58,
https://doi.org/10.1007/978-981-15-2545-2_2

9

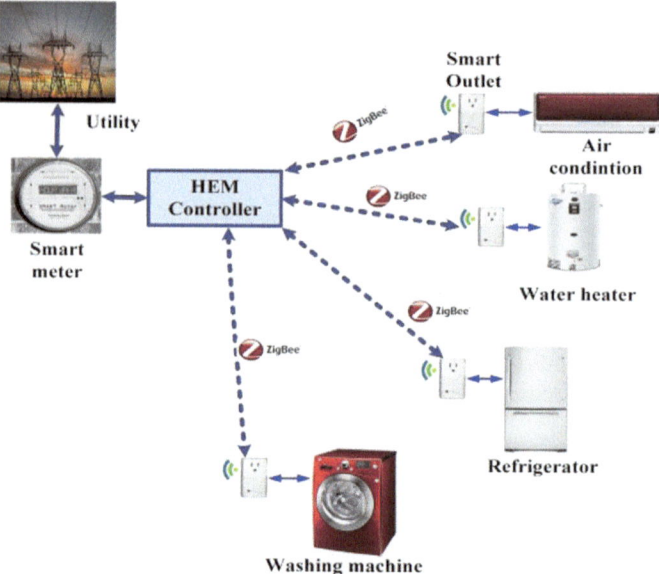

Fig. 1 Design of residential energy management system

In general, such system is installed in residential sector to monitor, control the energy usage by establishing a communication link with home appliances and receive the information to reduce the energy price [3, 4]. The application of demand response, load scheduling and shifting techniques have developed the existing smart energy management system to be more intelligent and robust [5]. Smart home energy management system (HEMS) includes control of different home appliances (air conditioners, refrigerators, lighting, electric vehicles, etc.) using sensors, actuators, smart loads and communication network [6, 7]. The architecture for energy management incorporating demand response signal from utility is shown in Fig. 1. In such architecture, the installation of smart meter set up the communication with grid and controls the appliances as per customer preferences. Such system consists of PC acting as centralized controller to exchange the flow of signals from home appliances and communicate with appliances using communication protocols like ZigBee. Home energy management controller receives the data regarding weather conditions in order to decide the customer preference. Renewable energy resources are connected to such controller to control the energy flow during peak load times.

2 Brief Idea on Energy Management System

Previously, the residential energy management system is designed based on microprocessor, and then, with growth of PCs, its performance gets enhanced. Different optimization algorithms have been developed to reduce the energy cost with reduction in its usage. Furthermore, some energy management systems have been developed while considering the surrounding conditions and different levels of its designing. Installation of appliance control interfaces between home appliances and Internet adaptors is done to control the entire network. To manage the lighting and switch sockets inside a room, the home energy management with infrared control is developed [1]. In 2012, existing HEMS is developed with demand response to lower the energy rate and energy usage [8].

Four appliances such as air conditioner, water heater, cloth dryer and electric vehicle are controlled as per user preferences. A smart home energy management controller with binary linear optimization is designed to allow the effective energy management for residential sector [9]. Authors have also discussed the stochastic dynamic programming with plug-in-electric vehicles in order to manage the power usage optimally. In this study, various mode is presented such as vehicle to grid, grid to vehicle and vehicle to home, and then, these modes are examined as per time-varying energy demand and energy cost [10].

Few researches are related to design of optimization algorithm for load scheduling while considering the battery storage size, dynamic prices and renewable sources to mitigate the entire energy consumption cost. Furthermore, the score role-based smart home energy management algorithm is designed to control the chosen home appliance for demand response events [11, 12]. In Turkey, the novel algorithm for home energy management is developed considering renewable energy sources [13]. In this study, the developed algorithm uses state of charge of battery and renewable resources to reduce energy cost tariff and schedule home appliances. Smart energy management system is also developed that includes dynamic variables (such as energy tariff, weather condition, appliance state, time of use). This system schedules the energy appliances by finding the link between load consumption patterns and power capacity [14]. Real-time energy control method with demand response tool is developed to schedule appliances in order to implement smart HEMS [15]. The state of charge of battery is determined using fuzzy controller and scheduling of loads during off-peak periods using rolling optimization approach.

In aforementioned techniques [16], the HEMS is developed without affecting consumer priority index. The smart energy management system is also developed using lookup table as per fuzzy logic and neural network [17]. Author has designed the fuzzy logic controller with renewable resources and storage battery for residential grid-connected micro-grids. In such study, the main objective is to reduce the power fluctuations while considering battery charging in limits [18]. Further, the artificial bee colony-based optimization technique is used to design HEMS for low price period [19]. An efficient control algorithm to manage air conditioner and water heaters in order to lower the power usage is discussed [20].

In commercial sector, different companies have developed residential smart energy management system to manage the energy usage during peak hours using solar and batteries. For example, Whirlpool Corporation has presented a residential power management controller to control the power usage within residents that exchange the information for on/off-peak time segments. Another company named as General Electric has also developed a smart system that controls and monitors the home appliances with mobile phones and Web applications [21]. Furthermore, Honda US has developed software to manage the home appliances operation within residents [22].

3 Demand Response Strategies in Residential Sector

Presently, the customer interest is bending toward demand response programs in order to mitigate the daily energy consumption with renewable sources or with effective scheduling of electrical load. Demand response modifies the energy consumption pattern of consumer as per variations in electricity cost during high peak hours.

In Europe and USA, demand response strategies have been largely implemented to modify the energy cost and usage level. In demand response programs, the customer can participate in three ways. In first case, the customer may reduce the

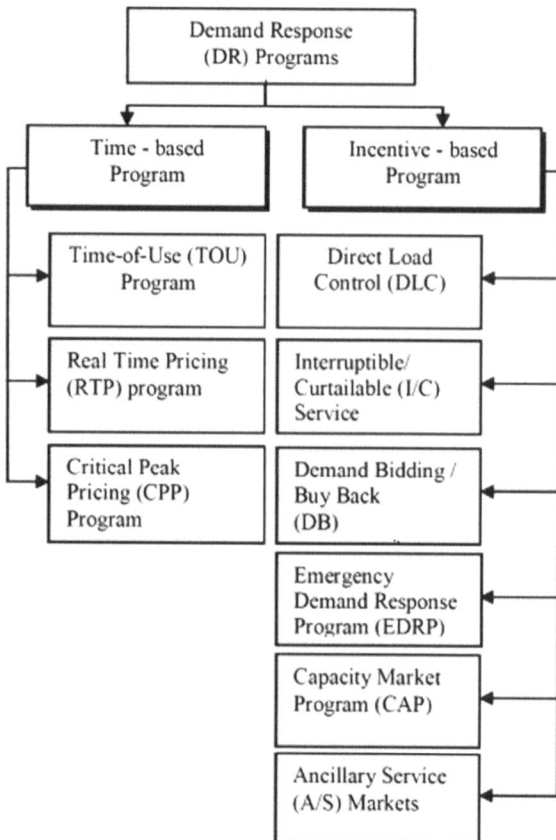

Fig. 2 Different demand response strategies

energy consumption for high peak hours. In second case, the customer may react to increasing energy cost by switching the energy usage from peak to off-peak hours. And, in third one, the customer may incorporate on-site generations to fulfill the customer demand. This kind of participation of customer in demand response programs can lead to saving in energy cost during peak times [23]. As per US Energy Department, the classification for residential demand response is shown in Fig. 2 [24].

3.1 Incentive-Based Demand Response

Such type of demand response is offering financial incentive to participating consumers and in response to this consumer is benefitted in discounted rate of energy cost. This aim is to shift the energy usage of consumers during peak load times [25]. As shown in Fig. 2, the incentive programs are classified in six categories as direct load control, demand building, interruptible services, emergency demand response, capital and ancillary service market. Remote controlling for common home appliances (water heater, air conditioner and lighting) by establishing the direct contract between utility and customers comes under the category of direct load control [26], whereas in case of interruptible services, the price-based contract is signed between utility and large industries to shift their operating appliances during high load times. Some bidding programs are also used where consumer has to set bid value voluntary on a day ahead and they are rewarded when actual energy saving reaches to required level. Further, there is no penalty cost for the consumers who fail to reduce their energy usage.

3.2 Price-Based Demand Response

Such type of demand response includes the financial benefits obtained by customers in response to reduced energy consumption. All these price-based programs provide benefits to the consumers by providing different tariff rates [27]. Depending on the need of consumers, the energy usage in different houses is modified as per real-time electricity cost during peak and off-peak hours. Such real-time pricing is done using time of use (TOU) pricing, real-time pricing (RTP) and critical peak pricing (CPP). In case of TOU, the different tariff rates are categorized as per different time frames and different weather conditions for a year. Suppose, during peak time, the tariff rate is high, and during off-peak hours, it is low, so it encourages the customers to shift their load toward low tariff rate. Further, in case of RTP also named as dynamic pricing approach, the price of electricity fluctuates at each hour and at each slot, whereas, in CPP, the consumers are rewarded depending on their level of control for energy consumption and load shifting on self basis. Another price-based program is peak time rebate pricing, where discount or rebate is awarded to

Table 1 Comparison for residential demand response programs

Demand response programs	Response time	Advantage	Disadvantage
TOU price program	Electricity rate varies for customers on hourly basis	During peak hours, tariff rate is high, and during off-peak hours, it is low, so it encourages the customers to shift their load toward low tariff rate	One tariff rate for all the consumers
RTP program	Electricity rate varies for customers on daily basis	Energy cost is reduced as per price change for a day or month	Continuous respond is required from customer end to reduce the energy bill
CPP program	Electricity rate varies any time at customer side	Discount notification is given to customer for short period	Shifting of home appliances is done for a period by customer
Direct load control program	Electricity rate varies any time at company side	Discount offers are given for curtailment of load by companies	Company requires the authorization from customer side to curtail the load
Interruptible service program	Electricity rate varies any time at customer side	Benefit of discounted price rate is based on respond from customer	Curtailment of load is for limited period
Bidding programs	Electricity rate varies any time at customer side	Special rebate benefits to consumer for load shifting	Curtailment of load is for limited period

consumers based on their reduced amount of energy consumption. The customer participation in demand response program is discouraged due to some uncertainties such as undefined quantity of load required by utility company for reduction and due to some more difficulties related to comfort level of consumers. The comparison among different demand response programs for residential sector is shown in Table 1.

4 Smart Technologies for Smart Homes

Smart home or home automation is an intelligent technology where the home appliances are Internet connected for their remote monitoring and controlling. It offers security, convenience, user comfort, reduction in carbon emissions along with energy savings. A part of Internet of things (IoT) and devices is connected together, collects the information from appliances and then performs automation actions as per user priority and minimized energy consumption. All these smart techniques set an organized schedule for the operation of appliances during peak hours as per demand response signal [28]. In the area of intelligent energy management system, various hardware models and control algorithms have been developed. Some researchers have developed power sockets in order to control the power consumption and turn it off when the power consumption is below some threshold value. Residential energy management system is also framed on the basis of rule-based algorithms where four loads are controlled as per user priority index. Furthermore, machine learning approaches with sensing techniques are designed to reduce the electricity billing cost. Some intelligent methods like real-time monitoring, real-time controlling and stochastic methods are used to reduce electricity billing cost [29]. In case of real-time controlling, the selected devices are controlled, whereas, in stochastic approach, the overall price value for energy usage is determined. Finally, for real-time control, the manageable loads are scheduled in real time and are displayed on real-time basis. In order to manage the switching of loads, software is developed that runs on Linux and written in Java, Python and hypertext markup language [30]. To limit the home power usage below set values and to control the selected appliance automatically, the intelligent residential energy management system is developed.

The main feature in intelligent energy management system is the integration of renewable resources with battery storage for energy consumption control. Some researchers have developed an embedded system to manage the energy management system by incorporating batteries and photovoltaic (PV).

5 Smart Sensors and Communication Protocols

In order to set up a communication link between home appliances and home energy management system, the wireless sensor technology is used. These information and communication technologies are used to design the load curtailment controller and energy management strategies. Appliances are integrated with different networks such as ZigBee, Wi-Fi and Bluetooth in order to receive signals from utilities within home. The highly preferable wireless technique for smart building is IEEE standard 802.11. This wireless setup link between devices and PC is the main key for intelligent energy management and home automation [31]. An intelligent energy management system is designed using ZigBee to offer intelligent services to consumers in a residential sector. Furthermore, cloud services are incorporated within energy management system using ZigBee communication in order to monitor the devices remotely and thereby reduce the energy consumption to 7.3%. An intelligent controller is also designed with ZigBee protocol to control home appliances as per dynamic pricing and customer priority. Power sensor nodes are used for ZigBee protocol to measure the power remotely and to control switching of appliances [32]. An energy management approach while considering energy resources, communication protocols, battery storage based on low energy Bluetooth is developed to establish the communication link between home appliances [33]. Due to various factors like Bluetooth range (limited to 10 m), network size and higher energy usage than ZigBee, the application of Bluetooth is limited [34]. Some of the advancements in information and communication technologies enhance the designing of home energy management system [35]. Although, Wi-Fi communication needs additional power and requires additional components such as routers to form complex infrastructure. So, the Wi-Fi enables smart plug design is developed to monitor and control residential appliances [36]. Currently available wireless networks such as Wi-Fi and Bluetooth are limited in its applications due to need of additional application layer gateway [37]. As shown in Table 2, the ZigBee is used in home automation due to suitable range of communication, low data rate, reduced energy consumption, reduced complexity, easy addition and removal of nodes and no impact on remaining network, higher nodes in the network, i.e., 65,535 [38]. To enable the monitoring and control of all load controllers such as lighting, Plug load and air conditioning loads using wireless and wired communication, the BEMOSS software platform is also discussed [39]. The data ranges of different communication technologies used are shown in Table 2. Different applications of data exchange protocols for communication depending on applications required are shown in Table 3.

Table 2 Data ranges of different communication technologies

Technology	Standard/protocol	Maximum data rate	Effective range
Ethernet	IEEE 802.3	10 Mbps–1 Gbps	Up to 100 m
Serial	RS-485	100 kbps–35 Mbps	Up to 1200 m
ZigBee	ZigBee	250 kbps	Up to 100 m
	ZigBee	250 kbps	Up to 1600 m
Wi-Fi	802.11x	2–600 Mbps	Up to 100 m
Bluetooth	802.15.1	1 Mbps	10 m

Table 3 Communications over data exchange protocols

Data exchange protocol	Applications	Allow communications over			
		Ethernet	Serial	Wi-Fi	ZigBee
1. BACnet (IP)	Building automation	×		×	
BACnet (MS/TP)			×		
2. Modbus (RTU)	Legacy device		×		
Modbus (TCP)	communications	×		×	
3. Web (e.g., XML, JSON, RSS/Atom)	Numerous applications	×		×	
4. ZigBee API	Building automation				×
5. OpenADR	Demand response	×		×	
6. Smart energy (SE)	Smart grid			×	×

6 Scheduling Techniques in HEMS

Scheduling of home appliances without affecting the customer's comfort level reduces the power consumption for HEMS. Reduction of energy demand and energy price with dynamic-based electricity tariffs in order to schedule the home loads during peak hours [40]. Implementation of optimal scheduling controllers, the customers may reduce their energy consumption while incorporating demand response approach and then switching the load from peak to off-peak hours. Home loads are categorized as scheduling and non-scheduling loads. A/C, water heater, electric vehicles, washing machine and clothes dryer come under the category of scheduling loads, whereas TV, lights, printers, oven, microwaves, etc., come under non-scheduling loads [41]. Different scheduling controller techniques are designed using rule-based technique, artificial intelligence and optimization techniques.

Rule-based technique creates if/then rules to manage the multiple appliances scheduling from higher priority to lower priority. This approach controls the higher energy cost with real-time pricing control. But this rule-based algorithm has certain limitations regarding unsuitability for system extension and inefficient to assign the large volume data, and it makes difficult controlling on real-time basis.

Further, the artificial intelligence approaches help to design appliance scheduling controllers for smart homes. These controllers are designed based on artificial neural network (ANN), fuzzy logic control (FLC), adaptive neuro-fuzzy interference system (ANFIS). ANN is an intelligent controller that depends on human thinking and adopts the computational algorithm to simulate the human brain. ANN replaces the simulation tools and acting as solution to problem of control and forecasting. An ANN-based thermal control approach is being developed to create comfortable thermal environment for domestic buildings [42]. Some studies use particle swarm optimization-based ANN to improve the learning rates with optimal number of neurons. Further, the artificial intelligence-based lightning search algorithm (LSA) is developed to forecast the switching state of home appliances, and this further enhances the learning rate and performance of ANN [43]. For energy management decisions, ANN-based distributed approaches are developed to reduce the energy cost and energy demand. Furthermore, FLC is designed to minimize the power usage and energy cost using four steps: fuzzification, defuzzification, rules-based and interference engine. This approach can handle linear and nonlinear systems and does not require mathematical model [44]. Author has also developed FLC-based day-ahead air conditioning scheduling to control the indoor temperature in respect of ambient temperature and energy price. In fuzzy systems, the input variables are considered as type of appliance and occupancy, whereas the output is considered to be the starting probability of each appliance within a minute. Scheduling controller for real time is designed based on fuzzy logic, where four appliances are taken in account with battery and PV [45]. Other AI controller for home energy management is ANFIS that schedules and manages the appliances for energy control. This controller presents many layers with no mathematical modeling [46]. The inputs are considered to be feedback signal from output, outside sensor and fuzzy subsystem. Further, authors have developed intelligent inference algorithm for ANFIS controller where the inference between appliances is improved and results in better performance than classic ANFIS [47].

Optimization technique determines the objective function while considering constraints and then calculates the most optimal solution to the problem. Different optimization techniques are used to schedule the appliance optimally as per feed-in-tariff, pricing schemes and comfortable level. Scheduling of energy-consuming loads to reduce the appliance's waiting time and energy cost with real-time pricing tariff [48]. Further, Lyapunov optimization technique is used to reduce the long-term energy rate that includes controllable and uncontrollable loads [49]. An optimization technique is developed for end customers to reduce the tariff rate by operating the appliances under demand response approach. Completely automatic demand response is significant for the home energy management. Results of optimization show that energy cost reduced to 22% by shifting the loads to off-peak hours. Further, optimal scheduling also achieved with mixed-integer nonlinear programming approach and game theory approach for energy saving. Presently, heuristic optimization is a stochastic algorithm approach used to solve optimization problem. Particle swarm optimization (PSO) technique is implemented for the optimal desirable points for the working hours of appliance [50]. Further, the binary PSO is also developed to create the optimal curtailment schedule of appliances, and it also schedules four controllable energy resources. Some researchers also implemented genetic algorithm approach with supervisory control to schedule residential loads and to reduce energy demand. A wind-driven optimization technique [51] is developed to reduce energy price rate and enhance the comfort level of consumers. Such proposed technique reduces the energy consumption up to 8.3% when compared with PSO. BPSO technique is also used for optimal scheduling of energy appliances. Results show that this BPSO-based heuristic approach is inefficient in terms of computational time [52]. According to aforementioned research, the heuristic approaches and mathematical methods both can be used to implement scheduling of home appliances. Although, such approach has drawback regarding computational time but it may be solved with optimization technique. Finally, the customer awareness regarding smart techniques is needed to realize for proper interaction between customer and intelligent home energy management.

7   Conclusion

This study has reviewed different technologies for the development of efficient residential energy management system. Demand response strategies and its programs are discussed in detail. Intelligent approaches used for the design of controller algorithm are also presented. Communication protocols related to wired and wireless communication such as Wi-Fi, Bluetooth, ZigBee, Serial and Ethernet are discussed in detail and also compared in respect of their operational range. Artificial intelligence techniques for the design of scheduling controller such as ANFIS, rule-based and FLC are also discussed in detail. Such study has also reviewed the mathematical and heuristic-based approaches for the scheduling and minimizing the energy price rate by switching of load from peak to off-peak hours. Its effectiveness is compared with computational time and complexity from other aforementioned techniques. The future trend is the development of self-learning artificial intelligence techniques for reducing the customer involvement for HEMS.

References

1. Shareef H, Ahmed MS, Mohamed A, Al Hassan E (2018) Review on home energy management system considering demand responses, smart technologies, and intelligent controllers. IEEE Access 6:24498–24509
2. Aman S, Simmhan Y, Prasanna VK (2013) Energy management systems: state of the art and emerging trends. IEEE Commun Mag 51(1):114–119
3. Theecoexperts.co.uk (2018) Home energy management systems: a comprehensive guide [online]. Available at: https://www.theecoexperts.co.uk/home-energy-managementsystems-a-comprehensive-guide. Accessed 15 Jan 2018
4. Beaudin M, Zareipour H, Schellenberg A (2012) Residential energy management using a moving window algorithm. In: 2012 3rd IEEE PES international conference and exhibition on innovative smart grid technologies (ISGT Europe). IEEE, pp 1–8
5. Amer M, El-Zonkoly AM, Aziz N, M'Sirdi NK (2014) Smart home energy management system for peak average ratio reduction. Ann Univ Craiova Electr Eng Ser 180–188
6. Li B, Hathaipontaluk P, Luo S (2009) Intelligent oven in smart home environment. In: International conference on research challenges in computer science, 2009. ICRCCS'09. IEEE, pp 247–250
7. Choi J, Shin D, Shin D (2005) Research on design and implementation of the artificial intelligence agent for smart home based on support vector machine. In: International conference on natural computation. Springer, Berlin, Heidelberg, pp 1185–1188
8. Han J, Choi C-S, Lee I (2011) More efficient home energy management system based on ZigBee communication and infrared remote controls. IEEE Trans Consum Electron 57(1)
9. Pipattanasomporn M, Kuzlu M, Rahman S (2012) An algorithm for intelligent home energy management and demand response analysis. IEEE Trans Smart Grid 3(4):2166–2173
10. Wu X, Hu X, Yin X, Moura SJ (2018) Stochastic optimal energy management of smart home with PEV energy storage. IEEE Trans Smart Grid 9(3):2065–2075
11. Di Giorgio A, Pimpinella L (2012) An event driven smart home controller enabling consumer economic saving and automated demand side management. Appl Energy 96:92–103
12. Squartini S, Boaro M, De Angelis F, Fuselli D, Piazza F (2013) Optimization algorithms for home energy resource scheduling in presence of data uncertainty. In: 2013 fourth international conference on intelligent control and information processing (ICICIP). IEEE, pp 323–328
13. Boynuegri AR, Yagcitekin B, Baysal M, Karakas A, Uzunoglu M (2013) Energy management algorithm for smart home with renewable energy sources. In: 2013 fourth international conference on power engineering, energy and electrical drives (POWERENG). IEEE, pp 1753–1758
14. Dittawit K, Aagesen FA (2013) On adaptable smart home energy systems. In: Power engineering conference (AUPEC)
15. Zhou S, Wu Z, Li J, Zhang X-P (2014) Real-time energy control approach for smart home energy management system. Electr Power Compon Syst 42(3–4):315–326
16. Missaoui R, Joumaa H, Ploix S, Bacha S (2014) Managing energy smart homes according to energy prices: analysis of a building energy management system. Energy Build 71:155–167

17. Shahgoshtasbi D, Jamshidi MM (2014) A new intelligent neuro–fuzzy paradigm for energy-efficient homes. IEEE Syst J 8(2):664–673
18. Arcos-Aviles D, Pascual J, Marroyo L, Sanchis P, Guinjoan F (2018) Fuzzy logic-based energy management system design for residential grid-connected microgrids. IEEE Trans Smart Grid 9(2):530–543
19. Zhang Y, Zeng P, Zang C (2015) Optimization algorithm for home energy management system based on artificial bee colony in smart grid. In: 2015 IEEE international conference on cyber technology in automation, control, and intelligent systems (CYBER). IEEE, pp 734–740
20. Hong Y-Y, Chen C-R, Yang H-W (2015) Implementation of demand response in home energy management system using immune clonal selection algorithm. In: 2015 IEEE congress on evolutionary computation (CEC). IEEE, pp 3377–3382
21. Vardakas JS, Zorba N, Verikoukis CV (2015) A survey on demand response programs in smart grids: pricing methods and optimization algorithms. IEEE Commun Surv Tutor 17(1): 152–178
22. John JS (2018) Is Europe ready for automated demand response? Online, greentechgrid, Tech Rep, Jan 2018
23. Ghazvini MAF, Soares J, Abrishambaf O, Castro R, Vale Z (2017) Demand response implementation in smart households. Energy Build 143:129–148
24. Federal Energy Regulatory Commission (2008) Assessment of demand response and advanced metering
25. Ghazvini MAF, Faria P, Ramos S, Morais H, Vale Z (2015) Incentive-based demand response programs designed by asset-light retail electricity providers for the day-ahead market. Energy 82:786–799
26. Lang C, Okwelum E (2015) The mitigating effect of strategic behavior on the net benefits of a direct load control program. Energy Econ 49:141–148
27. Asadinejad A, Tomsovic K (2017) Optimal use of incentive and price based demand response to reduce costs and price volatility. Electr Power Syst Res 144:215–223
28. Nanda AK, Panigrahi CK (2016) Review on smart home energy management. Int J Ambient Energy 37(5):541–546
29. Vivekananthan C, Mishra Y, Li F (2015) Real-time price based home energy management scheduler. IEEE Trans Power Syst 30(4):2149–2159
30. Fletcher J, Malalasekera W (2016) Development of a user-friendly, low-cost home energy monitoring and recording system. Energy 111:32–46
31. Keskin ME (2017) A column generation heuristic for optimal wireless sensor network design with mobile sinks. Eur J Oper Res 260(1):291–304
32. Peng C, Huang J (2016) A home energy monitoring and control system based on ZigBee technology. Int J Green Energy 13(15):1615–1623
33. Collotta M, Pau G (2015) A novel energy management approach for smart homes using bluetooth low energy. IEEE J Sel Areas Commun 33(12):2988–2996
34. Amzucu DM, Li H, Fledderus E (2014) Indoor radio propagation and interference in 2.4 GHz wireless sensor networks: measurements and analysis. Wirel Pers Commun 76(2):245–269
35. Khan AA, Razzaq S, Khan A, Khursheed F (2015) HEMSs and enabled demand response in electricity market: an overview. Renew Sustain Energy Rev 42:773–785
36. Wang L, Peng D, Zhang T (2015) Design of smart home system based on WiFi smart plug. Int J Smart Home 9(6):173–182
37. Ojuroye O, Torah R, Beeby S, Wilde A (2017) Smart textiles for smart home control and enriching future wireless sensor network data. In: Sensors for everyday life. Springer, Cham, pp 159–183
38. Abubakar I, Khalid SN, Mustafa MW, Shareef H, Mustapha M (2017) Application of load monitoring in appliances' energy management—a review. Renew Sustain Energy Rev 67:235–245

39. Pipattanasomporn M, Kuzlu M, Khamphanchai W, Saha A, Rathinavel K, Rahman S (2015) BEMOSS™: an agent platform to facilitate grid-interactive building operation with IoT devices. In: 2015 IEEE PES innovative smart grid technologies conference Asia (ISGT-Asia), 4–6 Nov 2015, pp 1–6

40. Marzband M, Alavi H, Ghazimirsaeid SS, Uppal H, Fernando T (2017) Optimal energy management system based on stochastic approach for a home microgrid with integrated responsive load demand and energy storage. Sustain Cities Soc 28:256–264

41. Zhou B, Li W, Chan KW, Cao Y, Kuang Y, Liu X, Wang X (2016) Smart home energy management systems: concept, configurations, and scheduling strategies. Renew Sustain Energy Rev 61:30–40

42. Yuce B, Rezgui Y, Mourshed M (2016) ANN–GA smart appliance scheduling for optimised energy management in the domestic sector. Energy Build 111:311–325

43. Liu Y, Yuen C, Yu R, Zhang Y, Xie S (2016) Queuing-based energy consumption management for heterogeneous residential demands in smart grid. IEEE Trans Smart Grid 7(3):1650–1659

44. Önder Z, Sezer SA, Çanak İ (2015) A Tauberian theorem for the weighted mean method of summability of sequences of fuzzy numbers. J Intell Fuzzy Syst Appl Eng Technol 28(3): 1403–1409

45. Wu Z, Zhou S, Li J, Zhang X-P (2014) Real-time scheduling of residential appliances via conditional risk-at-value. IEEE Trans Smart Grid 5(3):1282–1291

46. Premkumar K, Manikandan BV (2015) Fuzzy PID supervised online ANFIS based speed controller for brushless dc motor. Neurocomputing 157:76–90

47. Choi IH, Yoo SH, Jung JH, Lim MT, Oh JJ, Song MK, Ahn CK (2015) Design of neuro-fuzzy based intelligent inference algorithm for energy-management system with legacy device. Trans Korean Inst Electr Eng 64(5):779–785

48. Mohsenian-Rad A-H, Leon-Garcia A (2010) Optimal residential load control with price prediction in real-time electricity pricing environments. IEEE Trans Smart Grid 1(2):120–133

49. Guo Y, Pan M, Fang Y (2012) Optimal power management of residential customers in the smart grid. IEEE Trans Parallel Distrib Syst 23(9):1593–1606

50. Wang Z, Yang R, Wang L (2010) Multi-agent control system with intelligent optimization for smart and energy-efficient buildings. In: IECON 2010-36th annual conference on IEEE industrial electronics society. IEEE, pp 1144–1149

51. Rasheed MB, Javaid N, Ahmad A, Khan ZA, Qasim U, Alrajeh N (2015) An efficient power scheduling scheme for residential load management in smart homes. Appl Sci 5(4): 1134–1163

52. Mahmood D, Javaid N, Alrajeh N, Khan ZA, Qasim U, Ahmed I, Ilahi M (2016) Realistic scheduling mechanism for smart homes. Energies 9(3):202

Comparative Study of Treatment and Performance in Membrane Bioreactor and Sequencing Batch Reactor for Hospital Wastewater in Smart Cities

Nadeem A. Khan, Rachida El Morabet, Roohul Abad Khan, Sirajuddin Ahmed, Aastha Dhingra, Amadur Rahman Khan and Farhan Ali Adam

Abstract The hospital wastewater treatment is challenging, owing to complex constituents additional to conventional wastewater parameters. Many research works have evaluated sequencing batch reactor (SBR) and membrane bioreactor (MBR) performance but in terms of hospital wastewater treatment work is still lacking. This study performed a comparative study of sequencing batch reactor (SBR) and (MBR) in treating hospital wastewater. The SBR exhibited 88% removal efficiency in BOD5 removal exceeding MBR 78%, while in case of COD removal, SBR exceeds far well ahead with 86% removal efficiency as compared to 65% removal in case of MBR treatment. The removal efficiency is also validated by removal efficiency of total suspended solids (TSS) 87% and 83% and for mixed liquor suspended solids (MLSS) by 85% and 80% for SBR and MBR, respectively. The findings of the study aid in providing a comprehensive understanding, in order to select/adopt appropriate guidelines in selection of MBR and SBR applications for hospital wastewater treatment.

Keywords Hospital wastewater · Sequencing batch reactor · Membrane bioreactor · Efficiency

N. A. Khan (✉) · S. Ahmed · A. Dhingra · F. A. Adam
Department of Civil Engineering, Jamia Millia Islamia (Central University), Jamia Nagar, New Delhi 110025, India

R. E. Morabet
Department of Geography, Hassan II University of Casablanca, Casablanca, Morocco

R. A. Khan
Department of Civil Engineering, King Khalid University, Abha, Saudi Arabia

A. R. Khan
Faculty of Architecture, Aligarh Muslim University, Aligarh, India

© Springer Nature Singapore Pte Ltd. 2020
S. Ahmed et al. (eds.), *Smart Cities—Opportunities and Challenges*,
Lecture Notes in Civil Engineering 58,
https://doi.org/10.1007/978-981-15-2545-2_3

23

1 Introduction

Since last couple of decades, presence of hospital waste management has received more attention. The technology and modernization of society have rendered hospital to generate more and more waste each year [1]. Hospital waste originates from various hospital activities (medical and non-medical) viz. diagnosis, laboratory, emergency, first aid, operation, laundry, kitchen, etc. The resulting pollutants left after treatment from different medical treatments can easily reach thereby causing environmental pollution, which are affecting human and aquatic ecosystem [2]. Treatment of wastewater is a major concern since last century. However, hospital wastewater treatment is relatively recent in developing countries. The developing countries though have guidelines for wastewater discharge, specific guidelines for wastewater treatment are still lacking. Sequencing batch reactor and membrane batch rector are employed for wastewater treatment at various facilities [3].

The SBR provides adaptability in terms of sequence and cycle time but can be restricted in performance due to high variation in influent concentration [4]. However, recent advances combined biodegradation and membrane separation resulting in innovative technology termed as MBR [5]. The combined approach of two distinct processes rendered it as highly adopted process for treating complex wastewater [6–8]. The current scenario in the smart and newly developing cities is to adopt sustainable approach towards waster management. The complexity and the treatability of pharmaceutical effluent are quite new and lack proper treatment knowledge. Hence, will be area of recent concern nowadays.

The study aims to analyse comparative performance of SBR and MBR in treating hospital wastewater (HWW) in terms of conventional parameters. The analysis of conventional parameter can aid in revealing other constituents in wastewater affecting performance of treatment process.

2 Materials and Methods

2.1 Sampling of Wastewater

The samples were obtained onsite from treatment plant facility. The hospital is located in Delhi, NCR India. Table 1 presents the sampling details of wastewater samples and its storage duration for testing.

Table 1 Sampling detials with storage duration

Parameter	Technique of sampling	Volume of sample (ml)	Container	Max storage
BOD	Grab	1000	Glass	6 h/48 h
TSS	Grab	200	Glass	2 h
MLSS	Grab/Composite	200	Glass	7 days
COD	Grab/Composite	100	Glass	7 days/28 days
Alkalinity	Grab	200	Glass	24 h/14 days

The study was conducted on the working treatment plant facility at the hospital. The wastewater was analysed in terms of BOD, COD, alkalinity, TSS and MLSS. The TSS and MLSS were determined using loss of ignition method, alkalinity was determined using titration method, BOD was calculated using Winkler method and COD was obtained spectrophotometer.

3　Result and Discussion

The concentration of the parameters to be analysed was determined for the influent and effluent HWW for both SBR and MBR treatment facility. The results obtained are discussed in the section below.

3.1　SBR Treatment Results

The SBR treatment efficiency was analysed on the basis of concentration of parameters analysed for influent and effluent. Figure 1 presents the influent and effluent concentration of wastewater.

The sequencing batch reactor treatment efficiency was 77.29% for BOD, 86.14% for total suspended solids, 84.24% for mixed liquor suspended solids, 64% for COD and 28% for alkalinity. The low level of COD and alkalinity removal can be attributed to other contaminants present in hospital wastewater than in conventional wastewater.

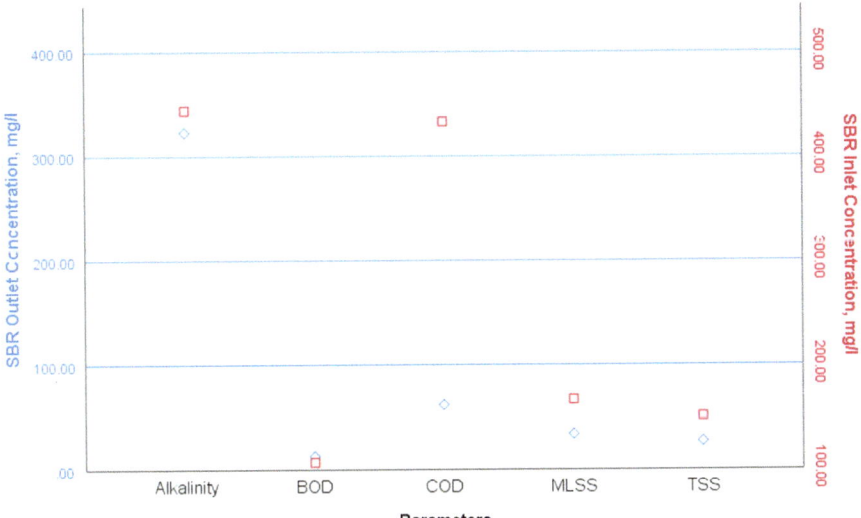

Fig. 1 SBR influent versus effluent concentration

3.2 MBR Treatment Results

The MBR treatment efficiency was also analysed on the basis of concentration of parameters analysed for influent and effluent. Figure 2 presents the influent and effluent concentration of wastewater. In membrane bioreactor removal, efficiency of 88.31 for BOD, 82.11% for TSS, 79.64% for MLSS, 85.68% for COD and 27.25% for alkalinity was achieved. The high alkalinity can be attributed to non-biodegradable waste in wastewater originating from various sources of hospital activities and operations.

3.3 Comparison of SBR Versus MBR

Even though MBR provides flexibility in operation but SBR surpasses it in removal efficiency percentage. In removal of BOD and COD, SBR exhibits higher removal percentage. However, in terms of TSS and MLSS, MLSS surpasses SBR. In alkalinity, removal efficiency both performs similarly with very low percentage of removal (Fig. 3).

The permissible concentration of various wastewater constituents was obtained from Central Pollution Control Board website. Table 2 presents effluent concentration of contaminants along with effluent concentration. It can be derived that the wastewater guidelines available are still very limited and cannot be successfully used in evaluating HWW treatment facility performance.

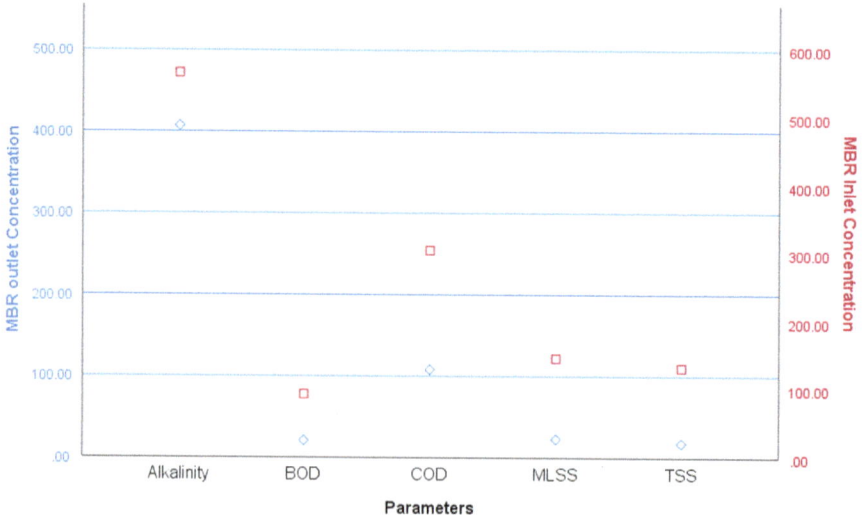

Fig. 2 MBR influent versus effluent concentration

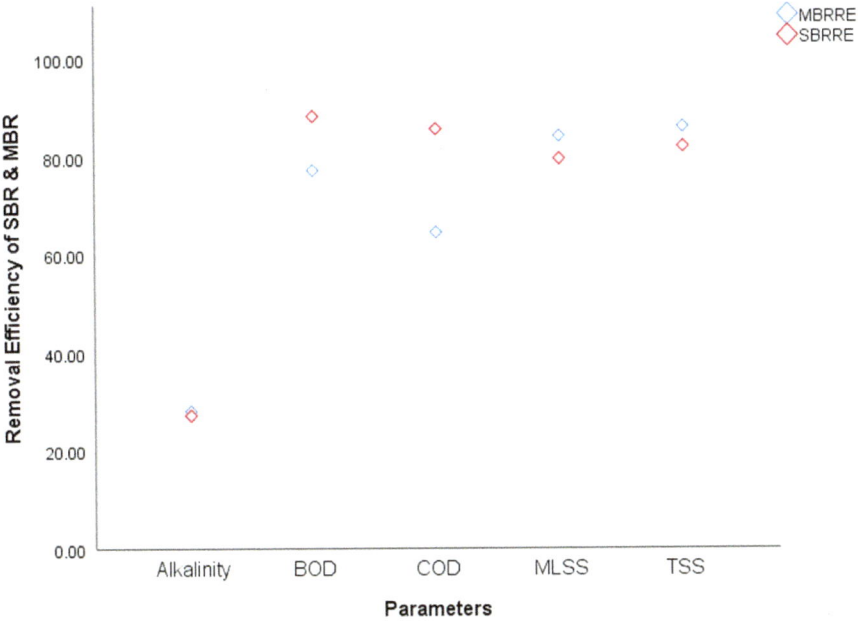

Fig. 3 Removal efficiency of SBR and MBR

Table 2 Effluent concentration of contaminants in SBR and MBR

S. no.	Parameter	SBR effluent	MBR effluent	Permissible standard
1	BOD	13	21	350
2	TSS	27	18	100
3	MLSS	34	23	–
4	COD	62	108	250
5	Alkalinity	323	407	–

4 Conclusion

In the comparative examination of MBR and SBR, Fadel et al. concluded that MBR surpasses SBR in treating high-strength leachate. However, the study concludes that in HWW treatment SBR surpasses MBR in treatment in terms of BOD and COD. Nonetheless, MBR provides better performance in removing TSS and MLSS from influent. The lower percentage of removal of alkalinity can be attributed to various chemical compounds ending in wastewater owing to various hospital activities. The study took into account only conventional wastewater parameters. However, if chemical compounds in HWW can be analysed and studied, they can provide a better understanding of treatment efficiency and can aid in wastewater treatment facilities designer to adopt optimized treatment process. Hence, full-scale wastewater treatment plant should account for non-biodegradable elements to reduce alkalinity and COD in the effluent concentration.

References

1. El-Gawad HAA, Aly AM (2011) Assessment of aquatic environmental for wastewater management quality in the hospitals: a case study. Aust J Basic Appl Sci 5(7):474–482
2. Kusuma Z, Yanuwiadi B, Laksmono RW (2013) Study of hospital wastewater characteristic in Malang City. Int J Eng Sci 2:13–16
3. Taylor P, Hashisho J, Hashisho J (2014) A comparative examination of MBR and SBR performance for the treatment of high-strength landfill leachate, pp 37–41
4. Hashisho J, Hashisho J (1969) Influenza: recommendations for immunization. U.S. public health service advisory committee on immunization practices. Ann Intern Med 71(3):617–618
5. Ahmed S, Dhoble YN, Gautam S (2012) Trends in patenting of technologies related to wastewater treatment
6. Grenoble Z et al (2007) Physico-chemical processes. Water Environ Res 79(10):1228–1296
7. Ahmed S (2008) Constructed wetland as tertiary treatment for municipal wastewater
8. Kumarathilaka P, Jayawardhana Y, Dissanayaka W, Herath I (2015) General characteristics of hospital wastewater from three different hospitals in Sri Lanka, December, pp 39–43

Design of Smart Lighting Control for the Built Environment

Richa Gupta, Saima Majid and Mohini Yadav

Abstract The inefficient use of electricity and the forgetting habits of humans related to manual controlling of lights results in wastage of electricity in both residential and commercial sector. The proposed design can get rid of this physical/manual switching and provides the energy-efficient environment. This lighting control design is programmable, cost effective and provides easy installation. It employs AT86C51 microcontroller with IR sensor modules for monitoring and controlling the LED light as per occupancy inside the room. Specifically, the developed design is used to monitor the entry and exit of occupants inside the room. It automatically senses the person's location and then displays the room occupancy with the glow of LED light. The LED light remains on as long as the person lies in the range of proximity of IR sensor; otherwise, it switches off. Classroom environment is considered as test bed for this analysis. The result analysis shows the effective and efficient usage of lights and thereby causes the energy saving with the rising energy demand. Keil microvision software is used in compiling the code.

Keywords Home automation · IR sensors module · Lighting control system · Microcontroller · Relay

1 Introduction

Electricity has now become a basic need for everyone. India is the world's third largest producer and third largest consumer of electricity [1] having higher energy generation capacity but lacks in insufficient framework to transform energy to end users. Around 300 million Indians are living in dark. In India, average electricity use (2017–2018) is 1149 kWh per capita. According to 2016 survey, 84.53% humans have access to electricity [2]. To ensure that everyone stays in light, it is

R. Gupta · S. Majid
Department of Computer Science and Engineering, Jamia Hamdard, New Delhi, India

M. Yadav (✉)
Department of Electrical Engineering, Jamia Millia Islamia, New Delhi, India

© Springer Nature Singapore Pte Ltd. 2020
S. Ahmed et al. (eds.), *Smart Cities—Opportunities and Challenges*,
Lecture Notes in Civil Engineering 58,
https://doi.org/10.1007/978-981-15-2545-2_4

high time to look forward to generate electricity as well as to reduce the energy wastage. Previously, the lighting control is based on manual on/off switching that leads to continuous power loss. The control may be done using light dependent resistors [3] with IR obstacle sensors and Arduino [4]. Meanwhile, different researches have been focused toward the significance of smart light systems [5–10]. However, home automation is the technique to smartly access the home appliances which are connected remotely to a network [11]. By smartly accessing, we mean any appliance which can be used with ease of convenience, can be secure, energy efficient and most importantly provides comfort. The system triggers events and performs the functions according to the need. Access to appliances can be from mobile using various technologies like Bluetooth, Wi-Fi, ZigBee, etc. As per the US market report, the global lighting and control system growth for the year 2017 is about USD 32.25 billion and is assumed to rise to a market value of USD 102.92 billion in 2024 [12]. This tremendous growth of smart lighting system over traditional lighting systems is beneficial for the growing modern infrastructure of commercial and residential sectors. Authors have also discussed the significance of lighting control for a specific building using dimmable control strategy and determine the relationship between internal and external luminance levels [13]. Keeping all this in mind, appropriate lighting control is a significant part for modern building in both residential and commercial sectors. Different controlling strategies [14] commonly used for modern building design are dimming, detecting human body presence, daylight harvesting and illuminance. Researchers have also discussed the method to assume the energy consumption in a building with daylight control and occupancy control [15]. This smart lighting technology leads to energy-efficient environment with high energy optimization. One of the major reasons behind the growth of smart lighting services is raising the demand of residential customers toward home automation. Smart lighting system is fragmented based on lighting source, connectivity and end users. Connectivity may be wired or wireless. Sensors as the name suggest 'senses' or 'detects' the changes in the environment. Information is mostly in the form of code/program, but sensors provide good actions. Sensors are the basic building block of home automation system. For developing smart energy buildings and to automate the surrounding environment, the machine learning and artificial neural networks are highly applicable. As an example, Amazon Echo or Google Home provides automation environment based on customer comfort and their priorities. The energy saving due to lighting control is divided into two aspects: The first one is related to dimming of artificial lights with daylight penetration, and the second one is the initial dimming while considering continuous illuminance level [16]. In 2007, the European government has introduced the standard EN15232, which provides the list of building automation and control system with some technical management functions to determine the impact of energy performance of buildings [17]. Authors have also presented the load prioritization technique for smart energy management [18] using microcontroller-based design and concept of approximately zero energy building using functional link neural network technique [19]. Authors have also discussed the lighting control strategies with daylight availability and their impact for efficient energy environment with experimental validation and case analysis [20]. The significance of this work lies with the fact of designing an energy-saving model with cost-effective approach to control the lighting system. This proposed design saves and conserves energy in such a way that lights switch on only when a movement of an object is detected at a certain distance specified by IR sensors. This paper designs this energy-saving model using AT86C51 microcontroller with IR sensor modules.

This study is organized in the following section as: Sect. 2 presents the design of the proposed automatic lighting control system. Section 3 presents the components required for hardware design. Section 4 presents the case study of the result. Section 5 presents the advantages and disadvantages of the proposed study and Sect. 6 presents the conclusion for the entire work.

2 Proposed Automatic Lighting System

In this study, the automatic lighting control system using 8051 microcontroller and IR sensors is designed to sense the nearby objects and then automatically switch on/off the LED lights.

At the input side of the module, two IR sensors are used to interface with microcontroller. This designing circuit includes IR sensor connected to one of the AT89C51 input pins on Port 2. All IR sensors work on 5 V supply given by AT89C51 microcontroller. Furthermore, both the sensors must be placed on the either side of the door and entrance of the room, and 5 V supply is also applied to relay. This design includes a connection between the AT89C51 microcontroller and a laptop. The connection is served to send the code regarding sensing of object from laptop to AT89C51 microcontroller which forms the brain of the circuit. At the receiving side, the microcontroller sends command to relay in the form of code. Relay signals the code and sends it to IR sensors. On receiving the command, IR transmitter transfers that IR signal within a specified range and at desired frequency to the IR receiver in such a way that IR receiver senses it and then forwards the command signal for the LED bulb to glow. If the reflecting surface absorbs the IR radiations, then there is no reflection, and the object is unable to be detected by sensor leading to no glow of bulb. The same occurs if the object is not present. A software program was developed for this design to perform various actions on the hardware. A Keil microvision software was used to compile the code.

3 Hardware Requirements

The components required in this proposed design of automatic lighting control system are given in Table 1. The internal connection established among different devices such as IR sensors, microcontroller and relay is shown in Fig. 1.

Table 1 Components required

S. no.	Components required	Specification
1	AT89C51 microcontroller chip	Completely static operation: 0 Hz to 24 MHz, three-level program memory lock, 128 × 8-bit internal RAM, 32 programmable I/O lines, two 16-bit counters
2	8051 development board	Programmable on-board Flash EPROM and ISP capabilities
3	2 infrared sensors	Vcc to the power supply 3–5 V DC
4	5 V, 4 channel relay module	5 V, 4-channel relay interface board with 15–20 mA driver current
5	Connecting wires	As per requirement

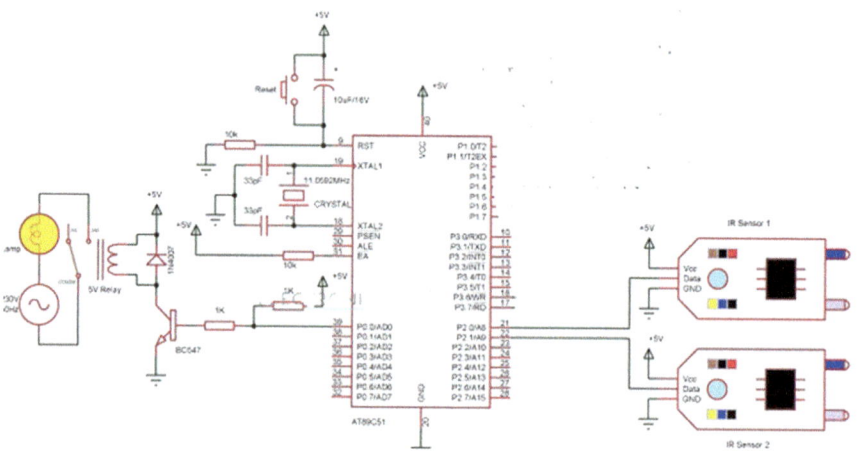

Fig. 1 Circuit diagram of the proposed system

As the object entered in the range of IR sensor 1, then the activation signal is sent to IR sensor 2. This signal sends the activation command to microcontroller used in this proposed design. This process is responsible for the glow of bulb. But, as the object leaves the room or lying outside, the IR sensor 1 rang, and then, bulb turns off. This process is presented in Fig. 2.

4 Result Analysis

Case 1 When an object enters within the range of IR sensor, sensor 1 detects the presence of object, and sensor 2 sends an activation signal to microcontroller. On receiving the command, microcontroller sends signal to relay, and it shows the turning on of LED light on relay as the output as shown in Fig. 3.

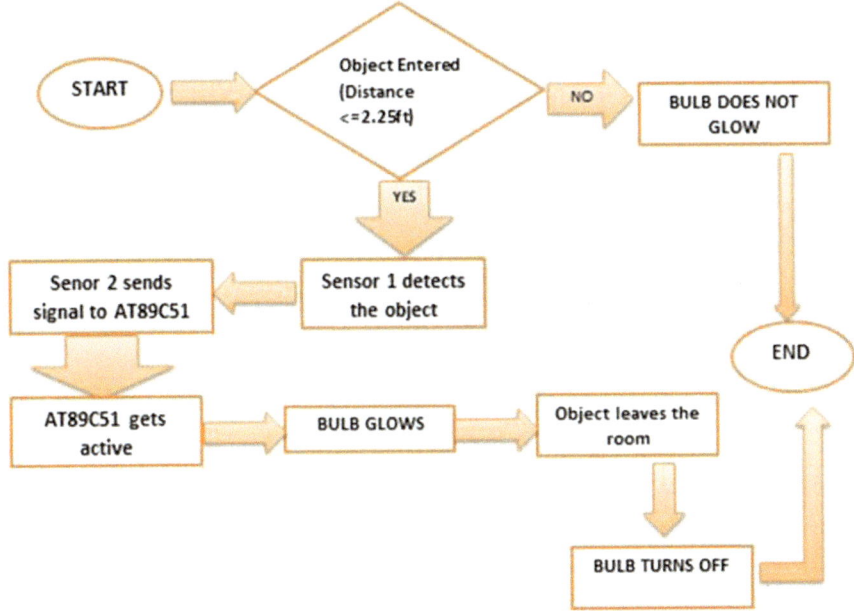

Fig. 2 Block diagram of the proposed technique

Fig. 3 Object entered inside the room

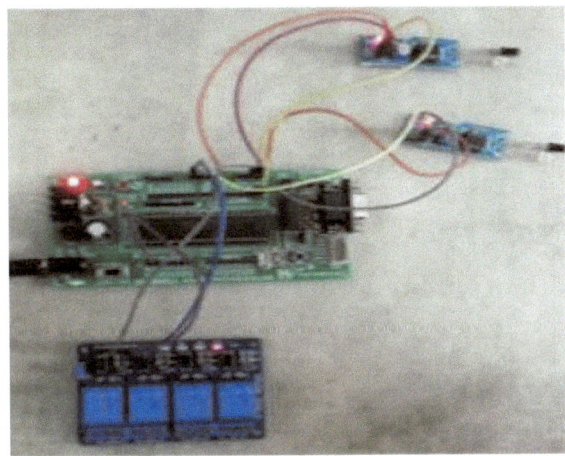

Case 2 When no object has entered or been detected within the range of 2.25 ft, then there is no glowing of light on relay as shown in Fig. 4 and thus signifying that there is no one inside the room and thereby saving electricity.

Fig. 4 No object inside the room

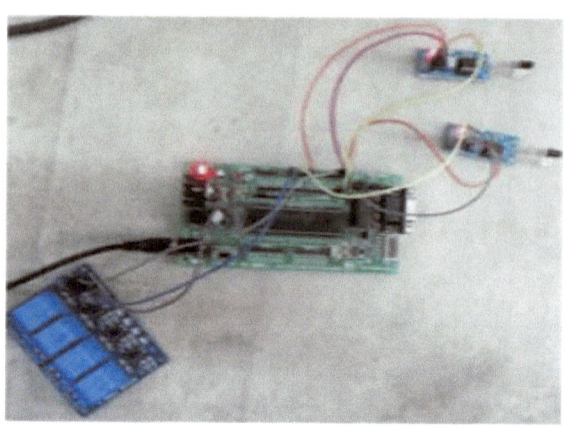

5   Advantages/Disadvantages of Proposed Design

5.1   Advantages

- Since the system is an automatic, no human action is needed to operate it. We have used microcontroller to do all the things.
- A more relaxing and comfortable living environment will also be made feasible by the system.
- The system will also take care of the careless mistake made by the majority of the home residents leading to reduction in the wastage of electricity.
- Simple, efficient and safest way to save energy.
- Power Consumption is much lower.
- System scalability.
- Easy extension—any change can be easily made in wireless, so extension is necessary.

5.2   Disadvantages

- Professional installation is required.
- Maintenance cost is high.

6   Conclusion

This proposed work has presented the design and implementation of an automatic lighting control system using 8051 microcontroller, to avoid manual switching of light. Since the nonrenewable energy resources are exhausting at higher speed, so it is the time to find its alternative. In order to save and conserve energy in an efficient manner, this study has presented the design to switch on the lights only when a movement of an object is detected at a certain distance; otherwise, it remains off. Keeping in view the long-term benefits, this project can be implemented at a larger scale. Implementation of circuit is simple, and also, the power consumed by the circuit is low because only few components are used in the circuit. As a future scope, this work may be expanded to many areas by not restricting to only home or classrooms. A security system can be added by adding an alarm system or a camera. In case, if any tragedy happens such as thefts, the owner receives a message. The system can be tested on different platforms such as Arduino. Dimming light feature can also be added to save a lot of energy and also can be used according to the darkness of the area.

References

1. Report (2018) Now, India is the third largest electricity producer ahead of Russia, Japan. Retrieved 26 March 2018
2. https://en.wikipedia.org/wiki/Electricity_sector_in_India
3. Jalan AS (2017) A survey on automatic street lightning system on Indian streets using Arduino. Int J Innov Res Sci Eng Technol 6(3):4139–4144
4. Louis L (2016) Working principle of Arduino and using it as a tool for study and research. Int J Control Autom Commun Syst 1(2):21–29
5. Jalan A, Hoge G, Banaitkar S, Adam S (2017) Campus automation using Arduino. Int J Adv Res Electr Electron Instrum Eng 6(6):4635–4642
6. Satyaseel H, Sahu G, Agarwal M, Priya J (2017) Light intensity monitoring & automation of street light control by IoT. Int J Innov Adv Comput Sci 6(10):34–40
7. Rath DK (2016) Arduino based: smart light control system. Int J Eng Res Gen Sci 4(2): 784–790
8. Akinyemi LA, Shoewu OO, Makanjuola NT, Ajasa AA, Folorunso CO (2014) Design and development of an automated home control system using mobile phone. World J Control Sci Eng 2(1):6–11
9. Shinde KP (2017) A low-cost home automation system based on power-line communication. Int J Creative Res Thoughts 5(3):20–24
10. Chunjiang Y (2016) Development of a smart home control system based on mobile internet technology. Int J Smart Home 10(3):293–300
11. Joy A, Thoppil AS, Alias BP, Kurup LS, Varghese R (2015) Microcontroller based room automation. Int J Adv Res Electr Electron Instrum Eng (IJAREEIE). ISSN: 2278-8875
12. Zion market research (2018) Available online: https://www.zionmarketresearch.com
13. Kaminska A, Ożadowicz A (2018) Lighting control including daylight and energy efficiency improvements analysis. Energies 11(8):2166
14. Borile S, Pandharipande A, Caicedo D, Schenato L, Cenedese A (2017) A data-driven daylight estimation approach to lighting control. IEEE Access 5:21461–21471
15. Larsen OK, Jensen RL, Antonsen T, Strømberg I (2017) Estimation methodology for the electricity consumption with daylight- and occupancy-controlled artificial lighting. Energy Procedia 122:733–738
16. Jung S-J, Yoon S-H (2018) Study on the prediction and improvement of indoor natural light and outdoor comfort in apartment complexes using daylight factor and physiologically equivalent temperature indices. Energies 11(7):1872

17. Favuzza S, Ippolito MG, Massaro F, Musca R, Riva Sanseverino E, Schillaci G, Zizzo G (2018) Building automation and control systems and electrical distribution grids: a study on the effects of loads control logics on power losses and peaks. Energies 11(3):667
18. Yadav M, Jamil M, Rizwan M (2018) Microcontroller based load prioritization technique in residential sector. In: IEEE international conference on power electronics, intelligent control and energy systems (accepted)
19. Yadav M, Jamil M, Rizwan M (2018) Accomplishing approximately zero energy buildings with battery storage using FLANN optimization. In: International conference on advances in computing, communication control and networking. IEEE (accepted)
20. Beccali M, Bonomolo M, Ippolito MG, Lo Brano V, Zizzo G (2017) Experimental validation of the BAC factor method for lighting systems. In: 2017 IEEE international conference on environment and electrical engineering and 2017 IEEE industrial and commercial power systems Europe (EEEIC/I&CPS Europe). IEEE, pp 1–5

Impacts of Urban Land Use Land Cover Pattern on Land Surface Temperature

Gupta Nimish, V. Banad Sudeep and H. Aithal Bharath

Abstract The impact of rising temperature due to climate change and its potential implications on human life in the past decade has become an issue of utmost significance. Rapid unplanned urbanization and its impact on land use change can be visualized spatially as well as temporally using remote sensing data combined with GIS. The present study computes the change in land surface temperature (LST) with the change in land use class. Landsat data for four different time periods from 1991 to 2017 was acquired to compute land use and LST using supervised learning and mono-window algorithm, respectively. Land use spatial patterns revealed that built-up increased from 1.85% in 1990 to 21.49% in 2017, and the vegetation reduced to half in this time period. LST was computed using modified emissivity method pointed out a rise in minimum temperature in the city, especially in the regions of recent development. Changes in peri-urban buffer zones and urban zones show a stark difference in temperature due to regions converted to open areas. The study points out that if the urban growth is allowed as usual, it would contribute largely in increased urban heat islands.

Keywords Land surface temperature · Land use · Single window · Urbanization

1 Introduction

Changes in weather and climate events, specifically an increase in warm temperature extremes, have been directly linked to human influence. The impact of rise in temperature due to climate change and its potential implications on human life in the past decade has become an issue of increased significance. IPCC report(s) indicated that last three decades have been warmer than any preceding decade since 1850 [1]. Out of the ten warmest years globally, nine are from 2007 to 2017, 2016

G. Nimish · V. B. Sudeep · H. A. Bharath (✉)
Ranbir and Chitra Gupta School of Infrastructure Design and Management,
Indian Institute of Technology Kharagpur, Kharagpur, West Bengal, India
e-mail: bharath@infra.iitkgp.ac.in

© Springer Nature Singapore Pte Ltd. 2020
S. Ahmed et al. (eds.), *Smart Cities—Opportunities and Challenges*,
Lecture Notes in Civil Engineering 58,
https://doi.org/10.1007/978-981-15-2545-2_5

being the warmest followed by 2015 and 2017 [2]. The rising temperature is closely associated with the rapid and unplanned urbanization. Unplanned growth leads to drastic land use (LU) changes and increased impervious surfaces [3].

Urbanization is one of the primary causes of the urban heat island phenomenon (UHI), where parts of urban areas experience higher temperatures as compared to the surrounding rural regions [4]. Urban region with a population exceeding one million or more is experiencing 1–3 °C warmer temperatures as compared to the surroundings [4]. Land surface temperature (LST) is one of the main indicators to assess UHI and can be defined as the radiative skin temperature of the land surface, as measured in the direction of the remote sensor [5]. LST is the main indicator of the surface energy balance of the earth, and it is an important input in many climate models, hydrological models and meteorological models [6].

LST data for a region is mainly collected from weather stations, thermometers mounted on automobiles and using IR sensors [7], but these methods are expensive and do not cover large spatial extent. The advent of satellite-based thermal infrared (TIR) with better spatial and temporal resolution has proven to be a cost-effective mechanism to acquire LST. LST retrieval from satellite-based TIR data is directly linked to the radiative transfer equation, and its history dates back to the 1970s [8]. Consequently, there are many algorithms and methods to retrieve LST [9]. Wan and Dozier [10] and Sobrino et al. [11] have proposed a generalized single-channel and split window method for retrieving LST and also listed various existing LST retrieval methods for Landsat data 8 TIR data. In Indian scenario, Mallick et al. [12] used Landsat 7 ETM + satellite data for estimation of LST over Delhi, and they also observed a strong correlation between LST and NDVI over different LU classes. Urbanization pattern of Coimbatore has been spatially visualized by Bharath et al. [13] using remote sensing data, and they also quantified the extent of urbanization using spatial metrics. But, most of these studies were limited to understand the city and its growth with changing conditions and not its buffer zone that forms an important aspect in surface dynamics. Therefore, the present study focuses on (i) understanding the impacts of land use change in the city and buffer (ii) retrieval of LST using satellite-based TIR data and understanding its relationship with altering land use.

2 Study Area

Coimbatore district is situated in the western part of Tamil Nadu on the banks of river Noyyal and has a population of 3.45 million. The urban population of the district is 75.7%, making it the third largest urban agglomeration in Tamil Nadu [14]. Coimbatore city is located between 10^0 11′ 27″ N–10^0 11′ 40″ N and 76^0 37′ 24″ E–77^0 31′ 55″ E and is one of the most industrialized cities in the state. The study area included Coimbatore Municipal City Corporation and additional 10 km buffer as shown in Fig. 1.

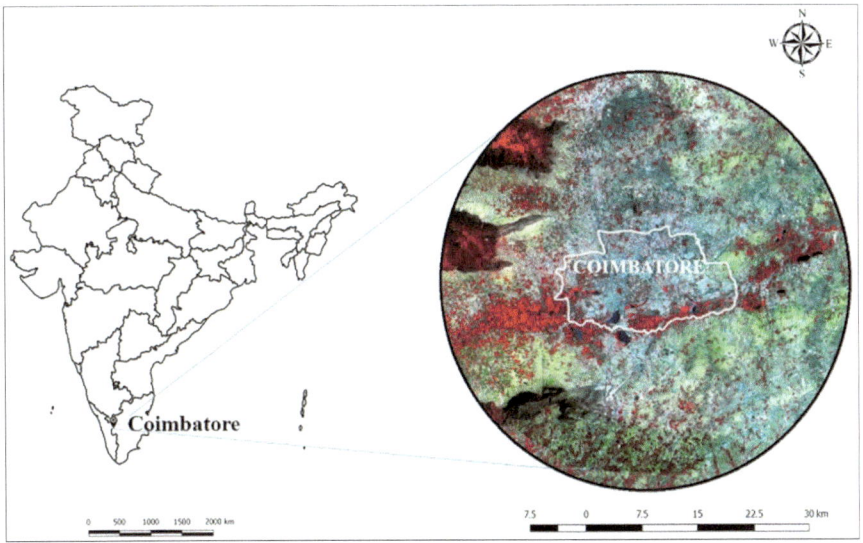

Fig. 1 Study area: Coimbatore and buffer region

3 Data

Data considered for the analysis is as listed in Table 1. Landsat data available in public domain was used for the analysis. Validation data was collected from the field and other sources available as listed.

Table 1 List of data used for study

	Data used	Purpose
1	Toposheets of 1:50,000 and 1:250,000 scale	Generate base layer and boundary layer maps
2	ISRO-Bhuvan and Google Earth	Used as secondary data and for creation of base layer
3	Landsat series (30 m): L-5 (TM), L-7 (ETM +) and L-8 (OLI)	Land cover and land use analysis
4	Landsat series(30 m) TIR data	LST retrieval
5	Global positioning system	Geo-correction and for validation data set

4   Method

To understand the spatio-temporal change in land use, remote sensing data (Landsat series) was acquired for the period 1991–2017 from United States Geological Survey (USGS) Earth Explorer. Figure 2 gives step-by-step account of the method followed and the software used.

4.1   Data Collection and Preprocessing

Administrative boundary of Coimbatore was digitized from the survey of India toposheets. Landsat data downloaded was geo-referenced and rectified, and it was cropped to the desired study area. An additional 10 km buffer from the city boundary was considered for the study. GPS points were acquired and used for geo-registration of remote sensing data.

4.2   Land Cover Analysis

Land cover analysis is the primary step to classify vegetation and non-vegetation class. It further helps in understanding the changes in the vegetation cover using the normalized difference vegetation index (NDVI).

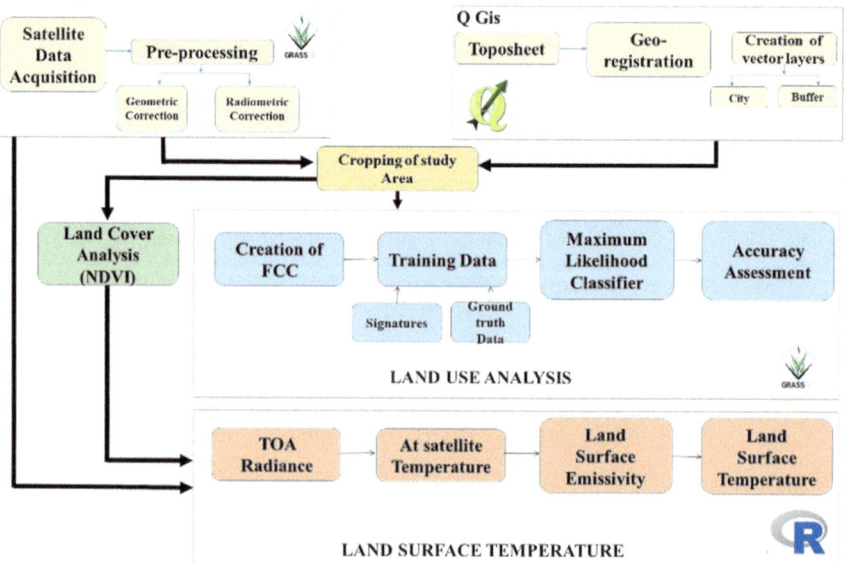

Fig. 2 Flowchart of the method adopted

$$NDVI = \frac{NIR - Red}{NIR + Red} \qquad (1)$$

NDVI values ranges from −1 to +1, and very low values indicate soil or barren areas of rock. Zero indicates water. Moderate values (0.1–0.3) indicate low-density vegetation, and higher NDVI values indicate thick vegetation. Calculation of NDVI is also important to find out the land surface emissivity (LSE) for different classes of land Use.

4.3 Land Use Analysis

Supervised classification technique was used for the land use analysis. Gaussian maximum likelihood classifier (GMLC) that classifies on the basis of the probability density function is more accurate and hence was used. The first step is creating false color composite (FCC) using near infrared (NIR), red and green bands as it helps in identifying heterogeneous patches in the study area. The next step is to create training data set. Training polygons should be well distributed throughout the study area and should cover at least 15% of the study area. The training polygons were digitized and converted into signatures of four different classes, namely built-up, vegetation, water and others. Table 2 lists different land use types categorized into four classes. Post classification, accuracy assessment was performed to estimate kappa statistics. The entire land use analysis was performed using GRASS GIS, an open-source software.

4.4 Retrieval Land Surface Temperature

Landsat 5 TM (thematic mapper) and Landsat 7 ETM+ (enhanced thematic mapper +) both have single TIR (thermal infrared) band, whereas Landsat 8 OLI sensor has two TIR bands. One of the bands (band11) is affected by stray light from outside the field of view; there is still anomaly on the amount of stray light entered, so we have

Table 2 Land use classification categories adopted

	Land use class	Land use type included in the class
1	Built-up	Residential, industrial and all the paved surfaces. The mixed pixels having built-up are included in this class
2	Vegetation	Forests, cropland, plantation and nurseries
3	Water	Tanks, lakes and reservoirs
4	Others	Open ground, quarry pits, rocks

Table 3 Single-channel LST retrieval formulae

	Formula	Description
1	$L_\lambda = (\text{Gain} * \text{DN}) + \text{Offset}$	L_λ = spectral radiance Gain = band specific multiplicative factor (obtained from metadata) Offset = band specific additive factor (obtained from metadata)
2	$T_B = K_2 / \ln((K_1/L_\lambda) + 1)$	K_1 and K_2 = calibration constants T_B = At-satellite brightness temperature (Kelvin)
3	$\text{LST} = (T_B/(1 + (\lambda T_B/\rho) * \ln(\varepsilon)))$ where, $\rho = (h * c/\sigma)$	ε = emissivity λ = mean wavelength of thermal band h = planks constant c = speed of light σ = Stefan Boltzaman constant

only one TIR band data even from OLI sensor. Keeping in mind, the above-mentioned constraints, single-channel method was used to retrieve LST from Landsat data.

The single-channel method involves the following steps and was performed using equations in Table 3 [15, 16].

- Conversion of standard product pixel values (DN) into TOA spectral radiance.
- Top of atmosphere brightness temperature (at satellite temp) from spectral radiance calculated.
- Emissivity is one of the important factors and must be known in order to estimate LST accurately from radiance measurements.
- Calculating land surface temperature which is the function of satellite temperature, wavelength of the emitted radiance and emissivity.

5 Results and Discussion

5.1 Land Use Analysis

Land cover maps were extracted through NDVI for the years 1991–2017. The land cover maps as shown in Fig. 3 indicate the vegetation cover in the region over the years. It can be seen that there has been a drastic decrement in vegetation cover from 1991 to 2017. To analyze further, land use analysis was performed.

Land use pattern analysis was performed for four time period as shown in Fig. 4 and the overall accuracy of classification were estimated to be 88%, 97%, 94% and 90% for 1991, 2001, 2009 and 2017, respectively.

The change in land use is as shown in Table 4, and it shows rise in built-up class from 1.87% in 1991 to 21.48% in 2017 resulting a proportional decrease in vegetation class. Vegetation in 2017 was 12.78% that is almost half when compared with 1991. Residential development in the northeastern part of city near Sri

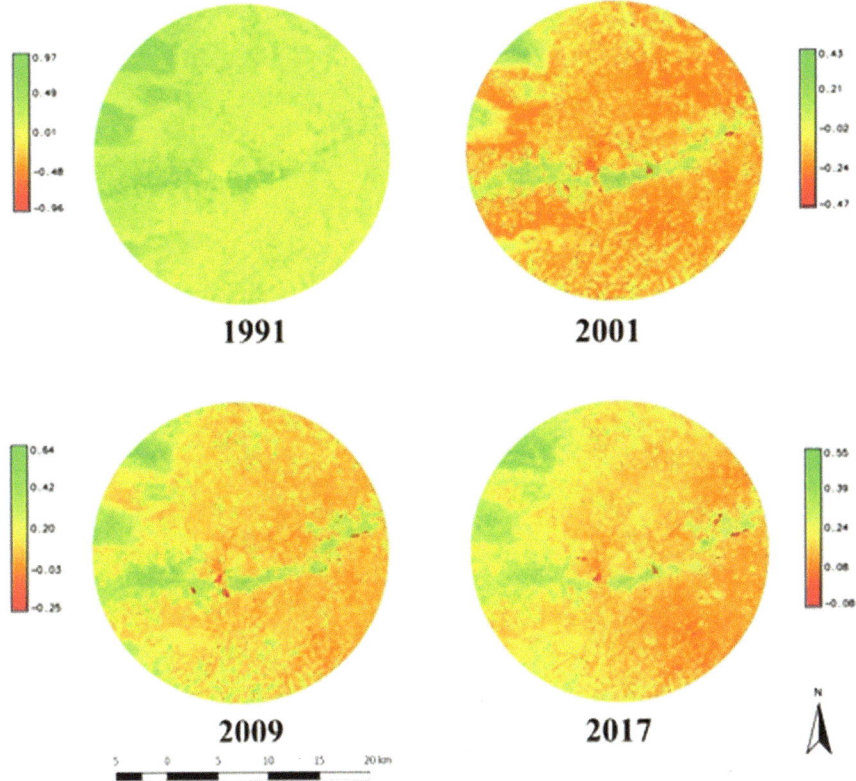

Fig. 3 Land cover for time period 1991–2017

Lakshmi nagar, Krishnarayapuram and also development near Periyakulam in the southern part of the city can be attributed to this high change in built-up class in the previous decade.

5.2 Land Surface Temperature Analysis

The LST maps for all the four time periods are shown in Fig. 5, it is evident that the LST (°C) is low (5–10 °C) in the northwest and southwest part of the study area, compared to the core city. One of the factors for this is the geographical location of the city. It is surrounded by Western Ghats on the west and reserve forests (Nilgiri biosphere reserve) on the north. From the LST maps, it can be visualized and inferred that the minimum temperature in the central part of the city is increasing.

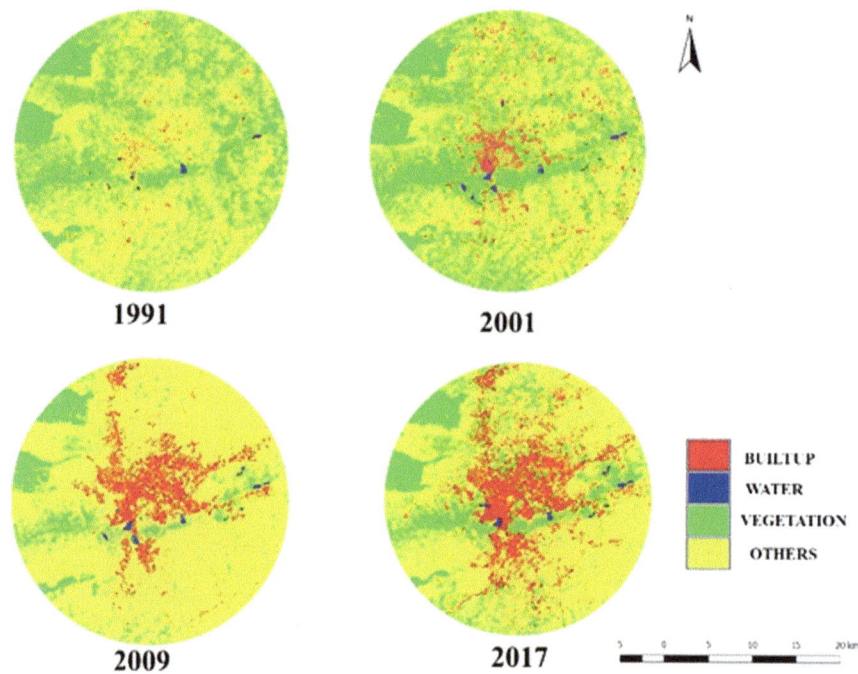

Fig. 4 Land use maps of Coimbatore for time period

Table 4 Land use statistics of Coimbatore

Land use class				
Year	Built-up	Vegetation	Water	Others
1991	1.87	25.21	0.32	72.61
2001	6.81	24.87	0.45	67.87
2009	13.27	20.66	0.39	65.67
2017	21.48	12.78	0.32	65.42

Blue color represents lower land surface temperature, due to the presence of vegetation and water that is prominently evident for 1991 and 2001. The intensity of the blue patch gradually decreases in the subsequent years. Change in land use class from vegetation and others to built-up is evident from the land use maps for the year 2017, and these changes had a significant impact on LST. The temperature in 2017 is in the range 36–40 °C in the northeastern part of the city. Table 5 lists year wise comparative assessment of all land use classes in the region.

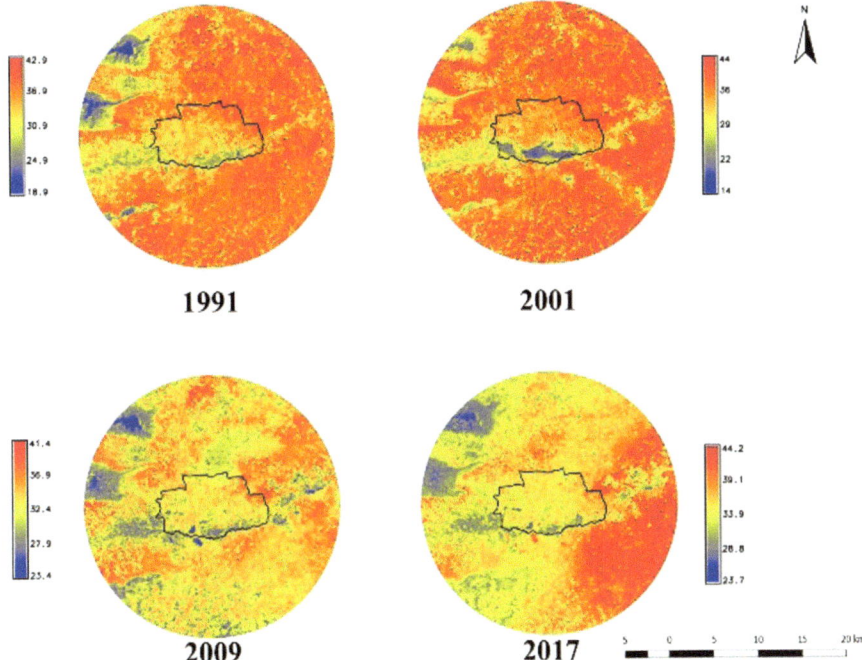

Fig. 5 LST maps for time period 1991–2017

Table 5 LST for different land use class across the time period

Year	Land use	Temperature (°C)		
		Minimum	Maximum	Mean
1991	Urban	22.01	42.90	36.06
	Vegetation	18.93	42.80	33.35
	Water	20.29	42.13	32.09
	Others	19.94	42.906	36.53
2001	Urban	14.20	43.48	35.87
	Vegetation	14.11	43.94	33.49
	Water	14.85	42.58	30.04
	Others	14.67	43.94	36.93
2009	Urban	23.91	40.96	33.71
	Vegetation	23.38	39.78	29.95
	Water	24.35	38.19	27.51
	Others	23.48	41.35	34.46
2017	Urban	26.90	43.97	35.98
	Vegetation	23.69	43.60	33.30
	Water	27.94	39.73	29.40
	Others	25.83	44.19	37.74

The emissivity of the impervious surface is less when compared to water body or vegetation, thus, radiance emitted by the impervious surface is less generating higher surface temperature. Table 5 shows the temperature variation across the different land use classes.

To understand the variation of surface temperature in the city for four time periods and to compare how the temperature in the center of the city and buffer (periphery) varies, boxplots were created for all land use classes as shown in Fig. 6. The plots indicated higher LST in urban areas when compared to other land use classes. To understand variation of LST within built-up class, separate boxplot for the city and the buffer were created that also provide information about the heat islands. The median of LST for the year 2017 turned out to be around 36 °C that is 2–3 °C higher than previous years, Fig. 7. The range of LST values experienced an increament (34–38 °C) for the city that over the previous time period was 32–36 °C. Figure 8 shows the plot of LST in the buffer; the higher values of median for 1991 and 2001 are due the presence of large open areas in the periphery (south-east) of study area that gets heated up quickly, also the surrounding built-up layer such as roads, residential and industrial patches shows higher LST values.

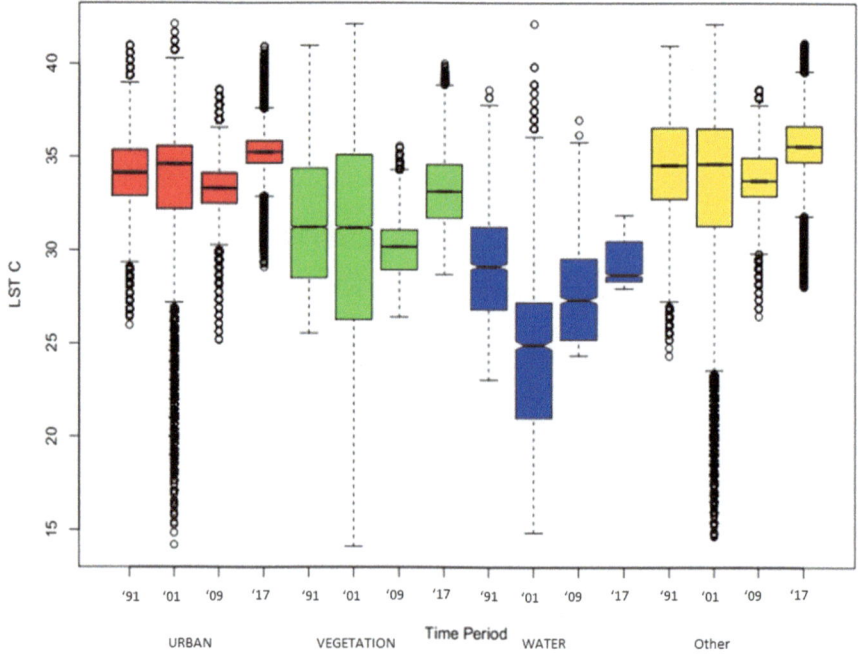

Fig. 6 Boxplot of LST for different land use classes for Coimbatore city

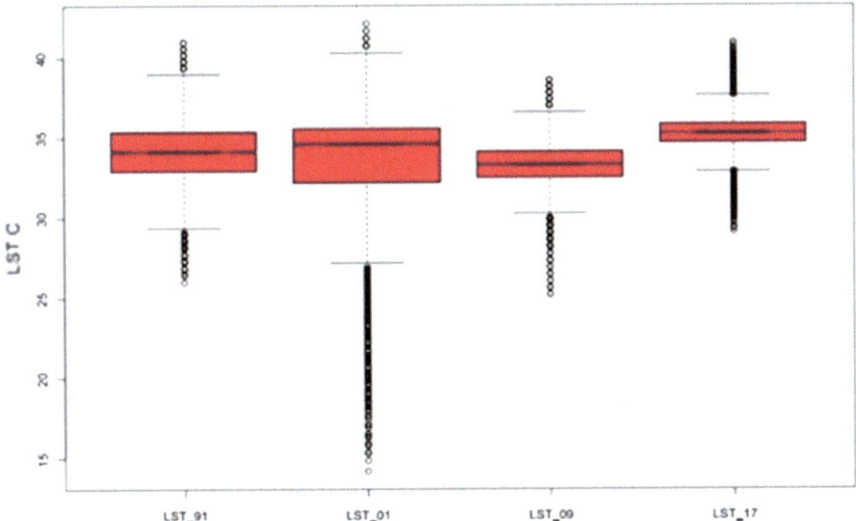

Fig. 7 Boxplot of LST for Coimbatore city

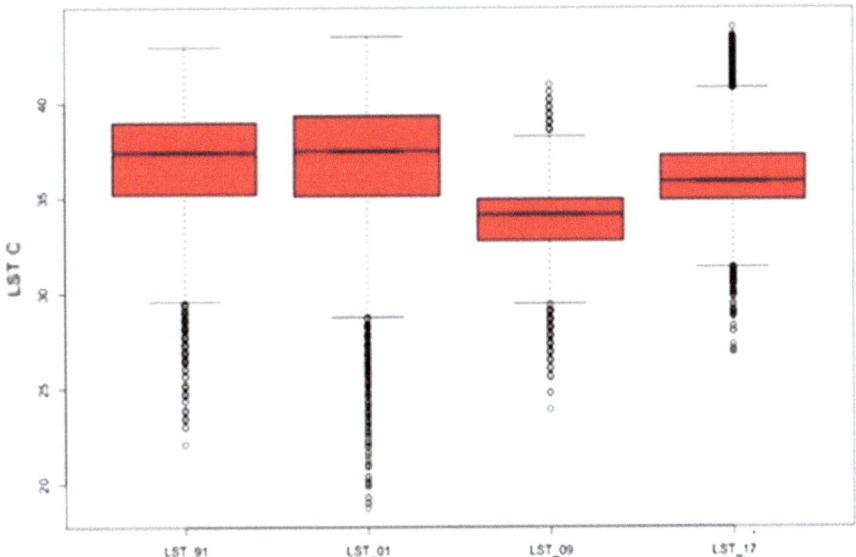

Fig. 8 Boxplot of LST for Coimbatore buffer

6 Conclusions

Land use analysis indicated loss of vegetation in the city of Coimbatore over the past two decades. The vegetation has decreased from 25.21 to 12.78% inferring to a direct result of increase in built-up area 1.87% in 1991 to 21.48% in 2017. The increase in built-up class has resulted in increased LST, especially in the densely urbanized areas. The spatiotemporal variation of LST can be easily visualized from the LST maps of the city and surrounding buffer area. LST analysis for various classes of land use clearly indicated higher surface temperature for urban areas. The minimum surface temperature of urban areas has seen a 4 °C rise. The boxplot of the LST for the city indicated a shift of LST range from 33-36 °C in 1991 to 36–38 °C in 2017. The LST values for the Coimbatore buffer region are in the range 34–36 °C, that is 2–4 °C lower than the core city. The present study clearly indicates rapid and unplanned urbanization that leads to high degree of changes in local climate and the land use. These play a vital role in maintaining the surface temperature of the region. Thus, a mixed land use without creating an imbalance in the land cover structure has to be maintained for sustainable development.

Acknowledgements We are grateful to (i) Science and Engineering Research Board, India (ii) Sponsored Research Cell, Indian Institute of Technology Kharagpur, (iii) Department of Science and technology, Government of West Bengal, for financial support to carryout research and (iv) Indian Institute of Technology Kharagpur, for infrastructural support.

References

1. IPCC (2014) Climate change 2014: synthesis report. Contribution of Working Groups I, II and III to the fifth assessment report of the Intergovernmental Panel on Climate Change Summary for Policymakers
2. NOAA (2018) 2017 was 3rd warmest year on record for US
3. Bhargava A, Lakmini S, Bhargava S (2017) Urban heat island effect: it's relevance in urban planning. J Biodivers Endanger Species 05:1–4. https://doi.org/10.4172/2332-2543.1000187
4. Oke TR (1997) Urban climate and global environmental change. Appl Climatol 273–287
5. Copernicus global land service. https://land.copernicus.eu/global/products/lst
6. Diamandi A, Oancea S, Alecu C (2010) Analysis of the land surface temperature estimated from different satellite sensors over Romania. Romanian Rep Phys 62:185–192
7. Zhang J, Wang Y, Wang Z (2007) Change analysis of land surface temperature based on robust statistics in the estuarine area of Pearl River (China) from 1990 to 2000 by Landsat TM/ETM + data. Int J Remote Sens 28(10):2383–2390. https://doi.org/10.1080/01431160701236811
8. McMillin LM (1975) Estimation of sea surface temperatures from two infrared window measurements with different absorption. J Geophys Res 80:5113–5117. https://doi.org/10.1029/JC080i036p05113
9. Li LZ, Tang BH, Wu H, Ren H, Yan G, Wan Z, Trigo IF, Sobrino JA (2013) Satellite-derived land surface temperature: current status and perspectives. Remote Sens Environ 15:14–37. https://doi.org/10.1016/j.rse.2012.12.008

10. Wan Z, Dozier J (1996) A generalized split-window algorithm for retrieving land-surface temperature from space. IEEE Trans Geosci Remote Sens 34(4):892–905. https://doi.org/10.1109/36.508406

11. Sobrino JA, Jiménez-Muñoz JC, Paolini L (2004) Land surface temperature retrieval from LANDSAT TM 5. Remote Sens Environ 90(4):434–440. https://doi.org/10.1016/j.rse.2004.02.003

12. Mallick J, Kant Y, Bharath BD (2008) Estimation of land surface temperature over Delhi using Landsat-7 ETM+. J Indian Geophys Union 12(3):131–140

13. Bharath HA, Vinay S, Ramachandra TV (2017) Characterization and visualization of spatial patterns of urbanisation and sprawl through metrics and modeling. Cities Environ 10(1):1–31

14. Census of India. http://censusindia.gov.in/

15. Pal S, Ziaul S (2017) Detection of land use and land cover change and land surface temperature in English Bazar urban centre. Egypt J Remote Sens Space Sci 20(1):125–145. https://doi.org/10.1016/j.ejrs.2016.11.003

16. Gaurav S, Shafia A, Bharath HA (2018) Urban growth pattern with urban flood and temperature vulnerability using AI: a case study of Delhi. IOP Conf Ser Earth Environ Sci 169:012092. https://doi.org/10.1088/1755-1315/169/1/012092

Experimental Investigation of RC Corrugated Beams Strengthened with FRP Sheets

Mohammed Rihan Maaze and Swapnil Patil

Abstract Steel corrosion in concrete is one of the main causes for the degradation of the structural members. Corrosion of steel causes internal damages in the reinforced structures which effects the durability and performance of the structure. In current investigation, total of 13 beams of size 1 m × 0.15 m × 0.15 m were casted and beams are subjected to accelerated corrosion, rates are varied incrementally by 5–15% and the beams were strengthened with carbon fiber-reinforced polymer (CFRP) sheets, glass fiber-reinforced polymer (GFRP) sheets, and CFRP plus GFRP sheets together and compared with controlled beam. Load-carrying capacity, flexural strength, ductility, and toughness are studied. Investigation shows that the beam strengthens with combination of CFRP and GFRP shows a better load-carrying capacity as compared to controlled beam; however, it can even achieve good results if bonding strength between sheets and beams is taken in consideration.

Keywords CFRP · Corrosion · Ductility · GFRP · Load-carrying capacity and toughness

1 Introduction

Corrosion of the steel is one of the main and serious problem facing in the infrastructure systems, many structures under moist environment have faced dispute loss and serviceability, and hence, there is a need to study the behavior of the corrugated members. Corrosion reduces the area of steel and, therefore, decreases the load-carrying capacity of members and corroded product occupies larger amount of steel compared to uncorroded steel which produces additional tensile

M. R. Maaze (✉)
Department of Civil Engineering, Nirma University, Ahmedabad, India
e-mail: rihanmaaz@gmail.com

S. Patil
Civil Aid Technoclinic Pvt. Ltd., Bangalore, India

© Springer Nature Singapore Pte Ltd. 2020
S. Ahmed et al. (eds.), *Smart Cities—Opportunities and Challenges*,
Lecture Notes in Civil Engineering 58,
https://doi.org/10.1007/978-981-15-2545-2_6

forces in the concrete which cause the expansion cracks and, thereby, concrete spalls. Extreme corroded RC structure inclines to fail due to forfeiture of bond splitting and bond this implies if corrosion can be stopped or deferred a certain degree of structural strength can be maintained [1, 2].

Rust of steel in structural member is a slow process as of the steel in the concrete is protected and reasonably it takes more time to undergo corrosion, and even with heavy moist condition, it is difficult to achieve and maintain the degree of corrosion [3]. Impressed current techniques have many advantages apart from saving time and money. The best advantage of this technique is the ability to control and vary the rate of corrosion which usually depends upon the resistivity, oxide concentration, and temperature [4]. In the present study, an attempt is made to achieve the corrosion rate by means of impressed current technique. Fiber-reinforced polymer sheets are seen rapidly used in the infrastructures in order to strengthen, repair or rehabilitate the concrete structures under corrosion [5]. Importance is due given to FRP sheets as it is having simple installation procedure, low cost, no clearance loss, and high strength-to-weight ratio. The use of CFRP fabric shows considerable low damage for the corroded beam specimen; much research has been done on beam wrapped with CFRP sheets subjected to freeze and thaw cycle [6] limited work is done with combination of different layers of FRP sheets. In the present study, different schemes were proposed to strengthen the corroded beams, and the corrosion rates are varied 5, 10, and 15% by impressed current techniques. The varied corroded beam is compared with the original conventional beam. The load-carrying capacity, failure mode, ductility, and toughness were presented in this paper.

2 Experimental Test Set up

Total of 13 reinforced beams was casted with a size of 1 m × 0.15 m × 0.15 m, three beams were corroded with 5% corrosion, three beams were corroded with 10% of corrosion, and three beams were corroded with 15% of corrosion. Corrosion is done by impressed current technique and corrosion rate is ensured by means of mass loss of the steel with respective percentage. Three strengthening schemes were proposed in the current study as shown in Table 1. Beams were designed with nominal two bars of 8 mm Fe 415 grade HYSD bars are placed in the bottom zone and two bars of 6 mm 316-type stainless steel were placed in the top compression as shown in Fig. 1. Additional tension reinforcement length of 100 mm is exposed as shown in Fig. 1 which acts as anode and a cathode is connected to a hollow stainless steel bar. Beams are placed individually in tank which consists of 5% NaCl upto bottom-third cover of the beam, i.e., 10 mm. Beams were tested under flexure with two-point loading the load deflection was recorded and studied.

Table 1 Mix proportion and compressive strength as per IS 10262: 2009 [7]

Sl. no	Mix	Cement (kg/m^3)	FA (kg/m^3)	CA (kg/m^3)	W/C	Super plasticizer % by weight of cement	Strength (N/mm^2)	Slump (mm)
1	M20	340	620	1225	0.5	0.2	28.66	75
2	M30	370	615	1135	0.45	0.2	38.25	70

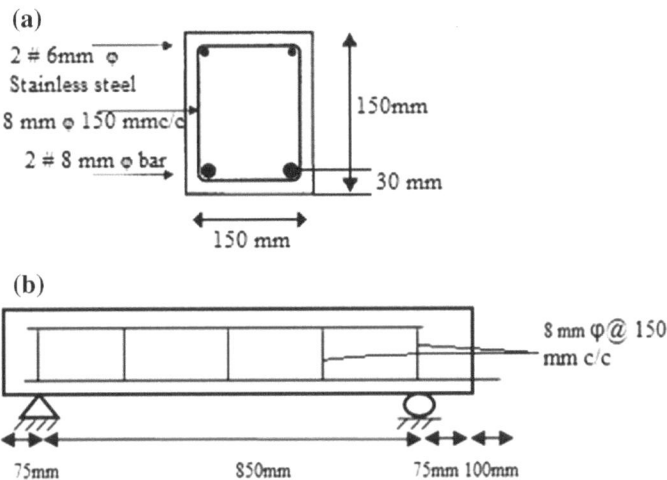

Fig. 1 Reinforcement detailing of the beam. **a** Cross section. **b** Longitudinal section

2.1 Materials Used

Ordinary portland cement of 53 grade cement was used. Locally available river sand conforming to zone iii used as fine aggregate (FA). Kapachi and Grit coarse aggregate (CA) of 2.7 specific gravity were used. Mix design is carried out as per the IS 10262: 2009 for M20 and M30 grade of concrete and Table 1 shows the mix proportion. Up to 75 mm M20, concrete is placed, and over that, M30 concrete is poured. CFRP and GFRP sheets were used to strengthen the beam. The properties of CFRP and GFRP are shown in Table 2. Nitowrap 30 is used as primer and Nitowrap 410 is used as saturant.

2.2 Resistivity Calculation

Electrical resistivity of concrete was measured by Resipod. Resistivity measurement can be used to assess the possibility of corrosion. Table 3 shows the resistivity of all beams. Electrical resistivity of concrete is less there will be less chance

Table 2 CFRP and GFRP properties

Properties	Nitowrap EP (CF)	Nitowrap EP (GF)
Weight of fiber (g/m^2)	450	920
Density of fiber (g/cc)	2.1	2.6
Fiber thickness (mm)	0.25–0.30	0.35
Fiber orientation	Unidirectional	Unidirectional
Tensile strength (N/mm^2)	4900	2300
Tensile modulus (N/mm^2)	240×103	7300

Table 3 Resistivity of beams

Beam designation	Resistivity (k Ω cm)	Risk of corrosion
B 0	112.5	Negligible risk
B 5	78.5	Low risk
B 10	38.6	Moderate risk
B 15	21.6	Moderate risk
B 5 Sc1	68.7	Low risk
B 5 Sc2	81.2	Low risk
B 5 Sc3	69.8	Low risk
B 10 Sc1	32.4	Moderate risk
B 10 Sc2	27.8	Moderate risk
B 10 Sc3	42.4	Moderate risk
B 15 Sc1	24.3	Moderate risk
B 15 Sc2	29.8	Moderate risk
B 15 Sc3	30.4	Moderate risk

of corrosion when it is more the likelihood of corrosion in the concrete will be more. Empirical test results analysis values are arrived and that can be used to determine the possibility of corrosion [8].

2.3 Strengthening Scheme

In the present investigation, three different types of schemes were proposed to study the behavior of the reinforced concrete corrugated beams as listed below and shown in Table 4.

Scheme 1: Long CFRP 850 × 150 mm is applied at the bottom surface of the beam longitudinally as shown in Fig. 2a.
Scheme 2: Long GFRP 850 × 150 mm is applied at the bottom surface of the beam longitudinally as shown in Fig. 2b and up to 40 mm in height sheets were wrapped in order to avoid the debonding.

Table 4 Strengthening schemes of RC beams

Sl. no	Beams	Corrosion (%)	Particular	Scheme
1	B 0	0	None	None
2	B 5	5	None	None
3	B 10	10	None	None
4	B 15	15	None	None
5	B 5 Sc1	5	1 long 850 × 150 mm CFRP sheets at the bottom	Scheme 1
6	B 5 Sc2	5	1 long 850 × 150 mm GFRP sheets at the bottom	Scheme 2
7	B 5 Sc3	5	1 long 850 × 150 mm CFRP sheets plus an U-shaped transverse GFRP sheet up to 100 mm height	Scheme 3
8	B 10 Sc1	10	1 long 850 × 150 mm CFRP sheets at the bottom	Scheme 1
9	B 10 Sc2	10	1 long 850 × 150 mm GFRP sheets at the bottom	Scheme 2
10	B 10 Sc3	10	1 long 850 × 150 mm CFRP sheets plus an U-shaped transverse GFRP sheet up to 100 mm height	Scheme 3
11	B 15 Sc1	15	1 long 850 mm CFRP sheets at the bottom	Scheme 1
12	B 15 Sc2	15	1 long 850 mm GFRP sheets at the bottom	Scheme 2
13	B 15 Sc3	15	1 long 850 × 150 mm CFRP sheets plus an U-shaped transverse GFRP sheet up to 100 mm height	Scheme 3

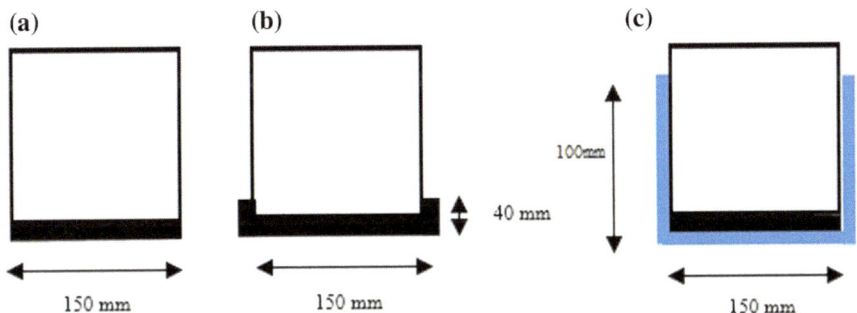

Fig. 2 Strengthening schemes of beams. **a** Scheme 1. **b** Scheme 2. **c** Scheme 3

Scheme 3: Long CFRP 850 × 150 mm is applied at the bottom surface longitudinally and in addition, GFRP sheet of 150 × 100 mm is wrapped as shown in Figs. 2c and 3, and subsequently, it will form a U hook which holds the premature failure by debonding.

B 0 Stands Beam with 0% corrosion, B 5 Sc1 state Beam corroded with 5% and strengthened with scheme 1, respectively.

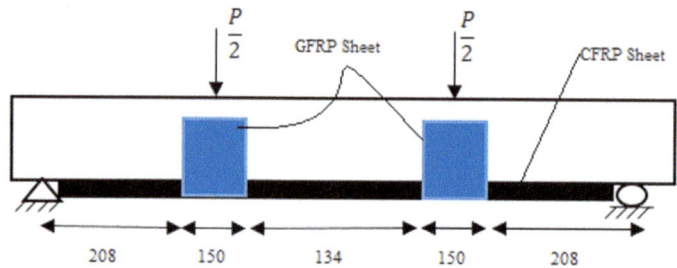

Fig. 3 Longitudinal section of scheme 3 strengthening method

2.4   Test Set up for Accelerated Corrosion

Impressed current technique is used to corrode the steel in the concrete, it also called as galvanostatic method. A constant current from DC source is applied to the steel present in the reinforced concrete to achieve the degree of corrosion. In present study, 4 amp power is supplied to the anode, steel bars will act as anode and stainless steel will act as cathode as shown in Fig. 4. Beams were placed in the pan and 5% of NaCl [1, 8] is added to acerbate the corrosion process. 1 amp consumes 1.04 g of iron using Faraday's law; the time for various degrees of corrosion were calculated and shown in Table 5.

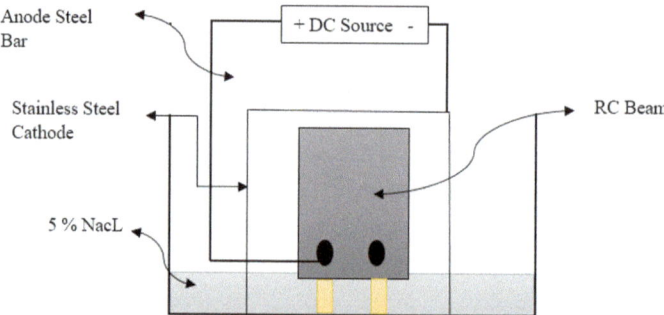

Fig. 4 Experimental set up for accelerated corrosion

Table 5 Time calculation to achieve various degrees of corrosion

Sr. No	Particulars	5%	10%	15%
1	Weight of steel in beams (g)	1105	1105	1105
2	Weight reduction (g)	55.25	110.5	165.75
3	Time required to corrode in hours	13.28	26.56	39.84

3   Results and Discussion

The beams were tested for flexure with two-point load as shown in Fig. 4. The load deflection curve for beams was studied and deflection is measured by dial gauge. Load deflection curves were compared with strengthened beam and conventional beams.

Figure 5a shows load deflection curve for controlled beams and corrugated beams with different degrees of corrosion, moreover it shows as increase in the percentage of corrosion the load carrying capacity is decreases and for B 15 shows the minimum load carrying capacity of the beams as the corrosion in the bar reduces the area of steel and which lead to decrease in the load carrying capacity of the beams. The ultimate strength of the B 5, B 10, and B 15 were 10, 20, and 40%

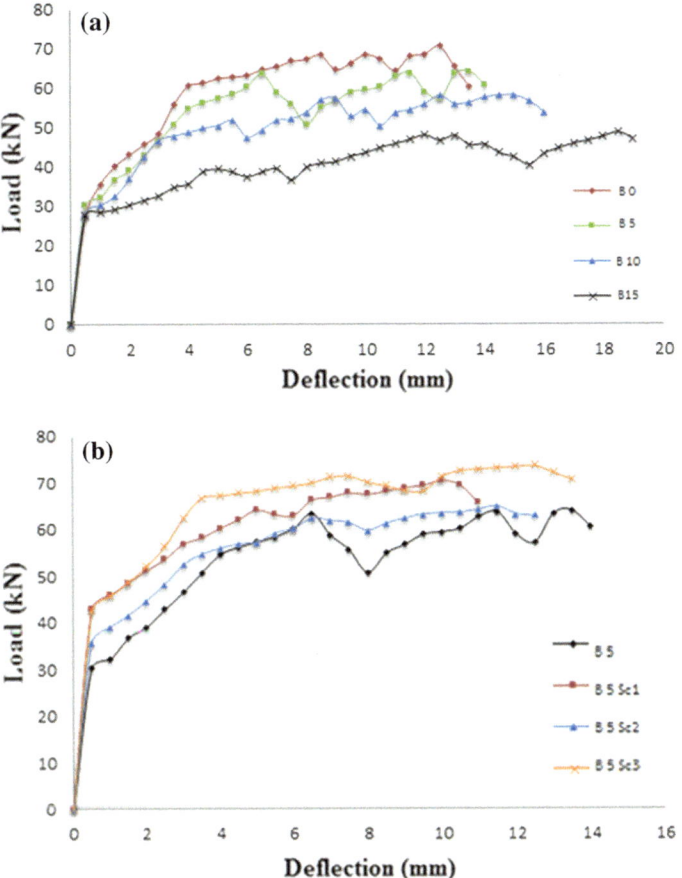

Fig. 5 Load deflection curve for beams. **a** Controlled beam with various degrees of corrosion. **b** 5% corroded beam with all three scheme

compared to control beam. Two longitudinal corrosion cracks were noted on the specimen before testing, each extending parallel to the length of the specimen. They were located on the bottom of the specimen, under the rebar and rust staining was noted along these cracks (Fig. 6). Generally, the corrosion cracks proceeded from the rebar to the soffit of the beam. No corrosion cracking was observed on the sides of the beam and beam fails under flexure as shown in Fig. 7b.

Figure 5b shows the load deflection curve for 5% corrosion and compared with three schemes, the ultimate strength of the B 5 Sc1, B 5 Sc2, and B 5 Sc3 were found 12, 6, and 14% compared to corroded B 5 beam; however, the specimen B 5 Sc1 failed due to debonding of CFRP sheet, B 5 Sc2 failed due to rupture of FRP sheets as shown in Fig. 7c, d, and e; moreover, the deflection in these beams were found 20, 10, and 8% compared to B 5 beam and in addition, the tensile strength

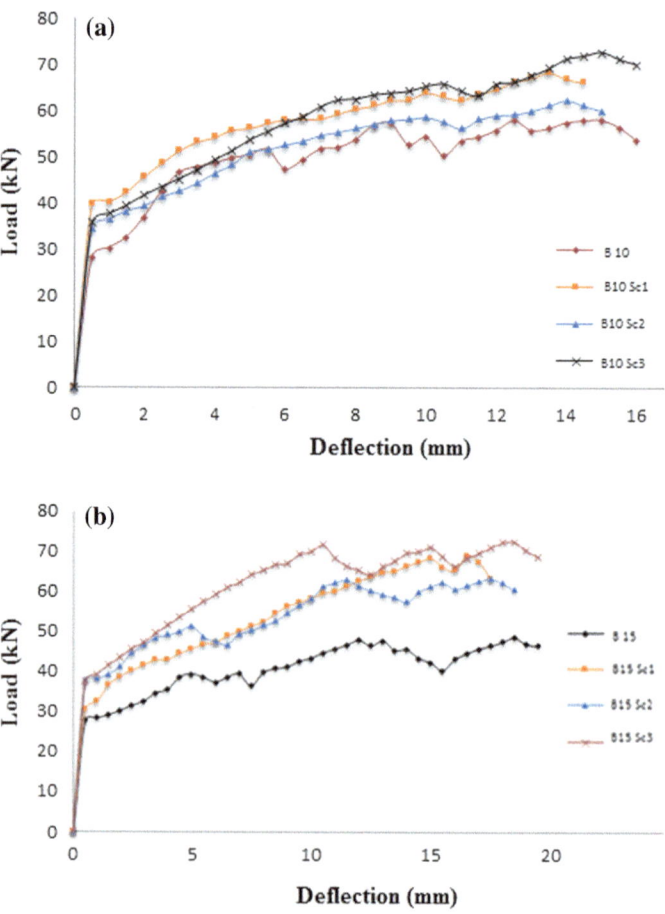

Fig. 6 Load deflection curve for beams. **a** 10% corroded beam with all three scheme. **b** 15% Corroded beam with all three scheme

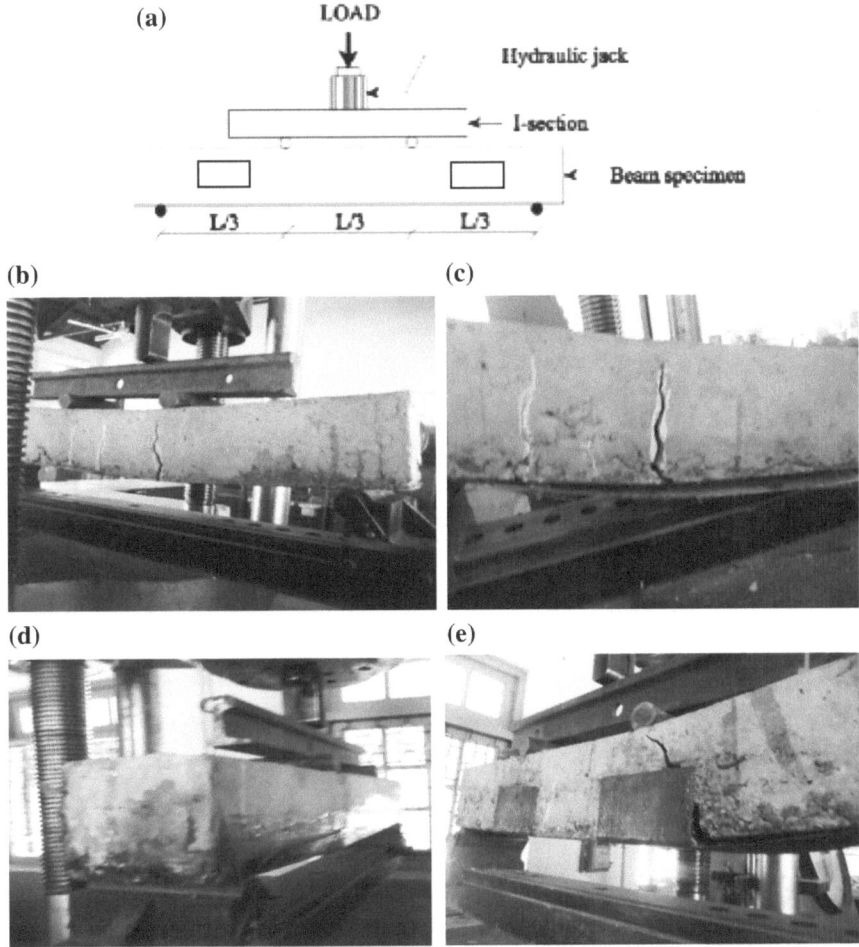

Fig. 7 **a** Experimental set up. **b** Flexure failure of controlled beam. **c** Debonding of scheme 1 beam. **d** Debonding of scheme 2. **e** Debonding and failure of scheme 3 beam

was found to be less in GFRP-coated beams as GFRP is having less tensile strength compared to CFRP. When used in combination, the ductility was increased and shown in Fig. 8.

Figure 6a shows the load deflection curve for 10% corrosion and compared with three schemes, the ultimate strength of the B 10 Sc1, B 10 Sc2, and B 10 Sc3 were found 20, 10, and 25% compared to corroded B 10 beam. Ultimate deflection of B 10 Sc1, B 10 Sc2, and B 10 Sc3 was 10, 5, and 2% less than the ultimate deflection of Specimen B 10. Yield and ultimate strength of specimens B 10 Sc1 and B 10 Sc3 are more than that of B 10 Sc2 and this is because the specimen B 10 Sc2 is wrapped only with GFRP sheet which has less tensile strength as compared to CFRP sheet and both CFRP and GFRP sheet together.

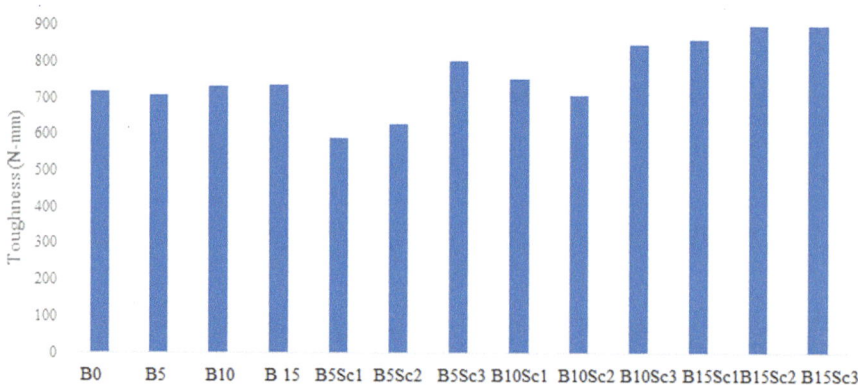

Fig. 8 Toughness of beam specimen

Figure 6b shows that ultimate strength of Specimens B 15 Sc1, B 15 Sc2, and B 15 Sc3 were 35, 25, and 45%, respectively, greater than the ultimate strength of Specimen B 15. Ultimate deflection of Specimens B 15 Sc1, B 15 Sc2, and B 15 Sc3 were 12, 5%, and 1.5%, respectively, less than the ultimate deflection of specimen B15. The load-carrying capacity of beam specimen B 15 Sc3 is more because it is wrapped with both CFRP and GFRP sheet and it undergoes large deflection, and it means that CFRP sheet prevents the brittle and sudden failure.

3.1 Toughness

Toughness is a measure of energy-absorbing capacity of the composite. Increased toughness means—improved performance in fatigue, impact and impulse loading and it also provides ductility, i.e., the ability to undergo larger shortenings before failure, it is often measured using a toughness index.

Trapezoidal rule is a piecewise quadratic approximation; it is generally more accurate compared to other methods. We take equal subdivisions and approximate f by a broken line of segments with end points $[a, f(a)]$, $[x1, f(x1)]\ldots[b, f(b)]$ on the curve 'f.' Then, the area under the curve 'f' between a and b is approximated by n trapezoid of areas,

$$1/2[f(a) + f(x1)]h, \; 1/2[f(x1) + f(x2)]h, \ldots 1/2[f(xn-1) + f(b)]h \qquad (1)$$

Taking their sum, we obtain trapezoidal rule

$$J = ab \int f(x)\mathrm{d}x \approx h[1/2f(a) + f(x1) + f(x2) + \cdots + f(xn-1) + 1/2f(b)]$$

$$(2)$$

where $h = (b-a)/n$, and xj, a, b are nodes.

Figure 8 shows the toughness of beam specimen for the above. It is observed that the beams strengthened with GFRP sheet shows high ductility index because the GFRP sheet has high ductility as compared to the CFRP sheet but the beams B 10 Sc2, B 10 Sc3, B 15 Sc2, and B 15 Sc3 show little increase in the ductility because of increase in corrosion rate.

3.2 Ductility Index

Ductility is a technical measure of the shortening of a structural member before it fails as per [9]. Ductility ratio can be defined as the ultimate deflection (δu) divided by the yield deflection (δy).

$$DI = u/\delta y \tag{3}$$

Figure 9 shows the ductility indices of all the beams which include control, control corroded, and strengthened with CFRP and GFRP sheets. The ductility indices of beam strengthened with scheme 1 for B 5 Sc1, B 10 Sc1, and B 15 Sc1 were 16, 43, and 40% less than the controlled beam B 0 and strengthened with scheme 2, B 10 Sc2 and B 15 Sc2 were 36 and 29% were less than controlled beam B 0, however, B 5 Sc2 was 4.4% greater than the controlled specimen B 0. Strengthened with scheme 3, B 10 Sc3 and B 15 Sc3 were 29 and 23.5% less than controlled beam B 0, although B 5 Sc3 was 25% greater than controlled specimen B 0.

It is observed that the beams strengthened with GFRP sheet shows high ductility index because the GFRP sheet has high ductility as compared to the CFRP sheet but the beams B 10 Sc2, B 10 Sc3, B 15 Sc2, and B 15 Sc3 show little increase in the ductility because of increase in corrosion rate.

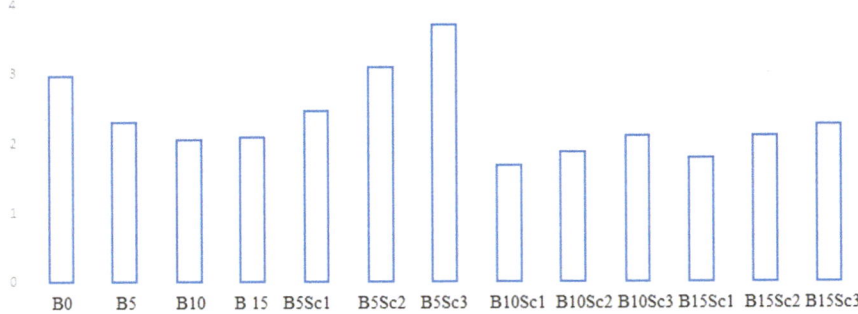

Fig. 9 Ductility index of all beams specimen

4 Conclusion

The following conclusion were drawn

(a) The degree of corrosion increases, the load-bearing capacity decreases. The 10, 20 and 40% variation is found for B 5, B 10 and B 15 corroded beams compared to the controlled beams. Specimens which are strengthen with schemes achieved the comparable strength with respect to controlled beam, however, the failure mode observed in strengthen beams is debonding of FRP sheets care should be taken to achieve more flexural strength.
(b) Beams strengthened with CFRP and GFRP, i.e., scheme 3 composites shown better strength compared to beam with scheme 1 and scheme 2 for all degrees of corrosion.
(c) Ductility indices for the 5% corrosion is found better as compared to 10 and 15% variation which shows the ability of materials undergo fracture, however, further precautions to be taken to improve the ductility for the higher degree of corrosion.
(d) Toughness of the all the corroded beams strengthened with scheme 3 found effective with in the range of 10% higher compared to controlled beam.

References

1. Al-Saidy AH, Al-Harthy AS, Al-Jabri KS, Abdul-Halim M, Al-Shidi NM (2010) Structural performance of corroded RC beams repaired with CFRP sheets. Compos Struct 92:1931–1938
2. Soudki KA (2006) FRP repair of corrosion-damaged concrete beams—Waterloo experience. In: Pandey M, Xie WC, Xu L (eds) Advances in engineering structures, mechanics & construction. Solid mechanics and its applications, vol 140. Springer, Dordrecht
3. Ahmed S (2009) Techniques for inducing accelerated corrosion of steel in concrete. Arab J Sci Eng 34:96–104
4. Austin SA, Lyons R, Ing MJ (2004) Electrochemical behavior of steel-reinforced concrete during accelerated corrosion testing. Corrosion 60(2):203–212
5. Maaddawy T, Chahrour A, Soudki K (2006) Effect of fiber-reinforced polymer wraps on corrosion activity and concrete cracking in chloride-contaminated concrete cylinders. J Compos Constr 10(2) April 2006
6. Mohammed T, Hamada H, Hasnat A, Al Mamun MA (2013) Corrosion of steel bars in concrete with the variation of microstructure of steel-concrete interface. J Adv Concr Technol 13:230–240
7. Bureau of Indian Standards (2009) Recommended guidelines for concrete mix design. (IS10262). New Delhi
8. Care S, Raharinaivo A (2007) Influence of impressed current on the initiation of damage in reinforced mortar due to corrosion of embedded steel. Cem Concr Res 37:1598–1612
9. Wang Y, Li J, Hamza AV, Barbee WT (2007) Ductile crystalline–amorphous nanolaminate. Proc Natl Acad Sci 104(27):11155–11160

Role of Particulate Matter on Air Quality Assessment of Delhi

Sanjoy Maji, Sirajuddin Ahmed, Santu Ghosh, Saurabh Kumar Garg and Tariq Sheikh

Abstract Ensuring clean air quality for the population is a priority consideration in many metro cities of the world particularly in middle- and low-income group South Asian cities. Urban population are exposed to various pollutants from vehicular and industrial sources. In a typical urban area, its population are exposed to more than 100 different chemical species. An extensive literature review has established the link between exposure to the classical pollutants and ill-health endpoints ranging from increases in asthma attacks, increases in acute bronchitis and decreased lung function to hospital admissions for respiratory-cardiovascular diseases and congestive heart failure. Communicating air quality through air quality index (AQI) system has become one of the major tools of air pollution information systems and is widely used for local and regional air quality management in many metro cities of the world. The metropolitan city of Delhi is the capital of India and is considered among one of the most polluted cities of the world. The study analyse the air quality for the City of Delhi, India with the help of AQI system proposed by EPA. A significant correlation was observed between air quality data with health data.

Keywords Air quality · Air quality index · Health effects · Respiratory mortality · Cardiovascular mortality

S. Maji (✉) · S. Ahmed · T. Sheikh
Faculty of Engineering and Technology, Jamia Millia Islamia (Central University),
Jamia Nagar, New Delhi 110025, India

S. Ghosh
Department of Biostatistics, St. Johns Medical College, Bangalore, India

S. K. Garg
AL-FALAH University, Dhauj, Faridabad, Haryana, India

© Springer Nature Singapore Pte Ltd. 2020
S. Ahmed et al. (eds.), *Smart Cities—Opportunities and Challenges*,
Lecture Notes in Civil Engineering 58,
https://doi.org/10.1007/978-981-15-2545-2_7

1 Introduction

Urban air pollution is recognized as a major threat to human health. In past few decades, research on air pollution health sciences has primarily focused on the linkages of daily mortality, hospital admission and visits to emergency department with fluctuation of air pollution level [1, 14, 23, 26]. The United Nations Environment Programme has estimated that globally 1.1 billion people breathe unhealthy air [25]. Estimates by the World Health Organization [28] put urban air pollution to be blamed for approximately 800,000 deaths and 4.6 million lost life-years each year around the globe. Worldwide many cities continuously monitor air quality scientifically designing ambient air quality monitoring network. The monitored pollutant concentrations are compared with national reference standards, which interpret air pollutants level as acceptable or not in comparison to the perspective reference standards. Air quality index (AQI) system is a different way of interpreting air pollution level which enables health risk communication of air pollution level for the common public. In AQI System, index values are proposed on the basis of potential health and environmental impacts of air pollutant concentrations. This is typically a numerical scale, intended to convey the likely severity of the adverse health effects at the monitored concentration levels. Health risk communication through AQI system serves several objectives, viz. to increase public awareness on likely health impacts of air pollution, to plan reduction in activities that aggravate air pollution, to aid sensitive groups (children, old age people and people with pre-existing cardio-respiratory diseases) to avoid exposure as well as assessing long-term pollution trends to increase awareness of the public health implications of air pollution.

1.1 Classification of Air Pollutants

Atmospheric pollutants have been classified according to their source and composition [4]. Followings are different ways of air pollutants classification.

1.1.1 Classification of Air Pollutants According to Origin

Primary pollutants: Pollutants which are directly emitted into the atmosphere and are found as such are called primary pollutants. e.g. nitrogen dioxide (NO_2), sulphur dioxide (SO_2), carbon monoxide (CO) and carbon dioxide (CO_2).

Secondary pollutants: Secondary pollutants are not directly emitted into the atmosphere rather they are derived from the primary pollutants due to chemical or photochemical reactions in the atmosphere are called secondary pollutants. e.g. peroxy acyl nitrate (PAN) and ozone (O_3).

1.1.2   Classification of Air Pollutants According to State of Matter

Gaseous pollutants: Pollutants which get mixed into the air and do not settle out. e.g. CO, CO_2, SO_2 and NO_2.

Particulate pollutants: These comprise of finely divided solids or liquids and often exist in colloidal state as aerosols. e.g. smog, dust, mist, fumes and spray.

1.2   Air Quality in Major Metropolitan Cities of the World

Ambient air quality is result of emission of primary air pollutants and meteorology-influenced atmospheric dispersion and transformation depending on ambient conditions. Some of the major factors of deteriorating air quality are increasing urbanization, growing vehicle use and increased energy demand. In many megacities, emission from vehicular sources is the primary causal factor of deteriorating air quality. Vehicular emissions are of particular concern since these are ground level sources and thus have the maximum impact on the general populations. Besides substantial CO_2 emissions, vehicular emissions give rise to different array of air pollutants: ozone precursors (CO, NOx, non-methane volatile organic compounds (NMVOC)), greenhouse gases (CO_2, CH_4, N_2O), acidifying substances (NH_3, SO_2), particulate matter (PM), carcinogenic species (poly-aromatic hydrocarbons (PAHs) and persistent organic pollutants (POPs)), toxic substances (dioxin, furans) and heavy metals. As per the estimates, the road transport alone emits around 13% of the global CO_2 emissions [15]. Ambient particulate matter concentration in some of the megacities of the world is shown in Fig. 1.

1.3   State of Air Quality—India

Like many other parts of the world, air pollution from motor vehicles is one of the most serious and rapidly growing problems in urban centres of India. A survey of 50 most highly polluted metropolitan cities of the world found 20 are in India [29]. Besides substantial CO_2 emissions, significant quantities of CO, HC, NOx, PM and other air toxins are emitted from these motor vehicles in the atmosphere, causing serious environmental and health impacts. Vehicles in major metropolitan cities are estimated to account for 70% of carbon monoxide (CO), 50% of hydrocarbon (HC), 30–40% of oxides of nitrogen (NOx) and 30% of particulate matter (PM) of the total pollution load of these cities [5].

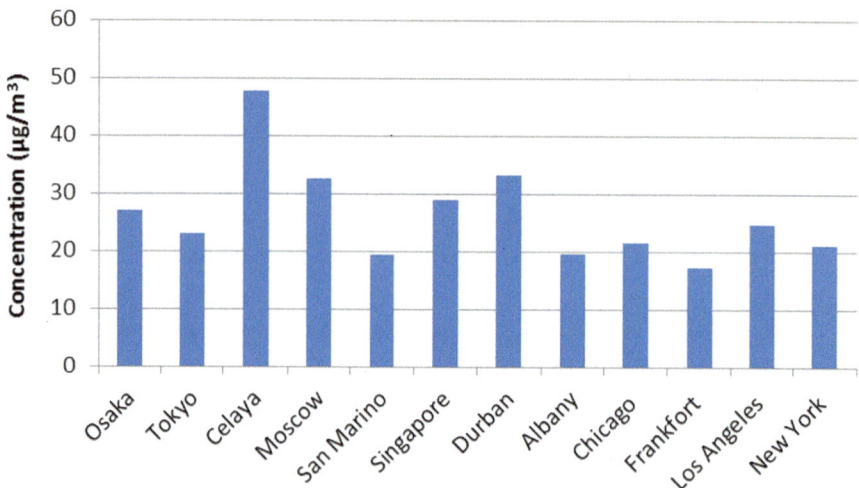

Fig. 1 Particulate air pollution concentration in some megacities of the world. *Source* www.who. int/phe

The rapid urbanization trend is likely to continue for the next two decades, and the majority of the country's population is expected to be living in cities within a span of next two decades. It is projected that about 40% of populations, i.e. about 600 million people would be living in urban areas by 2031 [22]. Simultaneously, number of cities with population of 1–10 million which is totalled at 32 in 2001 (as per Census of India, 2001) is estimated to increase to 85 by 2051 [22]. This urbanization trend together with industrialization, and economic growth is expected to lead to massive demand for road transportation [3]. The random urban development and unregulated increase of urban population have resulted in increased demands for transport, energy and other infrastructure which, in turn, have resulted in high levels of emissions. The national ambient air quality monitoring report published by CPCB, GoI [6] found almost all the metropolitan cities violate national ambient air quality standards (NAAQS) with respect to PM_{10} as well as NO_2. SO_2 concentration has been found to comply with NAAQS as well as WHO-AQG except Pune which violates the WHO-AQG. However, the capital city New Delhi violates NAAQS with respect to PM_{10} by a factor of 4. Air quality in some of the major metropolitan cities with respect to NAAQS and WHO-AQG [27] is shown in Figs. 2, 3 and 4.

1.4 State of Air Quality—Delhi

The national capital, Delhi, ranks among the worst polluted megacities of world [12]. Among the different sources of air pollutions, vehicular pollution accounts for 72% of total pollution load [13]. The vehicular population in Delhi has grown more

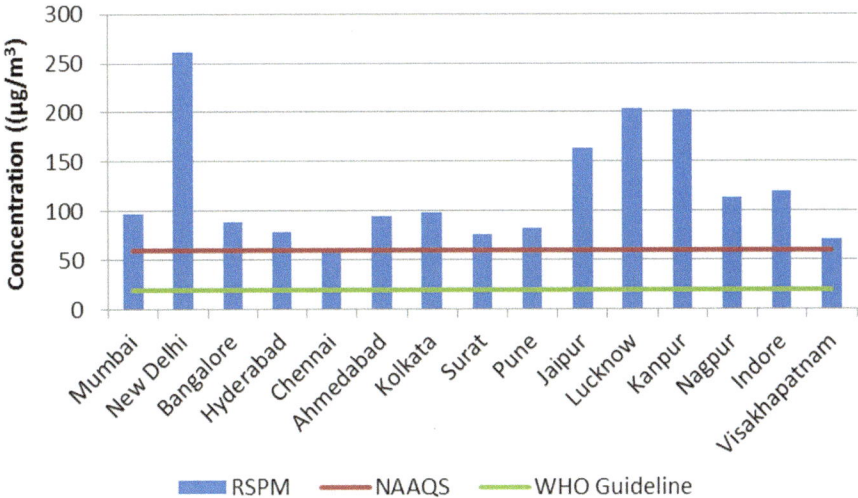

Fig. 2 RSPM concentration in some major metropolitan cities of India

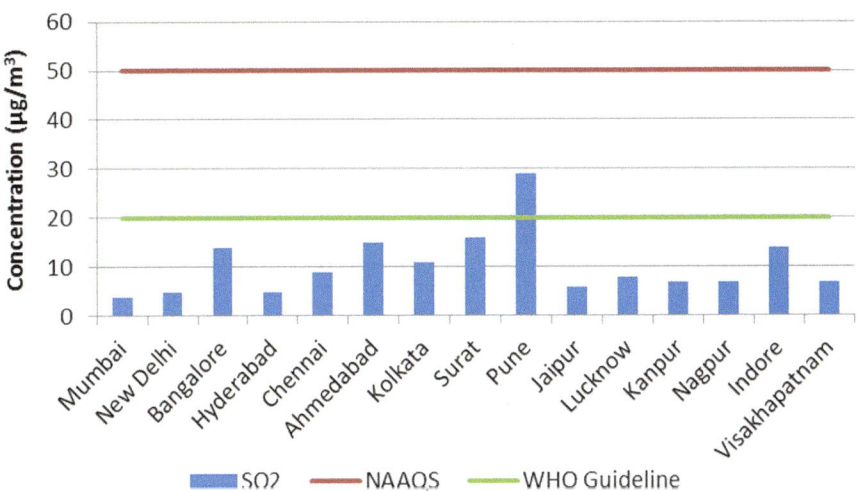

Fig. 3 SO_2 concentration in some major metropolitan cities of India

than 100% in the last decade (2000–2010), whereas petroleum and diesel consumption have grown by 400% and 300%, respectively, in the last decades (Delhi Statistical Abstract, various issues), Government of NCT of Delhi). Delhi alone accounts for about 8% of the total registered motor vehicles in India which is more than three other megapolitan (Mumbai, Kolkata and Chennai) taken together. In spite of several measures have been taken to control the vehicular pollution load in Delhi like introduction of emission standards, improvement of fuel quality,

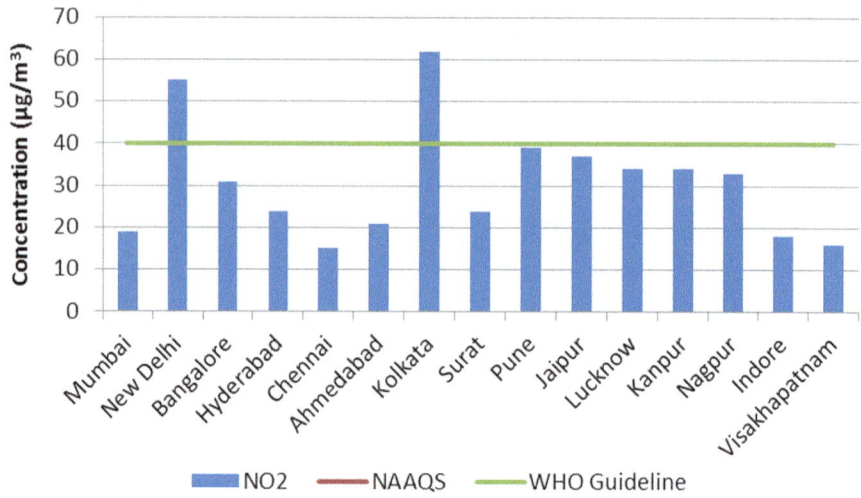

Fig. 4 NO₂ concentration in some major metropolitan cities of India

introduction of compressed natural gas (CNG) programme, phasing out old vehicles but still the problem of air pollution has not improved much in Delhi. Ambient air quality trend in different monitoring stations with respect to national ambient air quality standards and WHO-Air Quality Guideline [27] is shown in Figs. 5, 6 and 7. In all the stations, particulate matter concentration exceeds national ambient air quality standards by a factor of 4 or 5. Among the gaseous pollutants, though SO_2 concentration is found to vary very low bound but NO_2 concentration crosses national ambient air quality standards. Over the years, there is a decreasing trend for SO_2 but particulate matter and NO_2 concentration shows a gradual increasing trend.

2   The Historical Perspective of AQI

The National Ambient Air Quality Standards (NAAQS) forms the basis of AQI in USA [10]. It was started as an objective tool of NAAQS to be used by metropolitan areas with population of more than 350,000, voluntarily to report to the public on the status of air quality from its health point of significance. The USEPA-AQI considers air pollution data of five of the criteria pollutants, viz. particulate matter (PM_{10}, $PM_{2.5}$), sulphur dioxide (SO_2), ozone (O_3), nitrogen dioxide (NO_2) and carbon monoxide (CO) to regulate air quality in an urban area. Pollutant concentrations are integrated into a numerical range index of 0–500 which is further subdivided into six subindex classes corresponding to six categories of air quality based on their potential health significance. AQI scale values and their associated health effect statement are described in Table 1.

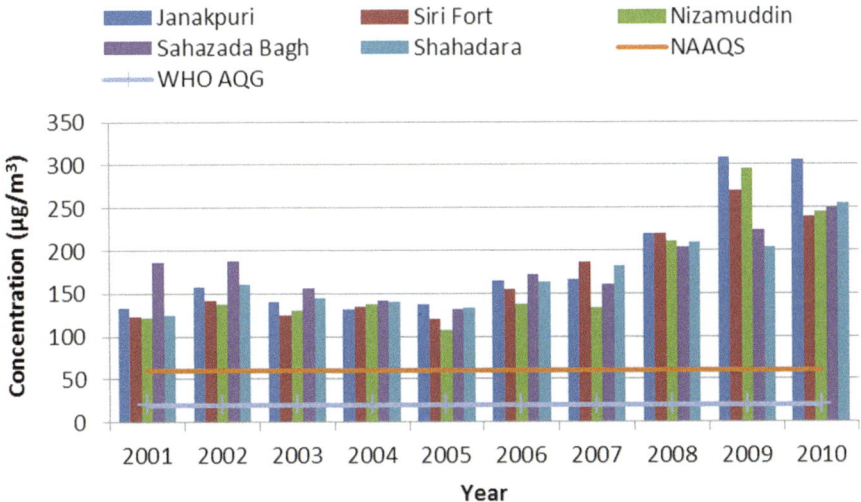

Fig. 5 Trend of RSPM concentration in air quality monitoring stations during 2001–2010 in Delhi

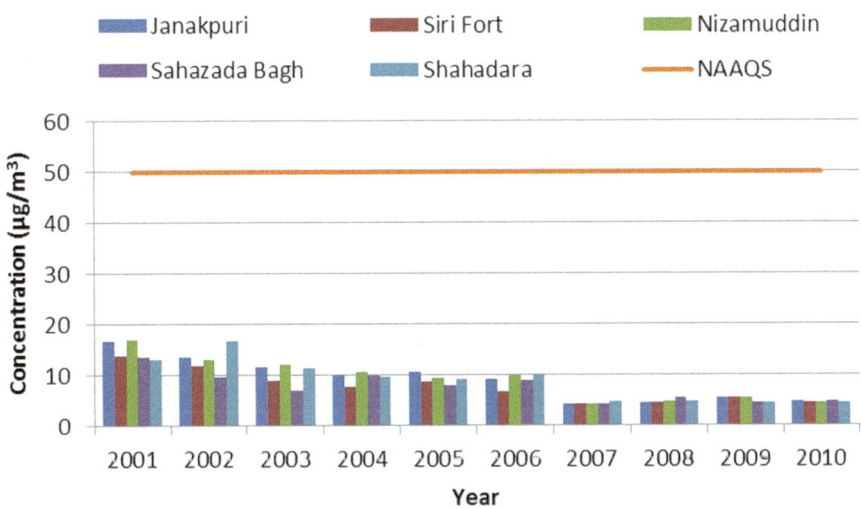

Fig. 6 Trend of SO_2 concentration in air quality monitoring stations during 2001–2010 in Delhi

AQI level 100 resembles health advisories as a precautionary measure, whereas AQI level 200 denotes "unhealthy" air for people with pre-existing cardio-respiratory diseases. At AQI level of 300, air pollution is "very unhealthy" for general population and Government Agencies to consider regulating industrial activities to limit operations. An AQI level of 300–500 resembles a

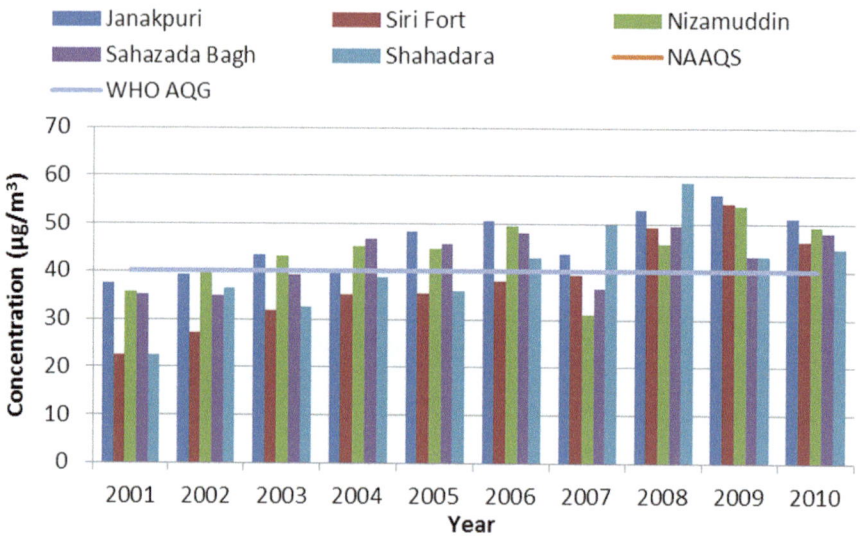

Fig. 7 Trend of NO$_2$ concentration in air quality monitoring stations during 2001–2010 in Delhi

Table 1 EPA air quality index values, levels of concern and health message

Air quality index (AQI) values	Levels of health concern	Health effect statement
0–50	Good	
51–100	Moderate	Sensitive individuals susceptible to respiratory symptoms
101–150	Unhealthy for sensitive groups	Increasing likelihood of respiratory symptoms in sensitive individuals; aggravation of heart or lung disease and premature mortality in sensitive group of population
151–200	Unhealthy	Increased aggravation of heart or lung disease and premature mortality in sensitive population; increased respiratory effects in general population
201–300	Very unhealthy	Sensitive population should avoid outdoor physical activity. General population should avoid prolonged exposure
301–500	Hazardous	Everyone should avoid all physical activity outdoors; sensitive group of population should remain indoors and keep activity levels low

Source [11]

"hazardous" category of air quality where air pollution generating industries, including transport system, require complete cessation. Pollutant-specific sub-indices for different ranges of air quality index (AQI) values have been described in Table 2. Each subindex is associated with a rage of pollutant concentration; on a given day, the pollutant concentration which gives the maximum AQI value is the AQI of that day.

3 Some Important AQI System

There is no universally accepted AQI system; different index systems are in vogue in different parts of the world. AQI system may be different with respect to selection of air pollutants, pollutant exposure matrix, AQI calculation technique as well as descriptors of the risk levels corresponding to index values which facilitates health risk communication.

CAQI (EU), 2008: To have a common index system for all the European cities, van den Elshout et al. [9] proposed common air quality index (CAQI) system. The CAQI has two different index systems, one is for traffic and other one is for city background concentration. The traffic index considers NO_2 and PM_{10} as the main criteria air pollutants with CO as a secondary component, whereas the city index comprises NO_2, PM_{10} and O_3 as primary criteria air pollutants with CO_2 and SO_2 as auxiliary component. The computation of CAQI is based on maximum operator concept likewise USEPA-AQI system though the choice of the classes for the CAQI is inspired by the European Commission legislation. In CAQI, the index scale has been formulated that resemble four categories of air quality ranging from "Very low" to "High" pollution category (Table 3).

Long-term air quality index (LAQx): Mayer et al. [20] have proposed long-term air quality index (LAQx) for Germany. In contrast to the daily air quality index system, long-term air quality index (LAQx) system considers "annual mean values" of benzene, NO_2, PM_{10} and SO_2 for the construction of index scale. Annual mean concentration of the pollutants has been subdivided into six classes as per German School grade system to obtain a reference to well-being and health of people.

Table 2 Breakpoints of USEPA-AQI

Index	8 h O_3 (ppm)	I hr O_3 (ppm)	24 h PM_{10} (μg/m3)	24 h $PM_{2.5}$ (μg/m3)	8 h CO (ppm)	1 h SO_2 (ppb)	1 h NO_2 (ppb)
0–50	0.000–0.059	–	0–54	0.0–12.0	0.0–4.4	0–35	0–53
51–100	0.060–0.075	–	55–154	12.1–35.4	4.5–9.4	36–75	54–100
101–150	0.076–0.095	0.125–0.164	155–254	35.5–55.4	9.5–12.4	76–185	101–360
151–200	0.096–0.115	0.165–0.204	255–354	55.5–150.4	12.5–15.4	186–304	361–649
201–300	0.116–0.374	0.205–0.404	355–424	150.5–250.4	15.5–30.4	305–604	650–1249
301–400		0.405–0.504	425–504	250.5–350.4	30.5–40.4	605–804	1250–1649
401–500		0.505–0.604	505–604	350.5–500.4	40.5–50.4	805–1004	1650–2049

Source USEPA, 2013

Table 3 Breakpoints of the CAQI of Europe

Index class	Grid	Traffic				City background					
		Mandatory pollutant			Auxiliary pollutant	Mandatory pollutant			Auxiliary pollutant		
		NO_2	PM		CO	NO_2	PM		O_3	CO	SO_2
			1-hour	24-hour			1-hour	24-hour			
Very low	0	0	0	0	0	0	0	0	0	0	0
	25	50	25	12	5000	50	25	12	60	5000	50
Low	26	51	26	13	5001	51	26	13	61	5001	51
	50	100	50	25	7500	100	50	25	120	7500	100
Medium	51	101	51	26	7501	101	51	26	121	7501	101
	75	200	90	50	10,000	200	90	50	180	10,000	300
High	76	201	91	51	10,001	201	91	51	181	10,001	301
	100	400	180	100	20,000	400	180	100	240	20,000	500
Very high	>100	>400	>180	>100	>20,000	>400	>180	>100	>240	>20,000	>500

Source [9]

AQI based on combined effects of air pollutants: Cairncross et al. [8] have proposed AQI system based on the relative risk of increased daily mortality associated with short-term exposure to common air pollutants: particulate matter (PM_{10}, $PM_{2.5}$), sulphur dioxide, ozone, nitrogen dioxide and carbon monoxide. The subindex values for each pollutant is based on published relative risk factors, in the range of air pollutant concentrations commonly experienced in urban areas. The final AQI is the sum of the normalized values of the individual indices for PM_{10}, $PM_{2.5}$, sulphur dioxide, ozone, nitrogen dioxide and carbon monoxide.

For synergistic effects of different pollutants, Murena [21] proposed a daily air quality index system by summing the ratios of the reference daily concentration of pollutant and the bottom breakpoint concentration corresponding to the pollutant category. However, the above-mentioned procedure could led to overestimation of the actual index since additive effects can be assumed if pollutants belong to the same category and have similar effects on human health. To overcome the problem (eclipscity), Kyrkilis et al. [16] proposed an aggregate AQI using weighting factor for aggregating the individual index of the pollutants:

$$I = \sum_{i=1}^{n} ((AQI_i)^{\rho})^{1/\rho}$$

where I = the overall air pollution index, AQI_i = is the AQI for a single pollutant i and $\rho = a$ constant.

The main objective of the air quality index system is to measure the air pollution with respect to its effects on the human health.

The link of air quality and ill-health impact has been worked by quite a few researchers for Delhi [2, 16–19, 24]. Therefore, an attempt has been made to analyse the air pollution status of Delhi through use of air quality index system and to correlate air pollution with sanitary data for the city of Delhi.

4 Materials and Methods

4.1 AQI Calculation

The mathematical calculation for calculating subindices was adopted after considering health criteria and subindex values of USEPA:

$$I = \frac{I_{high} - I_{low}}{C_{high} - C_{low}} (C - C_{low}) + I_{low}$$

where I = the (air pollution) index, C = the pollutant concentration, C_{low} = the concentration breakpoint that is $\leq C$, C_{high} = the concentration breakpoint that is $\geq C$, I_{low} = the index breakpoint corresponding to C_{low} and I_{high} = the index breakpoint corresponding to C_{high}.

The daily air quality index values for the city of Delhi for the year 2006, 2007, 2008 and 2009 were evaluated by the monitoring data of the three pollutants: the 24 h average concentrations of RSPM (PM_{10}), oxides of nitrogen (NOx) and sulphur dioxide (SOx). For calculating the air pollution index, the monitoring stations selected were Janakpuri, N. Y. School, Nizamuddin, Pitampura, Siri Fort and Town Hall. The air quality of all the above-mentioned monitoring stations is regularly monitored by the Central Pollution Control Board (CPCB) and National Environmental Engineering Research Institute (NEERI) under the National Air Quality Monitoring Programme (NAMP). The details of monitoring stations are presented in Table 4.

4.2 Population and Health Data

For the correlation study of the AQI values with the respiratory health symptoms of Delhi, the data on respiratory health status of Delhi were taken from the "Epidemiological Study on Effect of Air Pollution on Human Health (adults) Delhi [7]".

4.3 Results and Discussion

It was found that respirable suspended particulate matter (RSPM) is the main criteria pollutants in the determination of air quality index value. The average air quality value of the Delhi stands for "unhealthy for sensitive groups"; however, sometimes air quality index value also touches the "hazardous category". Different air quality categories observed at different air quality monitoring stations are shown in Fig. 8. Hazardous category of air quality was observed in Janakpuri, Nizamuddin and Town Hall monitoring station.

General trend of pollution during winter time (November–February) reaches maximum concentration. However, minimum index values are noticed during rainy season (July–September). Summer months (April–June) also show higher degree of pollution loads. High pollution loads during winter is due to reduced dispersion on

Table 4 Description of different monitoring station

Name of monitoring station	Area
Janakpuri	West Delhi
N. Y. school	South Delhi
Nizamuddin	Central Delhi
Pitampura	North Delhi
Siri fort	South Delhi
Town hall	North Delhi

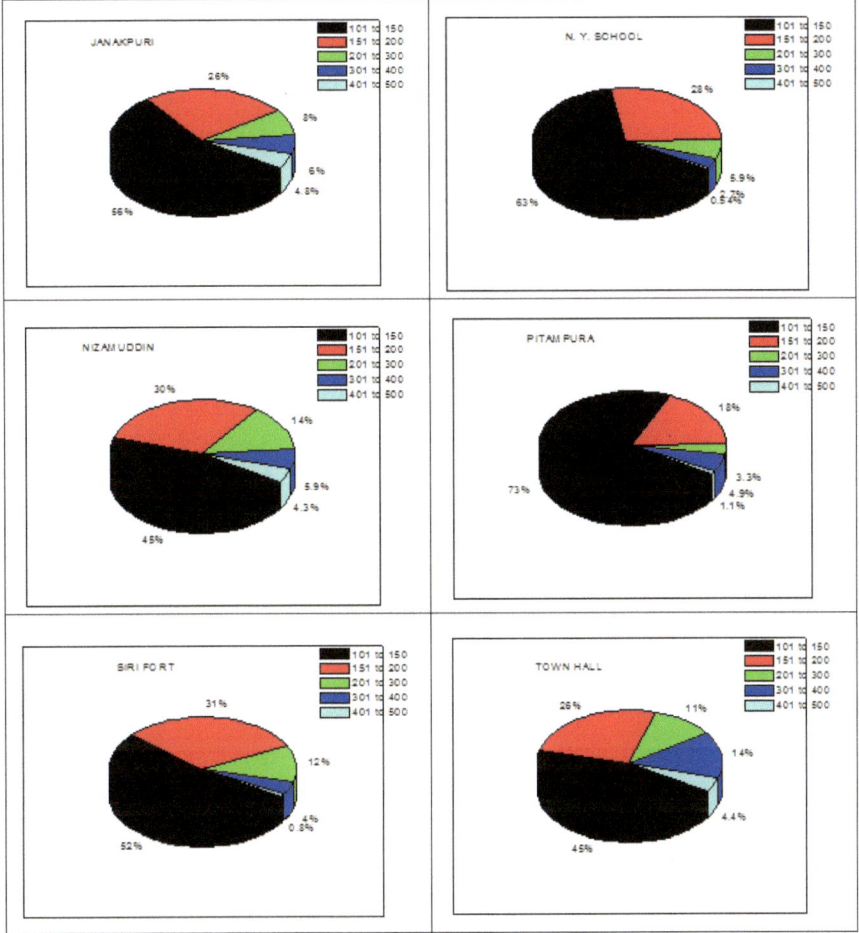

Fig. 8 AQI trend at individual monitoring station for the period 2006–2009

account of low wind velocity, whereas high pollution loads in summer can be attributed to dust storm and higher wind velocity. During monsoon period, because of large precipitation, low levels of pollution are observed. Temporal distribution of air quality is shown in Fig. 9.

Different respiratory health symptom shows a positive correlation with the AQI values. The lower respiratory symptoms (e.g. dry cough, cough with phlegm, chest discomfort, etc.) prevailing among the residents of Delhi shows a statistically significant relation (Table 5) with the AQI subindex values. The percentage of population with reduced lung function with AQI value also shows a positive correlation.

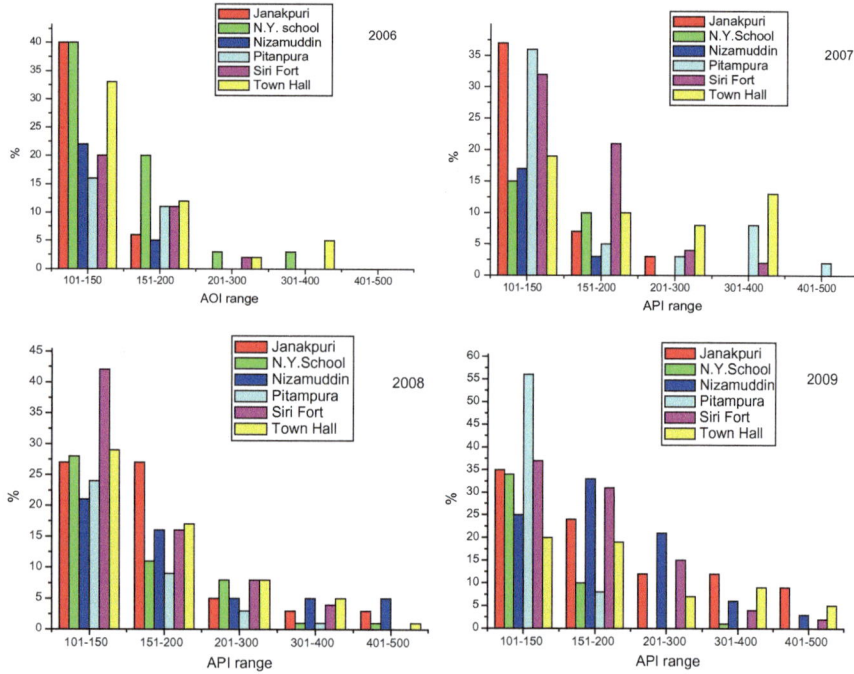

Fig. 9 Temporal distribution of air quality category during 2006–2009

Table 5 Prevalence (%) of lower respiratory symptoms in the population of Delhi

Station	RSPM Concn ($\mu g/m^3$)	API value	LRS (%)	Dry cough (%)	Cough with phlegm (%)	Chest discomfort (%)	Breathlessness (%)
Control	82.5	64.6	12.7	4.2	4.6	3.1	4.8
North	140.1	93	26.7	8.70	10.5	6.3	11.2
South	140.5	93	18.3	5.4	7.2	5.3	8.2
East	132.1	89	21.6	7.2	8	6.4	10.2
West	181.4	114	24.3	7.8	9.3	6.8	11
Central	186.9	117	23.0	8	9.5	6.8	10.1

Source [7]

5 Conclusion

The study has been conducted to show the air quality status of Delhi in relation to the EPA-AQI scale values and the inter-relationship of AQI subindex values with the respiratory health prevailing among the population of Delhi. Among the different criteria pollutants, we find SO_2 and NO_2 values are within the NAAQS standards limit; however, RSPM concentration varies widely among different stations in respect to different seasons as well as among the stations. Among the different stations, the AQI value of N. Y. School, Pitampura and Siri Fort varies from "unhealthy sensitive for group" to Unhealthy condition; it seldom reaches "hazardous" category. However, the AQI value at Town Hall and Nizamuddin touches the hazardous category range. The statistically significant association of the AQI value with the respiratory health symptom of the population of Delhi was observed.

References

1. Anderson HR, Spix C, Medina S, Schouten JP, Castellsague J, Rossi G, Bacharova L (1997) Air pollution and daily admissions for chronic obstructive pulmonary disease in 6 European cities: results from the APHEA project. Eur Respir J 10(5):1064–1071
2. Balakrishnan K, Ganguli B, Ghosh S, Sankar S, Thanasekaraan V, Rayudu VN, Caussy H (2011) Part 1. Short-term effects of air pollution on mortality: results from a time-series analysis in Chennai, India. Res Rep (Health Eff Inst) 157:7–44
3. Barnett, M (2011) New citizen forester program plans to grow tree Stewards. June 8. Online Available at http://nooga.com/152155/new-citizen-forester-program-plans-to-grow-tree-stewards. Accessed 30 Aug 2015
4. Bernstein JA, Alexis N, Barnes C, Bernstein IL, Nel A, Peden D, Williams PB (2004) Health effects of air pollution. J Allergy Clin Immunol 114(5):1116–1123
5. CPCB (Central Pollution Control Board) (2010) Programme Objective Series PROBES/136/2010 Status of the Vehicular Pollution Control Programme in India.
6. CPCB (Central Pollution Control Board) (2012a) National ambient air quality monitoring series NAAQMS/35/2011–2012. Available at www.cpcb.nic.in
7. CPCB (Central Pollution Control Board) (2012b) Epidemiological study on effect of air pollution on human health (adults) in Delhi. EHMS/01/2012. http://cpcb.nic.in/upload/NewItems/NewItem_188_Epidemiological_study_AP_Report.pdf
8. Cairncross EK, John J, Zunckel M (2007) A novel air pollution index based on the relative risk of daily mortality associated with short-term exposure to common air pollutants. Atmos Environ 41:8442–8454
9. van den Elshout S, Léger K, Nussio F (2008) Comparing urban air quality in Europe in real time: a review of existing air quality indices and the proposal of a common alternative. Environ Int 34:720–726
10. Environmental Protection Agency (EPA) (2009) Technical assistance document for the reporting of daily air quality—air quality index (AQI). EPA-454/B-09-010, Office of air quality planning and standards, Research Triangle Park, NC 27711
11. Environmental Protection Agency (EPA) (2013) Technical assistance document for the reporting of daily air quality—air quality index (AQI). Environmental protection agency, office of air quality planning and standards, Research Triangle Park, NC 27711
12. Gurjar BR, Jain A, Sharma A, Agarwal A, Gupta P, Nagpure AS, Lelieveld J (2010) Human health risks in megacities due to air pollution. Atmos Environ 44(36):4606–4613
13. Guttikunda SK, Goel R (2013) Health impacts of particulate pollution in a megacity—Delhi, India. Environ Dev 6:8–20
14. HEI International Oversight Committee (2004) Health effects of outdoor air pollution in developing countries of Asia: a literature review. Health Effects Institute, Boston, MA, (Special Report No. 15)

15. IPCC (2007) Transport and its infrastructure in climate change 2007: mitigation. Contribution of Working Group III to the fourth assessment report of the Intergovernmental Panel on climate change. http://www.ipcc.ch/pdf/assessment-report/ar4/wg3/ar4-wg3-chapter5.pdf
16. Kyrkilis G, Chaloulakou A, Kassomenos PA (2007) Development of an aggregate air quality index for an urban mediterranean agglomeration: relation to potential health effects. Environ Int 33:670–676
17. Maji S, Ahmed S, Siddiqui WA (2015) Air quality assessment and its relation to potential health impacts in Delhi,India. Curr Sci 109(5):902–909
18. Maji S, Ahmed S, Siddiqui WA, Ghosh S (2017) Short term effects of criteria air pollutants on daily mortality in Delhi, India. Atmos Environ 150:210–219
19. Maji S, Ghosh S, Ahmed S (2018) Association of air quality with respiratory and cardiovascular morbidity rate in Delhi, India. Int J Environ Health Res 28(5):471–490
20. Mayer H, Holst J, Schindler D, Ahrens D (2008) Evolution of the air pollution in SW Germany evaluated by the long-term air quality index LAQx. Atmos Environ 42(20):5071–5078
21. Murena F (2004) Measuring air quality over large urban areas: development and application of an air pollution index at the urban area of Naples. Atmos Environ 38:6195–6202
22. Planning Commission (2011) Recommendations of Working Group on urban transport for 12th five year plan. Planning Commission, Government of India, New Delhi
23. Pope CA, Burnett RT, Thurston GD, Thun MJ, Calle EE, Krewski D, Godleski JJ (2004) Cardiovascular mortality and long-term exposure to particulate air pollution epidemiological evidence of general pathophysiological pathways of disease. Circulation 109(1):71–77
24. Rajarathnam U, Sehgal M, Nairy S, Patnayak RC, Chhabra SK, Ragavan KV (2011) Part 2. Time-series study on air pollution and mortality in Delhi. Res Rep (Health Eff Inst) 157:47–74
25. UNICEF, United Nations Children's Fund & World Health Organization (2002) Children in the new millennium: environmental impact on health. UNEP/Earthprint.
26. WHO (2003) Health aspects of air pollution with particulate matter, ozone and nitrogen dioxide. Report on a WHO Working Group Bonn, Germany, 13–15 Jan 2003. WHO Regional Office for Europe, Copenhagen
27. World Health Organization (2006) WHO air quality guidelines for particulate matter, ozone, nitrogen dioxide and sulfur dioxide: global update 2005: summary of risk assessment
28. World Health Organization (WHO) (2014a) Burden of disease from ambient air pollution for 2012. World Health Organization
29. World Health Organization (2014b) WHO global urban ambient air pollution database (update 2016). http://www.who.int/phe/health_topics/outdoorair/databases/cities/en/. Accessed on 26 June 2015

Landslide Susceptibility Mapping Along Highway Corridors in GIS Environment

Sandeep Panchal and Amit Kr. Shrivastava

Abstract The hilly regions are developing at a very rapid rate. The anthropogenic activities due to road construction increase the instability of slopes along highways. Aim of this study is to prepare a landslide susceptibility map along State Highway 32. Landslide susceptibility maps along road section prove a good tool for effective mitigation and management of the landslide hazards. The parameters considered in this study are slope, aspect, elevation, drainage density, lithology, soil and distance from fault. Analytic hierarchy process (AHP) is used for evaluating various parameters and ranking them. Landslide Susceptibility Index (LSI) is calculated by using weighted linear combination (WLC) technique. The final landslide susceptibility map is divided into four categories from low to very high susceptibility zones. It is found that around 65% of the area lies under high and very high landslide susceptibility. The results of the study can be used by the urban planners, transportation planners and highway engineers.

Keywords Analytic hierarchy process · Landslides · Susceptibility mapping · Weighted linear combination

1 Introduction

Highways in a city play a very important role in its development. The transportation network is considered as lifeline of a city. Landslide is a common phenomenon along highways in the hilly regions. So, the disaster like landslides should be planned carefully in a smart city. Safety of transportation infrastructure is a matter of concern in the smart cities. The landslide susceptibility maps can play a major role in planning and mitigation of landslide events which take place along the highways. The landslide susceptibility maps are used for the safe and economical route planning of transportation infrastructure in a smart city.

S. Panchal (✉) · A. Kr.Shrivastava
Delhi Technological University, Delhi, India

© Springer Nature Singapore Pte Ltd. 2020
S. Ahmed et al. (eds.), *Smart Cities—Opportunities and Challenges*,
Lecture Notes in Civil Engineering 58,
https://doi.org/10.1007/978-981-15-2545-2_8

79

Many researchers have defined landslide in many ways. According to a research paper published by Cruden in 1991, a landslide is a downward movement of material under the effect of gravity. This was the first technical definition in the literature. Landslides are a type of "mass wasting", which denotes any down-slope movement of soil and rock under the direct influence of gravity (Highland; 2008). Highway construction projects involve a huge amount of cutting and filling work. Whenever a new road is constructed or an existed road is upgraded in hilly areas, the slopes along highways get disturbed. Due to this disturbance, there is a danger of slope failures. The mass movement of failed slopes is called landslides. Landslides are very common along highways and are responsible for economic losses [1]. A landslide along a highway section can block the traffic resulting in delay and discomfort to the passengers. Slope failures along highways are instant phenomenon that can even cause losses of life in adverse conditions [2].

The landslides are responsible for the huge economic losses which can be counted in monetary form. The losses due to landslides also include the traffic jams, delay in journey, minor injuries and deaths. These losses can't be calculated in the monetary form. The economic impact of landslides can be direct and indirect. The direct impact can be calculated easily. The debris and the failed material come to the road. The direct cost of cleaning and the maintenance of roads can be calculated. The cost of damaged vehicles and the accident cost can also be calculated. The landslides can cut off the mode of communication in the areas which are situated remotely. The houses which are situated in landslide-prone areas can be damaged partially or fully. The differential settlement of the foundations can take place. The cost of damage and cost of maintenance of affected infrastructure can be calculated. The triggering factors like earthquake and rainfall can increase the cost of damaged infrastructure.

Landslide susceptibility mapping is an important tool for planning the mitigations to the landslide hazards [3]. A landslide susceptibility map helps in understanding the chances of occurrence of landslides in different parts of the study area. The study area is divided in some parts on the basis of their susceptibility towards landslides [4]. A lot of methods are proposed by various researchers for mapping the landslide susceptibility of different areas. There are many qualitative, quantitative and semi-quantitative techniques available for landslide mapping. The direct methods of landslide susceptibility mapping are easy to implement. The methods which involve human expertise are known as expert methods [5]. The methods like analytic hierarchy process (AHP) and analytic network process (ANP) are improved form of human-driven approaches [6]. There are many data-driven approaches also. The mathematical models are used for forecasting the landslides. The statistical techniques give good results in case of landslide susceptibility mapping. The probabilistic approach is based upon the theory of probability which is considered good for landslide susceptibility mapping at regional level [7]. The methods like fuzzy logic and artificial neural network are new methods which can give highly accurate results [8].

Geographic information system (GIS) is an effective tool which is used to store, capture, analyse, query, manage and manipulate the data [9]. The GIS is used to show the cumulative effect of various factors. The main advantage of GIS is that the input can be varied to vary the output. So, it proves very effective tool for the planners. A landslide susceptibility map shows the cumulative effect of its causative factors. The parameters which are responsible for the occurrence of landslide event are slope, aspect, curvature, drainage and hydro-geological characteristics of the area. GIS proves very effective in finding the cumulative effect of these factors.

The proposed study area lies between 76°15′ and 76°35′ longitude and 31°28′ and 31°37′ latitude in hilly terrain of lower Shiwalik in Himachal Pradesh (H.P.) from Una to Bhota. The topography and geology of the area are such that it is prone to landslides. The proposed road section is upgraded from one lane to two lanes, recently. This disturbance in natural slope had made the slopes along roadside highly susceptible to failures. The road stretch is 67 km long, and the landslide susceptibility mapping is done along this road. The method used for the purpose of landslide mapping is analytic hierarchy process (AHP).

2 Parameters of Landslides

Seven parameters those influence the occurrence of landslides are considered in this study. These seven parameters are slope, aspect, plan curvature, drainage density, lithology/geology, soil type and distance from fault which are considered. Eighth layer is landslide inventory that is prepared by field survey. The landslide inventories give an idea of relation of occurrence of landslides and the causative factors. A landslide inventory can also be prepared from Google Earth. The field surveys can be done for preparation of event-based landslide inventories.

2.1 Lithology

Occurrence of landslides depends upon the lithology and geology of the area. The study area consists of mainly sandstones with shale, sandstone with clay and boulder conglomerate along the road section. A small area is covered with the alluvium soil near rivers. Lithology is given weightage 0.132 according to analytic hierarchy process (AHP). It is found according to the landslide inventory maximum number of landslides occurred on the sandstone with clay type of lithology. For observing the effect of lithology of the area on the occurrence of landslides, the study area is divided into four regions. Figure 1 shows lithology along SH-32.

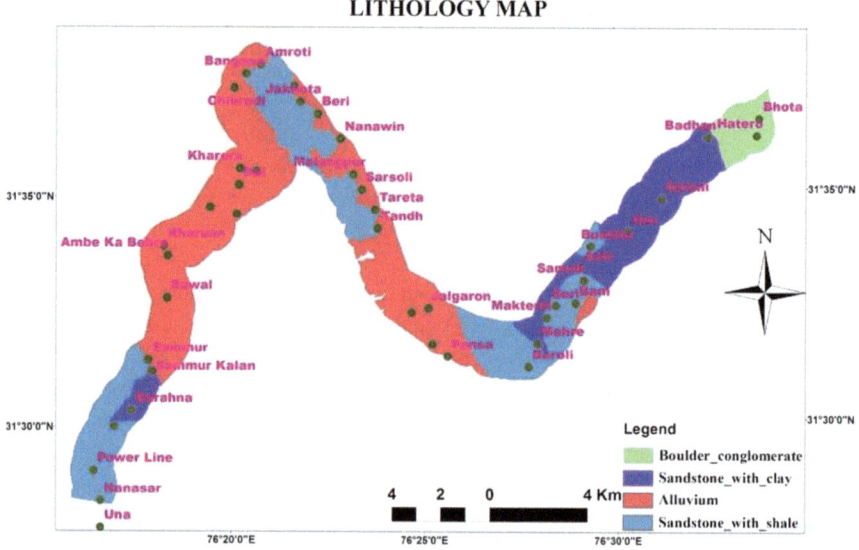

Fig. 1 Lithology characteristics

2.2 Soil Type/Geology

Occurrence of landslides also depends upon the soil characteristics of the particular area. The study area consists of mainly loamy soil with fine or coarse grains. Also, the area has some parts with sandy soil. Most of the landslides occur in the loamy soils. The loamy soils whenever come in contact of water, get weaker and tend to failure. Hence, the study area is divided into two type of soils, i.e. sandy and loamy. Maximum number of landslides took place in area with loamy soil. The AHP ranking of loamy soil is taken 0.8. Figure 2 shows the type of soil along SH-32.

2.3 Slope/Aspect

Slope angle is the most important factor that affects the slope stability. Slope angle may be defined as the rate of change of elevation w.r.t change in distance. So, landslide susceptibility of an area depends upon the slope angle (Fig. 3).

Aspect may be defined as the direction of steepest slope. Figure 4 shows the aspect along State Highway 32 (SH-32). The aspect affects the stability of slopes when these are subjected to high rainfall, weathering or other environmental factors. The slopes with aspect South (S), Southeast (SE) and Southwest (SW) direction.

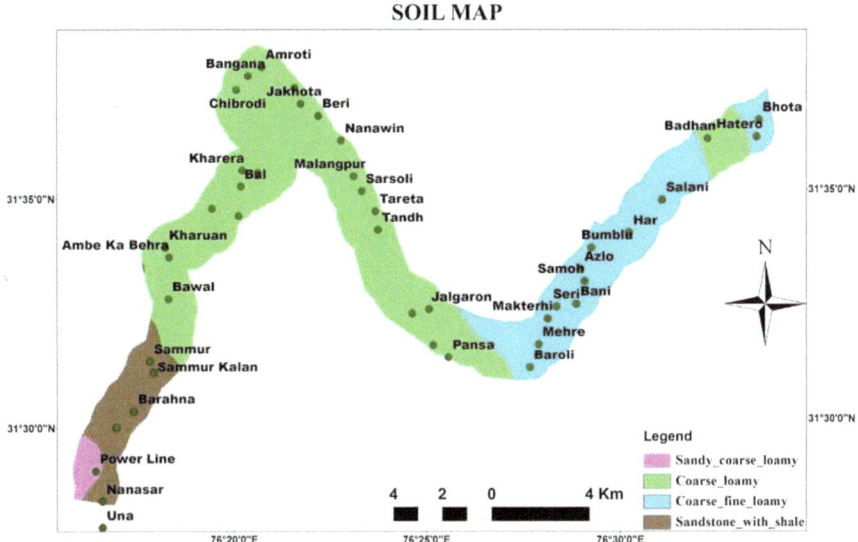

Fig. 2 Soil layer

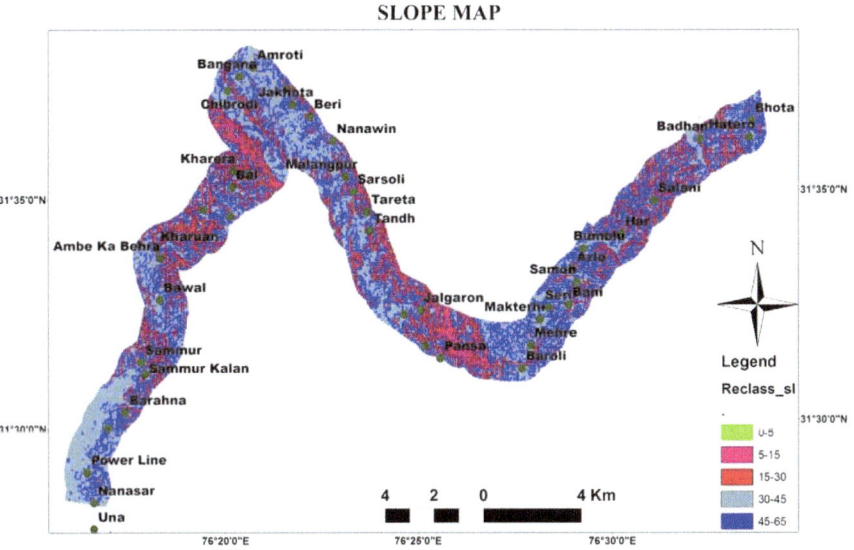

Fig. 3 Slope layer

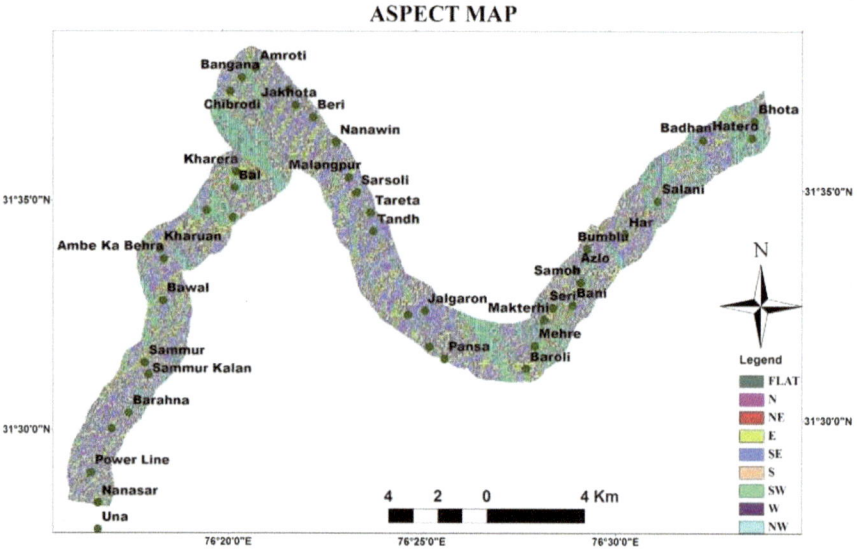

Fig. 4 Aspect layer

2.4 Curvature

Curvature is an important factor in mapping sensitivity to landslides. The curvature controls the soil's effect of water. Convex slopes are found to be more susceptible to failure after rainfall than convex slopes. The negative value of curvature represents the concave surface, and the positive values show convex surface. Zero value of curvature represents the flat surface.

Curvature layer is also prepared by using Spatial Analyst. Figure 5 shows the curvature along SH-32. The curvature can be of profile type or plan type.

2.5 Distance from Drainages

Landslide susceptibility of an area depends upon drainage characteristics of that particular area. Drainage characteristics are represented by drainage density. Drainage density may be defined as length of stream per unit area of the basin.

If the slopes are near to streams or the drainage density of the area is high, then the soil may get saturated due to the effect of these drainages. Such kind of saturated soil is more susceptible to failure.

Alluvium type of soil is found near the streams. Hence, alluvium is affected by the drainage characteristics of the area mostly. Figure 6 shows the drainage characteristics along SH-32.

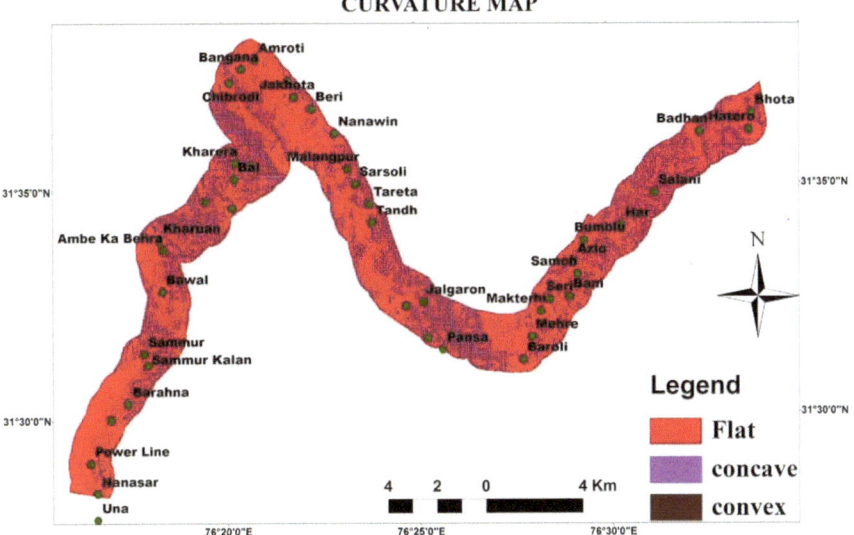

Fig. 5 Curvature layer

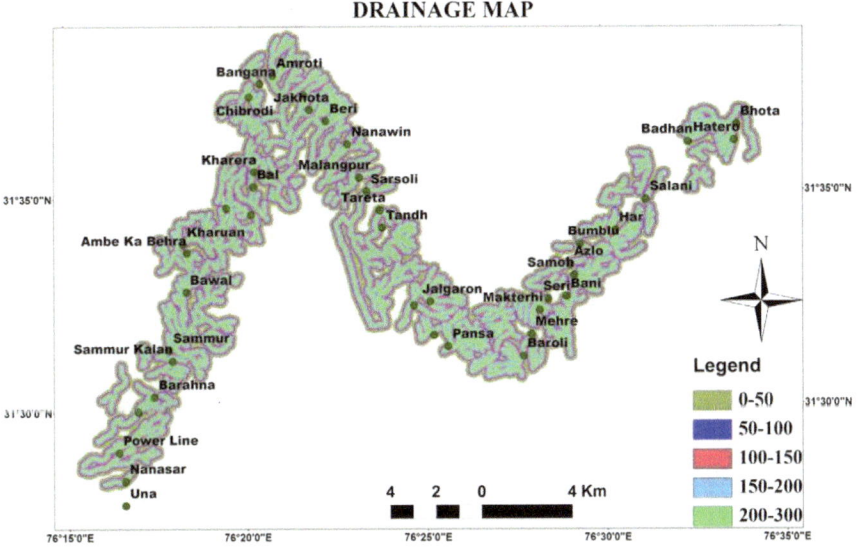

Fig. 6 Distance from streams

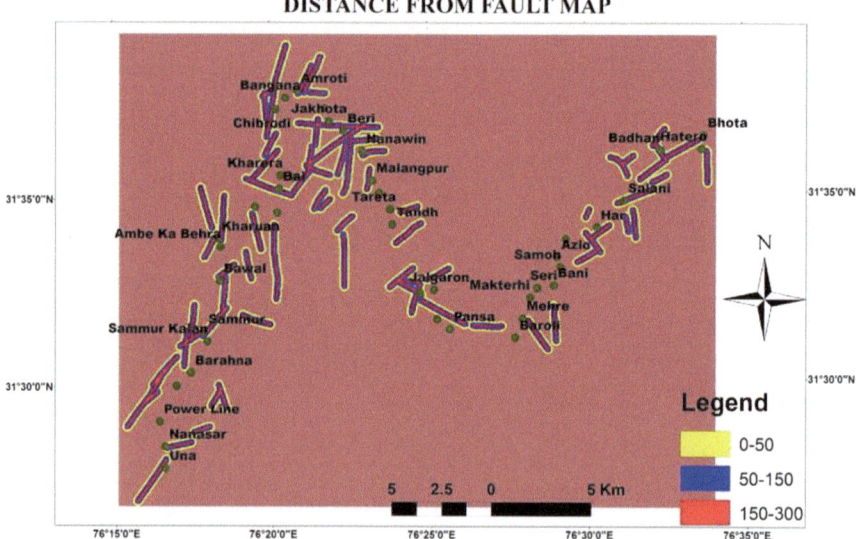

Fig. 7 Distance from faults

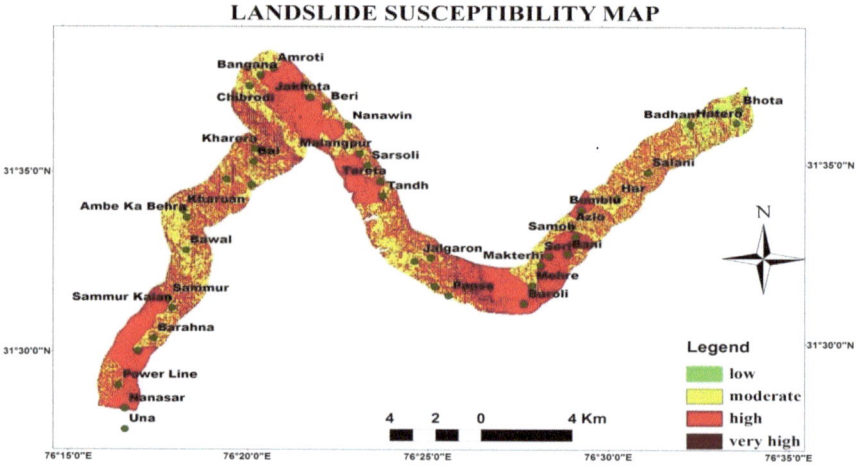

Fig. 8 Landslide susceptibility map

2.6 Distance from Faults

The distance of slopes from the faults also affects the stability of the slopes. The faults are symbol of weakness in the geological structure. Hence, these are very important to consider in the susceptibility mapping.

Faults are digitized on a map of scale 1:50,000. Buffering of fault layer is done to analyse the effect of faults on landslide susceptibility. The fault distance is divided into 250 m. Figure 7 shows the distance from faults.

3   Data Collection and Landslide Inventory

The basic principle of landslide susceptibility mapping is 'past is the key to future' [10]. So, the first step in landslide susceptibility mapping is to prepare a landslide inventory from the data of landslides occurred in the past. The data in this study is collected by field surveys. For the purpose of collection of data about landslides, a format is prepared. This format contains some characteristics of different landslides to recognize their properties required in landslide inventory. Landslide inventory consists of information like location of landslide, area of landslides, type of landslide, classification and date of occurrence of landslide if available. A photograph of each landslide is also taken with this information.

The location of landslide is defined by its longitude and latitude, that is, find out with the help of Global Positioning System (GPS). The areal extent of landslide gives information about the size of landslide. The type of slope failed is defined by the type of material it contains and the type of movement of the mass. For example, if the failed material has coarse particle about 80% and it has fall movement, then it is called 'Debris Fall'.

The maps of different layers are taken from Survey of India. The different topographic and hydrological maps are on a scale of 1:50,000 except soil map. The scale of the soil map is 1:2,200,000. The soil map available is of Himachal State. From this map, the study area is extracted.

A landslide inventory is an important part of landslide susceptibility mapping. In the present study, a basic level landslide inventory is prepared by field surveys. The landslide inventory includes information about longitude, latitude, type of landslide, classification of landslide, date of survey, date of occurrence and activity of landslide event. The location of the landslide is taken with the help of Global Positioning System (GPS). A format for the collection of landslide data for the inventory is prepared. A photograph is included with the landslide collection format for each landslide. In landslide inventory, about 20 landslides are recognized. Most of the landslides are found in the area where rock type is sandstone with clay.

4   Methodology

A landslide inventory is prepared by field surveys as described in Sect. 3. The proposed road section is digitized from the topographic map of the area. A buffer of 500 m each side of the proposed road is taken. Lithology, soil type, drainages and faults are digitized in GIS environment. From the topographic map, contours are digitized. These digitized contours are used to make Digital Elevation Model (DEM) of the area. Slope, aspect and curvature layers are extracted from this DEM. These layers are ranked with analytic hierarchy process (AHP) using online AHP calculator. These maps are overlaid according to their ranking and final map of landslide susceptibility map is derived.

Analytic hierarchy process (AHP) is used to rank the different criterions according to their effect on the occurrence of landslides [11]. The ranking for the main factors like slope, aspect, curvature, etc. is done by pairwise comparison in AHP calculator, and it is found that slope is the most important parameter. The different parameters are given a rank according to expert judgment on a scale from 1 to 9. After giving expert rankings, the parameters are compared and AHP rankings are calculated. If the value of consistency ratio is more than 10%, then the assigned rating is improved. If consistency ratio is less than 10%, then the ratings are consistent and can be used. The landslide susceptibility index (LSI) is calculated from the weightages of criteria and sub-criteria. The final map is based upon the landslide susceptibility index.

5   Results and Discussion

A landslide map shows the areas with different colour combination on the basis of their susceptibility towards the landslides. A landslide susceptibility map is created by overlaying the different layers in GIS environment. The map is divided into four categories low, moderate, high and very highly susceptible areas. The red and dark red colour shows high and very high susceptibility of area towards occurrence of landslides. The green and yellow colours show low and moderate susceptibility of area towards landslides. Figure 8 shows the final landslide susceptibility map.

It is found that around 65% area is under the high and very high susceptibility zone. The construction activities in such areas should be planned carefully to avoid the disaster. The retaining walls can be provided in the region of high susceptibility. The gabion walls are a suitable solution to stop the earth or rock mass. The vegetation can be provided on the slopes to reduce the landslide events.

6   Conclusion and Recommendations

Analytic hierarchy process (AHP) is a user perception-based method. The perception of expert is an advantage of this method. Hence, the ranking given in AHP depends upon the expertise of the user. Therefore, the accuracy of susceptibility maps prepared using AHP depends upon the realistic rankings given to the different parameters. The maps prepared using AHP are accurate enough at regional level. If more factors on which occurrence of landslide depends are considered, the accuracy of susceptibility maps can be increased. The present study is an attempt to prepare a susceptibility map at a preliminary level that can be extended to hazard and risk assessment. The results of the study can be used in planning and management by the road administrators.

References

1. Kumar R, Anbalagan R (2016) Landslide susceptibility mapping using analytic hierarchy process (AHP) in Tehri reservoir rim region, Uttrakhand. J Geol Soc India 87:1–16
2. Pandey VK, Sharma MC (2017) Probabilistic landslide susceptibility mapping along Tripti to Ghuttu highway corridor, Garhwal Himalaya (India). Remote Sens Appl Soc Environ 8
3. Demir G (2018) Landslide susceptibility mapping by using statistical analysis in the North Anatolian Fault Zone (NAFZ) on the northern part of Suşehri Town, Turkey. Nat Hazards 92:133–154
4. Othman AN, Naim WM, Noraini WM (2012) GIS-based multi-criteria decision making for landslide hazard zonation. Procedia-Soc Behav Sci 35:595–602
5. Ahmed B (2014) Landslide susceptibility mapping using multi-criteria evaluation technique in Chittagong Metropolitan Area, Bangladesh. Landslides 12:1077–1095
6. Ding Q, Chen W, Hong H (2016) Application of frequency ratio, weight of evidence and evidential belief function models in landslide susceptibility mapping. Geocarto Int 32: 619–639
7. Pradhan B (2011) Use of GIS-based fuzzy logic relations and its cross application to produce landslide susceptibility maps in three test areas in Malaysia. Environ Earth Sci 63:329–349
8. Lee S (2010) Landslide susceptibility mapping using an artificial neural network in the Gangneung area, Korea. Int J Remote Sens 28:4363–4383
9. Myronidis D, Papageorgiou C, Theophanous S (2016) Landslide susceptibility mapping based on landslide history and analytic hierarchy process (AHP). Nat Hazards 81:245–263
10. Highland M.L (2008) The landslide handbook: a guide to understanding landslides. Geological Survey, U.S., pp 112–120
11. Saaty TL (2008) Decision making with the analytic hierarchy process. Int J Serv Sci 1(1): 83–98
12. Cruden DM (1991) A simple definition of a landslide. Bull Int Assoc Eng Geol 43(1):27–29

Control Techniques to Optimize PV System Performance for Smart Energy Applications

Md. Ehtesham and Majid Jamil

Abstract Among the various renewable energy resources, photovoltaic (PV) power has emerged as one of the most promising source and has become the most attractive solution for smart energy in the form of a clean and green resource. As the PV panel characteristics are nonlinear in nature and vary greatly with the operating conditions, hence it becomes very much important to ensure that PV panels operate at the optimum point for which effective tracking algorithm has to be incorporated. Further, the power conditioning module plays a very crucial role for PV system interfacing, as it has not only to ensure system reliability and stability, but also to maintain the power quality simultaneously. In this paper, novel control techniques for all these challenges have been proposed. Here, a model-based (MB) technique has been applied that ensures effective tracking in dynamic conditions, where the environmental parameters change rapidly. Also, the control strategies for PV inverter have been designed based on the concept of active power filter (APF) that overcomes the limitations of commonly applied passive filtering scheme and provides a promising solution for eliminating harmonic distortions, thus improving the PV power profile. Hardware platform has also been used for experimental measurement of real-time parameters through sensors and data logger.

Keywords Photovoltaic (PV) · Maximum power point tracking (MPPT) · Power quality · Model-based (MB) MPPT · Active power filter (APF)

1 Introduction

The prolonged use of conventional sources over the years has led to some serious worldwide concerns like depletion of fossil fuel reservoirs, rapid increase in the fuel costs and above all the global climatic change. With the ever-growing demand of energy due to rapid trend of industrialization and economic development, it has been realized that there is the utmost invitation for exploiting smart alternatives.

Md. Ehtesham (✉) · M. Jamil
Jamia Millia Islamia, Jamia Nagar, New Delhi 110025, India

© Springer Nature Singapore Pte Ltd. 2020
S. Ahmed et al. (eds.), *Smart Cities—Opportunities and Challenges*,
Lecture Notes in Civil Engineering 58,
https://doi.org/10.1007/978-981-15-2545-2_9

Consequently, over the last two decades, there has been a tremendous spurt for deploying renewable sources of energy. Harvesting of solar energy by means of photovoltaic (PV) generation has shown the escalated trend recently because of its usefulness as an effective, inexhaustible and environmental friendly nature [1]. PV generation has many inherent advantages like that of the absence of moving parts, low maintenance cost, simplicity in operation, compact in area and abundance in availability. Due to the absence of moving parts, the hardware is very robust resulting in a long lifetime with very little maintenance.

However, with the increasing number of grid-integrated PV systems, there are various challenges that are being faced by the power system operators [2, 3]. Various standards and practices have been formulated globally for proper connection and operation of these systems [4], and their effects on the power system are being analyzed. As well known, the PV characteristics are highly nonlinear where the output power changes drastically with the operating conditions [5–7]. Therefore, MPPT algorithms are essentially incorporated which performs the task of optimizing the PV performance by tracking the operational point for maximum output power under different conditions. Many techniques for MPPT are found in the literature [8–10] each having their own merits and demerits on the basis of tracking speeds, accuracy, operating range, etc. But the major limitation commonly found in these techniques is their incapability to track the maximum power point under dynamic conditions where the environmental conditions vary rapidly [11, 12]. Their tracking response is found to be sluggish which leads to oscillations for the fast variations in temperature, insolation or partial shading. It has been found that a considerable amount of power loss occurs due to conventional MPPT convergence schemes applied for the measurements under real climatic conditions [13, 14].

Another major challenge that needs to be addressed for large-scale PV penetration is the threat to the quality of power being transferred. Issues like voltage rise, voltage fluctuations, harmonic distortions and power fluctuations are some of the common effects of widespread integrated PV systems [15, 16]. Generally, PV systems are interfaced through an inverter whose switches are connected and disconnected at discrete time instants to generate desired output voltage of specific magnitude and frequency. Such an output voltage in addition to its fundamental component contains a lot of undesired harmonic contents as well. Situation gets further worsened because of the increased application of nonlinear loads due to development in semiconductor technologies. The guidelines provided by IEEE standard 1547–2003 [17] are in general applicable to these PV systems. Traditionally, passive filters (PFs) were used for suppressing the current harmonics by connecting a low-pass filter consisting of typical LCL or LC configuration between PV inverter and utility grid. However, it has been found that PFs are very sensitive to the changes in source impedance or system frequency, and therefore, have some major limitations such as possibility of overloading and getting detuned for variations in load impedance [18, 19]. Furthermore, PFs are not suitable for unbalance conditions, and there always exists the chance of resonant problems with slight modifications in impedance [20].

Therefore, in order to overcome the deficiencies of the traditional approaches for resolving the above discussed challenges, this research work proposes novel control techniques. First of all, for effective MPPT in dynamic conditions, a model-based (MB) technique is incorporated which precisely estimates the parameters and allows quick jump in the operating point. Thus, it overcomes the limitation of drifting away of operating point from MPP under rapidly varying conditions. Environmental parameters have been measured through sensors like pyranometer, whereas for data recording and monitoring CR 1000 data logger has been used. Next, to mitigate the power quality issues, active power filter (APF) control scheme has been incorporated that basically injects the compensating component and cancels out the harmonic contents. Novel techniques have been designed for APF controller, where estimation of reference current has been carried out through Kalman filter and for generating switching pulses; adaptive hysteresis controller has been implemented. The effectiveness of APF scheme has been shown by simulated results, where the harmonic distortions have been brought much below to the acceptable level.

2 Modeling of System Configuration

Figure 1 shows the overall topology of the system with incorporated control schemes. Here, we will focus mainly on the control unit descriptions for MPPT and APF algorithm. Basically, maximum power point tracker is a power converter, which is interfaced in between PV panels and utility for extracting the maximum power output from PV panels. MPPT algorithm mainly controls the duty ratio (D) of the converter which results in the matching of actual load line (as viewed by

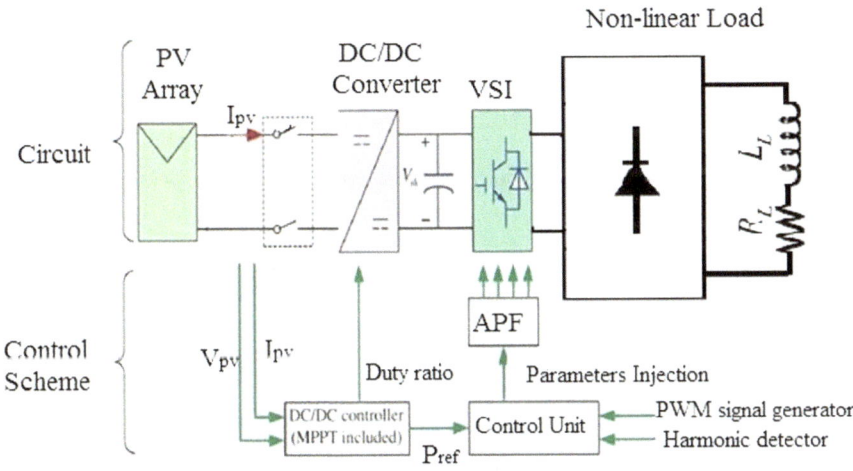

Fig. 1 Configuration of overall system with control schemes

PV panels) with the particular load that yields the maximum power. The detailed analysis of MB MPPT algorithm has been given in Sect. 2.1. The boost converter output is then fed to IGBT-based VSI with firing angles above 90°. The APF scheme functionally injects a compensating current, which are in opposite phase to that of harmonics component produced by that of nonlinear loads. Now, this compensating capability of APF is greatly influenced by technique employed for estimation of reference signals, and this particular task has been achieved by means of Kalman filter with a predictive loop. The hysteresis band is calculated optimally in order to achieve improved and efficient compensation with less distortion. The applied adaptive hysteresis technique takes account of both switching frequency and switching losses. Both these control algorithms are discussed in detail in Sect. 2.2.

2.1 MB MPPT Control Algorithm

The efficiency of PV panel is very much dependent on the converter switching and the potential of applied MPPT algorithm. Now, the most important aspect for application of MB MPPT technique is to accurately model the PV array and to precisely estimate the environmental parameters. Among the different PV models applied in the literature, single diode circuit is found to be the best in terms of the balance between simplicity and accuracy [21, 22], and hence, it is selected here as shown in Fig. 2. As series resistance corresponds to the losses due to current flow, thus for higher efficiency of the cell value of the series resistance should be relatively low; whereas, the value of shunt resistance is kept large, generally higher than 1 kΩ.

Using the standard equations for single diode model, the nonlinear I-V relationship for the PV cell can be deduced by subtracting diode current (I_D) and shunt current (I_{sh}) from photocurrent (I_{ph}) as given by Eqs. 1 and 2.

$$I = I_{ph} - I_D - I_{sh} \tag{1}$$

Fig. 2 Equivalent single diode PV circuit

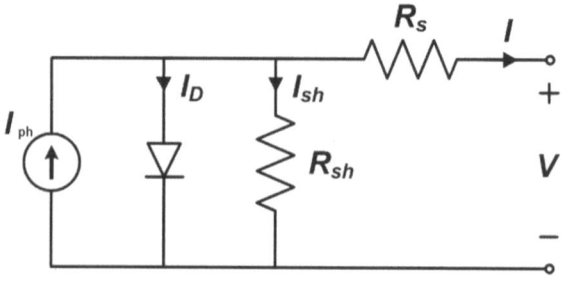

$$I = I_{ph} - I_o\left(\exp\left(\frac{V + R_s I}{\alpha V_t}\right) - 1\right) - \frac{V + R_s I}{R_{sh}} \qquad (2)$$

where I_o is diode reverse saturation current, α is the ideality factor, V is the output voltage, and V_t is thermal voltage. On the basis of above two equations, first photocurrent is evaluated which then leads to computation of panel output current as shown by simulated blocks in Fig. 3.

The MB MPPT technique essentially estimates the parameters with great precision, and for this very purpose sensors have been used. These measure the real working conditions for the installed 2 kW PV panels which have been shown in Fig. 4.

It is to be noted here that as any sort of sensor monitoring algorithm requires a data collection platform, therefore, here in this work CR 1000 data logger has been incorporated for this purpose. The real-time environmental conditions and their variations are continuously measured by the sensors which are then recorded by CR 1000 logger, and the readings are monitored in the display panel setup inside the lab. All the hardware elements are shown in Fig. 5.

Now, for the successful implementation of MB MPPT algorithm, a set of two Eqs. (3) and (4) are defined which characterize the PV module accurately in the region of MPP. Imposing these equations leads to fine tuning of the panel, and it is made to operate at MPP. Once voltage and current for maximum power operation is determined, then the panel operation at MPP is ensured by differentiating the power expression to null value as given below.

$$V_{mpp} = V_{oc} + V_T \ln\left(1 - \frac{I_{mpp}}{I_{sc}}\right) - I_{mpp}R_s \qquad (3)$$

$$\left.\frac{d(VI)}{dI}\right|_{I=I_{mpp}} = V_{mpp} + \frac{V_T I_{mpp}}{I_{mpp} - I_{sc}} - I_{mpp}R_s = 0 \qquad (4)$$

From the solution of above two equations and using the open-circuit (oc) and short-circuit (sc) parameters, the values of V_T and R_s are computed as per the Eqs. (5) and (6).

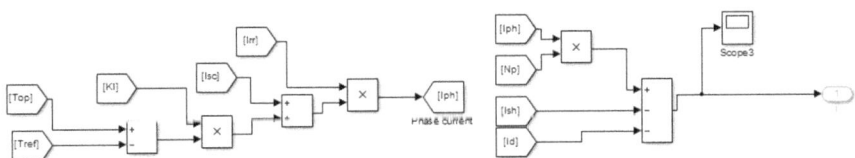

Fig. 3 Simulated blocks for computation of PV output current

Fig. 4 Installed PV panels

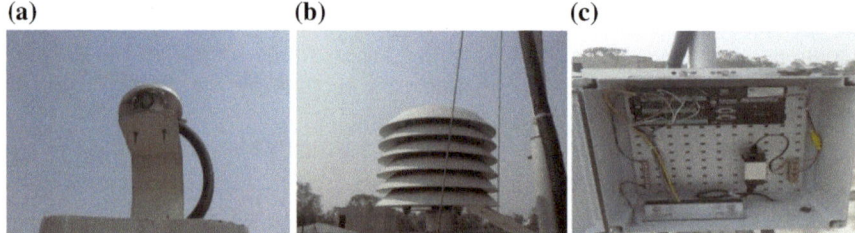

Fig. 5 a Pyranometer **b** humidity sensor **c** CR1000 data logger

$$V_{\mathrm{T}} = \frac{\left(2V_{\mathrm{mpp}} - V_{\mathrm{oc}}\right)\left(I_{\mathrm{sc}} - I_{\mathrm{mpp}}\right)}{I_{\mathrm{mpp}} + \left(I_{\mathrm{sc}} - I_{\mathrm{mpp}}\right)\ln\left(1 - \frac{I_{\mathrm{mpp}}}{I_{\mathrm{sc}}}\right)} \tag{5}$$

$$R_{\mathrm{s}} = \frac{V_{\mathrm{mpp}}}{I_{\mathrm{mpp}}} - \frac{2V_{\mathrm{mpp}} - V_{\mathrm{oc}}}{I_{\mathrm{mpp}} + \left(I_{\mathrm{sc}} - I_{\mathrm{mpp}}\right)\ln\left(1 - \frac{I_{\mathrm{mpp}}}{I_{\mathrm{sc}}}\right)} \tag{6}$$

To achieve high accuracy in MPPT, the panel is matched to its model to optimum possible extent. The panel is first tested at STC and then consequently insolation, temperature and shading are varied alternatively. The effect of variations in operating condition is monitored in display panel setup in the laboratory, and with the help of these readings *P-V* characteristic and MPP are investigated.

2.2 Active Power Filter Scheme

For effective implementation of APF scheme, designing proper control unit is one of the core aspects. It has been already mentioned that two main functions involved are the generation of reference current and switching pulses. Here, a predictive digital Kalman filter has been employed for estimating reference signals for compensation through APF scheme. This accounts for preventing the intermittent lagging nature of recursive filtering and enables to overcome the limitations like time delays or distorted waveforms. Further, in order to achieve improved and effective pulse generation scheme, adaptive hysteresis controller is applied which does not have a fixed hysteresis band and adaptively varies the band with respect to system parameter.

Estimation of the Reference Current Through Kalman Filtering

With slight modification and introduction of the predictive loop, lagging nature has been prevented for recursive Kalman filter. The digital filtering characteristic is governed by the following equation:

$$\hat{x}(k+1/k) = \emptyset \hat{x}(k/k-1) + G(k) \tag{7}$$

where $x(k)$ represents state variable vector, and state transition matrix is represented by $\emptyset$.

The expressions for (k) and $K(k)$ are given as follows:

$$G(k) = K(k)[z(k) - H\hat{x}(k/k-1)] \tag{8}$$

$$K(k) = \emptyset P(k/k-1)H^t[HP(k/k-1)H^t + R(k)]^{-1} \tag{9}$$

Thus, the prediction mean-squared error is obtained as:

$$P(k+1/k) = [\emptyset - K(k)H]P(k/k-1)\emptyset^t + Q(k) \tag{10}$$

For defining the state equations in terms of state variables, both frequency components, in-phase and in-quadrate, have been considered.

Generating Gate Pulses for APF Switching

The commonly applied hysteresis control methods are based on fixed band, they require various comparators for generating the required feeding values for their respective switching, and therefore, this work employs an adaptive hysteresis band (AHB) controller, capable of overcoming these limitations. Similar to that in HB controller, here also corresponding switch is turned-off whose band limit is crossed, to ramp the current within the band. The difference lies in the point that here in the adaptive controller, band modulation is done at discrete points of fundamental frequency cycle that maintains almost a constant switching frequency and also avoids the noises, hence reduces losses. The complete configuration of applied adaptive hysteresis controller is illustrated by blocks in Fig. 6, where f_c and L_f denote switching frequency and filter inductance, respectively.

Overall Parameters Injection for APF Control Scheme

The overall control strategy designed for implementation of APF scheme is illustrated in the Fig. 7. Here, in-phase frequency component for line voltage and

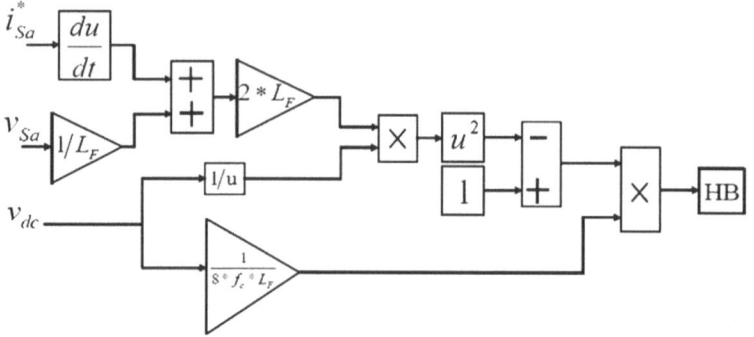

Fig. 6 Block diagram for adaptive hysteresis band calculation

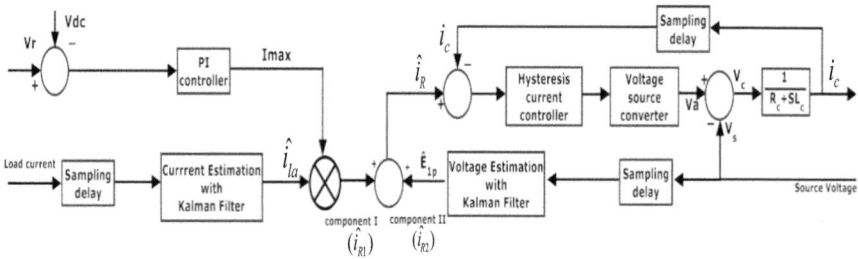

Fig. 7 Block diagram for incorporated controllers

nonlinear load current is generated by the Kalman filter, having time delay lower than that of the sampling interval. Then, the reference current is obtained using these frequency components. As shown in Fig. 7, Kalman filter generates the in-phase frequency components of line voltage and load current, with the time delay lower than that of sampling interval. Then, the reference current is obtained using these frequency components. Also, the block diagram shows that injecting current i_c is deduced as a voltage difference V_c on the link inductor with parameters L_c, R_c. Further, the voltage is obtained by processing converter voltage and source voltage as shown in block diagram.

3 Results and Discussions

This section presents the detailed performance analysis of proposed MB tracking algorithm and active power filtering scheme. First, the ability of MB technique to respond in dynamic conditions is assessed where MPP tracked for the rapid variations in environmental conditions is presented. Then, the effectiveness of applied APF technique in eliminating the harmonic distortion is demonstrated through the simulated results.

3.1 Analysis of MB Technique

Having estimated the parameters precisely as seen in the last section, now the tracking capability of proposed MB algorithm is analyzed under rapidly varying conditions. The tracked MPP is measured for large variations in the environmental conditions by dynamically changing solar insolation (100–1000 W/m²), temperature (12–45 °C) and partial shading (0–100%). First, the simulated results are obtained, then finally the real-time predicted MPP values are compared with the measured MPP through the plots as shown (Figs. 8, 9, 10, and 11).

3.2 Analysis of APF Scheme

Simulated results for active power filter compensation algorithm are obtained in MATLAB/Simulink platform. The in-phase frequency components generated by Kalman filtering are used for producing the reference current. The nonlinear load constituted with diode bridge rectifier has a resistance of 20 Ω and inductance of 10 mH. First, the source voltage waveform with peak value of 100 V is shown in Fig. 12. When this voltage is fed to the system of nonlinear load without any APF compensation scheme, then the load draws a highly distorted current as shown in Fig. 13. This distortion in current is also transferred to the source side, as source current happens to be same as load current, presently in the absence of filtering.

Hence, when the APF compensation strategy is not applied, the load current and hence the source current waveform come to be much distorted and far from being

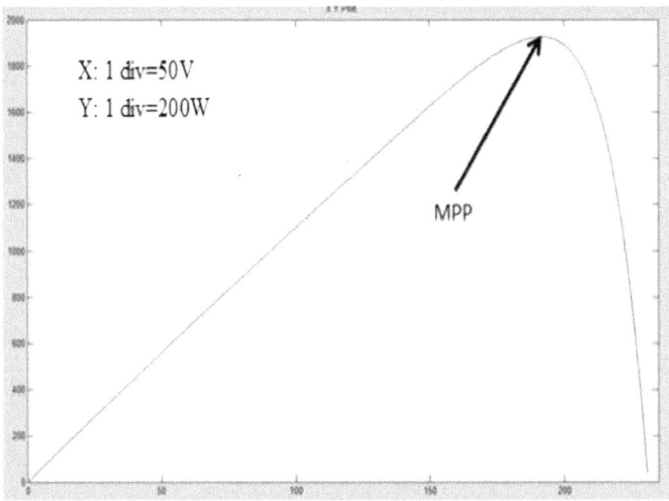

Fig. 8 MPP tracked at STC

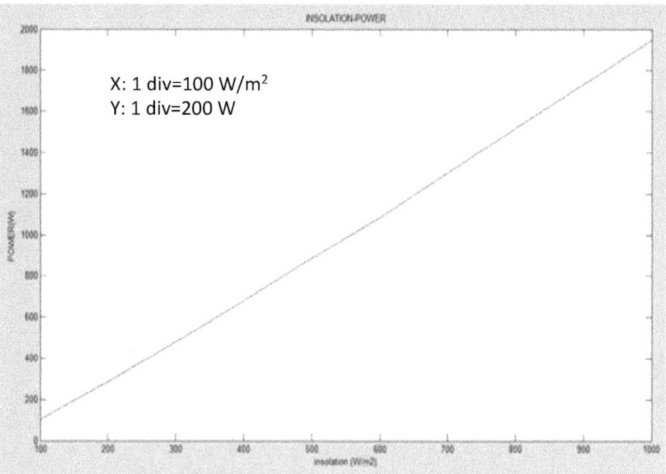

Fig. 9 MPP tracked with insolation

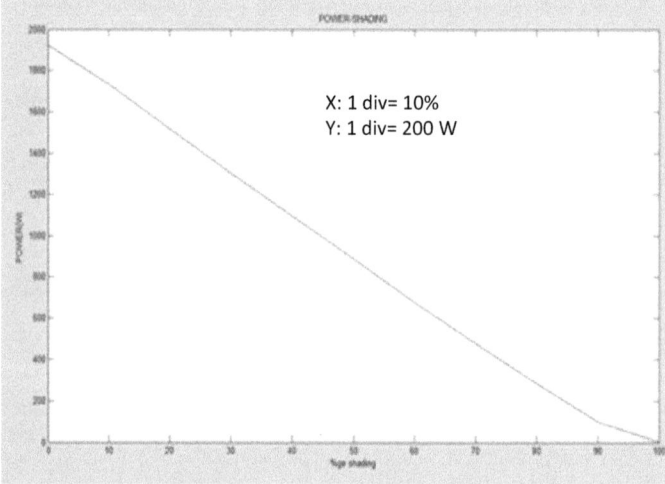

Fig. 10 MPP tracked with shading

sinusoidal. As shown by FFT analysis in Fig. 14, THD for uncompensated source current comes out to be around a high value of 29.3%.

Now, when the APF control scheme is applied, PI controller output is processed for the generation of signal $\hat{i}_{R1}$ as shown in Fig. 15, which is used to give one component of the reference current, and hence, it reduces the ripples in the current waveform. This generated compensation current is injected by the control algorithm, and it eliminates the distortions in source current, bringing it close to that of a

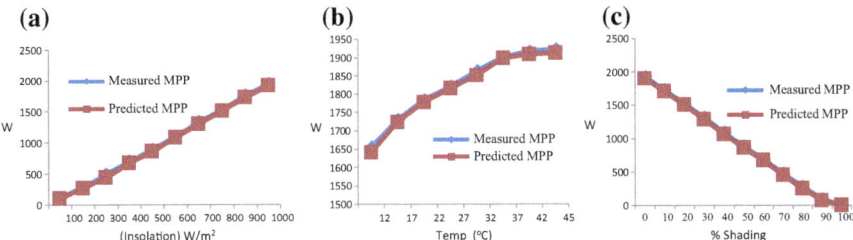

Fig. 11 Comparison of MPP tracked for variations in **a** insolation **b** temperature **c** shading

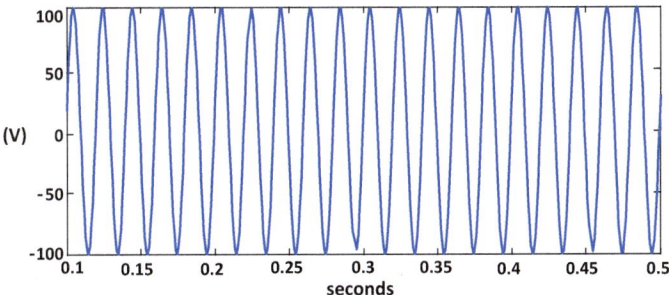

Fig. 12 Sinusoidal source voltage

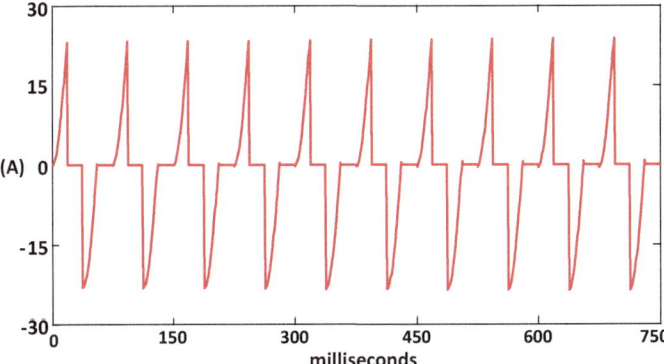

Fig. 13 Uncompensated source current

purely sinusoidal waveform. Figure 16 illustrates the effectiveness of APF compensation scheme where the waveforms of compensation current, compensated (filtered) source current and distorted load current are plotted on the same axis. It can be clearly seen that now the source current is not similar to load current, as it was the case before applied filtering; whereas, the transition in source current before

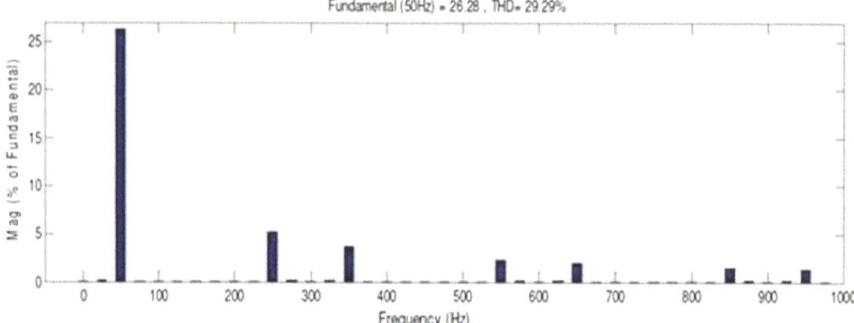

Fig. 14 FFT analysis for uncompensated current

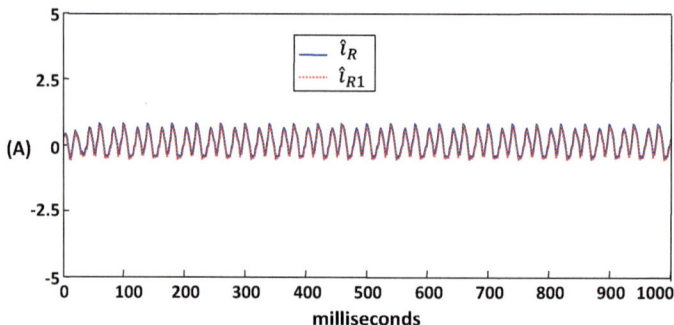

Fig. 15 Generated reference current

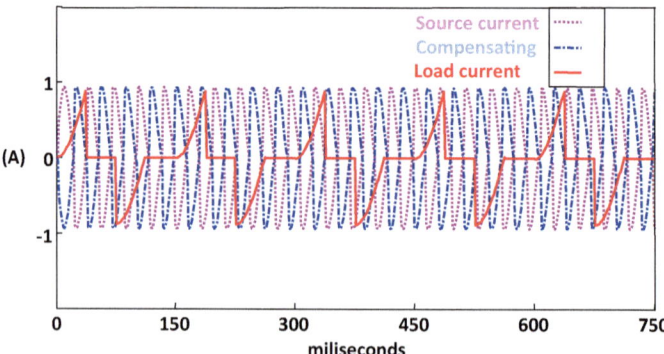

Fig. 16 Current waveforms, i_s, i_c and i_L

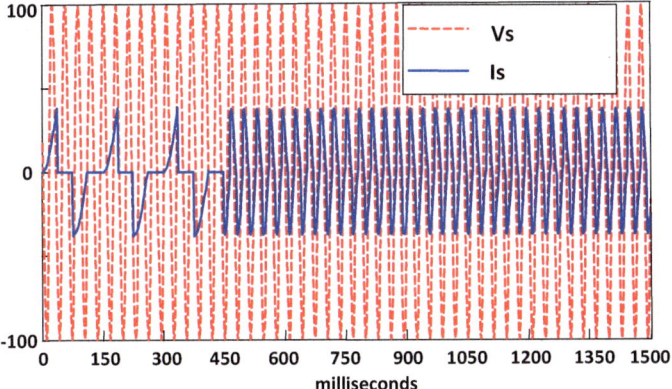

Fig. 17 Transition before and after filtering

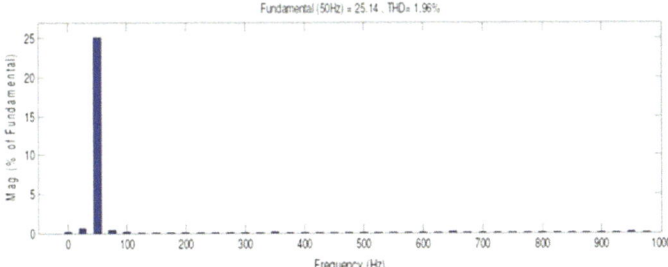

Fig. 18 FFT analysis of source current after filtering

and after compensation is shown in Fig. 17. Finally, FFT analysis of source current after APF compensation is shown in Fig. 18, where THD has been brought down to a very low value of 1.96%.

4   Conclusion

In this paper, some of the major challenges for large-scale penetration of PV systems have been addressed, and various control algorithms have been proposed for enhancing the PV performance for widespread application as a smart energy resource. A model based MPPT algorithm has been applied that efficiently works in the dynamic environmental conditions where the conventional MPPT techniques fail to perform satisfactorily. The environmental parameters are precisely estimated which is the most important feature of this MB algorithm. This allows fast jump of the array voltage under rapidly varying conditions, hence permitting dynamic response and thereby results in effective and efficient tracking. Further, novel techniques for application of APF in PV systems have been proposed for enhancement of power quality. Here, a predictive digital Kalman filter has been employed for estimating reference signals that prevented the intermittent lagging nature of recursive filtering and overcomes the limitations like time delays or distorted waveforms. Also, the adaptive hysteresis controller is applied for pulse generation, which does not have a fixed band and adaptively varies with system parameter. Thus, it takes account of both switching frequency and switching losses. The proposed APF scheme is found to be very effective for compensating the harmonics in the source side with distortion level limited to a low value of 1.96% only.

References

1. Xu G, Ortmeyer T (2017) Towards integrating distributed energy resources and storage device in smart grid. IEEE J Internet Things 4(1):192–204
2. Katiraei F, Aguero JR (2011) Solar PV integration challenges. IEEE Power Energy Mag 9(3): 62–71
3. Basak P et al (2012) A literature review on integration of distributed energy resources in the perspective of control and stability of microgrid. J Renew Sustain Energy Rev 16(8): 5545–5556
4. IEEE Standard 1547-A (2014) Standard for interconnecting distributed resources with electric power systems-amendment 1. IEEE 2014
5. Ngo T, Santoso S (2014) Grid-connected photovoltaic converters: topology and grid interconnection. J Renew Sustain Energy 6(3)
6. Batzelis EI, Georgilakis PS, Papathanassiou SA (2015) Energy models for PV systems under partial shading conditions: a comprehensive review. IET Renew Power Gener 9(4):340–349
7. Xiao W, Edwin FF, Spagnuolo G, Jatskevich J (2016) Efficient approaches for modeling and simulating photovoltaic power systems. IEEE J Energy Rev Photovolt 3(1):500–508
8. Esram T, Chapman P (2013) Comparison of PV array maximum power point tracking techniques. IEEE Trans Energy Convers 22(2):439–449
9. Verma D, Nema S, Dash SK (2016) MPPT techniques: recapitulation in solar PV systems. Renew Sustain Energy Rev 54:1018–1034
10. Quan D, Mao M, Duan P, Hu B (2017) An intelligent algorithm for MPPT in PV system under partial shading conditions. IEEE Trans Inst Meas Control 39(2):244–256
11. Hua C, Fang Y, Chen W (2016) Hybrid maximum power point tracking method with variable step size for PV systems. IET Renew Power Gener 10(2):127–132
12. Logeswaran T, Senthil A (2014) A review of maximum power point tracking algorithms for photovoltaic systems under uniform and non-uniform irradiances. Energy Procedia 54: 228–235
13. Tsang KM, Chan WL (2015) Maximum power point tracking for PV systems under partial shading conditions using current sweeping. Energy Convers Manag 93:249–258
14. Sayedmahmoudian M, Horan B, Stojcevski A (2016) Efficient MPPT using a new technique. Energies 9(3):1–18
15. Tummuru NR, Mishra MK, Srinivas S (2014) Multifunctional VSC controlled microgrid using instantaneous symmetrical components theory. IEEE Trans Sust Energy 5(1):313–322
16. Singh B, Arya SR (2014) Back-propagation control algorithm for power quality improvement using Dstatcom. IEEE Trans Ind Electron 61(3):1204–1212

17. IEEE recommended practice for utility interface of photovoltaic systems (2000) Technical report, IEEE Std.
18. Li D, Yang K, Zhu ZQ, Qin Y (2017) A novel series power quality controller with reduced passive power filter. IEEE Trans Ind Electron 64(1):773–784
19. Devassy S, Singh B (2017) Performance analysis of proportional resonant and adaline based solar PV integrated unified active power filter. IET Renew Power Gener 11(11):1382–1391
20. Hu H et al (2015) Potential harmonic resonance impacts of PV inverter filters on distribution systems. IEEE Trans Sust Energy 6:151–161
21. Laudani A, Riganti F, Salvini A (2014) Identification of one-diode model of PV modules from datasheet values. Sol Energy 108:432–436
22. Xiao W, Edwin FF, Spagnuolo G, Jatskevich J (2013) Efficient approaches for modeling and simulating photovoltaic power systems. IEEE J Photovolt 3(1):500–508

Age-Dependent Compressive Strength of Fly Ash Concrete Using Non-destructive Testing Techniques

A. Fuzail Hashmi, M. Shariq and A. Baqi

Abstract Non-destructive testing methods are important methods for civil engineers to monitor the structures periodically and ensure safety and serviceability without damaging the structures. In the present study, non-destructive testing (NDT) was carried out on plain and fly ash concrete mix using two techniques, i.e. ultrasonic pulse velocity and rebound hammer to evaluate the age-dependent compressive strength. The compressive strength of the specimen obtained using NDT was also compared with the experimental results. Laboratory investigations on pulse velocity were carried out by using PUNDIT and rebound number by using rebound hammer on 100×200 mm cylinders of plain and fly ash concrete at the ages varying from 28 to 180 days. The amount of fly ash replacement by cement was varied from 25 to 60%. The compressive strengths of the all the specimens have been obtained from the pulse velocity and rebound number reference charts. It has been observed that the compressive strength of fly ash concrete mix and plain concrete obtained by using PUNDIT are comparatively similar at all ages due to the homogeneous nature of fly ash concrete, whereas in case of rebound hammer, the strength gets reduced significantly for high-volume fly ash concrete.

Keywords Fly ash concrete · Non-destructive testing · Pulse velocity · Rebound hammer · Compressive strength

1 Introduction

Concrete is the most widely used material in the construction industry all over the world due to its good strength and durability. The continuous growth in the infrastructure has increased the demand for developing smart and sustainable material. The production of Portland cement consumes huge quantities of natural resources and releases large amount of harmful gases directly into the atmosphere which causes global warming and other climatic changes [1]. The utilization of fly

A. Fuzail Hashmi (✉) · M. Shariq · A. Baqi
Department of Civil Engineering, AMU, Aligarh, India

© Springer Nature Singapore Pte Ltd. 2020
S. Ahmed et al. (eds.), *Smart Cities—Opportunities and Challenges*,
Lecture Notes in Civil Engineering 58,
https://doi.org/10.1007/978-981-15-2545-2_10

107

ash in the concrete has significant effect in the construction industry. It improves the workability and durability of concrete, reduces overall cost of structure and water demand, minimizes drying and thermal shrinkage, enhances the durability to reinforcement corrosion, alkali silica expansion and sulphate attack [2]. The most important property of the concrete is the compressive strength as other properties such as flexural and tensile strength, modulus of elasticity all are closely related to it. It has a major role in evaluating the concrete of the existing building. Number of experimental studies has been carried out about compressive strength development of fly ash concrete. At early ages, the strength of fly ash concrete appears quite low but at the later ages, there is significant increase in the strength of fly ash concrete due to the pozzolanic reaction between fly ash particles and cement particles [3, 4].

Non-destructive testing (NDT) is an important method for civil engineers to evaluate the qualitative assessment of uniformity and homogeneity of concrete and the strength of existing building without damaging it. NDT identifies the cracks damages on the surface and interior of the materials without damaging and cutting. The core test is the effective method to know the quality of concrete in the existing structure but it is expensive and time consuming, therefore it is needed to carry out a calibration operation to relate the NDT measurements and compressive strength of concrete [5]. Various structures of concrete such as beams, columns, slabs, bridges, tunnels etc needs periodic investigation through NDT. In past decades, number of NDT methods has been developed such as infrared thermography, pulse-echo and impact echo method for the prediction of strength of concrete structures. Out of these, portable ultrasonic non-destructive digital indicating tester (PUNDIT) and rebound hammer are very common and widely used [6]. For the assessment and prediction of the quality of concrete, the use of only ultrasonic pulse velocity method is not appropriately enough to judge the concrete. So, other methods like the rebound number seem to be somehow more efficient in predicting the quality and strength of concrete under certain conditions [7]. Maria et al. [8] investigated the effect of ultrasonic and sonic wave propagation in the damaged RC frame affected by seismic action and observed that NDT give a brief idea about damages in the structures but for detailed results of completely damaged buildings, it is not suitable. Raffaele [9] did destructive and non-destructive tests on a building and found that rebound hammer is not useful for assessing the existing concrete structures, whereas the ultrasonic pulse velocity methods provide more satisfactory results. Shariq et al. [10] reported that the compressive strength of GGBFS concrete obtained through ultrasonic pulse velocity (UPV) is lower at all ages for different percentages replacement of cement. Mori et al. [11] developed a new NDT method to detect the defect and flaws in the concrete structures. The method is effective and useful based on vibrant response of concrete structures subjected to impact loading. Gholizadeh [12] reviewed the various NDT methods to evaluate the composite materials and described their advantages and disadvantages in the practical use.

This paper presents an experimental investigation into the age-dependent compressive strength of plain and fly ash concrete (having different content of fly ash) and compared the results with strength obtained from non-destructive testing methods by using PUNDIT and rebound hammer at the ages varying from 28 to 180 days.

2 Experimental Program

In this experimental investigation, the concrete mix was selected from the method of trial according to the guidelines of IS 10262 [13]. The experimental research was carried out for M20 mix. The test specimen was cast using cement, fly ash, coarse aggregate and fine aggregate. The properties of the materials such as fine aggregate, coarse aggregate, cement and fly ash used in the experimental investigation were confirmed as per the specification given in the Indian Standard Codes. The cement replacement by equal weights of 25, 40 and 60% by fly ash was done in plain concrete mixes to obtain fly ash concrete. The details of normal and fly ash concrete mixes are given in Table 1. The concrete cylinders of dimension 100×200 mm were cast by taking 0, 25, 40 and 60% partial replacement of ordinary Portland cement (OPC) with fly ash. Three samples were prepared for each percentage of fly ash. NDT has been carried out on these specimens by using rebound hammer and PUNDIT as shown in Fig. 1. The ultrasonic pulse velocity and rebound number measurements for normal and fly ash concrete were taken at the age of 28, 56, 90, 156 and 180 days. The samples were also tested in the compression testing machine to evaluate compressive strength at different ages for all percentages of fly ash. The specimens were kept at ambient temperature after 28 days curing until the time of test. Before taking the readings, the specimens were cleaned with dry cloth and rubbed with sand paper at both the ends. In the case of PUNDIT test, the lubricant was applied on the specimen at both the ends to make the surface smooth. Rebound hammer is basically dependent on the surface hardness, whereas PUNDIT is based on the ultrasonic pulse velocity. The repeated pulses with frequency of 20 kHz and above are generated by PUNDIT. The transmitter emitting ultrasonic pulses is placed on the free surface of concrete, whereas the receiver is placed at the other surface of concrete. The travel time of the ultrasonic waves through the concrete was noted down. Hence, by knowing the total length of travel path in concrete, the velocity of ultrasonic waves was obtained. Based on the value of pulse velocity, the condition of the structure can be evaluated.

Table 1 Fly ash concrete mix proportion

Per cent replacement of fly ash (%)	Cement (kg/m^3)	Fly ash (kg/m^3)	FA (kg/m^3)	CA (kg/m^3)	W/C ratio	Slump (mm)
0	400	0	850	1050	0.5	55
25	300	100				58
40	240	160				60
60	160	240				70

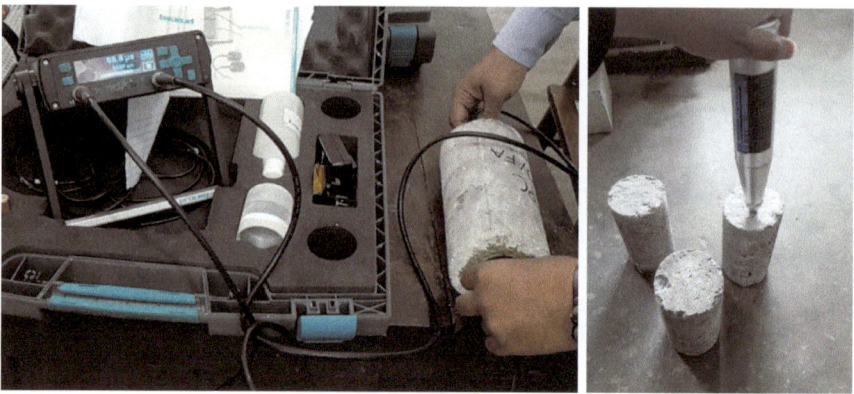

Fig. 1 Specimens under test using PUNDIT and rebound hammer

3   Results and Discussion

3.1   Compressive Strength Obtained Using PUNDIT

The variation of compressive strength of plain concrete and fly ash concrete is
shown in Table 2. It has been observed that the compressive strength obtained
using PUNDIT does not reduce drastically for high-volume fly ash (HVFA) concrete. At the age of 28 days, the compressive strength of fly ash concrete mixes
containing 25, 40 and 60% of fly ash is observed to be 98%, 93% and 85%,
respectively of plain concrete, whereas at the age of 180 days, the compressive
strength of fly ash concrete mixes containing 25, 40 and 60% of fly ash is observed

Table 2   Compressive strength obtained using PUNDIT

Fly ash (%)		Compressive strength at different ages (days)				
		28	56	90	156	180
0	Pulse velocity (Km/s)	4.22	4.30	4.37	4.45	4.47
	Comp. strength obtained from graph (MPa)	**20.5**	**23**	**26**	**28**	**30**
25	Pulse velocity (Km/s)	4.22	4.295	4.375	4.44	4.465
	Comp. strength obtained from graph (MPa)	**20**	**23**	**26.5**	**28**	**29.5**
40	Pulse velocity (Km/s)	4.21	4.295	4.35	4.42	4.425
	Comp. strength obtained from graph (MPa)	**19**	**22**	**25**	**27**	**28**
60	Pulse velocity (Km/s)	4.15	4.21	4.25	4.32	4.35
	Comp. strength obtained from graph (MPa)	**17.5**	**20**	**22**	**23.5**	**25**

to be 98%, 93% and 83%, respectively, of plain concrete. The above observation reveals that the compressive strengths of plain and fly ash concrete are not significantly differ at all ages for all substitution of cement with fly ash. Also, the compressive strengths of HVFA concrete are not reduced up to a greater extent due to the homogeneous nature of fly ash concrete.

PUNDIT determines the strength in terms of homogeneity, integrity, internal flaws and cracks of the material. The ultrasonic waves get scattered when it travels through cracks and flaws present inside the concrete. It takes more time to cover the same distance in cracked concrete than sound concrete which is free from surface cracks and internal defects. As a result of which the pulse velocity gets decreased and hence the strength of concrete gets reduced. The pulse velocity in concrete is normally affected by path length of the waves in the concrete, cross-dimension of the specimen which is being tested, existence of reinforcement steel and humidity of the concrete. However, the pulse velocity will not be affected by the profile of the concrete specimen, provided that the least lateral dimension (i.e. the dimension which is measured perpendicular to the wave propagation) is not as much of that the wavelength of the vibrating pulse. Fly ash concrete have good uniformity and durability, thats why the strength does not reduce much for high-volume fly ash concrete when evaluated using PUNDIT.

3.2 Compressive Strength Obtained Using Rebound Hammer

The variation of compressive strengths of plain concrete and fly ash concrete obtained using rebound hammer is given Table 3. It has been found that the strength obtained using rebound hammer reduced drastically for high-volume fly ash concrete. The strength development is quite slow in the case of HVFA concrete. This is due to the fact that the surface of fly ash concrete is not as hard as that of plain concrete. At the age of 28 days, the compressive strength of fly ash concrete mixes containing 25, 40 and 60% of fly ash is observed to be 87%, 81% and 65%, respectively, of plain concrete, whereas at the end of 180 days, the compressive strength of fly ash concrete mixes containing 25, 40 and 60% of fly ash is observed to be 84%, 76% and 72%, respectively. The rebound hammer basically measures the rebound of a spring-loaded mass striking in opposition to the surface of concrete. The test hammer hits the surface of concrete at a definite energy. The rebound of a hammer is dependent on the stiffness of the concrete and is calculated by the test equipment. The rebound number obtained from the hammer is used to find out the compressive strength of concrete with the help of reference chart. A lower rebound value indicates that the concrete has low strength and low stiffness as a result of which large energy is absorbed when a hammer strikes the surface of concrete. The hammer is placed at right angles to the flat and smooth surface of concrete at the time of conducting test. The readings in the rebound hammer is

Table 3 Compressive strength obtained using rebound hammer

Fly ash (%)		Compressive strength at different ages (days)				
		28	56	90	156	180
0	Rebound number	23	24	25	26	28
	Comp. strength obtained from graph (MPa)	**18.3**	**20**	**21.3**	**24.6**	**26.6**
25	Rebound number	20	23	24	25	26
	Comp. strength obtained from graph (MPa)	**16.0**	**18.67**	**20.0**	**20.6**	**22.6**
40	Rebound number	20	22	23	23	20
	Comp. strength obtained from graph (MPa)	**15.0**	**17.0**	**18.6**	**19**	**20.3**
60	Rebound number	14	20	20	21	23
	Comp. strength obtained from graph (MPa)	**12**	**16**	**16.5**	**17**	**19.3**

affected by the direction of the hammer, when the hammer is oriented upward, gravity increases the rebound distance of the mass, and vice versa for a test conducted on a floor slab. This method of testing does not give a straight measurement of the strength of the material. It simply gives a brief idea and indication of the surface properties of the concrete, it is appropriate for making comparisons among different samples.

3.3 Variation of Compressive Strength Obtained Using PUNDIT and Rebound Hammer

The compressive strength obtained using ultrasonic pulse velocity (UPV) is significantly high in comparison to the strength obtained using rebound hammer as shown in figs. For plain and 25% fly ash concrete, the experimental values of compressive strength are equivalent to the values obtained from PUNDIT, whereas for rebound hammer, compressive strengths are quite low at all ages as shown in Figs. 2 and 3. For HVFA concrete, the values of compressive strength obtained from PUNDIT are more at all ages in comparison to experimental values and values obtained from rebound hammer as shown in Figs. 4 and 5. The values of compressive strength obtained from rebound hammer remain low at all ages and for all replacement of fly ash. It is due to the porous nature of concrete matrix and slower rate of hydration which results in the reduction of compressive strength of concrete. For HVFA concrete, the strength obtained using rebound hammer is very low at initial ages because the surface is soft. But with respect to time due to the pozzolanic behaviour of fly ash the surface of the concrete hardened, and as a result of which the strength increases somehow. Thus, the readings of rebound hammer and

Fig. 2 Variation of strength of plain concrete

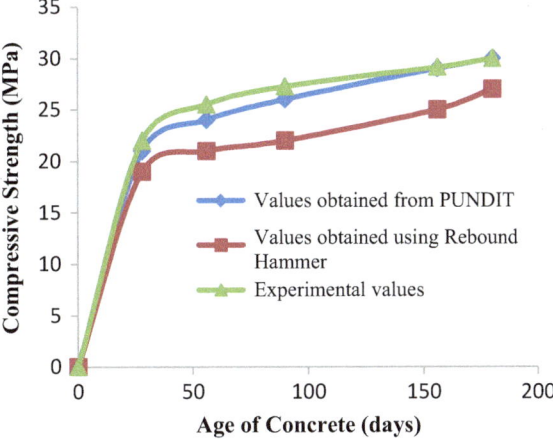

Fig. 3 Variation of strength of 25% fly ash concrete

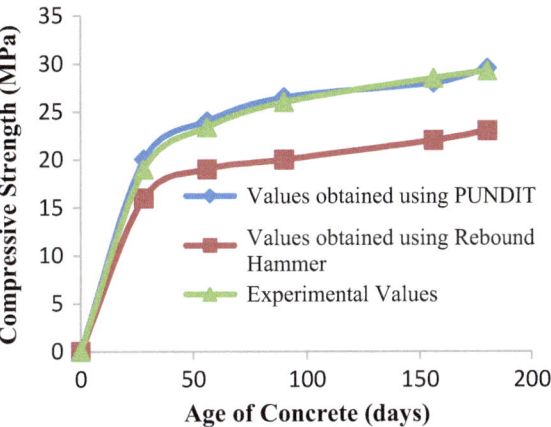

Fig. 4 Variation of strength of 40% fly ash concrete

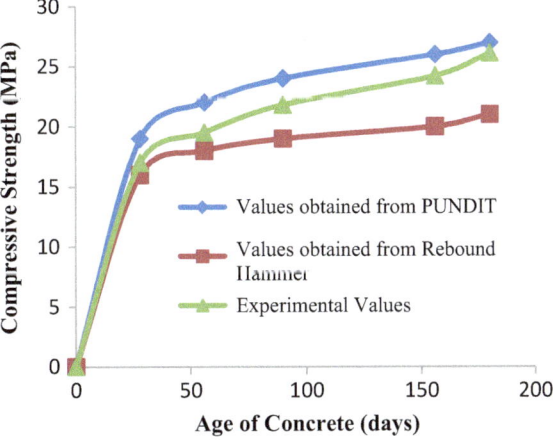

Fig. 5 Variation of strength of 60% fly ash concrete

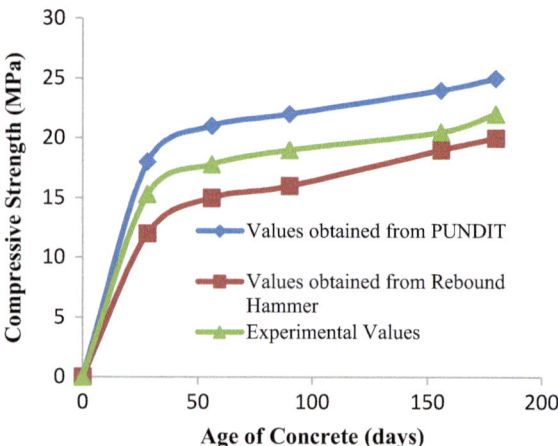

the hardness of the surface of concrete can be correlated with compressive strength of concrete. As far as PUNDIT test concerned, it gives higher compressive strength due to the good uniformity and durability of fly ash concrete. Ultrasonic pulse velocity (UPV) measures transit time which gives information about the uniformity of concrete, cracks, cavities and defects present inside the concrete.

4 Conclusion

The experimental investigation revealed that the compressive strength of fly ash concrete obtained using rebound hammer has been found to be lower than the compressive strength obtained using ultrasonic pulse velocity at all ages and for all percentage replacements of cement by fly ash. For high-volume fly ash concrete, the percentage variation is more at initial ages but at later ages, this variation gets reduced. Both of the NDT techniques, i.e. PUNDIT and rebound hammer do not assess the strength of concrete accurately. They just measure the surface hardness, cracks and internal defects in concrete. These are just indicative tests, other destructive testing is essential for the better evaluation of quality of concrete.

References

1. Bilodeau A, Malhotra VM (2000) High-volume fly ash system: concrete solution for sustainable development. ACI Mater J 97(1):41–48
2. Wang XY, Park KB (2015) Analysis of compressive strength development of concrete containing high volume fly ash. Constr Build Mater 98:810–819
3. Lam L, Wong YL, Poon CS (1998) Effect of fly ash and silica fume on compressive and fracture behaviors of concrete. Cem Concr Res 271–283

4. Poon CS, Lam L, Wong YL (2000) A study on high strength concrete prepared with large volumes of low calcium fly ash. Cem Concr Res 447–455
5. Youcef B, Said K, Khoudja AB (2018) Prediction of concrete strength by non-destructive testing in old structures: effect of core number on the reliability of prediction. MATEC web of conference
6. Kheder GF (1999) A two stage procedure for assessment of in situ concrete strength using combined non-destructive testing. Mater Struct 32:410–417
7. Hobbs B, Kebir MT (2007) Non-destructive testing techniques for the forensic engineering investigation of reinforced concrete buildings. Forensic Sci Int Elsevier 167:167–172
8. Polimeno MR, Roselli I, Luprano VAM, Mongelli M, Tati A, De Canio G (2018) A non-destructive testing methodology for damage assessment of reinforced concrete buildings after seismic events. Eng Struct, Elsevier 163:122–136
9. Pucinotti R (2015) Reinforced concrete structure: non destructive in situ strength assessment of concrete. Constr Build Mater, Elsevier 75:331–341
10. Shariq M, Prasad J, Masood A (2013) Studies in ultrasonic pulse velocity of concrete containing GGBFS. Constr Build Mater, Elsevier 944–950
11. Mori K, Spagnoli A, Murakami Y (2002) A new contacting non destructive testing method for defect detection in concrete. NDT&E Int, Elsevier 35:399–406
12. Gholizadeh S (2016) A review of non-destructive testing methods of composite materials. Struct Integr Procedia, Elsevier, pp 50–57
13. IS: 10262:2009. Recommended guidelines for concrete mix design. Bureau of Indian standards, New Delhi, India

Empowering Smart Cities Though Community Participation a Literature Review

Anika Kapoor and Ekta Singh

Abstract In the recent times, the concept of smart cities has gained momentum and has received considerable attention of urban planner, administrators, policymakers, etc., as a response to the complex problems associated with the unprecedented urbanization. However, citizen engagement or public participation is the key element in the smart-city concept to optimally reach the objectives. The intention of smart cities can only be met by making the communities smart and enabling the end-users and local people to involve in city governance. This article investigates the role of public participation in decision-making, transforming the same into a smart city in the real sense. The article also highlights the adaptable tools and techniques for effective public participation and limitations of this approach in the existing planning machinery in India. The paper facilitates a review of literature from varied research fields is accomplished to understand the role of community participation in smart-city context and to identify the different tools and techniques, empowering people to participate in city governance to make it smart.

Keywords Smart city · Smart community · Public participation · Local area planning

1 Introduction

Regardless of the discomforts, constraints, difficulty, and cost, particularly in intricate, heterogeneous and advanced twenty-first-century cities, smart cities as a concept, keep on drawing our attention as a relief. Smart cities in the absence of smart communities have failed to achieve the desired outcomes [21]. Uncontrolled urbanization with non-complimentary and inefficient urban infrastructure, a mismatch between demand and supply of resources and poor quality of life of the

A. Kapoor (✉) · E. Singh
Amity School of Architecture and Planning, Amity University, Noida, Uttar Pradesh, India
e-mail: esingh@amity.edu

© Springer Nature Singapore Pte Ltd. 2020
S. Ahmed et al. (eds.), *Smart Cities—Opportunities and Challenges*,
Lecture Notes in Civil Engineering 58,
https://doi.org/10.1007/978-981-15-2545-2_11

people in the urban areas are few of the driving forces for planners to introduce the concept of community participation in the planning process as an approach to facilitate local urban governance, thereby ensuring "smartness" of the cities.

The Wealth Report, 2012 associates the urban smartness to economic wealth. It elucidates that the cities in the developing economies will be able to participate in the global economy by 2050 because they are strategizing toward intelligent communities [9]. Community participation in decision-making for urban planning is based on the postulate that those who are affected by the decision should be given the right to get involved in the decision-making process. Community participation in planning and decision-making is not a novel concept. It is an ideology that seeks a common consensus over an issue of concern and is in practice since ages in varied forms and scales. Therefore, the paper intends to review existing literature on the subject in varied disciplines to understand the role of community participation in urban governance that anticipates cities to be "Smart". Paper, further strives to identify effective tools and techniques of involving people in the planning processes and highlight hurdles in adopting the public-participatory approach in urban governance.

2 Evolution of the Concept—Community Participation

Smart urban development is grounded on the concept of the smart city which is further reliant on the public or communities. The perceptions about public participation and its role in local urban governance, enabling the smart city, have been varied. Rousseau [20], a Genevan philosopher of the eighteenth century developed a political philosophy based on the concept of ideal community which influenced French revolution. He propounded the idea of the involvement of the common man in decision-making. He mentioned the concept of "General Will" which implies common good or public interest. He propagated that all citizens should participate for the general good; even if it is against their personal interests. But every time, the involvement seemed to be biased. His famous book "The Social Contract" quotes *"Man is born free and he is everywhere in chains"*. By this, he meant that even thoughts of the common man are influenced and imprisoned. According to him, participation has a psychological effect on the participants that ensures the continued relationships of individuals in the institutions and thus should be continued despite the hindrances. The word "Participation" became a part of the political vocabulary in the 1960s. Pateman [17] defined public participation as "a process where each individual member of the decision-making the body has equal power to determine the outcome of decision". In lieu of the changing political, social and technological world, Nabatchi [16] defined community participation "as the processes by which public concerns, needs, and values are incorporated into decision-making". This advocates that there is a drastic paradigm shift in the thought process in the last 5 decades where the role of common man has been redefined from a powerful entity in the decision-making process who has authority to decide his fate to the sheer end user, whose aspirations and demands are not necessarily determined in the process of distribution of goods or services. It has been realized that public participation is bedevilled with definitional problems [19].

Since the concept of citizen participation has a long lineage [5], some more literature has been reviewed to understand past, assuming some insights may be helpful to create a participatory framework for future to fulfil the objectives of the smart cities. It was discovered that the first written record of direct citizen participation came from the Greek city-states and one of its earliest expressions was in the Ecclesia of Athens [19] and now in the recent years, public participation in planning has resurged and is changing its role as it is determined by definition of problems, nature of prevailing planning paradigm, kind of knowledge base, ideology and technical know-how. Lane [12] observed that new technologies and approaches of governance have emerged and there is no systematic examination of the link between urban planning epistemology and public participation. Though, the ideology of public participation is very much present in all the eras of planning but was never effective in practice. This fact was realized much earlier with the revival of the participatory approach in the 1960s. Arnstein [2] rightly quoted that "the idea of citizen participation is a little like eating spinach and no one is against it in principle because it is good for you". She further elaborated that there is a contrast between a sheer void custom and belief of participation and having the effective participation which is required for meaningful development, especially in the era of smart cities.

3 Role of Community Participation

Another question in the mind of researcher could be that if public participation has been never effective in practice then why public participation in the decision-making process as a subject keeps on drawing the attention and interests masses. International Peacebuilding advisory Team [8] establishes the need for public participation as to sense the people's priorities, to develop a sense of citizenship, build trust in the authority, for good governance and effective implementation. Moreover, Nabatchi [16] summarized goals of citizen participation in public administration are to inform the public, explore an issue, transform a conflict, obtain feedback, generate ideas, collect data, identify problems, build capacity, develop collaboration and to make decisions. The relevance of the participatory approach to planning has been established globally and its need has been realized. In the Second World Urban Forum in Barcelona, it was quoted that "It became clear that citizen participation and stakeholder consultation, as practiced by some of the UN-HABITAT programs, were positioning urban planning at the cutting edge of the modern notion of good governance". In June 1992, the United Nations Conference on Environment and Development adopted *Agenda 21, the global action plan for sustainable development* [23]. Agenda 21, entitled "Local Authorities' Activities in Support of Agenda 21" states that all the problems addressed in the Agenda has its roots in local activities and can only be solved with the participation and cooperation of local authorities [1]. With the research, it is realized that community participation is a very deep-rooted concept, relevance and need of which is well established. As this approach has been practiced globally in various levels and scales, it is important to understand the ways to achieve it. The chronology of the research suggests that the definition of the concept has changed over the period of time and an approach that was useful at times is not always the best. One has to understand effective tools and techniques of involving people in the planning process and their evolution in the changing times to ensure validated inputs of the community in urban governance.

4 Effective Tools and Techniques

Literature has been reviewed to critically evaluate the different tools and techniques used in making public participation success in urban local governance. There is elaborate literature on the direct and indirect tool of participation. Indirect tools are nothing but an illusion to public participation and a make-belief. Selection of tools and techniques for participation depends upon the purpose of planning and intent of authorities to involve local people in the planning process. Krishnaswamy [11] stated that "Doing" participation effectively is more of an art than a technical skill that can be taught". According to him, the selection of tools varies according to desired outcomes, community profile and the social-political context, project size, budget, timeline, and resources allocated and skills of the team. Tools are also determined by the stage of planning to start from outreach to stakeholders, rationalizing common issues and goals, to evaluation and decision-making.

Tools are effective only when institutions are effective. George and Balan [6] stated that Local government can identify working groups (comprising of local people and government officials) for converting the ideas of the local people in respective areas which helps them in project formulation. Another perspective of the role of institutions is elaborated by Lowndes et al. [13] while advocating that Consumerist method is the most adopted means of public consultation (complaints and suggestions). Nearly half of the authorities used a focus group approach and very few used consultative innovations. The use of technology to extend participation and consultation is limited. They also pointed out that public participation is evident in the large urban area as against small urban areas or rural areas because of lack of resources and funds. They also analysed that political parties are not a significant factor and hardly impact participation initiatives. As against this point, *Frank Friesecke (2011)* stated that citizens often feel inadequately embedded in political decision-making. It is necessary to modernize public participation through optimizing formal participation processes, strengthening informal participation processes and more direct democracy.

"People believe to what they see and perceive". Maps are critical to making planners, policymakers, and residents for collection of knowledge, the establishment of boundaries, administration of municipality services, and empowerment of landowners. Maps are tools of planners that encourage public involvement. Warner [25] discussed that in developed cities and countries, technology is more widely available, affordable, and accessible, with a broader audience of citizens reachable through the Internet, smartphone applications and social media. In developing areas, participants in community-based research projects are often biased towards the educated, middle-class, and higher socio-economic status. *Richard* Kingston [10] also highlighted issues in adopting technology in public participation from people's perspective like IT Literacy, access to technology, GIS understanding, data copyright issues, and trust legitimacy. His analysis signified that though using technology in participatory planning is an efficient method but should not be replaced by traditional method rather should be used as complementary to the existing methods.

It is not only about how people will participate but also about how effective the participation is and what purpose it serves. Lowndes et al. [13] suggested that different participation approaches may be used to suit the needs of the area and type of organization. The review of the literature revealed that there are many tools and techniques but are ineffective in accomplishing the idea of public participation.

Many organization and agencies are working worldwide for establishing the legitimacy of public participation in the urban planning processes. *Global Report on Human Settlements 2009: Planning Sustainable cities* present global and regional trends in participatory urban planning and politics. The developed countries of Australia, Canada, the USA and Europe have introduced legislation for public participation and are providing technical and financial support for citizens to participate in public review processes. Many Latin American and Asian countries have realized the importance and need for participatory planning and have adopted a variety of measures to increase citizen participation and government accountability and responsiveness. Despite the vigorous efforts and numerous attempts, participatory approach to planning has failed to achieve its goals. The entire ideology is infected and diagnosis is difficult. Perceptions of experts and review of existing literature revealed the dilemma about non-acceptance of the approach in its full spirit.

5 Limitations of the Participatory Approach in Planning

Public Participation has been perceived differently since ages. There is a new dimension that combines participation with the local culture. Rezazade [18] stated that since public participation is a subjective phenomenon, it is critical to evolve an effective and adaptable philosophy to make participation a part of popular culture. Results of their analysis proved that economic and social factors have a large influence on citizen participation in urban development. Though, this judgment was made while evaluating the role of citizen participation in urban development in Nikshahar, Iran but it seems to be applicable on all the developing countries that are hesitant to involve people in the planning process.

Another limitation is associated with the perception of the local people. Lowndes et al. (2002) discovered from their research that people are more likely to relate to their and their community's immediate interests rather than the wider issues. People do not participate because of negative views about local authority, lack of awareness about opportunities to participate, lack of council response and social exclusion [13]. *Project Monitoring cell, the energy and research institute, India (2009)* found that the nomination process is biased and politically driven and citizens lack initiative in the development of their own area. In support, *Participatory Research in Asia, New Delhi (2010)* also advocated that literacy and confidence of the community to work along with are critical for participation and developed countries lack both.

This perception of people has been rightly built over a period of time. Zakaria and Ariga [28] stated that Bureaucrats and planners always dominate the planning process. Traditional leaders seem to have a very poor role in the participation process. Local people are only involved in the implementation stage. Role of citizens in the participation process is mostly seen as a receipt of a product. The government is taking advantage of people as a source of funds and cheap labour for projects that serve them without actually involving them in the decision-making. Lack of comprehensive and binding plans for the city leads to a lack of confidence in government agencies.

Participatory Planning seems to be a costly and time-consuming affair that discourages users. International Peace building advisory Team [8] highlighted resistances in the participatory approaches to public policy processes and governance. It stated that since these approaches are time and cost consuming, local people have no interest and they do not understand the issues. They are concerned about their daily needs only.

Another hurdle in the way effective public participation is the lack of a framework or model. Kayom et al. [26, 27] stated that public participation is meaningless, if very few urban community members were involved in physical planning due to lack of awareness that too only in the initial stage of the plan the formulation in the absence of the administrative hierarchy of government and formal strategies, policies, and guidelines for community participation. MacLaran et al. (2007) highlighted the limitations of Irish Urban Planning in facilitating community participation [14]. Lack of structure and resourcing of local authorities, conflicting interests within the area, advisory role of planners and not decision-making resulting in an inability to deliver promises to the community are few of the shortfalls of the system. While considering the case of Leipzig, Germany, Claude and Zamor [3], realized that there is no universal model that can be applied to all cities since each city has its own socio-economic, political and legal systems that provide a framework for workable solutions. Therefore, in the absence of a framework and model, it is difficult for authorities to implement public participation in urban governance. Despite resistance, India too realized the importance of public participation in decision-making for enhanced urban governance.

6 Participatory Planning in India

In India, the need for the participatory planning was realized after the 74th Constitutional amendment act, passed by the Indian Parliament in 1992. This act provided for autonomy to the Urban Local Bodies in urban India, through decentralization of the governance structure. This act designates Local bodies to be responsible for the planning, and implementation of local services. The 74th CAA requires the state governments to amend their municipal laws in order to empower ULBs "with such powers and authority as may be necessary to enable them to function as institutions of self-governance". 74th CAA stressed on the formation of ward committees and area sabhas. This has been taken forward in Master Plan of Delhi 2021, where the concept of Local Area Plans (LAP) has been introduced [15]. The local areas are defined as smaller areas with unique and uniform character and concerns. According to the LAP guidelines, it is mandatory to involve stakeholders to ensure optimum utilization of the resources for the creation of required infrastructure while planning for these areas. LAP refers to the plan of a ward; it is a qualitative and quantitative tool of assessment of public participation and satisfaction in the planning process. LAP states the strategy for the effective planning and sustainable development of a local area. INDO-USAID FIRE (D) Project, 2005 prepared guidelines for LAP preparation for enhanced local participation in urban governance.

Besides, there are few non-statutory steps taken by the Indian Government to encourage public participation. JNNURM (2012) requires certain reforms to be undertaken by states/cities in community participation, with the objective of institutionalizing citizen participation as well as introducing the concept of the Area Sabha in urban areas [7]. The larger objective is to involve citizens in municipal functions, e.g. setting priorities, budgeting provisions, etc. Recently, Smart Cities Mission Strategy (2015) has been introduced which requires the involvement of smart people who actively participate in governance and reforms. Citizen involvement is much more than a ceremonial participation in governance. The participation of smart people has been enabled by the Special Purpose Vehicle (SPV) through increasing use of ICT, especially mobile-based tools [22]. The newspaper article, *"The problem with smart cities", live Mint, 14 October 2016* states that "Apart from criticism on the quality of proposals and public participation, there were indications of a few cities hesitant to submit their proposals".

7 Discussion

Experts investigated factors that influence participatory planning and Claude and Zamor [3] realized that local spatial form in physical, economic, social, cultural and political dimensions of any city cannot be a result of globalization. There are many hidden factors that impact the development process. It is important to involve all those who will be affected by the decisions, to ensure fair and balanced urban development. He stated that "all stakeholders' values and concerns are legitimate and should be considered". Therefore, the factors affecting public participation are yet to be explored fully to understand the flaws in effective participatory planning to ensure the success of the smart cities vision.

This search of the framework has led to many investigations and analysis all over the world and it gave birth to concepts like SWARM Intelligence [24] or collective intelligence which has now emerged as an established science which states that collective efforts and competition may lead to an effective Consensus decision-making. It was in the 1990s that the experts and think tanks globally realized the role of the public is critical in development with the adoption of the African Charter for popular participation in development and transformation at the "International Conference on Popular participation in the recovery and the development process in Africa", Arusha, Tanzania. International Association for public participation was established in responding to rising global interest in public participation in 1990.

Roberts [19] correctly stated that "For the first half of the twentieth century, citizens relied on public officials and administrators to make decisions about public policy and its implementation. The latter part of the twentieth century saw a shift toward greater direct citizen involvement. This trend is expected to grow as democratic societies become more decentralized, interdependent, networked, linked by new information technologies, and challenged by "wicked problems".

8   Conclusion

The existing literature interprets the relationship between different perceptions and established works and highlights the significant contributions in the related field. In the process of reviewing literature, gaps in the previous researches have been identified. Since efforts for Public participation have not been able to achieve its deliverables, the success of smart cities remains a question. There is the need for research to investigate many veiled issues in the contemporary planning process, to address the issues pertaining to participatory planning thereby, establishing participatory planning as a qualitative and quantitative tool for assessing public satisfaction and participation in the planning process, thereby empowering smart cities to achieve the desired deliverables.

References

1. AGENDA 21 (1992) United Nations Conference on Environment and Development, Rio de Janerio, Brazil
2. Arnstein SR (1969) A ladder of citizen participation. J Am Inst Plan 35(4):216–224
3. Claude J, Zamor G (2012) Public participation in urban development: the case of Leipzig, Germany. J Public Adm Policy Res 4(4):75–83
4. Davidoff P (1965) Advocacy and pluralism in planning. J Am Inst Plan 31(4):331–338

5. Friesecke F (2011) Public participation in urban development projects—a German perspective, FIG working week 2011—bridging the gap between cultures Marrakech, Morocco, pp 18–22
6. George S, Balan PP (2011) People's participation in development planning in Kerela. In: Paper for India urban conference, Mysore, 17–20 Nov
7. Government of India. Ministry of Urban Development. JNNURM modified guidelines (submission for UIG). September 2006. Accessed 25 July 2012. http://www.india.gov.in/allimpfrms/alldocs/15518.pdf
8. International Peace building advisory Team (2015) Public participation and citizen engagement. Inter-peace, Geneva
9. Kinght F (2012) The wealth report 2012: a global perspective on prime property and wealth. Pure print group limited
10. Kingston R (2002) The role of e-government and public participation in the planning process. XVI Aesop Congress Volos, Greece
11. Krishnaswamy A (2012) Strategies and tools for effective public participation. FORREX Forum for Research and Extension in Natural Resources, Burnaby, Canada, pp 245–250
12. Lane MB (2005) Public participation in planning: an intellectual history. Aust Geogr 36 (3):283–299
13. Lowndes V, Pratchett L, Stoker G (2002) Trends in public participation: part 2—citizens' perspectives. Public Adm 79(2):445–452
14. MacLaran A, Clayton V, Brudell P (2007) Empowering communities in disadvantaged urban areas: towards greater community participation in Irish urban planning? Combat Poverty Agency Working Paper Series 07/04. ISBN: 978-1-905-48550-5
15. Master Plan for Delhi with the perspective for the year 2021
16. Nabatchi T (2012) Putting the "public" back in public values research: designing participation to identify and respond to values. Public administration review
17. Pateman C (1970) Participation and democratic theory. Cambridge University Press, Cambridge, MA
18. Raisi J, Miri GR, Rezazade MH (2015) Evaluation and analysis of the role of citizen participation in urban development (case study: the city of Nikshahar). Am J Eng Res (AJER). E-ISSN: 2320–0847 p-ISSN: 2320-0936. 4(4):74-78
19. Roberts NC (2004) public deliberation in an age of direct citizen participation. American Review of Public Administration 34(4):315–353
20. Rousseau JJ (1762) The social contract. translated by G. D. H. Cole, public domain
21. Satterthwaite D (1986) Urbanization and planning in the third world; spatial perceptions and public participation. Cities 3(2):170–171. https://doi.org/10.1016/0264-2751(86)90060-0
22. Sethi M (2015). Smart cities in India: challenges and possibilities to attain sustainable urbanisation #. Nagarlok 47:20. Indian Institute of Public Administration 0027-7584
23. United Nations General Assembly Draft outcome document of the United Nations summit for the adoption of the post-2015 development agenda. UN. Retrieved 25 Sep 2015
24. Wang J, Beni G (1988) Pattern generation in cellular robotic systems. In: Presented at the 3rd IEEE symposium on intelligent control. Arlington, Virginia, August 24–26
25. Warner C (2015) Participatory mapping: a literature review of community-based research and participatory planning. Social hub for community and housing faculty of architecture and town planning technion. Spring
26. Wilson K, Hannington S, Stephen M (2015a) Is human resources available to carry out physical planning in Uganda? Int Interdiscip J Sci Res. ISSN: 2200-9833, 2(2):18-32
27. Wilson K, Hannington S, Stephen M (2015b) The role of community participation in planning processes of emerging urban centres. A study of Paidha town in Northern Uganda. Int Ref J Eng Sci (IRJES), 4(6):61–71
28. Zakaria BI, Ariga T (2010) Community participation and urban development: evaluation of community participation practice in the sudanese capital region. In: 46th ISOCARP Congress 2010

Traffic Data Collection and Visualization Using Intelligent Transport Systems

Anurag Upadhyay, Asit Kumar and Varun Singh

Abstract Traffic conditions nowadays are in a grim situation caused by daily congestion and accidents. Thus, traffic state forecasting is considered as one of the most important traffic management techniques on roadway networks. Owing to financial and economic constraints, uses of sensors and cameras along the road are not a feasible option. Henceforth, probe vehicles equipped with GPS and other sensors are gaining prominence and are frequently used in developed countries to collect traffic data. In the probe vehicle concept, vehicles themselves are acting as roving traffic detectors, which are not bound to specific and fixed locations along the road infrastructure. In this paper, a sensor fusion model based on the extended Kalman filter and measurement inputs from a global positioning system (GPS) receiver and inertial measurement unit (IMU) sensors to improve absolute position estimation and to collect traffic data using ultrasonic sensors and dashcam has been presented. The proposed methodology has been tested for prevailing mixed traffic conditions in Prayagraj city. On the basis of the analysis of collected data, this paper presents a systematic solution to efficiently estimate the traffic state of large-scale urban road networks.

Keywords Intelligent transportation systems · Geographical information systems · Probe vehicle · Data collection · Kalman filter · Sensor fusion · Congestion prediction

1 Introduction

Collection of data is a core part of data engineering and geographical information systems (GIS). Setting up cameras and extrusive sensors, to record traffic data at each and every corner and intersection of road is not a feasible option. Hence, the concept of probe vehicles has emerged. They are mobile devices which are

A. Upadhyay (✉) · A. Kumar · V. Singh
Civil Engineering, Motilal Nehru National Institute of Technology Allahabad, Prayagraj, Uttar Pradesh 211004, India

© Springer Nature Singapore Pte Ltd. 2020
S. Ahmed et al. (eds.), *Smart Cities—Opportunities and Challenges*,
Lecture Notes in Civil Engineering 58,
https://doi.org/10.1007/978-981-15-2545-2_12

equipped with state-of-the-art sensors and collect positional, speed and accelerational data in real time.

Due a significant reduction in cost of these experiments, probe vehicles are gaining a firm ground in methods used to collect raw data.

The aim of this research is to provide a practical and simple solution to traffic data collection and hence support the development of driverless vehicles, congestion management, traffic control and route guidance.

2 Methodology

In this research, probe vehicles equipped with a variety of sensors were used to collect raw data, on which further analysis and visualization take place. Probe vehicles [1] give us real-time data, at a lower cost. If processing is done onboard, these could be developed to be used in self-driving autonomous cars.

2.1 Digital Map Design

In the GIS, the digital map comprises different types of spatial data in the form of layers.

When the spatial objects of one layer are changed, objects of other layers are not affected.

Open-source software QGIS Madeira 3.4 has been used for further data analysis. The recorded GPS data and related attributes are laid down on the world map. We defined the road network in the map so that we can identify the errors in GPS data.

As the data from GPS is geodetic in nature, we should convert it to a 2-D system, but considering a city and the area of travel in most cases would be less than 195.5 km^2, we have skipped Gaussian projection [2] and used Kalman filter to reduce errors.

2.2 Data Description

See Fig. 1 and Table 1.

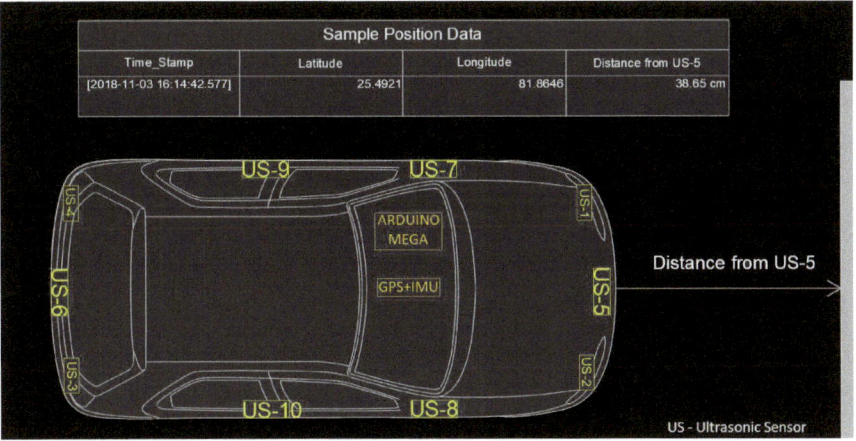

Sample Position Data			
Time_Stamp	Latitude	Longitude	Distance from US-5
[2018-11-03 16:14:42.577]	25.4921	81.8646	38.65 cm

Fig. 1 Probe vehicle layout

Table 1 Sample data

Sample data
Timestamp Latitude Longitude Ax Ay Az uS1 uS2 uS3 uS4 uS5 uS6 uS7 uS8 uS9 uS10 Theta

TimeStamp—Date and Time of recording of data
uSx—Stands for the distance to object from specified
Ultrasonic sensor
Ax—Acceleration in the specified axis
Theta—Angle of sensor

2.3 Description of the Devices Used for Analysis

Ultrasonic Sensor As the name indicates, ultrasonic sensors measure distance by using ultrasonic waves. The sensor head emits an ultrasonic pulse and receives the pulse reflected back from the target. Ultrasonic sensors measure the distance to the target by measuring the time between the emission and reception. A total of ten ultrasonic sensors of model HC-SR04 were used in the probe vehicle to measure the distance from all sides and to create an efficient two dimensional map around the vehicle.

Inertial Measurement Unit (IMUs) This combines inertial sensors such as linear accelerometers and rate gyroscopes, whose measurements are relative to inertial space, into one platform. By knowing angular rate and/or acceleration of an object, its change in position can be calculated by integrating the signal over time. IMU named MPU 6050 was used in the probe vehicle; it is a low power, low cost and high-performance device which combines three-axis gyroscopes and three-axis accelerometers with an onboard digital motion processing which processes complex six-axis motion fusion algorithms.

Fig. 2 Probe vehicle picture

Global Positioning System GPS is a satellite navigation system used to determine the ground position of an object. A GPS receiver combines the broadcasts from multiple satellites (minimum four) to calculate its exact position using triangulation. We have used ublox chips, which are dedicated entirely to determine the exact location.

Arduino Mega 2560 is a microcontroller board based on the ATmega2560. It has 54 digital input/output pins (of which 15 can be used as PWM outputs), 16 analog inputs, 4 UARTs (hardware serial ports), a 16 MHz crystal oscillator, an ICSP header. Arduino Mega acted like the brain of the whole system; along with fetching the data from other devices, it processed the data and further logged that data (Fig. 2).

2.4 Principles

Kalman Filter (Data Sheet and Map comparison) Kalman filter is a recursive filtering method for discrete data. The Kalman filter enables the estimation of past, present and future states of linear systems by using measurements in a fashion that minimizes the least mean squared error.

The Kalman filter is an algorithm that is widely used in sensor fusion applications, partially due to its computational efficiency [3]. The following equations describe the mathematical basis of the Kalman filter. Theory and equations of the Kalman filter are provided here (Fig. 3).

Fig. 3 Kalman filter
equations [4]

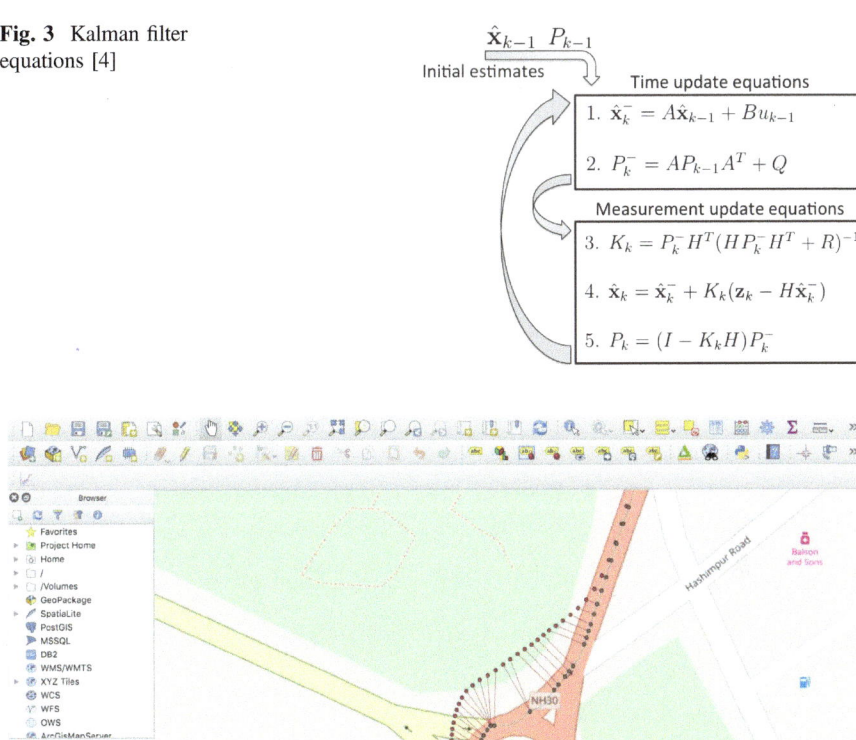

$$\hat{\mathbf{x}}_{k-1}\ P_{k-1}$$

Initial estimates

Time update equations

1. $\hat{\mathbf{x}}_k^- = A\hat{\mathbf{x}}_{k-1} + Bu_{k-1}$

2. $P_k^- = AP_{k-1}A^T + Q$

Measurement update equations

3. $K_k = P_k^- H^T (HP_k^- H^T + R)^{-1}$

4. $\hat{\mathbf{x}}_k = \hat{\mathbf{x}}_k^- + K_k(\mathbf{z}_k - H\hat{\mathbf{x}}_k^-)$

5. $P_k = (I - K_k H)P_k^-$

Fig. 4 Hubs and smoothing the data

Nearest Neighbor Analysis In order to achieve a relation between features of the spatiotemporal data, we projected the points on the road network using nearest neighbor analysis. We created hubs at small distances and took the projection of raw GPS locations on it so as to get smooth data for easier analysis. This step might induce some error at intersections of roads, or due to less distance between hub points [2]. After executing, this gave us the corrected data points, which could be used for traffic congestion prediction (Fig. 4).

Sensor Fusion: Different sensors to improve the quality of information such that the combined information becomes more valuable than the information would be from each individual sensor.

Sensor fusion across attributes involves several sensors that measure different attributes, e.g., acceleration, position and angular velocity of a vehicle. This type of sensor fusion increases the value of data by combining information from different sources, thus acquire data regarding different dynamics and aspects of an object [5].

3 Results

The results of trajectory analysis are summarized in Figs. 5, 6, 7, and 8.

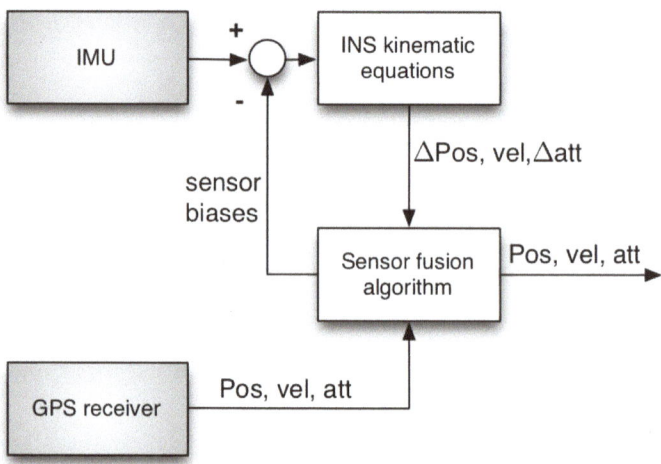

Fig. 5 Sensor fusion [6]

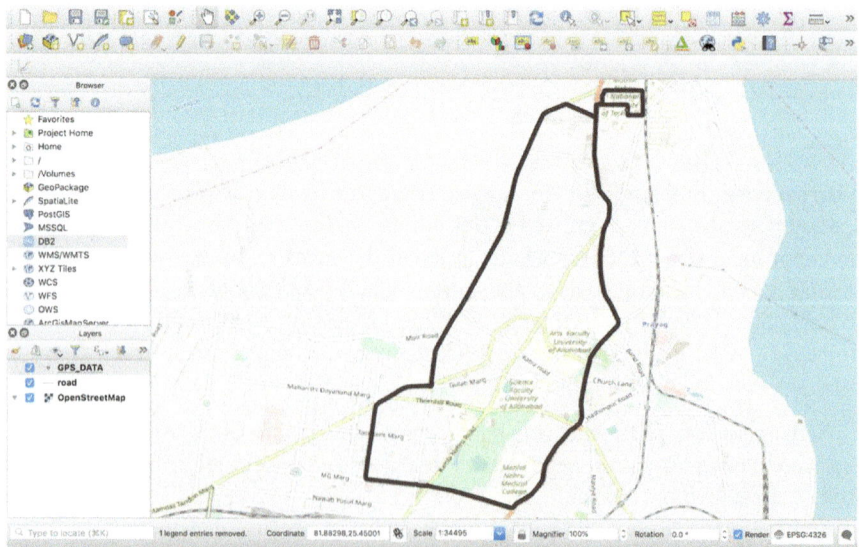

Fig. 6 GIS map 1

	fid	id	Time_stamp	xcoord	ycoord	altitude	ax	ay	az	theta	us1	us2
1	265	1	[2018-11-03...	25.4668432	81.8436943	23.367	-0.76	0.02	9.91	10	0	265.39
2	284	1	[2018-11-03...	25.466897	81.843723	24.059	-1.63	0.04	10.17	12	0	313.88
3	287	1	[2018-11-03...	25.4667472	81.843643	19.59	-1.34	0.61	10.89	6	0	58.86
4	286	1	[2018-11-03...	25.466793	81.8436673	21.648	-0.26	0.9	8.05	8	0	351.98
5	281	1	[2018-11-03...	25.46732	81.8439618	19.807	-0.76	-0.39	9.4	28	0	0
6	280	1	[2018-11-03...	25.4673642	81.8439873	19.75	-1.48	-1.04	10.87	30	0	68.54
												363.25
7	283	1	[2018-11-03...	25.4669522	81.8437518	24.637	-1.06	0.7	11	14	0	252.12
8	282	1	[2018-11-03...	25.4672218	81.8439067	23.541	-0.78	-0.6	8.79	24	0	0
9	277	1	[2018-11-03...	25.468814	81.8448262	24.806	-0.6	-0.34	8.6	98	0	120.54
10	276	1	[2018-11-03...	25.4688663	81.8448618	24.276	-1.11	-0.4	9.39	100	0	130.23
11	279	1	[2018-11-03...	25.468707	81.8447562	24.539	-0.93	0.61	10.55	94	0	74.16
12	278	1	[2018-11-03...	25.4687608	81.8447907	24.533	-0.98	0.02	10.32	96	0	0
13	273	1	[2018-11-03...	25.4690137	81.8449655	21.811	-0.8	-0.16	8.43	106	0	108.26
14	272	1	[2018-11-03...	25.4690593	81.8449988	21.855	-1.06	-0.13	11.18	108	0	101.46
15	275	1	[2018-11-03...	25.4689173	81.8448973	23.276	-0.86	-0.12	10.4	102	0	128.47
16	274	1	[2018-11-03...	25.4689663	81.8449318	22.743	-1.04	-0.35	10.61	104	0	121.95
17	269	1	[2018-11-03...	25.4691958	81.8451013	21.889	-1.4	-0.29	9.42	114	0	105.18
18	268	1	[2018-11-03...	25.4692417	81.8451355	21.913	-1.13	-0.17	9.72	116	0	99.49
19	271	1	[2018-11-03...	25.4691047	81.8450327	21.848	-1.34	0.95	9.71	110	0	106.3
20	270	1	[2018-11-03...	25.46915	81.8450672	21.957	-0.92	0.35	10.95	112	0	98.51
21	265	1	[2018-11-03...	25.469379	81.8452375	21.805	-1.36	0.09	10.26	122	0	98.72
22	264	1	[2018-11-03...	25.469425	81.845271	22.181	-0.53	0.15	10.15	124	0	105.25
23	267	1	[2018-11-03...	25.4692877	81.8451698	21.765	-0.78	0.34	9.29	116	0	102.72
24	266	1	[2018-11-03...	25.4693332	81.845204	21.791	-0.78	-0.18	11	120	0	106.86

Fig. 7 Ride data (live generated)

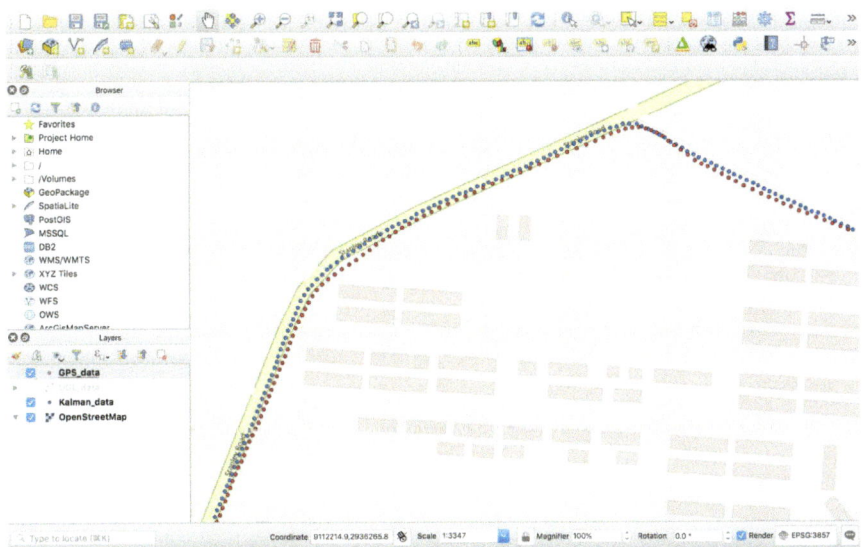

Fig. 8 Kalman versus non-Kalman data

4 Conclusion

It can be concluded that sensor fusion using Kalman filter helps in achieving positional accuracy. With the results available in the form of maps and data, analysis of the positional movement and speed patterns can readily be carried out. Using probe vehicle, traffic-related data can be collected concurrently in real time.

5 Future Applications

Methodology presented in this paper can be extended to include the

(i) Use of artificial neural networks and machine learning algorithms to train and generate better interpretations.
(ii) Use higher onboard processing techniques so that it could be used in self-driving cars.

References

1. Chen Y, Gao L, Li ZP, Liu YC (2007) A new method for urban traffic state estimation based on vehicle tracking algorithm. In: Proceedings of IEEE intelligent transportation systems conference. Seattle, WA, pp 1097–1101
2. Kong QJ, Zhao Q, Wei C, Liu Y (2013) Efficient traffic state estimation for large-scale urban road networks. IEEE Trans Intell Transport Syst 14(1):398–407
3. Kong Q-J, Li Z, Chen Y, Liu Y (2009) An approach to urban traffic state estimation by fusing multisource information. IEEE Trans Intell Transp Syst 10(3):499–511
4. Welch G, Bishop G (2006) An introduction to the kalman filter. Technical report, Department of Computer Science, University of North Carolina at Chapel Hill
5. Kong X, Xu Z, Shen G, Wang J, Yang Q, Zhang B (2016) Urban traffic congestion estimation and prediction based on floating car trajectory data. Futur Gener Comp Syst 61:97–107
6. Chen C (2008) Low-cost loosely-coupled GPS/odometer fusion: a pattern recognition aided approach. IEEE 1603–1608

Application of Renewable Solar Energy with Autonomous Vehicles: A Review

Mohammad Waseem, A. F. Sherwani and Mohd Suhaib

Abstract In the last two decades, automation and artificial intelligence improve accuracy as well as the quality of product and reduced processing time due to enormous advancement in the technology of robotics. At the present time, the most crucial factors in the modern world are scarcity of energy resources for global demand. Due to environmental issues and legislative straining, electrification is a glaring inclination to renovate performance and sustainability of the transportation system. Solar photovoltaic technology is an important research area to convert solar energy into useful electrical power. So far, robot extracts electrical energy stored in the batteries to run its mechanical, electrical and electronic devices to perform several tasks for industrial as well as commercial work. The robot can operate in a hazardous environment for a long duration of time without human assistant with high accuracy.

Keywords Hybrid vehicles (HVs) · Autonomous guided vehicles (AGVs) · Mars exploration rovers (MER) · Photovoltaic (PV)

1 Introduction

Energy crisis is the prime issue in the world at the present time as fossil fuels and uranium are the only available conventional energy resources. So, the prime resources of non-renewable energy are fossil fuel, petroleum and coal. Numerous injurious gasses such as carbon dioxide (CO_2) and nitrous oxide (NO_x), etc., are produced by the combustion of non-renewable resources of energy. These gaseous by-products result in damage of the ozone layer in the earth atmosphere and increment in global warming [1].

M. Waseem (✉) · A. F. Sherwani · M. Suhaib
Department of Mechanical Engineering, Faculty of Engineering and Technology, Jamia
Millia Islamia (A Central University), New Delhi 110025, India
e-mail: waseem159088@st.jmi.ac.in

© Springer Nature Singapore Pte Ltd. 2020
S. Ahmed et al. (eds.), *Smart Cities—Opportunities and Challenges*,
Lecture Notes in Civil Engineering 58,
https://doi.org/10.1007/978-981-15-2545-2_13

The most widely used technology in the existing transportation system is the internal combustion engine (ICE). The combustion of fossil fuel in ICE results harmful gases to the environment. Emissions of carbon-containing product from ICE vehicles are dominating the environmental issues pollution problems [2]. Due to several eco-friendly and non-polluting factors, the renewable resources of energy such as biomass, wind, geothermal and solar PV technology are getting more attention throughout worldwide [3].

The automobile industry is undergoing a revolution in designing new electrical platforms for vehicles to counter the sophistication involved with engine and carbon emission issues. Therefore, the alternate engine technology is needed to revamp the ICE vehicles. Electric vehicles are the alternate in place of ICE technology. Electric propulsion system not only diminishes the pollution issue but also conveys precision accuracy of power and easy vehicle handling. Electric vehicles are considered "green vehicles" to reduce carbon emission from transportation sector.

Hybrid electric vehicles are investigated by the automobile zone to lessen the application of ignition engine with integrating of electric motor/machine system, i.e. electric propulsion system. The proposed technology by the automobile sector has a lesser carbon emissions as compared to conventional engine [4]. Autonomous guided vehicles (AGVs) are the further advancement of technology to enhance the automation in industries for transferring the verities of material. In fact, AGVs are innovative electrical vehicle producing zero emissions at tail point [5].

Solar-powered robots/AGVs utilize renewable solar resources of energy. Renewable solar energy can be utilized directly or indirectly with the help of energy storage systems such as electric batteries, fuel cells and supercapacitors. Robots utilizing solar energy photovoltaic technology are the most promising new advancement in robotics and automation. These robots are self-generative (i.e. do not require an external power source) and save a great deal of energy. But, harnessing solar energy in robots has its own disadvantages. Solar power is very irregular, and very large panels are required to create small quantities of power. In this paper, applications of solar robotics system are reviewed.

2 Studies of Autonomous Guided Vehicles

Power security, comfortability and elasticity of AGVs are imperative subjects to use in automation industrial environment. Autonomous guided vehicles that can locate their position and choose the practicable path for execution on topography are well stated as "self-moving vehicle" or "unmanned guided vehicle". AGVs are getting more important in industrial as well as in transportation system to enhance the accuracy and quality of manufacturing with better time supervision. AGVs are considered as a catalyst in automation industries for developing and under developing countries to speed up manufacture and precision of product. Various path planning and navigation systems for AGVs are discussed below:

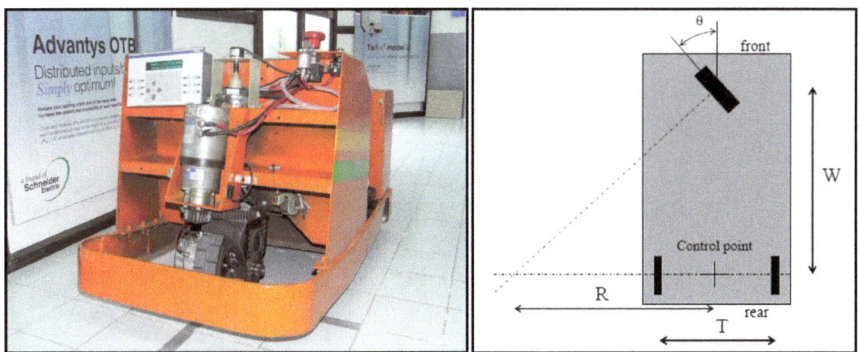

Fig. 1 Prototype and structure of AGVs [7]

Fuzzy and vector pursuit built nonlinear control methods for robust route following in industrial environment for AGVs are proposed by Villagra and Parez [6]. Butdee and Suebsomran [7] and Nayak et al. [8] have conferred vision-assisted image processing method for discontinuity path or an unclear line followed by AGVs as shown in Fig. 1. Several vision-based navigation methodologies have been proposed in the literature for [9–12]. Gulalkari et al. [13] proposed a Kalman filter-based navigation system for AGVs. Sahoo et al. [14] and Duinkerken et al. [15] designed and implemented proportional and proportional integral (PI) controllers.

3   Studies of Solar-Powered Electric Vehicles

There is a scarcity of natural resources to produce adequate electricity as compared to the demand by the people and industry in Bangladesh, even no electricity supply for the rural areas. Secondly, due to high population as well as conventional vehicles (auto rickshaw) that utilize fossils fuel results in pollution and traffic jam problems. Control and Application Research Group of BRAC University, Dhaka, Bangladesh, has implemented and evolved in two projects, namely solar battery charging station and torque sensor-based electricity-assisted rickshaw to resolve these major energy crises. Solar battery charging station implantation can help solar electrification process in the rural region as well as sustain the extra load of the grid in urbanized areas. The battery charged on the charging stations can be employed to power the efficient and environmentally friendly transportation system (rickshaw) which maintains the advance need while changing the dependency on carbon emission fuels by Faraz and Azad [16].

Shaha and Uddin [17] proposed renewable solar energy-based efficient model of an electric rickshaw with an improved range of driving, fast speed of the vehicle, enhanced lifetime and better travelling competency. A battery bank of the electric

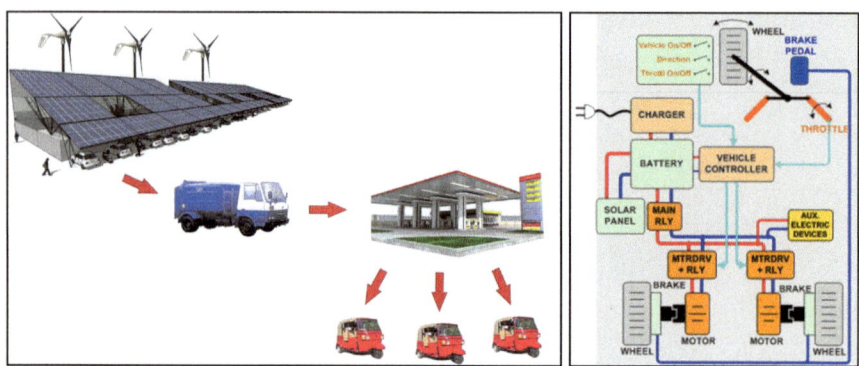

Fig. 2 Prototype and smart energy infrastructure and in-wheel drive technology for three-wheeler vehicle [17]

auto rickshaw is charged in two ways: first, by plug-in charging method and secondly, solar photovoltaic technology during an operation mode of the vehicle. Solar energy availability in the different cities of Bangladesh throughout the year is surveyed to support the construction of electrical auto rickshaw prototype. Fixed, installation and maintenance cost of the electric vehicle is estimated to compare the cost with a conventional auto rickshaw. ADVISOR software was employed to examine the hybrid electric vehicle performance, fuel economy and efficiency. Figure 2 shows the smart energy infrastructure and block diagram of in-wheel drive structure for three-wheeler vehicle.

Paudel and Kreutzmann [18] find out that environment pollution problems are mainly the outcomes of fossil fuel-based transportation. These factors are responsible for contributing pollution for economical and atmospherically sustainability, while one-third of worldwide energy is consumed by the US in transportation, taking a major part in CO_2 emissions. A hybrid tricycle design for a sustainable need of local commute is presented as an alternative means to revamp these factors and reducing energy consumption. Preitl et al. [19] solar hybrid vehicle (SHV) is an advancement form of a hybrid vehicle-mounted photovoltaic solar cells technology to utilized renewable energy as an alternate source of energy. A mathematical model has been developed for this HSV which consists of internal combustion engine, electric motor, solar panel and management unit for a vehicle to make power balance between powers of the electric generator, power obtained from PV panels, battery nominal power and electric power. Nonlinear and quasi-piecewise linear mathematical model for generator load torque, motor dynamics, generator behaviour and battery model was derived by assuming bilinear term as linear parameter varying system. Controllability and stability of design control system condition were studied, analysed and simulation performed to optimized fuel consumption of HSV.

Fang et al. (2015) [20] presented uninterrupted mechanical transmission (UMT) driveline control system for electrical vehicle to improve vehicle dynamics, economics and comfort performance of vehicle as government body emphasizing strict restriction on the usage of fossil fuel and their emissions due to an increase in atmospheric pollution, global warming and scarcity of oil resources. Fuzzy logic control (FLC) having an optimal control strategy at decision module layer of the controller has been executed in the supervisory strategy of a transmission system for targeting accurate shifting processes with the gearing module system. Sarkar et al. [21] proposed electric power system for a solar vehicle that utilizes the photovoltaic modules to harness solar irradiation. In the presented approach, they employed brushless DC motor for the solar electric vehicle due to better operating characteristics such as high torque, high efficiency, low inertia and need less space compared to DC motor.

4 Studies of Solar Robots

Various robots which utilized the solar power to perform different tasks discussed are summarized as below:

Satellite solar power station (SSPS) concept to meet the future energy demand in space based on photovoltaic conversion technology is represented by Glaser [22]. Design and performance analysis of Mars Exploration Rovers with a solar array to accumulate the excessive dust during the mission was presented by Stella et al. [23]. Lukic et al. [24] presented autonomous solar auto rickshaw having an electrical actuator to propel the vehicle and batteries charged by renewable solar energy that operates in an eco-friendly way to replace LPG and CNG-based conventional auto rickshaw.

Deor and Angel [25] presented the design and construction of an optimized charging system for Li-Po batteries with the aid of tracked solar panels for a VANTER robotic exploration vehicle. Cordova and Gonzalez [26] designed and experimented solar-powered underwater vehicle with an intelligent navigation system to measure the physical and chemical parameters of water quality. Riaz et al. [27] explained the design and fabrication of automated personal mobility vehicle (wheelchair) with retractable solar panels to help handicaps. Figure 3 shows the smart wheelchair and electrical circuit diagram.

5 Conclusion

Robots are advanced intelligent machines with the various synthesis of technology to alleviate the human endeavour and provide optimized output. AGVS are the mobile robots that automatically transport desired equipment/part to enhance automation in the logistic system have been studied. Electrical energy-powered vehicles are getting more attention in place of the conventional fuel-based

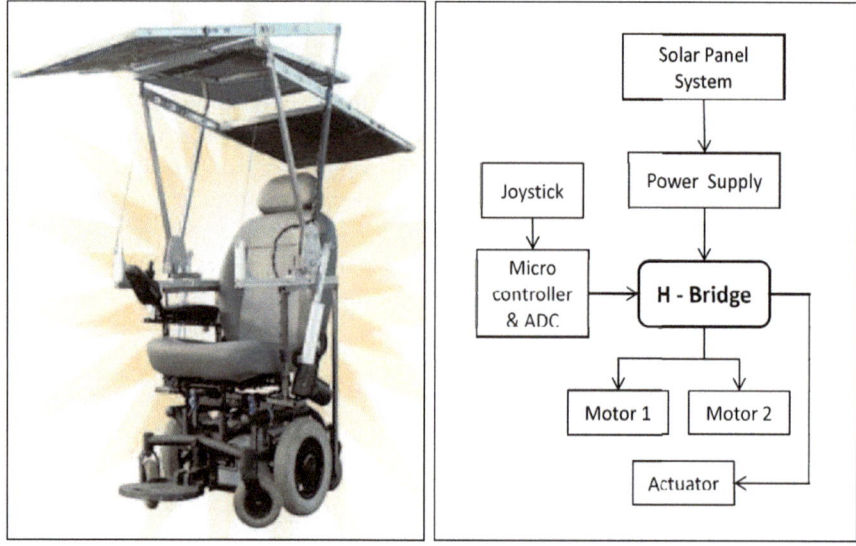

Fig. 3 Retractable solar wheelchair and its electrical circuit diagram [27]

transportation system to solve environmental issues such as pollution and global warming. One of the most reliable renewable energy resources is solar energy which is abundant in nature. In this paper, by combining these two concepts, various solar-powered vehicles are discussed and reviewed.

Acknowledgements We would like to acknowledge all the faculty members and associated staff of Mechanical Engineering Department, Jamia Millia Islamia, New Delhi, to support this research.

References

1. Masood B, Naqvi RAH, Asif RM (2014) Designing of a control scheme for the solar rickshaw in comparative study with conventional auto rickshaw. In: 2014 4th international conference on engineering technology and technopreneuship (ICE2T). IEEE, pp 324–329
2. Alahmad M, Chaaban M, Chaar L (2011) A novel photovoltaic/battery structure for solar electrical vehicles (PVBS for SEV). IEEE Veh Power Propuls Conf 1:1–4
3. Simaes MG, Franceschetti NN, Adamowski JC (1998) A solar powered electric vehicle. Appl Power Electron Conf Expo 1:49–55
4. Hannan MA, Azidin FA, Mohamed A (2014) Hybrid electric vehicles and their challenges: a review. Renew Sustain Energy Rev 29:135–150. https://doi.org/10.1016/j.rser.2013.08.097
5. Hossain S, Ali MY, Jamil H, Haq MZ (2010) Automated guided vehicles for industrial logistics—development of intelligent prototypes using appropriate technology. In: 2010 The 2nd international conference on computer and automation engineering (ICCAE). IEEE, pp 237–241
6. Villagra J, Herrero-Perez D (2012) A comparison of control techniques for robust docking maneuvers of an AGV. IEEE Trans Control Syst Technol 20:1116–1123. https://doi.org/10.1109/TCST.2011.2159794

7. Butdee S, Suebsomran A (2009) Automatic guided vehicle control by vision system. In: 2009 IEEE international conference on industrial engineering and engineering management. IEEE, pp 694–697

8. Nayak AA, Purniya DS, Pradhan GR, Sa PK (2012) Robotic navigation in the presence of static and dynamic obstacles. In: 2012 annual IEEE India conference (INDICON). IEEE, pp 952–955

9. Abe Y, Shikano M, Fukuda T, et al (1998) Vision based navigation system by variable template matching for autonomous mobile robot. IEEE Int Conf Robot Autom 952–957

10. Wang C, Wang L, Qin J, et al (2015) Development of a vision navigation system with fuzzy control algorithm for automated guided vehicle. In: 2015 IEEE international conference on information and automation. IEEE, pp 2077–2082

11. Wicing M, Kunemund F, Hec D, Rohrig C (2014) Hybrid navigation system for mecanum based omnidirectional automated guided vehicles. Jt Conference ISR 2014—45th Int symposium robot 2014—8th Gerenal Conference Robot ISR/ROBOTIK 2014, June 2, 2014 —June 3, 2014 663–668

12. Cucchiara R, Perini E, Pistoni G (2007) Efficient stereo vision for obstacle detection and AGV navigation. In: 14th international conference on image analysis and processing (ICIAP 2007). IEEE, pp 291–296

13. Gulalkari AV, Sheng D, Pratama PS, et al (2015) Kinect camera sensor-based object tracking and following of four wheel independent steering automatic guided vehicle using Kalman filter. In: 2015 15th international conference on control, automation and systems (ICCAS). IEEE, pp 1650–1655

14. Sahoo S, Subramanian SC, Srivastava S (2012) Design and implementation of a controller for navigating an autonomous ground vehicle. In: 2012 2nd international conference on power, control and embedded systems. IEEE, pp 1–6

15. Duinkerken MB, Lodewijks G (2015) Routing of AGVs on automated container terminals. In: 2015 IEEE 19th international conference on computer supported cooperative work in design (CSCWD). IEEE, pp 401–406

16. Faraz T, Azad A (2012) Solar battery charging station and torque sensor based electrically assisted tricycle. In: Proceedings of 2012 IEEE Global Humanitarian Technology Conference GHTC 2012, pp 18–22. https://doi.org/10.1109/GHTC.2012.62

17. Mulhall P, Lukic SM, Wirasingha SG et al (2010) Solar-assisted electric auto rickshaw three-wheeler. IEEE Trans Veh Technol 59:2298–2307. https://doi.org/10.1109/TVT.2010. 2045138

18. Paudel AM, Kreutzmann P (2015) Design and performance analysis of a hybrid solar tricycle for a sustainable local commute. Renew Sustain Energy Rev 41:473–482. https://doi.org/10. 1016/j.rser.2014.08.078

19. Preitl Z, Kulcsar B, Bokor J (2008) Piecewise linear parameter varying mathematical model of a hybrid solar vehicle. In: Proceedings of IEEE Intelligent Vehicles Symposium. Netherlands, pp 895–900

20. Fang S, Song J, Song H et al (2016) Design and control of a novel two-speed uninterrupted mechanical transmission for electric vehicles. Mech Syst Signal Process 75:473–493. https:// doi.org/10.1016/j.ymssp.2015.07.006

21. Sarkar T, Sharma M, Gawre SK (2014) A generalized approach to design the electrical power system of a solar electric vehicle. In: 2014 IEEE students' conference on electrical, electronics and computer science. IEEE, pp 1–6

22. Glaser PE (1977) The potential of satellite solar power. Proc IEEE 65:1162–1176. https://doi. org/10.1109/PROC.1977.10662

23. Stella PM, Ewell RC, Hoskin JJ (2005) Design and performance of the MER (mars exploration rovers) solar arrays. In: Conference record of the thirty-first IEEE photovoltaic specialists conference, 2005. IEEE, pp 626–630

24. Lukic S, Mulhall P, Emadi A (2008) Energy autonomous solar/battery auto rickshaw. J Asian Electr Veh 6:1135–1143
25. Deore T, Angal YS (2014) Optimization of battery charging system in solar-powered robotic vehicle using microcontroller 33–37
26. García-Córdova F, Guerrero-González A (2013) Intelligent navigation for a solar powered unmanned underwater vehicle. Int J Adv Robot Syst 10. https://doi.org/10.5772/56029
27. Riaz N, Aamir JB (2015) Electrical wheel chair with retractable solar panels. In: 2014 International conference on energy systems and policies, ICESP 2014

Influence of SCMs on Flow Properties of Self-compacting Mortar Made with Recycled Fine Aggregate

Monalisa Behera, A. K. Minocha, S. K. Bhattacharyya
and Md. Reyazur Rahman

Abstract The economic growth in India has led to the unprecedented infrastructure development, and as a result, the consumption of raw materials for the construction industry has grown at a very fast pace. Thus, they account for the significant contribution toward greenhouse emissions. In the current global scenario, the faster construction process has become the only mandate to fulfill the demand for infrastructure. However, the construction industry must follow the sustainability paramount to manage its environmental impact. Hence, this study has aimed to address these concerns by using recycled materials such as fly ash (FA), silica fume (SF), and recycled fine aggregate (RFA). The main aim of this paper is to investigate the effect of supplementary cementitious materials (SCMs) with recycled fine aggregate on the flow/rheological properties of fresh self-compacting mortar. Self-compacting mortar is made with cement and fine aggregate in which the cement was replaced with various proportions of FA and SF and the natural fine aggregate with RFA. The mortar mix composition has been finalized as per EFNARC guidelines. A total of twenty mixes is prepared by using different proportions of mineral admixtures and different types of sand for this experimental investigation. The flow properties of mortar are assessed in terms of slump flow through flow cone and in terms of V-funnel efflux time through V-funnel test. The flow tests are performed at varying water/powder (w/p) ratio and varying the dosage of superplasticizer (SP). The cube specimens of mortar are also tested at various ages to determine the strength. The influence of RFA on the flow properties of mortar was observed. It is noted that mortar made with RFA has shown a gradual decrease in the flowability due to its high water absorption and it demanded more SP dosage to maintain the flowability. For a better understanding of the effect of morphology of SCMs on the flow properties, the mortar flow properties were

M. Behera (✉) · A. K. Minocha · Md.Reyazur Rahman
CSIR-Central Building Research Institute, Roorkee, India

A. K. Minocha · S. K. Bhattacharyya
Academy of Scientific and Innovative Research, Ghaziabad, India

S. K. Bhattacharyya
Department of Civil Engineering, Indian Institute of Technology Kharagpur, Kharagpur, India

© Springer Nature Singapore Pte Ltd. 2020
S. Ahmed et al. (eds.), *Smart Cities—Opportunities and Challenges*,
Lecture Notes in Civil Engineering 58,
https://doi.org/10.1007/978-981-15-2545-2_14

investigated with binary and ternary combination of SCMs. From the result, it is observed that the self-compacting mortar can be prepared with RFA at ternary combination of binder (cement + SCMs) to achieve the desired flow characteristics at its 100% replacement level.

Keywords Recycled fine aggregate · Supplementary cementitious materials · Flow properties · Self-compacting mortar · Sustainability

1   Introduction

Natural aggregates have been used as construction materials from past centuries by all the civilization all over the globe. In the twenty-first century, we are at the peak of infrastructure development and urban expansion. In order to solve the environmental issues and to help the growing demand on natural resources, the Government of India has banned the use of natural river sand and put pressure on the construction sectors to use alternative to sand. Thus, one of the possibilities for reducing the environmental cost is by substituting the natural resources with recycled materials. Construction and demolition (C&D) waste is a heterogeneous material which may be composed of various types of wastes such as concrete rubble, bricks, tiles, and marbles. [1]. Faster construction process with sustainability goal will contribute toward smart city, which can be obtained by the assembly of the prefabricated building components or in place casting of the prefabricated components. Self-compacting concrete (SCC) is the mandate option to cast the prefabricated components with less time and economically. Because it is a new generation concrete, which is also known as a performance-based concrete because of its superior flowability under its own weight, it can pass through the congested reinforcement. The main challenges lie in SCC for increasing the flowability of particle suspension without the segregation or bleeding of the particle phases. It is a heterogeneous phase in which the surface chemistry plays an important role in balancing the flowability and stability. Hence, the stability is controlled by the surface-active admixture and the specific surface area of the fine content. However, SCC requires a very high binder/powder content and relatively high sand content to achieve the higher flowability. One of the demerits of SCC lies in its cost due to high cement content, high chemical admixture content, and also high sand content. Thus, one of the potential solutions to reduce the cost of SCC is by using SCMs to replace cement and by using recycled materials. Thus, the substitution of this powder with SCM and the natural sand with RFA will not only contribute toward sustainability, but also will reduce the environmental pollution caused due to the disposal of solid wastes. As SCC requires a more quantity of fine aggregate than coarse aggregate, the use of RFA can provide an opportunity to gain economy and the environmental benefits. The most common type SCMs used in concrete are FA, SF, and ground granulated blast furnace slag (GGBS), which are industrial by-products obtained from thermal power station, electric arc furnaces, and steel plant. These SCMs are used in concrete to reduce the effect of shrinkage, creep and to improve the durability of concrete at long term [2–4]. Use of locally available SCMs with RFA in producing SCC will improve its performance of mortar and concrete.

Most of the research so far confirmed that the RFA shows very high water absorption [5] value due to the adhered porous cement paste and it is having high fine content up to 30% [6]. As a result, it will provide a large number of water-filled pores around the RFA due to the absorption of water from the cementitious system. Thus, it will make the RFA particles heavier and it will lead to settle the particles from the cementitious substances if it is not stirred continuously. Hence, the optimization in the paste and mortar is necessary to keep the suspension in flowable condition in SCC.

The objective of this study is to evaluate the effectiveness of various SCMs in producing the flowable mortar incorporating RFA as replacement of NFA. In this study, mortar is prepared from both NFA and RFA, in which the cement is replaced with various percentages of FA and SF. These two SCMs were chosen, based on its particle shape (rounded for FA) in contributing toward high flowability and improving the strength with SF to compensate the negative effect of RFA on mortar and SCC properties. Mortar acts as a vehicle or suspension media for the workability properties of SCC. Hence, it is very essential to design and assess the properties of mortar as it provides flowability to the solid ingredients in SCC. Apart from that, the use of different types of SCMs plays a different role in the fluidity and it also affects the optimization of SP and viscosity modifying agent (VMA) dosage. In addition to this, the use of RFA in making SCC may affect the flow or rheological behavior of SCC due to its high water absorption value and its rough and porous surface structure. The rough and porous surface structure of the RFA will contribute more frictional forces to the free flow of mortar or concrete. An extensive study has been carried out at various combinations of SCMs with the replacement of cement and the effect of SP dosage on the flowability of mortar made with RFA was studied to optimize it in concrete. The flow properties of mortar are evaluated through the flow cone and V-funnel test. The SP dosage can significantly alter the apparent viscosity of cementitious system. However, its working principle still remains controversial in the different cementitious system. Hence, the SP dosage was optimized on the basis of saturation dosage in the corresponding mortar mix for its use in concrete.

2 Experimental Work

2.1 Materials Used

The basic materials used for self-compacting mortar are same as the traditional mortar such as cement, fine aggregate, and water in addition to mineral admixture and some chemical reagent. OPC 43 grade cement confirming to IS:8112 [7] has been used as the main binder having a specific gravity of 3.15. In this study, Class F FA (confirming to

IS:1727 [8] and IS:4032 [9], surface area: 4065 cm^2/gm) and condensed SF (confirming to IS:15833 [10], surface area: 16000 cm^2/gm) are used to replace cement having specific gravity of 2.25 and 2.16, respectively. The FA is used at various replacement levels, such as 20, 30, and 40%, and the SF is used at two replacement levels of cement such as 5 and 10%. The chemical composition of all the cementitious materials is shown in Table 1. The cement and the inorganic mineral additives are intergrinded in a ball mill to make them more fine and to blend with each other. The powder composition of all the mixes is shown in Table 2. Natural river sand confirming to zone II of IS:383 [11] is used as fine aggregate, and RFA confirming to zone II of IS:383 is used to replace NFA at 100% replacement level. The specific gravity of NFA and RFA are 2.67 and 2.10, respectively. The water absorption values of NFA and RFA are found to be 1% and 11.5%, respectively. A polycarboxylic ether-based SP is used as chemical admixture to reduce the water and to achieve the required fluidity by dispersing the cementitious particles.

2.2 Methodology

The w/b ratio of cement paste and mortar was finalized on the basis of w/b ratio at zero flow (β_P). Initially, the w/p ratio of the mix was decided with flow cone test from the paste study at various w/p ratios (by volume) such as 1.1, 1.2, and 1.3 for zero flow condition. For this, the chosen proportion of OPC and additives are mixed with selected w/p ratio and the flow is measured. The mix proportioning of self-compacting mortar was carried out in two phases. Initially, the minimum w/p ratio was considered for the paste study at zero flow condition. In the second phase, the mix proportioning of self-compacting mortar was carried out based on EFNARC guideline [12]. The mixing procedure was kept constant for all the mixes. After mixing, the tests were carried out on fresh mortar. The flowability and the viscosity of the mortar mixes were evaluated through mini slump cone and mini V-funnel tests. The SP was also optimized to achieve the desired self-compacting properties. The target values for slump flow and V-funnel time as per EFNARC guideline are 24–26 cm and 7–11 s, respectively. It was tried with both NFA and RFA separately for all the powder composition to achieve the target values by increasing the SP dosage gradually up to its saturation level. The major component in SCC is the fine fraction which includes the powder content and the fine aggregates. These fine fractions in combination with water provide the fluid media for the

Table 1 Chemical composition of cementitious materials

Binder	SiO$_2$	Al$_2$O$_3$	CaO	MgO	K$_2$O	Fe$_2$O$_3$	P$_2$O$_5$	Na$_2$O	SO$_3$
Cement	20.41	5.07	64.16	3.08	0.64	3.74	–	0.30	1.21
FA	54.23	30.57	2.38	1.02	0.20	1.59	0.30	0.35	0.24
SF	90.68	0.72	1.30	0.24	0.92	0.23	2.98	1.15	0.12

Table 2 Detail of powder composition in the mixes

Component	Mix-1	Mix-2	Mix-3	Mix-4	Mix-5	Mix-6	Mix-7	Mix-8	Mix-9	Mix-10
Cement	100	70	60	95	90	75	70	65	60	60
FA	0	30	40	0	0	20	20	30	30	35
SF	0	0	0	5	10	5	10	5	10	5

coarse aggregate to be suspended. Hence, the adjustment of admixture dosages was made in paste, mortar, and concrete and was carried out to validate a methodology for SCC mix design. Thus, by taking the combination of powder composition along with NFA and RFA into consideration, a total of 20 mixes have been tested for the flow properties of mortar. The control mix is made with only OPC and NFA. Finally, the flow behavior for all the mortar mixes has been compared with the control mix.

The compressive strength was determined on the cube specimens of size 50 mm × 50 mm × 50 mm, made with self-compacting mortar. The samples were casted into molds immediately after flow tests and cured in water till the testing of the specimen in UTM for various ages such as 7 days, 28 days, and 56 days. Casting and curing of samples are shown in Fig. 1. The strength was assessed to see the influence of SCMs along with RFA on the strength of mortar samples and to correlate the compressive strength development in concrete.

3   Results and Discussion

3.1   Paste Study

The spread of the paste with different w/p ratio is shown in Fig. 2. These values are obtained from the graphs plotted between w/p ratio and the relative slump flow. The graphs plotted for β_P determination and the values of β_P are shown in Fig. 3 and Table 3. The variation observed in β_P for various powder compositions can be due to the difference in the particle structure and their distribution in the system. The shape of particles plays a very crucial role in the flow behavior of any suspended medium. For example, the rounded shape of FA particles will impart a positive effect on the flowability as compared to the comparatively irregular shape of cement [13]. Similarly, finer the particle size, the more is the water demand due to more surface area and less is the flow as observed in the case of the powder compositions having SF. SF is much finer than the cement and has comparatively more rough physical structure than cement. Hence, finer particles are needed to maintain the viscosity of the mix, but these lead to lower flowability due to their higher specific area. This can be seen in the Mix-4 and Mix-5. The mix with 40% fly ash showed the lowest β_P as compared to other compositions. Mix-5 with 10% SF showed the

Fig. 1 Casting, curing, and testing of mortar specimens

highest β_P value. It is also observed that with gradual increase in FA content, the β_p values decreased and with the gradual increase in SF content, the subsequent β_p values increased in the mixes having both FA and SF content. It is due to the reduced interparticle friction with only FA and increased interparticle friction in FA and SF, respectively. Thus, the shape and distribution of all the cementitious substances will significantly influence the flowability of self-compacting mortar.

3.2 Mortar Flow Study

A total of two series of mortar has been prepared to investigate the effect of RFA over NFA along with various powder compositions. Each series contains ten mixes. The w/b ratio of all the mortar mixes is calculated from the β_P, obtained from paste study. Hence, it varies for all the mortar mixes. The water absorption values of aggregates are adjusted in the total water to achieve the desired flow and to compensate for the free water demand. The aim of this study is to achieve the target flow of mortar by varying the constituent materials.

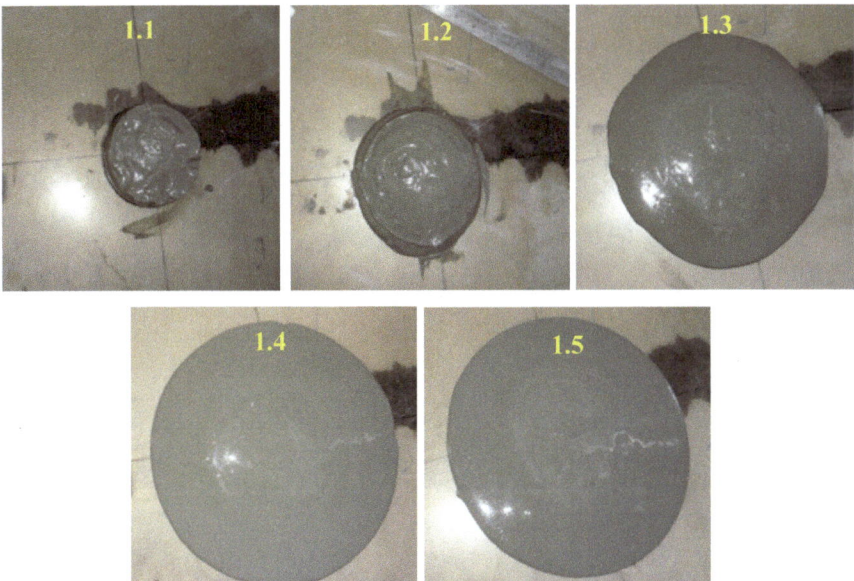

Fig. 2 Typical flow spread of paste with increasing *w/p* ratio

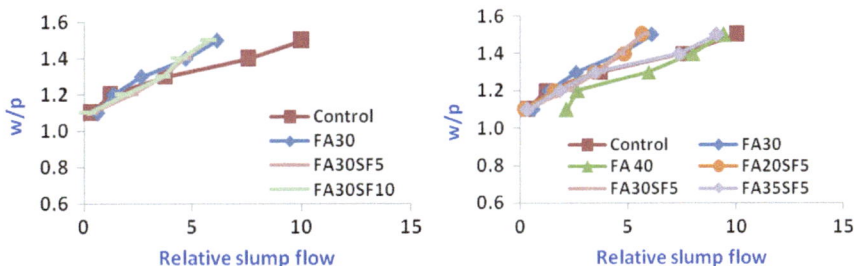

Fig. 3 Flow spread of different paste mixes wrt different *w/p* ratio

Table 3 *w/p* ratio of mixes for zero flow (β_P)

Mix	Mix-1	Mix-2	Mix-3	Mix-4	Mix-5	Mix-6	Mix-7	Mix-8	Mix-9	Mix-10
β_P	1.128	1.035	1.026	1.148	1.165	1.070	1.100	1.060	1.070	1.113

The sand content was taken as 50% of the total volume of mortar in accordance with EFNARC guideline. The weight of sand is different for the NFA and RFA series mortar due to the variation in the specific gravity of both the aggregates. The paste volume is taken as the remaining 50% of the total volume of the mix. The adjustment of SP dosage is carried out in mortar of various mixes depending on the *w/p* ratio. In the mortar flow study, the flow spread is recorded by increasing the SP dosage gradually until it attended a flow between 24 and 26 cm and a flow time of 7–11 s. In this manner, the SP dosage has been optimized for all the mixes as shown in Fig. 4. The optimized dosage of SP has been considered as the saturation dosage for this study. The flow diameter of the mortar is recorded in two perpendicular directions. The average diameter has been taken as the flow spread of the mortar. Finally, the flow spread has been calculated in terms of a relative slump. Similarly, the flow through V-funnel is recorded in terms of relative funnel speed. Segregation and bleeding were checked visually while performing the slump flow test during SP optimization. The results of the slump flow and V-funnel are shown in Table 4.

It can be observed in the Mix-4, Mix-5, Mix-7, and Mix-9 that the use SF leads to increase in the viscosity (v-funnel time) and decrease in the flow of material. Similar behavior is reported in the literature for SCC [14]. It is due to the higher fineness of SF which leads to the increase in interparticle friction. Thus, the optimum SP dosage has been suddenly increased in the mixes having 10% SF to achieve the desired flow. However, in case of RFA mortar series, the increment is quite high, which can be observed from the Fig. 5. It was also observed that the slump flow values of RFA mortar series were nearly equal to the slump flow values of NFA series mortar, and however, the v-funnel flow time of RFA mortar series is higher than the NFA mortar series. This difference noticed in the flow properties could be due to two reasons. Firstly, it is due to the porous nature of RFA, which leads to absorbing the available water instantly, thus leading to lower the free water from cementitious system. Secondly, it can be attributed to the increment in surface interaction between RFA and the binder which leads to more interparticle friction. The rough and porous structure of RFA led to cause hindrance in the free flow of mortar though v-funnel. It is due to the interlocking of particles with each other in the small opening of v-funnel that cause more friction and more time to pass through.

In case of NFA series, though the SP requirement increased with SF content, the influence of fly ash predominates over the influence of SF. From the observations, it can be concluded that for developing SCC with NFA, initial trials can be done with

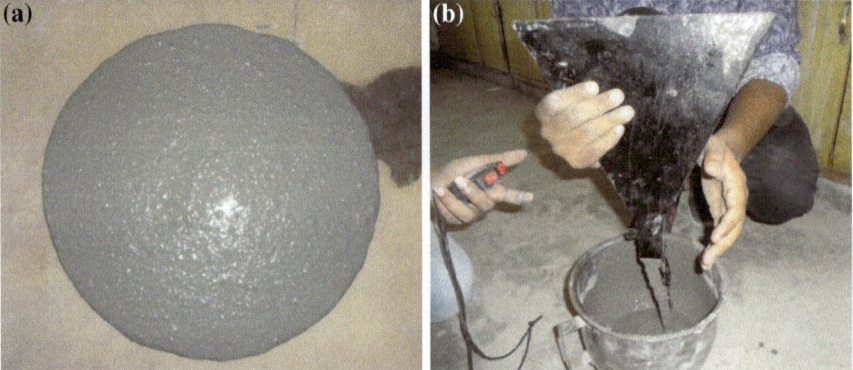

Fig. 4 **a** Mortar slump flow test and **b** mortar V-funnel test

Table 4 Flow spread and V-funnel time of NFA and RFA mortar

Mix name	NFA		RFA	
	Flow dia. (mm)	V-funnel time (sec)	Flow dia. (mm)	V-funnel time (sec)
Mix-1	271	7.00	255	9.37
Mix-2	248	9.43	251	8.38
Mix-3	246	8.20	255	9.84
Mix-4	251	7.94	274	8.12
Mix-5	247	7.85	255.5	8.28
Mix-6	249	10.00	255	10.97
Mix-7	258	10.70	255	11.50
Mix-8	253	10.94	259	11.00
Mix-9	255.5	9.70	257.5	11.25
Mix-10	261	7.22	256.5	11.43

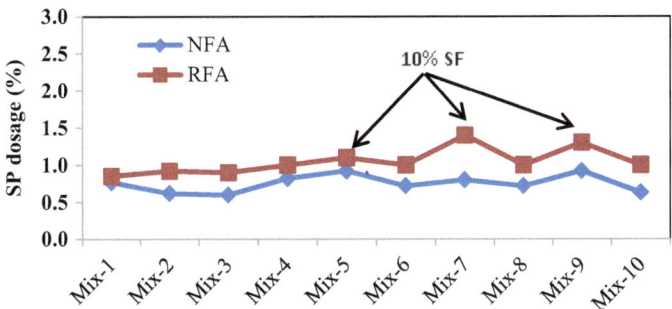

Fig. 5 Optimized SP dosage of different mixes

0.6–0.92% of SP dosage for various percentage of FA and SF. Similarly, for developing SCC with RFA, trials can be done with 0.85–1.4% of SP dosage. From the results of V-funnel test shown in Table 4, it is found that the use of SF increases the viscosity of the mix as it has increased the funnel passing time. Hence, we can replace or reduce the use of VMA and vice versa. Similarly, the flow increased and the flow time decreased with the gradual increase in FA content. It can be explained the same as in the paste study. Thus, FA is used to reduce the yield stress and the viscosity of the mixes. From visual inspection, it is also observed that the mix with 30 and 40% FA is more prone to bleeding and segregation than mixes having SF, confirming to the previous literature [15]. From RFA series of mixes, it is also observed visually that these mixes are less prone to bleeding and segregation due to the presence of more fine content in RFA. Thus, it increases the path of water movement, reducing the chances of segregation and bleeding.

3.3   Compressive Strength of Mortar

The compressive strength of all the mixes has been determined to evaluate the quality of concrete in its hardened state. The performance of self-compacting mortar has been determined on the same mixes of mortar that were used for determining flow diameter and flow time. Strength has been determined for various ages such as 7 days, 28 days, and 56 days and presented in Fig. 6. It can be seen from the graph, in case of NFA mixes, the highest strength at 28 days was achieved by Mix-5 having 10% SF followed by Mix-7 having 20% FA and 10% SF. Similarly, the 56 days strength showed that the maximum strength achieved by Mix-7 followed by Mix-9 with 30% FA and 10% SF. It is also found that the replacement of RFA leads to a decrease in compressive strength for all the ages. However, in case of RFA mixes, the maximum strength at 28 days and 56 days was achieved by Mix-5, followed by Mix-1. If more sustainability is taken into consideration, then it can be concluded that the Mix-9 having 30% FA and 10% SF performed better for both

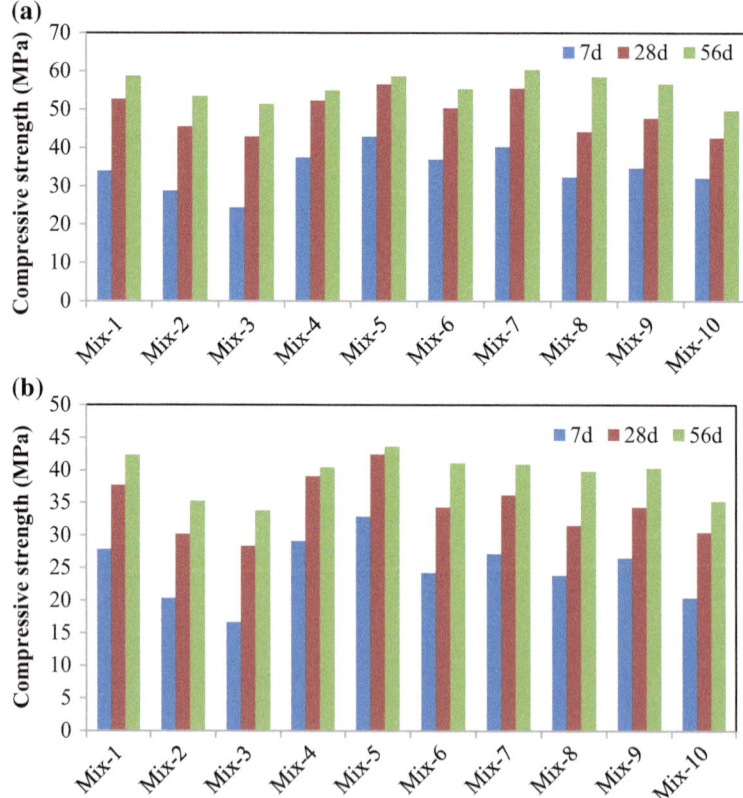

Fig. 6 Compressive strength of different mixes **a** NFA mixes and **b** RFA mixes

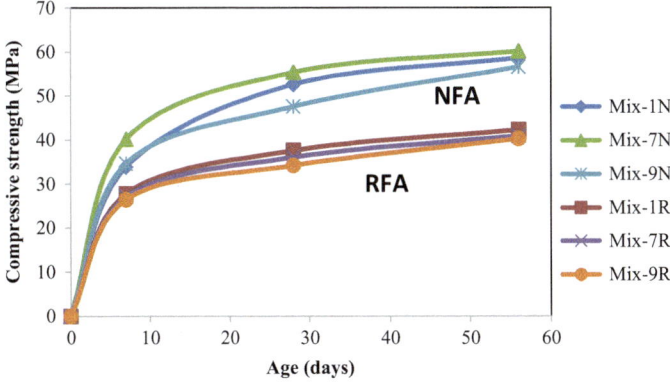

Fig. 7 Compressive strength gain with age in NFA and RFA mortar

NFA and RFA mixes. The SF led to the early strength development, and the FA led to the latter age strength development for both the series. The strength reduction at 56 days for RFA series is within 25.67–42% with respect to control mortar. The strength reduction of the mix with RFA and OPC was by 28% than the control concrete at 56 days. The early strength gain in the SF mixes is higher than the latter age strength gain for both the series. However, in the case of RFA series, the early day's strength was quite higher than the NFA mixes for all the mixes. This could be more due to the higher rate of hydration due to the more available moisture in RFA series and due to the presence of unhydrated cement particles in RFA which got hydrated at an early age. But, the ultimate strength gain was less than NFA series due to the poor quality of NFA than RFA. Thus, taking the three mixes combination with highest strength gain into consideration, such as Mix-1, Mix-7, and Mix-9, the strength results have been shown in Fig. 7 on a comparative basis. From this figure, it can be said that SCC can be made with RFA by taking a ternary binder combination. Thus, SCC can be tried with a combination of 30% FA and 10% SF.

4   Conclusions

This paper investigates the influence of different types of SCMs on the fresh properties of flowable mortar produced with RFA for its use in SCC. From this study, following conclusions can be drawn:

- Minimum water-to-powder ratio (β_P) is essentially needed to be determined for developing self-compacting mortar and concrete for determining the w/b ratio of mortar or concrete mixes while using a binary or ternary binder. Depending upon the mixes, the β_P value was found to be ranging from 1.026 to 1.165 by volume.

- The use of SCMs such as FA and SF will influence the flowability and viscosity of both the series of mortar, respectively. Hence, the use of cement, SP, and VMA can be optimized and reduced to obtain the desired flowability. From this study, it was found that for developing SCC with NFA, initial trials can be done with 0.6–0.92% of SP dosage for various percentages of FA and SF, while with RFA, this value was between 0.85 and 1.4%.
- Use of ternary binder performs better for developing the self-compacting mortar for achieving the required flowbility and segregation resistance than the use of binary binder for both NFA and RFA series. In the case of RFA mixes, the use of SF is to be optimized as it provides higher interparticle friction along with RFA. Thus, the surface structure of the constituent materials is one of the governing parameters to control or predict the flow behavior of self-compacting mortar or SCC.
- The use of SF increases the viscosity (v-funnel time) and decrease the slump flow of mortar due to the higher fineness which leads to the increase in interparticle friction. Also, the optimum SP dosage has been suddenly increased in the mixes having 10% SF to achieve the desired flow showing the more viscous nature of SF. However, by increasing the FA content, the flow value increased and the flow time decreased. It means that the viscosity decreases with the increase of the FA incorporation.
- From the compressive strength result, it was concluded that the combination of 30% FA and 10% SF gives most of the favorable results. For this combination, the strength reduction was 31%, which is similar to that with OPC and RFA. Hence, this combination can be used for the making of SCC mix for RFA.

Acknowledgements The paper forms part of CSIR R&D program (Government of India) and funded by CSIR-Central Building Research Institute, Roorkee. It is published with the kind permission of the Director, CSIR-Central Building Research Institute, Roorkee. The authors are also thankful to Burari recycling plant, New Delhi, for providing the raw materials such as RFA to carry out the research work.

References

1. Behera M, Bhattacharyya SK, Minocha AK, Deoliya R, Maiti S (2014) Recycled aggregate from C&D waste and its use in concrete—a breakthrough towards sustainability in construction sector: a review. Constr Build Mater 68:501–516
2. Malhotra VM, Zhang M, Read PH, Ryell J (2000) Long-term mechanical properties and durability characteristics of high-strength/high-performance concrete incorporating supplementary cementing materials under outdoor exposure conditions. ACI Mater J 97(5):518–525
3. Jianyong L, Yan Y (2001) A study on creep and drying shrinkage of high performance concrete. Cem Concr Res 31(8):1203–1206
4. Gedam BA, Bhandari NM, Upadhyay A (2016) Influence of supplementary cementitious materials on shrinkage, creep, and durability of high-performance concrete. J Mater Civ Eng 28(4):04015173

5. Evangelista L, Guedes M, de Brito J, Ferro AC, Pereira MF (2015) Physical, chemical and mineralogical properties of fine recycled aggregates made from concrete waste. Constr Build Mater 86:178–188
6. Zhao Z, Remond S, Damidot D, Xu W (2015) Influence of fine recycled concrete aggregates on the properties of mortars. Constr Build Mater 81:179–186
7. Bureau of Indian Standard (2013) IS 8112. Ordinary portland cement, 43 grade—specification, New Delhi, India
8. Bureau of Indian Standard (2004) IS 1727. Method of test for pozzolanic materials, New Delhi, India
9. Bureau of Indian Standard (2005) IS 4032. Method of chemical analysis of hydraulic cement, New Delhi, India
10. Bureau of Indian Standard (2003) IS 15388. Silica fume-specification, New Delhi, India
11. Bureau of Indian Standards (2016) IS 383. Coarse and fine aggregates for concrete—Specification, New Delhi, India
12. EFNARC (2005) The European guidelines for self-compacting concrete: specification, production and use. Self-Compacting Concrete European Project Group
13. Ahmaran MS, Christianto HA, Yaman IO (2006) The effect of chemical admixtures and mineral additives on the properties of self-compacting mortars. Cement Concr Compos 28:432–440
14. Benaicha M, Roguiez X, Jalbaud O, Burtschell Y, Alaoui AH (2015) Influence of silica fume and viscosity modifying agent on the mechanical and rheological behavior of self compacting concrete. Constr Build Mater 84:103–110
15. Jalal M, Pouladkhan AR, Ramezanianpour AA, Norouzi H (2012) Effects of silica nanopowder and silica fume on rheology and strength of high strength self compacting concrete. J Am Sci 8(4):285–288

WSN-Based Water Channelization: An Approach of Smart Water

Hrusikesh Panda, Hitesh Mohapatra and Amiya Kumar Rath

Abstract The existing water monitoring system of India is incapable to regulate the excessive water flow which leads to potential water wastage. The overpopulation, unrestrictive water supply, irregular monitoring, and lack of awareness are the major hurdles for the deployment of smart water system (SWS) in India. In this state of the art, we present a simulation-based work to monitor the water flow in accordance with the demand for water by several regions. In our experiment, we have designed a protocol which regulates water flow according to the priority value. The proposed **p**riority-based **w**ater **d**istribution (PWD) protocol executes through two phases. The initial phase assigns priority values to the geographical belts like an emergency belt (EB), residential belt (RB), and industrial belt (IB) by the utility center (UC). In the second phase, we install a *sensor-based smart valve* which is regulated by the UC corresponding to the priority values of several belts. This novice method is addressing one of the open challenges of smart water grid (SWG), i.e., data analytics over wireless communications.

Keywords Smart water · Water management · Wireless sensor network · Water distribution network · Water grid

1 Introduction

Priority basis water supply is an important issue which is being faced worldwide. The resources such as electricity, natural gas, and water need to be managed carefully to ensure future use and sustainability [1]. SWS is a new paradigm with ICT convergence for future generations [2]. The ICT-based SWS is expected to be helpful to the human to deal with various contexts such as scarcity of water, maintaining water quality, and proper distribution of water according to the command and demand of the situation [3]. In order to improve the quality of life, reduce the effects of human activities and the way in which people utilize water new

H. Panda (✉) · H. Mohapatra · A. K. Rath
Veer Surendra Sai University of Technology, Burla, Odisha, India

© Springer Nature Singapore Pte Ltd. 2020
S. Ahmed et al. (eds.), *Smart Cities—Opportunities and Challenges*,
Lecture Notes in Civil Engineering 58,
https://doi.org/10.1007/978-981-15-2545-2_15

157

technologies are being searched [4, 5]. People living in developed and underde-veloped countries are facing problems like non-availability of clean water and water at right time. To meet the demand for clean water by the growing population and urbanization has become a challenge at present [6]. During the process of addressing this issue, it invites many parallel challenges such as cost of manage-ment, water transmission, storage, distribution, and billing for consumption of water. The lifestyle of the people is changing rapidly. At the same time, the capacity of the people to pay has also increased. These attributes have certain negative effects on the use of water [7] (Fig. 1).

The term SWS does not have any generic or worldwide accepted definition, and hence, the definition of SWS is always dependent on the situation and the country structure. Every country and cities define the term SWS as per their own conditions and adopt a scientific and technological solution to focus on its different parts. According to the population index of 2018, India stands at the second position after China. As the population of India is proliferating, the water foundation should be prepared to take care of expanding water demand by considering the resource constraint as a factor. Secondly, to ensure good quality of water for clients round the clock, the conditioning of involved assets in water supply must be maintained. Parallelly, India is marching toward a developed country with the support of industrialization. The demand for water supply from both the sectors, i.e., RB and IB, invites many challenges for proper channelization and customer satisfaction.

2   Literature Review

The remote communication infrastructure and 24/7 sensing make WSN as a reliable tool for SWS. The WSN-based condition assessment retrieves more dynamic and real-time data. The existing water supply is controlled by a valve which is human operated. The intervention of human is an operation creates the problem of delay, mismanagement, and dependency. Hence, the better understanding of WSN with the existing water supply system allows us to plan, design, manage, and operate SWS efficiently [7]. The implementation of WSN also extended to monitor the water

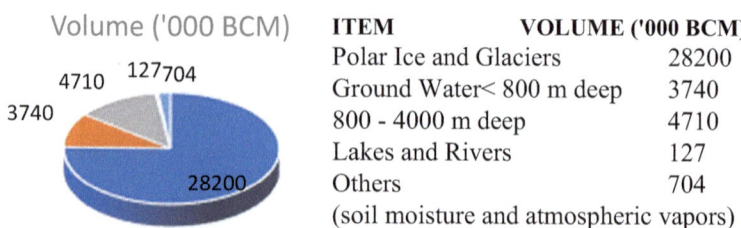

ITEM	VOLUME ('000 BCM)
Polar Ice and Glaciers	28200
Ground Water< 800 m deep	3740
800 - 4000 m deep	4710
Lakes and Rivers	127
Others	704
(soil moisture and atmospheric vapors)	

Fig. 1 Water availability status *Source* http://cwc.gov.in

quality in lakes and sea (Ex.MOBESEN Project) [8]. The WSN concepts also have been used for gauging the ion concentration level in the water through selective field effect transistors (ISFETs) and blind source separation (BSS) algorithm [9].

2.1 Residential

The water channelization process is increasingly drawing attention on the deployment and usage of SWS. The existing literature on smart water recognized several benefits over the traditional water supply system [10, 11]. The SWS provides the benefits of managing huge demand for water supply, water utilities, water billing, and water quality monitoring [12, 13]. As per the existing statistics, the per capita water consumption in India is estimated to 180 L and is distributed among various categories like personal hygiene (70 L), toilet flushing (40 L), dishes (25 L), laundry (20 L), drinking and preparation of food (15 L), and others (10 L) [14, 15]. However, the study says the average water consumption is nonlinear than linear. It seems the lack of importance of knowledge on usage of water in homes leads to heavy water wastage [16]. This field of study majorly focuses on the behavior and knowledge level of the end user. The term knowledge defines how, when, and by whom the water needs to be consumed for efficient water management [17, 18]. The lack of an efficient model needs intersection of sensor networking to operate on data basis.

2.2 Industrial

As per the water consumption statistics, industry stood second in the rank on water usage. The main source of water for the industrial sector is surface water and groundwater. The cost of supplying water is mainly decided by the administrator and by the government since the improper use of water becomes a normal practice [19]. Since the water supplied by the municipal water sources is not sufficient, ultimately industries are dependent on groundwater. (According to the ministry of water resources) [20].

2.3 Emergency

The effect of global warming largely influences on average rainfall. The irregular monsoon rains in most parts of India causes heavy rainfall with floods (Report by NDRF). After a natural calamity, constant supply of water for a long time is very necessary as per the situation concern. A new water supply system to facilitate disaster-resilient communities can be delivered from a traditional water distribution system by accommodating with WSN [19].

3 Proposed Work

In this section, we introduce a novice framework for water channelization monitoring. According to our proposed method, we divide the deployment geographical map into three specific zones such as IB, RB, and EB [20]. We have executed the simulation in Sambalpur district as our case study [21].

In each zone, we have deployed a fixed number of static sensor nodes which are responsible for data collection and transmission. Every SN is identified by SN_{id}. The traditional WSN architecture says that the SNs form clusters to maximize the lifetime of the network, where the cluster head (CH) monitors individual clusters. The CH is responsible for data aggregation and forwarding same to the base station (BS). In our proposed network, the UC acts as a base station which regulates the channelization task according to the data received from SNs.

PWD Algorithm

```
Inputs;
Vp, Zid = 1,2,3;
Number of zones = Zid ;
Water Channelization Graph = Gw ;
Initialization;
SNi = 1, 2, 3, ...,n ;
FORM : Cluster ;
MAP: SNid › → Zid;
READ: Vp by SN ;
REPORT: SN → BS;
INSTALL: SVs ;
READ: Vp by BS ;
Channelize: Water Supply ;
while Vp == 1 do
Redirect Water Supply ;
if Vp ≠ 1 then
Delay Amount of Time ;
else
Stop Water Supply ;
end
end
Repeat: Step 4 with Time instance Ti
```

3.1 Motivation

The motivation of work begins with a report produced by Associated Chambers of Commerce and Industry of India (ASSOCHAAM). This report revealed that the gap between supply and demand of water being nearly 1300 million liters. It is noted by PH division of Odisha that the state suffers a daily loss of nearly 150 million liters of drinking water because of non-uniform distribution of water among various regional belts. Secondly, the traditional water distribution network is operated through human intervention [22]. The lack of awareness among people regarding water scarcity leads to potential water waste [23, 24].

3.2 Methodology

Based on the specification and design requirement, the flow diagram is shown in Fig. 2. This flow diagram defines the overall functions to be performed by the system. This system contains three belts which are represented by SNs in the simulation environment. The distribution management unit is deployed with smart sensors which are operating on the basis of decisions taken by UC. The SNs or individual units of sensor network periodically report their status to the BS. Here, the status defines the priority value (V_p) of several zones according to their demands where V_P. When the priority value becomes 1 the UC redirects the water supply towards that zones subsequently to others according to their priority values. In Fig. 3, we have illustrated the flow diagram of the proposed model.

3.3 Results and Analysis

This paper mainly focuses to present the remote sensing-based water channelization system, only by considering few static SNs. This piece of work is simulated in Cupcarbon (SCI-WSN) simulator. Here, we assumed the UC availed with 24/7 water availability, the SNs are static in nature, as the SNs are domestically deployed there provided with time to time energy backup, and the SNs are in the coverage zone of sink node. Initially, the SNs are mapped with specific zones. Then, according to the command and demand of the situation, the V_p of several belts is updated with the time instance t_i. On the basis of data aggregated with the sink node or BS, the UC is taking decisions regarding channelization of water to a specific zone. Figure 5 illustrates the mapping of SNs with the zones (Z_{id}). The deployment of sensor node (SN) is illustrated in Fig. 4.

In the case of traditional water distribution network (WDN), the water gets supplied non-uniformly among several zones which leads to huge customer dissatisfaction. In Sambalpur zone, during the summer season, the water supply to the

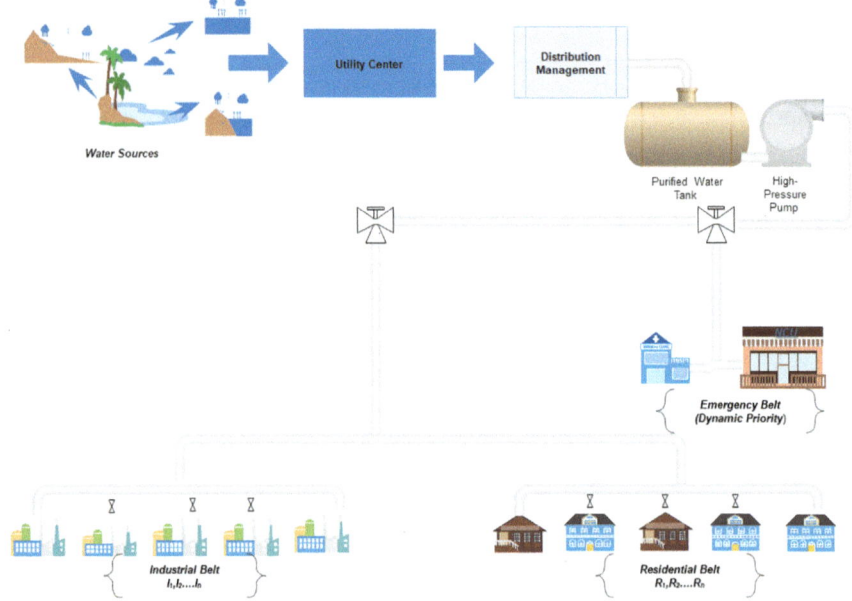

Fig. 2 Model of PWD system

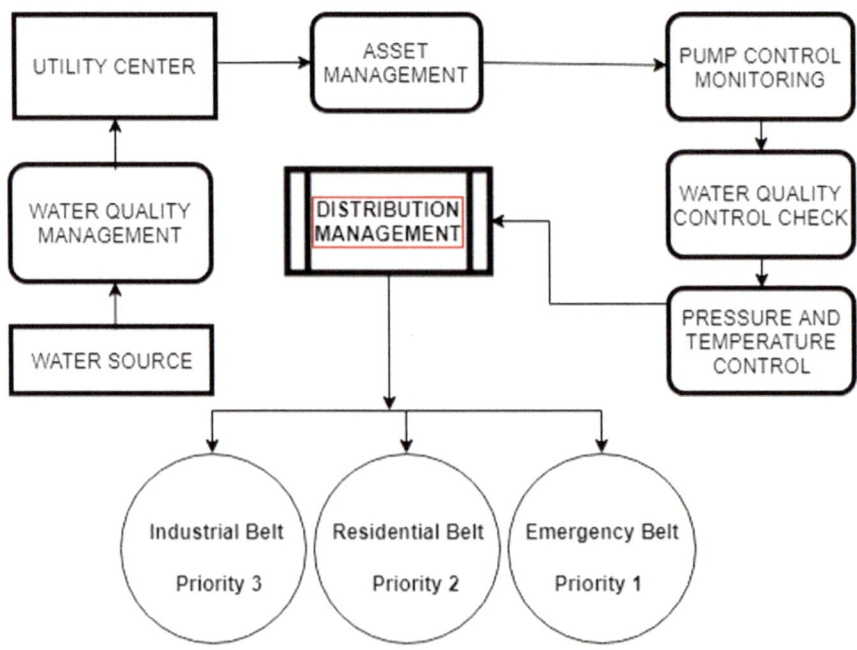

Fig. 3 DFD of proposed model

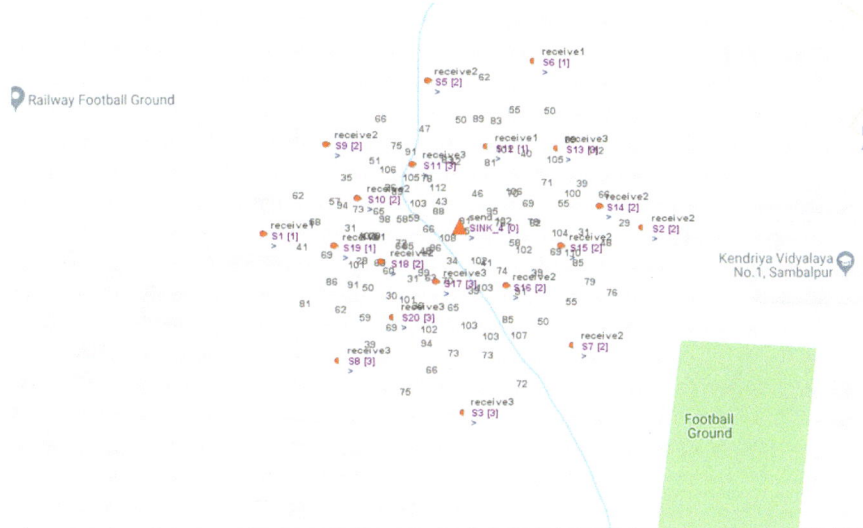

Fig. 4 Node deployment

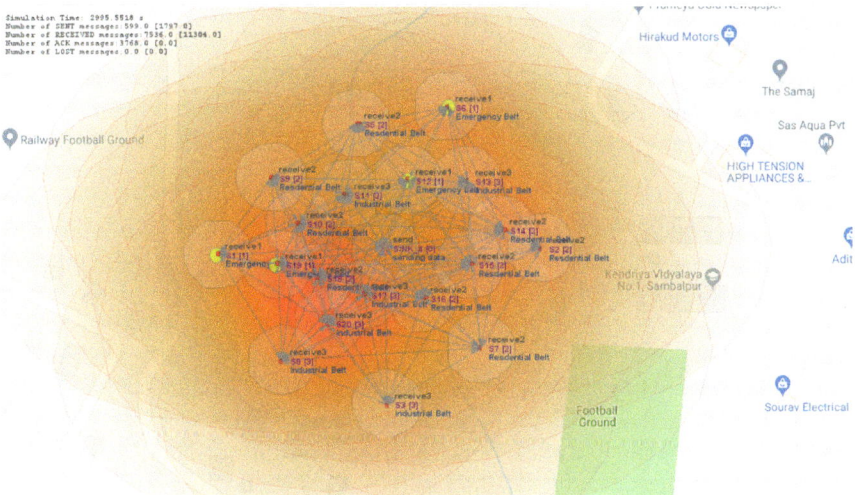

Fig. 5 Mapping phase

urban areas lasts for 13-15 min at maximum. Somehow, the existing WDN fails to reach up to the customer's expectation level. The result of such mismanagement costs in terms of revenue generation, customer satisfaction, and artificial scarcity of water. In Fig. 6, we have illustrated the uniform water supply to various zones according to the assigned time. In this figure, the Y-axis represents water resource

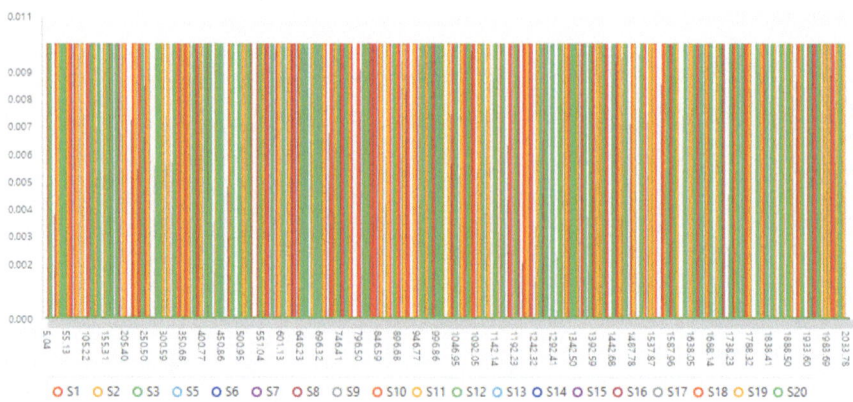

Fig. 6 Optimized water supply scheduling

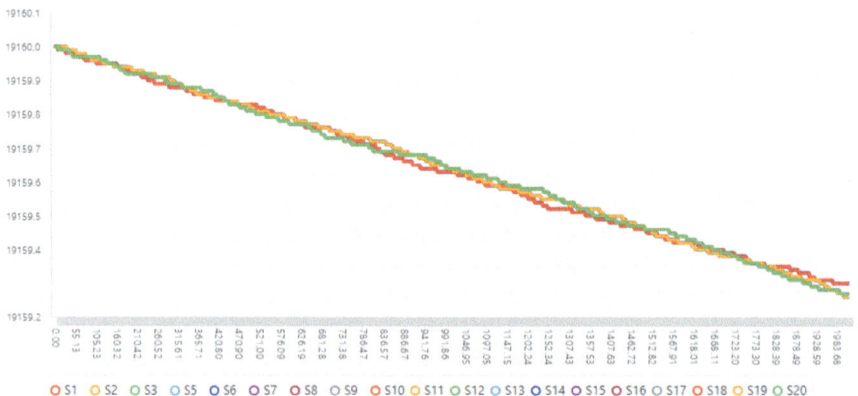

Fig. 7 Uniform water resource consumption

consumption rate, whereas the X-axis represents the SNs. Through our simulation, we have concluded that, when the water supply among various zones are uniform in nature, the sustainability os smart water lasts for long time and we can efficiently manage the water resource by avoiding surplus water supply to a specific zone. In Fig. 7, we present the uniform nonlinear water consumption rate.

4   Conclusion

In this state of the art, we have presented a WSN-based remote water channelization model. The water is a non-renewable resource; so, efficient resource management is needed. This model is implemented logically with SCI-WSN simulator where we have considered the district Sambalpur, Odisha, India as our area of study. The proposed system reduces human intervention as a controller for water channelization. The sensor-based smart valve is an automatic water channelizer. This piece of work can be extended by applying dynamic water demand from several zones.

References

1. Marais J, Malekian R, Ye N, Wang R (2016) A review of the topologies used in smart water meter networks: a wireless sensor network application. J Sens vol 2016, Article ID 9857568: 12. https://doi.org/10.1155/2016/9857568
2. Lin M, Wu Y, Wassell I (2008) Wireless sensor network: water distribution monitoring system. In: 2008 IEEE radio and wireless symposium, Orlando, FL, pp 775–778. https://doi.org/10.1109/RWS.2008.4463607
3. Yuanyuan W, Ping L, Wenze S, Xinchun Y (2017) A new framework on regional smart water. Procedia Comput Sci 107: 122–128, ISSN 1877-0509. https://doi.org/10.1016/j.procs.2017.03.067
4. Lim H, Kim W, Jung J (2018) Integrated water cycle management system for smart cities. In: 2018 2nd international conference on green energy and applications (ICGEA), Singapore, pp 55–58. https://doi.org/10.1109/icgea.2018.8356311
5. Philippe Gourbesville (2016) Key challenges for smart water. Procedia Eng 154: 11–18, ISSN 1877-7058. https://doi.org/10.1016/j.proeng.2016.07.412
6. Günther M, Camhy D, Steffelbauer D, Neumayer M, Fuchs-Hanusch D (2015) Showcasing a smart water network based on an experimental water distribution system. Procedia Eng 119. https://doi.org/10.1016/j.proeng.2015.08.857
7. Suresh M, Muthukumar U, Chandapillai J (2017) A novel smart water-meter based on IoT and smartphone app for city distribution management. In: 2017 IEEE region 10 symposium (TENSYMP), Cochin, pp 1–5. https://doi.org/10.1109/tenconspring.2017.8070088
8. Dallemagne P, Piguet D, Restrepo A, Sénéclauze M (2002) WSN mobility support for long term water monitoring
9. Bermejo S, Bedoya G, Parisi V, Cabestany J (2002, 5–8 Nov 2002) An on-line water monitoring system using a smart ISFET array. Paper presented at the IECON 02, IEEE 2002 28th Annual Conference of the industrial electronics society
10. Boyle T, Giurco D, Mukheibir P, Liu A, Moy C, White S, Stewart R (2013) Intelligent metering for urban water: a review. Water 5(3):1052–1108. https://doi.org/10.3390/w5031052
11. Sønderlund L, Smith JR, Hutton CJ, Zoran K, Savic D (2016) Effectiveness of smart meter-based consumption feedback in curbing household water use: knowns and unknowns. J Water Resour Plan Manag 142(12) Dec 2016
12. Britton T, Stewart R, O'Halloran KR (2013) Smart metering: enabler for rapid and effective post meter leakage identification and water loss management. J Clean Prod, 54: 166–176, Sept 2013. https://doi.org/10.1016/j.jclepro.2013.05.018
13. Stewart RA, Willis R, Giurco D, Panuwatwanich K, Ca-pati G (2010) Web-based knowledge management system: linking smart metering to the future of urban water planning. Australian Planner 47(2):66–74. https://doi.org/10.1080/07293681003767769
14. https://www.thehindu.com/features/homes-and-gardens/how-much-water-does-an-urban-citizen-need/article4393634.ece
15. http://water-purifiers.com/much-water-wasting-everyday/
16. Chena X, Yang S-H, Yang L, Chen X (2015) A benchmarking model for household water consumption based on adaptive logic networks. Procedia Eng 119:1391–1398. https://doi.org/10.1016/j.proeng.2015.08.998

17. Rathnayaka K, Malano H, Maheepala S, George B, Nawarathna B, Arora M, Roberts P (2015) Seasonal demand dynamics of residential water end-uses. Water 7(1):202–216. https://doi.org/10.3390/w7010202
18. Irons LM, Boxall J, Speight V, Holden B, Tam B (2015) Data driven analysis of customer flow meter data. Procedia Eng 119:834–843
19. Public Utilities Board Singapore (2016) Managing the water distribution network with a smart water grid. Smart Water. 2198–2619. https://doi.org/10.1186/s40713-016-0004-4
20. Shivashankar, Kumar RU, Roopesh N, Sihmar S, Ananthaswamy RA, Gowranga (2017) A smart water utilization technique based on voltammetric electronic tongues for domestic and industrial environment. In: 2017 2nd IEEE international conference on recent trends in electronics, information & communication technology (RTEICT), Bangalore, pp 1745–1749. https://doi.org/10.1109/rteict.2017.8256898
21. Alkandari A, Alnasheet M, Alabduljader Y, Moein SM (2012) Water monitoring system using wireless sensor network (WSN): case study of Kuwait beaches. In: 2012 second international conference on digital information processing and communications (ICDIPC), Klaipeda City, pp 10–15. https://doi.org/10.1109/icdipc.2012.6257270
22. Liu A, Giurco D, Mukheibir P (2015) Urban water conservation through customized water and end-user information. J Clean Prod. 112. https://doi.org/10.1016/j.jclepro.2015.10.002
23. Robles T et al (2015) An IoT based reference architecture for smart water management processes. JoWUA 6: 4–23
24. Mohapatra H, Rath AK (2019) Detection and avoidance of water loss through municipality taps in India by using smart taps and ICT. IET Wirel Sens Syst 9(6):447–457

Removal of Pb(II) from Industrial Wastewater Using of CuO/Alg Nanocomposite

Afzal Ansari, Vasi Uddin Siddiqui, M. Khursheed Akram, Weqar Ahmad Siddiqi and Shabana Sajid

Abstract Wastewater adversely affects humans and another animal including metal like Pb, As, Zn, Hg, and Cd in wastewater (domestic or industrial). These toxic metals affect human health and are a serious threat to the environment by the precipitation, adsorption, accumulation in the food chain and non-biodegradable nature, respectively. In the present study, treatment of industrial wastewater in terms of toxic Pb(II) removal was investigated by the using of copper oxide alginate (CuO/Alg) nanocomposite. The CuO/Alg nanocomposite was prepared by chemical reduction method in solution phase, and synthesized particles size were characterized by X-ray diffraction (XRD), transmission electron microscopy (TEM), scanning electron microscopy (SEM), and Fourier transform infrared spectroscopy (FTIR). The wastewater sample collected from WWTP of the local electroplating industry is located in Okhla Industrial Area, New Delhi. A series of experimental approaches have been used to remove Pb^{2+} from industrial wastewater with CuO/Alg nanocomposite, which includes sorbent mass, competitive ion, contact time, and SEM. The SEM image of CuO/Alg nanocomposite showed that particles had a sheet-like shape and mean diameter of about 18.09 nm. The test was performed under the batch condition to determine the adsorption rate and uptake at equilibrium from single component solution. The maximum uptake value of Pb^{2+} in single component solution was 118.40 mg/g from wastewater. The CuO/Alg nanocomposite identified as the most promising sorbent with an effective potential of removal of Pb^{2+} from wastewater is due to their high metal uptake.

A. Ansari (✉) · V. U. Siddiqui · W. A. Siddiqi
Department of Applied Sciences and Humanities, Faculty of Engineering
and Technology, Jamia Millia Islamia, New Delhi 110025, India

M. K. Akram
Applied Sciences and Humanities Section, University Polytechnic,
Faculty of Engineering and Technology, Jamia Millia Islamia,
New Delhi 110025, India

S. Sajid
Department of Chemistry, Gandhi Faiz-E-Aam College, Mahatma Jyotiba
Phule Rohilkhand University, Bareilly Uttar Pradesh, 242001, India

© Springer Nature Singapore Pte Ltd. 2020
S. Ahmed et al. (eds.), *Smart Cities—Opportunities and Challenges*,
Lecture Notes in Civil Engineering 58,
https://doi.org/10.1007/978-981-15-2545-2_16

Keywords Nanocomposite · Wastewater · Adsorbent · Treatment

1 Introduction

The contamination of water by heavy metal ions has become a serious environmental problem due to its toxicity and bioaccumulation tendencies. Across the world, surface and groundwater pollution by heavy metals is a major environmental issue [1, 2]. Heavy metal ions cause harmful effects on animals including humans and are toxic to living organisms through food chain transfer. Heavy metal ions can bind in a living organism to nucleic acids, proteins and small metabolites [3, 4]. These toxic heavy metal ions usually arise in mining industries, electroplating amenities, electric power industries, and other equipment manufacturing units and process waste streams from tanneries [1, 5, 6]. Many more environmental problems have been arising due to the release of toxic heavy metal ions in the natural environment and irrigation of agriculture area by the use of sewage water, these can be accumulated in the food chain due to their biodegradability and perseverance, thus may be a threat to human health [7–9]. In all heavy metals, Lead [Pb(II)] ion is a highly toxic substance, which can cause any adverse health effects including humans for exposure to organisms [10]. The level of exposure to Pb(II) ions in human can illation in lack of IQ level, learning inabilities, lack of attention shortfall, behavioral intricacies, undermined hearing and kidney failure [2, 11, 12]. Usually, children are more affected to lead ion toxicity than others as the lead ion concentration as per unit body weight (up to 40%) is much higher in children [13, 14].

Recently, the application of nanocomposite has become a fascinating area of research to remove pollutants from wastewater. The unique properties of nanocomposite are presenting distinct opportunities to remove toxic heavy metals in efficient and cost-effective approaches, and numerous nanocomposite has been used for this purpose [15–17]. Nanocomposite shows huge adsorption capacity mainly due to the more active sites for interaction with high surface area and metallic species [11, 12, 18, 19]. Metal oxide nanocomposites potentially provide more efficient and cost-effective wastewater treatment and processing techniques due to their size and removal efficiency [20, 21]. The present study aims to synthesis the CuO/Alg nanocomposite with the facile wet-chemical technique and to examine the removal efficiency of Pb(II) ions by the synthesized nanocomposite.

2 Materials and Methods

2.1 Chemicals Used

Copper nitrate trihydrate ($Cu(NO_3)_2 \cdot 3H_2O$), sodium hydroxide (NaOH), and sodium alginate [$(C_6H_8O_6)_n$] pure were obtained from Sigma Aldrich Co Ltd.

Double-distilled water was used for solution preparation. All the chemicals used in the study were analytical grade, and all the analysis was performed in triplicates.

2.2 Fabrication of CuO/Alg Nanocomposite

We have previously reported the synthesis of CuO/Alg nanocomposite [22] and briefly discussed here, as in this experimental part, pure copper oxide (CuO) has been synthesized by using copper nitrate trihydrate and sodium hydroxide as precursors. Two distinct solutions of copper nitrate trihydrate (1.0 M) and NaOH (8.0 M) were prepared in deionized water, respectively. Copper nitrate trihydrate (1.0 M) solution was stirred through dropwise (1 drop per 2 s) on a magnetic stirrer with sodium hydroxide (8.0 M) solution at 70 °C with 600 rpm till gel formation. The mixture was cooled at room temperature and separated by centrifuge at 800 rpm with deionized water for washing. The obtained powder was dried at 70–80 °C for 24 h and got a dry product of nanocomposite. For CuO/Alg nanocomposite, the addition of sodium alginate in copper nitrate trihydrate (1.0 M) solution was used.

2.3 Characterization Techniques

An advanced X-ray diffractometer (Rigaku) was used for structural analysis of the synthesized product by using monochromatic Cu K_{α} radiation in the 2θ angular range of 20°–80°. The morphology of the samples was studied by using the transmission electron microscope (TEM) (JEOL, Tokyo, Japan) at 200 kV and scanning electron microscope (SEM) (JSM 6510LV JEOL, Tokyo, Japan). Fourier transform infrared (FTIR) (Bruker, Tensor 37) was used for functional group analysis.

3 Results and Discussion

3.1 X-Ray Diffraction (XRD) Analysis

The highest peak in the result of XRD analysis corresponds to CuO/Alg (Fig. 1). The average particle size of synthesized sample can be determined through the results obtained from XRD analysis. In the peaks of the diffraction patterns, CuO/Alg showed a monoclinic structure, and Miller indices have been identified with JCPDF card no. 89-5895. The lattice framework parameters were determined from XRD data and the average crystallite sizes were calculated approximately 18.09 nm

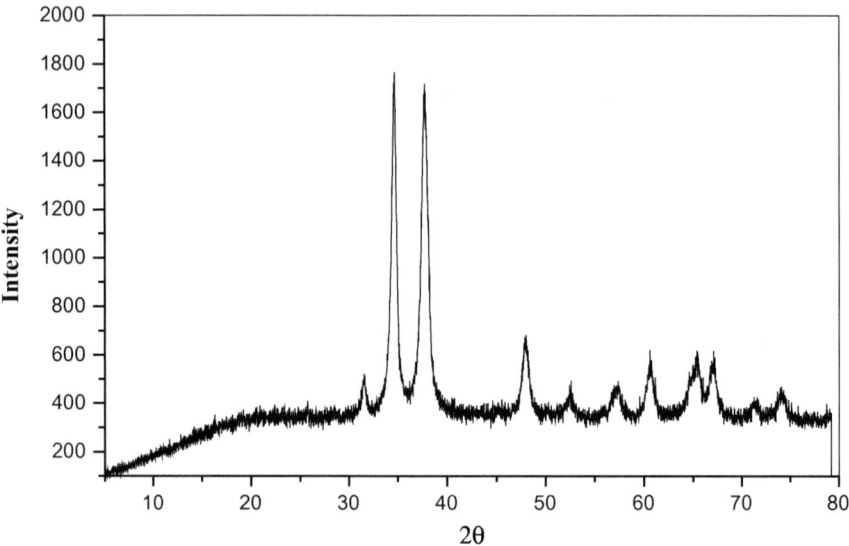

Fig. 1 XRD pattern of CuO/Alg nanocomposite

by the using of Debye–Scherrer formula. The latticework parameters are estimated from XRD data, $a = 0.46$ nm, $b = 0.34$ nm, and $c = 0.50$ nm.

$$D = 0.9\lambda \ / \ \beta\cos\theta \tag{1}$$

where β is full-width half maxima of the peak, θ is angle, and λ is x-ray wavelength of XRD patterns.

3.2 Fourier Transform Infrared Spectroscopy (FTIR) Analysis

The FTIR spectra with the KBr pellet technique were examined in solid phase in the range of 400–4000 cm^{-1}. CuO/Alg nanocomposite treated at 80 °C is shown in Fig. 2. The FTIR spectra exhibit three vibrations in the sample occurring at approximately 416, 498, and 599 cm^{-1}, which can be associated with the vibrations of CuO, affirming the fabrication of pure CuO/Alg nanocomposite. The three absorption peaks at 599, 498, and 416 cm^{-1} are the characteristic peak of copper oxide. The absorption peaks of 599 and 498 cm^{-1} can be assigned to high-frequency Cu-O vibration stretching. The peak perceived at approximately at 3419 and 1020 cm^{-1} is observed due to the adsorption of water on the metal surface and corresponds to O–H stretching and deformation frequency, respectively. The peak at 1633 cm^{-1} is due to the C=O bond which may be the reason for the absorption of CO_2 by the KBr because of KBr is highly CO_2 adsorbent [7].

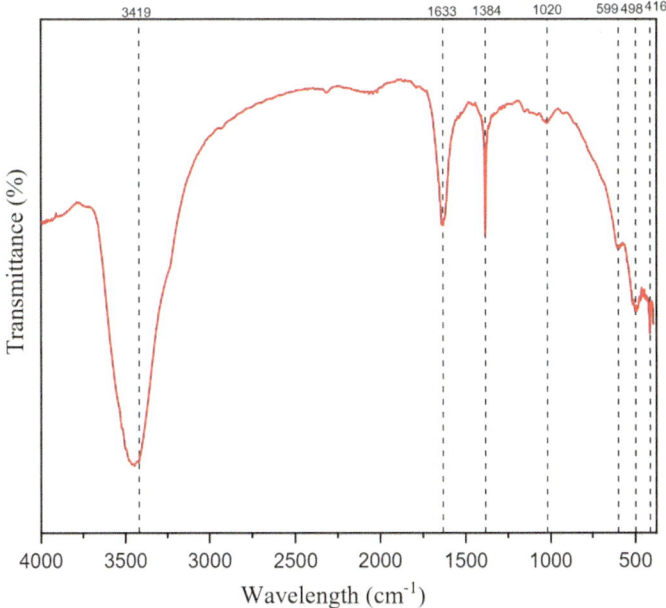

Fig. 2 FTIR spectra of CuO/Alg nanocomposite

3.3 Transmission Electron Microscopy (TEM) Analysis

The surface morphology and size of the prepared CuO/Alg were examined by TEM as shown in Fig. 3. It can be clearly noticed that the synthesized nanocomposite contained significant nanosheets with size around 11 nm with some agglomeration. Finally, the particle sizes obtained from TEM are well measured by the XRD peak width.

Fig. 3 TEM images of CuO/Alg nanocomposite

3.4   Scanning Electron Microscope (SEM) Analysis

Based on SEM images, the nanocomposite is predominantly composed of CuO/
Alg. In this study, the changes in surface morphology of the prepared CuO/Alg
nanocomposite due to adsorption of Pb(II) ions was studied by the SEM images as
shown in Fig. 4. The SEM image of nanocomposite reflects a homogeneous distribution of circular shaped nanoparticles with irregular distribution. The particles
were distributed irregularly after adsorbing the lead ions.

3.5   Removal of Pb(II) from Wastewater

The removal of lead is performed by the batchwise at room temperature. For the
effective removal of heavy metal ions such as Pb(II), the CuO/Alg nanocomposite
was found fit in the presence of feasible ions because there is no requirement of
adjusting pH and oxidation. Based on the result of SEM analysis, the particle size of
the sample was found approximately 50–70 nm. The removal of Pb(II) with different CuO/Alg dosage (2–16 g/L) was studied. The optimum dose of the
nanocomposite for solutions containing 16 g/L and lead removal efficiency was
114.2 mg/g with 120 min of contact time (Fig. 5). The maximum removal efficiency of lead was 118.4 mg/g at contact time 120 min, and maximum percentage
removal was observed 92.2% with a CuO/Alg dosage of 8.0 g/L (Fig. 6).

The removal of Pb(II) ions is also shown in Fig. 4b which was increased
gradually with increasing contact time and CuO/Alg dose. The short stability time
was reported by other researchers for adsorption of heavy metals ions on
nanocomposite [6, 8, 23]. Similar behaviors have been reported by many authors
[10, 12, 19] for the uptake of heavy metal ions from aqueous solution by different
metals' adsorbent. At lower contact time and dose, the percentage removal of Pb(II)

Fig. 4 a SEM image of CuO/Alg. **b** SEM image of CuO/Alg after removal of lead

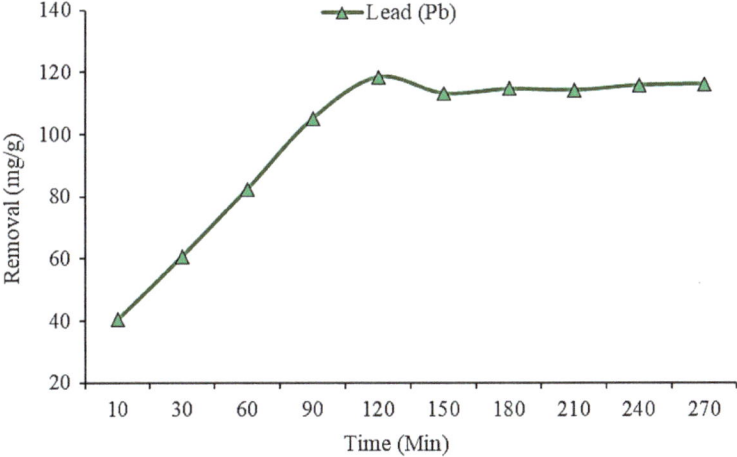

Fig. 5 Effect of contact time on Pb(II) sorption by CuO/Alg, at 25 °C

was low for CuO/Alg nanocomposite because protons in large quantities compete with positive metal ions for the adsorption sites. Therefore, it obstructs the surface of metal ions to reach the functional group. Consequently, the percentage of lead ion [Pb(II)] removal may decrease at low doses of nanocomposites [9, 11, 15]. This is in contrast to other traditional porous adsorptions, which have to be absorbed through other spreading stages [4]. This result is promising because contact time plays a major role in economic viability for waste treatment plants .

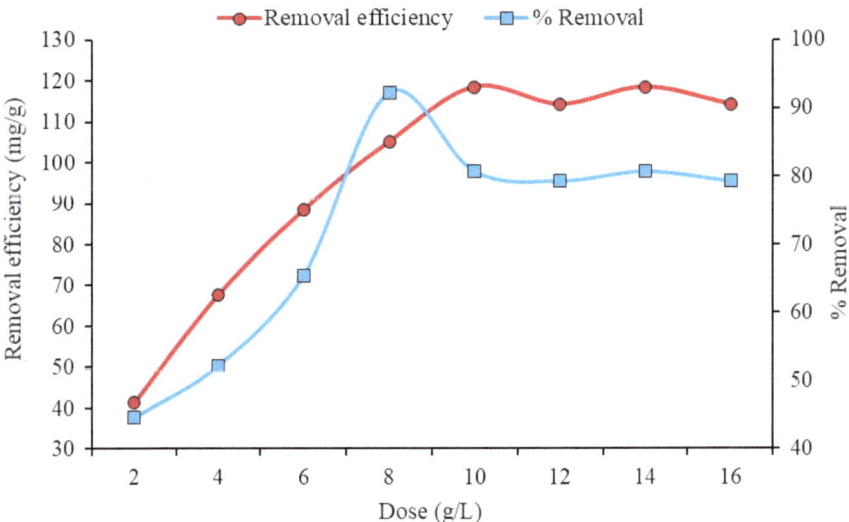

Fig. 6 Effect of dosage for Pb(II) sorption by CuO/Alg at 25 °C

4 Conclusion

It was concluded from the present study that the CuO/Alg nanocomposite was synthesized by wet-chemical reduction method in solution phase. The XRD spectrum of CuO/Alg exhibits that the particle size was approximately 18.09 nm which well agreed with the XRD data. The study of surface morphology of the nanocomposite was done in the context of SEM, TEM, and FTIR spectroscopy techniques. The CuO/Alg nanocomposite was successfully experimented to remove Pb(II) from industrial wastewater. The removal process was conducted in the acidic and basic environment with a contact time of 10, 30, 60, 90, 120, 150, 180, 210, and 240 min. The obtained data indicated that the removal efficiency of Pb(II) ion was higher and nanocomposite was found to be dependent on experimental conditions. From the reported observation, CuO/Alg has no commercial value and is a good, inexpensive source of the readily available biomaterial. It can be concluded that this can be positively used as a cost-effective and environment-friendly nanocomposite for the removal of Pb(II) ions from wastewater.

Acknowledgements The author, Afzal Ansari gratefully acknowledges for the financial assistance in terms of "Non-NET fellowship" by University Grant Commission (UGC), New Delhi. Further, the authors are also grateful to the Department of Applied Sciences, Faculty of Engineering and Technology, Jamia Millia Islamia, New Delhi, for providing the experimental facility.

References

1. Farghali AA, Bahgat M, Enaiet Allah A, Khedr MH (2013) Adsorption of Pb(II) ions from aqueous solutions using copper oxide nanostructures. Beni-Suef Univ J Basic Appl Sci 2:61–71. https://doi.org/10.1016/j.bjbas.2013.01.001
2. Heidari A, Younesi H, Mehraban Z (2009) Removal of Ni(II), Cd(II), and Pb(II) from a ternary aqueous solution by amino functionalized mesoporous and nano mesoporous silica. Chem Eng J 153:70–79. https://doi.org/10.1016/j.cej.2009.06.016
3. Liu X, Hu Q, Fang Z, Zhang X, Zhang B (2009) Magnetic chitosan nanocomposites: a useful recyclable tool for heavy metal ion removal. Langmuir 25:3–8. https://doi.org/10.1021/la802754t
4. Hussain M (2014) Synthesis, characterization and applications of metal oxide nanostructures. Linköping University Electronic Press, Department of Science and Technology, Linköping University
5. Yin P, Xu Q, Qu R, Zhao G, Sun Y (2010) Adsorption of transition metal ions from aqueous solutions onto a novel silica gel matrix inorganic–organic composite material. J Hazard Mater 173:710–716. https://doi.org/10.1016/j.jhazmat.2009.08.143
6. Jiang M, Wang Q, Jin X, Chen Z (2009) Removal of Pb(II) from aqueous solution using modified and unmodified kaolinite clay. J Hazard Mater 170:332–339. https://doi.org/10.1016/j.jhazmat.2009.04.092
7. Munagapati VS, Yarramuthi V, Nadavala SK, Alla SR, Abburi K (2010) Biosorption of Cu (II), Cd(II) and Pb(II) by Acacia leucocephala bark powder: kinetics, equilibrium and thermodynamics. Chem Eng J 157:357–365. https://doi.org/10.1016/j.cej.2009.11.015

8. Uheida A, Iglesias M, Fontàs C, Hidalgo M, Salvadó V, Zhang Y, Muhammed M (2006) Sorption of palladium(II), rhodium(III), and platinum(IV) on Fe3O4 nanoparticles. J Colloid Interface Sci 301:402–408. https://doi.org/10.1016/j.jcis.2006.05.015
9. Ruparelia JP, Duttagupta SP, Chatterjee AK, Mukherji S (2008) Potential of carbon nanomaterials for removal of heavy metals from water. Desalination 232:145–156. https://doi.org/10.1016/j.desal.2007.08.023
10. Mohapatra M, Anand S (2007) Studies on sorption of Cd(II) on Tata chromite mine overburden. J Hazard Mater 148:553–559. https://doi.org/10.1016/j.jhazmat.2007.03.008
11. Gupta VK, Agarwal S, Saleh TA (2011) Synthesis and characterization of alumina-coated carbon nanotubes and their application for lead removal. J Hazard Mater 185:17–23. https://doi.org/10.1016/j.jhazmat.2010.08.053
12. Gebru KA, Das C (2017) Removal of Pb (II) and Cu (II) ions from wastewater using composite electrospun cellulose acetate/titanium oxide (TiO 2) adsorbent. J Water Process Eng 16:1–13. https://doi.org/10.1016/j.jwpe.2016.11.008
13. Hassan KH, Mahdi ER (2016) Synthesis and characterization of copper, iron oxide nanoparticles used to remove lead from aquous solution. 04: 730–738
14. Moezzi A, Soltanali S, Torabian A, Hassani A (2017) Removal of lead from aquatic solution using synthesized iron nanoparticles. Int J Nanosci Nanotechnol 13:83–90
15. Hu J, Chen G, Lo IMC (2006) Selective removal of heavy metals from industrial wastewater using maghemite nanoparticle: performance and mechanisms. J Environ Eng 132:709–715. https://doi.org/10.1061/(ASCE)0733-9372(2006)132:7(709)
16. Afkhami A, Conway BE (2002) Investigation of Removal of Cr(VI), Mo(VI), W(VI), V(IV), and V(V) Oxy-ions from industrial waste-waters by adsorption and electrosorption at high-area carbon cloth. J Colloid Interface Sci 251:248–255. https://doi.org/10.1006/jcis.2001.8157
17. Yang M, He J, Hu X, Yan C, Cheng Z (2011) CuO nanostructures as quartz crystal microbalance sensing layers for detection of trace hydrogen cyanide gas. Environ Sci Technol 45:6088–6094. https://doi.org/10.1021/es201121w
18. Türker AR (2007) New sorbents for solid-phase extraction for metal enrichment. Clean–Soil, Air, Water 35: 548–557. https://doi.org/10.1002/clen.200700130
19. Ghorpade A, Ahammed MM (2017) Water treatment sludge for removal of heavy metals from electroplating wastewater. Environ Eng Res 23:92–98. https://doi.org/10.4491/eer.2017.065
20. Engates KE, Shipley HJ (2011) Adsorption of Pb, Cd, Cu, Zn, and Ni to titanium dioxide nanoparticles: effect of particle size, solid concentration, and exhaustion. Environ Sci Pollut Res 18:386–395. https://doi.org/10.1007/s11356-010-0382-3
21. Al-Saad KA, Amr MA, Hadi DT, Arar RS, Al-Sulaiti MM, Abdulmalik TA, Alsahamary NM, Kwak JC (2012) Iron oxide nanoparticles: applicability for heavy metal removal from contaminated water. Arab J Nucl Sci Appl 45:335–346
22. Siddiqui VU, Khan I, Ansari A, Siddiqui WA, Akram K (2018) Advances in polymer sciences and technology. Springer Singapore, Singapore
23. De Gisi S, Lofrano G, Grassi M, Notarnicola M (2016) Characteristics and adsorption capacities of low-cost sorbents for wastewater treatment: a review. Sustain Mater Technol 9:10–40. https://doi.org/10.1016/j.susmat.2016.06.002

Effect of Elevated Temperature on the Residual Compressive Strength of Normal and High Strength Concrete

Ateequr Rehman, Amjad Masood, Sabih Akhtar and M. Shariq

Abstract Experimental test has figured out to investigate behavior of remaining compressive strength concrete prepared using normal and high strength grades. For the same purpose, cube-shaped and cylindrical-shaped specimens of concrete material were casted and consecutively subjected to heating and cooling condition in laboratory-controlled environment. A hold period of three hours was provided to impart heating–cooling phenomenon inside the electrical furnace at four different set of temperatures. The elevated temperatures chosen for the present compressive behavior study are 200, 400, 600 and 800 °C. Strength was also determined at ambient environment for the purpose of comparing the effects of thermal loads on behavior of strength. Significant loss in compression strength has been visualized by plotting the curves for different set of concrete mixes at various temperatures. The compositeness of concrete results as one of the main reason in controlling strength loses inculcated due to elevated temperature. The outcomes of the current experimental work are termed useful while understanding key mechanical characteristics of concrete under the effect of overburdened thermal loads.

Keywords Residual compressive strength · Thermal loads · Concrete · Mechanical behavior · Experimental techniques

1 Introduction

Among several natural hazards, fire has its own peculiar nature which attacks building construction. The buildings undergo deterioration when exposed continuously against fire due to its high temperature. The buildings have been designed so as to provide protection to life of living being as well as to cause minimum deterioration when undergoes through these types of hazard. Prime importance has been given to increase evacuation timing for escaping the people when the building

A. Rehman (✉) · A. Masood · S. Akhtar · M. Shariq
Department of Civil Engineering, Aligarh Muslim University, Aligarh, India

© Springer Nature Singapore Pte Ltd. 2020
S. Ahmed et al. (eds.), *Smart Cities—Opportunities and Challenges*,
Lecture Notes in Civil Engineering 58,
https://doi.org/10.1007/978-981-15-2545-2_17

is subjected to fire. In recent times, the use of high strength concrete (HSC) in civil construction has been rigorously boomed while designing multi-story buildings especially in skyscraper, where structural resistance against fire has been taken as one of major design perceptions while discussing the safety of living beings. The phenomenon of concrete spalling is an another major concerns while dealing higher strength concrete members that has often been developed due to the presence of low ratios of water–cement in the concrete mixes. While reviewing the literature, it has been observed that various researchers all around the globe had observed the degrading pattern of various mechanical properties of concrete prepared using different ingredient's mix proportions when divulged to severe temperature ranges. One of the many contributions has been provided by Saad et al. [1] who studied the consequence of thermal loads on physical and mechanical characteristics of concrete mixed with silica fume. In another investigation, Phan and Carino [2] analyzed and compared various mechanical and physical characteristics of concrete having normal and higher strengths at exalted temperatures. After making on tedious experimental investigations, he found that the differences in strength characteristics are more arrested when the temperature amplitudes are in spectrum of 25 and 400 °C. It has been also observed that strength reduction in HSC had shown at a higher slope when compared that of NSC. Another study carried out by Kodur [3] investigated the HSC's strength and durability performances followed up with the comparison with that of NSC's when used as a structural member. Moreover, Luo and Chan [4] discussed the effects of higher thermal loads on the remaining or say, residual strength of pore structure in various high performance concrete (HPC) contrasting with that to NSC. Other investigations in obtaining the performances of mechanical properties such as residual Young's modulus of elasticity, flexural behavior and compressive strength for various concrete mixes under fire had been performed by Cülfik and Özturan [5]. The temperature spectrum chosen in study to impart the thermal effects on specimens ranges between 300 and 900 °C while using higher strength mortars specimens. Authors visualized higher contraction in magnitudes of concrete flexural strength than compressive strengths at exalted temperatures. Similar study has been performed by Li et al. [6] who investigated the degrading pattern of compressive performance, split tensile performance along with the bending strengths of NSC and HSC at exalted temperatures and successively concluded with plots showing decrements in relative strength of concrete at different fire exposure conditions. Phan and Carino [7, 8] determined the mechanical and physical characteristics of concrete at higher thermal loads by performing the experimentations under stressed and unstressed conditions. The investigation done by Xiao and Gert [9] revealed significant reduction in residual strength at 400 °C consecutively expanded to 75% loss at 800 °C fire exposure. In another study, Kodur and Phan [10] reported that the fire strength performance in concrete structures has dependencies on fire timing, its types, heat evolved and the location and area affected. While dealing with exalted thermal load, another key finding on remaining compressive strength of concretes of different grades along with its consecutive mechanical performance after adding different percent of silica fume and different water–cement ratios (w/c) had been investigated by Behnood and Ziari [11]. Authors identified that the concrete mixed with silica fumes shows lesser reduction in strength depended characteristics than the ordinary concrete up to 600 °C. Fire effect on pozzolanic mixed concrete composites had been studied by Alidoust et al. [12], whereas Hachemi [13] reported decrement in mass and density of concrete while studying fire-resistant behavior of NSC and HSC under exalted temperatures.

The present study investigates the characteristic variation of NSC and HSC when subjected to thermal loading at five different temperatures, hence evaluating the residual strength and corresponding fire-resisting performance of various concrete mixes. The future objectives may include in obtaining the mathematical relations of strength at any future point of time with respect of having dependencies on fire characteristics and ingredient percentages uses for preparing mixes. The study also aids the future designers in developing appropriate solutions against fire effect HSC members that may help in prolonging the structural life and mitigations purposes.

2 Experimental Program

2.1 Material Properties

Ordinary Portland cement of Grade-43 has been utilized for preparing the concrete mixes throughout the current investigation having its cements properties conforming to Indian codal provision, IS-4031 [14] (see Table 1). Microsilica procuring Grade 920 U (i.e., content of silica >92%) along with river sand as fine aggregate acquiring modulus of fineness 3.27 mixed with well-graded crushed granite stone pieces of 20 mm (max) and sp. Gravity of 2.69 g/cm^3 was used for experimental program. The physical characteristics of both the fine and coarse aggregates are conformed to Indian codal provision, IS-383 [15]. For the purpose of hydrating the mix content, the potable water free from hazardous or deleterious elements has been used having recommendation to IS: 456 [16]. The admixture, Glenium 51 has been mixed to provide workability of HSC concrete mixes.

2.2 Proportioning of NSC and HSC Mix

Sticking to the guidelines and clauses mentioned in Indian code IS: 10262 [17], two different mixes were prepared. The NSC mix for obtaining characteristics compressive strength of magnitude 20 MPa and HSC mix prepared for obtaining characteristic strength of magnitude 60 MPa were prepared applying the hit and trial methodologies. The quantity of superplasticizers was taken as 2% to weight of

Table 1 Physical characteristics of cement utilized in preparing concrete specimens

Characteristics	Results obtained	Specified value (as per IS: 8112–1989)
Specific gravity(g/cc)	3.14	3.15
Soundness (mm) (Le Chatelier's test)	1.0	10 (max)
Autoclave expansion (%)	0.056	0.8 (max)
Normal consistency (cement % by weight)	31	30
Setting time of cement (Min)		
(1) Initial	140	30 (min)
(2) Final	360	600 (max)
Compressive strength (MPa)		
3 days	24	23
7 days	33	33
28 days	43	43

cement for the purpose of obtaining cement workability. The detailed proportioning of various mixes has been tabulated after performing several experimental trials shown in Table 2.

2.3 Casting and Testing Procedure

The 28th day compressive strength of specimens prepared utilizing NSC and HSC mixes has been obtained experimentally by testing cubes of size 150 mm × 150 mm × 150 mm and cylinders of size 150 mm × 300 mm sticking to codal provisions IS: 516 [18]. For the purpose of mixing the ingredient, firstly, the coarse aggregate was placed inside the electricity-driven mixer followed up by adding 1/3rd of water quantity. The remaining ingredients were followed up inside the mixer as fine aggregates, cementitious materials, appropriate quantity of silica fumes, and then, the remaining quantity of water maintains a gradual mixing procedures. The specimens inside the molds were demolded after 24 h and are

Table 2 Proportioning of concrete mixes obtained after working out several trails

Mix	Cement (kg/m^3)	Fine aggregate (kg/m^3)	Coarse aggregate (kg/m^3)	W/C ratio	Silica fume (kg/m^3)	Super plasticizer (% by weight of cement)
NSC	350	630	1235	0.50	–	–
HSC	600	624	1050	0.34	49.8	2

cured using potable water for 28 days after which these NSC and HSC specimens are subjected to testing at ambient and exalted temperatures for different regimes of heating–cooling cycles.

2.4 Heating and Cooling Regime

The electric-controlled furnace depicted in Fig. 1 was utilized for procuring experimentation observations for concrete cubes and cylinders at exalted thermal loads. The furnace includes the heating elements owing the heating rates 5 °C per min fixed from two inside sideways while having refractory lining on all six faces. The furnace dimensioning has been done in such a way so as to provide internal clear spaces of about 1000 mm × 760 mm × 510 mm (length × width × height). A hole having dia. of 20 mm has been provided in front of the furnace to release fumes. This hole apart of providing prevention from blasting also has been utilized in placing thermocouples inside the furnace. The current electric furnace comprises a programmable microprocessor that acts as temperature controller for the instrument. The temperature inside the furnace has the rating of 1150 °C (max) that is designated standardized using fire curve of stand fires.

Fig. 1 Electric-based furnace used in the experimental study for controlling temperature environment

3 Results and Discussion

3.1 Effect of Temperature on Residual Concrete's Compressive Strength

The performance of NSC and HSC concrete mixes at ambient and higher temperatures has been plotted as shown in Fig. 2 and 3. The residual strengths of NSC and HSC specimens have been observed about 28.66 and 61.38 MPa, respectively, at ambient temperature, and it decreases gradually with increasing thermal loads. The figures shown below perpetuate that NSC specimens observed the least strength at all temperatures regimes. With increments in thermal loads, the minimal losses in residual strength under compression have been observed in this mix. NSC cubes have shown a gradual fall in strength of 82.79, 55.3 and 19.81% strength, at 200, 400 and 600 °C temperature, respectively. Moreover, in case of cylinder-shaped specimens, the strength at these temperatures was observed to acquire lower values when comparatively analyzed to cube-shaped specimens. On the other hand, the strength of HSC cubes was found decreasing gradually from 87.29, 64.09, 45.19 and 18.80% at 200, 400, 600 and 800 °C temperatures, respectively. Significant amount of decrement in strength of specimens using HSC mix has been observed while increasing temperature up to 400 °C, which is marginal in case of NSC. At 400 °C, both the cube and cylinder compressive strength in HSC specimens is approximately same. Due to development of extensive cracks in HSC specimens, the both cubes and cylinder strengths have been observed near about (Figs. 4 and 5). After 400 °C, though the strength of HSC has observed elevated magnitudes to temperature of 800 °C, the compressive strength of NSC at 600 °C is very low in the cubes. These specimens did not depict any value of strength after this temperature. No strength beyond 600 °C in NSC cylinders has been observed in present experimentation work.

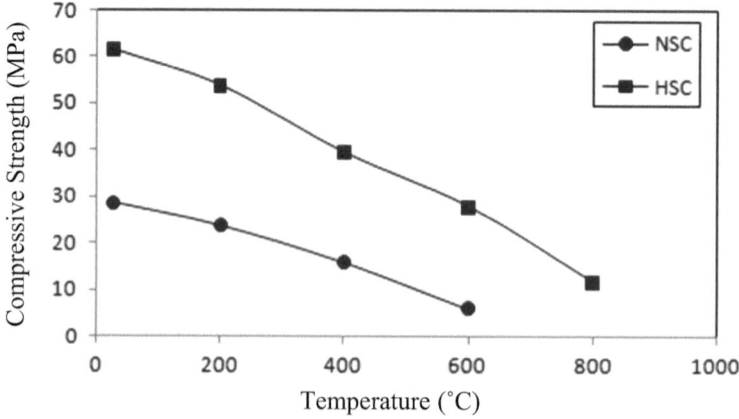

Fig. 2 Residual cube compressive strength for different concrete mixes at exalted thermal loads

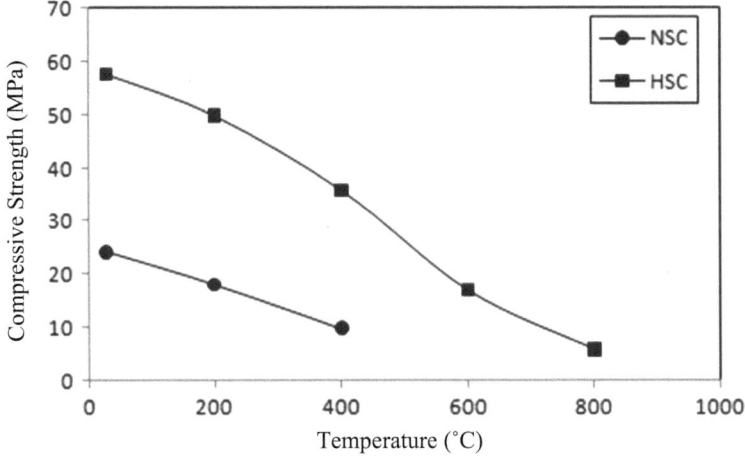

Fig. 3 Residual cylinder compressive strength for different concrete mixes at exalted thermal loads

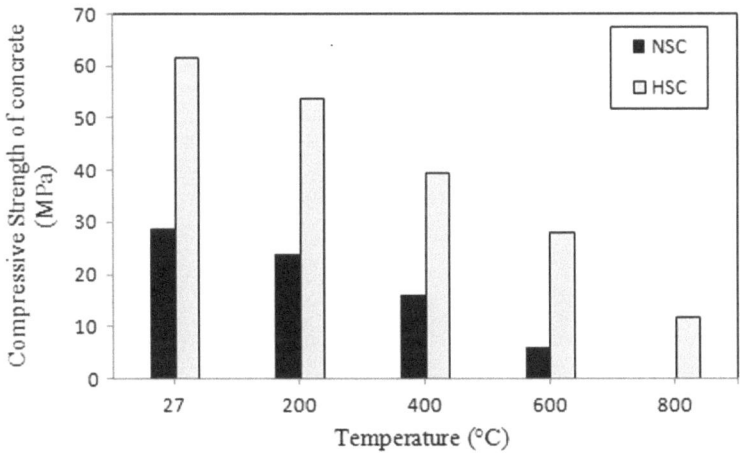

Fig. 4 Residual cube compressive strength in MPa for various concrete mixes at exalted temperature

3.2 Relative Compressive Strength of Concrete Mixes

Relative compressive strength of cubes and cylinders of all the concrete mixes is shown in Figs. 6 and 7. In these figures, the negative effect of temperature on the strength of NSC and HSC is clearly visible. Comparison of the relative strength of these concrete mixes shows that the loss of relative residual strength is marginal up to 600 °C, whereas it is substantial beyond this temperature.

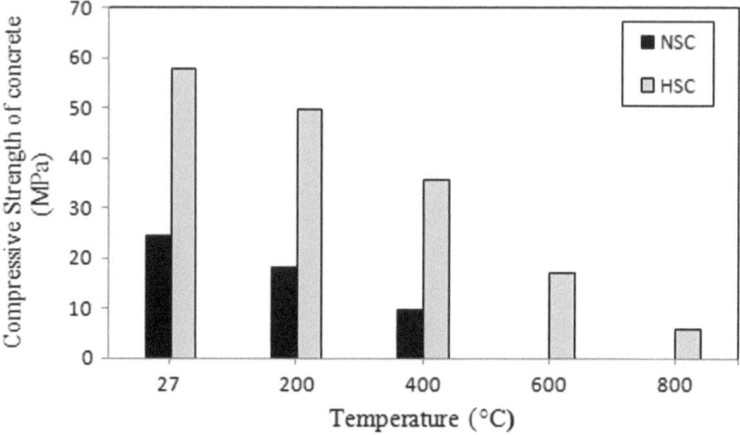

Fig. 5 Residual cylinder compressive strength in MPa for various concrete mixes at exalted temperature

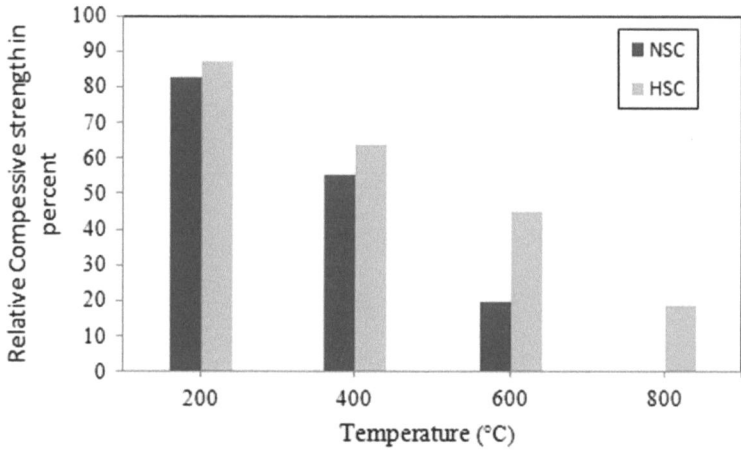

Fig. 6 Relative cube compressive strength in normalized percentage for various concrete mixes at exalted temperature

With increasing temperature, NSC has shown the maximum reduction of relative strength. At 200 °C, the relative strength of this mix has decreased nearly to 82.79% of its strength at ambient temperature. The maximum loss of strength in NSC is observed at 400 °C. The relative strength of this mix is only 19.81% at 600 °C. Between all the mixes which have been taken in to consideration in this study, NSC has shown the maximum differences between the relative compressive strength of cubes and cylinders at all the temperatures. At 200 °C, the difference between the relative compressive strength of NSC and HSC cubes and cylinders is

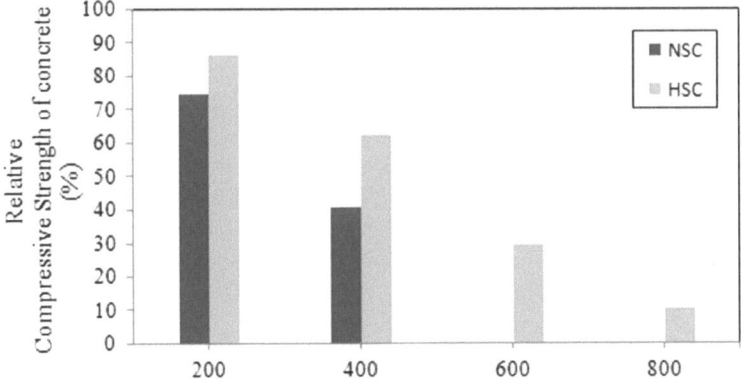

Fig. 7 Relative cylinder compressive strength in normalized percentage of various concrete mixes at exalted temperature

about 8 and 1%, respectively. The same difference is about 14.86 and 2.08% for NSC and HSC at 400 °C, respectively. At 600 and 800 °C, the relative compressive strength of HSC cubes and cylinders (which is the only concrete mix that has shown some strength at the same temperature) is about 16% and 9%, respectively.

4 Conclusions

The experimental investigation has been carried out efficiently under lab-controlled conditions to obtain the results evaluating the mechanical performance of concrete behavior under overburdened thermal loads. The point-to-point conclusions drawn after visualizing plots, carrying out the statistical analysis and hence evaluating thermal behavior of concrete of different mixes are given below.

- The grades of mixes used for obtaining normal strength (NSC) and high strength (HSC) in concrete are 28.66 and 61.38 MPa, respectively.
- Exposure of concrete to higher thermal loads reduces the compressive strength effectively. Further, the losses in strength of concrete mixes have been found varying directly with respect to temperatures.
- In a temperature ranging between 27 and 400 °C, HSC pertains a higher loss in compression strength when compared to that of specimen having normal strength.
- It has been also observed that above 400 °C, the specimens that were prepared using HSC lose a significant amount of its compressive strength when compared with specimens tested at ambient temperature conditions. The mean loss in strength at 600 °C has been evaluated about 80.15% in specimens prepared using NSC.
- The strength reduction in the temperature of 27–400 °C is minimal in all mixes.

- In the similar manner, an average loss of about 81.2% strength at 800 °C has been evaluated using statistical formulas.
- Specimens prepared using higher strength concrete mixes have been found more valuable when exposed to severe thermal loading conditions when compared to the other normal strength mixes. One of many reasons may include lower magnitude of losses in its relative compressive strength up to 400 °C temperature while studying the present two-phase study. The study will help the future researcher in field of investigating mechanical characteristic of concrete mixes under elevated thermal loading environmental conditions in obtaining precise estimation of failure strengths.

References

1. Saad M, Abo-El-Encini SA, Hanna GB, Kotkatat MF (1996) Effect of temperature on physical and mechanical properties of concrete containing silicafume. J Cem Conc Res 26 (5):669–675
2. Phan LT, Carino NJ (1998) Review of mechanical properties of HSC at elevated temperature. J Mater Civ Eng 10(1):15502
3. Kodur VKR (1998) Performance of high strength concrete-filled steel columns exposed to fire. J Civ Eng 25(6):975–981
4. Luo X, Sun W, Chan YN (2000) Residual compressive strength and microstructure of high performance concrete after exposure to high temperature. J Mater Structr/Materiaux et Constr. 33:294–298
5. Cülfik MS, Özturan T (2002) Effect of elevated temperatures on the residual mechanical properties of high performance mortar. J Cem Conc Res 32:809–816
6. Li M, Qian C, Sun W (2003) Mechanical properties of high-strength concrete after fire. J Cem Conc Res 35:2192–2198
7. Phan LT, Carino NJ (2000) Fire performance of high strength concrete: Research needs. In: Proceeding national institute of standards and technology, reprinted from the advance technology in structural engineering, Philadelphia
8. Phan LT, Carino NJ (2003) Code provisions for high strength concrete strength-temperature relationship at elevated temperatures. J Mater Struct/Materiaux et Constr 36:91–98
9. Xiao J, Gert K (2004) Study on concrete at high temperature in China-an overview. J. Fire Safety 39:89–103
10. Kodur VK, Phan L (2007) Critical factors governing the fire performance of high strength concrete systems. J Fire Safety 42:482–488
11. Behnood A, Ziari H (2007) Effects of silica fume addition and water to cement ratio on the properties of high-strength concrete after exposure to high temperatures. J Cem Conc Compo 30(2):106–112
12. Alidoust O, Sadrinejad I, Ahmadi MA (2007) A study on cement-based composite containing polypropylene fibers and finely ground glass exposed to elevated temperatures. J World Acad Sci 34
13. Hachemi S (2014) Evaluating residual mechanical and physical properties of concrete at elevated temperatures. World Acad Sci Eng Technol 86, Publication/9997531
14. IS: 4031 (Part 1 to 15) (1999) Indian standard methods of physical tests for hydraulic cement. Bureau of Indian Standard, New Delhi
15. IS: 383 (1970) Indian standards specification for coarse and fine aggregates from natural sources for concrete. Bureau of Indian standards, New Delhi, India

16. IS: 456 (2000) Indian standard code for design of plain and reinforced concrete structures. Bureau of Indian Standard, New Delhi, India
17. IS: 10262 (1982) Indian standard recommended guidelines for concrete mix design. Bureau of Indian Standard, New Delhi, India
18. IS: 516 (1959) Indian standard methods of tests for strength of concrete. Bureau of Indian Standard, New Delhi, India

Planning for Low-Carbon Smart Cities in India

Athar Hussain and Alpana Gupta

Abstract A human greed is never ending which gives rise to social as well as environmental externalities. In the past decade, India's extraordinary urbanization has analogous growth in primary energy demand. With urban per capita scenario, commercial energy prompts three times higher energy demand than rural areas because urban areas are the foundation of energy and CO_2 emission giving rise to climate change. This paper will first review the traditional practices and approaches in the context of the low-carbon cities and related climate-resilient cities initiatives, as development strategies for addressing and highlighting urbanization challenges. An attempt has been made through this study to explore the major root causes and factors of climate change and variable ideas of low-carbon resilient cities. The article is an exploratory type, in which different practices worldwide for a low-carbon and resilient city model have been incorporated.

Keywords Urbanization · Low-carbon smart city · Greenhouse gases · Climate change · Resilience

1 Introduction

Urbanization is an important challenge facing today's developing countries. By 2030, 60% of the population will be responsible for urban residents with 60% of the population living in cities with 41% of the population of India and 87% of the population of the US urban centers being the key players in greenhouse gases (GHGs) [1]. Space agglomeration of activities leads to urban growth, while population, economic

A. Hussain (✉)
Department of Civil Engineering, Ch Brahm Prakash Government Engineering College, Jaffarpur, New Delhi 110073, India

A. Gupta
Master of Environmental Planning, School of Planning and Architecture, An Institution of National Importance, M.H.R.D., Government of India, Bhopal 462030, India

© Springer Nature Singapore Pte Ltd. 2020
S. Ahmed et al. (eds.), *Smart Cities—Opportunities and Challenges*, Lecture Notes in Civil Engineering 58, https://doi.org/10.1007/978-981-15-2545-2_18

activities and spatial models and built environments are the same concentrate. They are therefore posturized with increased risks as a result of floods, heat waves, the rise in sea level and other risks that many treaties are expected to complicate with climate change mitigation and adaptation. An international environmental treaty was adopted on May 9, 1992, and in Rio de Janeiro from June 3 to 14, 1992 the United States Framework Convention on Climate Change, (UNFCCC) [2]. They are aimed at "stabilizing GHG emissions to cut down emissions which rises due to anthropogenic interference." It sets non-binding greenhouse gas emission limits for individual countries and does not contain mechanisms for enforcement.

The concentration of CO_2 has increased since pre-industrial phase ranging from 280 to 379 ppm in 2005, according to India's National Action Plan on Climate Change. The elevation from sea level is between 0.18 and 0.59 m. That will have immense impact on freshwater level, productivity, flooding of coastal areas and increasing chart of diseases (National Action Plan on Climate Change, p. 13). Besides this, Indian economy has increased by 7–8% on average, with a large amount of confrontation in cities such as urbanization, migration openings, slums and squatters, infrastructure demands, road congestion, vehicular pollution, depletion of groundwater, depletion of non-renewable energy and alternative modeling and simulation that cities need. In order to address growing environmental concerns, the Global Environment Facility (GEF) Trust Fund was recognized on the eve of the Rio Earth Summit of 1992. Thirty-nine donor countries contribute GEF funding and are reviewed every four years. The World Bank is the trustee of the GEF and the mobilization of GEF funds.

1.1 Concept of Smart City and Climate Change

It is classified as an urban area with different types of electronic sensors for the efficient management of resources. The Smart Cities Mission is a new government of India initiative designed to foster economic growth and improve the quality of life of the population through local involvement and technological development.

Smart connections are connected and satisfy the needs of citizens by providing a range of links, including sustainable transport, online access, technological advancement (smart transport) and social inclusion. Innovative, enterprise-friendly and effective collaboration are promoted by smart economics which offers high paid jobs and enriched living standards. Smart people form a foundation for smart cities that have access to knowledgeable partnerships which embrace technology and innovation.

Smart living links people to good health, education and security. Management deals with sophisticated allocation of resources, such as safe drinking water, infrastructure and other resources, under budgetary constraints. The smart environment is a magnificent pillar that bridges or balance planning growth with protected resources (Fig. 1).

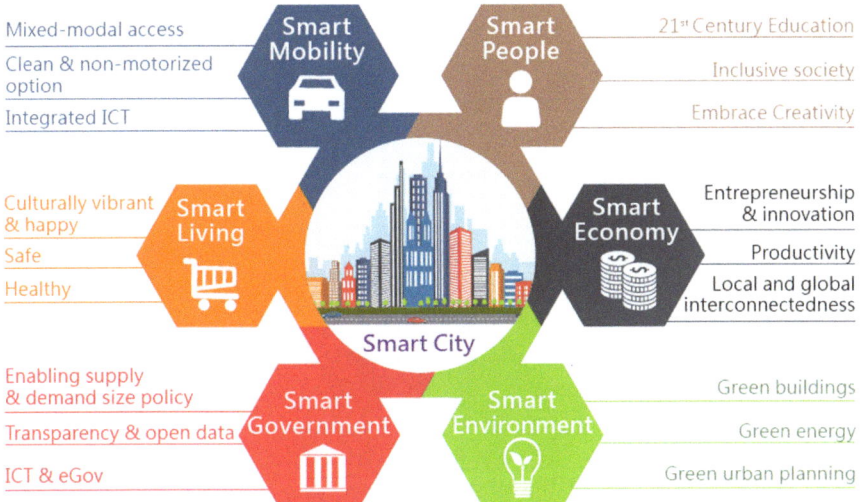

Fig. 1 Components of smart city [3]

Climate change refers to global climatic changes, a sudden apparent shift between the middle and the late twentieth century, largely attributed to the increased carbon dioxide produced by the increased use of fossil fuels.

1.2 Smart City and Benefits [4]

The smart cities facilitate the integration of public administration systems and processes and create transparency in the provision of better decision-making information. It is also useful for optimal use and resource allocation. The project also raises the population's satisfaction as it enables greater participation by civil society through the inclusion of technological instruments that help monitor public services, inform citizens and interact with the municipality in tackling particular urban issues. Singapore's National Secretariat for Climate Change (NCCS, 2012) enforces that it takes time to implement adaptation indications. The concept of risks and impacts of climate change for public health, energy demand and biodiversity should be identified and understood before supplementing policy initiatives which are useful in developing and implementing adaptive measures in the city.

The Government of Singapore has therefore encouraged and formulated a resiliency framework in the next 50–100 years that will promote and enhance its efforts to protect Singapore from the anticipated adverse effects of climate change. The framework is shown in (Fig. 2), and this is part of the national framework

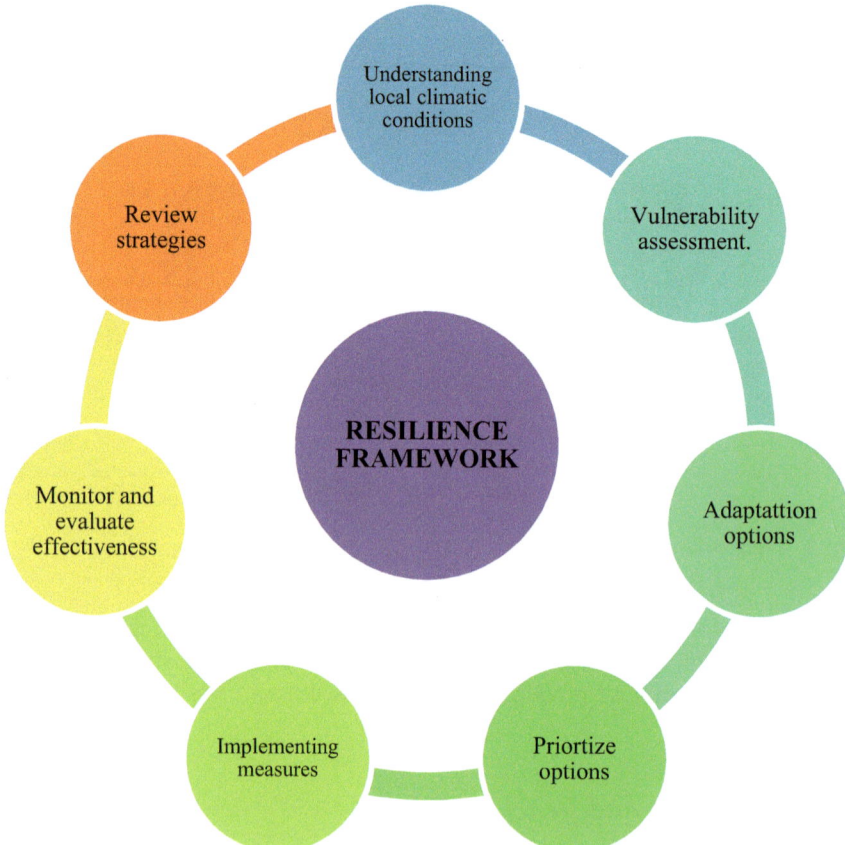

Fig. 2 Resilience framework [5]

category. The framework represents the process for the Singapore Government to adapt to climate change. It covers the understanding of prevailing local climate conditions and evaluates vulnerability, uncertainty and impacts of climate change in order to identify adaptation options [5]. The options are then weighted, and the application prioritized. The options are properly evaluated, and alternatives are chosen. This is a platform for reviewing the strategy that constitutes the basis for local climate understanding. The Nanyang Technological University's Future Resilient Systems (FRS) research group reviews and adopts this approach to making cities resilient that involves dynamic interaction between infrastructures and social behavior.

2 Smart Cities Are Climate-Resilient

The world is developing, and by 2050, there will be over 6 billion people living in cities that are going to cross the threshold. The effects and threatening effects of climate change on the cities and risks to floods, droughts and heat waves increase. In addition, the IPCC refers to the insecurity and risk-facing India. The US-based World Resources Institute echoes historical and catastrophic events like the 2005 Mumbai floods and the recent paradoxical Srinagar disaster. When India blinds to the global intelligent city club, it will create work to integrate planning interventions and strong remodeling for the urban development ministry to help smart cities develop [6]. The new projects in India have enormous effort and potential. But local stakeholders and even the state plan to restructure and simulate cities in today's situation without taking into account the consequences of climate change. This is the right time to find alternatives to combat the terrain scenario. Smart cities are expected to lighten urban stagnation by restoring urban areas in line with the objectives of the smart city concept and remove them from the environmentally unsustainable areas.

2.1 Approaches Using Landscape and Nature-Based Adaptation

Natural solutions reduce the risk of flooding, droughts and urban heat, provide services to ecosystems and improve cities' live capacity. The soil–water green system should be considered as the basis for landscape adaptation. Co-creation with governments and local stakeholders is strongly needed when designing and implementing natural solutions. Adaptive circular cities and water-smart cities [7] established the cities simultaneously contribute to climate change mitigation, adaptation and resource efficiency. A transition from drained cities to water-smart cities is therefore needed to (re)design cities to restore the natural drainage ability of cities. Intelligent combinations of technical, civil engineering and nature-based climate adaptation solutions with the transition to water-smart city will generate great business and innovation opportunities.

2.2 Global Initiative for Low-Carbon Cities

There have been global initiatives in various sectors to improve quality of life and boost economic growth. On September 4, C40 City Climate Leadership Group (C40) and Siemens have reflected on 29 towns and 37 projects. For its tremendous efforts to promote green mobility, Bogota received urban transport awards [8]. Launched in 2000, the City's Rapid Transit Bus system achieved reductions in emissions of more than 350,000 ton a year. Various sustainable approaches were adopted by shifting to hybrid and electric buses to cut down the emission by the end of year 2024.

Melbourne builds sustainable buildings that are energy-efficient. This is evident with the joint involvement of building managers with energy-efficient construction and water efficiency design. Tokyo has been at the queue for a program to reduce CO_2 emissions from large commercial and industrial buildings, which was launched in April 2010. Over 1100 facilities participated in a 13% reduction in emissions during their first year, a further 10% reduction was achieved the following year, bringing to more than 7 million tons of CO_2 total emissions reductions achieved. For the first time, Singapore has been known for the electronic price and congestion tolls for congestion and emission reduction.

2.3 Need for Low-Carbon City in India

The worst conditions prevail in Mega polis (Delhi and Mumbai). Mumbai is vulnerable to flooding, according to an article in 2012 which highlighted on climate change which is posing great threats to cities and also ranked Mumbai Sixth amongst 20 port cities. In November 2012, an article appeared in Times of India, [9] which stating that climate change and growing development are leaving Mumbai vulnerable to flooding, damage from high storm winds that rank Mumbai sixth in a list of 20 port cities worldwide. Mumbai is largely affected by the absorption of solar radiation by concrete, which activates the urban heat island effect. It reduced the permeability and additional stress on the creepy old drainage system was imposed by its high population density. The Mega Polis such as Mumbai, Delhi should be properly planned and managed. However, the urban low-carbon solution to combat climate change is becoming increasingly necessary. The article addresses all issues which highlight the gaps by means of urbanization scenarios, greenhouse gas emissions and CO_2 emissions causes. The concept of low-carbon cities is based on a Malaysian case study which reduces greenhouse gas (GHG) criteria and identifies carbon-reduction parameters using similar models for Indian cities.

2.4 Causes of Carbon Dioxide Emissions

In the transport sector, CO_2 emissions are driven by motors. Carbon emissions in corporate world is mainly through the running of offices and shops 24 h, which leads to increase in energy consumption. In the housing sector, areas have been expanded as a result of the growth of nuclear families and single houses, the development of IT and increased size of household appliances, etc. The number of households is increasing [10]. Buildings and structures are also accumulating that are not energy-efficient or [10] low-carbon, since energy efficiency is lower than convenience, comfort and economy low-carbon emitting, since energy efficiency has lower priority compared to convenience, comfort and economy.

3 Indian Scenario of Urbanization

India accounts for 286 million people in 2001 and gains the second-largest urban population country [11]. According to the Indian registrar general and census officer, the urban population in India will increase by 38% during the next 25 years to reach 534 million by 2026 [12]. The problem is that the government is not capable of bridging the gap between large demands and supplies and that the mega projects will consume a large share of energy use. Thus, when low-carbon cities are simulated by energy-efficient buildings that encourage public transport in cities, a solution is required that will otherwise exacerbate the current urbanizing rate in future. India is the third-largest emitter of CO2 in absolute emissions. These emissions will continue to increase as the economy expands. India, as is the case in the usual business scenario, will increase its CO_2 emissions five times before 2050 while already relying heavily on fossil fuels. ICLEI South Asia., 2012.

3.1 Low-Carbon City

It can be defined as a city consisting of societies using sustainable green technology, green practices and relatively low carbon or GHG emissions in comparison with current practice [13], in order to prevent adverse impacts on climate change. The adoption of the Low-Carbon City Program was initiated by a resolution, which led to a strong desire for the 2009 National Energy Conference. The targets formulate low-carbon communities in both municipalities and counties over two years, six low-carbon towns over five years and four low-carbon towns, each of which will be completed by 2020 in the north, central, south and east of Taiwan [14]. Renewable energy, energy conservation, low carbon occupancy and green mobility were the major indices of low carbon development (Tables 1 and 2).

3.1.1 Keetha, Malaysia

Malaysia Low-Carbon City Framework formulates the need to study the local government plan on the ground and to implement it, which will initiate a reduction of carbon emissions to preserve the balance of the environment. The urban

Table 1 Low-carbon community promotion initiative [14]

Renewable energy and energy conservation	Low-carbon buildings	Green mobility
Renewable energy planning • Solar thermal energy • Bio-mass energy	**Low-carbon building management** Water conservation and waste reduction	**Low-carbon modes and sustainable modal choices** • Bicycle (electric bike) • Public transport system • E-rickshaws and other low-carbon modes
Energy conservation LED and compact fluorescent light installation	**Green building material** Evaluation of the lifecycle (production, process, use, demolition, disposal)	

Table 2 Indices planned for low-carbon smart cities in Taiwan include the following: [14]

Classification	Rural	Urban	Solid waste recycling rate	Wastewater recycling rate	Rural type
Major indices	Fossil fuel consumption	Greenhouse gas emissions			
Secondary indices	Ratio of total solid waste recycled			Ratio of total wastewater recycled	
Supplementary measures	Coping with local adverse conditions				
Low-carbon city classification	Renewable energy ratio increase		Conserving Energy		

environment performance criteria describe the extent and boundaries of urban planning by site selection, which includes different factors such as the urban base spring which focuses on the concept of infill development and the control of growth outside the green space boundary. KeTTHA, [13] which will initiate on reducing the carbon emissions, thus maintains the atmospheric balance. Urban transportation indicates the shift to low-carbon modes and other high transit capacities and encourages people to travel without motorization. Urban infrastructure provides wastewater management strategies. For green parks, outdoor areas, recycled water is suitable. Green building or energy efficient buildings have been adopted which made use of recycled materials. For example, in many government office buildings fly-ash has been used as green material as they are light weight. Green buildings also incorporates energy efficient appliances (5 star appliances) which leads to less energy consumption.

3.1.2 Programme Level

The Environment Program of the United Nations (UNEP), with a total budget of 2.49 million pounds, is now supporting low-carbon transport to India. There are also a number of models for the urban emission. The SIM-Air Model (simple interactive models for improved air quality) is an integrated air pollution analysis tool that ranges from estimating emissions to pollution impacts for a given scenario. The other VAPIS model is the vehicle emission calculator to match and compare the emission inventory, including the emissions factor database repository [15]. Smart CART is an important model, which includes a simple carbon analysis calculator along a road corridor and a flexibility to expand to other pollutants. The air quality index, which is generally used in many cities around the world and designed to plug your data into AQI estimates in real time or forecast, is also an important parameter. Atoms Dispersion Model (The Atmospheric Transport Modeling System serves as a (FORTRAN language based) simplified Lagrangian dispersion model to generate transfer matrices for multiple source and multiple pollutant types; for direct input to the SIM-air model. *Fugitive emissions of dust by road vehicles are symbolic and simple to use V-Dust calculator* (urban emissions, Webpage).

3.1.3   Need for an Indicator

No management without measurement: While the Kyoto Protocol has elaborated on the methods to ensure allowable reduction of SGM, it remains for individual countries to decide how these mechanisms are to be used and what individual goals will they set at home. In 1995, the IPCC Working Group 13 concluded that there is an apparent human influence on the global environment. The **GHG Indicator** allows companies to meet this need.

4   Comprehensive Climate and Smart City Planning Model

Smart cities with resilience to climate and low carbon can be described as digitally linked to all sectors and features. It covers everything in connection with sustainable, resilient, circular, efficient and urban connectedness [16]. Climate change mitigation as well as adaptation goals are included at every step of planning process in urban governance.

4.1   Connectivity of Climate Change Actions and Plans

The increasing number and responses to climate change have identified the need for an integrated (3Cs) carbon-based model for smart planning that incorporates climate change mitigation policy goals at every stage of the process.

4.2 Benefits for Cities in Climate Change Action

There are at least four reasons why climate action is in the cities' best interests. First, there are very high costs of inaction. Urgent steps to guide building codes and practices, density and connectivity infrastructure will be necessary in rapidly growing cities. Secondly, green action's co-benefits often cover more than cost. Thirdly, the adoption of such a major global cause helps cities to position themselves in a group of leaders, access and learn through information and technology. Fourth, the best way for small and poor cities to access the best experience available from around the world is to take up and share global products and practices.

Cities are also good pilots of climate action and have key climate change responsibilities. Through screening of infrastructure and transport investment, financial, partnerships, energy suppliers regulation, cities can promote green growth. Practical approaches climate risk evaluation/city vulnerability study, which shows how the particular area is susceptible to hazards, is taken into consideration. Analytical hierarchical process should include adaptation and mitigation strategies. Landscape and nature-based adaptation strategies majorly focus on the adaptive practices of that area.

(i) Climate workshops/design workshops—The participation of parties involved can include advanced urban planning and design solutions.
(ii) The tool to support naturally oriented adaptation strategies focuses on developing cities in order to combat shocks and stress. In a situation, it is better to adapt rather than reduce it.
(iii) Modeling of urban heat stresses may very well be explained where the floor design is less permeable, absorbing heat that causes the surrounding heat waves to block.
(iv) Green health check concentrates on several parameters such as green mobility, green design and efficient construction.
(v) Water-sensitive urban design includes sensors for water saving to control the excess flow of water.

4.3 Climate and Disaster-Resilient Smart Cities

The smart city mission, which is India's largest urban development government effort, should unequivocally focus on urban climate resilience. The first 20 approved smart city plans are compared in a TARU Leading Edge study, and there is no explicit focus on climate change in any of the 20 documented city plans. The study was entitled "Climate Change and Disaster Resilience" in Indian Cities: the preparation of city governments and funding arrangements for development across the various sectors, as opposed to urban initiatives, and the sites examined are the municipalities of Mumbai, Kolkata, Chennai, Surat, Indore, Kochi, Guwahati, Bhubaneswar, Aizawl and Panaji [17]. For example, climate resilience can be classified as a cross-cutting issue but does not form part of a plan.

4.4 Indices for Smart City Planning

The ecological stress framework suggests a new approach to evaluating community socio-ecological vulnerability to disasters. In order to study the diversity of ecological stress in these communities, remote sensing data are needed in combination with household surveys that document emotions and behavior and social practices. The City Prosperity Index needs to be introduced in the smart cities planning process, which states that resilience depends not only on physical assets but also on policies and social capital.

4.5 Mainstreaming Disaster Resilience for Sustainable Development of Cities in India

4.5.1 Case Study of Guwahati and Shillong

The case study emphasized mainly the need for disaster resilience to be integrated into city development plans. The HIGS Framework incorporates the four components of which relate to cities: hazard, infrastructure, governance and economic status), and GIS spatial analysis was used to carry out vulnerability and haze assessments in the cities of Guwahati and Shillong. The main component of the organization is governance and institutional framework [18]. The presence of multiple agencies which lead to issues of coordination, irresponsibility and inefficiency in service delivery generally causes a lack of governance in cities. Deficit in infrastructure is the result of growing greed and population growth. Sewage, solid waste and drainage infrastructure in both towns are poorly developed and out of date to cope with current urban demand. These are thus the indicative measures to incorporate the resilience plan into development plans, to protect both people's and nature's interests.

5 Smart City and Low-Carbon Digital Connections [16]

The preparatory models provide spatially and strategically urban growth while maintaining GHG emissions under control and reducing the pace toward climate change sensitivities, improving quality of life, in particular through the use of

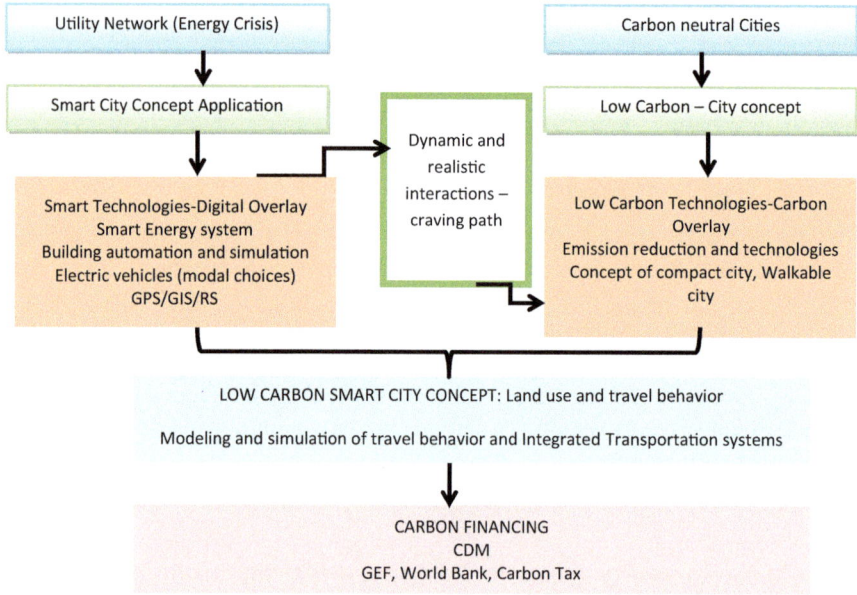

Fig. 3 Conceptual framework for low-carbon smart city

state-of-the-art automation and the development of Internet control technology [16]. With the help of growing ICT applications in specialized and growing areas, such as transport, land-use planning, energy, water, waste management and so on, the complex, spatial Web of a low-carbon smart city can be integrated in an integrated holistic approach via networking and information sharing (Fig. 3).

The concept of carbon overlay is a quantitative index. There are many green building ratings for efficient construction for each category. The results for every LEED loan depend on the carbon footprint analysis of the LEED building [16]. Carbon footprint is complementary to the construction and operation of gasoline gases, including:

- Building systems energy consumption;
- Embedded water emissions related to electricity used in water geysers;
- Embedded solid waste emissions
- Embedded material emissions indicating emissions from materials manufacture and transport.

Table 3 Government Initiatives under smart city

Policy/mission	Key focus area
INDC	Development of green mobility, climate-resilient urban areas
National sustainable habitat mission	Sustainable transport, waste management and energy-efficient construction
Smart city mission	To develop 100 smart cities
Atal Mission for rejuvenation and urban transformation	Revitalize 500 cities (basics, climate and green spaces policies)

5.1 National Policies Focusing on Sustainable Urban Development [19]

NDC majorly reforms cities into climate resilient to combat risks and uncertainties. There should be a proper corporate and official sector in each state to deal with climate agenda and plan according to the development goals. Various organizations are formed (5.1) that deal with climate hazards but are not planning accordingly; there is a need to have decentralized planning with an effective scale. Various initiatives were taken by GoI under smart city umbrella by setting up decentralized organizations for dealing or developing city as smart or eco-friendly (Tables 3 and 4).

Table 4 JNNURM, smart cities mission and AMRUT

	JNNURM [3, 20]	Smart cities mission [21, 22]	AMRUT [20]
Number of cites	65	100	500
Period	7 years (2005–2012)	5 years (2015–2020)	5 years (2015–2020)
Criteria	Class A cities will include populations over 4 millions per census for 2001), seven class B cities will include populations between 1 to 4 million per 2001 census), 28 class C cities will include religious/historic towns, and 30 cities will be included in class A cities	Cities should meet the following requirements: master plan, GIS maps, public service online, IT-based platform, etc.	Towns with 1 lakh population, including some towns on the main river, some capital cities and major towns in hilly areas, islands and tourist areas
Central Government outlay	Allocation of Rs 66,085 crores	Allocation of Rs 48,000 crores	Allocation of Rs 50,000 crores

(continued)

Table 4 (continued)

	JNNURM [3, 20]	Smart cities mission [21, 22]	AMRUT [20]
Reform objectives	Transparent budgeting, planning and accountable management	Movability, provision of infrastructure, healthy environment and IT connectivity	Project-based approach: funding allocation, bye-laws, urban local authorities and central planning
Role of center	Central authority should sanction projects	City challenge competition	State shall present the central agency with plans

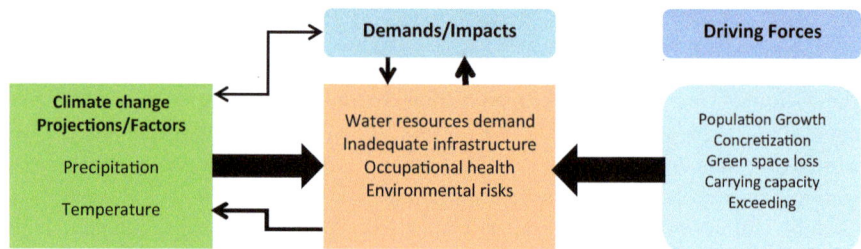

Fig. 4 Climate change vulnerability assessment

5.2 Vulnerability Assessment

Firstly, vulnerable factors affecting the specified location or area should be known. Vulnerability assessment can be done through remote sensing and GIS application. Various thematic and info graphics map can be prepared for better understanding of hot spots. Detailed analysis and inferences can be drawn, based on that policies can be framed and implemented by knowing the targeted groups. Vulnerability assessment can be best quoted from the climate change projection factors (Extreme temperature/precipitation) which will pose occupational as well as environmental hazards (Fig. 4).

5.3 Mainstreaming Climate Change Resilience in Urban Development Plan

Migration, slum growth, poor housing finance, poor local government, unexpected growth and restrictive zoning regulations all take into account India's urban challenges. For efficient resource-based planning, resource-based plans need to be incorporated into development plans as well as sufficient governance. Governance

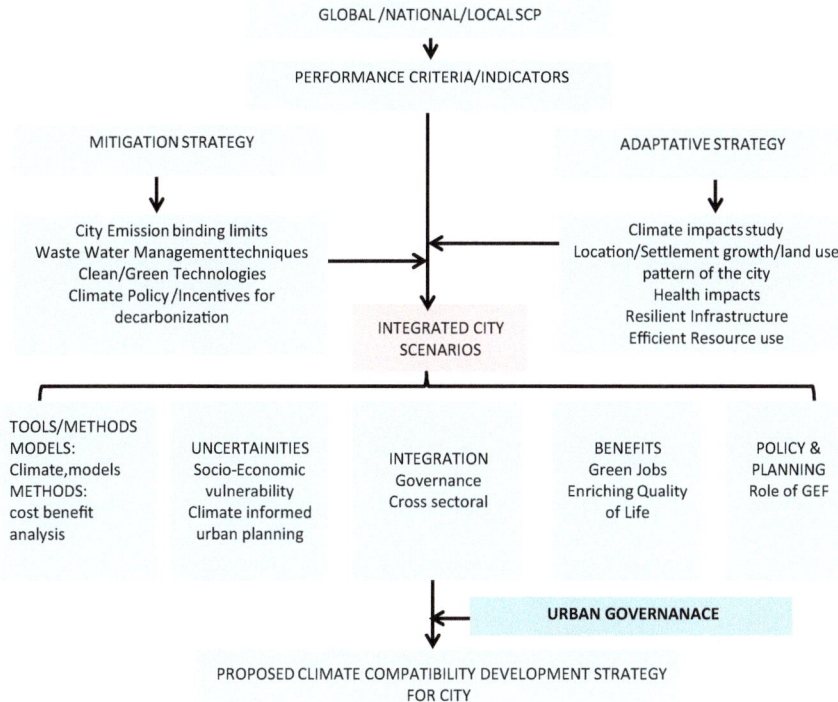

Fig. 5 Framework for climate change resilience in development plans

is an art of government leadership. The main components of governance are responsible, self-governing, provision, (Table 2) regulation, enabling and partnership. UNDP has listed different good governance parameters: participation, rule of law, transparency, responsivity, alignment, equity, efficiency, accountability, strategic vision (Fig. 5). Performance criteria are established for both mitigation and adaptation strategy. Climate compatible city development plan can be integrated with urban governance to check on performance indices.

5.4 Resilience Strategies

The capacity of a socio-ecological system to absorb shocks and stresses is defined as climatic resilience and in order to maintain the role that climate change places in the pace of external stresses [9] and to bridge the gap between adaptation and mitigation and evolve into more desirable configurations that improve the sustainability of the system (Fig. 6; Table 5). Resilience strategies can be adopted to combat environment risks such as proper land use planning, urban governance,

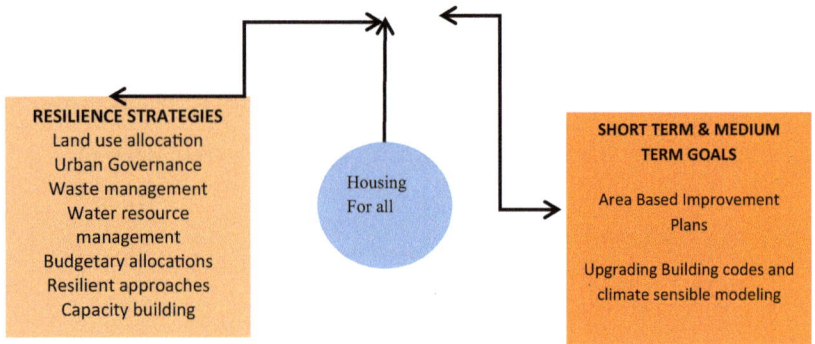

Fig. 6 Resilience strategies [21, 22]

Table 5 Urban governance approaches

Self-governing	Planning of buildings according to climate-resilient adaptation
Governing through enabling	Local governments should encourage local stakeholders and private actors to contribute their share in the pay of adaptation
Governing by provisioning	Warning systems and emergency planning
Governing by authority involving	Urban planning and other influential roles of local government

upgrading building codes, climate sensible modeling and capacity building. Good Governance approaches serves as the backbone for proper functioning of perspective plan.

6 Limitations of Climate Smart Cities

In these days, climate cities are always a big concern. The concept of adaptation and mitigation is much more comprehensible. The fourth evaluation of the IPCC shows that the urban resilience framework is emergent by the end of 2050. The key issue is the conflict of interest arising from the specificity of the project being undertaken in consultation with different stakeholders. In order to achieve stability, proper allocation and fair distribution of the budget should be made between all sustainable sectors. Megapolis as well as metropolis are planning according to Master plans. But in great need of Master plan, settlements and their are growing at a very fast pace, which needs to be checked upon. Master plan must incorporate a sustainable vision of growth of city in eco-friendly manner. The plan should be strengthened, focusing on public involvement and adaptation and mitigation goals in order to enhance the town's resilience.

7 Conclusions

In India, disaster and climate resilience framework can be integrated into the cities' development plans, with missions such as Smart City, AMRUT and HRIDAY. Geospatial and remote sensing techniques can be used to make city climates resilient, allowing for vulnerability and hazard assessment. Capacity building, the appropriate institutional framework and programs (decentralized planning) are needed for better coordination of cities at all levels to cope with climate change. In order to tackle current urbanization challenges, it is also necessary to integrate the low-carbon smart city concept.

References

1. Sperling J (2013) Exploring health outcomes as a motivator for low carbon city development: implications for infrastructure interventions in Asian cities. Habitat Int 37:113–123
2. Lankao PR (2012) Governing carbon and climate in the cities: an overview of policy and planning challenges and options. Eur Plan Stud 20(1):7–26
3. Kapur A (2013) Performance audit on JNNURM, Comptroller and Auditor General of India. Accountability Initiative (Research and Innovation for Government Accountability)
4. Smart City (2018) Retrieved from https://blogs.iadb.org/ciudadessostenibles/2016/07/11/smart-city-2/
5. European Partnership for Innovative Cities within an Urban Resilience, EPICURO - Outlook (2018): SWOT Analysis Final Report, European Union Civil Protection and Humanitarian Aid
6. Sriraj K (2015) Smart cities are climate-resilient. Retrieved from https://www.dailypioneer.com/2015/columnists/smart-cities-are-climate-resilient.html
7. Wageningen climate solutions Glascow (2017) A guide to Wageningen University & Research activities. ECCA Conference
8. Ten Cities Tackling Climate Change (n.d.) Retrieved from https://www.smartcitiesdive.com/ex/sustainablecitiescollective/10-cities-tackling-climate-change/178136/
9. Chinmayi S (2012) Climate change poses grave threat to Indian cities, Times of India, TNN Nov 12, 2012. Retrieved from http://articles.timesofindia.indiatimes.com/2012-11-12/global-warming/35067898_1_climate-change-mumbai-oecd-study
10. Allaoua Z (2013) Guide to climate change adaptation in cities. The International Bank for Reconstruction and Development, World Bank
11. Vaidya C (2009) Department of Economic Affairs, Ministry of Finance, GOI, working paper no 4/2009-DEA
12. Population projections for India and states 2001–2026 (n.d.) Retrieved from https://nrhmmis.nic.in/UI/Public%20Periodic/Population_Projection_Report_2006
13. Kettha (2011) Low carbon cities framework and assessment system. Green Tech Malaysia and Malaysian Institute of Planners
14. Towards low carbon cities in Taiwan (2009). Retrieved from http://unfccc.saveoursky.org.tw/2011cop17/images/cadiis/book-dowload/06_Towards_Low_Carbon_Cities_in_Taiwan.pdf
15. UNEP Risø Centre EC (2013) Promoting low carbon transport in India. United Nations Environment Programme
16. Kim K-G (2018) Planning models for climate resilient and low-carbon smart cities: an urban innovation for sustainability, efficiency, circularity, resiliency, and connectivity. Springer International Publishing

17. Climate disconnect in India's smart cities mission (2017) Retrieved from https://indiaclimatedialogue.net/2017/10/11/climate-disconnect-indias-smart-cities-mission/
18. Climate and disaster resilient smart cities (2015) Integrated Research and Action for developement, New Delhi
19. Pathak M (2015) Mainstreaming climate change resilience in urban development plan: case of Ahmedabad city
20. Union Cabinet approves Atal Mission for Rejuvenation and Urban Transformation and Smart Cities Mission to drive economic growth and foster inclusive urban development (2015). Retrieved from Press Information Bureau http://pib.nic.in/newsite/PrintRelease.aspx?relid=119925
21. Draft concept note on Smart Cities Scheme (2014) Ministry of Urban Development, Lok Sabha Secretariat Parliament Library and Reference, Research, Documentation and Information Service
22. Prasad N (2009) Climate resilient cities: a primer on reducing vulnerabilities to disasters, World Bank

Comparative Cost Analysis of MMFX Bars in Indian Scenario

Virendra Kumar Paul, Salman Khursheed and Md. Asif Akbari

Abstract In this research work, an attempt has been made to do a comparative study of MMFX bars in place of Fe-500 bars which are used extensively in building construction industry these days. The average tensile strength of which vary from 620 to 1030 MPA as compared to the tensile strength of conventional Fe-500 bars which is in the range of 500 MPA. As in the case of normal Fe-500 bars, the main issue was of rusting hence to overcome this high chromium content is mixed for making the MMFX bars. The chloride content is also taken four times higher in case of MMFX bars than the normal carbon steel bars. An attempt has been made to check the savings in quantity of steel by providing reinforcement with higher capacity as compared to Fe-500. At the same time, a cost model is also tried to work out to check its overall feasibility as per Indian condition. The findings of this report will benefit the cost consultants as well as researchers who have keen interest to implement and encourage the use of smart materials for building smart cities.

Keywords MMFX bars · Fe-500 bars · Smart materials for smart cities · New construction materials · Emerging material · Cost comparison · Optimization in RCC structure cost

1 Introduction

The main goal is to have a sustainable economic growth by improving the life of citizen. For this, the terms sustainable cities, smart cities or new technology communities are coined. There are many components of a smart city like energy, governance, environment, mobility and building and services. The use of advanced

V. K. Paul · S. Khursheed (✉) · Md. A. Akbari
Department of Building Engineering and Management,
School of Planning and Architecture, New Delhi, India
e-mail: salman.khursheed@spa.ac.in

V. K. Paul
e-mail: vk.paul@spa.ac.in

© Springer Nature Singapore Pte Ltd. 2020
S. Ahmed et al. (eds.), *Smart Cities—Opportunities and Challenges*,
Lecture Notes in Civil Engineering 58,
https://doi.org/10.1007/978-981-15-2545-2_19

material as one of the parameters is for judging the smart city network [1]. As per [2] the smart building is defined as a component of smart city and for that focus is given on the selection of innovative materials. As per [3] apart from sustainable environment, smart communication spaces, smart devices and smart cities also refer to a city using smart material. The smart material refers to a material having superior strength, ductility, toughness, initial and life cost efficiencies, ease of manufacture and application [4]. As per [5] smart structure is defined as the key component of smart city which apart from fulfilling technical criteria must also fulfil economic, sustainability and environment criterion. Therefore, there is a need to look beyond conventional steel and RCC framed structure and focus on those advanced construction technologies and materials which have better properties in terms of durability, sound insulation, earthquake resistance, strength and at the same time the project can be completed in lesser time with cost feasibility. As per [6] in his research titled "Smarter Material for Smart Cities" there are eleven smart materials defined, TMT bars was one of those. It is manufactured primarily from recycled scrap. Reinforcement with this higher capacity could provide various benefits by reducing member cross sections and reinforcement quantities, leading to savings in material, shipping, and placement costs. MMFX2 rebar can be found in the world's best-engineered construction projects. Since 2002, this revolutionary steel product has been specified in public infrastructure and public/private development projects throughout the USA, Canada, Mexico and the Middle East.

2 Methodology

The methodology involves exploratory research on MMFX bars. To achieve the objective a building is modelled on ETABS 2016 using MMFX as well as Fe-500 bars as per case study. The ETABS 2016 model is then analysed for various structural properties and a relationship is defined among structural parameters of design and cost modelling after taking values of the various loading conditions according to the case study model. Then, at last a cost model is obtained from the relationship established.

3 Comparison with Stainless Steel Rebars

MMFX bars were invented at the University of California by Prof. Gareth Thomas [7]. The organisation based in USA, producing MMFX bars is the MMFX Steel Corp [8]. The MMFX bars have higher strength, fatigue resistance and ductility as compared to other high-strength bars because of its microstructure laminated lathe structure. Because of these changes at microstructure level MMFX bars has longer service life as well as lower construction cost. In terms of corrosion resistance, MMFX is similar to stainless steel. Comparing service life span both MMFX bars and SS rebars gives 100 years of service life but cost-wise MMFX is not even half of the cost of SS rebars if comparatively similar strength bars is selected as per American scenario [9]. As per Virginia Transportation Research Council (VRTC), the MMFX products are designated as the most cost-effective solution [10].

Further, MMFX rebar can be handled just like conventional steel bars, without the special handling requirements associated with SS bars. For example, MMFX is not considered dissimilar to carbon steel for galvanic corrosion purposes. Therefore, MMFX does not need to be isolated from carbon steels in construction, as SS bars do.

The drawback to reinforcing steel is its susceptibility to corrosion. One of the potential alternative to this problem is MMFX bars. MMFX bars do not corrode and come in many different forms that lend themselves to both exterior application for rehabilitation of existing RC columns and use as internal reinforcement to extend initial design life. As per [11] MMFX bars reduce the vulnerability of reinforcement to corrosion and works as an effective high-strength bar. The MMFX bars as compared to the conventional steel when used in an experiment provided better crack control after the yielding [12]. MMFX bars have the yield strength of approximately 828 N/mm^2 [13].

4 Structural Modelling of six Storey Commercial Block

The case study of a building considered is a super structure with the plan as shown in Fig. 1 Structural plan of building considered for case study. The building is modelled using ETABS 2016 having dimensions of 19 × 13.5 m and it is of six storey (G + 5). The building in the study considered is a commercial block with column, beam and slab type framed structure with shear walls. Since the effect of MMFX bars is being checked, the building is first modelled using Fe-500 conventional bars and then using MMfx bars. The building has non-uniform grid in both the directions. There are three different types of column which are 22 in numbers along with three different types of beams. The shear wall is provided in the lift core.

The structural model of building is prepared, analysed and designed in ETABS 2016 version. The load combination for different types of forces is applied on the structure. The member sizes and details are fixed by analysing the structure. Analysis is performed and results are compared for base shear, support reactions, storey drift, mode shapes, etc. The following criteria are assumed while analysing the building:

- The floor diaphragm is assumed to be rigid.
- Dynamic analysis to be performed using response spectrum method.
- All dimensions are in m, unless otherwise specified.
- The size of the framing plan is 19 m × 13.5 m.
- The framing plan is assumed to be Special Moment Resisting Frame (SMRF) (Table 1).

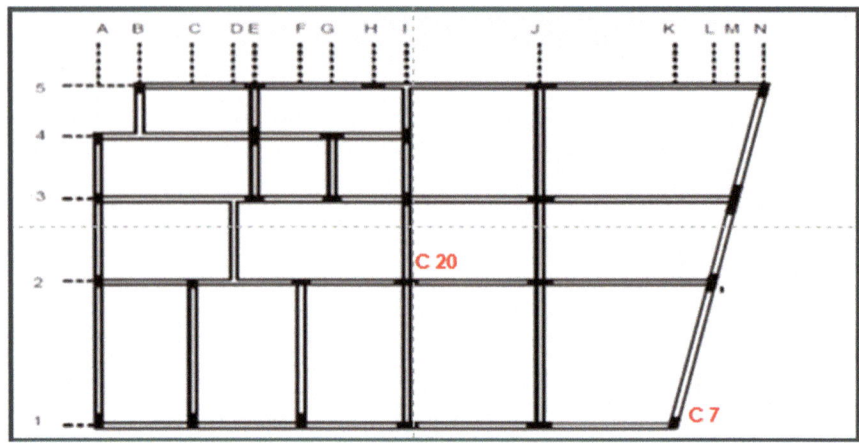

Fig. 1 Structural plan for building considered for case study showing columns C7 and C20

Table 1 Structural modelling of building

Modelling details of building		Remarks
Area	13.5 m × 19 m	
Storey details	Ground + 5 storeys	
Frame type	SMRF (Special Moment Resisting Frame)	Table 9, IS:1983:2016, part-I
Height of basement	5 m	
Height of ground storey	3 m	
Height of typical storey	3 m	
Purpose	Mercantile	IS:875:1987 part -II
Foundation support system	Fixed	
Seismic zone	Zone-IV	IS:1983:2016, part-I
Zone factor (z)	0.24	IS:1983:2016, part-I
Type of soil	Medium	IS:1983:2016, part-I
Damping ratio	5%	IS:1983:2016, part-I
Response reduction factor (R)	5	IS:1983:2016, part-I
Importance factor (I)	1	IS:1983:2016, part-I
Time period in X direction	0.319 s [clause 7.6.2, IS:1893: 2002, part-I]	IS:1983:2016, part-I
Time period in Y direction	0.378 s [clause 7.6.2, IS:1893: 2002, part-I]	IS:1983:2016, part-I

(continued)

Table 1 (continued)

Modelling details of building		Remarks
Material properties		
Grade of reinforcement (longitudinal/main)	Fe-500	MMFX grade 100 (690 MPA) for second building
Grade of reinforcement (shear/ties)	Fe-415	
Concrete		
Grade of concrete (beam, slab, staircase)	M25	
Grade of concrete (column, shear wall)	M30	
Density (M25)	25 KN/Cum	
Density (M30)	30 KN/Cum	
Poisson's ratio	0.20	
Structural members		
Thickness of slab	160 mm	
Thickness of sun-shade	125 mm	
Thickness of staircase	250 mm	
Thickness of shear wall	115 mm and 230 mm	
Thickness of shear wall	230 mm	
Dimension of beam	230 mm × 500 mm	
	300 mm × 600 mm	
	300 mm × 450 mm	
Dimension of column	230 mm × 400 mm	
	230 mm × 600 mm	
	300 mm × 600 mm	
Dead load intensities		
For 230 mm thick wall	12.34 KN/m	IS:875:1987 part-I
For 115 mm thick wall	6.82 KN/m	IS:875:1987 part-I
For parapet wall	5.14 KN/m	IS:875:1987 part-I
For slab	5.24 KN/sqm	IS:875:1987 part-I
For terrace	6.64 KN/sqm	IS:875:1987 part-I
For staircase	10.54 KN/sqm	IS:875:1987 part-I
Imposed load		
On every floor except terrace	5 KN/sqm	IS:875 part-II
On terrace	1.5 KN/sqm	IS:875 part-II

The modelling design and analysis are done using ETABS 2016 software. The building is analysed and designed for conventional Fe-500 and MMFX bars. The effect of brickwork is also taken into account along with the portal frame.

The storey and grid systems are defined and then the material properties of different items taken in the structural modelling followed by the load cases, then the earthquake resistant design parameters.

5 Modelling Results

It deals with both the structural comparative analysis was done using ETABS 2016 as well as cost comparative analysis of building modelled using Fe-500 and MMFX grade 100 bars. The below-mentioned graphs are showing storey drifts when the rebar is changed for Fe-500 to MMFX grade 100 rebars.

It is observed from Fig. 2 that storey drift for building with Fe-500 is more than building modelled using MMFX bars. The changes in the properties of MMFX bars in comparison to Fe-500 may be the reason behind this (Figs. 3 and 4).

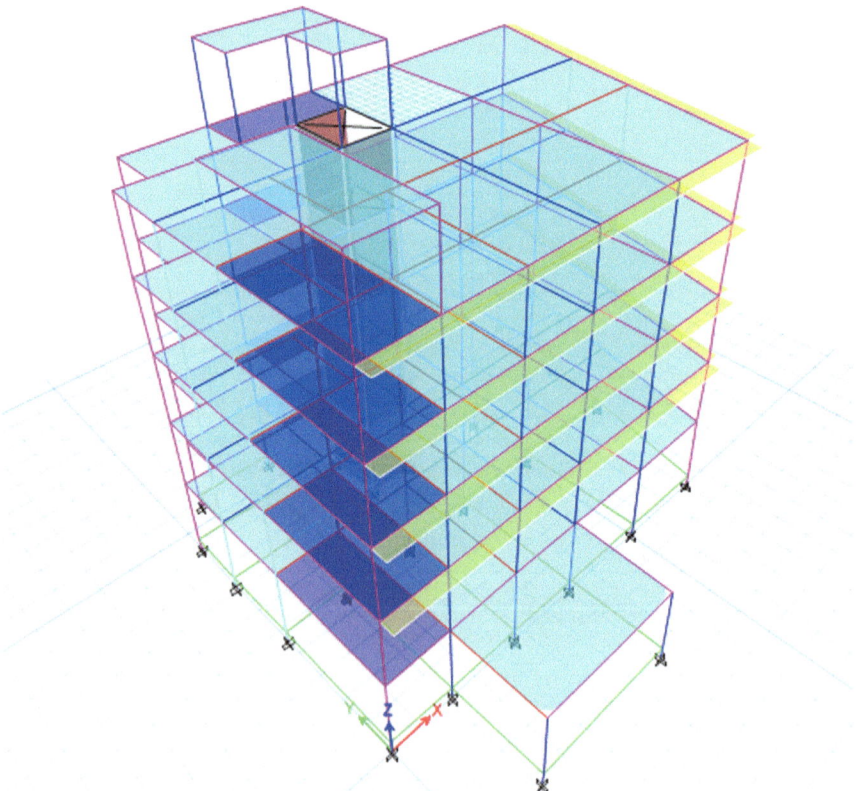

Fig. 2 3D view of commercial block modelled in ETABS 2016

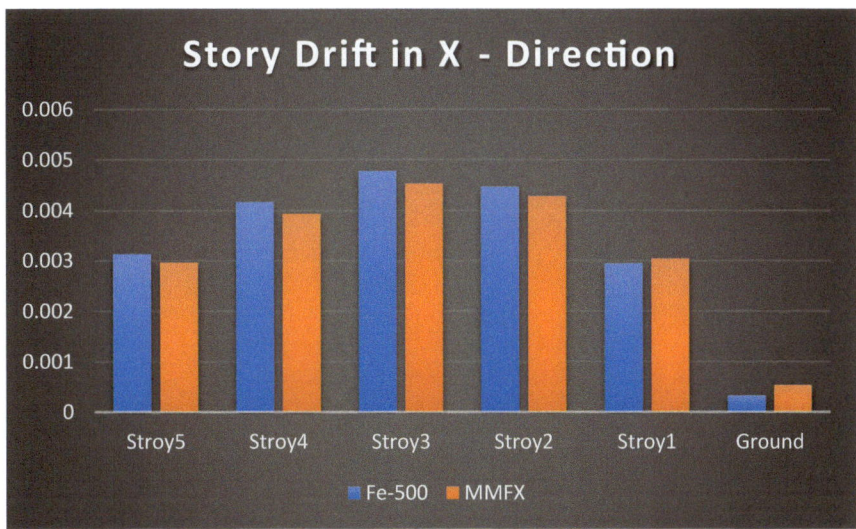

Fig. 3 Representation of storey drift in X direction when the rebar is changed for Fe-500 to MMFX grade 100 rebars

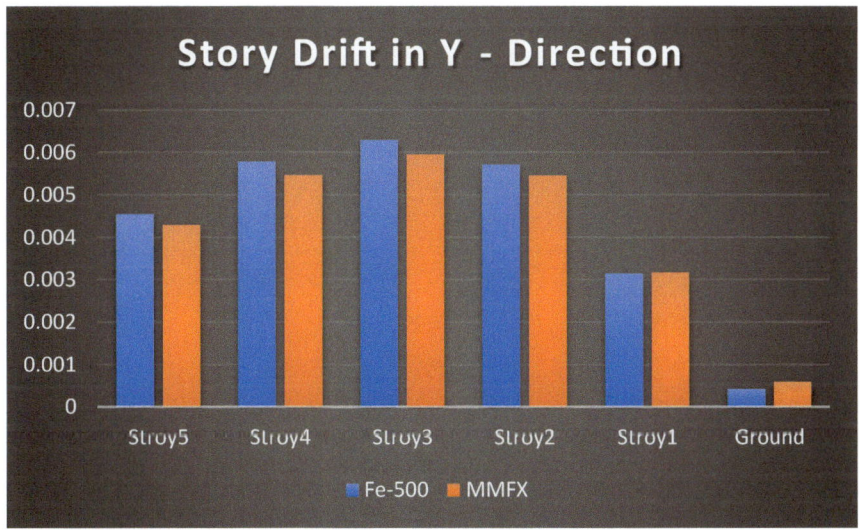

Fig. 4 Storey drift in Y direction for Fe-500 and MMFX grade 100 rebars

The same happened with storey drift in Y direction, because of the changes in the material properties of MMFX bars as compared to conventional FE_500 bars (Figs. 5 and 6).

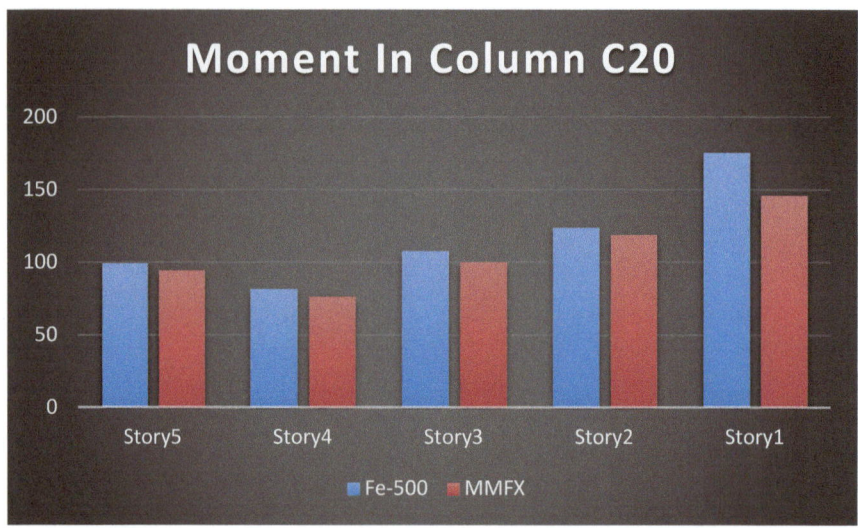

Fig. 5 Percentage variation of moment in column C20 (middle column)

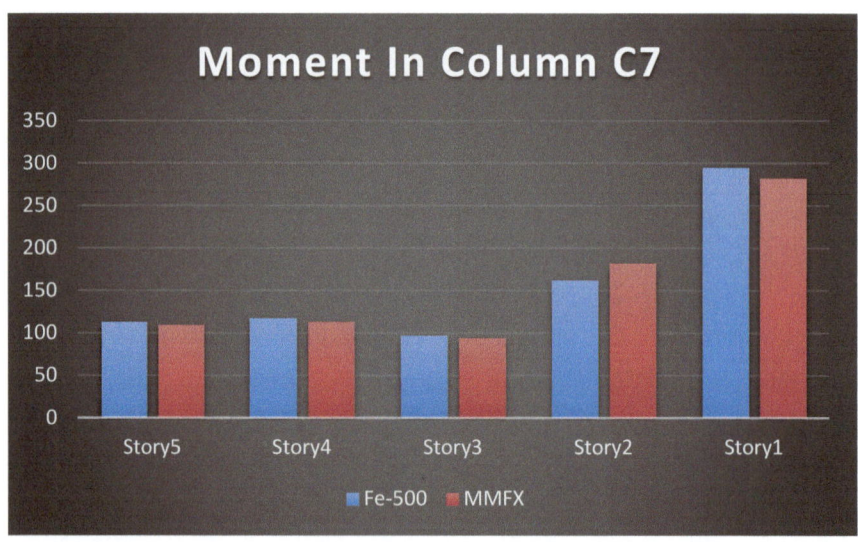

Fig. 6 Percentage variation of moment in column C7 (front right corner)

As per the architectural plan shown in Fig. 1, the middle column (C20) and the front corner column (C7). The variation in moment is about 5–17% on different stories for the same column modelled using Fe-500 and MMFX bars. The variation in moment is due to the different properties of Fe-500 and MMMFX bars (Table 2).

Table 2 Weight of steel obtained in building modelled using Fe-500

Members	Weight (kg)
Beam (longitudinal reinforcement)	
Subtotal	18,557
Beam (shear reinforcement)	
	2.94 kg/m
Subtotal (shear reinforcement)	2682
Total reinforcement in beam	21,539
Column (longitudinal reinforcement)	
Subtotal (longitudinal reinforcement)	8409.45
Column (shear reinforcement)	
Average weight of shear reinforcement per metre length of column	13.5 kg/m
Subtotal (shear reinforcement)	4333.5
Total reinforcement in column	12,742.95
Total weight of reinforcement used in building with conventional bars (Fe-500)	34,281.95

6 Reinforcement Quantity Calculation and Comparison

The area of reinforcement calculated by ETABS is further modified after the detailing done by using conditional formatting function of MS-Excel and the actual area which is got at the end is used for further calculation of cost. The same procedure is repeated for building modelled both using Fe-500 bars as well as MMFX bars. The total quantity of reinforcement thus obtained is summarised in Table 3 as follows in and the summary of total quantity if reinforcement for MMFX bars is made in Table 4.

Table 3 Weight of steel used in building modelled using MMFX grade 100 bars

Members	Weight (kg)
Beam (longitudinal reinforcement)	
Subtotal	16,082
Beam (shear reinforcement)	
	2.94 kg/m
Subtotal (shear reinforcement)	2682
Total reinforcement in beam	18,764
Column (longitudinal reinforcement)	
Subtotal (longitudinal reinforcement)	6280.55
Column (shear reinforcement)	
Average weight of shear reinforcement per metre length of column	13.5 kg/m
Subtotal (shear reinforcement)	4333.5
Total reinforcement in column	10,614.05
Total weight of reinforcement used in building with conventional bars (Fe-500)	29,378.05

Table 4 Cost of importing MMFX rebars at Delhi location

Particulars	Quantity	Unit
The weight obtained from structural detailing	19,684	kg
Rate in Dubai per ton	47,034	Rs.
Approximate fright charge (via sea) [14]	68,058.36	Rs.
Fright charge per ton	3402.918	Rs.
Total cost at Kolkata port/ton	50,436.92	Rs.
Import duty at Kolkata port for 20 ton [15, 16]	14,276	Rs.
Import duty per ton	713.8	Rs.
Total cost in India per ton	51,150.72	Rs.
Transportation cost to Delhi/ton [17]	3971.7	Rs.
Total cost at Delhi location/ton	55,122.42	Rs.
Total cost at Delhi location/Kg	56.12	Rs.

Table 5 Cost comparison of Fe-500 and MMFX bars

Sl. No.	Particulars	Unit	Quantity	Rate/Kg	Cost
1	Fe-500	Kg	29,948.5	33.56	1,005,070
2	MMFX	Kg	22,362.6	56.12	1,254,986

7   Cost Comparison

The rate of the Fe-500 bars is taken for Delhi location from site with the brand name of TATA Tiscon TMT Steel Bars. The rate is ₹33,560/ton, which is ₹33.56/Kg. Similarly, the rate for MMFX bars is taken from the site negotiated rate at Dubai location which is 2600 Dirhams/ton.

Excluding Fe-415 bars which are used for shear reinforcement, as per Table 5, the total quantity of Fe-500 and MMFX grade 100 bars are as follows.

The cost of MMFX bars is higher than the cost of Fe-500 by 25%. Hence, the benefit–cost ratio comes out to be negative which shows non-feasibility for such scale projects.

8   Inferences

The following inferences are drawn from the study done in this research:

- The MMFX bars have superior strength and corrosion resisting property than conventional Fe-500 bars due to its nanostructural arrangement.
- The exploratory study shows that there is no such Indian Standard code which defines the use of MMFX bars in Indian context.
- The density of MMFX bars is slightly more than that of Fe-500 bars (negligible).

- The minimum yield strength of MMFX bars used in this case is 690 MPA and that of Fe-500 is 500 MPA. The former is 38% more than the later one. This results in stiffer structure.
- The quantity of reinforcement obtained in the modelling done using MMFX bars is 29,378 kg and the quantity of reinforcement consumed in building modelled using Fe-500 bars is 34,281 kg. The quantity of reinforcement saved in the modelling done using MMFX bars approximately 16.69%.
- The cost of reinforcement obtained in the modelling done using MMFX bars is INR 1005069 and the cost of reinforcement obtained in building modelled using Fe-500 bars is INR 1254986. The cost difference of 25% is seen as per the study.

9 Conclusions

As seen from the above results, the MMFX bars are better for storey drifts in both X and Y direction. The case study considered in this research is of a mercantile building with six storey. Even after saving 16.69% of reinforcement as compared to Fe-500 bars, the import cost shows that the cost of reinforcement is approximately 25% higher if we opt for MMFX rebars. The possible reasons for this higher cost are that, in this case, the building is considered is of six storey only, and hence the lesser amount of reinforcement is used as compared to tall buildings in terms of total weight of reinforcement. The fright charge from Dubai to Kolkata and the transportation cost from Kolkata to Delhi may have decreased if we would have ordered the same MMFX bars in bulk against the present ordered quantity of just 20 ton. As the code does not takes into consideration of MMFX bars, this study is a kind of way forward for the use of MMFX bars in Indian context as an innovative material for building smart cities but can not be treated as recommendation for use of MMFX bars.

References

1. Lopes IM, Oliveira P (2017) Can a small city be considered a smart city? Procedia Comput Sci 121:617–624
2. Venkat Reddy P, Siva Krishna A, Ravi Kumar T (2017) Study on concept of smart city and its structural components. Int J Civ Eng Technol 8:101–112
3. Apurva S, Tailor S, Rastogi N (2017) Smart materials for smart cities and sustainable environment. J Mater Sci Surf Eng 5:520–523. Doi: 10.jmsse/2348-8956/5-1.5
4. Elattar SMS (2013) Smart structures and material technologies in architecture applications. Sci Res Essays 8:1512–1521. https://doi.org/10.5897/SRE2012.0760

5. Agrawa Y, Saxena R, Gupta T, Sharma RK (2017) Sustainable structures for smart cities and its performance evaluation. Int Res J Eng Technol 4
6. Mishra V, Gandhi N, Desani P, Mehta D (2016) Smarter material for smart cities. Glob Res Dev J Eng. e-ISSN: 2455-5703
7. Hung HQ, Duc Cong N, Anh NMT (2017) Bending behaviour of concrete beams reinforced with MMFX steel preventing the corrosion. Vietnam J Constr, 56th Year
8. Praveen S, Rajesh S, Akhil S (2016) A study on applications of nanotechnology in civil engineering. Asian Rev Civ Eng 5:36–41
9. Report CA (2016) Reinforcing steel comparative durability tourney consulting group, LLC
10. Weyers RE, Sprinkel MM, Brown MC (2006) Summary report on the performance of epoxy-coated reinforcing steel in Virginia. Virginia Transportation Research Council, Charlottesville
11. Malhas F, Mahamid M, Rahman A (2016) Serviceability performance of concrete beams reinforced with MMFX steel. Int J Mod Eng 17(1):50–59
12. El-Tahawy R, Hassan TK, Hamdy O, Rizkalla S (2009) Flexural behavior of concrete beams reinforced with high strength steel reinforcing bars. In: Paper presented at the 13th international conferences on structural and geotechnical engineering, Cairo, Egypt
13. Chajes MJ, Richardson DR, Wenczel GC, Liu W (2005) MMFX rebar evaluation for I-95 service road bridge 1-712-B. Delaware Center for Transportation, University of Delaware, Newark, Delaware
14. Rates (Online) WF (2013). http://worldfreightrates.com/en/freight
15. Cybex (2018) Cybex. http://www.cybex.in/indian-custom-duty/Ferro-Alloys-Hs-Code-7202.aspx
16. www.census.gov (2017). https://www.census.gov/foreign-trade/schedules/b/2018/c72.html. 2018
17. www.truckbhada.com (2018). http://www.truckbhada.com/CalculateFreight

Feasibility Investigation of Energy Storage Systems of Hybrid Power System and Its Benefits to Smart Cities

Abid Hussain Lone, Tahleela Navid and Anwar Shahzad Siddiqui

Abstract The objective of this research is to design and model a hybrid system using photovoltaic (PV) and battery, integrated with grid system using hybrid optimization model for electric renewables (HOMER) software. The model is tested using three different battery types: lead acid (LA), vanadium redox (VR), and zinc bromine (ZnBr) flow batteries. It has already been mandated that 10 per cent of the smart cities energy requirement should come from solar energy. Solar rooftops can go a long way in providing a clean and green living environment in these smart cities. Solar energy can be harnessed by installing solar panels that can reduce our dependency on non-renewable sources of energy, but due to the intermittent nature of solar energy and its variability, the need for battery energy storage becomes mandatory to stabilize the operation of such hybrid systems and to store the surplus energy produced and supply it at a time of insufficiency. These battery types are compared in terms of system sizing, economy, technical performance, and environmental stability. A case study of an Institutional area of New Delhi is carried out to compare the performance of individual batteries. The results demonstrate that the hybrid system using ZnBr batteries is the most favourable choice. Using this configuration, the economic parameters, including net present cost (NPC) and levelized cost of energy (LCOE), are found to be lowest and emissions into the atmosphere are found to be optimum.

Keywords Flow battery · Lead–acid battery · Optimization · HOMER

1 Introduction

Solar energy can play a big role in creating smart cities. Solar applications, such as solar street lights, solar water heaters, and solar rooftop, will help to make the buildings energy efficient and green in these smart cities. Switching to renewable resources like solar energy from fossil fuels [1] needs an extensive survey of the

A. H. Lone (✉) · T. Navid · A. S. Siddiqui
Department of Electrical Engineering New Delhi, Jamia Millia Islamia, New Delhi, India

© Springer Nature Singapore Pte Ltd. 2020
S. Ahmed et al. (eds.), *Smart Cities—Opportunities and Challenges*,
Lecture Notes in Civil Engineering 58,
https://doi.org/10.1007/978-981-15-2545-2_20

available potential of different natural resources in the area to be electrified. According to the Ministry of New and Renewable Energy (MNREL) report 2016–17 [2], the state of New Delhi (Latitude: 28 °59′ N, longitude: 77 °59′ E) considered for the present study, has a solar potential of 2050 MW. Therefore, solar energy is considered as a reliable resource for generating electricity. Since solar energy is intermittent in nature and it varies with cloud cover, the use of a standalone PV system for serving the base load may not be a reliable solution that is why PV power is considered as unstable source. Hence, to stabilize the operation of such hybrid systems, batteries are incorporated in them. An assessment of hybrid systems with battery energy storage systems (BESS) was discussed in [3]. This paper presents the life cycle cost-benefit analysis of BESS using different types of batteries in a hybrid system. The main purpose of the hybrid system with BESS proposed is to reduce the economic parameters as much as possible. The incorporation of a battery bank makes the control operation more practical [4]. Batteries can be compared on the basis of their cost, cycle life, replacement, and, most importantly, their safe disposal. Five different battery types lead acid (LA), zinc bromine (ZnBr), sodium-based, nickel-based, and vanadium redox (VR) are normally used in renewable energy systems. Amongst these batteries, the sodium-based are not considered in this study as they need an extra system to ensure a high operating temperature, thereby increasing their high annual operating cost [5]. Similarly, nickel is a toxic material, whose decomposition may lead to environmental hazards [6]. Table 1 shows a technical comparison, for the remaining three battery types: LA, ZnBr, and VR. These batteries are used individually with the designed system and are compared on the basis of economics, technical performance, and environmental effects. Energy density of a battery is a measure of how much energy the battery can store, in a given size or mass. Therefore, a battery having higher energy density can power a load for longer duration than the one with a low energy density and the same physical size or mass. Its units are in Wh/kg or Wh/litre. A battery's depth of discharge (DoD) is the percentage of the battery that has been discharged relative to the overall capacity of the battery. If a battery is charged and discharged more frequently, its lifespan gets reduced. It is generally not recommended to discharge a battery entirely, as that dramatically shortens the useful life of the battery. If a battery is discharged

Table 1 Technical comparison of batteries

Parameter	LA battery	ZnBr battery	VR battery
Unit cell voltage (V)	2	1.8	1.15–1.55
Open circuit voltage (V)	12	100	49–57
Energy density (Wh/litre)	60–110	35	20
Cycle life (cycles)	500–2000	>10,000	>10,000
Life span (years)	5–15	10–15	20-30
Depth of discharge (%)	70–80	100	100
Round trip efficiency (%)	85	75	75–80

regularly at a lower percentage amount, it will have more useful cycles than a battery frequently drained to its maximum DoD. The ratio of energy put into the battery to the energy retrieved from the battery is the round trip efficiency expressed in per cents (%). The higher the round trip efficiency, the less energy loss due to storage, the more efficient the system as a whole.

2 Description of Parameters

2.1 Load Profile

A hostel building namely J&K hostel in Jamia Millia Islamia, New Delhi is considered for case study. It includes 217 rooms, 2 dining halls, 1 reading room, 1 common room, and 2 warden offices. The details of different types of loads, their ratings and quantity commonly used in this hostel are listed in Table 2. The daily average hostel load is 3113 kWh with a random variation of 10% while the peak load is 426 kW with a random variation of approx. 8%. Average load profile of weekdays and weekends is shown from Figs. 1, 2, 3, and 4. There is a huge variation in load profile because a minimum load was recorded during the vacation period, whereas in peak winter and summer days, load rapidly increases up to 426 kW.

Table 2 Approximate daily electric load details of hostel

Loads	Rating (W)	Usage (hours/day)	Quantity
Tube lights	20	8	788
Water cooler[a]	775	24	28
AC[a]	1667	8	3
Laptop	65	2	650
Exhaust fan	40	24	131
Iron	1000	1	220
Phone charger	3.5	2	650
Washing machine	500	3	56
CFL	7	16	896
Induction heater	2000	2	2
Refrigerator[a]	150	24	2
Fans	80	16	480
Room heater[a]	2000	8	2
Geyser[a]	2000	5	28
Air cooler[a]	200	8	150
TV	70	2	1
Halogen	200	12	7

[a]Seasonal loads

Fig. 1 Approximate load variation (hot weekdays)

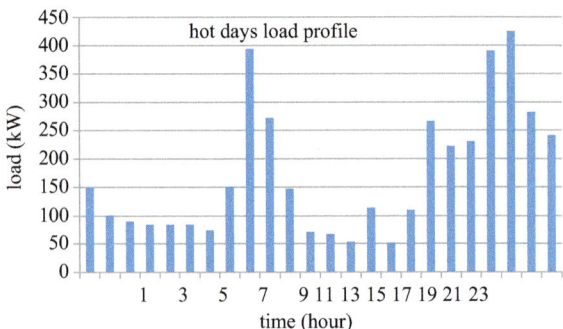

Fig. 2 Approximate load variation (hot weekends)

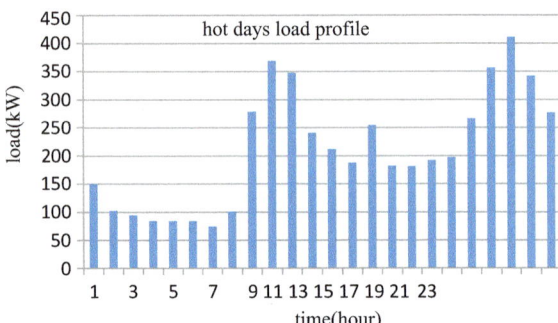

Fig. 3 Approximate load variation (cold weekdays)

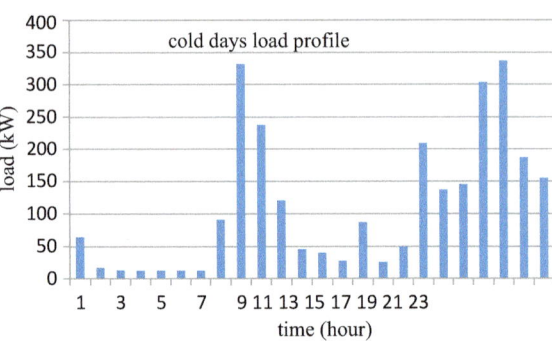

Fig. 4 Approximate load variation (cold weekends)

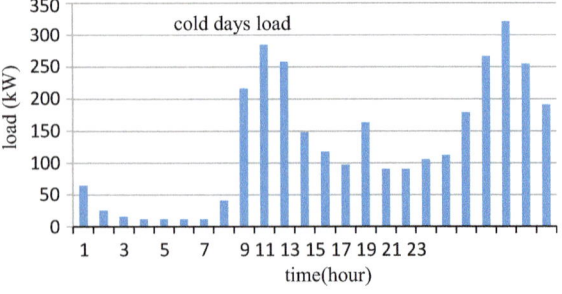

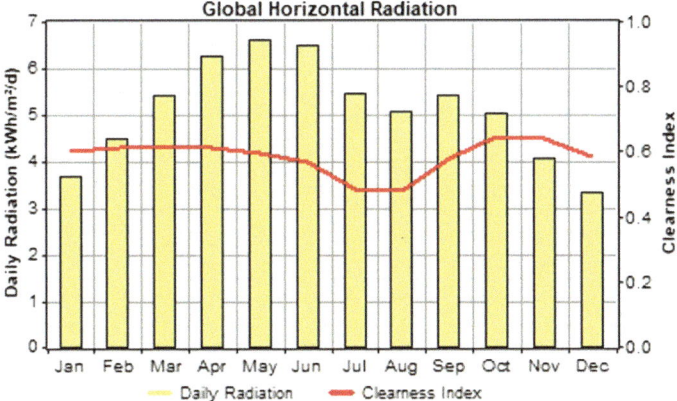

Fig. 5 Annual solar radiation and clearness index for New Delhi, India

2.2 Solar Radiation

The solar resource input for the various months throughout the year was obtained from the Internet via HOMER software [11] by providing the latitude, longitude, and time zone information [7]. Indian Meteorological Department (IMD) measures and provides solar radiation data along with other climatic parameters over various locations across the country. Figure 5 shows the solar radiation data inputs of New Delhi along with the clearness index of the solar irradiation. The clearness index is the fraction of the solar radiation that is transmitted through the atmosphere to strike the surface of the Earth. It is a dimensionless number ranging between 0 and 1. The clearness index has a high value under clear and sunny conditions and a low value under cloudy conditions. The solar radiation for each month was obtained where it was found that the maximum solar radiation was found for the month of May with daily radiation of 6.616 kWh/m^2/d, whereas the minimum radiation was found for the month of December with daily radiation 3.33 kWh/m^2/d. The average radiation throughout the year was 5.108 kWh/m^2/d.

2.3 Hybrid System Components, Their Specifications and System Constraints

Photo Voltaic System: A 440 kW PV system is chosen for the system. The parameters considered for PV are: lifetime: 20 years, ground reflectance: 20%, and scaled annual average 5.11 kWh/m^2/d. The capital cost, replacement cost, and operating and maintenance (O&M) cost considered for the PV system are shown in Table 3.

Table 3 Cost of equipments used in hybrid system

Equipment	Size (kW)	Capital INR/ kW	Replacement INR/ kW	O&M INR/ kW
Diesel generator	280, 120, 50	36,000	29,000	21,600
Solar PV	440	51,000	51,000	11,972
LA battery	1156 Ah	21,900	21,900	100
ZnBr battery	1000 kWh	54,000	54,000	0
VR battery	350 kWh	45,600	45,600	0
Converter	420	20,999	20,999	0

Converter: A converter of rating 420 kW is used in this modelling having an efficiency of 90% and 85% for inverter and rectifier, respectively, and a life span of 25 years.

Battery Bank: Commercially available battery models, like Surrette S6CS25P LA, VR, and ZnBr batteries are considered one by one for hybrid system simulation. Cost summary of different batteries used is shown in Table 3. For optimal operation of batteries, the cycle charging (CC) dispatch strategy is set. The CC strategy is a dispatch strategy, whereby whenever a power source (generators, storage bank or grid) needs to serve the primary load, it operates at full output power. Surplus electrical production goes towards the lower-priority objectives in order of decreasing priority like serving the deferrable load, charging the storage bank, or serving the electrolyzer.

System Constraints:
Annual interest rate = 6%
Plant working life span = 25 years, maximum annual capacity shortage = 0%

Grid:
Cost of energy (COE)   INR 8.4/kWh
Sellback price   INR 4/kWh
Carbon dioxide (CO_2) emission   630 g/kWh
Sulphur dioxide (SO_2) emission   2.74 g/kWh
Nitrogen oxide (NO) emission   1.34 g/kWh

The price listed in Table 3 is taken as the unit price of the equipments used in the proposed system for simulation. It is done by collecting data from local market [8–10, 14].

3 Modelling and Simulation Results

The optimization and technical results and cost details for the two different models, that is, Generator-PV-Battery connected system and Grid-PV-Battery connected system are compared when three different types of batteries (LA, ZnBr, and VR) are used in above two models and the emissions are recorded.

3.1 Economic Analysis

Economics play an integral role both in HOMER's simulation process, wherein it operates the system so as to minimize total NPC, and in its optimization process, wherein it searches for the system configuration with the lowest total NPC.

After performing a rigorous simulation process, the most feasible and optimal configuration is taken for modelling the system.

(1) *Generator-PV-Battery*:

The model and optimal size and optimal cost of different components used in this hybrid configuration is shown in Fig. 6, Tables 4 and 5, respectively.

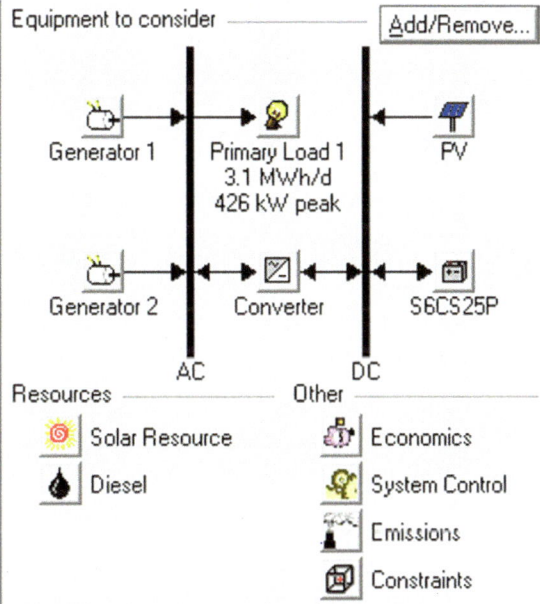

Fig. 6 Schematic of proposed hybrid PV and generator system with battery storage

Table 4 Optimal size of different components

Model	PV (kW)	Diesel generator s (kW)	Battery (kWh)	Converter (kW)
Gen-PV	450	350, 100	–	230
Gen-PV-LA	450	260, 100	763	240
Gen-PV-ZnBr	450	240, 105	5500	400
Gen-PV-VR	450	260, 105	350	250

Table 5 Optimal cost for different configurations

Model	NPC (Million) INR	LCOE (INR) INR	O&M (Million) INR	IC (Million) INR
Gen-PV	358.61	24.67	277.73	43.98
Gen-PV-LA	304.389	20.951	234.604	43.358
Gen-PV-ZnBr	259.14	17.82	190.042	49.39
Gen-PV-VR	297.906	20.51	225.46	50.01

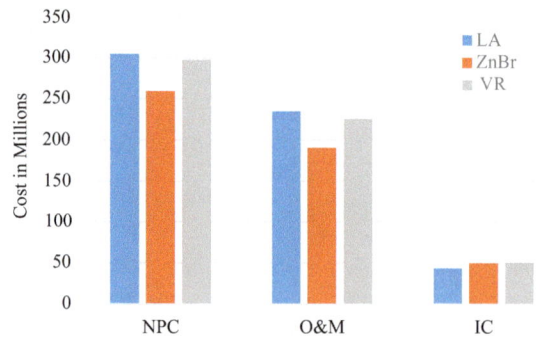

Fig. 7 Comparison of economics of Gen-PV-Battery hybrid systems

Table 6 COE with different types of batteries

Parameter	Gen-PV-LA	Gen-PV-ZnBr	Gen-PV-VR
COE (INR)	20.91	17.82	20.51

The configuration using ZnBr batteries was found to have the lowest total NPC of 259.14 million INR and the least LCOE of 17.82 INR/kWh in this configuration as shown in Fig. 7 and Table 6 respectively.

(2) *Grid-PV-Battery:*

Optimal size, cost and model of different components in this hybrid system obtained after many simulations are shown in Tables 7 and 8 and Fig. 8 respectively.

In this configuration also, using ZnBr batteries, the system has the lowest total NPC of 80.08 Million INR and the least LCOE of 4.31 INR/kWh.

Table 7 Optimal size of different components

Model	PV (kW)	Grid (kW)	Battery (kWh)	Converter (kW)
Grid-PV	450	450	–	290
Grid-PV-LA	440	300	1387	320
Grid-PV-ZnBr	450	400	1000	330
Grid-PV-VR	440	350	200	320

Table 8 Optimal cost for different configurations

Model	NPC (Million) INR	LCOE (INR) INR	O&M (Million) INR	IC (Million) INR
Grid-PV	103.46	5.621	69.143	29.039
Grid-PV-LA	97.8	5.402	56.58	33.54
Grid-PV-ZnBr	80.06	4.31	54.015	21.63
Grid-PV-VR	83.867	4.45	52.51	26.637

Fig. 8 Schematic of proposed hybrid PV and grid system with battery storage

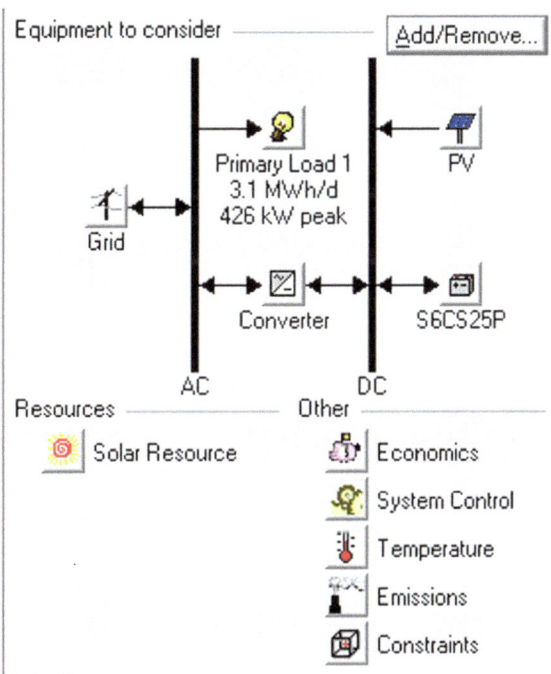

Table 9 COE with different types of batteries

Parameter	Grid-PV-LA	Grid-PV-ZnBr	Grid-PV-VR
COE (INR)	5.402	4.31	4.45

Table 10 Emission from two different hybrid system configurations

Pollutant	Emissions (kg/year) Generator-PV-Battery	Emissions (kg/year) Grid-PV-Battery
Carbon dioxide	743,664	288,472
Carbon monoxide	1836	0
Unburnt hydrocarbons	203	0
Particulate matter	138	0
Sulphur dioxide	1493	1251
Nitrogen oxides	16,379	612

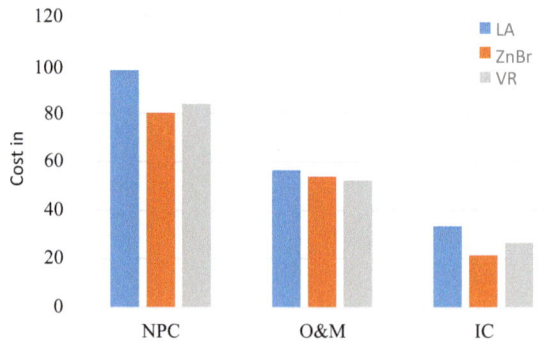

Fig. 9 Comparison of economics of Grid-PV-Battery hybrid systems

3.2 Environmental Effects

The major factors causing environmental hazards in the present hybrid systems are emissions from the diesel generators. The various harmful gases emitted from the two chosen configurations are shown in Tables 9 and 10. Also a comparison was made for this hybrid configuration for different types of batteries as indicated in Fig. 9.

4 Conclusion

Two different configurations are compared for supplying electricity to a modelled load in this research work comprising of diesel generator, solar PV, and battery, and another configuration consisting of grid, solar PV, and battery. Three different types of batteries (lead acid, zinc bromine and vanadium redox) are used [12]. Numerous alternatives of feasible hybrid systems are obtained from the HOMER software. Amongst the three different battery types (LA, VR, and ZnBr) used in the study, the hybrid system PV-Grid using ZnBr batteries has been recommended for the case study site. This configuration is found to offer the most compact arrangement, requiring a nominal PV system capacity of 450 kW, and a battery bank rating of 1000 kWh for the considered load. Furthermore, the hybrid system using ZnBr batteries is the most economical configuration, with the lowest NPC of INR 80.06 M and the least COE of 4.31 INR/kWh. At present time, the cost of energy for the grid-connected system is Rs 8.4/kWh, which is expected to increase with time. Energy storage systems are emerging as technologies that can increase the local management of the microgrid electricity network in the smart cities by ensuring energy is delivered effectively and efficiently at microlevels [13].

References

1. Zhang C, Zhou K, Yang S, Shao Z (2017) On electricity consumption and economic growth in China. Renew Sust Energy Rev 76:353–368
2. https://mnre.gov.in/annual-report
3. Lasseter R, Erickson M (2009) Integration of battery based energy storage element in the CERTS microgrid. Report U.S. Department of Energy, Madison, WI
4. Wang Y, Tan KT, So PL (2013). Coordinated control of battery energy storage system in a microgrid. In: 2013 IEEE PES Asia- Pacific power and energy engineering conference (APPEEC)
5. Kawakami N, Iijima Y, Fukuhara M, Bando M, Sakanaka Y, Ogawa K (2010) Development and field experiences of stabilization system using 34 MW NAS batteries for a 51 MW wind farm. In: IEEE 2010 International symposium on industrial electronics, Bari, Italy. New York, NY, USA: IEEE. pp 2371–2376
6. Lacerda VG, Mageste AB, Santos IJ, Da Silva LH, Da Silva MD (2009) Separation of Cd and Ni from Ni-Cd Batteries by an environmentally safe methodology employing aqueous two-phase systems. J Power Sources 193:908–913
7. NASA Surface Meteorology and Sola Energyi(Online). Available: http://www.nasa.gov
8. https://india.alibaba.com/index.html
9. https://my.indiamart.com/
10. www.rollsbattery.com
11. Getting Started Guide for HOMER version 2.1, April 2005
12. Tharani KL, Dahiya R (2017) Choice of battery energy storage for a hybrid renewable energy system. Turk J Electr Eng Comput Sci (elk-1707-350) 26(2):666–676
13. Wang C (2008) Power management of a stand-alone wind/photovoltaic/fuel cell energy system, energy conversion. IEEE Trans 23(3):957–967
14. http://www.kirloskar-electric.com/gens.shtml

Investigating IoT Middleware Platforms for Smart Application Development

Preeti Agarwal and Mansaf Alam

Abstract With the growing number of Internet of Things (IoT) devices, the data generated through these devices is also increasing. By 2030, it has been predicted that the number of IoT devices will exceed the number of human beings on earth. This gives rise to the requirement of middleware platform that can manage IoT devices, intelligently store and process gigantic data generated for building smart applications such as smart cities, smart health care, smart industry and others. At present, market is overwhelming with the number of IoT middleware platforms with specific features. This raises one of the most serious and least discussed challenges for application developer to choose suitable platform for their application development. Across the literature, very little attempt is done in classifying or comparing IoT middleware platforms for the applications. This paper categorizes IoT platforms into four categories, namely publicly traded, open-source, developer-friendly and end-to-end connectivity. Some of the popular middleware platforms in each category are investigated based on general IoT architecture. Comparison of IoT middleware platforms in each category, based on basic, sensing, communication and application development features, is presented. This study can be useful for IoT application developers to select the most appropriate platform according to their application requirement.

Keywords IoT · Middlewares · Analytics · Protocols · Platforms

1 Introduction

The advancement in sensor, actuator, computing and storage technologies has given rise to era of "Internet of Things" (IoT). The emergence of smaller and cheaper interoperable wireless devices over low-powered wireless medium has made the communication among these devices and human possible. These wireless devices

P. Agarwal (✉) · M. Alam
Jamia Millia Islamia, New Delhi, India
e-mail: malam2@jmi.ac.in

© Springer Nature Singapore Pte Ltd. 2020
S. Ahmed et al. (eds.), *Smart Cities—Opportunities and Challenges*,
Lecture Notes in Civil Engineering 58,
https://doi.org/10.1007/978-981-15-2545-2_21

can foster development of many smart applications in various domains such as smart home, smart traffic, smart healthcare, smart city, smart agriculture and smart logistics [1].

IoT devices sense the surroundings. If a large number of IoT devices is connected, then they are going to sense a large number of events, generating massive data. This data can be either structured, unstructured and semi-structured. The data can be generated at constant rate or time triggered. For analysis of such kind of data, it is essential to be stored resourcefully. Different IoT vendors worldwide are coming with different middleware platforms to support application development requirement. These IoT middleware platforms mainly sit between actual deployed sensors and applications. They consist of a set of functionalities such as device interaction, management, data storage, and processing which required to build smart application. Broadly, IoT platforms can be divided into the following four broad categories [2]: publicly traded, open source, developer friendly and end-to-end connectivity platforms.

Finding the most appropriate IoT middleware for an application development is the major challenge faced by the developers today [3]. The functionalities provided by different middleware vendors are almost similar, but they differ mainly in underlying technologies. Services provided by different IoT vendors mainly include data acquisition, device management, data storage, security and analytics [4]. Selecting the right platform according to the detailed analysis of application requirement is one of the necessary steps in application development [5].

This paper presents some of the most popular platforms in each of the four categories and examines their features according to the general IoT architecture. This study can help the application developers in choosing the right platform according to their need. Section 2 briefly describes some studies carried related to IoT platforms. Section 3 presents the general architecture of IoT platform. Section 4 presents categorization of IoT platforms. Section 5 gives comparisons of different features of IoT middleware layer-wise and provides tools and technologies used in each layer. Finally, Sect. 6 discusses criteria for choosing the right platform according to the application requirement, and Sect. 7 gives the concluding remarks.

2 Literature Review

In this section, we will discuss studies carried out in the IoT platform. Ray [6] studied twenty-six platforms based on ten parameters (application development, device management, system management, heterogeneity management, data management, analytics, deployment, monitoring, visualization and research).

Farahzadi et al. [7] discussed various challenges of storage, management and aggregation of IoT data in middleware platforms. They presented a comparison of platforms on the basis of overall characteristics such as adaptability, connectivity, context management, energy efficiency, flexibility, interoperability, maintainability, platform portability, quality of service, real-time tasks, resource discovery, reusability, security and privacy, transparency and trustworthiness.

da Cruz et al. [8] classified IoT platforms and IoT cloud platforms based on the application enablement platforms, application development platforms and device management platforms.

Singh and Kapoor [9] studied different hardware modules and their integration with different IoT platforms. This study focused only on hardware devices making it insufficient for application development.

Mineraud et al. [10] evaluated open-source platforms. Gap analysis is done to improve platforms and provide business opportunities.

Scott and Östberg [11] presented a comparative study of open-source IoT middleware platforms. This comparison was restricted to the scalability and reliability measure. Machorro et al. [12] carried out a comparative study of platforms from industry point of view and discussed case studies with challenges. Fortino et al. [3] discussed an outline of middleware for smart objects and smart environments, compared them with specific requirements and identified in the literature.

All the above-mentioned studies have discussed parameters that give the overview of different platforms but do not compare layer-wise features of platforms from application development perspective. Our investigation divides the task of smart application development into layers as described in Sect. 3 and provides layer-wise features of identified platforms. Data analytics is the heart of application development in IoT cloud environment. A number of applications can be developed using big data analytics in cloud environment [13]. Task scheduling and load balancing are required to improve the efficiency of platforms [14]. In the next section, we present the general architecture of IoT platform.

3 IoT Middleware Platform General Architecture

This section describes the basic elements of IoT middleware platform reference architecture [5]. All elements of this architecture are described in the bottom-up manner as shown in Fig. 1.

- **Sensors**: Sensor consists of a hardware component capable of acquiring information about physical environment. This acquired information is transmitted in the form of electrical signals to the connected devices. Devices connectivity can be either wired or wireless.
- **Actuator**: It is a hardware component, which receives command in the form of electrical signals from connected device and performs some kind of physical action. Like sensor, it can also be connected to the device either in a wired or wireless manner.
- **Device**: A device is a hardware component consisting of processor and storage. It is connected to sensors and actuators. By the help of software, it can establish connection to IoT integration middleware.

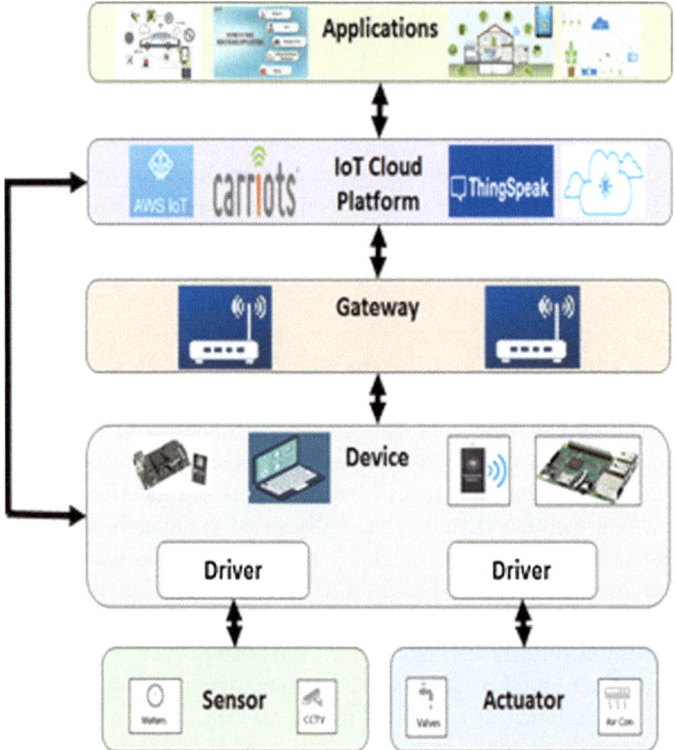

Fig. 1 General IoT architecture

- **Gateway**: Sometimes gateway is required to connect the device to IoT platform. Gateway is an interface that provides technologies and mechanism to interconnect between communication technologies and protocols. All devices can access gateway if they are IP enabled. Gateway is also able to store, filter and process received data before sending to cloud.
- **IoT Platform**: Main responsibilities of this layer are:

    a. It integrates data received from different kinds of connected devices.
    b. Process the received data.
    c. Control devices.
    d. Provide received data to various applications.

IoT cloud platform can also directly communicate with the device if both are using compatible technologies and protocols.

IoT cloud platform layer is also responsible for providing functionalities such as time series database or graphical dashboards, aggregation and utilization of data received from devices. Mostly, IoT platforms are accessed through HTTP-based REST APIs.

- **Application**: Applications are built on top of various IoT cloud platforms to provide services to some real-life scenario such as smart cities, smart health care, smart industry.

4 Categorization of IoT Middleware Platforms

Various IoT middleware platforms can be categorized into the following four categories, namely publicly traded IoT cloud platforms, open-source IoT cloud platforms, developer-friendly IoT cloud platform and end-to-end connectivity IoT cloud platform as shown in Fig. 2. This section describes various popular platforms in each of these categories:

- **Publicly Traded IoT Middleware Platforms**: This category consists of platforms developed and maintained by large publicly traded companies such as AWS IoT platform [15], Microsoft Azure IoT Hub [16], IBM Watson IoT Platform [17], Google IoT Platform [18] and Oracle IoT Platform [19].
- **Open-source IoT Middleware Platforms**: This category consists of platforms that provide data management services under open licenses such as Kaa [20] and ThingSpeak [21].
- **Developer-Friendly IoT Middleware Platforms**: This category of platforms is developer friendly and can be easily integrated with Arduino, Raspberry, etc., to

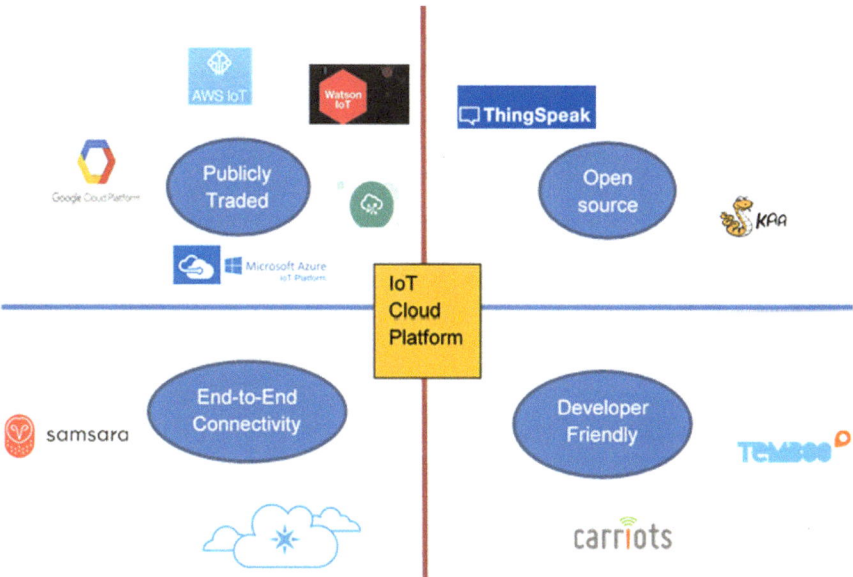

Fig. 2 Categorization of IoT middleware platform

develop users' applications. Some of the platforms belonging to this category are Carriots [22] and Temboo [23].

- **End-to-End Connectivity IoT Middleware Platforms**: Platforms designed based on supplied hardware and required solution such as Samsara [24] and Particle Cloud [25].

5 Comparison of Features of IoT Middleware Platforms

IoT platforms in each of the above-mentioned category are compared based on the basic features, sensing features, communication features and application development features.

5.1 Basic Features

The basic features of the IoT middleware platforms include open-source or open API, deployment model of cloud, availability, data format supported, programming languages supported and pricing model. Different platforms support different features. The application developers can select a platform according to their requirements. The basic features of platforms are provided in Table 1. We have used "Y" in tables to show support of certain feature, and the absence of "Y" shows lack of support.

5.2 Sensing Features

The sensing features of the IoT middleware platforms include support for multi-device, heterogeneous devices and hardware compatibility of platform for sensing environmental information. Sensing features are provided in Table 2; "Y" is used to show support of certain feature, and the absence of "Y" shows lack of support.

5.3 Communication Features

Communication features include communication of IoT sensed data for storage and processing. Major features compared here are: gateways, protocols, security mechanism. Communication technologies used in platforms play a major role in

Table 1 Basic features of IoT middleware platforms

IoT middleware	Open source/open SDK	Deployment type	Availability (24 * 7)	Data format supported	Programming languages support	Pricing
Publicly traded platform						
AWS IoT platform	Open-source SDK	PaaS, IaaS	Y	JSON	Java, C, NodeJS, Javascript, Python, SDK for Arduino, iOS, Android	Pay when execute your own written functions
Microsoft Azure IoT Hub	Open-source API	IaaS	Y	JSON	.NET, UWP, Java, C, NodeJS, Ruby, Android, iOS	Pay according to number of devices and messages per day
IBM Watson IoT platform	Open-source SDK	PaaS, IaaS	Y	JSON, CSV	C#, C, Python, Java, NodeJS	Pay according to number of devices, data traffic and data storage
Google IoT platform	Open API	PaaS, IaaS	Y	JSON	Go,Java, .NET, Node.js, php, Python, Ruby	Priced per MB
Oracle IoT platform	Open-source SDK	PaaS	Y	CSV, REST API	Java, Javascript, Android, C, iOS	Subscription based
Open-source platform						
Kaa	Open SDK	IaaS	–	REST API, JSON	Java, C, C ++	Free
ThingSpeak	Open source	PaaS	–	ThingSpeak API, JSON, XML	MATLAB	Free
Developer-friendly platform						
Carriots	Open-source API	PaaS	Y	XML, JSON, REST API	Java	Subscription based
Temboo	Open-source API	PaaS	–	Excel, CSV, XML, JSON	C, Java, Python, iOS, Android, Javascript	Subscription based
End-to-end connectivity platform						
Samsara	Open API	–	–	JSON	–	Paid services
Particle cloud	Open source	PaaS, SaaS	–	CSV	Javascript, particle js	Free access for the first 100 devices after that paid per device

Table 2 Sensing features of IoT middleware platform

IoT middleware	Multi-device support	Heterogeneous device support	Hardware compatibility
Publicly traded platform			
AWS IoT platform	Y	Y	Broadcom, Marvell, Renasas, Texas Instruments, Microchip Intel
Microsoft Azure IoT Hub	Y	Y	Intel, Raspberry, FreeScale, Texas Instrument
IBM Watson IoT platform	Y	Y	ARM mbed, Texas Instruments, Raspberry Pi, Arduino Uno
Google IoT platform	Y	Y	Raspberry Pi
Oracle IoT platform	Y	Y	Raspberry Pi, iMX6 sabrelite
Open-source platform			
Kaa	Y	Y	Udoo, Samsung Artik, Raspberry Pi, Intel edison
ThingSpeak	Y	Y	Arduino, Particle photon, ESP8266 wifi, Raspberry Pi
Developer-friendly platform			
Carriots	Y	Y	Arduino, Raspberry Pi, Nanode, BeagleBone
Temboo	Y	Y	Texas Instrument, Arduino, Samsung Artik
End-to-end connectivity platform			
Samsara	–	–	–
Particle cloud	Y	Y	Electron, Photon, Raspberry Pi

selecting the right platform for application developers. In case application requires to connect different networks together than gateway support is must. The major communication protocols supported by IoT devices are: MQTT, CoAP, HTTP, WebSockets. Different security features supported by platforms can be encryption, authentication, authorization, auditing and scope for user facilities. Depending on the security requirement of the application, the developer can choose the platform. Table 3 presents communication features; "Y" shows the presence of certain feature, and the absence of "Y" shows absence of features.

5.4 Application Development Features

The support for various application development technologies is shown in Table 4. Application development technologies deployed on the platform are considered the heart of that middleware. In Table 4, we have discussed the support for

Table 3 Communication features of various IoT middleware platform

IoT middleware	Gateway	Protocols	Security				Scope for user defined policies
			Encryption	Authentication	Authorization	Auditing	
Publicly traded platforms							
AWS IoT platform	–	HTTP, MQTT, Websockets	Y	Y	Y	Y	–
Microsoft Azure IoT Hub	Y	HTTP, AMQP, HTTP	Y	Y	Y	–	Y
IBM Watson IoT platform	Y	MQTT	–	Y	Y	–	–
Google IoT platform	–	MQTT, HTTP	–	Y	–	–	–
Oracle IoT platform	Y	REST APIs	–	Y	Y	–	–
Open-source IoT platforms							
Kaa	Y	MQTT, CoAP	Y	–	–	–	Y
ThingSpeak	–	MQTT	Y	–	–	–	–
Developer-friendly platforms							
Carriots	Y	MQTT	–	Y	Y	–	–
Temboo	Y	HTTP, MQTT, CoAP	–	–	–	–	–
End-to-end connectivity platforms							
Samsara	–	HTTP	–	–	–	–	–
Particle cloud	–	HTTP	Y	Y	Y	Y	–

Table 4 Application development support features in IoT middleware platform

IoT middleware	Support for application development							Technologies used
	M2M application	Real-time analytics	Machine learning	Artificial intelligence	Analytics	Visualization	Event and reporting	
Publicly traded platform								
AWS IoT platform	–	Y	Y	Y	Y	Y	Y	AWS Lambda, Amazon Kenisis, Amazon Machine learning, Amazon Dynamo DB, Amazon CloudWatch, AWS CloudTrail
Microsoft Azure IoT Hub	–	Y	Y	–	Y	Y	Y	Azure CosmosDB, Azure Tables, SQL database
IBM Watson IoT platform	–	Y	Y	–	Y	Y	Y	Cloudant NOSQL DB
Google IoT platform	–	Y	Y	–	Y	Y	Y	Google's BigData tool, Riptide IO, BigQuery, Firebase, PubSub
Oracle IoT platform	–	Y	–	–	Y	Y	Y	NoSQL Database
Open-source platform								
Kaa	–	–	Y	–	Y	Y	Y	NoSQL, Cassandra, Hadoop and MangoDB
ThingSpeak	Y	Y	–	–	Y	Y	Y	MATLAB, dashboard
Developer-friendly platform								
Carriots	Y	–	–	–	Y	Y	–	NoSQL BigDatabase
Temboo	–	–	Y	–	Y	Y	Y	Microsoft Power BI, Google BigQuery
End-to-end connectivity platform								
Samsara	–	–	–	–	Y	Y	–	–
Particle cloud	–	–	–	–	Y	–	Y	IFTTT

technologies such as M2M applications, real-time analytics, machine learning, artificial intelligence, analytics, visualization and event reporting. "Y" shows the support for particular features, and the absence of "Y" means lack of support. Real-time analytics refers to the application of analytical algorithms on streaming data, whereas analytics refers to the application of analytical algorithms on historic data.

6 Criteria for Selection of the Right IoT Middleware Platform

The most challenging task for an application developer in IoT is choosing the right platform. Each platform has certain specific features and services. The selection depends on certain criteria which are discussed below:

- **Availability**: Availability and stability are important parameters for application requirement. For example, smart healthcare application requires patient medical data to be monitored continuously (24 * 7). Therefore, application related to patient data monitoring needs to choose a platform with 24 * 7 availability, whereas in case of smart industry, timings can be restricted to limited hours. As shown in Table 1, most of the publicly traded platforms satisfy availability requirements.
- **Deployment type**: Open-source platforms allow platforms to be managed by the application developer according to their need; in contrast, this facility is not provided by commercial IoT platforms such as AWS IoT and IBM Watson. Professionals at the middleware provider end manage the commercial platforms. It depends on requirement of application to be developed, whether developer needs to keep the flexibility of platform management on his job part or wants to relax itself by levering this responsibility on vendor. Most of middleware platforms use cloud technology for storage purpose. According to the resource requirement, developer can choose among different cloud deployment models— Infrastructure as a Service (IaaS), Platform as a Service (PaaS) and Software as a Service (SaaS).
- **Pricing Model**: Different vendors adopt different pricing models. Some support pay as you execute, some pay per storage, some pay according to the number of connected devices and some are subscription based. For small-scale applications, some platforms provide free limited storage. Particle cloud provides free access up to 100 devices. AWS IoT charges only when function is executed on stored data. Microsoft Azure charges based on number of devices and messages. This can be suitable for applications where there is a variance in the number of devices and messages every time. IBM Watson charges on the basis of storage required. Kaa and ThingSpeak are open sources and free, good for personalized applications. For applications with constant storage rate and devices, subscription-based platforms are good such as Carriots and Tembo.

- **Support for Required Hardware**: A number of IoT boards like Arduino Yun, Photon, Raspberry Pi, etc., are available in market. Each supports different standards and features. These boards mainly differ on the basis of processor, GPU, clock speed, size, RAM, memory, support for different programming language and price [9]. Choosing the right one close to requirement is important. AWS IoT has the highest number of compatible hardware devices.
- **Security Requirement**: Levels of required security vary in different applications, as banking application requires high security as compared to smart health care. Different platforms provide different levels of security. Some platforms even allow the application developers to implement their own security algorithms. Microsoft Azure IoT and Kaa platform provide flexibility to users to implement their own security policies. AWS IoT and Particle Cloud provide the highest security features.
- **Type of Communication Protocol Support**: Multiple communication protocols are supported by IoT devices. Some are lightweight, and some are secure [26]. Which protocol to choose depends on the application requirement. CoAP is similar to HTTP but is lightweight, therefore more suitable for mobile applications. MQTT is also lightweight and supports broker concept, making it good for limited bandwidth applications [27]. CoAP is good for multicast and broadcast.
- **Storage Technologies Used**: Mainly vendors use cloud as their storage. Different storage and processing technologies on top of cloud support different type of analytics. As per processing requirement of data, cloud with different storage technologies can be selected. AWS IoT supports largest number of storage technologies.
- **Type of Analytics Supported**: IoT applications usually require either real-time/streaming data or historic data for the application development [28]. Choosing the right technology for applying analytics on the kind of data generated by IoT device in the application is another important factor. ThingSpeak and Kaa support M2M applications. AWS IoT, Microsoft Azure IoT, Google IoT, Oracle IoT and ThingSpeak support real-time analytics on streaming data. AWS IoT also supports artificial intelligence.

7 Conclusion

In this paper, we have discussed the basic features, sensing features, communication features and support for the application development features. This information can help the IoT application developers in selecting appropriate platforms according to their application need. Basic features include support for open-source/open SDK/ open API, cloud deployment model used, availability, data format supported, programming languages supported and pricing model of various platforms. Sensing features tells about the capability to sense information from multi-devices, heterogeneous devices, compatibility with hardware. Communication features include support for gateways, different protocols supported, security features such as encryption, authorization, authentication and provision for the extension of security policy. In application development features, we discussed about the support of tools for the application development. Different applications require different kind of tools. Lastly, we have discussed criteria for selection of appropriate platform and the IoT middleware platforms satisfying it. This investigation can help application developers in choosing platform according to tools required in their application.

References

1. Razzaque MA, Milojevic-Jevric M, Palade A, Clarke S (2016) Middleware for Internet of Things: a survey. IEEE Int Things J 3(1):70–95
2. Postscapes (2019) IoT cloud platform landscape. 2019 vendor list [Online]. Available at: https://www.postscapes.com/internet-of-things-platforms/. Accessed 7 Feb 2019
3. Fortino G, Guerrieri A, Russo W, Savaglio C (2014) Middlewares for smart objects and smart environments: overview and comparison. In: Internet of Things based on smart objects. Springer, Cham, pp 1–27
4. Al-Fuqaha A, Guizani M, Mohammadi M, Aledhari M, Ayyash M (2015) Internet of Things: a survey on enabling technologies, protocols, and applications. IEEE Commun Surv Tutor 17 (4):2347–2376
5. Guth J, Breitenbücher U, Falkenthal M, Leymann F, Reinfurt L (2016) Comparison of IoT platform architectures: a field study based on a reference architecture. In: Cloudification of the Internet of Things (CIoT). IEEE, pp 1–6
6. Ray PP (2016) A survey of IoT cloud platforms. Future Comput Inf J 1(1–2):35–46
7. Farahzadi A, Shams P, Rezazadeh J, Farahbakhsh R (2018) Middleware technologies for cloud of things: a survey. Digital Commun Netw 4(3):176–188
8. da Cruz MA, Rodrigues JJP, Al-Muhtadi J, Korotaev VV, de Albuquerque VHC (2018) A reference model for Internet of Things middleware. IEEE Int Things J 5(2):871–883
9. Singh KJ, Kapoor DS (2017) Create your own internet of Things: a survey of IoT platforms. IEEE Consum Electron Mag 6(2):57–68
10. Mineraud J, Mazhelis O, Su X, Tarkoma S (2016) A gap analysis of Internet-of-Things platforms. Comput Commun 89:5–16
11. Scott R, Östberg D (2018) A comparative study of open-source IoT middleware platforms
12. Machorro-Cano I, Alor-Hernández G, Cruz-Ramos NA, Sánchez-Ramírez C, Segura-Ozuna MG (2018) A brief review of IoT platforms and applications in industry. In: New perspectives on applied industrial tools and techniques. Springer, Cham, pp 293–324
13. Khan S, Shakil KA, Alam M (2018) Cloud-based big data analytics—a survey of current research and future directions. In: Big data analytics. Springer, Singapore, pp 595–604
14. Ali SA, Alam M (2016) A relative study of task scheduling algorithms in cloud computing environment. In: 2016 2nd international conference on contemporary computing and informatics (IC3I). IEEE, pp 105–111
15. Amazon Web Services, Inc (2019) IoT applications & solutions. What is the Internet of Things (IoT)? AWS [Online]. Available at: https://aws.amazon.com/iot/. Accessed 7 Feb 2019
16. Azure.microsoft.com (2019) IoT Hub. Microsoft azure [Online]. Available at: https://azure.microsoft.com/en-in/services/iot-hub/. Accessed 7 Feb 2019

17. Ibm.com (2019) IBM Watson Internet of Things (IoT) [Online]. Available at: https://www.ibm.com/internet-of-things. Accessed 7 Feb 2019
18. Google Cloud (2019) Google cloud IoT—fully managed IoT services. Google Cloud [Online]. Available at: https://cloud.google.com/solutions/iot/. Accessed 7 Feb 2019
19. Cloud.oracle.com (2019) Internet of Things. Oracle cloud [Online]. Available at: https://cloud.oracle.com/iot. Accessed 7 Feb 2019
20. Kaa IoT platform (2019) Kaa enterprise IoT platform [Online]. Available at: https://www.kaaproject.org/. Accessed 7 Feb 2019
21. Thingspeak.com (2019) IoT analytics—ThingSpeak Internet of Things [Online]. Available at: https://thingspeak.com/. Accessed 7 Feb 2019
22. Altairsmartworks.com (2019) Altair SmartWorks. Home [Online]. Available at: https://www.altairsmartworks.com/. Accessed 7 Feb 2019
23. Temboo.com (2019) IoT: Temboo [Online]. Available at: https://temboo.com/iot. Accessed 7 Feb 2019
24. Samsara.com (2019) Samsara. Internet-connected sensors [Online]. Available at: https://www.samsara.com/. Accessed 7 Feb 2019
25. Particle (2019) Particle company news and updates [Online]. Available at: https://www.particle.io/. Accessed 7 Feb 2019
26. Ammar M, Russello G, Crispo B (2018) Internet of Things: a survey on the security of IoT frameworks. J Inf Secur Appl 38:8–27
27. Hejazi H, Rajab H, Cinkler T, Lengyel L (2018) Survey of platforms for massive IoT. In: 2018 IEEE international conference on future IoT technologies (future IoT). IEEE, pp 1–8
28. Mohammadi M, Al-Fuqaha A, Sorour S, Guizani M (2018) Deep learning for IoT big data and streaming analytics: a survey. IEEE Commun Surv Tutor

Thermal and Optical Investigation of Lime Mortar for Repetitive Thermal Loading

Sumedha Moharana and Venkata Vishala

Abstract Lime-based mortar interlayer plays an important role in structural and architectural construction. Lime mortar undergoes repetitive changes in temperature when it is exposed to its service environment. Change in temperature has a significant effect on the thermal deformation of cement mortar and even affects directly the durability of century decade monuments and structure under extreme regions. This paper deals with thermal deformation of lime mortar, which is the prime constituent of cement and the oldest construction material by measuring its deformed geometrical configuration. In the first part, it is aimed to investigate the thermal deformation of mortar cube, its changes in length and mass of the specimens continuously measured during the heating and cooling cycles within the range of temperatures of -20 to $70\ ^\circ$C. The second part of this paper aims to investigate the pore formation and pore size distrubution in lime and lime mortar sample using optical microscopic images. Although a careful visual inspection can provide very good accuracy information, it has inherent limitations in assessing the structural integrity, and also, visual inspection can provide limited information to a damaged facility. In this paper, a detailed pore distribution analysis has done using digital image processing tool for better interpretation of thermal damages in lime mortar. The above-proposed tool can be used as aided tool for retrofitting and conservation of many architectural heritages.

Keywords Digital image processing · Pore size distribution · Lime · Mortar · Interfacial transition zone · Thermal deformation

S. Moharana (✉) · V. Vishala
Department of Civil Engineering, Shiv Nadar University, Dadri 201314, India
e-mail: sumedha_maharana@snu.edu.in

V. Vishala
e-mail: vv434@snu.edu.in

© Springer Nature Singapore Pte Ltd. 2020
S. Ahmed et al. (eds.), *Smart Cities—Opportunities and Challenges*,
Lecture Notes in Civil Engineering 58,
https://doi.org/10.1007/978-981-15-2545-2_22

245

1   Introduction

Mortar is one of the oldest building materials, which is used to construct large structures from small components such as bricks, blocks and stone. If lime is used as a binder, it is known as lime mortar. Lime has been an important component of mortars since many years, millennia in fact. Lime mortar provides unique benefits, namely water retention and air entrainment. Its water retention is very useful for resistance to rain penetration. Lime mortars have high levels of flexural bond strength [1]. Lime mortars also can be used to minimize the potential for water penetration, increase the durability and compressive strength, and also lime mortar provides uniform performance characteristics in the field.

The change in volume that occurs as water crystallizes is known as freezing–thawing cyclic process. This process occurs within the mortar during freezing just before its liquefaction during melting. Lime mortars have excellent freeze–thaw durability in the field and in freeze–thaw testing, regardless of the degree of air entertainment [2].

In the presence of water, exposure of mortars to low temperatures and frost can lead to freeze–thaw damage, causing the mortar to fail or burst. One of the ways to minimize this damage, which is very important, is carbonation of lime mortar Freezing of water held in capillary pores is understood to generate inward tensile pressures in mortar [3]. The deleterious effect of freeze–thaw cycling is mitigated by deliberately introducing millions of tiny air voids into the material, known as entrained air voids. Also, climate change will affect the presence and duration of freezing and consequently impact any freezing mechanisms that occur [4–6].

In many tropical countries, where there is a high intense rainfall where the materials used in masonry might be saturated for longer periods, there is a high risk of binder leaching and consequent deterioration. Damage due to freeze–thaw deterioration can take various forms, and the most common is micro-cracking. This damage potentially progresses individually within the structural member and also reduces the strength and serviceability [7]. Factors that affect pore properties play an important role in freeze–thaw deformation. Keeping this in mind, there is a high importance of need to study deterioration of lime mortar [8].

Lime mortar has wide applications including masonry works such as brick masonry, plastering walls and columns. Cement mortar and asphalt mortar undergo repetitive changes in temperature when it exposed to the atmosphere, i.e. changes in temperature have a significant effect on the thermal deformation of cement mortar. Many ancient structures are constructed by various materials which include lime mortar. In many western countries in the world, for exterme cold in winter and croaching summer, the repetative variation of temperature has corrosive effect in building materials [9, 10]. This creates cracks, patches and cavities. In the colder parts of the world, there is a corrosive effect on the structure due to rainfall. Plastering starts falling because of freeze–thaw cycles. When the snow melts, water flows into the cracks or gaps of the structure; as when water freezes to ice, it occupies more volume than that of water. During nights, when temperature goes down, the water, which entered the cracks, freezes which causes expansion, weakening and crumbling of some of the material.

The application of digital image processing tool in civil engineering structures and material characterization has significantly increased among researchers. This study also attempts the pore distribution analysis using digital image processing tool through MATLAB. The analysis done on the microscopic images of the lime mortar samples is used to establish mineralogical and chemical profiles of the samples and to verify the results of the experimental field methods. **This approach can be used as an aid for nondestructive evaluation and continuous monitoring of historical monuments and their material characterization. Also, pore analysis through digital image processing tools always stands for different utilities and smart tools for urban development and smart cities.**

2 Experimental Investigation

Experimental Investigation
Lime mortar has prepared by mixing lime with sand in 1:3 ratio, and an optimal amount of water is added for the preparation of samples. The casting of lime and lime mortar cylinderical sample (length = 73 mm and dia = 22 mm) has been done for experimental procedure (see Figs. 1 and 2). These moulds have kept for air curing for 20 days. The samples were kept in woven. The temperature rises and lowers repetitively at the rate of 1 °C/min at the range 23–70 °C. Weight and volume changes of specimens were recorded. The process was repeated for 40 cycles of thermal loading.

Before putting them lime and lime mortar sample in repetative thermal loading, the basic physical characterization of lime based mortar has done in accordance with IS codal guidlines. The following experiments have been conducted.

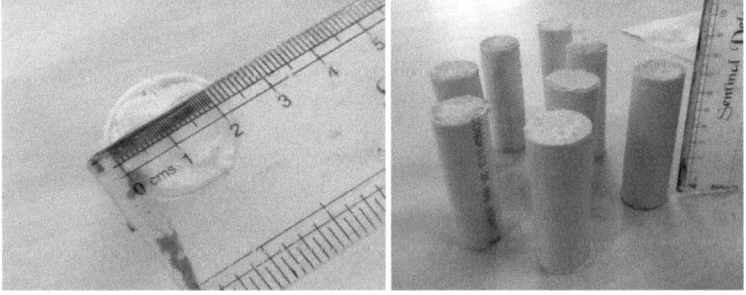

Fig. 1 Lime paste kept in cylindrical mould of diameter: 22 mm and length: 73 mm

Fig. 2 Lime paste in cubical
mould

Table 1 Physical parameters
of lime sample

Tests	Results
Consistency	32%
Initial setting time/final setting time	38 and 363 min
Soundness	4 mm

  i. Consistency of lime
 ii. Initial and final setting time of lime
iii. Soundness of lime.

The results are tabulated in Table 1.

Experimental Thermal Analysis

From the experimental results, it can be observed that the length, volume and
weight parameters are decreasing as the number of cycles is increasing. This change
is due to the change of temperature and loss of water content in the sample. Loss of
moisture from the moulds would lead to strains, since moisture is an important
factor of frezze–thaw deterioration. Free water in the sample exists only in pores so
that porosity is closely related with the evaporation of water amount. Free water in
the sample exists only in pores so that porosity is closely related with the evaporation of water amount.

From the above figures, it can be observed that the length and volume parameters are decreasing as the number of cycles is increasing. The change in these
parameters is due to the change of temperature and loss of water content in the
sample. Loss of moisture from the moulds would lead to strains, since moisture is
an important factor of freeze–thaw deterioration. Samples with larger pores permit

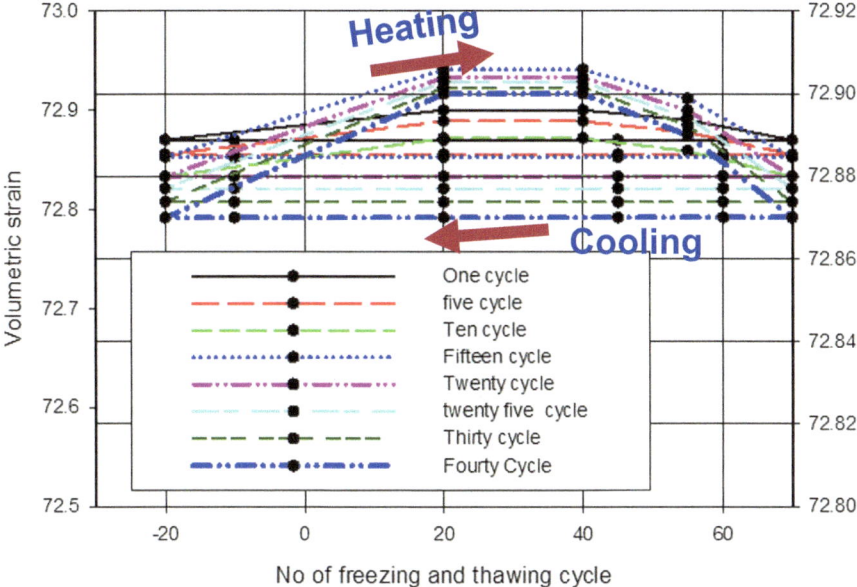

Fig. 3 Volumetric strain in lime mortar cubes with respect to thermal loading

larger water absorption in saturated condition and larger evaporation of water in drying process accordingly. Free water in the sample exists only in pores so that porosity is closely related with the evaporation of water amount. The changes in length and volume can be seen more during dry cycles (see Figs. 3, 4 and 5). Figure 7 shows the strain temperature behaviour of a sample from cycle 1 to cycle 40.

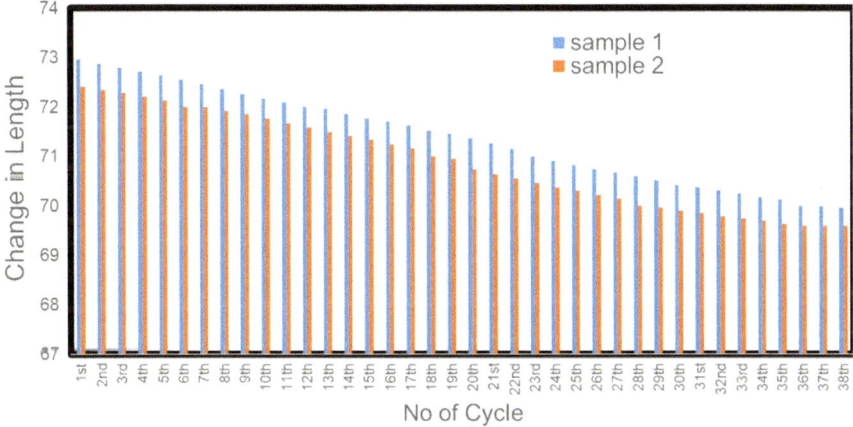

Fig. 4 Change in length in lime mortar cylinder with respect to thermal loading

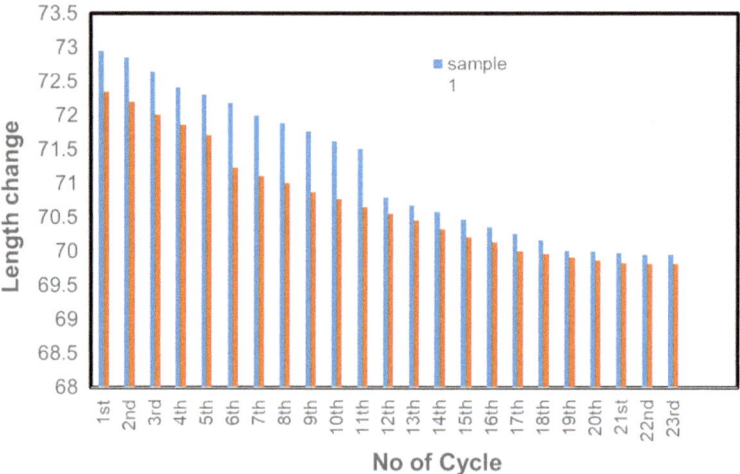

Fig. 5 Change in length in lime mortar cubes with respect to thermal loading

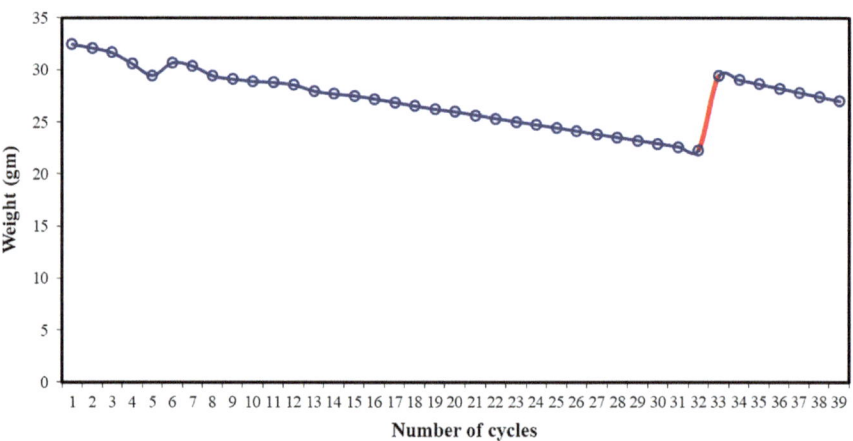

Fig. 6 Change in weight in lime mortar cubes with respect to thermal loading

Figure 6 represents the weight variation in the lime mortar sample (cube) for cycles of thermal loading. The strain during the thawing was higher compared to freezing using dry cycles; the main difference can be seen in 0 °C during wet cycles showing that the feature is present because of freezing of the water around the sample.

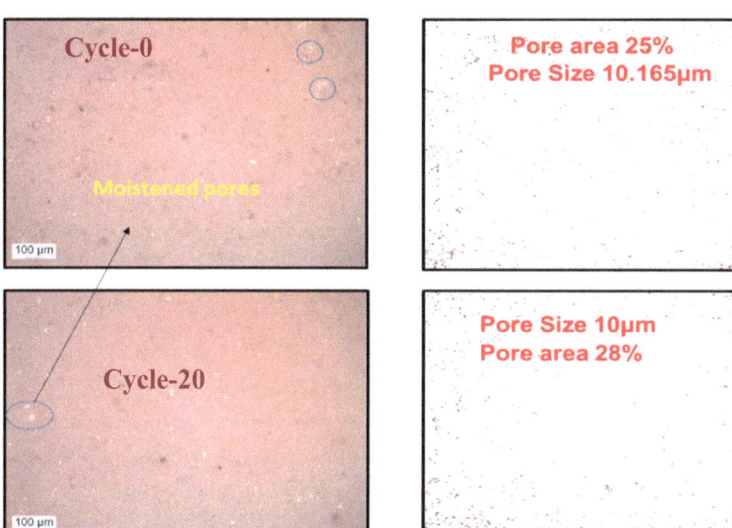

Fig. 7 Lime cubes (for first 0 and 20 cycles of thermal loading)

3 Optical Study of Mortar/Lime Samples After Repetitive Thawed-Freeze Action

Optical microscope is a useful method to observe fracture surfaces of thin sections that can be obtained from the samples. Through optical observation, several changes due to freeze and thaw exposure can be noticed. The total pore volume and the pore size distribution have a **major** influence on the freeze–thaw durability of a mortar. Quantitative optical microscopy methods can be used to determine the aggregate size distribution of the mortar which are very useful when only a limited edition amount of sample is present. This method should only be applied to mortars with a rather uncomplicated composition [10]. An optical microscope was used at $20\times$ magnification in an attempt to investigate the macro- and micro-structural changes of the sample due to freezing and thawing cycles. For the examination of lime and lime mortar samples with optical microscope, thin section samples were prepared. These samples were allowed to dry for two days for setting. The dried samples were then investigated with an optical microscope under reflected light. The position was marked on the sample to observe the changes in the pores.

4 Pore Distribution Analysis Though Optical Image Processing

In this section, for different types of samples (lime mortar) and for number cycles of thermal loading, pore sizes and their distribution have been studied through digital image processing algorithm. Figures 7, 8, 9 and 10 represent pore population in the lime mortar sample under various freezing and thawing cycles. The exposed top surface of both lime and lime mortar sample optical scanning area of sample (591.623 μm × 778.351 μm) is optically studied and compared with before and after thermally exposed. The percentage of pores and size of lime mortar sample remain the same for the first few cycles but reduced to 3% after 25 cycles. The pore size and its distribution significantly increased after 40 cycles of freezing and thawing. The pore diametre (micron) and its distribution over cycles of freezing and thawing are plotted in Figs. 11 and 12 for enhancement of this study The above plot (Figs. 11 and 12) represents the increase in pore sizes and their distribution over scan area over various cycles of heating and cooling. For lime sample, after few cycles of freezing and thawing, the pore size does not increase much, but there is a significant increase in their distribution for small size pores with respect to increase in thermal loading cycles (see Fig. 11). For lime mortar samples, the size of pores is larger, and their distribution increases with thermal cycles. The addition of sand for lime mortar further produces the calcium silicate hydrate which further induces increase of pore size after 35 and 50 cycles (see Fig. 12).

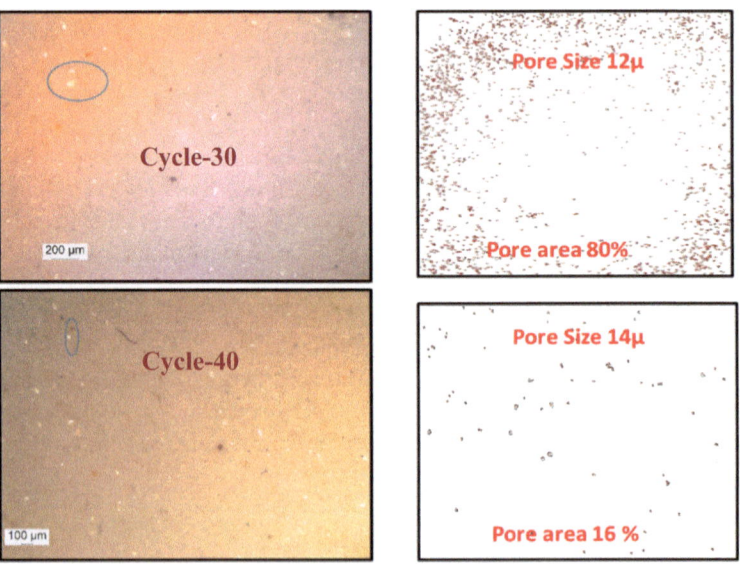

Fig. 8 Lime cubes (for first 30 and 40 cycles of thermal loading)

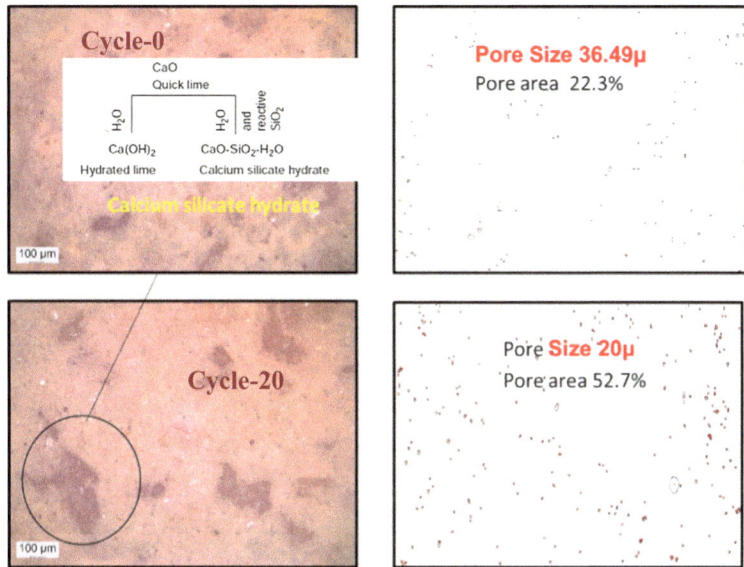

Fig. 9 Lime mortar cubes (for first 0 and 20 cycles of thermal loading)

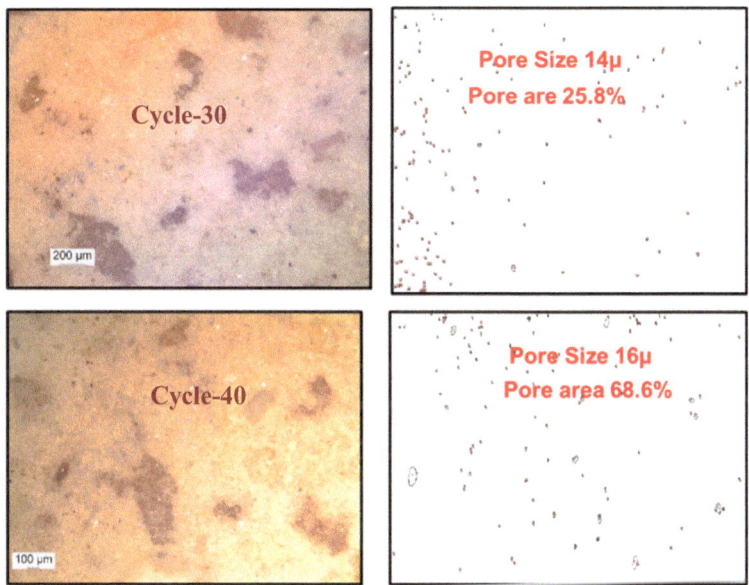

Fig. 10 Lime mortar cubes (for first 30 and 40 cycles of thermal loading)

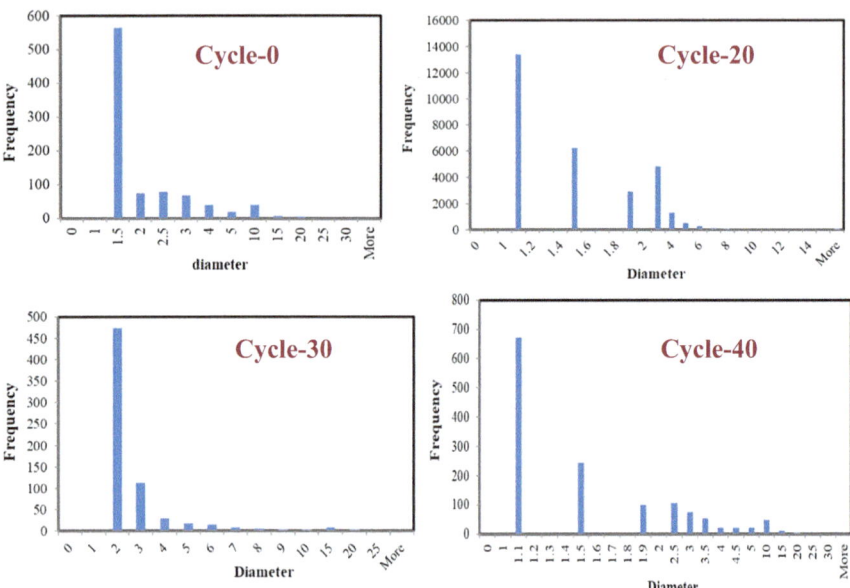

Fig. 11 Distribution of pores in lime sample for different cycles of thermal loading

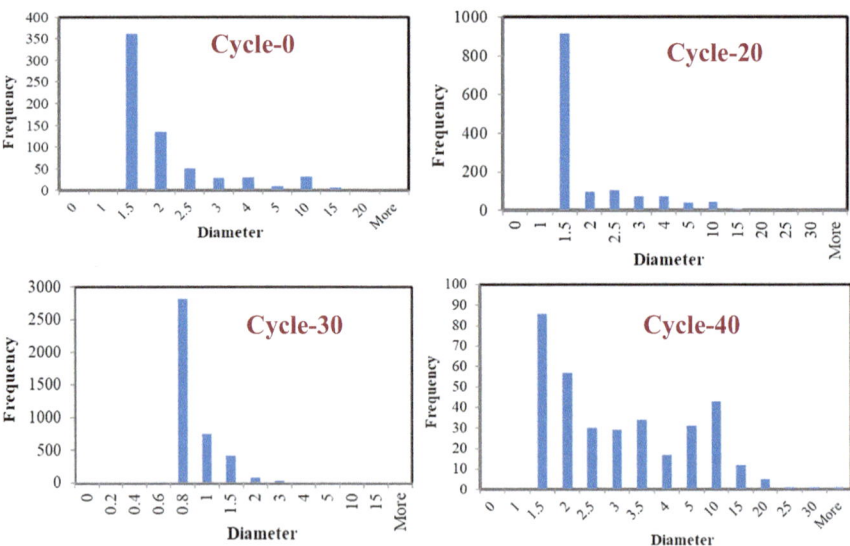

Fig. 12 Distribution of pores in lime mortar sample for different cycles of thermal loading

5 Conclusions

This paper presents the thermal deformation of lime and lime mortar for deterioration due to freezing and thawing. Freezing and thawing cycles have done repeatedly on different samples of lime and lime mortar. They were freeze and thawed till the volume and length parameters remained constant for higher number cycles. Optical microscopic study of the lime and lime mortar samples have studied further to investigate the pore sizes and their occurrence for the given laboratory samples. Using digital image processing simulation, a detailed analysis has been made to find the exact pore sizes and their distribution for lime and lime mortar sample for different cycles of thermal loading. From the experimental results and pore analysis, this method can be utilized for nondestructive evaluation and characterization of old age monuments for their restoration.

References

1. Costigan A (2013) An experimental study of the physical properties of lime mortar and their effect on lime-mortar masonary. Thesis. Trinity College (Dublin, Ireland). Department of Civil, Structural and Environmental Engineering, p 470
2. Passa DS, Sotiropoulou AB, Pandermarakis ZG, Mitsopoulos GD (2012) Thermal and drying cyclic loading for cement based mortars and expanded polystyrene foam layers. Appl Mech Mater 204–208:3648–3651
3. Bryan M, Callan EJ, Mather K, Dodge NB (1953) Laboratory investigation of certain limestone aggregates for concrete. U.S. Army Corps of Engineers, Waterways Experiment Station Technical Memorandum 6–371
4. Speyer RF (1994) Thermal analysis of materials, vol 7. Marcel Dekker, pp 165–175
5. Wunderlich B (1990) Thermal analysis, 1st edn. Academic Press, pp 279–293
6. Schneider U (1986) Properties of materials at high temperatures—concrete. RILEM, pp 101–106
7. Wells LS, Clarke WF, Levi EM (1948) Expansive characteristics of hydrated limes and the development of an autoclave test for soundness. J Res Natl Bur. Stan 41
8. Zhou XL, Xie YJ, Zheng KR, Zhang S, Cai FL (2013) Thermal deformation of cement-asphalt mortar under repetitive heating and cooling. In: Advanced materials research, vols 639–640, pp 304–308
9. Botas SMS, Rato VM, Faria P (2010) Testing the Freeze/Thaw cycles in lime mortar. In: 2nd historic conference HMC2010 and RILEM TC 203-RHM final workshop. Prague, Czech Republic, 22–24 Sept 2010
10. Lagrou D, Dreesen R, Broothaers L (2004) Comparative quantitative petrographical analysis of Cenozoic aquifer sands in Flanders (N Belgium): overall trends and quality assessment. Mater Char, pp 317–326

Assessment of Municipal Solid Waste Management in Jammu City: Problems, Prospects and Solutions

Adil Masood and Kafeel Ahmad

Abstract The study reports about the improper and poor municipal solid waste management system in the city of Jammu and highlights other issues like lack of public education, scientific techniques, machinery and legislation in the area that solicit immediate attention. The main aim of this research is to conduct a SWOT analysis (strength, weakness, opportunities and threats) of the municipal solid waste management system (MSWM) in Jammu city and propose an indicative solid waste management (SWM) plan by analysing both internal and external factors of SWOT. The city produces a whopping 350–400 MT (metric tons) of solid waste daily at a rate of 0.45 kg/cap/day, with contributions from domestic, institutional, commercial and street sweepings as 0.3, 0.03, 0.10 and 0.02 kg/cap/day, respectively. The operational efficiency of Jammu Municipal Corporation (JMC) is 50% and at present, only 18 wards have the facility of door-to-door collection of waste. Out of the total waste dumped by the JMC, the ragpickers collect only 20 MT of waste for recovery, reuse and recycling purposes. This study, therefore, not only attempts to improve the MSW problem in the city but also emphasizes the need of smart waste management systems that would further help in the development of a more habitable and environmentally friendly smart city.

Keywords Solid waste management · SWOT · Generation rate · Jammu city

1 Introduction

India, a home to a billion people is having the seventh largest economy in the world in terms of GDP and fourth largest in terms of purchasing power parity (PPP) [1]. At present, there are three megacities in the country and this number will creep to

A. Masood (✉) · K. Ahmad
Faculty of Engineering and Technology, Jamia Millia Islamia University,
New Delhi 110025, India

© Springer Nature Singapore Pte Ltd. 2020
S. Ahmed et al. (eds.), *Smart Cities—Opportunities and Challenges*,
Lecture Notes in Civil Engineering 58,
https://doi.org/10.1007/978-981-15-2545-2_23

six by 2021 [2]. With rapid industrialization and increasing per capita waste generation rate, the municipal solid waste has become a critical issue to be taken care of for a smart and livable city. A formal waste management system can be considered as a key element of smart city. The smart city concept, therefore, may induce the growth of smart services in the waste management sector for a city like Jammu. Many major Indian cities are grappling with the issue of waste management and presently no city has a foolproof segregation system in place with 100% efficiency [3]. Moreover, poor waste management has heavily contributed to the growing problem of air pollution and most of the cities, including Delhi, experience poor air quality conditions throughout the year [4]. The waste collection efficiency for major metro cities in India varies between 70 and 90% and for smaller cities and towns; this efficiency drops to a value of less than 50% [5]. The per capita waste generation rate has seen a small rise from 0.44 kg/day in 2001 to 0.5 kg/day in 2011 [5]. The municipal solid waste (MSW) generation rates for different states and union territories has been shown in Fig. 1. For the city of Jammu, the per capita waste generation is 0.45 kg/capita/day and the total solid waste generated is around 350–400 MT/d (metric tons per day) [7]. Presently, due to lack of a formidable waste management plan, the waste is heaped up across the street corners and other low-lying areas or strewn around nooks and vacant spaces [8]. JMC carries out the duty of waste collection, transport and disposal in the area, but the lack of adequate technical, managerial, administrative and financial resources of managing urban solid waste restricts the efficient functioning of the municipal body. JMC manages a mere 50% of the total MSW generated in the city, forcing them to dispose of the rest of the waste unscientifically without any processing [7]. Our study addresses the current issue of waste management in the Jammu municipal region and besides this highlights the importance of the door-to-door collection, segregation and processing of municipal solid waste. SWOT analysis has been adopted in our study to ascertain the existing situation, shortcomings and opportunities related to the municipal solid waste management (MSWM) in the region. Hosts of other researchers have also utilized SWOT analysis as a formidable tool to study and examine the different domains of waste management. For example, Mbuligwe [9] had carried out SWOT analysis of the already established SWM system of the Dar Es Salaam city. His analysis called for further improvement in waste collection and disposal practices in the city. Raharjo et al. [10] conducted a SWOT analysis for developing strategies such as formulating a local regulation, in order to enhance the potency of solid waste bank for overall development of the local waste management in Padang city, Indonesia. On similar note, Verma et al. [11] pointed out via SWOT analysis that the role of the private sector is critical in terms of establishing an effective MSW management in Ho chi Minh city. They also proposed certain recommendations for the ongoing waste management process adopted in the city based on their SWOT analysis. Certain studies, Aich and Ghosh [12] and Suman et al. [13] reported the use of SWOT analysis in the selection of optimal technology and for processing and disposal of MSW. These kinds of studies are missing in the

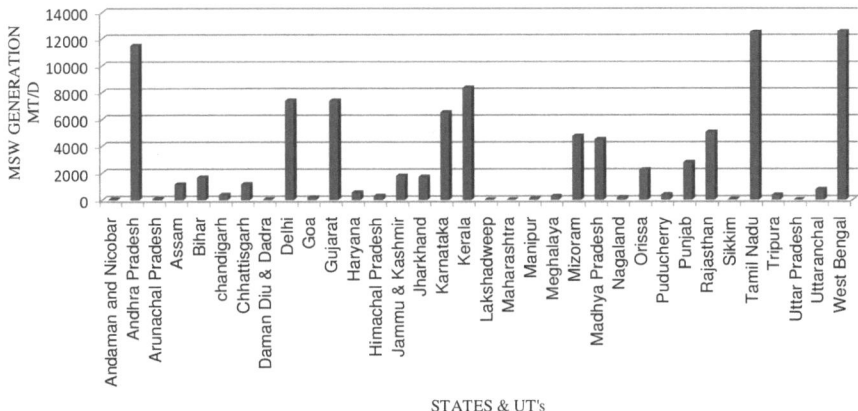

Fig. 1 Municipal waste generation for states and union territories from 2009 to 2012 in MT/D [6]

Jammu context and, therefore, this study will be a valuable addition to the MSWM literature. The paper has been divided into five sections. Section 1 presents an overview of the present status of solid waste management in India during the last decade and in addition to this discusses scholarly literature on SWOT analysis. Section 2 provides a comprehensive description of the state of waste management in Jammu city and provides an insight into the existing practices of waste collection, storage, transport and disposal. The methodology for SWOT analysis of MSWM in Jammu is explained in Sect. 3. Section 4 presents a strategy based on decisive measures that can be used to address the MSWM issues in the region. Finally, in Sect. 5, the main conclusions are drawn.

In contrast to the studies mentioned above, our study has presented a case of SWOT analysis, which pivots around the following objectives:

(i) Assessment of the status quo with respect to collection, storage, transportation and disposal activities for the MSWM in city of Jammu.
(ii) Propositions for critical strategies and guidelines for modelling an indicative waste management framework for the city.

Moreover, our study presents a brief review of the pragmatic solutions, which require implementation for developing a methodical solid waste management system in the region. The study represents one of the first attempts of using SWOT analysis as a tool to delineate the urban waste sector of Jammu and provides the basis for further development of effective ideas on sustainable management of MSW in Jammu city. In addition to this, our study also aims at promoting awareness towards scientific practices involving reduction, reuse and recycle of MSW empowering encouragement to the competent authorities, researchers to work towards the improvement of waste management system in the city.

2   Materials and Methods

2.1   Geographic and Demographic Profile of the Area

Jammu, popularly known as the 'City of Temples', lies on the uneven ridges of Shivalik hills between the geographical coordinates of 74° 24′–75° 18′ E longitude and 32° 50′–33° 30′ N latitude [14]. It is approximately 600 km (370 miles) from the national capital, New Delhi. With an average elevation of 327 meters, the city is bounded by Udhampur district in the north and north-east, Kathua district in the east and south-east, Pakistan (Sialkot) in the west and POK (Bhimber) and Rajouri district in the north-west [15]. The city has a subtropical and humid climate with temperatures reaching high forty's (degree Celsius) in summers and dropping to sub-zero levels in winters [40]. The city spreads around the Tawi riverbanks, with new neighbourhoods located on the left bank and the old city located on the right. The population of Jammu has been recorded as 612,163 based on a census in 2011 with a representation of 52.7% as males and 47.3% of females. In addition to this, the city has been divided into 71 sanitary wards (Fig. 2) at the municipal level for the execution of various administrative activities.

2.2   Municipal Solid Waste Management in Jammu City

2.2.1   Current status of MSWM

JMC has been performing the task of upkeep and maintenance of civic amenities including solid waste management in the city. Covering an area of 112 km^2, JMC

Fig. 2 Jammu city wardwise map (NOT to scale)

comprises of 71 wards, 3 zones and 2 divisions for the execution of various activities at the field level. The organizational structure of JMC is delineated into a well-defined structure as per the municipal corporation act 2000 as shown in Fig. 3. The structure is divided into 3 levels, viz. the corporation council, the standing committee and the municipal commissioner. The mayor, who is elected to the post for a term of one year, has all the executive powers vested in him and is solely responsible for the implementation of the resolutions passed by the corporation council. The JMC produces 350–400 MT/d (metric tons per day) of solid waste. Just to make the arithmetic simpler and the perspective clearer, each person who has the right to vote in the city is adding 600 grams to the urban solid waste every day. Most residents in the city resort to burning and illegal off-site disposal of solid waste, resulting in unhygienic conditions and breeding grounds for disease vectors. The solid waste is usually dumped outside in open plots and around street corners. A mere 29% of the households use private bins and utilize the services of local garbage collector for disposal of garbage, 22% of the residents dispose of their waste in community bins and the remaining percentage of residents simply burn their waste in open plots [16]. The role of JMC in terms of managing biomedical waste is ambiguous and the associated waste generating institutions are unequipped to handle and dispose of such waste [17]. With a total budget of 1.7 Million US$ for SWM and a staff of 1202, JMC proves to be inefficient in terms of technical, operational, functional and financial expertise [18]. Most of the municipal bodies in India shell out approximately US$ 30/tones for MSWM. About 60–70% of this sum is spent on collection, 20–30% on transportation and fewer than 5% on final disposal [19].

2.2.2 MSW Sources, Composition and Characteristics

Throughout the region, the principal source of solid waste generation has been the local residents, anaz mandi, vegetable market, sweet shops, restaurants, hospital,

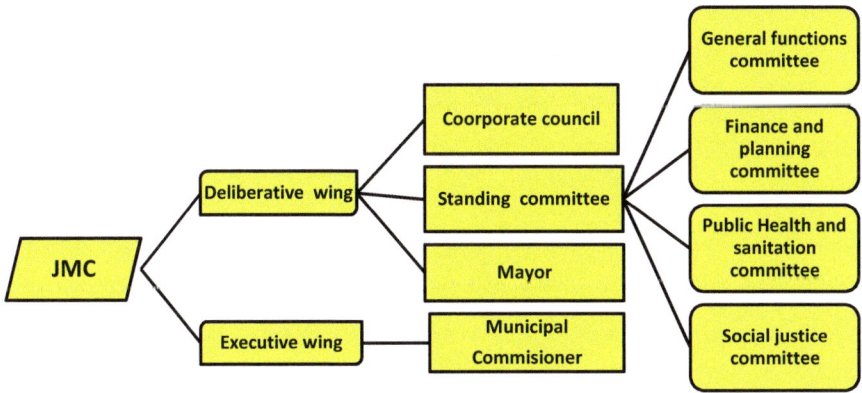

Fig. 3 JMC organogram

dispensaries, domestic and stray animals, shops and commercial establishments. Jammu city generates a whopping 350–400 metric tons (MT) of solid waste daily with a per capita generation rate of around 0.45 kg/capita/day [20]. The composition of the waste largely depends on a wide range of factors like food structure, culture, lifestyle, climate, economic development and local landscaping [21]. The bulk density for MSW in Jammu city is found to be 500 kg/m^3, which is almost 42% more than the national average value of 350 kg/m^3 [22]. The solid waste generated in the city is mainly composed of the MSW with some proportion of hospital, industrial, construction and demolition waste [7]. In this study, SWM data was collected from 25 randomly selected houses through data collection surveys and direct measurements (weight and volume) during the study period of 30 days. The data shows that the MSW has the potential for reuse, recycle and recovery. Moreover, composition analysis of the total MSW generated in the region indicates that the biodegradable fraction present in the total MSW is 60% and the non-biodegradable fraction is 40% (recyclable) (Fig. 4). A breakdown of the physical composition of MSW for Vidhata Nagar, Jammu city, has also been shown in a similar study [23] (Table 1). It is observed from the composition analysis (Fig. 4), that organic waste in the form of fruit, food and vegetable has the most considerable share in the MSW composition (40%) and has the maximum potential for composting, followed by inert waste, plastic, paper and textile. The findings are in accordance with those of Xiao et al. [25], Visvanathan et al. [26], Bandara et al. [27], Yousuf and Rahman [28], who reported the solid waste composition in most of the Asian cities to be highly biodegradable and largely composed of organic fraction constituting mainly paper, leather, wood, rubber, plastic and textile. The quantity of waste recovered in the form of paper and plastic in Indian cities is lower in comparison to the cities of the northern hemisphere, USA (65%) and Western Europe (48%) [29]. The chemical characteristic of MSW in Jammu city shows an increase in moisture content. The reason behind this may be due to the presence of a high proportion of unprocessed vegetable waste. Developing cities generally tend to have high organic content and a low energy value. The high moisture content and low calorific values make the waste a prime candidate for biochemical conversion technologies like waste to energy (WTE) and composting

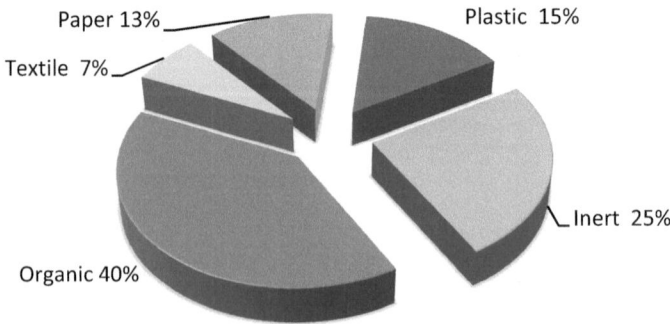

Fig. 4 MSW composition in Jammu city [24]

Table 1 Overall composition of solid waste for Jammu city (Vidhata Nagar) [23]

Waste category	Gross domestic waste generated (Kg) (25 houses)	Waste generated (kg/c/d)	Waste generated (kg/house/d)	Gross monthly waste generation (kg) (25 houses)	Gross waste generation (Kg/house/month)
Biodegradable					
Vegetable waste	11.00	0.096	0.440	330.00	13.20
Food waste	15.75	0.076	0.630	472.50	18.90
Fruit waste	11.50	0.100	0.460	345.00	14.00
Textile waste	0.324	0.002	0.013	9.74	0.38
Paper waste	7.58	0.064	0.300	227.9	9.10
Total	46.154	0.338	1.843	1385.14	55.58
Non-biodegradable					
Plastic waste	7.10	0.06	0.284	213.00	8.52
Metal waste	1.400	0.012	0.056	42.00	1.70
Glass waste	8.50	0.006	0.34	255.00	10.2
Total	17.00	0.078	0.68	510.00	20.42
Inert waste					
Stone, dust, Hair, wax	0.324	0.002	0.0128	9.72	0.288

Table 2 Physical and chemical characteristics of MSW in Jammu city [7, 18]

Parameters	Value
Compostables (%)	40–45
Recyclables (%)	21.08
C/N ratio	26.79
HCV (Kcal/kg)	1782
Average bulk density (kg/m^3)	500
Moisture (%)	40

[6]. For the city of Jammu, the organic content of MSW is around 40–45% and the calorific value is 1782 kcal/kg. A lower C/N ratio of 26.79 has also been reported which indicates a rapid process of waste decomposition (Table 2) [18]. Analogous trends have been observed in other studies carried out in this part of the world. Sharma

[8] conducted a study for top Paloura, Jammu, based on the problems, prospects and challenges of managing MSW in the region. She found out that the organic content of MSW in the region was 69.02%. A study conducted by Kumar and Singh [15] for Kathua district, and Jammu showed correlating trends in terms of biodegradable and non-biodegradable composition. The biodegradable component of MSW in their study was 70.62% and the non-biodegradable component was 25%, respectively.

2.2.3 Collection, Storage and Transportation Practices

- *Primary collection and storage of MSW*: Generally considered as the first stage in the waste management process where the role of external factors is put into use. This task of gathering waste from one house to another is performed by waste collectors, especially in developing areas, and is referred to as the primary collection. Shortage of workers, vehicles, and garbage bins has led to the failure of cleanliness drive in the region. The segregation of municipal solid waste in organic, inorganic and recyclable material is not possible at the source, i.e. the household level. Ragpickers do spread out a part of the waste for sorting and collection purposes [7]. At present, only 18 wards have the facility of door-to-door collection of waste, which covers around 25% of the registered houses in the JMC area. Due to inadequate primary collection and lack of awareness amongst the local inhabitants, a major share of the waste is dumped in nearby vacant plots or streets. Workers sweep the roads, lanes and bylanes, collect the waste, load it into the handcarts and transport it to the nearest temporary collection centres. In the city, there are around 81 temporary collection centres. Use of regular handcarts is still prevalent in the city and in some areas. Containerized handcarts with four bins each of 30 Liters capacity have been recently taken into the service [7]. Due to lack of stringent rules and regulations, these workers fill up their handcarts for only 3–4 times a day resulting in heaps of waste left unattended and littered. Road sweepings and waste removal cover around 60–75% of the total municipal area and the remaining 25% is covered by door-to-door collection. At present, JMC has a strength of 2000 workers and 57 sanitary inspectors, engaged in effective execution of waste management works in the corporation. The current waste management system administered in the city is below par as per the MSW rules 2000 [7]. There are 453 temporary collection points provided in the urban limits of the city. These temporary waste collection points are of four different types, viz. the open depots with capacities ranging from 20 to 30 m^3 (masonry structures), dumper placer bins, refuse collector (RC) bins and skips along roadsides. The municipal administration in the coming years is planning to convert these collection points into transfer stations provided with compaction, segregation and processing technology in order to formalize the solid waste management system in the city [30].

- *Secondary collection and storage*: Solid waste handling in the city is rather poor due to lack of experience and expertise, resulting in environmental pollution and health hazards. The residents often dump their waste in nearby low-lying areas, wide roadsides, street corners or a spot adjacent to public offices treating them as a receptacle of waste. Systematic and scientific methods are still not followed in the region, resulting in a waste collection efficiency of a mere 50% [7]. In some of the wards, the waste collected from the roadsides and low-lying areas is brought to temporary collection centres and stored in collection bins. Waste is then lifted manually or mechanically with the help of front-end loaders onto the tipper trucks. This step is essential in terms of the solid waste management process in the area and is referred to as the secondary collection. JMC is having both hauled and stationary container systems at its service. Various motorized and manual vehicles used for waste collection and transport by JMC are hoist trucks (2–10 m^3), tipper trucks, manually loaded dumpers and dumpers filled with the help of pay loaders. Further details of the waste management machinery in use by the JMC have been shown in Table 3.

2.2.4 Disposal of MSW

People dwelling in and around the Jammu city region are unacquainted with scientific waste management techniques; therefore, they are dumping their waste on roads, open plots, nooks and street corners. With time, these dumping sites transform into breeding grounds for rodents and other animals. Lack of roadside vats, dumpers and waste collection bins forces people to dump their waste at a common site whose selection is decided haphazardly. Waste segregation practice at the

Table 3 Details of machinery and equipments owned by JMC [7]; On-site survey

S. no.	Items	Specifications	Available	Required	Deficit
1.	Dumper placer	–	7	16	9
2.	Front-end loader	–	2	7	5
3.	JCB (Backhoe loaders)	–	3	7	4
4.	Road sweeping machines	–	2	2	0
5.	Battery operated tricycle	–	–	221	221
6.	Tata ace tipper	1.5 m^3	–	43	43
7.	Auto three wheeler tipper	1 m^3	–	35	35
8.	RC bins (blue)	1 m^3	60	–	–
9.	RC bins (blue)	6.5 m^3	0	3	3
10.	Dumper placer bins (yellow/ green)	4.5 m^3	340	178	162
11.	Open storage sheds (collection points)	20–25 m^3	53	–	–

disposal stage has progressively deteriorated to an extent that the solid waste is now directly thrown into the water drains. This treacherous practice not only clogs the drain, but also triggers the problem of waste handling to a level where environmental implications may be hazardous to human health [31]. A cattle menace also prevails in the region as the animals, mostly feed on the dumped domestic waste. These conditions have evolved due to lack of community bins, workforce, transport and disposal sites [7, 23]. Failure to identify a permanent landfill site to dump the MSW has consequently resulted in an unpremeditated and unscientific dumping of waste not only in the deep trenches created in the forest region around the city but also around the Tawi riverbed. This situation has led to surface and groundwater pollution in the region [16, 32]. At present, JMC is having two temporary dumping sites situated at Bhagwati Nagar and Bandurakh. About 350–400 MT of waste is daily disposed at the dumping site through unsystematic and haphazardous dumping and about 20 MT is collected by the ragpickers and others for material recovery, reuse and recycling purposes. All the operations of SWM in JMC are executed under four heads—sweeping, collection, transfer and disposal as shown in Fig. 5. The Bhagwati Nagar dumping site spreads across 60 kanal of land (30,351 m^2) and is subjected to uncontrolled, continuous and haphazardous dumping of waste, which has led to the choking of Tawi river bed. The deteriorating situation around the river bed has resulted in a direct intervention from the National Green Tribunal and the Jammu high court who have directed JMC to develop integrated waste management schemes for the city and clear out the catchment area of the river.

2.3 Methodology

A series of field surveys were carried out to assess the status of solid waste management in the city during the months of January 2015 to February 2016. Separate questionnaires were designed for the staff of JMC and for the households and waste pickers. A total of 50 staff members of the corporation, 40 ragpickers and 10 waste dealers were invited for the structured interviews. The extracted information from these interviews was further utilized as a data source for our study. The collected data also helped to identify the strengths, weaknesses, opportunities and threats of the local waste management process. An acronym to strength, weakness, opportunities and threats (SWOT) is a tool for preliminary stages of decision-making that addresses the internal attributes and external factors affecting an organization, process, project or a company [38, 41]. In the present study, SWOT analysis has been conducted in order to evaluate the MSWM scenario based on semi-structured interviews with stakeholders in JMC. Correlating SWOT studies [9, 11, 33, 34] based on similar domains have been adopted as a reference in performing the SWOT analysis of our research. Some of the research questions derived from the SWOT analysis are presented in Table 4.

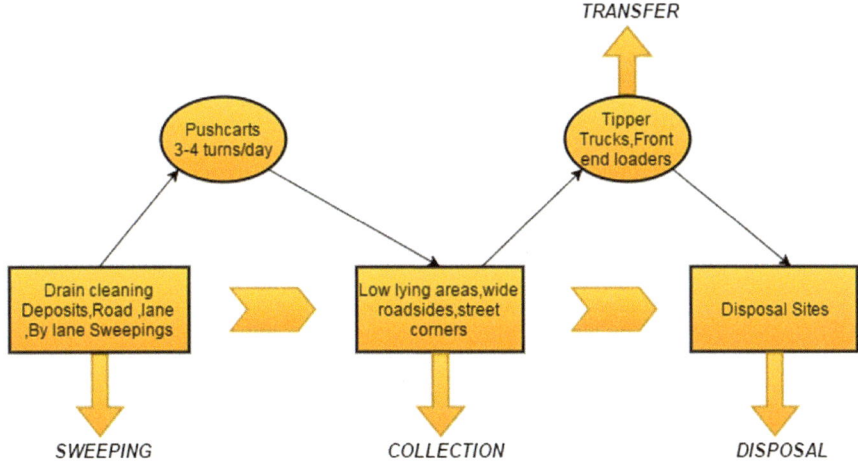

Fig. 5 Flow chart for current SWM scheme implemented by JMC

Table 4 SWOT analysis details

Factors	Questions
Strengths	What are the factors that facilitate the promulgation of waste management and minimization by the JMC?
	What are the driving agents that are enabling Jammu to be a strong competitor in terms of SWM and establish a formal waste management framework?
Weaknesses	What factors hinder the progress of waste management in the city?
	What is not done efficiently?
	What elements could be improved in promoting MSWM?
	Are there any major frailties in the system?
Opportunities	What likely benefits may occur?
	What reforms are observed in socio-economic patterns, lifestyle and economic standards of the project beneficiaries?
	What changes are observed in the usual practices and available techniques of waste management?
Threats	What external obstacles do the MSWM face?
	Is the current state of MSWM posing a threat to the environment?
	Do stakeholders/private owners show interest in supporting waste management program?

3 Result and Discussion

3.1 SWOT Analysis for MSWM System in Jammu City

Strengths

S1: State Government's realization of the dilapidated waste management system in the city, succeeded by a series of counteractive measures taken to mitigate this situation. Some of these pronounced measures are:

- Frequent visits by government officers to severely affected areas.
- Sanctioning of funds for waste management-related activities.
- Imposing fines at uncontrolled illegal dumping.
- Installation of environmentally sound machinery like Bioneer machines for waste processing.

S2: Periodic initiatives taken by JMC like anti-polythene drive and thematic cleanliness drive under the name of 'Where is my dustbin' involving shopkeepers, tours & travel and resident welfare associations as stakeholders.

S3: Identification of landfill site.

S4: Willingness of the state government to involve private partners for dispensing advisory services to the JMC.

S5: Approval of integrated SWM facility as per SWM rules 2000.

S6: Encouragement to night scavenging.

S7: Workshops organized to spread awareness on integrated SWM by JMC have caught the attention of pollution control boards, safai karamcharis union, sabzi mandi and industries association.

Weaknesses

W1: No waste segregation at source due to lack of environmental awareness amongst the local people.

W2: Technology spread at a sluggish pace and low level of private sector investment.

W3: Lack of adequate experience and expertise in JMC.

W4: Waste collection efficiency is a mere 50%, due to a shortage of human resource and machinery.

W5: Non-availability/Identification of land for SWM facility.

W6: No practical solution for waste collection, recycling and energy recovery.

W7: Central waste treatment and other facilities like landfills are absent.

W8: Lack of biomedical waste collection leading to open dumping in and around the Tawi river bed.

W9: No investment for enhancing the waste valourization process.

Opportunities

O1: Subsidiary benefits associated with improved waste management system like employment opportunities and income generation.

O2: Non-recyclable waste can be utilized as refused-derived fuel (RDF) in cement factories and power plants.

O3: An efficient SWM system would create a formal synergistic network between rag pickers, itinerant waste buyers (IWB), small enterprises middleman (SEM) and WWD (whole sale waste Dealers).

O4: Waste processing techniques like composting and vermicomposting have proved to be propitious to agricultural activities.

O5: Job creation in services sector, which is second to agriculture in terms of employment opportunities in this part of the country.

O6: High organic fraction in MSW would develop potential for waste to energy or compost schemes.

O7: JMC has been organizing the 'clean India' campaign drive under the banner of Swachh Bharat Abhiyan, which has provided opportunities to unemployed youth, social & non-governmental organizations (NGO) workers to train and participate in the awareness programs.

Threats

T1: Brisk economic growth and elevated living standards have proliferated waste generation in the city.

T2: Unscientific waste management and disposal are the principal cause of ground water contamination because of leachate production.

T3: Natural water resources, like river Tawi in the region, are under great stress due to pollution, encroachment and siltation.

T4: In spite of the potential benefits associated with the waste management scenario, the private sector is still ambivalent on the point of investing.

T5: lack of public–private government alliance.

T6: Lack of scientific set-up to manage, collect and transport the hazardous waste, leading to open dumping.

3.2 Proposed Strategy for SWM

A strategy for waste management is formulated with due consideration of the observations made and the results drawn out during the SWOT analysis (Table 5).

Table 5 Proposed waste management strategy [18, 39]

S. no.	Waste management process	Further clarification	Compliance criteria/proposed action
1.	Waste generation at source	• Domestic • Road sweepings • Commercial • Institutional	

2.	Collection of waste		• House to House collection of MSW(via community bin method)
			• Collection of waste from slums, illegal colonies, commercial areas (fish market, slaughter house)
			• Source specific quantification of waste by:
			  – Load count analysis
			  – Weight volume analysis
			  – Material balance analysis

3.	Transportation		• Proper maintained vehicles with provision of waste cover • Route optimization for minimum exposure to public • Separate class of vehicles for slaughterhouses, meat processing units and other animal products-based units

4.	Recovery/ processing of MSW		• Use of methods/technologies like Bio-processing to make use of the waste and minimize the burden on a landfill
			• Testing facility provision for compost quality assessment
			• Use of composting and vermicomposting for waste stabilization
			• Recovery of resources and energy via
			  – Incineration
			    ➡ Energy recovery
			  – Pelletization

5.	Disposal		• Disposal site improvement/Up gradation as per MSW rules 2000 • Check or complete prohibition of land filling of non-biodegradable, toxic and other waste, which may be hazardous in nature • Mixed waste land filling to be avoided if waste is not suitable for processing Landfill to be constructed as per MSW rules 2000

3.3 Current Issues to Solid Waste Management in Jammu City

3.3.1 Increase in Solid Waste Generation Due to Population

The growing population in Jammu is a grave matter of concern in terms of present MSW management scenario in the city. A constant rise in population calls for better management of MSW because of poor sanitation and scarcity of land [35]. The population growth of Jammu city is on the rise and looking at the current growth rate, the city is rapidly moving towards becoming a mega city by 2047. As an outcome of this steady rise in population, a proportional effect on the available resources as well as on the waste generation levels has been observed in the region [36]. The consumption capacity of the city residents is on the surge due to the socio-economic shift developed over the last few years. Validating the claims of constant population growth in the region, decadal growth rate of the Jammu city since 1961 is shown in Table 6.

The rapid increase in population, uncontrolled urbanization and rise in living standards has led to the generation of a colossal volume of waste in the city, a major share of which is left unattended. The solid waste generation rate and its projection for the coming years have been shown in Fig. 6.

3.3.2 Lack of Transfer Station and Scientific Disposal Practices

Transfer stations act as an intermediate station lying between the collection and final disposal sites, where the municipal solid waste is dropped off by smaller capacity vehicles and then loaded onto larger capacity vehicles from where the waste is hauled to a processing or a final disposal site [37]. Jammu city faces an acute problem of unscientific and indiscriminate disposal of garbage. Issues like lack of awareness and public participation towards scientific waste disposal have existed in the region. The current scenario calls for a decentralized low-cost waste management system where a provision for recycling, material recovery, transfer, collection and disposal facilities has been made. It is proposed that at least two transfer stations, each with a capacity of 800 ton per-day, must be provided for all

Table 6 Decadal growth rate of Jammu city	Year	Population	Population increase	Percentage increase
	1961	102,738	–	–
	1971	157,708	54,970	53.50
	1981	214,737	57,029	36.16
	1991	369,960	155,223	72.28
	2001	549,791	179,831	48.60
	2011	576,195	26,404	4.80

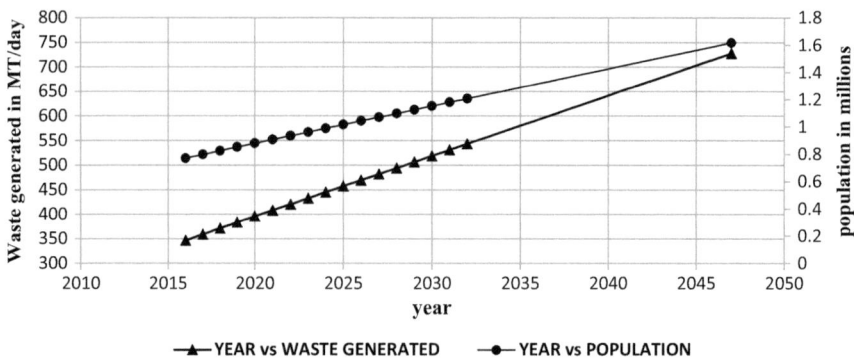

Fig. 6 Temporal waste generation trends with respect to population

71 wards which may increase the efficiency of the vehicles and also assist in developing a methodical solid waste management scheme for this area [7].

3.3.3 Lack of Effective Number of Vehicles and Containers for Efficient SWM

Due to lack of financial investments in the waste management sector, JMC often faces a shortage of vehicles, storage containers and other machinery. At present, the machinery and vehicles in use for sweeping, collection, transportation and disposal are inadequate in number and also in terms of serving the purpose of SWM effaciously [7].

4 Conclusion and Recommendations

This research is an attempt to interpret the existing solid waste management scenario in the city of Jammu. With a total budget of 1.7 Million US$ for SWM and a staff of 1202, JMC proves to be inefficient in terms of technical, operational, functional and financial expertise [18]. The existing unscientific disposal of waste near the ecologically sensitive zone of the river Tawi must be immediately abandoned, as it is leading to environmental complications in the region. In order to revive the MSWM scenario in the region, the state government has invited advisory services for MSW disposal through waste to energy (WTE) means in Public–private partnership (PPP) mode. The core set of reasons due to which the winter capital is still striving for a scientific and an efficient waste management system are delays in completion of government funded waste management projects, failure to identify a landfill site, non-compliance with MSW rules 2000, poor public participation in waste management activities and lack of resources and workforce. SWOT analysis executed in our study focuses on some major opportunities associated and threats that need to be mitigated in the current system. Based on the trends of the waste collection, transportation, processing, disposal action and other observations made via SWOT analysis, an indicative waste management plan has also been proposed which can be further employed as a decision support tool. The data and the outcomes presented in this study are important in terms of smart management of MSW and will, therefore, allow planners, researchers and JMC officers to work towards transforming the city of Jammu into a successful smart city.

The following are some of the pragmatic solutions that have been proposed for each level of the waste management process.

- Implementing separate door-to-door MSW collection scheme. At present, door-to-door collection covers a mere 25% of the total municipal area. The situation calls for encouragement through extensive awareness and public participation programs. Use of handcarts, battery operated tricycles and Tata Ace (1.5 m^3) should assist in execution of this door-to-door collection scheme.
- Executing waste segregation process at the household level. A set of twin bins of 15 Liters capacity may be allocated to each household for advocating the collection of dry and wet waste in a phased manner.
- Inflating the number of community bins (4.5 m^3) and refuse collector bins (1 m^3) at areas and narrow bylanes which have become open dumping spots for the residents.
- Increasing the frequency of waste collection and discouraging manual handling of waste by persistent use of front-end loaders and refuse collectors at waste collection spots.
- Introducing stringent measures like spot fining at commercial areas to restrain littering.
- Enforcing strict regulations to prevent illegal dumping of MSW in deep trenches around the banks of river Tawi in order to control air and water pollution in this eco-sensitive region.
- Exploring alternatives for a new landfill site, where suitable arrangements like bottom liner, lightning, fencing, weighbridge, leachate collection and gas venting are kept in provision.

References

1. Myers N, Kent J (2003) New consumers: the influence of affluence on the environment. Proc Natl Acad Sci 100(8):4963–4968
2. Taubenböck H, Wegmann M, Roth A, Mehl H, Dech S (2009) Urbanization in India-spatiotemporal analysis using remote sensing data. Comput Environ Urban Syst 33 (3):179–188
3. Joshi R, Ahmed S (2016) Status and challenges of municipal solid waste management in India: a review. Cogent Env Sci 2(1):1139434

4. Akhtar A, Masood S, Gupta C, Masood A (2018) Prediction and analysis of pollution levels in Delhi using multilayer perceptron. In: Data engineering and intelligent computing, pp 563–572
5. Annepu RK (2012) Sustainable solid waste management in India. M.Sc. thesis, New York, pp 1–189
6. TERI (2015) Municipal solid waste generation in India (State—wise) ENVIS Centre on Renewable Energy and Environment Municipal Solid Waste Generation in India. http://terienvis.nic.in/index3.aspx.aspx?sslid=4110&subsublinkid=1348&langid=1&mid=1
7. JMC (2016) Detailed draft report, municipal solid waste management scheme for JMC. Jammu. http://environmentclearance.nic.in/writereaddata/Online/TOR/0_0_23_Feb_2016_1752557731PFR.pdf
8. Sharma A (2015) Generation, composition and management of solid waste in top Paloura, Jammu (J&K). Int J Env Sci 6(2):213
9. Mbuligwe SE (2012) Solid waste management strategies and practices that work: a developing country perspective. Int J Env Waste Manage 10(2–3):201–221
10. Raharjo S, Matsumoto T, Ihsan T, Rachman I, Gustin L (2017) Community-based solid waste bank program for municipal solid waste management improvement in Indonesia: a case study of Padang city. J Mater Cycles Waste Manage 19(1):201–212
11. Verma RL, Borongan G, Memon M (2016) Municipal solid waste management in Ho Chi Minh City, Viet Nam, current practices and future recommendation. Proc Env Sci 35:127–139
12. Aich A, Ghosh SK (2016) Application of SWOT analysis for the selection of technology for processing and disposal of MSW. Proc Env Sci 1(35):209–228
13. Suman S, Khan MK, Pathak M (2015) Performance enhancement of solar collectors—a review. Renew Sustain Energy Rev 49:192–210
14. Khanna P (2015) Physico-chemical parameters of groundwater of Bishnah, district Jammu, India. Proc Natl Acad Sci India Sect B Biol Sci 85(1):121–130
15. Kumar A, Singh S (2013) Domestic solid waste generation—a case study of semi-Urban area of Kathua District, Jammu, J&K, India. Int J Sci Res Publ 3(5):1–5
16. WAPCOS (2006) City development plan for Jammu city, Sectorwise details (Chapter 3)
17. Biomedical (2016) GOI, Ministry of Environment Forest and climate change notification. Gazette of India Part II, http://mpcb.gov.in/biomedical/pdf/BMW_Rules2016.pdf
18. CPCB (2006) Central pollution control board, Assessment of status of municipal solid waste management in metro cities and state capitals, pp 1–30. http://cpcb.nic.in/status-of-implementation-of-solid-waste-rules/
19. Parihar RS, Ahmed S, Baredar P, Sharma A (2017) Characterisation and management of municipal solid waste in Bhopal, Madhya Pradesh, India. Proc Inst Civ Eng Waste Resour Manage 170:95–106
20. Daily Excelsior (2014) Union Govt sanctions solid waste management projects for capital cities. http://www.dailyexcelsior.com/union-govt-sanctions-solid-waste-management-projects-for-capital-cities
21. Wang X, Geng Y (2012) Municipal solid waste management in Dalian: practices and challenges. Front Env Sci Eng 6(4):540–548
22. Khalil N, Khan M (2009) A case of a municipal solid waste management system for a medium-sized Indian city, Aligarh. Manage Env Qual 20(2):121–141
23. Rakesh J (2015) Status of household solid waste generation in Vidhata Nagar, Bathindi. Jammu. Indian Streams Res J 5(7):1–9
24. SWM Rules (2016) Jammu and Kashmir state integrated solid waste management strategy 11 (1). http://jkhudd.gov.in/pdfs/policy%20swm.pdf
25. Xiao Y, Bai X, Ouyang Z, Zheng H, Xing F (2007) The composition, trend and impact of urban solid waste in Beijing. Environ Monit Assess 135(1–3):21–30
26. Visvanathan C, Tränkler J, Kuruparan P, Basnayake BF, Chiemchaisri C, Kurian J, Gonming Z (2005) Asian regional research programme on sustainable solid waste landfill management in Asia. In: Proceeding tenth international waste management and landfill symposium. Sardinia

27. Bandara NJ, Hettiaratchi JP, Wirasinghe SC, Pilapiiya S (2007) Relation of waste generation and composition to socio-economic factors: a case study. Environ Monit Assess 135(1–3):31–39

28. Yousuf TB, Rahman M (2007) Monitoring quantity and characteristics of municipal solid waste in Dhaka City. Environ Monit Assess 135(1–3):3–11

29. Chattopadhyay S, Dutta A, Ray S (2009) Municipal solid waste management in Kolkata, India—a review. Waste Manag 29(4):1449–1458

30. Hazra T, Goel S (2009) Solid waste management in Kolkata, India: practices and challenges. Waste Manag 29(1):470–478

31. Sharma R (2014) Composition and management of solid waste in ward no. 39, Shiv Nagar Extn Jammu Env Sci 4(9):237–239

32. Pattnaik S, Reddy MV (2010) Assessment of municipal solid waste management in Puducherry (Pondicherry), India. Resour Conserv Recycl 54(8):512–520

33. Srivastava PK, Kulshreshtha K, Mohanty CS, Pushpangadan P, Singh A (2005) Stakeholder-based SWOT analysis for successful municipal solid waste management in Lucknow, India. Waste Manage 25(5):531–537

34. Yuan H (2013) A SWOT analysis of successful construction waste management. J Clean Prod 39:1–8

35. Matsunaga K, Themelis NJ (2002) Effects of affluence and population density on waste generation and disposal of municipal solid wastes. Earth Eng Cent Rep, 1–28

36. Saxena A, Oraon R (2016) Municipal solid waste management in Bareilly. Int J Tech Res Appl 4(3):51–56

37. Ramachandra TV (2011) Integrated management of municipal solid waste. Env Secur Hum Anim Health 30:466–484

38. Jain S, Pant P (2010) Environmental management systems for educational institutions: a case study of TERI University, New Delhi. Int J Sustain High Educ 11(3):236–249

39. Kumar S, Bhattacharyya JK, Vaidya AN, Chakrabarti T, Devotta S, Akolkar AB (2009) Assessment of the status of municipal solid waste management in metro cities, state capitals, class I cities, and class II towns in India: an insight. Waste Manag 29(2):883–895

40. Anita B (2012) History of Jammu region through archaeological evidences early and early medieval period. In: Early and early medieval period, pp 1–367

41. Mor S, Kaur K, Khaiwal R (2016) SWOT analysis of waste management practices in Chandigarh, India and prospects for sustainable cities. J Environ Biol 37(3):327

A Review of Lake City Tehri as Smart City Tehri

Tripti Dimri, Shamshad Ahmad and Mohammad Sharif

Abstract A smart city aims at sustainable economic development of the existing city for providing better quality of life with judicious use of natural resources by participation from the common people. The quality of life can be improved by applying smart technological solutions to the day–to-day life activities and ultimately enhance the quality of living in the city. The Smart Cities Mission in India was launched on 25 June 2015, for five-year duration in its first phase. The mission targets to transform 100 cities into smart cities by 2020. The paper mainly focuses on studying and analysing the process involved in the Smart Cities Mission of India for New Tehri city in the state of Uttarakhand, India.

Keywords Smart Cities Mission · New Tehri · India · Urban planning

1 Introduction

A smart city refers to a city that is planned in an impeccable manner and provides basic infrastructure to its inhabitants in a reliable manner. The inhabitants of the smart city are expected to utilize the available facilities and infrastructure in a cost-efficient manner. During the last decade, the term smart city has become quite popular with several governments adopting the idea of smart city in a comprehensive manner. Basically, a smart city aims at judicious use of available resources by its inhabitants. Smart cities, if planned properly, shall lead to alleviation of poverty and eliminate adverse impacts of urbanization.

T. Dimri (✉) · S. Ahmad · M. Sharif
Department of Civil Engineering, Jamia Millia Islamia, New Delhi, India
e-mail: sahmad8@jmi.ac.in

M. Sharif
e-mail: msharif@jmi.ac.in

© Springer Nature Singapore Pte Ltd. 2020
S. Ahmed et al. (eds.), *Smart Cities—Opportunities and Challenges*,
Lecture Notes in Civil Engineering 58,
https://doi.org/10.1007/978-981-15-2545-2_24

2   Indian Smart Cities Mission

The Smart Cities mission is a flagship programme of the Government of India that is aimed at providing a high quality of life to its citizens. With this aim in mind, the Indian government initiated the execution of the smart city concept in over 100 cities in India. At present, considerable progress has been made in providing a reasonably good-quality infrastructure in these selected cities. The long-term focus of the smart city concept is on sustainable development of the society through resource management. Particular emphasis has been laid on the management of energy and water infrastructure in these cities. After the completion of the first phase of infrastructure creation in these cities, the Indian government plans to widen the scheme to a greater number of cities. A smart city aims at providing housing opportunities to all, creating and expanding existing road networks not only for vehicles and public transport but also for pedestrians and cyclists. It is anticipated that under the Smart Cities Mission, the existing cities will function more efficiently and offer an improved quality of life for its people, attract greater investments and generate higher gross domestic product (GDP). For example, in transport sector, the use of smart solutions in traffic management system can reduce commute time and cost of travel, thereby affecting the productivity of the people. Water management in the city can be improved by using smart metering and wastewater recycling.

The smart city initiative of Government of India focuses on sustainable and inclusive development of the existing cities by application of technology-driven solutions in the fields of energy management, water and waste management and e-governance and other citizen services. It also aims at preserving and developing open spaces such as parks and playgrounds to enhance the quality of life of citizens and reduce the urban heat effect to promote eco-balance. Application of smart solutions makes infrastructure and services better by use of technology and information [1, 2, 5, 10].

There are other schemes and programmes of central and state governments which if work as a complimentary scheme with smart cities, will result in comprehensive development of the infrastructure in the concerned city. Similarly, there are schemes like Swachh Bharat Mission (SWM), National Heritage City Development and Augmentation Yojana (HRIDAY), Digital India, Skill Development, Pradhan Mantri Awas Yojana, etc., which if work as a complimentary scheme right from planning stage will lead to overall physical, institutional, social and economic infrastructure development of the concerned potential smart city. Infrastructure is the foundation for the development and improvement of any existing city as smart city. This includes smart buildings, smart mobility and transport, smart energy, smart water management, smart waste management and smart health care. The long-term aim of all the government policies and schemes is the attainment of all these smart solutions. Such a smart city is well prepared for any kind of natural calamity and disaster and is capable of safeguarding its people and infrastructure in best possible way [1–3, 5, 7, 11].

The overall purpose of this mission is to enhance the quality of life of the inhabitants using technology-based solutions which will derive economic growth in long run. By local area development, the existing localities will be transformed into better planned ones and new areas will be developed around the cities to accommodate the expanding population of the cities. This comprehensive development will lead to enhance the quality of life for the inhabitants of the city. Comprehensive development also promotes mixed land use in the area. This includes planning for unplanned areas having variety of compatible activities and makes land usage more efficient by allowing land use activities close to each other. This can be adapted by having flexibility in land use [1–3, 5, 7, 11].

The strategy adopted in smart cities mission for local area development includes city improvement (area about 500 acres), city renewal (area about 50 acres) and city extension (area about 250 acres). For Himalayan regions, the above-prescribed area for different developmental models is reduced to half of its value. City improvement includes retrofitting of existing built-up area to make it more efficient and livable as per the smart city criterion. This will be done in consultation with the people residing in the city and will vary from city to city depending on their current conditions. City renewal includes redevelopment of existing built-up area and use mixed land use to co-create new layout with enhanced infrastructure (Fig. 1).

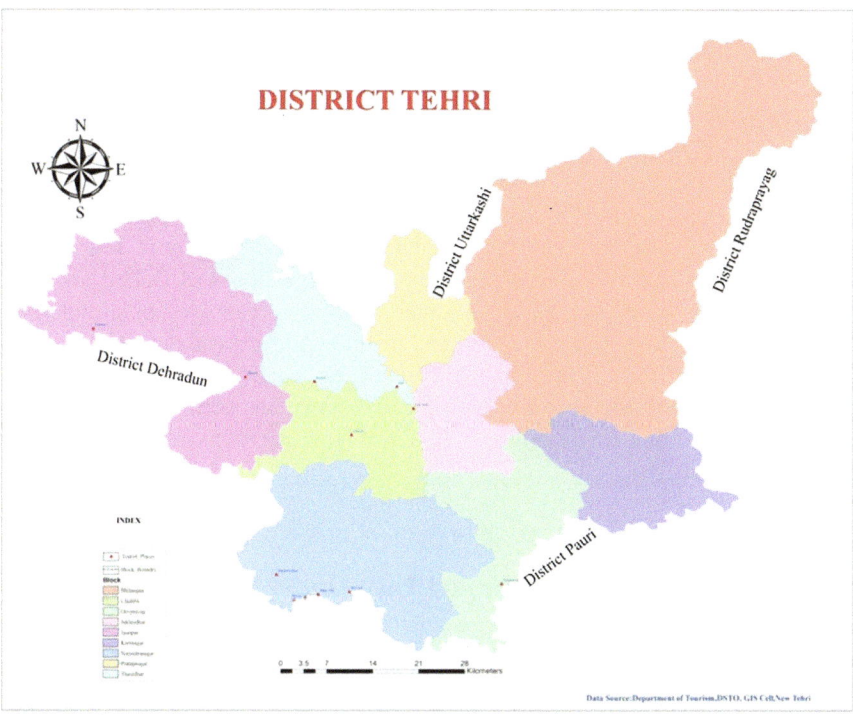

Fig. 1 Map of Tehri district (*Dept. of Tourism, GIS cell, New Tehri*)

The smart city proposal of each shortlisted city is to include either of the three above-mentioned developmental methods. The understanding of these developmental methods by the decision-makers at different levels is very crucial for the success of the mission. The smart cities mission requires an active participation of smart people in governance and reforms. It is for the first time the Ministry of Urban Development has used challenge or competitive approach to select cities for funding and implementation of the area-based developmental methods. Some consultancy firms have been shortlisted region-wise by the government to assist the cities to participate in smart city challenge. The mission covered 100 cities for tenure of five years (2015–16 to 2019–20) thereafter evaluating it upon incorporating the learning from the mission. The cities are selected from each state based on equitable criteria which give equal weightage to urban population and towns. Based on these criteria, the state of Uttarakhand has one city (Dehradun) to be developed as smart city under this mission [1, 5, 7].

The present review is to study and evaluate the different criteria which will qualify New Tehri as a smart city in the state of Uttarakhand. New Tehri qualifies to be in the list of one of the planned cities of India and only planned city of Uttarakhand but still needs to be considered for inclusion into the Smart Cities Mission.

3   Lake City Tehri

New Tehri (30° 22′ 48″ N, 78° 28′ 48″ E) is the only planned city of Tehri Garhwal District in the Indian state of Uttarakhand situated at an elevation of 1750 m above msl. New Tehri has 11 wards under its jurisdiction from Vidhi Vihar to Vishwakarma Puram (Koti colony). The old Tehri town is located at the confluence of river Bhagirathi and Bhilangana. After the construction of Tehri dam in 2006, this town was totally submerged and the population was shifted and rehabilitated to New Tehri town which is a modern town spread over an altitude from 1550 to 1950 m. New Tehri is considered as the world's most successful rehabilitation programme. It has now emerged as a very beautiful hill station garnering new avenues for tourism sector in the state. The town has seen several protests in the past by environmental activists, who fear adverse impacts of dams on the ecology of the region.

4   Selection Process of a Smart City Under the Mission

Every city competing to qualify for being selected as smart city needs to undergo two stages in the selection processes. The authorities have already indicated the number of cities from each state which they will be developing as smart cities. For example, the state of Uttarakhand can have only one city which can be developed as smart city. The first stage of qualification is intra-state where potential cities in the state compete based on current development scenario. The city scoring highest in this is recommended for second round from that state.

In the second stage, the qualified potential cities prepare a proposal. The proposal includes one of the above-mentioned models which will be adopted by the city for its development. It also includes the strategy and framework for implementation of the proposed model and its cost-effectiveness. The local development bodies should also provide the phase-wise timeframe in which the model is implemented. The credibility of implementation is also analysed using last three-year data of the city. The Ministry of Urban Development reviews and evaluates the submitted proposals by a committee of experts from national and international organizations and institutes. The cities qualifying in this start implementing their proposal keeping in line with the time frame. The cities which are not selected are also provided with suggestions for improvement in their proposal. Such cities might be included in smart city at a later stage after they implement the suggested improvements into their development strategy. The different steps in selection of smart cities are given in Fig. 2.

The preparation of smart city proposal is a challenging work, and a state requires technical assistance support from consulting firms and handholding agencies. Consulting firms are the agencies which will help the states to prepare smart city proposal as per the financial rules and guidelines. Handholding agencies are the national and international institutes and organizations which will help the states to prepare smart city proposal as they have experience in the field of smart city

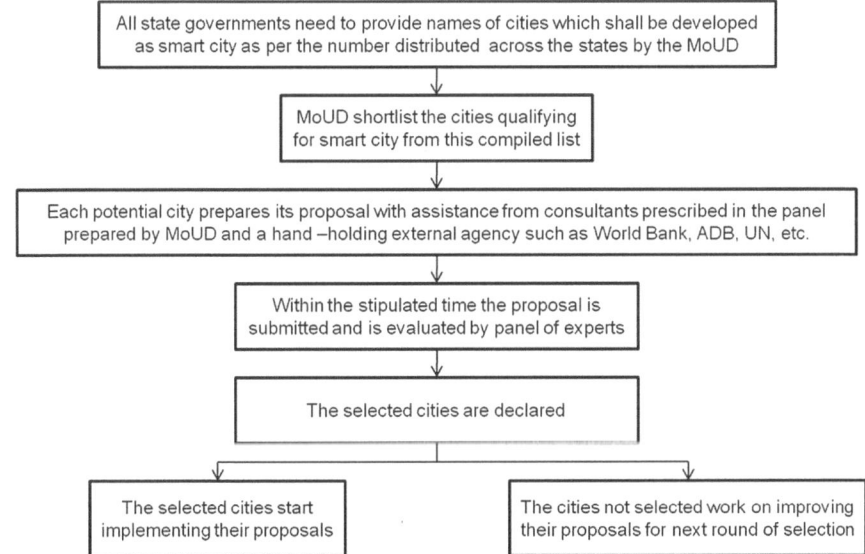

Fig. 2 Flowchart showing different steps in selection process of smart cities [1]

development. For the purpose of proper implementation of this mission, Special-Purpose Vehicles (SPVs) are created in each proposed city. These SPVs play a key role in the evaluation of smart city projects. This entire project under Smart Cities Mission is funded by central government agencies [4, 6, 8, 9].

5 Smart City Tehri

Upon studying and following the guideline prescribed under Smart Cities Mission, the following conclusions were made for Tehri city. New Tehri city is one and only planned city in the state of Uttarakhand, and this can be considered an advantage in developing Tehri as a smart city in near future. A city-wide concept plan depicting the vision, mission and key challenges faced by the city needs to be identified. This will give a clear view to the decision-makers on how city visualizes itself in five years. Keeping this in mind, the city aims at improving the living standard of people by applying smart solutions using latest technology to infrastructure improvement, energy management, water and wastewater management. The smart solution enables the use of mobile and telecommunication for address and solution of any situation. As per the census 2011 report, the total population of New Tehri city has increased about 2.35% from the last census. New Tehri city was developed after the construction of Tehri Dam as its rehabilitation area. It is a well-planned city with planned housing societies with all the major amenities like school, colleges, hospitals, banks, etc. Being a hydropower site, the city has a 24-h water and electricity supply.

Being self-sufficient in water and power supply sectors, it has also developed itself into a major centre of tourist attraction. It attracts hundreds and thousands of tourists which boosts its tourism industry. To tackle the problem of waste management, this comes with the tourism sector; the government policies are such that they promote tourism with sustainable development and preserve the natural environment of the city. Though the city has faced major protests during the construction of Tehri dam, it was severely impacting the nearby flora and fauna and in turn the weather, but now with strict policies of the government, the situation is under control. The major portion of revenue is generated from power and tourism sectors in the city. Both these sectors are also providing new avenues for the employability of youth in nearby area as well. Various other schemes of the government such as JnNURM, SWM, Digital India and Pradhan Mantri Awas Yojana have also been successfully implemented in the city. They serve as supporting schemes to SCM. With the construction of all-weather roads in the state of Uttarakhand, the city of New Tehri will get benefitted in a way that it will have a better connectivity with other parts of the country.

Based on the guidelines given by Smart Cities Mission, Tehri city is scoring 65 and 55 marks in the prescribed criteria in the first stage of Smart Cities Mission. These scores are obtained after analysing the existing service levels, institutional systems and capacities, self-financing capacities and past track record of development and reforms in the city. The format of these forms and score cards are provided in the guidelines of Smart Cities Mission by Ministry of Urban Development as annexure 3. These forms are submitted by state government bodies to the Ministry of Urban Development for evaluation. In the light of above-stated information, it is suggested to have a mix of City Improvement (Retrofitting) and Pan-city development model in the proposal of smart city Tehri. Pan city is an additional feature provided in the guidelines where smart solutions are applied to the existing city-wide infrastructure. A very good example for this is implementation of intelligent traffic management system which will reduce the average commuting time which in turn will reduce the cost of travel and boost the tourism industry in the area. The Ministry of Urban Development evaluates the smart city proposal for the city, and if passed, it also regularly monitors the implementation of the proposal.

6 Conclusions

Smart city concept is concerned more with progressive development of the city rather than rating a city for development. An attempt was made in this paper to critically evaluate New Tehri City based on the guidelines of Smart Cities Mission of Government of India.

There had been improvement in the living standard of the people, sustainable and economic development in the Tehri city in the last three years. Transportation condition has improved and is expected to improve further after the construction of all-weather roads in the state. Housing, safety and water and wastewater management situation has improved and is constantly improving in the area. The administration of the city has improved by implementation of stricter government policies. It can improve further by the use of information and communication technology (ICT) in the city. The overall development of the city towards a smart city is possible only by the active participation of people residing in the city. The media and technology should empower the citizens and derive them to achieve smart city status for their city.

Smart city concept and evaluation of New Tehri city presented in this paper require further research, alterations and improvements.

References

1. Aijaz R (2016) Challenge of making smart cities in India. In: ASIE visions, vol 87. IFRI
2. Anand A, Sreevatsan A, Taraporevala P (2018) An overview of the smart cities mission in India. Report, Center for Policy Research, India
3. Bhattacharya S, Rathi S, Patro S, Tepa N (2015) Reconceptualising smart cities: a reference framework for India. Report. Center for Study of Science, Technology and Policy, India

4. Chourabi H, Nam T, Walker S, Gil-Garcia JR, Mellouli S, Nahon K, Pardo TA, Scholl HJ (2012) Understanding smart cities: an integrative framework. In: Proceedings of the annual Hawaii international conference on system sciences, pp 2289–2297. https://doi.org/10.1109/HICSS.2012.615

5. Government of India, Ministry of Urban Development (MoUD) (2015) Smart cities mission statement & guidelines. Report

6. Gupta K, Hall RP (2017) The Indian perspective of smart cities. In: Smart city symposium, Prague, Oct 2017. https://doi.org/10.1109/SCSP.2017.7973837

7. Jawaid MF, Khan Saad A (2015) Evaluating the need for smart cities in India. Int J Adv Res Sci Eng 8354(4):991–996. https://doi.org/10.1016/j.ijrobp.2008.08.040

8. Municipal Corporation Bhopal (2016) The smart city challenge stage 2 smart city proposal Bhubaneswar, vol 7, p 113. https://doi.org/10.1017/CBO9781107415324.004

9. Ojo A, Dzhusupova Z, Curry E (2016) Exploring the nature of the smart cities research landscape. https://doi.org/10.1007/978-3-319-17620-8

10. Parise A (2019) Smart city strategies: management of resources and qualities. Easy chair preprint, no. 731

11. Selvakanmani S (2015) Smart city—the urban intelligence of India. Int J Res Appl Sci Eng Technol (IJRASET) 3(VI):302–307. https://doi.org/10.1111/eva.12239

Forward Osmosis (FO)—Exploring Niche in Various Applications: A Review

S. Dhiman and N. Ahsan

Abstract As industrialization and urbanization are escalating, so is the water resources scarcity and pollution problem as well as energy demand. Urbanization and development have been a great threat to the water resources and play a major role in water pollution problem if not planned in a sustainable way. Nowadays, more focus is on water recycling and reuse. Forward osmosis (FO) is an emerging technology providing a great alternative approach compared to conventional water/ wastewater treatment techniques. FO works on the principle of differential osmotic pressure of feed solution, FS (low osmotic potential) and draw solution, DS (high osmotic potential) with no hydraulic pressure and low fouling. FO has been applied in various industrial applications, viz. food and beverage, textile, oil and gas and pulp and paper, etc. Many FO membranes and DSs of desired characteristics have been developed and studied. But, still, there are many challenges and issues that need to be considered and resolve. This paper provides the information on the FO process, membrane fouling, DS concept and applicability of the FO/FO-hybrid technology in the various sectors.

Keywords Urbanization · Water pollution · Forward osmosis · FO membrane · Draw solution · FO applications

1 Introduction

Sanitation, public health and waste management are few of the main aspects of the smart cities. Public should have access to safe and adequate water supply [1]. This needs to be done in a sustainable manner with the treatment technologies having minimal environmental footprints. With the development of smart cities, urbanization is at a great pace. Rapid urbanization poses a great challenge on the wastewater management and treatment, especially in Asian and African region.

S. Dhiman (✉) · N. Ahsan
Department of Civil Engineering, Faculty of Engineering and Technology,
Jamia Millia Islamia, New Delhi 110025, India

© Springer Nature Singapore Pte Ltd. 2020
S. Ahmed et al. (eds.), *Smart Cities—Opportunities and Challenges*,
Lecture Notes in Civil Engineering 58,
https://doi.org/10.1007/978-981-15-2545-2_25

This will ultimately reflect in strictness of effluent discharge standards. The current world population is around 7.7 billion which is expected to grow over 9 billion by 2050 [2]. The growing population will make this task even more difficult. Wastewater can be a good source of resource recovery if managed properly. More focus should be on recycle and reuse rather than treat and discharge. As per Central Pollution Control Board (CPCB), presently India produces sewage of approximately 61,754 million litres per day (MLD) of which around 60% is discharged untreated into the streams/rivers. And, the sewage generated is around 80% of the water supplied [3]. To meet increasing water needs worldwide, research is ongoing on recovering/reclaiming water of the required quality from wastewater or contaminated water sources.

Membrane filtration technology is viable in the wastewater treatment. Commonly used membrane filtrations are: ultrafiltration (UF), microfiltration (MF), nanofiltration (NF), pressure retarded osmosis (PRO) and reverse osmosis (RO). These pressure-driven processes have been applied combined or separately for wastewater treatment [4–8]. Forward osmosis (FO) is another membrane technology that has come into the light recently for wastewater treatment and is being studied extensively.

Forward osmosis is defined as the movement of solvent molecules across a selectively semi-permeable membrane from a region of low solute concentration (feed solution, FS) to the higher solute concentration region (draw solution, DS), driven by the action of differential osmotic pressure. The process will continue until the concentration of the solutions on the both sides reaches an equilibrium stage. FO follows the second law of thermodynamics which states that solvent molecules tend to move naturally from a region of lower concentration to higher concentration to equilibrate overall chemical potential [9]. FO process requires no pressure, whereas RO involves hydraulic pressure for purification of wastewater. The generalized equation for pressure and osmotic pressure-driven processes is given by Darcy's law (Eq. 1) [10].

$$J_w = A \times (\sigma \cdot \Delta\pi - \Delta P) \tag{1}$$

where J_w is the water flux (L/m^{2}h), A is water permeability constant (L/m^{2}hPa), σ is reflection coefficient (−), $\Delta\pi$ is differential osmotic pressure across the membrane (Pa) and ΔP is the applied pressure (Pa). The reflection coefficient σ is taken as 1 for the membrane achieving 100% solute rejection. For RO process, the applied pressure is more than differential osmotic pressure ($\Delta P > \Delta\pi$), and for PRO, osmotic pressure is more than applied pressure ($\Delta P < \Delta\pi$). For FO, ΔP is zero ($\Delta P = 0$).

Growing interest in FO is because of its many positive points: involves low or no hydraulic pressure; high rejection of contaminants, salt and particulate matter; simple and low strength equipment required and low membrane fouling property than RO [11, 12]. FO has found its applications in various sectors, for instance, food processing [13], oil and gas [14], textile [15, 16], landfill leachate [17, 18] and municipal wastewater treatment including sludge dewatering [19, 20].

This paper reviews the FO process and FO membrane development, and challenges have also been addressed. The importance of the DS selection and how the concentration of DS effects the system has been discussed. This paper enlightens the applications of FO in various sectors and the works need to be done to achieve better results.

2 FO Membrane Development and Challenges

Earlier in 1970s, RO membranes were used in the FO process, and the obtained water flux was much lower as RO membranes have thick support layer and contribute to concentration polarization (CP) problem. Later, cellulose triacetate (CTA) membrane came into the picture developed by Hydration Technologies Incorporation (HTI), having thin support layer and thickness less than 50 μm [10]. CTA membranes are widely used. CTA membranes are resistant to chlorine, adsorption of minerals, oils and fatty acids. These membranes can also withstand thermal, chemical and biological degradation [12]. Commercial thin film composite (TFC) membranes are also available. These are superior to CTA membrane because of good permeability and use over broader pH range (2–12). Later, robust TFC membranes were formed through phase inversion with lesser support thickness and through interfacial polymerization. But, the properties that contribute to these structural parameters can also cause mechanical fragility [21]. Recently, TFC aquaporin biomimetic membranes [22] and thin film inorganic (TFI) membranes [23] have also been studied and have shown great prospects in terms of high water flux and solute rejection.

Asymmetric FO membranes possess two different layers: active layer (AL) and support layer (SL). The AL is generally dense and SL is porous. SL provides mechanical support. There can be two cases for positioning the AL of the membrane: AL facing feed side (AL–FS mode) and AL facing draw solution (AL–DS mode). Zhang et al. [24] found that in AL–FS mode, there was a rapid decline in water flux at first and became gradual later. But, in case of AL–DS, at first, flux decline was slow, and then, the flux decreases rapidly in comparison with AL–FS. The water flux of AL–FS mode exhibited 7.5% higher flux than AL–DS and found to be better.

FO involves various membrane-related phenomena which contribute to lower flux. The observed flux is much lower than the expected as the differential osmotic pressure across membrane is lower than the difference in the bulk osmotic pressure of FS and DS (Fig. 1). For AL–FS mode, which is preferred in FO, as the water from AL enters the porous support layer, the DS in support layer gets diluted. This is known as dilutive internal concentration polarization (ICP).

During FO, a layer of solute particles gets build up on the membrane active surface (FS–AL mode) and results in fouling and ultimately lower water flux [11]. This problem is much severe in pressure-driven processes such as RO. Fouling layer developed in pressure-driven process is irreversible and much compacted

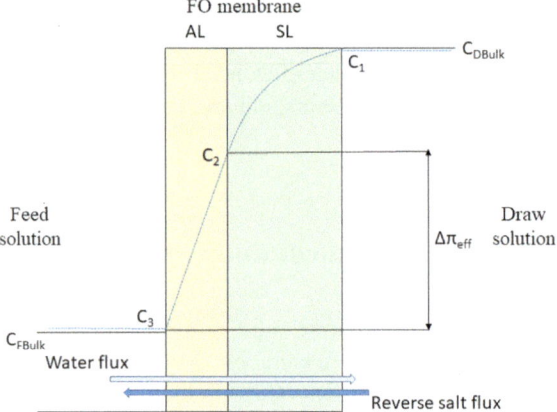

Fig. 1 Illustration of effective driving potential across membrane in FO, AL facing feed side. C_{FBulk} and C_{DBulk} are the bulk osmotic potential of feed and draw solution, respectively. C_1 and C_3 are the osmotic potentials at the DS–SL and FS–AL interface, respectively. C_2 is the osmotic potential at the AL–SL junction and $\Delta\pi_{eff}$ is the effective driving force

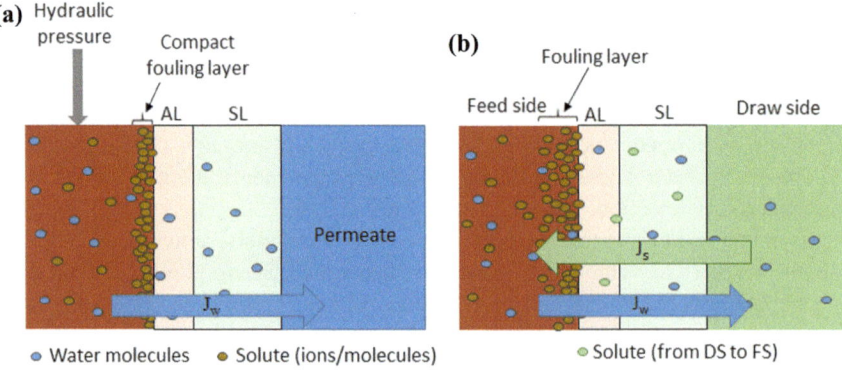

Fig. 2 Illustration of fouling behaviour in **a** pressure-driven membrane processes and **b** differential osmotic pressure-driven membrane processes

compared to osmotic driven process (Fig. 2). They also increase the maintenance of the system and deteriorate the membrane as well. Fouling may occur because of: organic fouling (due to the presence of organic matter, proteins, fats, etc.); inorganic fouling (due to the crystallization of the less soluble salts after saturation point); biofouling (due the development of the microbial film on the membrane surface) and colloidal fouling (due to the colloidal particles present in the effluent).

Membrane reusability is another important aspect to consider in FO process. Membranes with usage are prone to fouling that reduces the performance by a significant amount. The separation and interaction of organic/inorganic particles in an aqueous solution are dependent on the interaction of colloidal matter with the surface of the membrane [25]. Fouling in FO membrane is usually reversible in nature, and water flux recovery of up to 100% can be achieved after chemical cleaning [16]. For good performance, membrane with low structural parameters (S) is preferred [21]. Membrane structural parameter reflects the thickness of the support layer, reverses solute rejection capability and porosity and is given by Eq. 2.

$$S = K \times D = (t_s \times \tau)/\varepsilon \tag{2}$$

where S is known as structural parameter, K is the resistance to solute diffusion coefficient, D is the diffusion coefficient of the reverse solute, t_s is the thickness of support layer, τ is the tortuosity and ε is the porosity of the membrane. Low S-value is preferred which requires high porosity and less thickness and is a very difficult task to achieve in membrane development field. To minimize internal concentration polarization (ICP) and attain maximum water flux, the resistance to draw solution diffusion must be reduced. A good FO membrane should possess the qualities of: a thin support layer; hydrophilicity of active layer; low fouling property and high chemical and mechanical strength to withstand long-term operations.

3 Draw Solution Selection

FO is based on the differential osmotic pressure across the membrane, making the choice of suitable DS a vital aspect in the FO. So, more focus should be on the type of DS used in FO. Various types of DSs have been used. Sea water is easily available and the most commonly used DS [12]. For concentrating sucrose in sugarcane, sea bittern was used as a DS [26]. NaCl was used for wastewater treatment to recover phosphorous [27]. Different salts like $MgCl_2$, NaCl, $CaCl_2$ and KCl have been used as DSs. Till now, around 40% of work has been done with NaCl as DS [10, 12].

For wastewater treatment, Achilli et al. [28] suggested that magnesium chloride ($MgCl_2$) is the most ideal DS. As magnesium chloride is a divalent ion, so there will be less solute passage. Also, the osmotic pressure of $MgCl_2$ is more than NaCl at the same concentration [10].

Fertilizers as DS in FO is beneficial as it eliminates the DS regeneration after FO with the diluted DS being used for fertigation; this saves energy and hence cost [29]. Urea (NH_2CONH_2) is the most commonly used fertilizer. The limitation in this case is the low osmotic pressure, and the final concentration of fertilizer has to be according to the cropping season and soil nutrient condition [30]. In case of fertilizers, the use of wastewater as FS gives better flux results ($\approx 80\%$ higher) than saline or brackish feed [31].

The DS must have high osmotic pressure than the FS for FO to work properly. The osmotic pressure of the DS also depends on the concentration being used. The relation between the osmotic pressure and the concentration of the DS is expressed by the Morse equation (Eq. 3) [32].

$$\pi = iMRT = i(n/V)\,RT \tag{3}$$

where π is the osmotic pressure of the solution (Pa), i is the van't Hoff factor, M is the molarity of the solution (mol/L), R is the universal gas constant ($8.3145\ \mathrm{J\ K^{-1}\ mol^{-1}}$) and T is the absolute temperature (K).

The DS must also have small molecular weight and low viscosity [9]. As the diffusion coefficient is inversely proportional to the molecular mass and viscosity, DS with high mass will lead to low diffusion and ultimately to concentration polarization. There are two types of solute diffusion in FO: forward (FS–DS) and reverse (DS–FS). Reverse salt diffusion into the feed decreases the overall chemical potential difference between the FS and DS and ultimately reduces the water flux. Reverse salt flux is defined by Fick's law (Eq. 4).

$$J_s = B \cdot \Delta c \tag{4}$$

where J_s is the reverse salt flux, B is the solute permeability coefficient and Δc represents the solute concentration difference across the membrane.

DS should possess the characteristics of: high solubility and high degree of dissociation to be a good quality of DS; non-toxic and free from impurities as their presence reduces the osmotic potential of the solution by disturbing the solvent structure and easy regeneration steps or the possibility of using as it is after FO.

4 Applications in Various Fields

4.1 Municipal Wastewater

Conventional wastewater treatment involves biological processes which consume a lot of energy and have a substantial environmental footprint as they produce gases (N_2 and CO_2) which can result into greenhouse gas (N_2O) [33]. With increasing demand of water and depletion of water resources at a great pace, more focus has been given on the wastewater recycling, reuse and less energy-intensive treatment processes. Wastewater is also being explored for resource recovery [34, 35].

Hey et al. [20] studied direct membrane filtration (DMF) and direct forward osmosis (DFO) from energy perspective for municipal wastewater as feed using NaCl and sea water as DS and membrane was aquaporin membrane. AL–FS mode was used in the process. The results showed both DMF and DFO can be used as treatment method for low-scale wastewater treatment plants. It was concluded that the pre-treatment (e.g. coagulation–flocculation) can be avoided by FO technology using aquaporin membrane in the process. Treatment process including FO yielded more CH_4 gas compared to DMF and conventional treatment practices. Illustration of the experimental set-up used in FO process is given in Fig. 3.

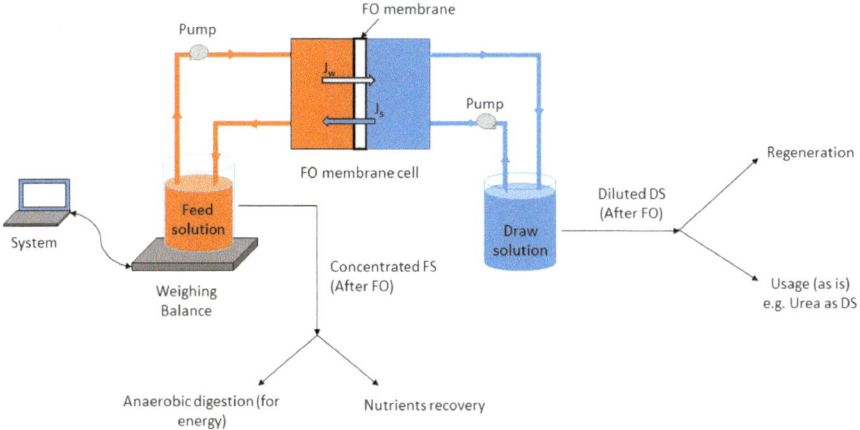

Fig. 3 Schematic representation of FO experimental set-up

Zhang et al. [24] concentrated municipal wastewater by FO to use concentrated feed as energy recovery resource. Chemical oxygen demand (COD) concentration was achieved up to 300%, and observed water flux was in the range of 3–7.4 L/m^{2}h. The flux decline was contributed mainly because of cake layer formation on the AL of membrane (AL–FS mode) and drop in the differential osmotic pressure due to the dilution of DS.

FO was studied by Hey et al. [36] to assess its suitability for wastewater treatment in the absence of biological treatment for small treatment plants and analysed the effects of the pre-treatment techniques (micro sieving and microfiltration) prior to FO. Two flat sheet TFC membranes were used in the study (aquaporin and HTI technology), and 2M NaCl was used as DS with AL–FS mode. Use of pre-treatment before FO exhibited stable and increased water flux. Aquaporin showed less decline in water flux (25%) from initial compared to HTI membrane (43%) as fouling, and concentration polarization was experienced more in case of HTI membrane. Also, rejection of more than 96% was achieved for biochemical oxygen demand (BOD), total phosphorus (TP) and total nitrogen (TN) for both the membranes with or without pre-treatment.

4.2 Textile Industries

Textile industries consume a lot of freshwater, and also, effluent is of very complex nature having high COD, dyes, surfactants and reagents [37]. One of the challenges is the treatment of non-biodegradable reactive dyes having high residence time when exposed to the environment.

Korenak et al. [16] used real textile wastewater as feed in FO using NaCl and MgCl$_2$ as DS. Also, mixtures of concentrated dyes were tested as DS to eliminate the need for regeneration step for DS after FO. The concentrations of dyes as DS were kept high initially and diluted in FO to a certain amount to be used again in the dyeing sector after FO. MgCl$_2$ proved to be more efficient as DS in terms of reverse salt flux (0.04 g/m^{2}h) compared to NaCl (2.9 g/m^{2}h) and dyes (up to 5.7 g/m^{2}h). 100% rejection for dye was achieved. COD was rejected up to a greater extent (>95%), and rejection of (>99%) was achieved for total dissolved solids (TDS) and total suspended solids (TSS). The feasibility study using FO in textile industry showed that the volume of the influent for the municipal treatment plant from the industry can be reduced to half, and the rest half can be incorporated into dye mixing or reused in other maintenance work.

The possibility of using polyacrylamide (PAM) as DS was investigated for the treatment of dye wastewater and compared with the KCl as DS using polyamide-based TFC membrane [38]. PAM was proven more efficient in terms of water flux stability and membrane fouling. PAM also showed lower reverse flux (<0.009 g/m^{2}h) compared to KCl (>0.2 g/m^{2}h). Increase in water flux was proportional to temperature (5.09 L/m^{2}h at 45 °C) because of decrease in kinematic viscosity and increase in diffusion of water across membrane.

A hybrid system involving FO and coagulation–flocculation (FO–CF) was studied by Han et al. [39] for treating synthetic textile wastewater. Textile feed was concentrated using FO with TFC membrane and NaCl as DS. Dye was removed from the concentrated feed using coagulation–flocculation. This will reduce the amount of chemicals needed in CF process. 95% removal of dye was achieved. Overall flux decline was also not much, and flux of 12 L/m^{2}h was obtained with corresponding water recovery of 90%.

4.3 Produced Water Treatment

Gas and oil recovery operations involve by-product water called produced water. It contains a huge amount of TDS and can range up to 150 g/L [40] and also has organics which are recalcitrant in nature and cannot be treated using biological treatment.

Bell et al. [41] treated produced water during oil and gas exploration having high TDS using CTA and polyamide TFC membranes. Treatment study of the produced water was performed for three weeks. 1M NaCl was used as DS. Rejection of 90% was achieved for hydrophobic compounds by both the membranes. Results showed that CTA membrane performed much better overall than TFC membrane in terms of high water flux and lower reverse flux. Despite using antifouling TFC membrane (more smoother, neutrally charged and hydrophilic nature), it exhibited more fouling than CTA membrane. High fouling concluded that the pre-treatment of the complex wastewater streams needs to be done to make FO more efficient.

Coal seam gas (CGS) produced water containing low TDS was treated by Chun et al. [42] using bench-scale FO (AL–FS mode) and PRO (AL–DS mode) system. Polyamide TFC and CTA membranes were used with NaCl as DS. Cross-flow cell of dimensions (8 cm × 6 cm × 0.2 cm) was used. FO mode showed much consistency in terms of water flux. CTA-FO showed only 13% flux decrease compared to 15% in TFC-FO, whereas flux decline for CTA-PRO (35%) and TFC-PRO (55%) mode was much greater. Flux decline in FO experiments was experienced due to the external and internal concentration polarization. FO rejected dissolved salts up to 96%, and it was concluded that for low foulant CGS streams, FO can be a good alternative pre-treatment method.

FO-hybrid system has also been investigated for the produced water treatment. Sardari et al. [43] studied electrocoagulation (EC) (2 and 1 min reaction time) along with FO to treat produced water by hydraulic fracturing process using NaCl and ammonium bicarbonate (NH_4HCO_3) as DS. CTA membrane (HTI) was used in FO process. EC helped in removing around 70% of the TSS, organic carbon and turbidity and minimizing the fouling during FO operation. Water recovery was around 45% (NaCl as DS) and 70% (NH_4HCO_3 as DS). As the concentration of the DS increased, reverse solute flux also increased and contributed in the membrane fouling and concluded that prediction of membrane fouling behaviour is very difficult and complex because of many factors involved. For long-term operations, Coday et al. [44] concluded that fouling depends on foulant–foulant interactions once the cake layer formation occurs and not the membrane type and can be controlled by selecting suitable chemical cleaning reagent for membrane cleaning.

4.4 Distillery Wastewater

Distilleries use a large amount of freshwater. In India, annual wastewater generation from distilleries is around 40.4×10^4 million litres [45]. Effluents from distilleries have a dark colour, high TDS and can contain complex organic contaminants which are non-biodegradable in nature [46].

Singh et al. [22] investigated the treatment of distillery wastewater using FO to evaluate the rejection of melanoidins (phenolic compound having antioxidant properties), COD and TDS. Biomimetic aquaporin membrane (Aquaporin A/s) was used with $MgCl_2 \cdot 6H_2O$ as DS. Rejection of melanoidins, antioxidant activity and COD was 90, 96 and 84%, respectively, with water flux of 6.3 L/m^2h using synthetic melanoidins (10% v/v) as FS. Whereas using real distilleries wastewater as FS, water flux obtained was much lower (2.3 L/m^2h) with similar rejection range for melanoidins, antioxidant activity and COD. 24 h study using FO was conducted which yielded water recovery around 70%, much higher than RO (40%). Transport of molecules was observed from FS to DS which can deteriorate the DS and can create problem in the regeneration steps or as it is further use of DS.

4.5 Groundwater Treatment

The presence of heavy metals in the groundwater is a big concern. Leaching of contaminated soil under the acid rain action can pollute the aquifer underground up to a great extent [47]. There is a threshold limit for every element before it shows an adverse effect on the human body, making assessment and treatment of heavy metals very important [48].

FO study was performed to remove divalent heavy metals (Cd^{2+}, Zn^{2+}, Pb^{2+} and Cu^{2+}) using thin film inorganic (TFI) membrane with NaCl as DS [23]. Study showed that TFI membranes are able to withstand chemical and thermal (20–70 °C) changes efficiently. They are applicable in a wide range of pH (5–9). The process yielded water flux of 69 L/m^2h and rejection of all the studied metals up to 94% at an initial concentration of 200 mg/L in the FS.

Groundwater pollution due to arsenic (As) is a very serious problem, and upon ingestion, it gets absorbed by the gastrointestinal (GI) tract and leads to complicated health issues [49]. Mondal et al. [50] conducted a study to remove As^{5+} ($Na_2AsO_4 \cdot 7H_2O$ as source in the study) from groundwater using FO and analyse the effect of other ions/elements present in the stream on the removal efficiency. CTA membrane (moderate hydrophilic and slightly negatively charged) by HTI was used in the process and glucose as DS which will reduce the need of regeneration steps as it can be a good alternative for an emergency drink in serious situations. 95.5% rejection of As^{5+} was obtained which was reduced by approximately 15% in the presence of the phosphates, whereas humic acid showed a positive impact on the As^{5+} rejection (99.5–99.3%) due to the fouling occurred which acted as fluid-like loose layer. The presence of silica showed no significance on the rejection efficiency.

4.6 Leachate Treatment

Landfill is the most commonly used method for waste disposal in India [51]. Leachate is generated by the action of the rainfall mixing with the waste and percolates into the soil and tends to contaminate groundwater. Leachate is comprised of dissolved organic matter, inorganic elements/ions, heavy metals and xenobiotic compounds (from household waste) [52].

Iskander et al. [17] discussed the possibilities of integrating bioelectrochemical system (BES) with FO. The ammonia recovered in treatment of leachate can be used as DS. Landfill leachate treatment was studied by Dong et al. [18] with membrane bioreactor (MBR) effluent as feed in FO. Membrane used was CTA and DS was NaCl. Rejection of COD, TP and ammonia was 98.6, 96.6 and 76.9%, respectively. Also, chemical cleaning achieved better flux recovery (98.9%) compared to hydraulic cleaning (88.9%).

4.7 Food and Beverage Industry

Fruits juices are a great source of energy, vitamins and essential nutrients. They need to be concentrated by processes such as thermal vacuum treatment for long-term storage, usage and reduce the transportation cost. These processes can deteriorate taste, nutritional value and affect the colour of juices [53].

Sant'Anna et al. [54] used FO process for the concentration of jaboticaba juice which is good source of vitamins, mineral and antioxidants. NaCl was used as DS, and membrane used was CTA. FO showed good results in terms of retaining the organoleptic properties of juice and suggested to use it as an alternative in food and beverages industry.

Tuna cooking and processing industry produces a large amount of cooking juice with a high concentration of protein. Khongnakorn and Youravong [55] used FO and were able to increase protein concentration in concentrated juice feed by 9%, but low corresponding flux was obtained due to the fouling and increased osmotic concentration of feed with time, lowering the driving force.

An et al. [56] studied a hybrid system of FO and membrane distillation (FO–MD) for concentrating apple juice using TFC polyamide membrane for FO and poly (vinylidene fluoride-co-hexafluoropropylene) for MD. Potassium sorbate was used a DS to help avoiding bacterial growth in apple juice feed by the action of reverse salt flux of DS. FO was able to retain most of the nutritional value of the apple juice with potassium sorbate concentration of 0.45 mg/L, well below the standard level in food industry (1.0 mg/L).

5 Conclusions

Depletion of water resources along with water pollution problem has become a great problem in the urban world. Water demand is increasing day by day. Smart cities need to have wastewater treatment facilities incorporating resource recovery methods in the system and generate revenue. FO is one of the technologies that has been emerging as a less energy-intensive alternative for wastewater treatment and showing advantages like less prone to fouling and great contaminants rejection. Properties of the membrane surface and selection of the DS are of great importance for the FO process to achieve better results. FO has been incorporated with many other techniques for wastewater treatment and resource recovery. Commercialization of recovered resources needs to be done effectively. Overall, FO has shown great prospects in a few industrial applications and continue exploring the niche in others as well.

References

1. Smart Cities Mission, Ministry of Housing and Urban Affairs, Government of India. http://smartcities.gov.in/content/innerpage/what-is-smart-city.php
2. Koop SHA, Leeuwen CJ (2015) Assessment of the sustainability of water resources management: a critical review of the city blueprint approach. Water Resour Manage 29:5649–5670. https://doi.org/10.1007/s11269-015-1139-z
3. Central Pollution Control Board (2015) http://www.sulabhenvis.nic.in/Database/STST_wastewater_2090.aspx
4. Bunani S, Yörükoğlu E, Yüksel Ü, Kabay N, Yüksel M, Sert G (2015) Application of reverse osmosis for reuse of secondary treated urban wastewater in agricultural irrigation. Desalination 364:68–74. https://doi.org/10.1016/j.desal.2014.07.030
5. Sahinkaya E, Sahin A, Yurtsever A, Kitis M (2018) Concentrate minimization and water recovery enhancement using pellet precipitator in a reverse osmosis process treating textile wastewater. J Environ Manage 222:420–427. https://doi.org/10.1016/j.desal.2012.03.015
6. Linares RV, Li Z, Yangali-Quintanilla V, Ghaffour N, Amy G, Leiknes T, Vrouwenvelder JS (2016) Life cycle cost of a hybrid forward osmosis—low pressure reverse osmosis system for seawater desalination and wastewater recovery. Water Res 88:225–234. https://doi.org/10.1016/j.watres.2015.10.017
7. Coskun T, Debik E, Kabuk HA, Demir NM, Basturk I, Yildirim B, Temizel D, Kucuk S (2016) Treatment of poultry slaughterhouse wastewater using a membrane process, water reuse, and economic analysis. Desalin Water Treat 57:4944–4951. https://doi.org/10.1080/19443994.2014.999715
8. Yordanov D (2010) Preliminary study of the efficiency of ultrafiltration treatment of poultry slaughterhouse wastewater. Bulg J Agric Sci 16:700–704
9. Ge Q, Ling M, Chung T-S (2013) Draw solutions for forward osmosis processes: developments, challenges, and prospects for the future. J Membr Sci 442:225–237. https://doi.org/10.1016/j.memsci.2013.03.046
10. Cath TY, Childress AE, Elimelech M (2006) Forward osmosis: principles, applications, and recent developments. J Membr Sci 281:70–87. https://doi.org/10.1016/j.memsci.2006.05.048
11. Holloway RW, Childress AE, Dennett KE, Cath TY (2007) Forward osmosis for concentration of anaerobic digester centrate. Water Res 41:4005–4014. https://doi.org/10.1016/j.watres.2007.05.054
12. Lutchmiah K, Verliefde ARD, Roest K, Rietveld LC, Cornelissen ER (2014) Forward osmosis for application in wastewater treatment: a review. Water Res 58:179–197. https://doi.org/10.1016/j.watres.2014.03.045
13. Garcia-Castello EM, McCutcheon JR, Elimelech M (2009) Performance evaluation of sucrose concentration using forward osmosis. J Membr Sci 338:61–66. https://doi.org/10.1016/j.memsci.2009.04.011
14. Zhao S, Minier-Matar J, Chou S, Wang R, Fane AG, Adham S (2017) Gas field produced/process water treatment using forward osmosis hollow fiber membrane: membrane fouling and chemical cleaning. Desalination 402:143–151. https://doi.org/10.1016/j.desal.2016.10.006
15. Dutta S, Nath K (2018) Feasibility of forward osmosis using ultra low pressure RO membrane and Glauber salt as draw solute for wastewater treatment. J Environ Chem Eng 6:5635–5644. https://doi.org/10.1016/j.jece.2018.08.037
16. Korenak J, Hélix-Nielsen C, Bukšek H, Petrinić I (2019) Efficiency and economic feasibility of forward osmosis in textile wastewater treatment. J Clean Prod 210:1483–1495. https://doi.org/10.1016/j.jclepro.2018.11.130
17. Iskander SM, Brazil B, Novak JT, He Z (2016) Resource recovery from landfill leachate using bioelectrochemical systems: opportunities, challenges, and perspectives. Bioresour Technol 201:347–354. https://doi.org/10.1016/j.biortech.2015.11.051

18. Dong Y, Wang Z, Zhu C, Wang Q, Tang J, Wu Z (2014) A forward osmosis membrane system for the post-treatment of MBR-treated landfill leachate. J Membr Sci 471:192–200. https://doi.org/10.1016/j.memsci.2014.08.023

19. Wang Z, Zheng J, Tang J, Wang X, Wu Z (2016) A pilot-scale forward osmosis membrane system for concentrating low-strength municipal wastewater: performance and implications. Sci Rep 6:21653. https://doi.org/10.1038/srep21653

20. Hey T, Bajraktari N, Davidsson Å, Vogel J, Madsen H, Helix-Nielsen C, Cour Jansen J, Jonsson K (2017) Evaluation of direct membrane filtration and direct forward osmosis as concepts for compact and energy-positive municipal wastewater treatment. Environ Technol 39:1–39. https://doi.org/10.1080/09593330.2017.1298677

21. Shaffer DL, Werber JR, Jaramillo H, Lin S, Elimelech M (2015) Forward osmosis: where are we now? Desalination 356:271–284. https://doi.org/10.1016/j.desal.2014.10.031

22. Singh N, Petrinic I, Hélix-Nielsen C, Basu S, Balakrishnan M (2018) Concentrating molasses distillery wastewater using biomimetic forward osmosis (FO) membranes. Water Res 130:271–280. https://doi.org/10.1016/j.watres.2017.12.006

23. You S, Lu J, Tang CY, Wang X (2017) Rejection of heavy metals in acidic wastewater by a novel thin-film inorganic forward osmosis membrane. Chem Eng J 320:532–538. https://doi.org/10.1016/j.cej.2017.03.064

24. Zhang X, Ning Z, Wang DK, Costa JCD (2014) Processing municipal wastewaters by forward osmosis using CTA membrane. J Membr Sci 468:269–275. https://doi.org/10.1016/j.memsci.2014.06.016

25. Howe KJ, Ishida KP, Clark MM (2002) Use of ATR/FTIR spectrometry to study fouling of microfiltration membranes by natural waters. Desalination 147:251–255. https://doi.org/10.1016/S0011-9164(02)00545-3

26. Mondal P, Tran ATK, Bruggen BV (2014) Removal of As(V) from simulated groundwater using forward osmosis: effect of competing and coexisting solutes. Desalination 348:33–38. https://doi.org/10.1016/j.desal.2014.06.001

27. Huang L-Y, Lee D-J, Lai J-Y (2015) Forward osmosis membrane bioreactor for wastewater treatment with phosphorus recovery. Bioresour Technol 198:418–423. https://doi.org/10.1016/j.biortech.2015.09.045

28. Achilli A, Cath TY, Childress AE (2010) Selection of inorganic-based draw solutions for forward osmosis applications. J Membr Sci 364:233–241. https://doi.org/10.1016/j.memsci.2010.08.010

29. Xiang X, Zou S, He Z (2017) Energy consumption of water recovery from wastewater in a submerged forward osmosis system using commercial liquid fertilizer as a draw solute. Sep Purif Technol 174:432–438. https://doi.org/10.1016/j.seppur.2016.10.052

30. Phuntsho S, Shon HK, Hong S, Lee S, Vigneswaran S, Kandasamy J (2012) Fertiliser drawn forward osmosis desalination: the concept, performance and limitations for fertigation. Rev Environ Sci Bio/Technol 11:147–168

31. Chekli L, Kim Y, Phuntsho S, Li S, Ghaffour N, Leiknes T, Shon HK (2017) Evaluation of fertilizer-drawn forward osmosis for sustainable agriculture and water reuse in arid regions. J Environ Manage 187:137–145. https://doi.org/10.1016/j.jenvman.2016.11.021

32. Devia YP, Imai T, Higuchi T, Kanno A, Yamamoto K, Sekine M, Le T (2015) Potential of magnesium chloride for nutrient rejection in forward osmosis. J Water Resour Prot 7:730–740. https://doi.org/10.4236/jwarp.2015.79060

33. Rodriguez-Caballero A, Aymerich I, Poch M, Pijuan M (2014) Evaluation of process conditions triggering emissions of green-house gases from a biological wastewater treatment system. Sci Total Environ 493:384–391. https://doi.org/10.1016/j.scitotenv.2014.06.015

34. Bergmans BJC, Veltman AM, Loosdrecht MCM, Lier JB, Rietveld LC (2014) Struvite formation for enhanced dewaterability of digested wastewater sludge. Environ Technol 35:549–555. https://doi.org/10.1080/09593330.2013.837081

35. Singh N, Dhiman S, Basu S, Balakrishnan M, Petrinic I, Helix-Nielsen C (2019) Dewatering of sewage for nutrients and water recovery by Forward Osmosis (FO) using divalent draw solution. J Water Process Eng 31:100853. https://doi.org/10.1016/j.jwpe.2019.100853

36. Hey T, Zarebska A, Bajraktari N, Vogel J, Hélix-Nielsen C, Cour Jansen J, Jönsson K (2017) Influences of mechanical pretreatment on the non-biological treatment of municipal wastewater by forward osmosis. Environ Technol 38:2295–2304. https://doi.org/10.1080/09593330.2016.1256440

37. Holkar CR, Jadhav AJ, Pinjari DV, Mahamuni NM, Pandit AB (2016) A critical review on textile wastewater treatments: possible approaches. J Environ Manage 182:351–366. https://doi.org/10.1016/j.jenvman.2016.07.090

38. Zhao P, Gao B, Xu S, Kong J, Ma D, Shon HK, Yue Q, Liu P (2015) Polyelectrolyte-promoted forward osmosis process for dye wastewater treatment—exploring the feasibility of using polyacrylamide as draw solute. Chem Eng J 264:32–38. https://doi.org/10.1016/j.cej.2014.11.064

39. Han G, Liang C-Z, Chung T-S, Weber M, Staudt C, Maletzko C (2016) Combination of forward osmosis (FO) process with coagulation/flocculation (CF) for potential treatment of textile wastewater. Water Res 91:361–370. https://doi.org/10.1016/j.watres.2016.01.031

40. Abualfaraj N, Gurian PL, Olson MS (2014) Characterization of marcellus shale flowback water. Environ Eng Sci 31:514–524. https://doi.org/10.1089/ees.2014.0001

41. Bell EA, Poynor TE, Newhart KB, Regnery J, Coday BD, Cath TY (2017) Produced water treatment using forward osmosis membranes: evaluation of extended-time performance and fouling. J Membr Sci 525:77–88. https://doi.org/10.1016/j.memsci.2016.10.032

42. Chun Y, Kim S-J, Millar GJ, Mulcahy D, Kim IS, Zou L (2017) Forward osmosis as a pre-treatment for treating coal seam gas associated water: flux and fouling behaviour. Desalination 403:144–152. https://doi.org/10.1016/j.desal.2015.09.012

43. Sardari K, Fyfe P, Lincicome D, Wickramasinghe SR (2018) Aluminum electrocoagulation followed by forward osmosis for treating hydraulic fracturing produced waters. Desalination 428:172–181. https://doi.org/10.1016/j.desal.2017.11.030

44. Coday BD, Almaraz N, Cath TY (2015) Forward osmosis desalination of oil and gas wastewater: impacts of membrane selection and operating conditions on process performance. J Membr Sci 488:40–55. https://doi.org/10.1016/j.memsci.2015.03.059

45. Chowdhary P, Raj A, Bharagava RN (2018) Environmental pollution and health hazards from distillery wastewater and treatment approaches to combat the environmental threats: a review. Chemosphere 194:229–246. https://doi.org/10.1016/j.chemosphere.2017.11.163

46. Krishnamoorthy S, Premalatha M, Vijayasekaran M (2017) Characterization of distillery wastewater—an approach to retrofit existing effluent treatment plant operation with phycoremediation. J Clean Prod 148:735–750. https://doi.org/10.1016/j.jclepro.2017.02.045

47. Zheng S-A, Zheng X, Chen C (2012) Leaching behavior of heavy metals and transformation of their speciation in polluted soil receiving simulated acid rain. PLoS ONE 7:1–7. https://doi.org/10.1371/journal.pone.0049664

48. Sridhar SGD, Sakthivel AM, Sangunathan U, Balasubramanian M, Jenefer S, Mohamed Rafik M, Kanagaraj G (2017) Heavy metal concentration in groundwater from Besant Nagar to Sathankuppam, South Chennai, Tamil Nadu, India. Appl Water Sci 7:4651–4662 (2017). https://doi.org/10.1007/s13201-017-0628-z

49. Abdul KSM, Jayasinghe SS, Chandana EPS, Jayasumana C, Silva PMCSD (2015) Arsenic and human health effects: a review. Environ Toxicol Pharmacol 40:828–846 (2015). https://doi.org/10.1016/j.etap.2015.09.016

50. Mondal D, Sanna Kotrappanavar N, Reddy AVR, Ghara K, Maiti P, Upadhyay S, Ghosh P (2015) Four-fold concentration of sucrose in sugarcane juice through energy efficient forward osmosis using sea bittern as draw solution. RSC Adv 5 (2015). https://doi.org/10.1039/c5ra00617a

51. Joshi R, Ahmed S (2016) Status and challenges of municipal solid waste management in India: a review. Cogent Environ Sci 2:1139434. https://doi.org/10.1080/23311843.2016.1139434

52. Kjeldsen P, Barlaz MA, Rooker AP, Baun A, Ledin A, Christensen TH (2002) Present and long-term composition of MSW landfill leachate: a review. Crit Rev Environ Sci Technol 32:297–336 (2002). https://doi.org/10.1080/10643380290813462

53. Rastogi NK (2018) Chapter 13—Reverse osmosis and forward osmosis for the concentration of fruit juices. In: Rajauria G, Tiwari BK (eds) Fruit juices. Academic Press, pp 241–259 (2018). https://doi.org/10.1016/b978-0-12-802230-6.00013-8
54. Sant'Anna V, Gurak PD, Vargas NS, Silva MK, Marczak LDF, Tessaro IC (2016) Jaboticaba (*Myrciaria jaboticaba*) juice concentration by forward osmosis. Sep Sci Technol 51:1708–1715 (2016). https://doi.org/10.1080/01496395.2016.1168845
55. Khongnakorn W, Youravong W (2016) Concentration and recovery of protein from tuna cooking juice by forward osmosis. J Eng Sci Technol 11:962–973
56. An X, Hu Y, Wang N, Zhou Z, Liu Z (2019) Continuous juice concentration by integrating forward osmosis with membrane distillation using potassium sorbate preservative as a draw solute. J Membr Sci 573:192–199. https://doi.org/10.1016/j.memsci.2018.12.010

Smart Construction: Case of '3-S' Prefab Technology for Sustainable Mass Housing

Mamata R. Singh and S. D. Naskar

Abstract India has witnessed rapid urban transformation in the recent decades, and the ongoing 'Housing for All' and 'Smart Cities' mission of the Central Government has added new thrust to search for alternate technologies. Considering huge housing shortage in EWS and LIG segments, India's construction industry needs to evolve and start embracing smart construction technologies for mass housing addressing sustainable and inclusive development. The use of industrialized manufacturing using prefab technologies for smart construction has today emerged as a preferred and promising alternative to the conventional in situ construction practices to ensure speed, safety, strength and sustainability. This paper deals with the case of a mass housing project for EWS and LIG category in Narela, Delhi, using an emerging technology branded as '3-S' prefab technology for smart construction. Needless to say, such prefab technology for smart construction would play a key role in meeting Government's crucial missions of 'housing for all' and 'smart cities'. Use of such modern smart construction technologies would thus strengthen Government's commitment towards inclusive and sustainable development.

Keywords Affordable housing · Prefab technology · Smart cities · Smart construction · Sustainable development

1 Introduction

As per the previous study, there was around 24.7 million house shortage in India on an average [1]. Most of the shortage (99%) was for the category of lower-income group (LIG) and economically weaker section (EWS) [2]. However, situation is no

M. R. Singh (✉)
Directorate of Training and Technical Education, Govt. of NCT of Delhi, Pitampura, Delhi, India

S. D. Naskar
M/s B. G. Shirke Construction Technology Pvt. Ltd., New Delhi, India
e-mail: sdnaskar@shirke.co.in

© Springer Nature Singapore Pte Ltd. 2020
S. Ahmed et al. (eds.), *Smart Cities—Opportunities and Challenges*,
Lecture Notes in Civil Engineering 58,
https://doi.org/10.1007/978-981-15-2545-2_26

different today. Considering huge shortage of affordable houses which is one of the basic necessities of human being, Government of India unveiled its ambitious plan of 'Housing for All' under the scheme Pradhan Mantri Awas Yojana (PMAY) launched on 1 June 2015. As per this mission, the target is by 2022, when independent India celebrates its 75th Independence Day, every Indian family should have a house of his/her own. Under this scheme, 2 crore affordable houses will be built in selected cities and towns using sustainable construction methods for the benefit of the urban poor population covering entire urban areas in the country [3]. For fulfilling this ambitious mission of Government, such massive requirement of housing particularly for EWS and LIG categories cannot be met with conventional outdated methods. Apart from being costly, these methods to a larger extent have outlived their technical and commercial usefulness since they on their own cannot meet massive housing demand and supply imbalance prevailing in the country.

Another recent notable initiative undertaken by the Government is 'Smart City Mission' which is an urban renewal and retrofitting programme by the Government of India to develop 100 citizen-friendly cities across the country from 2017 to 2022 that provide sustainable infrastructure and give a decent quality of life to its citizens with the application of 'Smart' solutions [4]. The focus is on sustainable and inclusive development, and the idea is to look at compact areas and create a replicable model which will act like a lighthouse to other aspiring cities. The key infrastructure elements of a smart city include affordable housing especially for the poor and sustainable environment.

Smart construction would serve as the smartest solution towards fulfilling the Government's missions of 'Housing for All' and 'Smart Cities'. Smart construction deals with building design, construction and operation that through collaborative partnerships make full use of digital technologies and industrialized manufacturing techniques to improve productivity, minimize whole life cost, improve sustainability and maximize user benefits [5]. This paper deals with the case study of smart construction through '3-S' industrialized manufacturing techniques of prefabrication for an affordable mass housing project in Narela, Delhi. This case study would serve as one of the best practice smart construction model for several upcoming smart city projects under consideration by the Government.

2   Case Study

Delhi Development Authority (DDA) has been at the forefront of providing affordable housing since long using prefab technology. In response to Prime Minister's ambitious mission of 'Housing for All', DDA invited the tender for the greenfield project consisting of construction of 24,660 LIG and 4855 EWS houses at Narela, Delhi, using prefab technology on 'Design Build Lump Sum Turnkey' basis. After due process of tendering, the project was awarded to M/s B. G. Shirke Construction Technology Pvt. Ltd. (BGSCTPL) for execution using their tried, tested and proven 3-S prefab technology.

2.1 Project Brief

The salient features of the project [6] are briefly indicated in Table 1.

LIG units and EWS units both have two rooms, kitchen, WC and bath and a sit out/balcony with designated areas confirming to Delhi building by-laws and IS: 8888. For LIG units, the building blocks are conceived generally in a G+11 to G +12 floors with varying number of modules of 8–10 dwelling units per floor, and then carefully combined to form a composite cluster court building with adequate number of lifts and staircases. The total building height (up to the roof slab) is approx. 39.0 m for LIG units. The plinth height varies from 450 to 900 mm depending on site slope, contours, etc. For EWS units, the building blocks are conceived with modules of four dwelling units per floor for each stair, and coupled to form building module of 40 and 60 dwelling units per block with five floors walk up format. The blocks are completed module wise and placed judiciously to form cluster.

Table 1 Project brief

S. No.	Titles	Description
1	Name of Client	DDA, Delhi
2	Name of Architect	BGSCTPL
3	Name of Construction Agency	BGSCTPL
4	Type of Tender	Lump sum turnkey contract
5	Type of Structure/No. of stories	Prefab RCC framed structure, G + 4 for EWS and G + 11 and G + 12 for LIG
6	Number of Pockets/Sites	12 nos.
7	Total Built-up Area	15,320,042.4 ft^2
8	Built-up Area of 1 Dwelling Unit	LIG 546.39 ft^2 EWS 380.24 ft^2
9	Defect Liability	3 years
10	Minimum Carpet Area of 1 Dwelling Unit	LIG 35 m^2 EWS 25 m^2
11	No. of Dwelling Units	LIG 24,660 EWS 4855
12	Cost per Dwelling Unit	LIG Rs. 878,299/- EWS Rs. 600,352/-
13	Period of Completion	36 months
14	Awarded Contract Value	2624.28 Cr.

2.2 Project Execution

The executing agency of the project M/s B. G. Shirke Construction Technology Pvt. Ltd. (BGSCTPL) is Integrated Management Systems (IMS) accredited organization, which covers ISO 9001:2008, ISO 14001:2004 and OHSAS 18001:2007 for quality, environment, health and safety, respectively. The project was executed by Shirke on lump sum turnkey basis including soil investigation, preparation of architectural, structural and services drawings and designs, taking permissions/approvals from statutory authorities on single-point responsibility as against outdated system of item rate or item rate percentage tenders. The conventional system of executing housing projects based on item rate or item rate percentage tenders has several inherent flaws; as before commencing, the work a number of formalities (site investigation, appointment of consultants, tender preparation including drawings and specifications, statutory permissions, etc.) are to be completed by the client department. Also, during execution, quantities and specifications vary, and many extra items crop up due to interpretation of drawings and designs provided by the client department. Therefore, the final cost is very much higher than the estimated/tendered price, and hence, the term 'Lowest' price is misleading. In conventional tenders, the product is non-existent, whereas with the prefab system, the products are with ISI/BIS marks and can be inspected for the quality with cost.

3 Smart Construction and Mechanization Adopted for Mass Housing

Considering the volume of work to be executed in 3 years, a state-of-the-art onsite factory was established for manufacturing precast components using '3-S' prefab technology for smart construction, a brand name developed and perfected by Shirke after years of strenuous research and development supplemented by extensive field trials for its effective implementation in construction of mass housing projects. The factory, which is biggest automatized prefab factory in Asia, is laid out over 25,200 m^2 area with storage/stacking area of 46,000 m^2 for precast components [6].

3.1 Industrialized Manufacturing for the Project

The industrialized manufacturing system branded as '3-S' (S-Strength, S-Safety, S-Speed and Sustainability) for smart construction is also recommended by Building Material and Technology Promotion Council (BMTPC) after extensive study of technical specifications and performance characteristics since it fulfils end-users' ultimate need of owning a dream house, which is strong, safe and available in short time at affordable price and is sustainable.

One of the most unique features of the project is the high degree of mechanization and automation achieved at par with European standards. Right from dispatch of concrete from the batching plant through an overhead concrete carrying capsule/shuttle which delivers concrete to the 'Comcaster' which in turn lays the concrete into the column/wall/slab/moulds which already have the required reinforcement steel in place. Cutting and bending machines are used for fabrication of reinforcement. Batching plant is used for mixing of concrete, while concrete pumps and transit mixers are used for transport and placing of concrete.

An automated and programmable reinforcement cage manufacturing unit (refer Figs. 1 and 2) [6] capable of handling 150 tons of steel per day has also been installed. The entire structural component manufacturing facility is supported by high-quality compaction and vibration stations, surface finishing equipments and high-capacity overhead cranes, lifting and tilting stations, etc. After concreting of the precast components in robotic moulds, components undergo an accelerated curing process in specially designed curing chambers which ensures that the concrete components achieve their design strength which in turn enables handling of the precast components within a short period of 8–12 h. Besides this, accelerated curing of beam is achieved through hot water pipes concealed in the beam moulds. This ensures de-shuttering of moulds in shortest possible time, thereby leading to optimum utilization of moulds. Further, the components are lifted, shifted and erected at site by totally mechanical means (refer Figs. 3 and 4). A total of 60 tower cranes and 2 mobile cranes accompanied by a large number of small cranes (hydra) were pressed in service only for erection of precast components. The overall production capacity [6] of this factory is as under:

Fig. 1 Reinforcement processing plant

Fig. 2 Stacking of reinforcement cages

Fig. 3 Erection work in progress

Fig. 4 Erection of precast components in progress

Fig. 5 Manufacturing of precast wall panels

(a) Precast columns = 132 nos/day
(b) Precast beams = 520 nos/day
(c) Precast wall panels = 250 nos/day (refer Fig. 5)
(d) Precast slabs = 463 nos/day (refer Fig. 6)
(e) Precast rakers = 9 nos/day
(f) Lift core units = 16 nos/day

Fig. 6 Manufacturing of
precast slabs

(g) Precast parapet walls = 35 nos/day

(h) Miscellaneous components, viz. precast panels, chajjas, etc. = 136 nos/day.

3.2 '3-S' Prefab Technology for Smart Construction

Under 3-S prefab technology [7] for smart construction, all the structural components are pre-engineered and manufactured in factories/site factories with objective
quality control resulting into dimensional accuracy, correctness in spacing of
reinforcement, uniform protective cover, full maturity of components and assurance
on design strength due to use of design mix concrete having minimal water–cement
ratio which ultimately results into durable structure. The constituent components are
standardized for mass manufacturing in factories and thus universalized in order to
bring them under the purview of specification of Bureau of Indian Standard
(BIS) and/or specialized manufacturers' norms. This 3-S system differs from other
prefab systems as this is an open componentized system that imparts total flexibility
to accommodate every possible architectural planning and aesthetic application for
entire range of dwellings, i.e. small one room to luxurious HIG tenements in
single-storey to high-rise framed structures with minimal modular constraints. After
adoption of modular planning by construction industry, 'off-the-shelf' prefab
structural components can also be made available under this system.

The 3-S prefab technology consists of in situ foundation with conventional
methods and superstructure frame in precast dense reinforced cement concrete
(RCC) consisting of hollow core columns, structural walls, beams, stairs, floor/roof
solid slabs, lintels, parapets and chajjas. AAC blocks are used for partition walls.

Fig. 7 Manufactured and ready to shift precast half-slabs with lattice girders

Hollow core columns are erected above cast-in situ substructure, over which beams are integrated in the column notches followed by erection of slabs. Reinforced screed/lattice girder is provided on slab to have monolithic construction (refer Fig. 7). Structural continuity and robustness are achieved through wet jointing using dowel bars/continuity reinforcement of appropriate diameter, length and configuration placed at connections and filling the in situ self-compacting concrete in hollow cores of columns, beam and slab tops in order to ensure monolithic, continuous, resilient, ductile and durable behaviour (refer Fig. 8). Precast auto-claved aerated cellular (AAC) reinforced Siporex slabs and Siporex blocks [8] are manufactured in permanent factory, and other structural components of super-structure like columns, beams, wall panels, slabs, staircases, lintels, chajjas, etc., are manufactured at precast factories established at site under stringent quality control. A view of completed buildings is shown in Fig. 9.

4   Appraisal of Smart Construction Adopted

The 3-S prefab technology for smart construction was also subjected to third-party appraisal by many reputed organizations with the help of field and laboratory tests. The outcome of such evaluation [9] is briefly as follows:

(a) City Industrial Development Corporation (CIDCO) has carried out load test to check the safety and stability of the structure by loading the structure and found that the structural behaviour was most satisfactory.
(b) Tests were carried out by Indian Institute of Technology, Mumbai (refer Figs. 10 and 11), and they have certified that the joints fully established the

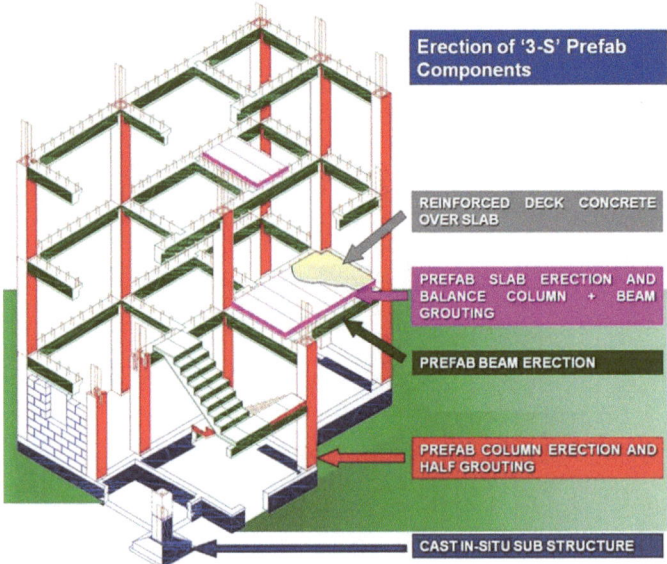

Fig. 8 Model depicting 3-S prefab system

Fig. 9 A view of completed buildings for LIG

Fig. 10 Performance evaluation test on full building by IIT, Mumbai

Fig. 11 Test on column beam joint by IIT, Mumbai

behaviour in the elastic range with adequate safety margins; absence of any separation cracks or any structural distress in the joints; adequacy of the bare portal to offer resistance to horizontal forces; ultimate load is on the higher side and ductility ratio is more than what is specified and required; joints of the beam column connections have behaved as monolithic as designed.

(c) TOR Steel Research Foundation of India has carried out the tests and concluded that there is no distress feature in any of the joints and assembly of precast units is safe for resisting the loads for which they are designed for.

(d) Tests were carried out by Civil Engg. Department of Stanford University, and it is certified that the design calculations and detailing of the structure are such that for vertical loads, seismic loads and wind loads, the buildings should provide safe and desired performance.

(e) Central Building Research Institute (CBRI), Roorkee, has also carried out tests and experimental results on full-scale building structure (refer Fig. 12) and established the desired performance and behaviour of '3-S' prefab technology for smart construction under all design load conditions including seismic (zone IV) for high-rise buildings. CBRI has also certified that protective treatment given to steel reinforcement in Siporex is quite effective compared to corrosion of steel in normal conventional concrete.

(f) Wind Tunnel Study of Tall building by Structural Engineering Research Centre, Chennai, also provided satisfactory results.

In addition to the third-party appraisal, the structure and design also confirms to the provisions of relevant Indian codes and standards [6] as follows:

(a) The structure conforms to the provisions of IS 1893 for earthquake resistant structural design for earthquake zone IV. Besides, the IS code provisions on progressive collapse stipulated vide IS 15916 and 15917 have also been incorporated in the structural design and detailing.

Fig. 12 Test on full-scale building by CBRI, Roorkee

(b) All structural floor elements are detailed for 1-h fire rating, and all vertical load bearing structural elements are detailed for 2-h fire rating as per relevant applicable provisions of IS:456 and IS:1642. Accordingly, member sizes and cover to reinforcement are maintained.
(c) Fire-fighting installations of the housing schemes are as per latest National Building Code (NBC) and CPWD specifications. The proposals obtained the Clearance from Delhi Fire Service (DFS).
(d) Provision of space and areas for LIG and EWS units confirm to Delhi building bye laws and IS: 8888.

5 Sustainable Benefits of Smart Construction Adopted

Conservation of national resources, maintaining ecological balance, inclusive and sustainable development in line with Government's missions is a national necessity. These were achieved in the present case study through [6, 9].

5.1 Saving in Construction Materials

(a) Reduction in dead weight of structural components with high strength-to-weight ratio and better performance under the seismic loads.
(b) Substantial reduction of concrete and steel and other materials per unit of built-up area due to use of prefab construction.
(c) Negligible wastage of materials
(d) Elimination of plaster due to form finished precast units such as slab and wall panel.

5.2 Use of Green Materials, Green Features and Resource Conservation

(a) Use of industrial waste like fly ash, GGBS in concrete.
(b) Total elimination of timber/wood for both shuttering and doors including frames due to use of reusable steel moulds and prefabricated GI door systems.
(c) Use of lightweight AAC blocks for interior walls offering thermal and acoustic insulation along with enhanced fire protection and also provides calmer and quieter interiors to the occupants.
(d) Rain water harvesting system comprising of recharge pits with grease trap chambers.

(e) Minimum air, water and noise pollution during construction due to industrialized manufacturing.
(f) Use of fully cured and matured components considerably reduces water consumption.
(g) Use of recycled water from STP for flushing.
(h) Use of Siporex which is non-toxic, environmentally safe material
(i) Negligible generation of construction debris.

5.3 Cost Benefits

(a) High speed of construction thereby resulting in early completion and occupancy with accompanying benefits of reduced financing cost, saving in interest on investment, early return on investment, saving in escalation and establishment cost.
(b) Due to use of precast structural members, cycle time required for each floor is reduced substantially resulting in cost-saving.
(c) Siporex (AAC) material having very good heat and sound insulation property, thereby resulting in considerable reduction in recurring cost of energy.
(d) Siporex material having very high fire resistance, thereby resulting in lesser fire insurance premium cost.
(e) Reduction in fixed and escalation cost due to timely completion of projects.
(f) Due to turnkey contract, saving in planning and design fee.
(g) Reduction in dead weight results in saving in foundation and framework cost.
(h) Saving in cost due to elimination of slab, wall panel plaster.
(i) Cost-saving in maintenance due to quality construction.
(j) Minimum dependence on skilled manpower and hence cost-saving.

5.4 Quality Benefits

(a) Quality is ensured automatically as structural units have BIS (ISI) marks and are manufactured in permanent/site factories.
(b) Objective quality control hence less or almost nil rework/rejection.
(c) Dense concrete avoiding porosity, due to pre-casting in horizontal position results in assured quality, thereby enhancing the durability.

6 Conclusion

There is huge shortage of affordable houses to the tune of 95% of total requirements [9] in the country particularly in LIG and EWS segments. Such massive requirements cannot be met with conventional construction methods. 3-S prefab technology for smart construction, a brand name developed and perfected by Shirke, has undergone the test of time in addition to rigorous tests and appraisals conducted at test houses of national and international repute. The 3-S system is thus, tried, tested and proven to suit the Indian climatic, seismic, rainfall and wind conditions. In addition to speed and several other sustainable benefits, there is cost-saving of about 30–40% if tangible and intangible financial benefits are to be quantified [9] due to technological advantages.

The uniqueness of the smart construction in the project is amply brought out right from inception itself by inviting lump sum turnkey tender, setting up a state-of-the-art onsite precast manufacturing unit, achieving full mechanization/automation, incorporating sustainable features and delivering on both parameters of quality and time. The 3-S (S-Strength, S-Safety, S-Speed and Sustainability) prefab technology for smart construction would thus go long way in mitigating the housing shortage through industrialization of civil engineering, and Shirke has already taken lead on this front as is evident through the case study. This case study has been able to exhibit several unique features and demonstrate global best practices by introducing cutting-edge technology which makes this project a model and trendsetter that can be replicated to fulfil the Government's missions of 'Smart City' and 'Housing for All'.

Use of prefab construction has reportedly helped in saving almost 60% of time in completion of projects as compared to the conventional construction methods [10]. To summarize, it can be said that the industrialized manufacturing system using prefab technology such as 3-S for smart construction is the best solution to meet huge demand supply gap particularly in LIG and EWS categories and to ensure sustainable and inclusive development. Such smart construction technology would also offer an economy of scale and hence reduced cost, greater speed and green construction [11].

Acknowledgements The authors wish to acknowledge the support extended by Shirke's top officials in providing necessary inputs for shaping this paper.

References[1]

1. National Urban Housing and Habitat Policy (2007) Ministry of Urban Employment and Poverty Alleviation, Government of India
2. Roy UK, Roy M, Saha S (2009) Energy optimization through open-industrialised building system in mass housing projects. J Indian Build Congr
3. Pradhan Mantri Awas Yojana (PMAY)—Housing for All (Urban), Ministry of Housing and Urban Affairs, Government of India. https://pmaymis.gov.in/
4. Smart Cities Mission, Ministry of Housing and Urban Affairs, Government of India. http://smartcities.gov.in/content/
5. Smart Construction—a guide for housing clients, Construction leadership council (2018). T02072156476E, construction.enquiries@beis.gov.uk, www.constructionleadershipcouncil.co.uk
6. *Mass affordable/social housing projects with precast technology-the way forward
7. *'3 S' prefab building technology
8. *Siporex; The wonder building material of the world. A profile from Siporex India Pvt. Ltd.
9. *Tried, tested and proven 3-S prefab technology: a solution for affordable mass housing in India
10. An article (2018) Is India's construction industry ready to build smart cities?, Deccan Chronicle, Published on 20 Nov 2018. https://www.deccanchronicle.com/technology/in-other-news/201118/is-indias-construction-industry-ready-to-build-smart-cities.html
11. Affordable Housing (2018) A whitepaper by CBRE and FICCI

---

[1]*Unpublished Report as obtained directly from B. G. Shirke Construction Technology Pvt. Ltd.

Spatio-temporal Analysis of Urban Air Quality: A Comprehensive Approach Toward Building a Smart City

Kanika Taneja, Shamshad Ahmad, Kafeel Ahmad and S. D. Attri

Abstract Observing and understanding atmospheric processes and impacts of air pollution in urban areas is essential for a sustainable and healthy city. Knowledge of temporal and spatial distribution of air quality at the landscape scale is essential for regional climate control and urban planning. For this purpose, spatial analysis of atmospheric aerosol concentrations and air pollution data by geo-statistical methods is carried out. The present study is focused on the megacity Delhi and its surrounding areas which are one of the most polluted urban regions in the world. The relationship between land-use structure and satellite retrieved AOD has been analyzed for the first time over Delhi, and the impacts of topography on the aerosol distribution have been discussed. It is observed that the major part of central and east Delhi showing built-up area depicted highest aerosol concentrations including high particulate matter (PM_{10} and $PM_{2.5}$), whereas the surrounding areas around Delhi mostly having agricultural farmlands show lower values of the same. Further, application of stochastic modeling technique in analyzing the future trends of aerosol optical properties was performed using the Box–Jenkins autoregressive integrated moving average (ARIMA) model. After rigorous evaluation, the ARIMA $(1, 0, 0) \times (0, 1, 2)_{12}$ was identified as the best predictive model for the time series under study, suggesting a simplistic modeling technique for determining the future values of AOD. This can provide a comprehensive approach that state and local authorities can use to deal with environmental protection issues.

Keywords Air quality · Geo-statistical · ARIMA · Delhi

K. Taneja (✉) · S. Ahmad · K. Ahmad
Jamia Millia Islamia, New Delhi 110025, India

S. D. Attri
India Meteorological Department, New Delhi, India

© Springer Nature Singapore Pte Ltd. 2020
S. Ahmed et al. (eds.), *Smart Cities—Opportunities and Challenges*,
Lecture Notes in Civil Engineering 58,
https://doi.org/10.1007/978-981-15-2545-2_27

1 Introduction

The pace of urbanization in the developing country like India is unprecedented. The effective and efficient management of rapid urbanization is the need of the hour. It has become essential for the urban cities to maintain their three key pillars, i.e., social, environmental and economic sustainability. According to the United Nations (UN) urban and rural population datasets [1], the migration from rural to urban areas has been growing seemingly over the past few years. A significant proportion of population lives in cities with growing infrastructure and industrial development. This has led to exceeding limits of air quality index for several air pollutants like particulate matter, ozone, nitrogen dioxide, etc. These pollutants pose a severe health and environmental impacts. In order to deal with these environmental issues, development of real-time air quality management system has become an essential component of any smart city. As defined by Giffinger et al. [2], a smart city is the one with six major components which include smart economy, smart transportation, smart environment, smart citizens, smart life and smart management. The most important and often neglected component of all is the environment.

Both economic and population growth in the urban areas have resulted in the use of large number of vehicles, industries and energy consumption. This has led to the unforeseen impact on the surrounding atmosphere. The urban air pollution is becoming pervasive with increasing incidents of foul air, smog and low visibility. In order to completely understand the air quality conditions, these pollutants must be accurately monitored at different locations in the city. For this purpose, various pollution sensors are established by city officials to measure and monitor the real-time pollutants level in the air at different locations to get good spatial coverage. Most of the smart cities have adopted a system of integrated network of sensors which is termed as Internet of Things (IoT) [3]. The IoT devices can be used to generate large amount of data by collecting and exchanging information. This big data can be further used to enhance air quality management systems and forecasting purposes [4]. The data collected from sensors can be utilized for data analytics, prediction and finally generating alert messages in case of severe air quality conditions. This form of observation of concentration trend of atmospheric pollutants would allow the city officials and urban planners to detect potential emergency situations and suggest appropriate mitigation measures to support urban development.

For a smart city initiative, it is essential to develop a system that not only monitors air quality but also provides pollution forecasts at a regional level. For establishing sensors at specific locations in the city, it is essential to know the spatial distribution of these air pollutants on the basis of the data collected from satellite and ground-based instruments.

The present study attempts to integrate the spatial distribution of major pollutants in the megacity of Delhi and surrounding regions with its topography using geo-statistical tools. This paper also addresses a unique time series modeling method to forecast the aerosols concentration in the atmosphere.

2   Site Description

The present study pertains to Delhi, an urban megacity in the heart of Indo-Gangetic Plain. It lies between latitude of 28.59° N and longitude of 77.22° E at 213 m above mean sea level. The region has a mixed land-use pattern with a population of over 16 million people as per Census of India, 2011. The city is one of the world's highly polluted cities, due to rapidly increasing number of vehicles, thermal power plants and an increasing number of industrial units [5]. Regulatory monitoring of air pollutants in the city is administered by Central Pollution Control Board (CPCB). The Continuous Ambient Air Quality Monitoring Stations (CAAQMS) produce air quality data that is regularly monitored and controlled by CPCB, Delhi Pollution Control Committee (DPCC) and India Meteorological Department (IMD) [6]. Figure 1 shows the study location with Delhi and adjoining parts of Uttar Pradesh and Haryana.

3   Data Resource

3.1   Satellite Data

For spatial distribution, the data on aerosol optical depth (AOD) at 550 nm wavelength was retrieved from Moderate Resolution Imaging Spectroradiometer (MODIS) instrument aboard the Terra satellite. The MODIS instrument measures radiance at 36 wavelengths, ranging from 0.4 to 14.5 mm with nadir on ground spatial resolutions between 250 and 1000 m [7]. It covers the Earth surface with a swath of 2330 km thereby daily, with equatorial crossing local time of 10:30 am.

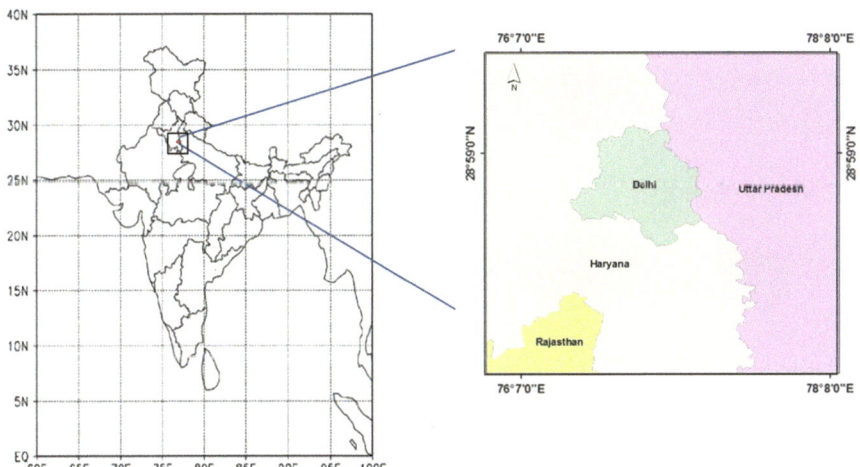

Fig. 1 Location map of the study area

The MODIS aerosol product was acquired from the NASA Goddard Earth Science Distributed Active Archive Center (DAAC) for the period of 2004–2016 (https:// ladsweb.modaps.eosdis.nasa.gov/). The retrieval of AOD at 550 nm was performed at 10×10 km^2 pixel size. Further details on AOD retrieval were referred to studies by Remer et al. and Levy et al. [8, 9]. These studies also reported the expected error or uncertainty over land to be $\pm$ (0.05 + 0.15 $\times$ AOD). Various validation studies have been conducted between MODIS aerosol products with ground-based instruments like that of AERONET and handheld MICROTOPS-II sun photometers over the Indo-Gangetic Basin [10, 11], rendering the satellite datasets suitable for further aerosol research.

3.2 LULC Data

For integration of aerosol data with the topography of the study area, the land-use/ land cover (LU/LC) map of the present study area was collected from the ISRO Bhuvan Web Services (http://bhuvan.nrsc.gov.in). The LULC mapping was collected on 1:250,000 scale using multi-temporal AWiFS datasets, portraying major land cover classes.

3.3 Ambient Air Quality Data

The other set of data on ambient air quality in Delhi NCR was collected from the 40 stations established in the city by CPCB, DPCC and IMD. Some of the air quality parameters measured at these stations are particulate matter (PM$_{2.5}$/PM$_{10}$), nitrogen dioxide (NO$_2$), sulfur dioxide (SO$_2$), ammonia, carbon monoxide (CO) and ozone (O$_3$), along with the daily meteorological data. The PM$_{2.5}$/PM$_{10}$ data from these stations has been considered for the present study.

4 Methodology

4.1 Spatial Variability of Aerosol Concentration W.R.T. LULC

The satellite data on AOD (550 nm) acquired from MODIS onboard Terra satellite was averaged on monthly and annual basis. The annual averaged spatial data on AOD was overlaid as contours on the LULC map of the study region using the ArcGIS software.

Further, the ground -based air quality data from 40 stations in the Delhi NCR was recorded for each day and averaged annually. The collected data of ambient air quality at observation sites was interpolated in GIS environment while importing the geographic locations of sampling points. The concentrations of $PM_{10}/PM_{2.5}$, NO_x were analyzed spatially after interpolation through inverse distance weightage (IDW) technique.

Based on the values of the study variables at observed locations, the IDW method predicts the values of these variables at unobserved locations, while spatially correlating the variables [12]. The interpolated values were then represented as contours and integrated with the LULC maps.

4.2 Time Series Analysis

Based on the objective of observing the past behavior of the time series to predict the future trend of the variable, the Box–Jenkins autoregressive integrated moving average (ARIMA) modeling method [13] is used in the present study. A simple ARIMA (p, d, q) model consists of p, d and q orders denoting autoregression, differencing and moving average, respectively. Considering the seasonality in time series of satellite retrieved AOD, a multiplicative ARIMA denoted as ARIMA $(p, d, q) \times (P, D, Q)s$ is used for the study, where P, D, Q represent seasonal autoregressive, differencing and moving average orders, respectively, and 's' denotes number of seasons. A seasonal ARIMA (p, d, q) $(P, D, Q)s$ model can be denoted by the following equation:

$$\phi_p(B)\,\Phi_P(B^s)\,(1-B)^d Y_t = \theta_q(B)\,\Theta_Q(B^s)\,\varepsilon_t \qquad (1)$$

where B denotes the backshift operator, ε_t is the error variable, Φ and ϕ are the seasonal and nonseasonal autoregressive parameters; Θ and θ are the seasonal and nonseasonal moving average parameters, respectively. The ARIMA methodology consists of a four-step iterative process which includes identification of a stationary time series, estimation of parameters followed by a diagnostic check and finally forecasting of future values. The detailed procedure of ARIMA modeling technique is mentioned in Taneja et al. [5].

5 Results and Discussion

5.1 Spatial Variation in AOD

The AOD (550 nm) data retrieved from MODIS was averaged for each month for 2011–2016. Figure 2 depicts the spatial distribution of monthly averaged AOD

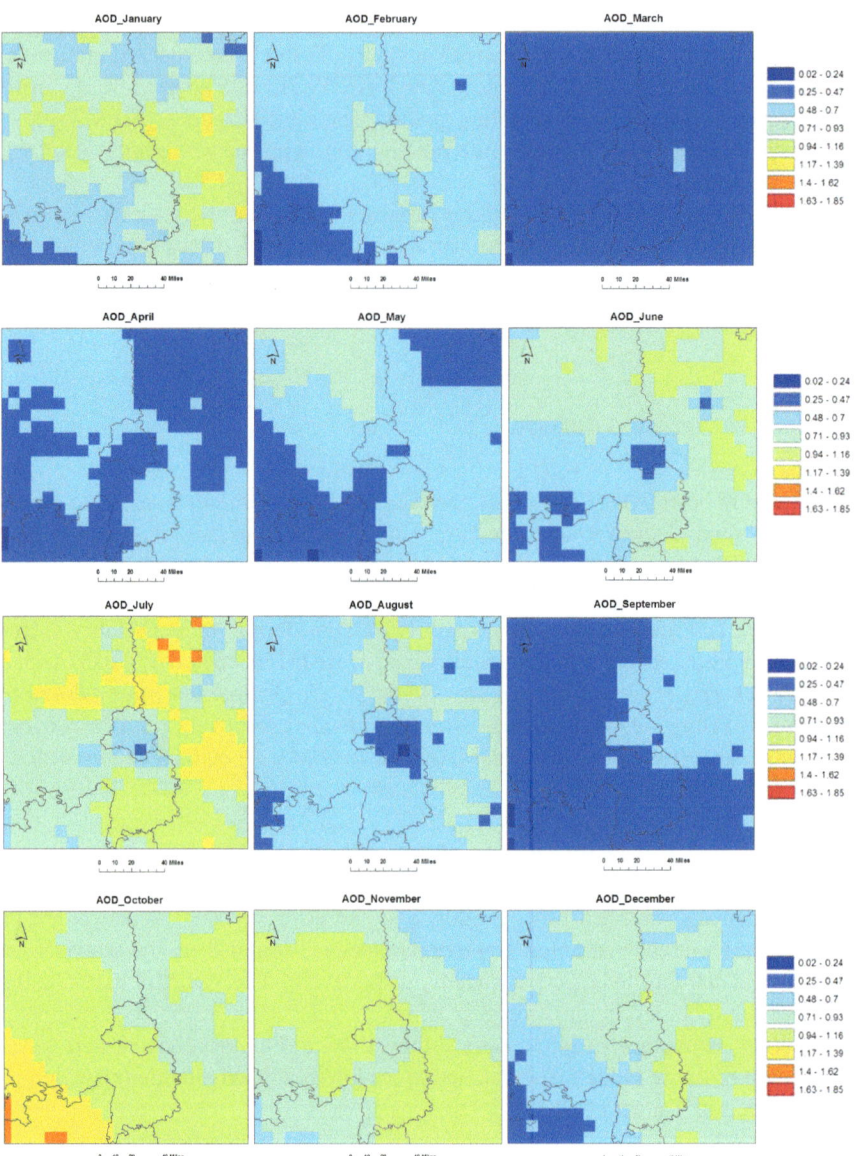

Fig. 2 Monthly spatial variation in AOD over Delhi NCR

over the study region. In winter months, especially January, AOD shows high concentration (0.7–1.1) in Delhi. The North, central and east part of Delhi are observed to have higher AOD (0.94–1.16). The region beyond Delhi towards U. P. shows higher AOD. The minimum temperature in this month of the year falls below 10 °C. This region is mostly inhabited by small villages where people burn leaves and wood stocks during this season, leading to higher aerosol content in the local atmosphere. Additionally, this region is covered with dense fog during this season. In February, AOD is reduced to a range of 0.5–0.9 almost throughout Delhi, followed by March month that is characterized with the offset of winters when the temperatures start rising. The fog is scattered and thus, AOD falls below 0.25 throughout Delhi and surroundings.

The pre-monsoon season (April, May, June) is characterized by dust storms called 'Aandhi' coming from Northern Pakistan, Afghanistan and Peninsular regions. The AOD starts to increase in April (>0.5) as the dust particles approach from northwestern regions. May is characterized by higher AOD (>0.7) with dust reaching Delhi from the northwestern part. The wind speed is higher in this season resulting in lifting of loose and dry soil particles from the barren land in the surrounding regions of Delhi. Monsoon hits Delhi in the end of June or the first week of July. AOD tends to be more during this month due to hygroscopic growth of aerosols due to increase in relative humidity. However, AOD reduces to around 0.5–0.7 in August and September, indicating the withdrawal of rainfall.

The post-monsoon months (September, October, November) are characterized by higher AOD again (>0.7) in and around Delhi. This is attributed to biomass burning activities in the northwestern direction from Delhi in Punjab and Haryana. The kharif crops after post-harvesting are burnt by farmers to get their agricultural land ready for the next crop. Also, during October, November months, festive activities of Dussehra and Diwali occur which lead to increase in aerosol load in and around Delhi. December marks the onset of winters in North India with temperatures falling to 10 °C. During this month, AOD is between 0.7 and 1.0 throughout Delhi.

5.2 Aerosol Concentration W.R.T. LU/LC

The LU/LC map of Delhi and surrounding area (1:250,000 scale) of 2015–16 was acquired from the Web Map Service offered by Bhuvan-Indian Geo-Platform of ISRO. The AOD data is displayed as raster and then converted to contour plots. The resulting product is overlaid on the LU/LC map of the same rectangular coordinates. It is evident from the resulting image (Fig. 3) that most parts of Delhi (central, south and east Delhi) are covered in red representing the built-up area that covers both residential and industrial activities. These areas show correspondingly high AOD contour (>0.7). On the contrary, the area surrounding Delhi in Haryana and U.P. show comparatively less AOD than the core region of Delhi, as they are characterized by agricultural farm lands with kharif/rabi crops and mostly plantations. This indicates that the smoke emitted from industrial developments and dust from the construction sites are responsible for high concentration of AOD in these regions. However, the agricultural regions on the outskirts of the city depict lower AOD (<0.7) due to the presence of more plantations and lesser concrete areas.

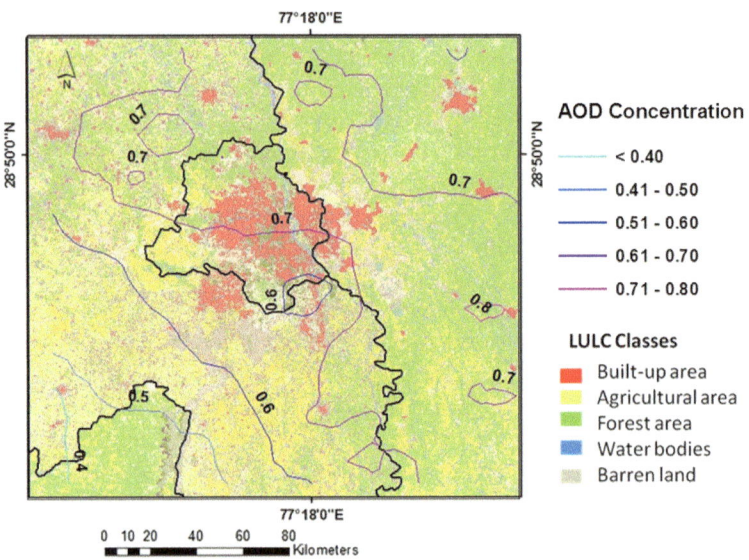

Fig. 3 Spatial relationship between AOD concentration and LU/LC over Delhi and surroundings

Due to the use of satellite pixel data, the spatial variation of AOD cannot be adequately estimated at regional level. In order to have a clear understanding of the relationship of aerosol characteristics with LULC, ground based data on PM_{10} and $PM_{2.5}$ were utilized for the study.

The CPCB data for PM_{10} and $PM_{2.5}$ collected from 40 sites in Delhi NCR was integrated as contours with LULC map of the study region (Fig. 4). It was observed that areas in East Delhi including Anand Vihar and Ghaziabad are highly polluted with the PM_{10} levels in severe and emergency category. This can be attributed to various reasons like the presence of railways and bus station, waste dumping ground in Gazipur, entry-exit point for many interstate vehicles and construction activities going on in this part of the city. The highest PM_{10} (>300 μgm^{-3}) and corresponding $PM_{2.5}$ (>190 μgm^{-3}) values were observed in North Delhi region around Narela, Alipur, Bawana, Mundka, Jahangirpuri. The high concentration of $PM_{10}/PM_{2.5}$ can be due to the smoke released from the small and large-scale industries present in these areas. This corresponds to the high AOD concentration in North and East Delhi, justifying the inter-relationship between aerosol concentration and topography of the study area.

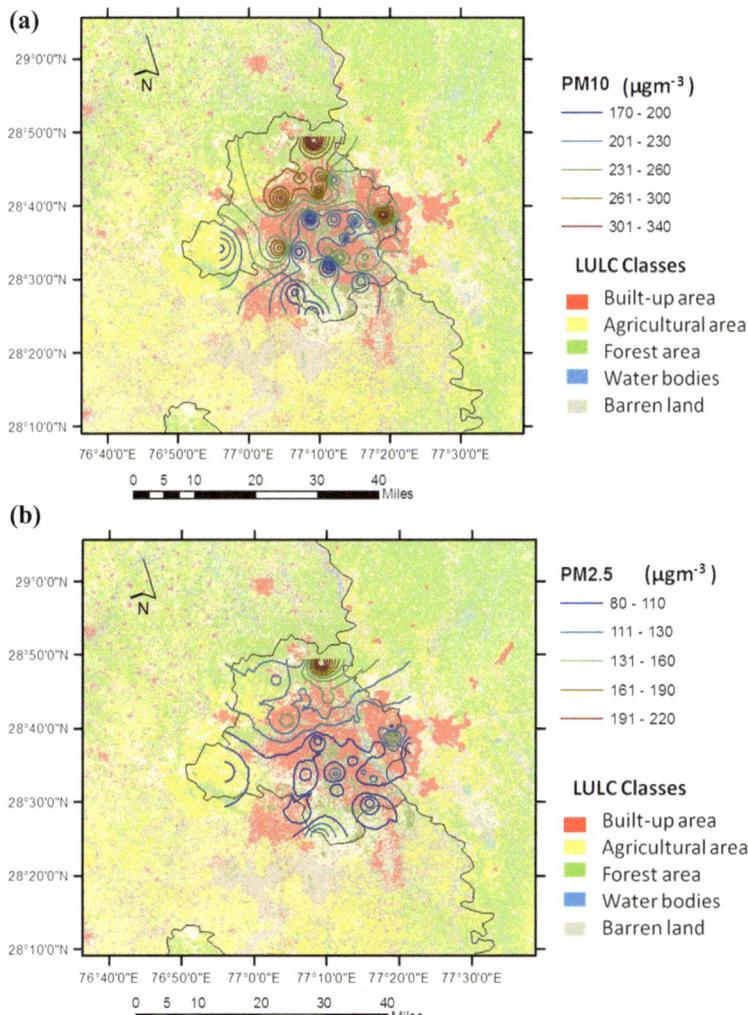

Fig. 4 Spatial relationship between **a** PM$_{10}$ and **b** PM$_{2.5}$ concentrations and LU/LC over Delhi NCR

5.3 Time Series Analysis and Forecasting

On the basis of the past trends in the MODIS AOD data collected for 2004–2014, the future trends in the time series could be identified using ARIMA modeling method. The iterative procedure of model selection, estimation of parameters, diagnostic checking and forecasting is detailed in Taneja et al. [5]. On trying several combinations for the model parameters, few models were tentatively identified to simulate the AOD time series. The maximum likelihood estimates of the parameters

Table 1 Model statistics of ARIMA $(1, 0, 0) \times (0, 1, 2)_{12}$ model

Fit statistic	Mean
Stationary R-squared	0.533
R-squared	0.674
RMSE	0.128
MAPE	16.905
MaxAPE	92.497
MAE	0.095
MaxAE	0.345
Normalized BIC	−3.941

for the identified models were determined based on stationary R-square; R-square; root mean squared error (RMSE); mean absolute error (MAE); mean absolute percentage error (MAPE) and normalized Bayesian information criteria (BIC).

A model with the lowest value of normalized BIC was considered as the best fit model that could be further used to generate the forecasts. Normalized BIC is a score based upon the mean square error and includes a penalty for the number of parameters in the model and the length of the series. Out of all the tentatively selected models, ARIMA $(1, 0, 0) \times (0, 1, 2)_{12}$ was identified as the best fit model with lowest normalized BIC value of −3.941. Table 1 shows the model statistics of the selected ARIMA model.

The best fit ARIMA model was used further to compute the future values of AOD till the year 2018 with a 95% level of confidence. The accuracy of the forecasted values for 12 months of 2014 has been already reported in Taneja et al. [5]. The residuals between predicted and actual values were well within ±0.1 limit for the whole year, thereby rendering the model accurate for further forecast. In this paper, the same model has been utilized to predict extended forecast values till 2018. Figure 5 depicts the time series plot of AOD with satellite observed values

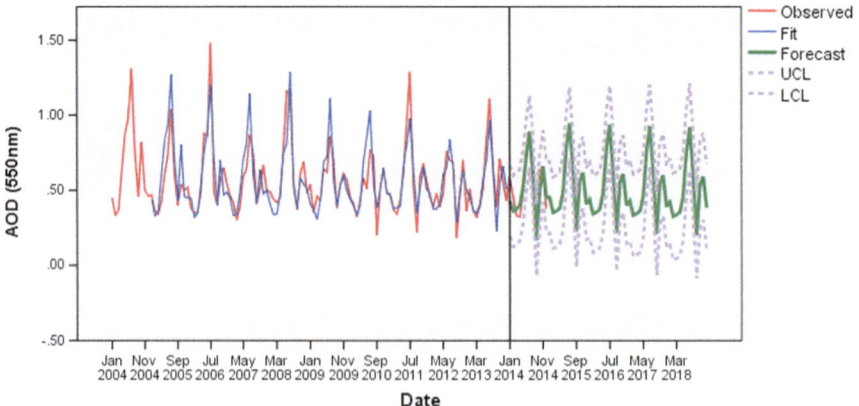

Fig. 5 Time series plot showing the observed and modeled output of AOD for the best fit model

and modeled values along with the forecasted values within the 95% confidence limit. The forecast values till 2018 are shown at the 95% confidence level (green line).

6 Summary and Conclusion

In order to have a comprehensive approach toward building a smart city, the air quality of Delhi and surrounding areas was monitored and forecasted. The spatial and temporal variations in satellite retrieved AOD along with ground-based observations of PM_{10} and $PM_{2.5}$ and LU/LC were examined, and their relationships were studied. The spatial distribution of AOD over the study region exhibits significant spatial clustering of higher concentrations of AOD and corresponding PM_{10} and $PM_{2.5}$ in the built-up area in the city. On the contrary, lowest concentrations were observed over the surrounding agricultural regions during 2011–2016. The analysis also revealed that the aerosol concentration has increased over the years owing to LU/LC changes in the regions where there are more anthropogenic activities, as compared to lower concentrations in areas with dense forests and less LU/LC changes. This impact of topography on the aerosol distribution can be further used by the state and local authorities for urban planning and better air quality management.

The study also presents the stochastic behavior of satellite AOD data over Delhi. Based on the features of AOD time series, an ARIMA model has been identified and evaluated to simulate and reproduce the monthly AOD till 2018. Based on the evaluation criteria of Box–Jenkins theory, ARIMA $(1, 0, 0) \times (0, 1, 2)_{12}$ model was identified as best fit model for forecasting of future AOD values. It is observed from the study results that few aspects of the past trends in time series continue to influence the future values of AOD. Thus the geo-statistical tools integrated with time series modeling techniques prove to be essential for air quality management. This can provide a comprehensive approach that the state and local authorities can use to deal with environmental protection issues.

Acknowledgements MODIS data were obtained from the Level 1 and Atmosphere Archive and Distribution System (LAADS) at Goddard Space Flight Center (GSFC), (http://ladsweb.nascom. nasa.gov/data/). The authors acknowledge the CPCB online data system (https://app.cpcbccr.com/) for providing ambient air quality data. The LU/LC map from the Web Map Service offered by Bhuvan-Indian Geo-Platform of ISRO is also acknowledged.

References

1. United Nations: world population prospects: the 2012 revision, highlights and advance tables (2013)
2. Giffinger R (2007) Smart cities ranking of European medium-sized cities. Centre of Regional Science, Vienna UT
3. Atzori L, Iera A, Morabito G (2010) The Internet of Things: a survey. Comput Netw 54 (15):2787–2805. https://doi.org/10.1016/ j.comnet.2010.05.010
4. Mahajan S, Chen LJ, Tsai TC Short-term PM2.5 forecasting using exponential smoothing method: a comparative analysis. Sensors 18(10):3223. https://doi.org/10.3390/s18103223
5. Taneja K, Ahmad S, Ahmad K, Attri SD (2016) Time series analysis of aerosol optical depth over New Delhi using Box–Jenkins ARIMA modeling approach. Atmos Pollut Res 7(4):585–596. https://doi.org/10.1016/j.apr.2016.02.004
6. Central Pollution Control Board: Ambient Air Quality Data of Delhi-NCR (2018). http://www.cpcb.nic.in
7. de Meij A, Lelieveld J (2011) Evaluating aerosol optical properties observed by ground-based and satellite remote sensing over the Mediterranean and the Middle East in 2006. Atmos Res 99(3): 415–433. https://doi.org/10.1016/j.atmosres.2010.11.005
8. Remer LA, Kaufman YJ, Tanré D (2005) The MODIS aerosol algorithm, products, and validation. J Atmos Sci 62(4):947–973. https://doi.org/10.1175/JAS3385.1
9. Levy RC, Remer LA, Dubovik O (2007) Global aerosol optical properties and application to Moderate Resolution Imaging Spectroradiometer aerosol retrieval over land. J Geophys Res Atmos 112:D13210. https://doi.org/10.1029/2006JD007815
10. Prasad AK, Singh RP (2007) Changes in aerosol parameters during major dust storm events (2001–2005) over the Indo-Gangetic Plains using AERONET and MODIS data. J Geophys Res Atmos 112:D09208. https://doi.org/10.1029/2006JD007778
11. Tripathi SN, Dey S, Chandel A (2005) Comparison of MODIS and AERONET derived aerosol optical depth over the Ganga Basin, India. Ann Geophys 23(4):1093–1101. https://doi.org/10.5194/angeo-23-1093-2005
12. Kumar A, Krishna AP (2018) Assessment of groundwater potential zones in coal mining impacted hard-rock terrain of India by integrating geospatial and analytic hierarchy process (AHP) approach. Geocarto Int 33:105–129. https://doi.org/10.1080/10106049.2016.1232314
13. Box GEP, Jenkins GM, Reinsel GC (eds) (1994) Time series analysis: forecasting and control. Prentice Hall, New Jersey

Utilities and Services in Smart Cities: A Case Study of Jaipur City

Anuja Sharma, Gautami Tyagi, Geeta Saha and Kakoli Talukdar

Abstract Smart City Mission is an intervention to enhance the urban livelihood with various steps taken by the collaboration of government and special purpose vehicles towards its implementation. The goal is to study the policies and their implementation in the development of Jaipur under this mission. The objective of this study is to observe and analyze the status and effects of the convergence of various bodies in the implementation of the various projects under Smart City Mission and critically analyze the status progress and concerns of the groundwork of projects in the study area. This paper covers a field trip analysis of the Jaipur walled city within the limits of the stretch Surajpole to Chandpole where the ongoing projects are observed and their immediate interactions are analyzed under categories of mobility, infrastructure, and heritage. We used a qualitative interviewing approach with the project heads of the various implementation bodies to understand the challenges faced and their vision towards the end of this mega project. The study shows that implementing "advanced and innovative solutions" while being unable to benefit people on all spheres of life is not a smart solution towards what the mission stands for.

Keywords Smart City · Policies · Jaipur · ICT (information and communication technology) · SPV (special purpose vehicle)

1 Introduction

Smart City interventions in Jaipur can be seen in the cityscape today when you go around the walled city where construction zones are being created due to the ongoing development projects under the Smart City Mission. The importance here is given to the redevelopment, retrofitting and majorly through information technology interventions such as online apps and modern surveillance facilities throughout the city.

A. Sharma (✉) · G. Tyagi · G. Saha · K. Talukdar
Faculty of Architecture and Ekistics, Jamia Millia Islamia, New Delhi 110025, India

© Springer Nature Singapore Pte Ltd. 2020
S. Ahmed et al. (eds.), *Smart Cities—Opportunities and Challenges*,
Lecture Notes in Civil Engineering 58,
https://doi.org/10.1007/978-981-15-2545-2_28

329

After the independence of India in 1947, Jaipur was made the capital city for the state of Rajasthan. The civil authorities and democratically elected candidates run the state. Millions of domestic and international tourists came to relive its royal past.

1.1 Policy Framework of Rajasthan

The Government of Rajasthan is a signatory to the Urban Reform Incentive Fund (URIF) scheme of central government. The policy document attempts to address the urban sector Issues (such as land, planning, finance).

The policy is divided into the following two groups:

Infrastructure and Services

Urban Transport, Water and Sanitation, Solid Waste Management, Housing, and Drainage.

City competitiveness

Economy and Investments and Heritage and Culture.

1.2 Agencies Responsible for Planning, Implementation, Operation, and Maintenance in Different Sectors

See Fig. 1.

1.3 Timeline of Jaipur Policies Development

See Fig. 2.

1.4 Analysis of Building Bye-Laws for Walled City of Jaipur, 1970

The walled city is still facing the risk of losing its heritage to prohibited structures and encroachment. Illegal construction is flourishing and residents are not following the permissible height. Moreover, the state government is not keen on making any strict guidelines. The amendments proposed in the building bye-laws to preserve and maintain the rich heritage of the city are stinging dust till date. There is a need to reexamine the existing building bye-laws and framework to authenticate the ongoing process without negotiating on conserving heritage and livability.

Sector	Planning	Implementation	Operation and Maintenance
Landuse / Master Plan/ Building Byelaws	JDA	JDA, JNN	JDA, JNN
Water Supply	JDA, PHED, JNN, RUIDP, RHB, PriDev	JDA, PHED, JNN, RUIDP, RHB, PriDev	JDA, PHED, JNN, RHB, PriDev
Sewerage	JDA, PHED, JNN, RUIDP, RHB, PriDev	JDA, PHED, JNN, RUIDP, RHB, PriDev	JDA, PHED, JNN, RHB, PriDev
Roads/ Bridges/ flyovers/ RoB/Multilevel Parking	JDA, JNN, RHB, PWD, RUIDP, NHAI	JDA, JNN, RHB, PWD, RUIDP, NHAI	JDA, JNN, RHB, PWD, NHAI
Traffic Control and Management Systems	JDA (JTB), JP, RTO	JDA (JTB), JP, RTO	JDA (JTB), JP, RTO
City Public Transportation	RSRTC, PriOper	RSRTC, PriOper	RSRTC, PriOper
Street Lighting	JDA, JNN, RHB, PWD, RUIDP	JDA, JNN, RHB, PWD, RUIDP	JDA, JNN, RHB, PWD
Storm Water Drainage	JDA, JNN, RHB, ID, RUIDP	JDA, JNN, RHB, ID, RUIDP	JDA, JNN, RHB, ID,
Solid Waste Management	JNN, RHB	JNN, RHB	JNN, RHB
Parks / Playground/ golf course/ beautification of road intersections/ urban forest	JDA, JNN, ID, FD	JDA, JNN, ID, FD, PDCOR	JDA, JNN, ID, FD
Air, water and noise pollution Control	JNN, RSPCB, PHED,	JNN, RSPCB, PHED,	JNN, RSPCB, PHED,
Slum Development	JNN, JDA	JNN, JDA	JNN, JDA
Urban Poverty Programme	JNN	JNN	JNN
Housing for EWS	JNN, JDA, RHB	JNN, JDA, RHB	JNN, JDA, RHB
Public Conveyance	JDA, JNN	JDA, JNN	JDA, JNN
Heritage Building Conservation	JDA, JNN, RUIDP, AD, PHED, DD, PriOwn	JDA, JNN, RUIDP, AD, PHED, DD, PriOwn, SpSo	JDA, JNN, AD, PHED, DD, PriOwn, SpSo

Fig. 1 Agencies responsible for planning, implementation, operation, and maintenance in different sectors. Source : Line Department Survey, LASA, 2006

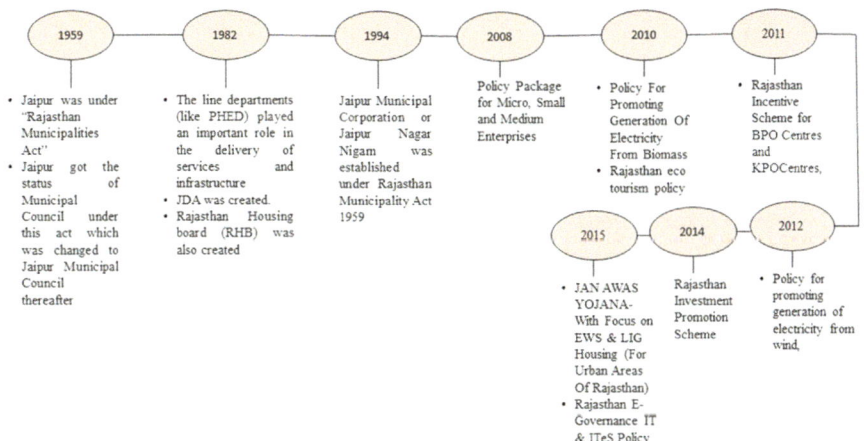

Fig. 2 Timeline of Jaipur policies development. Source : Author

Therefore, suggested bye-laws should be comprehensive, inclusive, participatory, and controlling in approach rather than mere prohibitory.

2 Smart City Mission in Jaipur

Smart City Mission is the one body that will follow the guidelines of all the major policies for infrastructure, mobility, tourism, and livelihood development. The Jaipur SPV is known as Jaipur Smart City Ltd. that was formed on 12th March 2016.

The vision for Jaipur as a Smart City has emerged from exchanged views of people from all walks of life and prepared city documents, like City Development Plan, Master Development Plan of Jaipur city, and self-assessment surveys relating to its city profile, etc.

Convergence of Policies within Smart City Mission

- National urban livelihood mission
- Atal Mission for Rejuvenation and Urban Transformation (AMRUT)
- Jawaharlal Nehru National Urban Renewal Mission (JNNURM)
- National Tourism Policy
- National Heritage City Development and Augmentation Yojana (HRIDAY)
- Digital India Programme.

The following Smart City projects are analyzed within the limits of the stretch Surajpole to Chandpole under the categories of Heritage and Tourism, Mobility, and Infrastructure:

- Smart Heritage and Tourism Precinct: Adaptive Reuse of Heritage Structure, Façade Improvement of Bazaar.
- Smart Mobility: Public Bicycle Sharing.
- Smart and Sustainable Civic Infrastructure: Smart Classroom, Smart Toilet, Smart Road, Multilevel Car Parking, Mobile Application, Video Surveillance, Interactive Kiosk, Environmental Sensor, Smart Light.

2.1 Smart Heritage and Tourism Precinct: Adaptive Reuse of Heritage Structure

Rajasthan School of Arts, a Building with rich historical value, is located in a unique context on the busy Kishanpol Bazaar near Ajmeri Gate. The museum building was originally the Residence of Pandit Shivdeen. In 1857, donated to be converted into the Madarsa-e-Hunari or Institute of Arts under the patronage of the king. In 1886, the name was changed to Maharaja School of Arts and Crafts. In 1988, it was set up and succeeded as Rajasthan School of Arts. In 2016, School was shifted to a new campus.

The project envisages a restoration of the complete building including damaged wall, ceiling floors, redoing lime plaster on damaged portions and finally giving a

Table 1 SWOT analysis of adaptive reuse of Rajasthan School of Art

Strength	Weakness	Opportunity	Threat
Physical restoration Heritage awareness	Previous identity of the structure as an art school is completely lost due to restructuring of the building's utility	Retrofitting Revival of old craftsmanship	Loss of authenticity

Source : Author

finishing touch over the surface with Khamira. Work also involves restoration of architectural elements like stone columns, arches, brackets, parapets, jalies, stone flooring, and khat work on ornamental plaster surface (Table 1).

2.2 Smart Heritage and Tourism Precinct: Façade Improvement of Bazaar

The major work planned in this project is restoring the façade which has gone haywire due to a lot of unplanned development in the walled city area. The main work which is being done in the façade improvement is scrapping, repairing doors and windows, stucco painting around door and windows, araish work, etc. These are age-old techniques that were used while the Jaipur city was laid down.

The façade improvement initiative is executed on the ground in the stretch from Chand pole to Kishan pole/Bari chaopar but there is no work done beyond the base plasterwork after Bari Chopar till Suraj pole.

The façade improvement initiative takes care of the façade treatment and also the aesthetic view of the street which involves the above-mentioned components. There is no control over encroachment. Verandah spaces are still filled with the outdoor units of AC's and toilet structures with overhanging electric wires in a haphazard manner (Table 2).

2.3 Smart Mobility: Public Bicycle Sharing

National Public Bicycle Scheme (NPBS) was introduced by the Ministry of Urban Development (MOUD) in December 2012. As part of the NPBS, the toolkit was prepared for MOUD by a team from the Institute for Transportation and Development Policy (ITDP) (Table 3).

Table 2 SWOT analysis of facade improvement of bazaar

Strength	Weakness	Opportunity	Threat
Physical restoration Heritage awareness	Time-consuming Creates hindrance in the daily routine of the public	Enhanced tourism Revival of old craftsmanship and construction technique	Loss of authenticity

Source : Author

Table 3 SWOT analysis of public bicycle sharing

Strength	Weakness	Opportunity	Threat
An economical and convenient mode of transport Reduces pollution level	Cycling to public transport is unsafe, Consumes time and effort For efficiency, there is no supporting infrastructure yet on all the streets	Promotes physical activity within the community, leading to improved health of overall community Flexible to be disassembled and reassembled to another location	Negligence of traffic rules can result in accidents May be difficult to shift people's commuting patterns

Source : Author

2.4 Smart and Sustainable Civic Infrastructure: Smart Classroom

Jaipur Smart City Limited is striving to enhance the learning outcomes of school children using diverse techniques. One of the techniques is to empower teachers by training them to use multimedia content and interactive techniques.

Jaipur Smart City Limited has set up smart classrooms in the selected Government Institutions of the ABD area of Jaipur including the supply of equipment, installation, training, and maintenance of the complete smart classroom system.

There is only one classroom having whiteboard facilities in Maharaja Girls High School. There were no proper seating arrangements in the classroom and no computer system in the computer lab. More focus was given on the façade improvement of the building rather than facilities to enhance education (Table 4).

2.5 Smart and Sustainable Civic Infrastructure: Smart Toilet

Smart toilets are purposed for the conveyance for the tourist and the daily commuters of the market area (Table 5).

2.6 Smart and Sustainable Civic Infrastructure: Smart Road

Under this project, the designated parking space on the side of roads, E- rickshaw lanes, footpath for walking on the side of the street, and demarcated vending zones would create extra space so that the movement of traffic is smooth (Table 6).

Table 4 SWOT analysis of smart classroom

Strength	Weakness	Opportunity	Threat
Enhance the interest of student in learning using digital techniques	Insufficient effort upon implementation	Improve quality of teaching and learning	The intent of the project will be lost if not executed as per necessary

Source : Author

Table 5 SWOT analysis of smart toilet

Strength	Weakness	Opportunity	Threat
Improve urban environment and station status	Lack of public awareness	Maximum coverage of sanitation facilities	Waiting in a queue create crowding, inconvenience

Source : Author

Table 6 SWOT analysis of smart road

Strength	Weakness	Opportunity	Threat
Lively and comfortable spaces in the bazaar for better active and passive interaction with the surrounding Reduction in commuting time Non-motorized transport reduce pollution	During the construction phase, there is always traffic congestion		

Source : Author

2.7 Smart and Sustainable Civic Infrastructure: Multilevel Car Parking

The Multilevel car parking (MLCP) project is undertaken by Jaipur Smart City Ltd. to ease out the congestion in the marketplace. The traffic has increased manifold and the walled city area which is a retrofit area has the limitation of expanding hence the idea of multilevel car parking has been planned for this area.

The project was started during mid of 2017 and proposed to be completed within 2 years. But currently, as seen in pictures, excavation work is in the process and it can be clearly visible that the process is surely going to take extra time as compared to the proposed area (Table 7).

Table 7 SWOT analysis of multi-level car parking

Strength	Weakness	Opportunity	Threat
Reduced congestion in commercial area safe and comfortable mode of transportation Greater sense of security	There is a greater construction cost per space It is not recommended for high peak hour volume facilities	Better business opportunities in the commercial plaza as the facility will attract more consumers Reduction in vehicular emissions	Degradation of air quality during construction

Source : Author

2.8 Smart and Sustainable Civic Infrastructure: Mobile Application

This project is introduced to highlight all the problems related to oneself using Government introduced app. As mobile applications enable the citizens to report the issues and grievances related to the streets in the city and other issues related to the government departments. All the reported issues are accounted for and the one gets assured support from the concerned department.

The use of ICT deviates from the past and municipalities lack the requisite skills and knowledge to adopt the technology.

1. Lack of awareness in local people about the use of this mobile application.
2. Many features of application do not work properly, e.g., booking bus tickets, making payments, selecting parking lots, etc.
3. The service quality and interface of the app are poorly maintained and crashes frequently.

MOBILE APPLICATIONS		ABOUT THE APP	ANALYSIS
JAIPUR SAMADHAN APP		THIS APP IS UNDER JAIPUR NAGAR NIGAM TO REACH OUT DIRECTLY TO THE CITIZENS. USING THIS APP THE CITIZENS OF JAIPUR CAN REGISTER COMPLAINS AND PROBLEMS RELATED TO JAIPUR NAGAR NIGAM. THEY CAN ALSO CHECK THE LIVE STATUS OF COMPLAINS POSTED. USERS CAN ALSO VIEW COMPLETE HISTORY OF COMPLAINS DONE BY THEM TILL DATE.	THIS APP IS NOT USER- FRIENDLY. IT IS ONLY MADE FOR ANDROID PHONES ONLY. INTERNET ERROR CAMES OFTENLY
'NUEVAS GPS'		TO TRACK HOOPERS IN EVERY WARD OF THE CITY. LIVE TRACKING OF THE GARBAGE COLLECTION VEHICLE CAN BE CONDUCTED THROUGH THIS APP AND ON THE OFFICIAL WEBSITE. PEOPLE CAN ESTIMATE WHAT TIME THE HOOPER WILL BE AT THEIR DOORSTEP FOR COLLECTING GARBAGE.	THERE IS NO AWARENESS OF THIS APP AMONG PEOPLE. GOVT. IS NOT TAKING ANY INITIATIVES TO MAKE PEOPLE AWARE OF THIS APP AND NO FOLLOW UP OF GARBAGE PICKING IN SEVERAL LOCALITIES
RAJASTHAN SAMPARK		RAJASTHAN SAMPARK IS AN INTEGRATED PLATFORM FOR READRESSAL OF GRIEVANCES RELATED TO DELIVERY OF GOVERNMENT SERVICES.	THIS APP CAN ONLY BE USED IN ANDROID PHONES. THE GOVERNMENT DOES NOT RESPOND TO PROBLEMS QUIKLY.
BHAMASHAH WALLET		THE GOVERNMENT AIMS TO REVOLUTIONIZE DIGITAL PAYMENTS BY OFFERING A UNIQUE, SECURE AND USER-FRIENDLY ONLINE PAYMENT PLATFORM.	NOT WORKING
RAJ MAIL		ITS FREE EMAIL SERVICE FOR EVERY CITIZEN OF RAJASTHAN BY RAJASTHAN STATE GOVERNMENT	THIS APP ONLY WORK FOR ANDROID PHONES. IT IS EASILY WORKABLE ON DESKTOP THAN MOBILE PHONES.
E-SAKHI APP		AN INNOVATIVE DIGITAL LITERACY PROGRAM INITIATED BY DOITC, RAJASTHAN TO ENHANCE GIRL POWER OF AGE 18-35 YEARS.	
BHAMASHAH YOJNA		THE OBJECTIVE OF THE BHAMASHAH YOJANA SCHEME IS FINANCIAL INCLUSION, WOMEN EMPOWERMENT AND EFFECTIVE SERVICE DELIVERY.	NOT WORKING . SLOW DELIEVERY

2.9 Smart and Sustainable Civic Infrastructure: Video Surveillance

In order to record incidents that occur throughout the city, and to closely monitor respective areas IP-based Surveillance Solutions have been installed at key. It also provides a live video feed to the Jaipur Police control room, leading to faster response times and improved success rates, allowing local people and tourists alike to feel safer in Jaipur.

Advanced cameras (noise sensitive) have been placed at tourist places, parks, hospitals, entry/exit points, and monuments in Jaipur. JDA and DOITC (Department of Information Technology and Communication) are responsible for installing CCTV cameras in Jaipur. The feedback from all these cameras from JDA and DOITC is monitored by the Police Control command center (Table 8).

2.10 Smart and Sustainable Civic Infrastructure: Interactive Kiosk

Jaipur's Smart City kiosk was planned by JDA as a broad build-in web 2.0 created as an interactive system where users can access a diverse range of content such as location, mobile recharge, city connectivity, etc. on a 42 in. touch display.

Main features of Interactive Kiosks are Recharge facilities, Movie booking, Ticket booking, and online tourism information.

Interactive Kiosks have been planned to be installed at Albert Hall, Amber fort, Zoo, Jantar Mantar, and Hawa Mahal. JDA has also plans to put these Kiosks at 17 other public places in Jaipur. Interactive Information Kiosks features city information and location services. These Kiosks will also provide facilities like device charging, information on train schedules, status of reservations, flight information, etc. (Table 9).

Table 8 SWOT analysis of video surveillance

Strength	Weakness	Opportunity	Threat
Controlling public safety and mass events Public and tourists safety		Optimize efficiency and response time of emergency services Provide surveillance in public spaces and secure public administration transactions	

Source : Author

Table 9 SWOT analysis of interactive kiosk

Strength	Weakness	Opportunity	Threat
Offer instant customer service	Lack of public awareness	Improved efficiency Saves cost and time	Longer waiting time due to queue formation

Source : Author

Table 10 SWOT analysis of environmental sensor

Strength	Weakness	Opportunity	Threat
Provides air quality status report Improve living conditions Reports data on PM_{10}, SO_2, and NO_2	Lack of public awareness	Planning considerations Better health prospects Improve the environment	

Source : Author

2.11   Smart and Sustainable Civic Infrastructure: Environmental Sensor

Environmental Sensors have been placed at 32 key locations in the city. It provides air quality status reports in real-time. It provides necessary data to Jaipur authorities including RSCPB to handle pollution and problems caused by climatic conditions. This data is further used to increase green belt coverage in specific areas or to take planning considerations/decisions.

Policy reform based on these information and then their implementation would be the effective next steps to improve the air quality (Table 10).

2.12   Smart and Sustainable Civic Infrastructure: Smart Light

2200 Smart light fittings have been placed under Smart lighting project on a 2 km stretch. It includes automated dimming and illuminance (upon traffic movement).

This can be used for better traffic planning by traffic analytics that monitors car and pedestrian traffic conditions (Table 11).

Table 11 SWOT analysis of smart light

Strength	Weakness	Opportunity	Threat
Providing light as and when required Better for vehicle and pedestrian traffic conditions Upgraded to the level of intelligent lighting Energy saving		Optimizing the power consumption to the maximum Allows the authority to have a planned or scheduled dimming cycle via a web application	

Source : Author

3 Conclusion

We've found that the Smart City project had been designed on the basis of smart visions that implement "advanced and innovative solutions" for providing public services by using ICT.

We can state that the Smart City Mission is lacking in providing or uplifting the basic needs of society. Along with attaining smart technologies, and providing smart heritage, mobility, and infrastructure, Smart City vision must also have considered the basic principles of a city which accounts for fair accessibility of infrastructure that can preserve the best of the past while promising creativity and innovation in the development of a sustainable future.

The study highlighted that a large percentage of local people and tourists are still not connected to the network due to the Lack of IT infrastructure and digital illiteracy. In order to incorporate ICT in the existing infrastructure and create smart utilities, the basic infrastructures such as water distribution, public transport, and waste management have to be worked upon with minimum quality standards.

In a historical city like Jaipur, the new developments and new design should fulfil the general principles of a city. The process and method of developing any smart project in such a place should be able to connect new design with its historic setting. An approach from one facet alone is doubtful to be successful.

As per our study tour in and around the walled city of Jaipur and interviewing the concerned department officials, we came to the following conclusion.

Jaipur Smart City Limited (JSCL) is taking up projects mentioned under the Smart City Proposal (SCP) approved by the Government of India. JSCL is taking up projects independently as well as part of the convergence from line departments to integrate various civic bodies with Jaipur Municipal Corporation (JMC) and Jaipur Development Authority (JDA).

After meeting the project heads of various organizations, it is reflected that there is no consideration upon certain key area, namely rural area and population growth with time.

Again Smart City proposals had been largely limited to people with access to the Internet, thus leaving out the poorer segments of the population, and that, by not requiring Smart City plans to address the root causes of poverty and discrimination, the initiative was unlikely to create more inclusive and human rights based urbanization.

Acknowledgements We are thankful to Ar. Mohammad Saquib and Ar. Mohit Dhingra (Jamia Millia Islamia) for their guidance during the course of this research.

Bibliography

1. Ashutosh ATP (2015) Preparing smart city proposal
2. Manju Y (2017) Rajasthan tourism: problems and government policies

3. Jaipur Municipal Cooperation (2008) Chapter-11, Urban Governance and Institutional Framework; City Development Plan for Jaipur
4. The Smart City Challenge Stage 2: Smart City Proposal Jaipur. New Delhi, India: Ministry of Urban Development, Govt. of India, 2016; City Development Plan for Jaipur (2008)
5. https://www.jscljaipur.com/abd-area-projects.html

Techno-Economic Feasibility Analysis of Hybrid System

Tahleela Navid, Abid Hussain Lone and Anwar Shahzad Siddiqui

Abstract This paper presents an overview of the techno-economic feasibility analysis of hybrid renewable energy system (HRES) of interest that can be used in smart cities. A case study of an institutional area of New Delhi is carried out to compare the various configurations of HRES using hybrid optimization model for electric renewable (HOMER) software and choosing the best configuration among them on the basis of various parameters which are prerequisite for the development and sustainability of a smart city. These configurations were considered in two different modes, grid-connected and the isolated mode, and were compared on the basis of economic parameters like calculation of cost of energy, net present cost, and life cycle cost of the plant, technical aspects, and the impact on the environment. The HRES model proposed here serves a load demand of 3113 kWh/day. After performing the load data analysis and simulation, the results showed that the grid–PV–battery hybrid system configuration could bring financial benefits, improve the power reliability and energy security and lead to less greenhouse gas emissions and thus reducing the cities' impact on the environment.

Keywords Smart city · Microgrid · Optimization · HOMER

1 Introduction

According to the worldwide smart grid initiative, the future electricity network must be flexible, accessible, reliable, and economic. This is also mandated by the Sustainable Energy Authority of Ireland (SEAI) and European Electricity Grid Initiative (EEGI). Research on various configurations of microgrid (μG) system is being carried out in order to meet these objectives and to reduce greenhouse gas (GHG) emission, particularly with high penetration of renewable energy sources.

T. Navid (✉) · A. H. Lone · A. S. Siddiqui
Department of Electrical Engineering, Jamia Millia Islamia New Delhi, New Delhi, India

© Springer Nature Singapore Pte Ltd. 2020
S. Ahmed et al. (eds.), *Smart Cities—Opportunities and Challenges*,
Lecture Notes in Civil Engineering 58,
https://doi.org/10.1007/978-981-15-2545-2_29

Depending on the resource availability, geographical locations, load demand, and existing electrical transmission and distribution system, μG can be either connected to the grid or can work in an islanded mode. Renewable energy technologies will contribute a lot in achieving the objectives of the smart cities' mission to make them zero-polluting and self-sustaining cities. Over last two decades, India is witnessing unparalleled transformation from rural to urban living. It has been mandated that 10 percent of the smart cities' energy requirement should come from solar energy [1]. Use of techno-economic analysis and feasibility study of a new site, as suggested in [2, 3], was a revolutionary step to solve the site selection for a better performance of microgrid. The increase in demand for quality power and depleting sources of conventional fuels has brought into focus the potential of renewable resources. Considering the renewable potential of India, it becomes important to study and develop methods for checking the techno-economic feasibility of the hybrid renewable energy system (HRES). The socioeconomic and the environmental implications associated with the use of fossil-fueled generators can be very alarming [4].

The Hybrid Renewable Energy System is widely used in recent times because it combines various power sources to optimize each sources strength while compensating for the short comings of others [5]. Compared to systems comprising of a single energy source, the HRES allows improving the system efficiency, reliability of the power supply, and reduces the energy storage requirements. It is essential to ensure that microgrid projects are technically reliable, financially feasible, and socially sustainable [6]. Furthermore, μG can reduce environmental pollution and global warming by utilizing low-carbon technology [7]. A microgrid can be of the form of a shopping center, industrial park, or college campus. Microgrid can be categorized into two types: (i) off-grid (ii) grid-connected. In this paper, various topologies of microgrid are analyzed to find out the best possible configuration in terms of achieving various desirable objectives like reduction in CO_2 footprints, better economical configuration, and other environmental issues. The benefits of grid-connected or isolated μG with storage have also been identified [8].

2 Objective of This Study

Analysis is done to carry out the following objectives:

   i. To scale up the deployment of HRES in all those areas of the country, where it is deemed the best option for electrification.
  ii. To reduce the purchase of energy from grid.
 iii. To provide a reliable, economical, and optimized power system in the form of microgrid.

iv. To help the government of India in realizing a faster adoption of microgrids for the development of smart cities.

v. To make a city zero-carbon emission city by maximizing the utilization of renewable energy resources.

3 Description of Parameters

3.1 Load Profile

A hostel building, namely J&K hostel in Jamia Millia Islamia, New Delhi, is considered for case study. It includes 217 rooms, dining halls, reading room, common room, and warden office. The details of different types of loads, their ratings, and quantity commonly used in this hostel are listed in Table 1. The daily average hostel load is 3113 kWh with a random variation of 10% while the peak load is 426 kW with a random variation of approximately 8%. Average monthly load variation (weekdays and weekends) is shown in Figs. 1 and 2. There is a huge variation in load profile because a minimum load was recorded during the vacation period, whereas in peak winter and summer days, load rapidly increases up to 426 kW.

Table 1 Approximate daily electric load details of hostel

Loads	Rating (W)	Usage (h/day)	Quantity
Tube lights	20	8	788
Water cooler[a]	775	24	28
AC[a]	1667	8	3
Laptop	65	2	650
Exhaust fan	40	24	131
Iron	1000	1	220
Phone charger	3.5	2	650
Washing machine	500	3	56
CFL	/	16	896
Induction heater	2000	2	2
Refrigerator[a]	150	24	2
Fans	80	16	480
Room heater[a]	2000	8	2
Geyser[a]	2000	5	28
Air cooler[a]	200	8	150
TV	70	2	1
Halogen	200	12	7

[a]*seasonal loads*

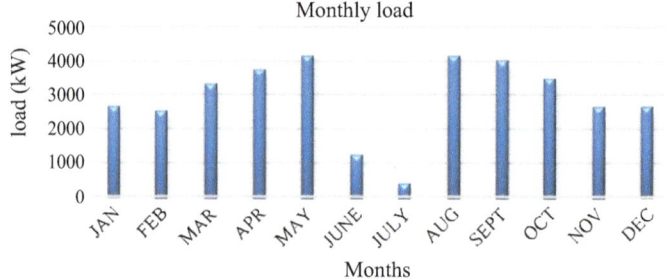

Fig. 1 Approximate monthly load variation (weekdays)

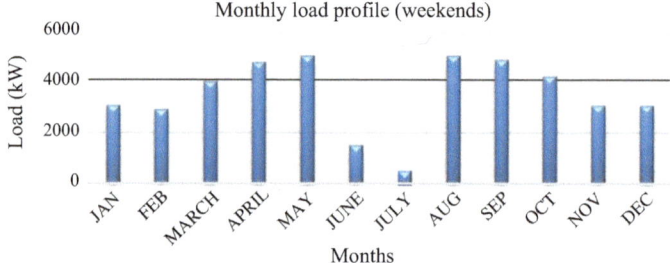

Fig. 2 Approximate monthly load variation (weekends)

3.2 Solar Radiation

The solar radiation and clearness index input for the various months throughout the year were obtained from the Internet via HOMER software by providing the latitude, longitude, and time zone information. Indian Meteorological Department (IMD) measures and provides solar radiation data along with other climatic parameters over various locations across the country. Figure 3 shows the solar radiation data inputs of New Delhi along with the clearness index of the solar irradiation. The clearness index is the fraction of solar radiation that is transmitted through the earth. It ranges between 0 and 1.

The clearness index has a high value under clear and sunny conditions and a low value under cloudy conditions. The solar radiation for each month was obtained, where it was found that the maximum solar radiation was found for the month of May with daily radiation of 6.616 kWh/m^2/d, whereas the minimum radiation was found for the month of December with daily radiation 3.33 kWh/m^2/d. The average radiation throughout the year was 5.108 kWh/m^2/d.

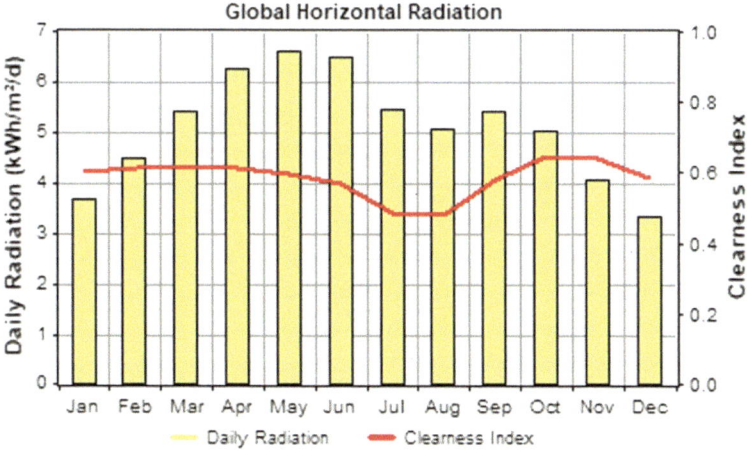

Fig. 3 Annual solar radiation and clearness index for New Delhi, India

3.3 Wind Resource

The wind energy is an important source of sustainable energy resource worldwide. The energy contained by the wind depends on the density and velocity. The need to integrate the renewable energy like wind and solar into power system is to make it possible to minimize the environmental impact of the conventional power plants. The overall capacity of all wind turbines installed worldwide by the end of 2017 reached 5,39,291 MW [9]. As the ratio of installed wind capacity to the system load increases, the required equipment needed to maintain the AC grid stable increases, forcing an optimum amount of wind power in a given system. Therefore, design of individual components of microgrid containing wind turbine must be sized properly. In this modeling, 10 kW DC wind turbine is used. The power curve and cost curve for wind turbine are shown in Figs. 4 and 5, respectively.

The lifetime taken as 15 years, and hub height is 25 m for the wind turbine considered. Figure 6 shows wind resource for a given simulation. The daily average wind speed measured at 25 m height is 4.5 m/s.

Fig. 4 Power curve of a wind turbine

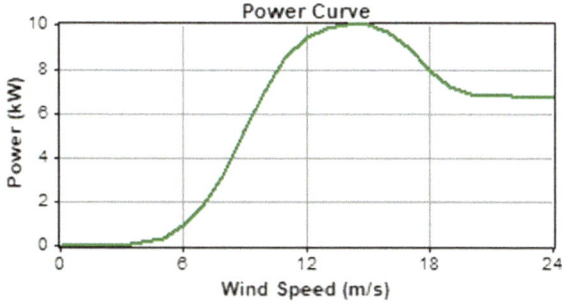

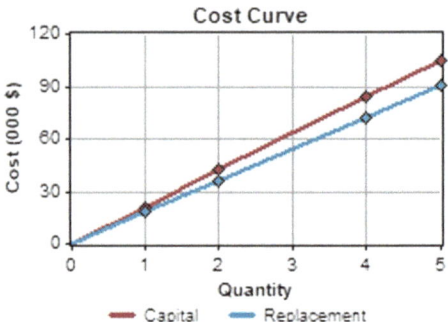

Fig. 5 Cost curve of a wind turbine

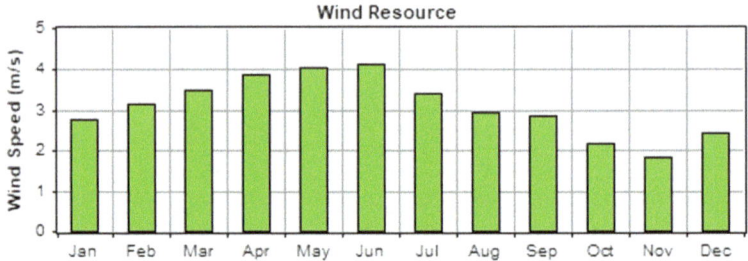

Fig. 6 Wind resource

3.4 Diesel Generators

Diesel engines are considered less environmental friendly with more emissions but are still the most popular fuel-type generators, mainly used for off-grid generation. Low installed capacity, high shaft efficiency, suitable for start–stop operation, and high exhaust heat are some of the advantages of combustion diesel generators.

For an uninterruptable power supply, in this modeling three generators of ratings 280, 120, and 50 kW are used. Figure 7 shows the cost curve of diesel generator rated for 10 kW. The price of diesel considered here is 65 rupees/L.

3.5 Hybrid System Components, Their Specifications, and System Constraints

Photo Voltaic System: A 440 kW PV system is chosen for the system. The parameters considered for PV are lifetime: 20 years, ground reflectance: 20%, scaled annual average 5.11 kWh/m^2/d. The capital cost, replacement cost and

Table 2 Cost of equipments used in hybrid system

Equipment	Size (kW)	Capital (INR/kW)	Replacement (INR/kW)	O&M (INR/kW)
Wind turbine	10	1,53,300	1,31,4000	3650
Diesel generator	280, 120, 50	36,000	29,000	21,600
Solar PV	440	51,000	51,000	11,972
Battery	1156 Ah	21,900	21,900	100
Converter	420	20,999	20,999	0

operating, and maintenance cost (O&M) considered for the PV system are shown in Table 2.

Converter: A converter of rating 420 kW is used in this modeling having an efficiency of 90 and 85% for inverter and rectifier, respectively, and a life span of 25 years.

Battery Bank: Commercially available battery model, Surrette 6CS25P lead–acid, is considered for hybrid system simulation.

System Constraints:

Annual interest rate	6%
Plant working life span	25 years
Maximum annual capacity shortage	0%

Grid:

Cost of energy (COE)	INR 8.4/kWh
Sellback price	INR 4/kWh
Carbon dioxide (CO_2) emission	630 g/kWh
Sulfur dioxide (SO_2) emission	2.74 g/kWh
Nitrogen oxide (NO) emission	1.34 g/kWh

The price listed in Table 2 is taken as the unit price of the equipments used in the proposed system for simulation. It is done by collecting data from local market [10–13].

4 Modeling and Simulation Results

The following configurations are analyzed for technical and economic feasibility of the system.

After performing a rigorous simulation process, the most feasible and optimal configuration is taken for modeling the system.

4.1 Results

(1) *Diesel generator* (270, 120, and 60 kW):

Here, it is assumed that diesel generator is the only power source for meeting the whole demand. In this case, three generators of rating 270, 120, and 60 kW are used to meet the peak and base load demand. This is the worst-case scenario where the COE, NPC, OC, and CO_2 emissions are having the high values of INR 26.64, INR 387.24 M, INR 29.025 M/year, and 1075,848 kg/year, respectively. So it is not desirable to meet the whole demand only from diesel generators in terms of economic and environmental economic aspects. Average monthly electricity production from considered generators is shown in Fig. 9.

(2) *Utility grid*:

In this case, it is assumed that the whole load demand is met by the utility grid and the grid is available to supply the power at all the time. The COE, NPC, OC, and CO_2 emissions in this case are INR 8.32, INR 120.85 M, INR 9.45 M/year, and 7,18,102 kg/year, respectively.

(3) *Diesel generator* (350, 100 kW), *PV* (450 kW):

In this case, it is assumed that there is no grid supply and load demand which is fulfilled by the combinations of diesel generator and renewable energy sources, i.e., through solar PV system. The COE, NPC, OC, and CO_2 emissions obtained in this case are INR 24.67, INR 358.6 M/year, 24.61 M/year, and 880,595 kg/year, respectively.

(4) *Diesel generator* (260, 100 kW), *PV* (450 kW), *LA battery bank* (763 kWh):

In this case, here it is assumed that there is no grid supply and load demand which is fulfilled by the combinations of diesel generator and renewable energy sources and battery storage, i.e., through solar PV system. The COE, NPC, OC, and CO_2 emissions obtained in this case are INR 20.44, INR 304.7 M/year, 20.42 M/year, and 743,664 kg/year, respectively.

(5) *Grid, PV* (440 kW), *and battery bank* (1387 kWh):

In this case, the combinations of grid, solar PV, and wind turbine systems are considered for supplying power to the expected load. The COE, NPC, OC, and CO_2 emissions obtained in this case are INR 5.621, INR 103.39 M, 5.816 M/year, and 307,203 kg/year, respectively. If a battery bank of 250 kWh is added in this case, then the values of COE, NPC, OC, and CO_2 emissions will change to INR 5.402, INR 97.8 M, INR 5.026 M/year, and 315,825 kg/year, respectively.

(6) *Grid, PV* (420 kW), *wind turbine* (10 kW), *and battery bank* (694 kWh)

In this case, the combinations of grid, solar PV, and wind turbine system are considered for supplying power to the expected load. The COE, NPC, OC, and CO_2 emissions obtained in this case are INR 6.059, INRi 109.32 M, 6.14 M/year, and 334,958 kg/year, respectively.

5 Discussion

According to the optimization results, the optimal combination of hybrid system components as shown in Fig. 8 are a 450 kW PV array, 1387 kWh Surrette 6CS25P battery, and a 320 kW converter with a cycle charging (CC) strategy of load following. The total NPC, IC, LCOE, and CO_2 emission for such a hybrid system are 103.39 M, 33.5 M, 5.621 INR, and 307,203 kg/year. The technical and economical details of all the configurations of the hybrid system from the optimization process are shown in Tables 3 and 4. For the selected location, we have an average wind speed of 4.5 m/s which makes the installation of wind turbine technically not feasible.

Figures 9 and 10 illustrate the distribution of monthly average electricity produced in kW by the combinations of diesel generators (case 1) and of optimal combination of PV–grid–battery.

Table 3 Optimal size of different components

Model	PV (kW)	Diesel generator (G1, G2, G3) (kW)	Battery (kWh)	Converter (kW)	Grid(kW)
Generators	–	270,120,60	–	–	–
Gen–PV	450	350,100	–	230	–
Gen–PV–battery	450	260–100	763	240	–
Grid	–	–	–	–	1000
Grid–PV–battery	440	–	1387	320	300
Grid–PV–wind turbine–battery	420	–	694	270	350

Table 4 Optimal cost for different components

Model	LCOE (INR)	Initial capital (IC) (INR in millions)	NPC (INR in millions)	O&M (INR in millions)	CO_2 emissions (Kg/year)
Generators	26.64	16.2	387.24	29.025	10,75,848
Gen–PV	24.67	43.98	358.6	24.61	8,80,595
Gen–PV–battery	20.44	43.43	304.7	20.42	7,43,664
Grid	8.32	0	120.85	9.45	7,18,102
Grid–PV–battery	5.621	33.55	103.39	5.816	3,07,203
Grid–PV–wind turbine–battery	6.059	30.81	109.32	6.14	3,34,958

Fig. 7 Cost curve of 270 kW
diesel generator

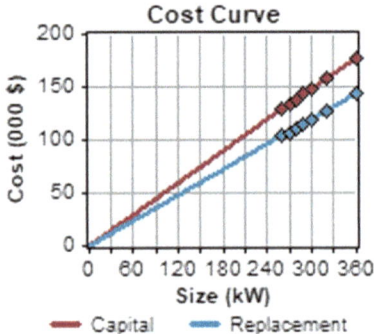

Fig. 8 Schematic of
proposed hybrid PV and grid
system with battery storage

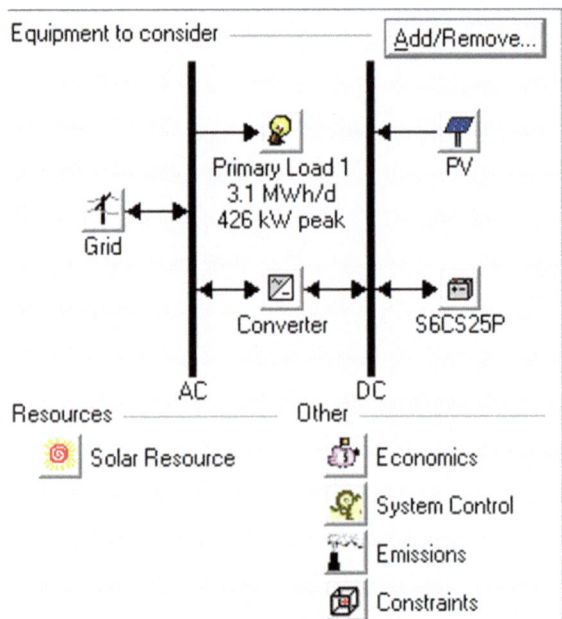

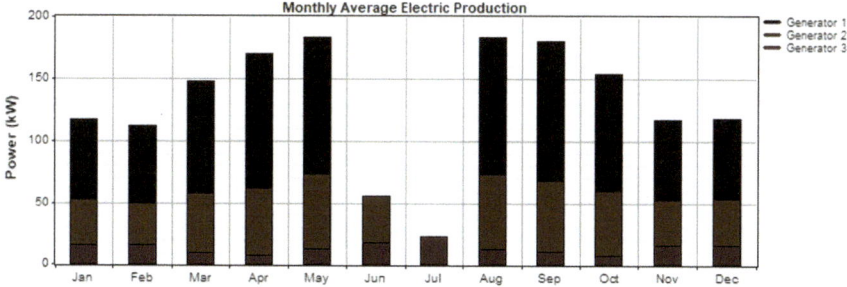

Fig. 9 Average monthly electricity production (case 1)

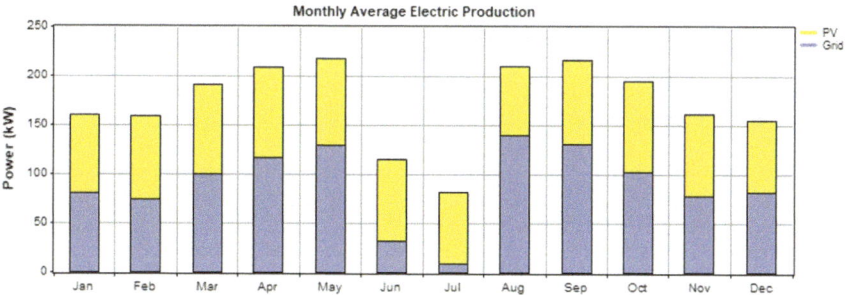

Fig. 10 Average monthly electricity production (case 5)

6 Conclusion

In order to make the zero-polluting and self-sustaining cities, renewable energy technologies will contribute a lot in achieving objectives of the smart cities' mission. Cities are now home to more than half of the world's population and account for almost two-thirds of global energy use and three-quarters of energy-related carbon emissions [14]. With about 20% of their energy supply currently from renewable, their potential for scaling up local renewable supplies is without doubt. Since the buildings and transport are the two largest energy consumers in cities, these are identified as priority areas for action in order to scale up rapidly and substantially. Therefore, technical and economic feasibility analysis of a HRES is a preliminary step in moving toward the development of smart cities. Educational institutes have a responsibility to become a role model for the nation to save energy and promote the optimization and can be a good platform to raise awareness and promote energy saving. For different locations and system constraints, these microgrid topologies may differ, according to the resources available. Based on a large number of simulation results obtained from HOMER software, combinations of grid, solar PV, and battery are found to be the most optimal case for the specified load. In this case, the values of COE, NPC, OC, and CO_2 emissions are found to be less with respect to other considered topologies.

References

1. https://www.eprmagazine.com/green-zone/solar-power-the-future-smart-cities/for
2. Chatterjee A, Rayudu R (2017) Techno-economic analysis of hybrid renewable energy system for rural electrification in India. In: 2017 IEEE innovative smart grid technologies—Asia (ISGT-Asia). Auckland, pp 1–5
3. Das G, De M, Mandal KK, Mandal S (2018) Techno-economic feasibility analysis of hybrid energy system. In: 2018 Emerging trends in electronic devices and computational techniques (EDCT). Kolkata, pp 1–5
4. Okundamiya MS, Emagbetere JO, Ogujor EA (2014) Assessment of renewable energy technology and a case of sustainable energy in mobile telecommunication sector. Sci World J 2014:1–13, article ID 947281
5. Okundamiya MS, Akpaida VOA, Omatahunde BE (2014) Optimization of a hybrid energy system for reliable operation of automated teller machines. J Emerg Trends Eng Appl Sci 5:153–158
6. Zhou W, Lou C, Lu L, Yang H (2010) Current status of research on optimum sizing of stand-alone hybrid solar-wind power generation systems. Appl Energy 87:380–389
7. Wies R, Johnson RA, Agarwal AN, Chubb TY (2005) Simulink model for economic analysis and environmental impact of a photovoltaic with diesel-battery system for remote villages. IEEE Trans Power Syst 20(2):692–700
8. Yan-Hua L, Nan Z, Xu Z (2012) Research on grid connected/ islanding smooth sharing of microgrid based on energy storage. In: IEEE Power System Conference. Newzealand
9. http://wwindea.org
10. https://india.alibaba.com/index.html
11. https://my.indiamart.com/
12. www.rollsbattery.com
13. http://www.kirloskar-electric.com/gens.shtml
14. Adaramola MS, Paul SS, Oyewola OM (2014) Assessment of decentralized hybrid PV solar-diesel power system for applications in Northern part of Nigeria. Energy Sustain Dev 19:72–82

Review of dSPACE 1104 Controller and Its Application in PV

Tanushree Bhattacharjee, Majid Jamil and Abdul Azeem

Abstract The increase in penetration of renewable energies or clean energies in power sector and their integration with existing conventional grids generates the need for design and implement good controller circuits in its hardware model. This paper is having a discussion over dSPACE 1104 controller circuit and its use in a solar system. dSPACE controller board overview with its operation has been given in detail. The controlling action of the controller board is being shown here. dSPACE DS1104 card and its use of RTI library to design the control system that can connect a simulated model with the controller board has been shown. Lastly, solar inverter controller circuit with controlling operation and connection of inverter circuit with dSPACE controller circuit has been given that can be used in future implementation of dSPACE 1104 controller circuit in a PV simulated model hardware prototype. From this paper, the readers will get to know about dSPACE 1104 controller board and its idea of implementation in any simulated prototype model.

Keywords dSPACE 1104 controller · RTI library · PV inverter controller · Control system

1 Introduction

The penetration of renewable energies and upgradation of conventional plants are getting increased day by day. Among various resources of energy, PV is increasing in demand due to its availability and economic factors [1]. While using solar energies for the generation of power, there is a very important factor to care of and that is inverter system. As the generated power is DC, always we need a good inverter circuit to convert it in useful AC form [2]. The solar energy generation is lacking in its efficiency in respect of generation and power conversion. To improve

T. Bhattacharjee (✉) · M. Jamil · A. Azeem
Electrical Engineering Department, Jamia Millia Islamia,
Jamia Nagar, New Delhi 110025, India

© Springer Nature Singapore Pte Ltd. 2020
S. Ahmed et al. (eds.), *Smart Cities—Opportunities and Challenges*,
Lecture Notes in Civil Engineering 58,
https://doi.org/10.1007/978-981-15-2545-2_30

the efficiency by reducing losses of energy conversion through inverter controlled switching of inverters are very important [3].

The controlling action in solar inverter circuit can be incorporated by the use of good controller circuits so that its switching can get controlled by using different switching schemes like sine pulse width modulation or SPWM switching, space vector pulse width modulation or SVPWM scheme, etc. With the use of PWM techniques, the inverter switching signal's duty cycles will get controlled as per requirement [4]. Any type of controller circuits can be modeled using MATLAB Simulink software and can test its result outputs before using any real-time controller circuits in PV model [5]. The dSPACE controller application is increasing in automobile industries and power industries because dSPACE controller is having the facility of directly connecting any summation model with real-time hardware circuit by converting it into C codes [6]. Actually, the dSPACE 1104 controller while get connected to any system it will convert the system or PC with the ability to develop rapid control prototypes or RCP models. These prototypes can get implemented in real time using the controller board which is a Power PC microprocessor-based system having digital input/output or I/O interfaces [7].

This paper is focused on discussion over dSPACE 1104 controller board, its software libraries, and its application in solar inverter circuits. In Sect. 2, the dSPACE 1104 controller circuit with its circuit diagram and controlling operation has been discussed in detail. The dSPACE 1104 controller is having a real-time interface or RTI library that will be added with the existing MATLAB library to connect simulation model with the hardware part as described in Sect. 3. Also, the ControlDesk software that is used in controlling action of hardware has been shown. Section 4 contains solar inverter control system operation and implementation of dSPACE controlling logics with the inverter simulation model and its hardware prototype. In Sect. 5, conclusion has been drawn from all the above studies with its possible future outcomes.

2   dSPACE 1104 Controller

2.1   Overview

In real-time hardware models, the dSPACE controllers with digital controlling action are using rapidly in robotics, aerospace, automobiles, power and research industries. This extensive use of dSPACE controller is because of the reason that the inverter controller simulation model can get directly connected to the hardware circuit through this controller [8].

The controller board with its master connector 37-pin ribbon cable is shown in Fig. 1. It can be seen that dSPACE 1104 model is having 8 ADC and DAC ports, one master digital I/O or PWM switch and a slave I/O that we can use as per our

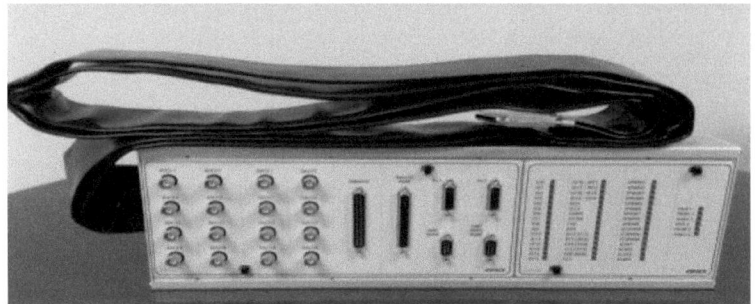

Fig. 1 dSPACE 1104 controller board and 37-pin connector ribbon

requirements. The master PPC is a PowerPC 603e microprocessor running at 250 MHz and slave subsystem with DSP Texas Instruments TMS320F240 DSP that is running at 20 MHz [9].

2.2 Controlling Action

The block diagram of dSPACE controller has been shown in Fig. 2. In the figure, it is shown that how the signals are going to travel from PC through controller to the inverter circuit.

The inverter outputs will get sensed by sensors, and the signal will feed into the dSPACE controller by ADC channels as shown in Fig. 2 [10]. The controller is also connected with the PC that is having the simulation model and MATLAB library with RTI and also ControlDesk software for executing controlling actions. From the simulation model, the dSPACE controller will get all the reference values and controlling power. The PWM signals generated from the controller board will then be transferred through the digital I/O or PWM pin to the inverter for performing controlled switching of its IGBTs [11]. The controlling action and signal flow can be also seen in Fig. 3 [12].

3 DS 1104 Card

3.1 RTI Library

The DS1104 card of the dSPACE controller needs to get inserted inside the CPU of PC, and it is having RTI library that will be added with the MATLAB library to get used while building the control system block for connecting the simulation model with controller board. There are various blocks present in the RTI library like DS1104ADC_Cx for analog to digital conversion, DS1104DAC_Cx for digital to

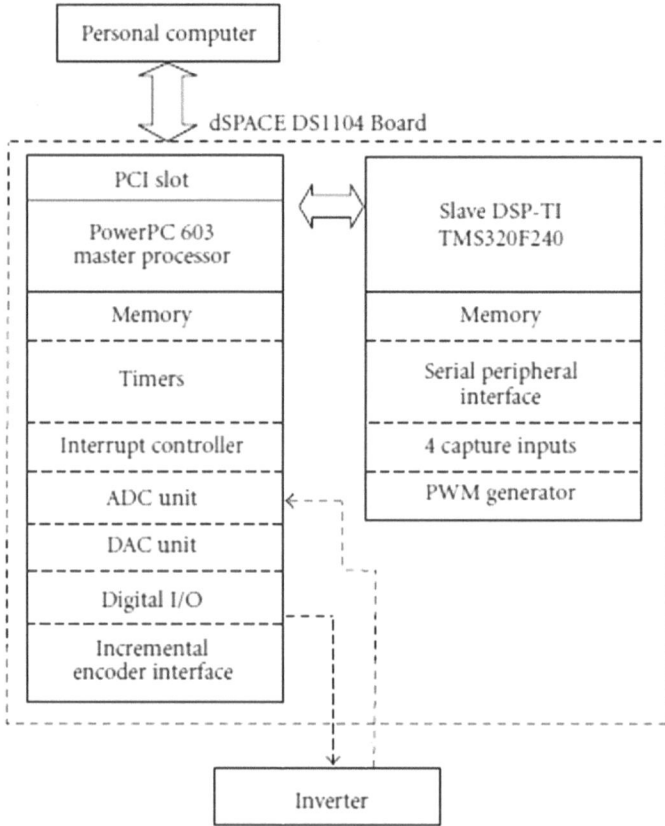

Fig. 2 dSPACE 1104 controlling block diagram

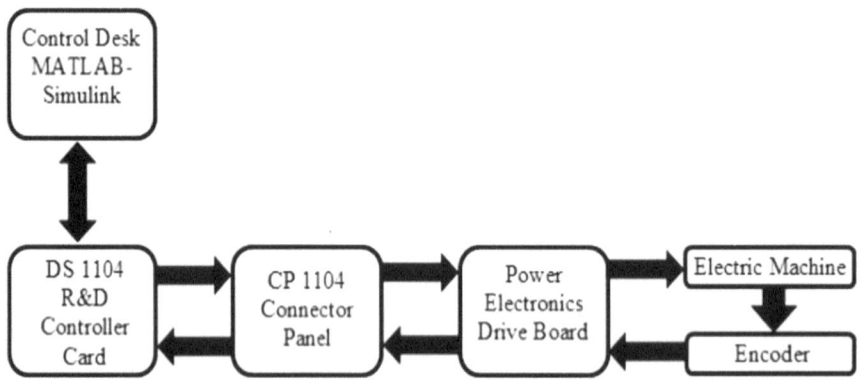

Fig. 3 Control step blocks

analog conversion, DS1104BIT_OUT_Cx for output port to connect dSPACE controller board, and DS1104SL_DSP_PWM for PWM signal generation. The I/O blocks of dSPACE make the auto-conversion of Simulink model into C code using its real-time workshop function or RTW [13]. The RTI library in MATLAB with all its function blocks has been shown in Fig. 4.

So, the PC used here will have MATLAB Simulink for simulating the complete solar inverter controller model in software. Also, it will have RTI library blocks that will be used to simulate the control subsystem of dSPACE controller for connecting simulation model and hardware. Lastly ControlDesk software for executing all the controlling operations like changing of parameters and control logics. To get an overview, all the software windows have been shown in Fig. 5 [14].

3.2 Control System

The dSPACE 1104 controller board needs to get connected with the simulation model for doing the control operations, and for that, a control system block needs to

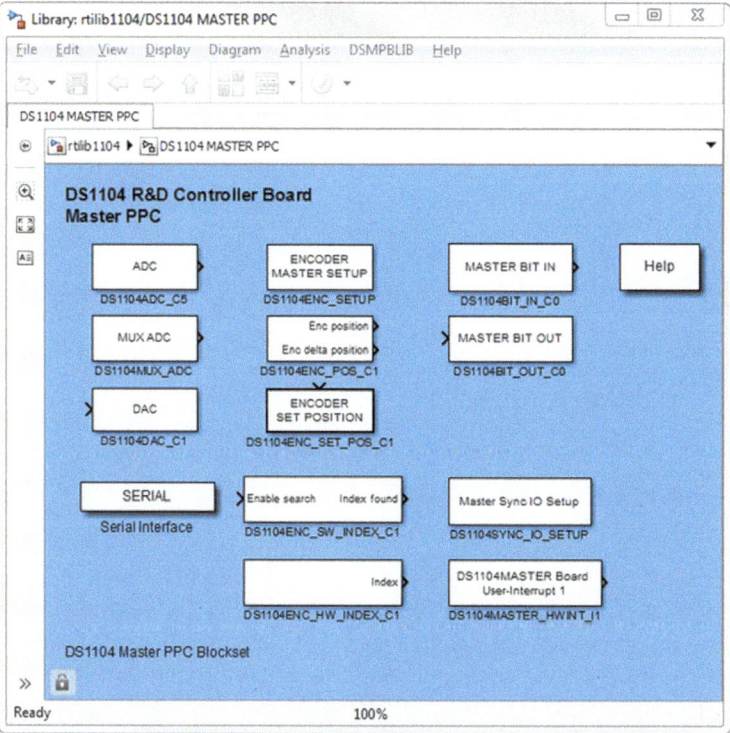

Fig. 4 RTI block sets of DS1104 card

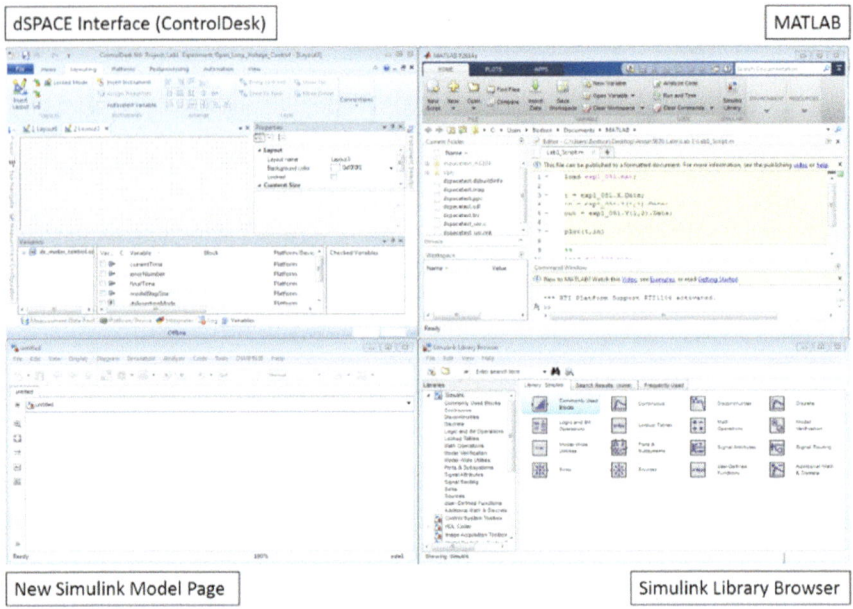

Fig. 5 Interfacing of MATLAB Simulink, RTI, and ControlDesk software

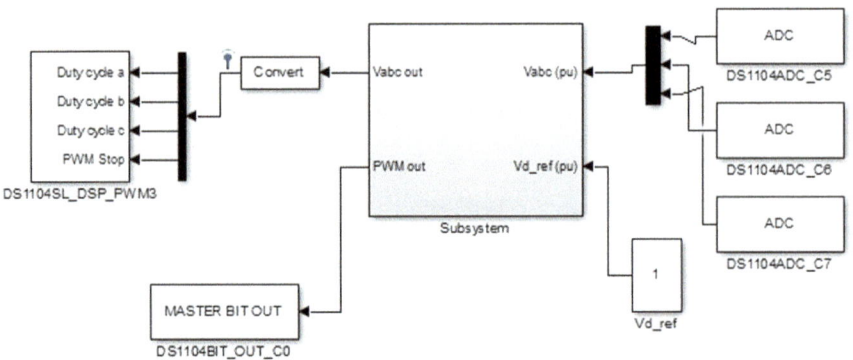

Fig. 6 Control system simulation from RTI

simulate using RTI library of DS 1104 card. A control system block can get simulated as shown in Fig. 6. The blocks used from RTI library are DS1104ADCs; these are used to convert the analog signals coming from inverter inside the controller and will get converted into digital form for performing the sampling needed for PWM generation. DS1104SL_DSP_PWM is used for PWM signal generation, and DS1104_OUT_CO can be used for switching of booster circuit in the PV model for increasing efficiency of operation [15].

4   Solar Inverter Controller and dSPACE 1104 Connection

4.1   Solar Inverter Controller Operation

To know the controlling operation to perform using external hardware controllers, first need to know about the simulation controller model that needs to simulate using MATLAB for getting reference signals that will be used in generating PWM signals. The block diagram representation of a PV system controlling operation has been shown in Fig. 7. The solar energy from solar array will get transferred to a booster circuit to get most of the outputs, then that will feed to inverter circuit and filtered from harmonics using a filter circuit. The DC link voltage and output from the filter circuit will get sensed using sensors and given to controller circuit where PWM signals will get generated and given to the inverter circuit back for controlling the switching of IGBTs and also to booster for controlling switching of its IBGT to. The final output after controlling will be harmonics free and stable which gets supplied to the loads [16].

Now, the controller which can be PI, PID, or PR or any other type can be formed as per the requirement will perform controlling operation as shown in Fig. 8. The abc form sensed signal of the inverter output will be transformed in dq or controller readable format and compared with reference signals then given to PI (PI have been taken here as an example) and after performing the controlling operation it will be again transformed in abc format. This signal will be compared with a very high-frequency triangular carrier wave to generate the SPWM signals and supplied to inverter and booster for switching operation [17]. The PLL blocks will give synchronization the controlling signals to 50 Hz.

4.2   Applying dSPACE 1104 Controller in PV

The dSPACE controller will be fed by inverter filtered outputs like voltage and current signals at inverter end and also PV voltage and current ratings from booster circuit by sensing through sensors as shown in Fig. 9. These signals will be

Fig. 7 Solar inverter operation block diagram

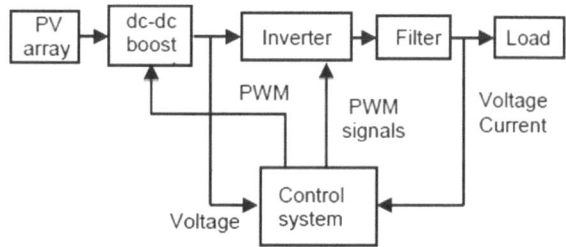

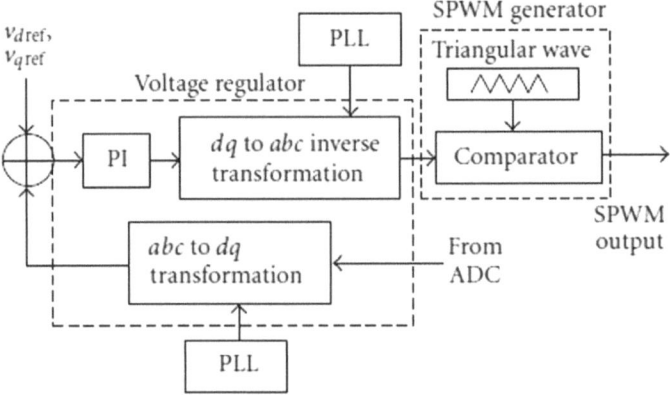

Fig. 8 Controlling operation of solar inverter controller

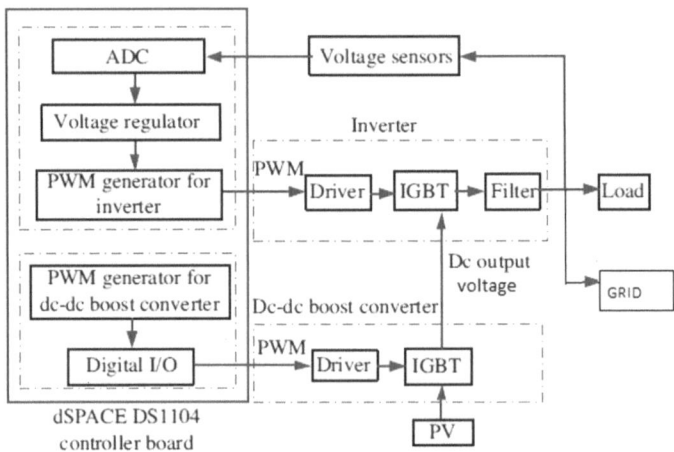

Fig. 9 dSPACE 1104 controller in PV application

transferred through ADC ports of the dSPACE board. Then, dSPACE will do the controller operation by using ControlDesk software and simulation blocks to generate the SPWM signals needed for IGBT switching. These signals will be sent back to the inverter circuit by master PWM digital I/O pins connecting to the inverter IGBTs and slave digital I/O pins are connecting with booster IGBTs. Here, the dSPACE controller circuit is working at a very low voltage around 10 volts while the inverter or power circuit will work at higher voltages (e.g., 400 V) depending on its capacity. So, there must be an isolation between the two circuits and that is why an optocoupler will be connected in between inverter and controller circuit [11].

The controller output will deliver to the load and grid circuit connected to it that forms a microgrid. With advanced communication networking, this microgrid system can get approached in smart grid operations. In upgradation of conventional cities into smart city, smart grid can play an important role in automated power delivering and information traveling operations.

5 Conclusion

The basic idea about dSPACE 1104 controller circuit and its application in solar or other energy sectors has been given in this paper. The dSPACE controller board with its controlling operation has been shown with block diagrams. Then, its DS1104 card with its RTI library has been discussed with diagrams. Also, its software part that is ControlDesk software and MATLAB simulation model formation or control system with RTI blocks to connect the simulation with hardware has been depicted here. The solar system with inverter, booster, controller, and load circuit operation has been discussed. The application of dSPACE controller in the simulation model hardware prototype with its connections has been given in detail with all diagrams.

In the future, the simulation model can be formatted in MATLAB and its prototype can be developed with dSPACE 1104 controller and solar inverter circuits. The result output can be analyzed and load or any grid can get supplied with this model. The prototype connected with grid will give smart grid operations in the development of a smart city.

References

1. Blaabjerg F, Chen Z, Kjaer SB (2004) Power electronics as efficient interface in dispersed power generation systems. IEEE Trans Power Electron 19(5):1184–1194
2. Sanjeevikumar P, Rajambal K (2008) Extra-high-voltage DC-DC boost converters topology with simple control strategy. Model Simul Eng 2008:6
3. Subiyanto AM, Hannan MA (2010) Photovoltaic maximum power point tracking controller using a new high performance boost converter. Int Rev Electr Eng 5(6):2535–2545
4. Sibi Raj PM, Rashmi MR (2015) Reduction of common mode voltage in three phase inverter. In: 2015 International conference on advancements in power and energy (TAP energy), pp 244–248. IEEE 2015
5. Selvaraj J, Rahim NA (2009) Multilevel inverter for grid-connected PV system employing digital PI controller. IEEE Trans Industr Electron 56(1):149–158
6. Sefa I, Altin N, Ozdemir S, Demirtas M (2008) dSPACE based control of voltage source utility interactive inverter. In: SPEEDAM 2008 international symposium on power electronics, electrical drives, automation and motion, pp 662–666. IEEE 2008
7. Vijayakumari A, Devarajan AT, Devarajan N (2012) Design and development of a model-based hardware simulator for photovoltaic array. Int J Electr Power Energy Syst 43(1):40–46

8. Azharuddin SM, Vysakh M, Thakur HV, Nishant B, Babu TS, Muralidhar K, Paul D, Jacob B, Balasubramanian K, Rajasekar N (2014) A near accurate solar PV emulator using dSPACE controller for real-time control. Energy Procedia 61:2640–2648

9. Salam Z, Soon TL, Ramli MZ (2006) Hardware implementation of the high frequency link inveter using the dSPACE DS1104 digital signal processing board. In: IEEE international, power and energy conference, 2006. PECon'06, pp 348–352. IEEE 2006

10. Gadekar A, Virulkar VB (2014) Effective dSPACE inverter controller for PV application. In: 2014 IEEE students' conference on electrical, electronics and computer science (SCEECS), pp 1–5. IEEE 2014

11. Atkar D, Udakhe PS, Chiriki S, Borghate VB (2016) Control of seven level cascaded H-bridge inverter by hybrid SPWM technique. In: 2016 IEEE international conference on power electronics, drives and energy systems (PEDES), pp 1–6. IEEE 2016

12. Hannan MA, Abd Ghani Z, Mohamed A (2010) An enhanced inverter controller for PV applications using the dSPACE platform. Int J Photoenergy

13. dSPACE DS1104 (2008) Hardware installation and configuration and controldesk experiment guide. Paderborn, Germany

14. ECE 5671/6671—Lab 1 (2016) dSPACE DS1104 control workstation and Simulink tutorial, pp 1–38

15. Ghani ZA, Hannan MA, Mohamed A (2013) Simulation model linked PV inverter implementation utilizing dSPACE DS1104 controller. Energy Build 57:65–73

16. Alkhazragi MS, AL-Shamaa NK (2017) Cascaded H-bridge multilevel inverter using SPWM and MSPWM strategies. Int J Eng Res Appl 7(6), (Part-2):14–20

17. Avci E, Ucar M (2017) Analysis and design of grid-connected 3-phase 3-level AT-NPC inverter for low-voltage applications. Turk J Electr Eng Comput Sci 25(3):2464–2478

India's Lethal Informal E-waste Recycling: A Case Study of Delhi and NCR Region

Athar Hussain, Sanjay Kumar Koli, Rajdeep Tripathi and Suneel Pandey

Abstract In India, there is a very less awareness about the formal recycling of electronic wastes. The recycling is taken into account informally to make profit in terms of money. The current study has been carried out in order to examine the recovery potential from residues collected from e-waste sites at Delhi NCR location. All these recycling works are done informally and unorganized manner. Informal recyclers adopt techniques such as pyro-metallurgy (using heat), hydro-metallurgy (using acid) and electro-metallurgy (using current). After extraction of precious metals, the remaining residues are not properly dumped into landfill sites. These residues are just thrown on lands without any precautions. However, in these remaining residues, still there is a quantity of precious metals that are affix. Au is one of the prime precious metals which is found to be present in satisfactory amount in all the samples of e-waste. However, of all the metals, the lowest concentration of cobalt metal has been observed, and highest concentration of copper metal has been observed in all e-waste samples of the study. The key to success in terms of e-waste management is to develop eco-design devices, properly collect e-waste, recover and recycle material by safe methods, dispose of e-waste by suitable techniques, forbid the transfer of used electronic devices to developing countries and raise awareness of the impact of e-waste. A national scheme such as EPR is a good policy in solving the growing e-waste problems.

Keywords Electronic waste · Printed circuit board · Heavy metals

A. Hussain
Department of Civil Engineering, Ch. Brahm Prakash Government Engineering College Jaffarpur, New Delhi 110073, India

S. K. Koli (✉)
Department of Environmental Engineering, Ch. Brahm Prakash Government Engineering College Jaffarpur, New Delhi 110073, India

R. Tripathi
Department of Civil Engineering, Gautam Buddha University, Noida, Uttar Pradesh, India

S. Pandey
Environment and Waste Management Division, TERI, New Delhi, India

© Springer Nature Singapore Pte Ltd. 2020
S. Ahmed et al. (eds.), *Smart Cities—Opportunities and Challenges*,
Lecture Notes in Civil Engineering 58,
https://doi.org/10.1007/978-981-15-2545-2_31

1 Introduction

Due to gigantic increment in population and advancement in technology day by day, the waste generation is surging tremendously throughout the globe. The human population is expected to reach 11.18 million by the end of 2100 (U.N. Population Division). The global waste generation rate is expected to reach 2.2 billion tons per day by 2025. In spite of having so many advanced treatment technologies available in the world, most of the waste ends up in the landfill, and it creates a major problem not as an environmental aspect but health and development aspect as well [7]. E-waste is a generic term encompassing various forms of electrical and electronic equipment (EEE) that are old, end-of-life electronic appliances and have ceased to be of any value to their owners [15]. E-waste stream is increasing by 4–5% every year [5]. E-waste is a heterogeneous mixture of materials, its composition increasing rapidly and continuously [3]. Globally, 41.8 million metric tonnes of e-waste was generated in 2014. These electronic items are the hardware part of the computer [4]. India is the fifth biggest producer of e-waste in the world discarding 1.7 million tons (MT) of electronic and electrical equipment in 2014 [6].

Apart from Moradabad (U.P.), this is one of the major sites for recycling of e-waste in North India. Huge mountains of e-waste are formed day by day because of change in lifestyle of people, GDP growth of country and using of new electronic product in the market, and this creates a huge problem in developing as well as in developed countries. Kumar et al. describe that the population does not have any relation with the e-waste production, while GDP has direct correlation with e-waste production. In developed countries, e-waste constitutes 1–2% of the total solid waste generation. The growth rate of discarded electronic waste is high in India since it has emerged as an information technology giant and further aggravated due to modernization of lifestyle. E-waste is considered to be carcinogenic and hazardous waste [10].

Informal scrap networks have historically generated income by collecting, extracting and selling recyclable materials [2]. Inappropriate disposal of e-waste led to the human and environmental damage, along with the high value of harmful component materials. In patients, ten times more lead concentration has been found in blood steams [8]. Basically, copper, silver, gold and platinum encourage e-waste material recovery, and open burning to extract metals is widely adopted [1], and heavy metals are ultimately accumulated in the roots of the soil [9]. Gold, silver and palladium can be recovered with a high level of efficiency in the refining process. Copper and precious metal smelting works are linked to the material value of the PCBs supplied. A precise analysis data is available for different types of assembled PCBs.

The physical composition of e-waste is very diverse and contains over thousands of different substances, which falls under organic and inorganic fractions. Heavy metals form a significant part of inorganic fraction accounting for 20–50% and may affect the nervous system, blood and kidney [11]. These fractions consist of hazardous metallic elements like lead, cadmium, chromium, mercury, arsenic and selenium and precious metals like silver, gold, copper and platinum. More efficient recovery systems could be developed for the conservation of precious or valuable metals from e-waste.

Considering the above mentioned-facts in view, a study was being planned in order to assess the e-waste management strategy and to incorporate all of the major parameters of e-waste recovery, their uncertainties and interactions into an analysis of recovery of useful materials components. This includes analysing available separation and recovery technologies. The major objectives of the present study have been undertaken in order to assess the material recovery situation in informal sector of recycling in Delhi NCR region. Also, evaluating the concentration of precious and other metals from residues of PCBs generated after informal processing from Delhi NCR location is also a part of the study. The framework of environmentally sound management of e-waste residue has also been included in the study.

2 Materials and Methods

2.1 Study Area

In the present study, the Delhi NCR region comprising of e-waste sites located at Tila Sahbazpur, Loni (Rajiv Nagar, Mandoli), Ghaziabad and Old Seelampur, Delhi, has been surveyed and selected for the present study. At these sites, both collection and recycling procedures have been carried out informally by unprofessional workers. It has also been informed that most of the persons involved in recycling business are migrants from Bihar ($\approx$90%). E-waste is transported from institutions and cities across the country at two main clusters of Seelampur and Mayapuri.

2.2 Sample Collection, Preparation and Analysis

Burnt and discarded PCB samples of recycled e-waste were collected from the study site. The random grab sampling technique has been used for the collection of e-waste residue samples from the site. PCBs thrown by workers were collected from sites after and processing carried out by leaching process to recover all the precious metals undertaken in the study. The seven samples were collected of burnt and discarded PCBs. Pretreatment includes several steps as oven drying, crushing from cutter, grinding through mixer, sieving and sample weighing. PCB samples were dried in the oven at a temperature about 60–70 °C for 2–4 h and crushed by cutter into smaller pieces. The samples collected from different sites were being labelled and designated as s5, s6, s7, s8, s9, s10 and s11.

All collected samples were shredded into smaller pieces converted into powdered form by grinding through mixer. After grinding, the samples were passed through 60 mm size sieve. The analysis and testing of all the samples were being carried out in triplicate, and after digestion, 100 mL of each sample was collected for further study. The metal concentration has been determined using atomic absorption spectrophotometer (AAS). The analysis and stepwise calculations of the obtained results in the present study are shown in Fig. 1.

The stepwise calculations are as:

(i)   Metal in new PCB (g/kg of PCB) (A)
(ii) Metal in PCB after copper (Cu) recovery (g/kg of PCB) (B)
(iii) Metal discarded in environment (g/kg of PCB) (C)

Metal recovered in the second phase after the first phase of copper (Cu) removal has been calculated as PCB $(\%) = \frac{A-B}{A} \times 100$. Similarly, metal recovered in the third phase as compared to phase I after being burnt and discarded has been calculated using equation PCB $(\%) = \frac{A-C}{A} \times 100$

2.3   E-waste Survey on Selected Sites of Delhi/NCR

The information technology (IT) industries and related organizations have expanded electronics marketing in Delhi and its NCR regions broadly over the past few years. The constant upgradation of software has reduced the lifespan of personal computers (PCs) to about ten years [12]. As a result, a large number PCs are auctioned in Delhi. E-waste generated over years is algebraically directly proportional to the growth of IT companies in the city. Delhi produces about 9.5% of total e-waste generated in India. Delhi is the second largest city generating e-waste in India after Bangalore [13]. Besides the computer assembly hub operating in the city, the traders also receive huge quantities of PCs from different parts of country.

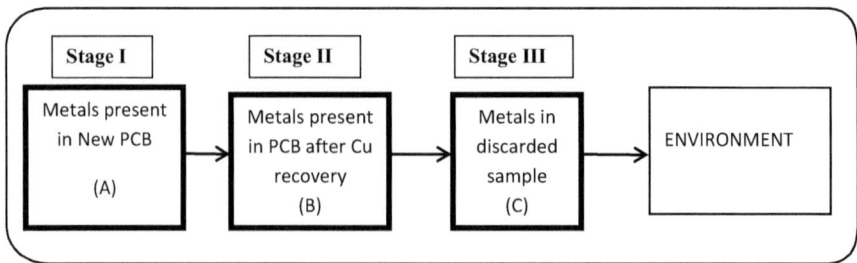

Fig. 1 Stepwise calculation for recovery rates of metals from PCB samples

Scrap dealers sell certain components like PCBs and integrated circuits (IC) directly to dismantlers for dismantling. Although, there are many sites at major locations in Delhi/NCR where e-waste get recycled and material recovery is done. The major e-waste generating sites undertaken in the present study have been discussed and surveyed accordingly.

3 Results and Discussions

The analysis of collected samples of gold and zinc has been carried out, and the result has been summarized in Table 2 and in Fig. 2. From Fig. 2, it can be observed that the maximum Au has been recovered from s11 as compared to the minimum value of sample from site s5.

Gold (Au) is a noble metal and highly precious in nature. It has high quantity of PCBs from e-waste. The highest Au concentration of 46.19 mg/L has been observed in sample s11. However, mean concentration of 32.4 mg/L Au has been observed in all the samples of the present study. Zinc (Zn) metal is most commonly used as an anticorrosion agent. A widely used alloy that contains zinc in brass, where copper is alloyed with from 3 to 45% zinc, depending upon the type of brass is to be manufactured. In the present study, the concentration of Zn in all samples ranges from 18 to 74 mg/L. The present investigation has also been carried out in order to determine the concentration of valuable metals present in the e-waste (Fig. 3).

An attempt on recovery potential of valuable metals in collected seven samples (PCB) from the Loni site of NCR region has also been undertaken in the present study. Samples are categorized into three categories comprising of sample number s5, s6, s7, s8, s11 as PCBs while sample number s9 as discarded and burnt after informal recovery of metals on site and sample number s10 having new PCB

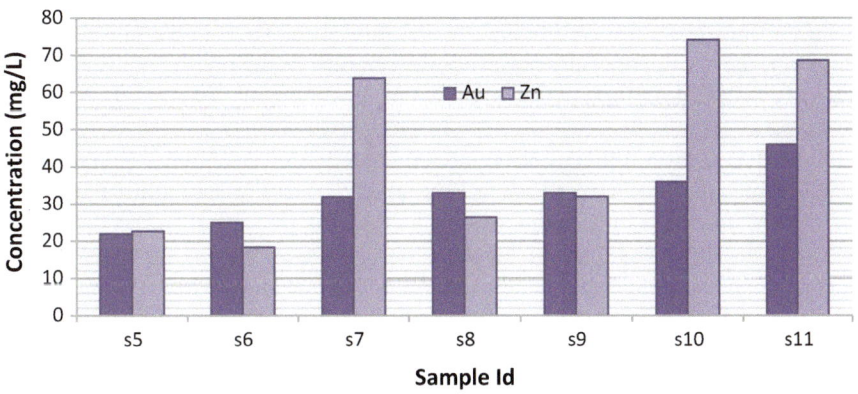

Fig. 2 Concentration of Au and Zn in samples from different locations of study area

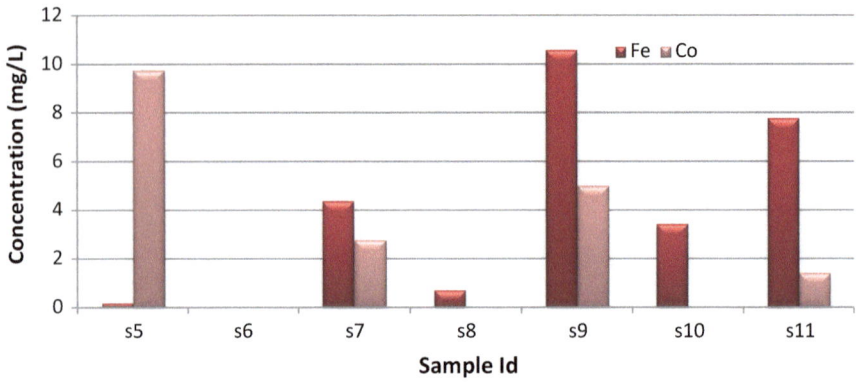

Fig. 3 Concentration of Fe and Co in samples from different locations of study area

Table 1 Metals concentration in e-waste as analysed in the present study

Sample id	Concentration (mg/L)									
	Ag	Al	Au	Fe	Cr	Ni	Pb	Zn	Co	Cu
s5	636	16,950	22	0.19	111	13	1465	22	9.76	1096
s6	746	321	25	0	11	30	713	18	0	2179
s7	6523	13,112	32	4.39	107	157	2175	63	2.78	2940
s8	791	280	33	0.72	8.86	16	679	26	0	2112
s9	16	16,899	33	10.6	547	21	782	32	5.01	34.5
s10	37	5110	36	3.43	113	51	2099	74	0	19
s11	29	10,671	46	7.78	297	14	377	68	1.42	2383
Average	1254	9049	32	3.87	170	43	1184	43	2.71	1538

collected and PCB after Cu recovery as summarized in Table 1. Ferrous concentration has also been observed in all the e-waste samples. The concentration of Fe in sample s6 has been observed below the detection level (bdl) in the present study. However, the mean concentration of all the samples in the present study has been observed to be 4.52 mg/L. The highest concentration of 10.64 mg/L has been observed in sample s9. Most part of the produced cobalt is utilized in preparing alloys and is used in different areas in manufacturing of electronic equipments.

Therefore, the produced cobalt is consumed in most of the alloys being produced. The foremost application of cobalt is as the free metal, in production of certain high-performance alloys. However, in the present study, the concentration of cobalt in most of the samples has been found to be below the detection limit. The maximum cobalt concentration has been observed in sample s9. Silver (Ag) is another major precious metal that has also been analysed in the present study. The average concentration of silver in all the samples has been observed to be 415.35 mg/L. However, it has also been observed that Ag can be recovered on large

Table 2 Metals concentration (g/kg) from e-waste as per category of PCB

Metals (g/kg)	New PCB collected (phase I)	PCB after Cu recovery (phase II)	PCBs discarded and burnt after informal recovery (phase III)
Ag	0.016	0.037	0.482
Al	16.90	5.110	7.740
Au	0.03	0.036	0.032
Fe	0.01	0.003	0.003
Cr	0.55	0.112	0.108
Ni	0.02	0.051	0.047
Pb	0.78	2.099	1.818
Zn	0.03	0.074	0.046
Co	0.01	0.00	4.653
Cu	0.03	0.019	1.788

scale from e-waste. The maximum aluminium (Al) concentration of 16,950 mg/L has been observed in the sample s5 taken in the present study. However, the maximum gold (Au) concentration of 46 mg/L has been assessed in sample s11 in the present study. Au is one of the prime precious metals that is found to be present in satisfactory amount in all the samples of e-waste. The highest chromium (Cr) concentration of 297 mg/L has been observed in the sample s10 of the study area.

The average concentration of 170 mg/L of chromium has been assessed in all samples of the study area. Nickel (Ni) is another important metal generally used for electroplating and metal coating by most of the metal manufacturing industries. The highest concentration of 157 mg/L of Ni has been observed in sample s7 as compared to all other samples. Lead (pb) is another important metal which is toxic heavy metal and can be recovered 100% if proper recovery method is applied. The highest Pb concentration of 2175 mg/L has been observed in sample s7 in the present study. However, the average lead concentration of 1670 mg/L has been found in all samples. The highest zinc (Zn) concentration of 74 mg/L has been observed in sample s10 of the study area. However, of all the metals, the lowest concentration of cobalt metal has been observed, and highest concentration of copper metal has been observed in all the e-waste samples of the study.

The load-based metal concentration at phase I, phase II and phase III from PCB samples has been determined and summarized in Table 3. In phase I, the maximum concentration of aluminium has been found to be 16.9 g/kg, and minimum concentration in phase I has been found to be 0.01 g/kg of metals Fe and Co. In phase II, the maximum Al metal concentration of 5.11 g/kg is observed, and no Co metal is detected in the phase II of the study. However, the maximum AL metal concentration of 7.74 g/kg has been observed, while minimum concentration of 0.003 g/kg of Fe is detected in phase III of study. The precious metals Ag and Au have also been recovered in all the phases of the study. The Au concentration of 0.03, 0.036 and 0.032 g/kg has been obtained in phase I, II and III of the study,

Table 3 Concentration of metals in all the samples analysed in the present study

Category	Sample id	Concentration (mg/L)									
		Ag	Al	Au	Fe	Cr	Ni	Pb	Zn	Co	Cu
Discarded and burnt PCBs	s5r1	665.31	17,637.5	25.31	0.06	117.37	13.12	1579.94	28.35	13.04	1218.81
	s5r2	579.23	16,735.7	22.46	bdl	109.92	14.46	1413.68	19.47	13.63	1032.19
	s5r3	661.99	16,475.9	16.91	0.32	8.89	11.72	1401.14	20.09	2.61	1037.31
	s6r1	712.89	282.2	19.53	bdl	10.38	30.97	719.83	20.01	bdl	2277.95
	s6r2	758.29	371.4	25.51	bdl	13.96	31.99	702.61	17.17	bdl	2206.54
	s6r3	765.71	308.8	28.94	bdl	98.67	27.67	717.32	17.97	bdl	2051.03
	s7r1	563.67	13,811.1	29.86	3.62	112.79	139.41	2098.75	59.84	bdl	3081.46
	s7r2	756.55	12,710.1	33.17	4.31	109.58	182.92	2204.39	58.32	bdl	2961.89
	s7r3	637.86	12,813.9	32.94	5.26	6.34	148.04	2221.88	73.23	2.78	2777.09
	s8r1	909.08	255.6	31.53	0.55	10.73	18.01	730.97	29.35	bdl	2112.61
	s8r2	627.82	267.8	32.83	0.99	9.52	13.61	646.84	25.33	bdl	1820.28
	s8r3	837.33	317.1	34.06	0.62	534.40	17.74	658.61	24.29	bdl	2401.64
New PCBs	s9r1	14.80	16,635.8	32.25	10.35	564.23	17.23	790.19	30.34	4.40	46.41
	s9r2	16.87	16,852.7	33.90	11.13	541.40	24.42	816.04	32.90	4.68	33.54
	s9r3	17.11	17,207.3	33.50	10.44	113.78	22.14	740.50	32.81	5.96	23.20
PCB after Cu recovery	s10r1	36.24	5128.7	33.33	3.09	102.17	76.97	2246.09	66.86	bdl	26.62
	s10r2	37.12	5203.4	35.06	2.84	120.85	30.86	2109.75	91.55	bdl	12.39
	s10r3	36.74	4997.0	40.69	4.36	328.73	44.09	1941.88	64.11	bdl	16.66
Discarded and burnt PCBs	s11r1	33.37	11,167.0	47.84	8.24	258.79	13.13	3229.51	72.45	bdl	2550.88
	s11r2	28.84	9988.8	45.74	7.46	302.54	15.25	4641.54	67.88	1.82	2448.24
	s11r3	25.63	10,859.0	44.99	7.63		14.24	3458.82	65.64	1.03	2149.96

respectively. However, the concentration of Ag has been found to be 0.016, 0.037 and 0.482 g/kg in phase I, II and III of the present study, respectively. However, in phase III, when PCBs are discarded and burnt after informal recycling, the maximum and minimum concentration of Al and Fe was observed as 7.74 and 0.003 g/kg in the e-waste samples (Table 2).

3.1 E-waste Situation at Major Sites in Delhi/NCR

3.1.1 Loni Study Area

Motherboards are processed at unconfined industrial area of Mandoli at a place called Tila. The area is owned by local landholders, and they give their land on rent to the recyclers. Dismantling and refurbishment of e-waste take place at five locations in Delhi as Shastri Park, Seelampur, Turkmangate, Mustafabad and Mayapuri. These cluster organizations tender and collect e-waste. Workers undertake testing of the collected materials, and if found suitable, the same is sent for refurbishment, and if they are not economically refurbishable, the same is sent for extracting metals and other valuable components. Materials that are recovered include ABS plastic, motherboard, tin, picture tube (99% refurbished), silver and gold. Also, it has been reported that the recycling efficiency is low due to unavailability of proper technology and remains only in extent of 30–40%. Wires are being burnt to take metal out of wires. There is no proper system and management for collected e-waste. Moreover, based on survey at site, the recyclers in Loni buy @ 400/kg and extract about Rs. 900/kg else there shall be no feasibility to run e-waste recycling business.

3.1.2 Old Seelampur Study Area

Trucks carrying e-waste unload the waste at site from 4 am–12 pm. The cluster is almost 20 years old. There are about 300 traders working in old Seelampur cluster. These traders buy e-waste and sell the same to the processors of the waste for recovering metals which have immediate monitory value. The tubes from the monitors are sometimes usable and used to make televisions. Unlike old Seelampur, Mayapuri cluster deals more in waste of electric motors. The e-waste collected includes as computer boards (Rs. 40–200 per kg), CFL (Rs. 50 per kg), chargers, fan motors, wires, cooler kits, monitors (Rs. 500–1000 per PC) and batteries, cameras, etc. Waste is bought to these traders in PVC bags and usually also common types of waste per bag. The mode of transportation is railways, trucks and transport systems on sharing basis. Waste from institutions and also even the e-waste collected from recognized recyclers many a time are sold among these traders for processing. The rates of e-waste fluctuate as market rates of metals vary. The most expensive e-waste is motherboard of computers which even has recoverable gold in it. The turnover of an e-waste trader has been informed as 5 tonnes per month, and economically, it is found varying from Rs. 18,000 per month to more than Rs. 500,000 per month.

3.2 Major Observations at E-waste Generation Sites of Study Area

3.2.1 Loni Site

Cluster has recycling units for rubber (tyres, shoe sole, etc.). The cluster also has recycling unit of glass and produces thread for flying kites (Manja). Some small workshops heat the motherboard of PCs, computer and other electronic instruments by using a domestic LPG cylinder. The varying components are removed. Motherboards after removing components still have metals. These boards are bought, stacked and open burned. Upon cooling, the metals like copper, iron and brass are segregated manually by cheap labour. It has also been observed that these motherboards are burnt during night hours (maybe to avoid attention of authorities). Acid used for recovering after use is dumped on ground without any treatment. The motherboards after all the possible recovery are also dumped randomly at convenient locations in a haphazard manner. Parts like resistors and capacitors are handled by other kinds of recyclers. These parts are first grinded and then heated at very high temperature to get rid of the plastics. The metals then secured are copper, aluminium, iron and brass. These pieces of metals are sorted through hand picking, sieving and magnets. Recyclers specialize in recycling of different parts of motherboard. Motherboards are heated to remove parts like resistors, capacitors, etc.

3.2.2 Old Seelampur Site

No specific land or cluster is allotted by the government to the site. No facilities, campaigning and awareness are initiated by the government to improvise the e-waste management process. It is also observed that there are lakhs of people who are employed through this business, and these are mostly labour class.

4 E-waste Recovery at Sites of Study Area

4.1 Loni Site

Material recovery in recycling units is highly generic and rudimentary in nature. The workers work in closed dingy areas. They handle acid, caustic solution and other health hazardous chemicals. The workers are thus exposed to several toxic elements. Further, on oversight at site of Loni, it is investigated that motherboards from expensive equipment like magnetic resonance imaging (MRI) and CT scan machines have high quantities of precious materials. These motherboards are not processed in India but are sent to a company in Belgium. It is claimed that the company can recover 98% of the precious metals from the motherboards. According to site workers, 18 tons of motherboard, 2.5 kg of gold (24 carat), 170 kg of silver and 1.75 kg of platinum are recovered. On an average, each kg of motherboard contains three pieces, which is sold the Belgian company at Rs. 5000–6000.

4.2 Old Seelampur Site

The e-waste collected is directed to the informal recovery sites where expensive metals like gold, silver, copper, brass, aluminium and shoulder are recovered from the waste. The unrecovered materials from the e-waste such as mica boards and IC parts land up in dump sites after processing. The informal trading and recovery sites are located around residential areas. Thus, the recyclers face issues with the area and money issues. The informal recyclers need recognition from the government apart from allotment of space and basic technology to safely recover material.

4.3 E-waste Recycling

Based on the latest demonstration, eco-friendly procedures such as biotechnological approaches which seem to be a valuable tool for recovery of precise metals (gold, silver and platinum) are observed. The presence and restoration of precious metals in electronic waste even in small quantities such as gold and silver influence more than 50% of the economic value of recycling business. The PCBs in the present study have been recovered by using simple mechanical and chemical procedures. Density and magnetic separation method have been used in order to collect, disassemble, pulverization and separation of PCBs. However, in the present study, 98.82% purity of $CuSO_4$ hydrate and $Al_2(SO_4)_3$ hydrates were recovered from the PCBs by using the chemical recycling method.

The scrap dealers generally adapt an environmentally friendly dismantling technique in order to recover valuable materials and to minimize the adverse effects of hazardous materials contained in CRT and PCB's scrap. Therefore, the useful and hazardous materials can be manually separated and recovered by using this eco-friendly dismantling process. Thereafter, the retrieved materials are sent for particular treatment facilities accordingly depending upon the characteristics of that particular material.

Furthermore, from recycling of printed circuit boards for the recovery of valuable materials, the effective use of high-temperature pyrolysis is accrued out. However, silver, gold, palladium and platinum are the recovered precious metals present in waste PCBs having a clear tendency to form solutions with the main metallic constituents of waste PCBs, namely copper, tin and lead. The effective concentration has been observed due to the high affinity between copper and these precious metals. Also, a small percentage of 5–20% of valuable metals has been detected in the nonmetallic part after heat treatments in the temperature range of 800–135 °C.

5 Conclusion

As per the experimental work carried out in the present study, it has been observed that recovery rate of copper metal in phase II is estimated to be 46% by weight of e-waste as compared to phase I. Moreover, recovery rate of aluminium, gold, iron and chromium from PCBs (after burnt and discarded in phase III) has been found to be 54.19%, 2.89%, 68.96% and 80.3% by weight, respectively, as compared to phase I. Due to rapid growth of informal sector recycling, recovery or extraction of reusable components like as ceramics, polymers and metals from PCBs is processed in informal sector of Delhi NCR. However, the challenge is to adopt efficient process with maximum recovery of materials and minimal loss. The local people are suffering from health concerns as nitrogen dioxide (NO_2) and mercury (Hg) fumes might be hazard to workers since it can damage the respiratory systems. The threats recognized on e-waste recovery sites are the produced fumes while recovery process, untreated effluent and dumped solid waste. Hence, there should be a proper management and precaution systems for safety of labours at e-waste recycling sites.

References

1. Awasthi AK, Zeng X, Li J (2016) Comparative examining and analysis of e-waste recycling in typical developing and developed countries 35:676–680. https://doi.org/10.1016/j.proenv.2016.07.065
2. Borthakur A, Govind M (2017) How well are managing e-waste in India: evidences from the city of Bangalore. Energy Ecol Environ 2(4):225–235. https://doi.org/10.1007/s40974-017-0060-0
3. Chancerel P, Meskers CEM, Hagel C, Rotter VS (2009) Assessment of precious metal flows during preprocessing of waste electrical and electronic equipment 13(5). https://doi.org/10.1111/j.1530-9290.2009.00171.x
4. Debnath B, Roychoudhuri R, Ghosh SK (2016) E-waste management—a potential route to green computing. Procedia Environ Sci 35:669–675. https://doi.org/10.1016/j.proenv.2016.07.063
5. E-waste monitor (2014)

6. Koli SK, Hussain A, Ahmed A (2018) Status Electron Waste Manag India 5(2):75–80
7. Kumar S, Bhattacharyya JK, Vaidya AN, Chakrabarti T, Devotta S, Akolkar AB (2009) Assessment of the status of municipal solid waste management in meto cities state capitals, class I cities, and class II towns in India : an insight 29, 883–895. https://doi.org/10.1016/j.wasman.2008.04.011
8. Kush A, Arora A (2013) Propos Solut E-waste Manag 2(5):7–10. https://doi.org/10.7763/IJFCC.2013.V2.212
9. Panwar RM, Ahmed S (2018) Assessment of contamination of soil and groundwater due to e-waste handling 114(1):166–173
10. Rakib A, Ali M (2014) Full length research paper electronic waste generation : observational status and local concept along with environmental impact 2(10):470–479
11. Sivaramanan S (2013) E-waste management. Dispos Its Impacts Environ Abstr 3(5):531–537
12. Tran CD, Salhofer SP (2018) Processes in informal end—processing of e-waste generated from personal computers in Vietnam. J Mater Cycles Waste Manage 20(2):1154–1178. https://doi.org/10.1007/s10163-017-0678-1
13. Wath SB, Chakrabarti PSDT (2011) E-waste scenario in India, its management and implications 249–262. https://doi.org/10.1007/s10661-010-1331-9
14. U.N. Population Division
15. United Nations Environment Programme (2007)

Feasibility of Aquatic Plants for Nutrient Removal from Municipal Sewage in Smart Cities

Mohd. Najibul Hasan, Abid Ali Khan, Sirajuddin Ahmed, Henna Gull, Mohammed Sharib Khan and Beni Lew

Abstract An attempt was made to investigate the removal of nitrogen and phosphorous from municipal sewage using four aquatic plants (two emergent plants— *Typha latifolia and Phragmites australis* and two floating plants—*Eichhornia crassipes and Lemna gibba*). Batch studies were carried out in five reactors. Each batch reactor was having an effective volume of 49L. All batch reactors were fed with municipal sewage. The NH_4-N, NO_3-N and PO_4-P concentrations were measured at an interval of three days. Results of this study indicate that the highest removal efficiencies of NH_4-N were observed as 80% and NO_3-N and PO_4-P were 75% using the emergent plant (*Typha latifolia*) at an hydraulic retention time (HRT) of 21 days. The final value of treated effluent NH_4-N, NO_3-N and PO_4-P concentrations were found to be 7.5, 1.48 and 3.0 mg/L, respectively. The primary cause of the removal of nutrient from municipal sewage using *Typha* could be the presence of vigorous roots of this plant providing an expanded surface for microbial growth compared to other aquatic plants.

Keywords Municipal sewage · NH_4-N · PO_4-P · Aquatic plants · Dissolved oxygen

1 Introduction

The concept of aquatic plants-based treatment system is one of the smart ways to reduce the burden of pollution of water bodies and could be a possible way to reuse or achieve disposal standards for the municipal wastewater generated from smart city community. Lacking access to clean water and sanitation could be a common problem affecting human health throughout the smart cities like developing town.

Mohd. Najibul Hasan (✉) · A. A. Khan · S. Ahmed · H. Gull · M. S. Khan
Department of Civil Engineering, Jamia Millia Islamia (Central University),
Jamia Nagar, New Delhi 110025, India

B. Lew
ARO, Volcani Center, Bet Dagan, Israel

© Springer Nature Singapore Pte Ltd. 2020
S. Ahmed et al. (eds.), *Smart Cities—Opportunities and Challenges*,
Lecture Notes in Civil Engineering 58,
https://doi.org/10.1007/978-981-15-2545-2_32

Centralized wastewater system, energy and cost-intensive technologies are ineffective to resolve the complex water-related problem [1]. There is a need to investigate and implement alternative treatment system for the current wastewater technology [2]. In the field of wastewater treatment, aquatic plants are one of the low-cost treatment concepts which have not been recognized yet for smart cities. The natural systems for the treatment of low-strength municipal wastewater are still one of the most economical and sustainable technologies widely used around the world due to its simple operation and user friendly [3, 4]. In general, these treatment concepts were used in a variety of geometrical shapes and sizes like constructed wetlands (CWs), lagoons waste stabilization ponds (WSPs) with or without aquatic weeds and frequently used for the treatment of municipal wastewater. This type of system is also effective to reduce the large amounts of nonpoint source pollutants that occurred by the rainfall and rain washing the village grounds and fields [5]. Researcher also reported that constructed and natural wetlands are often used for municipal wastewater effluent, i.e., single-residence septic tank effluent and at large municipal wastewater plants due to low cost of treatment [3, 6]. Various simultaneous processes like filtration, sedimentation, adsorption, bioconversion, combination of nitrification/denitrification, uptake by microorganism, wetland microphyte and microbial processes were occurring in natural systems for the removal of nutrients. The selection of aquatic plant in constructed wetlands should be based on strong vitality, pollution resistant, survive in aquatic and wet environment, diseases and pests [7]. Enormous studies are found in the literature on the performance evaluation of the constructed wetlands (CWs) based on geometry like dimension, area and substrate used. The role of plants is equally important but limited studied observed on fate of organic and nutrients removal for the treatment of municipal sewage [8].

An attempt was made to investigate the nutrient removal using four widespread aquatic plants, two emergent plants and two floating plants from municipal wastewater and feasibility for smart city.

2   Materials and Methods

2.1   Reactor Configuration

Five batch reactors were used for the present study. Each reactor was made of plastic material having cylindrical shape of diameter 35 cm and height of 51 cm. The effective volume of each reactor was 49 L. Various types of plants, i.e., emergent plants (*Typha Latifolia* and *Phragmites Australis)* and floating plants (*Lemna gibba* and *Eichhornia crassipes)*, were introduced in four different batch reactors. For emergent plants, the density of plant was kept six, while for floating plants, planting coverage area was maintained half of the sewage surface. One batch reactor was treated as control without any plant.

Three layers of different strata/media, i.e., gravels (20–40 mm), at the bottom followed by fine gravels (1–2 mm) and sand mixed with mud (0.0039–0.065 mm) above it were made in each batch reactor that was obtained from vicinity of the campus of university. The gravels, coarse sand and silt were thoroughly washed with tap water until the supernatant solution was clear. The emerging plants, i.e., *Typha latifolia* and *Phragmites australis*, were procured from nursery, while floating plants, i.e., *Lemna gibba* and *Eichhornia crassipes*, were collected from Okhla barrage at Yamuna River near Kalindi Kunj, New Delhi. The collected aquatic plants were washed to remove dirt and other adhesive materials introduced in the reactors. The sewage was brought to laboratory from the sewer pumping station Batla House, New Delhi.

2.2 Experimental Protocol

Experimental setups (batch reactors) were operated at the ambient temperature varied from 20 to 32 ± 5 °C, with a light intensity of 1500 Lx during day time. Initially, all the batch reactors were filled with municipal sewage and kept for fifteen days for the stabilization of plant. Later, sewage in the reactors was replaced with fresh municipal sewage. The present study was performed for 45 days. The depth of the municipal sewage above the top surface was maintained constant by keeping 60 mm height avoiding the evaporation intervention. During the course of the study, this height was maintained by adding freshwater in the reactors. The porosity of the reactors was found to be 40% which was calculated after the stabilization of plants in the reactors.

Analytical Procedure
Samples were analyzed to determine the concentration of ammonia (NH_4^+ -N), nitrate (NO_3^- -N) and phosphate (PO_4-P) at an interval of three days. All samples were analyzed in triplicate according to the Standard Methods for Examination of Water and Wastewater [9].

3 Results and Discussions

3.1 Emergent Plant (Typha latifolia and Phragmites australis)

The influent and effluent concentrations and percent removal for NH_4-N, NO_3-N and PO_4-P were monitored in the batch studies planted with *Typha latifolia* and *Phragmites australis* with an interval of three days.

NH$_4$-N and NO$_3$-N

Figures 1 and 3 show the variation of ammonium concentration in reactors planted by *Typha latifolia and Phragmites australis and Temporal variation of NO$_3$-N by Typha latifolia and Phrgamites australis respectively*. The result indicates that there is a significant reduction in NH$_4$-N and NO$_3$-N. The reduction in NH$_4$-N reached to 7.5 and 11 mg/L from 38 mg/L in reactor planted by *Typha latifolia and Phragmites australis*, respectively, and similarly, NO$_3$-N reduced to 1.48 and 1.76 mg/L from the initial value of 6 mg/L. The regression coefficient R^2 = 0.85–0.89. Removal of ammonia and nitrate was influenced by temperature that varied from 20 to 38 °C. During start-up, the removal efficiency of NH$_4$-N and NO$_3$-N was significantly higher with both types of plants. From Figs. 1 and 2, removal of NH$_4$-N was observed about 75–80% and NO$_3$-N about 71% at HRT of 21 days by *Typha latifolia and Phragmites australis* planted reactor and 21–45 days, varied from 71 to 90%. The graph shows less variation in removal after 21 days, the reasons is dead lives or plant litter decomposition. Figure 2 shows the variation in removal of NH$_4$-N and NO$_3$-N in control reactor. Results were statistically found significant, and the percentage removal of nitrogenous pollutants in planted reactors is significant (p > 0.05) (Fig. 2).

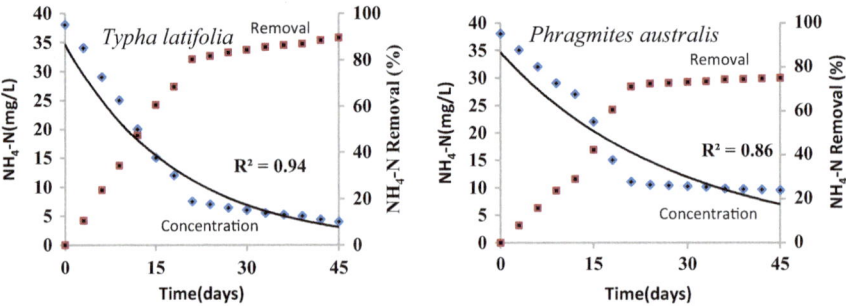

Fig. 1 Temporal variation in NH$_4$-N by *Typha latifolia* and *Phrgamites australis* planted reactor

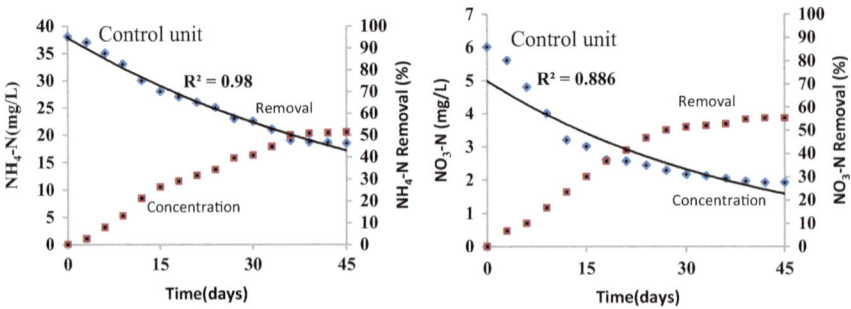

Fig. 2 Temporal variation of NH$_4$-N and NO$_3$-N by control reactor

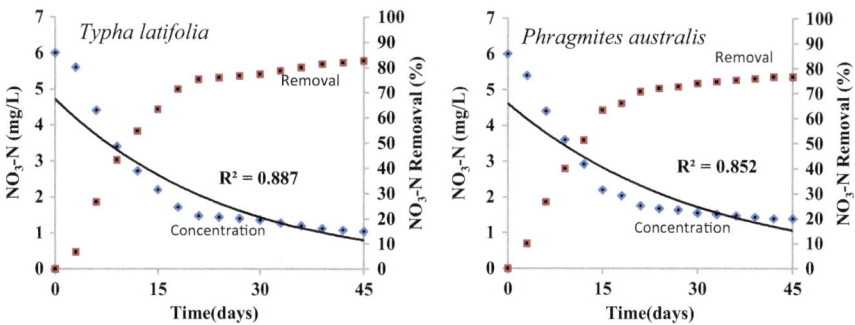

Fig. 3 Temporal variation of NO$_3$-N by *Typha latifolia* and *Phrgamites australis* planted reactor

PO$_4$-P

Figure 4 shows the variation of phosphorous removal in batch reactors having *Typha latifolia and Phragmites australis*; graph trends show 21 days retention time is an appropriate period for satisfactory removal. The removal of PO$_4$-P was observed greater than 70% and reduces to 3 mg/L from the initial concentration of 12 mg/L at an HRT of 21 days. The removal efficiencies of PO$_4$-P achieved in present study at an HRT of 21 days were 75 and 73% by *Typha latifolia and Phragmites australis* planted reactors were higher than the removal reported by [10] against of 50% removal efficiency in constructed wetland vertical type flow. The higher removal of PO$_4$-P might be due to the high density of plant and high HRT. The highest removal may also be due to the PO$_4$-P adsorbed in the media bed, rather than in the plant as sand used in this study contain oxides of Fe, Al and Ca. These minerals could enhance the phosphorous retention due to chemical adsorption and precipitation in wetlands. Other possible cause of PO$_4$-P removal might be the adsorption in algal and microorganisms that utilize phosphorous as an essential nutrient and contain phosphorous in their tissues. Results of this study were well

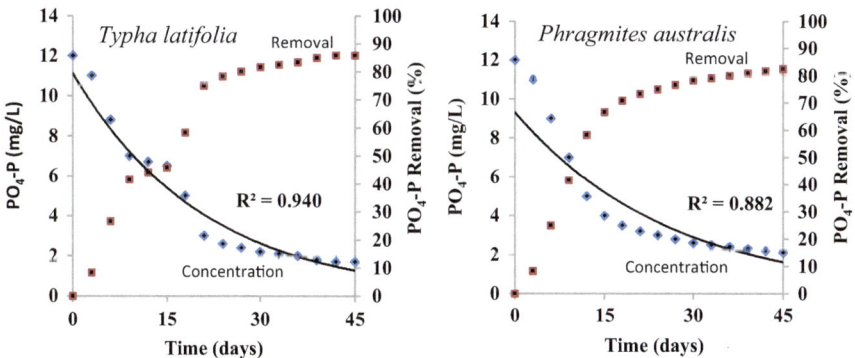

Fig. 4 Temporal variation of PO$_4$-P by *Typha latifolia* and *Phrgamites australis* planted reactor

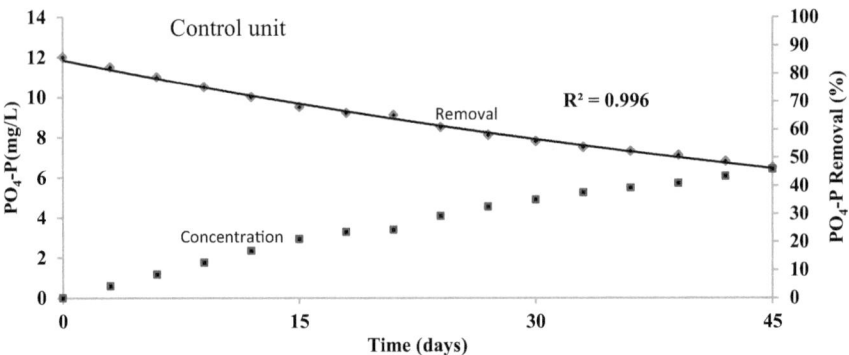

Fig. 5 Temporal variation of PO_4-P by control reactor

supported by one-way ANOVA with significant difference ($p < 0.05$) for removal of *Typha latifolia and Phragmites australis*-based reactor. The removal by *Typha latifolia*-based reactor was higher compared to *Phragmites australis*-based reactor. The overall phosphorous removal efficiency of *Typha latifolia*-based reactor was observed significantly good based on comparison with control reactor (Fig. 5).

3.2 *Floating Plants (***Eichhornia crassipes** *and* **Lemna gibba***)*

In the batch studies planted with *Eichhornia crassipes* and *Lemna gibba,* the influent and effluent concentrations along with percent removal for NH_4-N, NO_3-N and PO_4-P were analyzed.

Variation of NH_4-N and NO_3-N
Figures 6 and 8 shows the variation of NH_4-N and NO_3-N concentration in batch reactors planted two different plant species, viz. *Lemna gibba* and *Eichhornia crassipes*. Results show the removal of NH_4-N and NO_3-N was higher in *Lemna gibba*-based batch reactor compared with *Eichhornia crassipes*. The NH_4-N removal efficiency in *Lemna gibba*-based batch reactor at HRT of 21 days observed 68%, and an initial concentration of 38 mg NH_4-N/L reduced to 12 mg NH_4-N/L with same reactor. The removal of NO_3-N was 68%, and the final concentration was observed as 1.92 mg/L (Fig. 7).

Results were well supported by statistically evaluation using single-way ANOVA and found that the removal of NH_4-N and NO_3-N significant ($p > 0.05$) difference from control reactor.

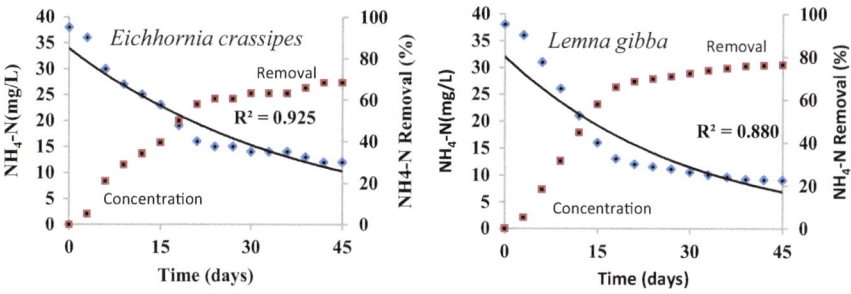

Fig. 6 Temporal variation in NH₄-N by *Eichhornia crassipes* and *Lemna gibba* planted reactors

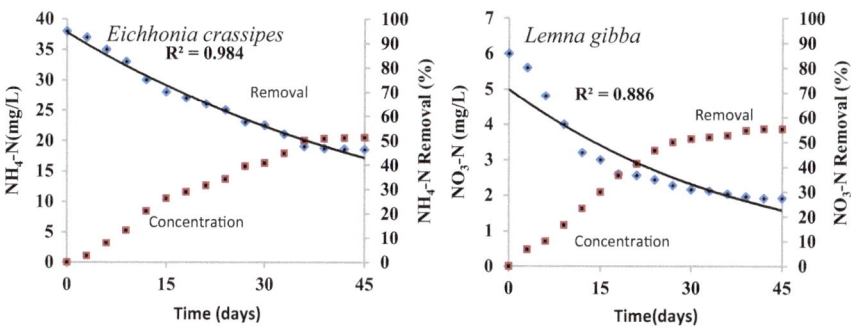

Fig. 7 Temporal variation of NH₄-N and NO₃-N by control reactor

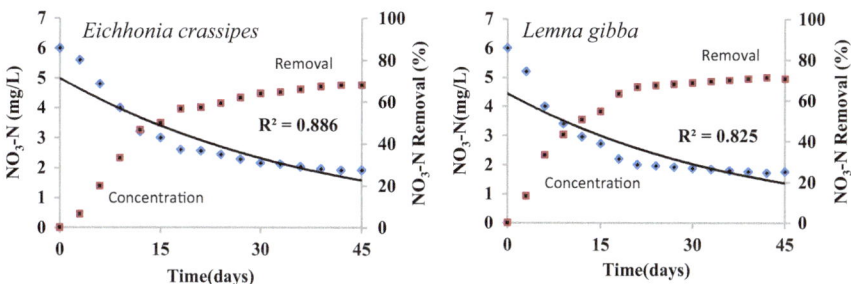

Fig. 8 Temporal variation of NO₃-N by *Eichhornia crassipes* and *Lemna gibba* planted reactor

Variation of PO4-P

Figure 9 shows the variation of phosphorous removal by *Lemna gibba* and *Eichhornia crassipes*. Initially, the PO_4-P concentration of municipal sewage was 12 mg/L and at retention time of three weeks. The concentration of PO_4-P reduced to 4.4 and 5.7 mg/L in batch reactors having *Lemna gibba* and *Eichhornia crassipes*, respectively, with regression coefficient of $R^2 = 0.89$–0.92. The removal

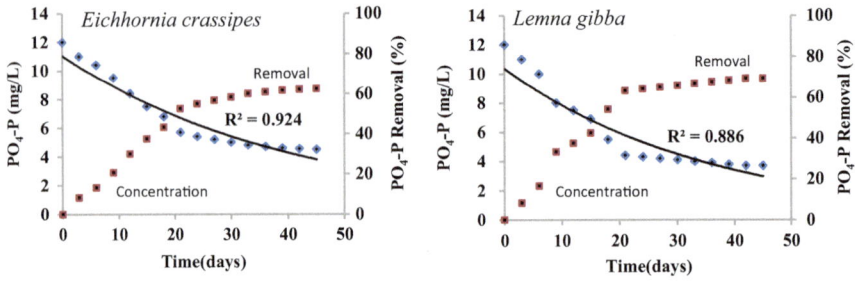

Fig. 9 Temporal variation in PO$_4$-P by *Eichhornia crassipes* and *Lemna gibba* planted reactors

efficiency was achieved as 63 and 53% by *Lemna gibba* and *Eichhornia crassipes*. The trend of nutrient removal during retention period of 21–45 days observed scantly due to dead leaves and plants.

4   Conclusions

Batch studies investigated the comparison of the performance of emergent and floating plant for the removal of nutrients from municipal sewage. The highest removal of nutrients in terms of NH$_4$-N, NO$_3$-N and PO$_4$-P was observed in Typha latifolia plant species at an HRT of 21 days.

- The highest removal of the NH$_4$-N was 80, and NO$_3$-N and PO$_4$-P were 75%, and the final treated effluent concentration was reduced to 7.5, 1.48 and 3.0 mg/L, respectively.

Results inferred that the aquatic weeds are the promising option for the treatment of municipal sewage for small communities due to lower capital cost, easy operation and maintenance and highest removal in nutrients. However, further studies are required at demonstration scale in order to upscale the technology.

5   Acknowledgement

Authors are highly thankful to the Department of Science and Technology and UGC, Government of India to provide the financial support to carry out this study.

References

1. Zhang DQ, Jinadasa KBSN, Gersberg RM, Liu Y, Ng WJ, Tan SK (2014) Application of constructed wetlands for wastewater treatment in developing countries–a review of recent developments (2000–2013). J Environ Manage 141:116–131
2. Ahmed S, Dhoble Y, Gautam S (2012) Trends in patenting of technologies related to wastewater treatment. Available at SSRN 2148918
3. Ahmed S, Popov V, Trevedi RC (2008) Constructed wetland as tertiary treatment for municipal wastewater. In: Proceedings of the institution of civil engineers-waste and resource management. Thomas Telford Ltd., vol. 161(2), pp. 77–84
4. Tomenko V, Ahmed S, Popov V (2007) Modelling constructed wetland treatment system performance. Ecol Model 205(3–4):355–364
5. Liu H, Dai ML, Liu XY, Ouyang W, Liu PB (2004) Performance of treatment wetland systems for surface water quality improvement. Huan jing ke xue = Huanjing kexue 25 (4):65–69
6. Vymazal J (2010) Constructed wetlands for wastewater treatment: five decades of experience. Environ Sci Technol 45(1):61–69
7. Ziqiang A, Jie Z, Guiqun P, Jiaqi F, Cheng J, Jihai X (2017) Plant selection of constructed wetlands for treatment of piggery wastewater. Meteorological and Environmental Research 8 (2):85
8. Hasan MN, Khan AA, Ahmad S, Lew B (2019) Anaerobic and aerobic sewage treatment plants in Northern India: Two years intensive evaluation and perspectives. Environmental Technology and Innovation, 100396
9. American Public Health Association (APHA) (2005) Standard methods for the examination of water and wastewater, 21st edn. American Public Health Association, Washington DC
10. Zurita F, De Anda J, Belmont MA (2009) Treatment of domestic wastewater and production of commercial flowers in vertical and horizontal subsurface-flow constructed wetlands. Ecol Eng 35(5):861–869

Modeling Security Threats for Smart Cities: A STRIDE-Based Approach

Malik Nadeem Anwar, Mohammed Nazir and Adeeb Mansoor Ansari

Abstract With the rapid advancement in IOT devices and networking technologies in recent years, the concept of smart cities has emerged as an important paradigm. A smart city is a heterogeneous network of ubiquitous sensors, along with intelligent processing and control systems. A smart city monitors the residents in real time and facilitates its inhabitants with intelligent services in terms of health, transportation, energy, governance, etc. Thus, security and privacy have become inevitable concerns for such systems. Therefore, ensuring security and privacy in smart cities is essentially needed in order to enhance trust and confidence level among citizens, thereby increasing their participation and trust with the services. STRIDE, a popular threat modeling technique developed by Microsoft, is widely used for modeling security threats in a complex system. In this paper, authors have made an attempt to model security-related threats in smart cities using the STRIDE methodology, categorize them accordingly, and suggest possible countermeasures for each threat.

Keywords Smart cities · STRIDE · Threat modeling · Data-flow diagram (DFD) · Countermeasures

1 Introduction

Smart cities can be defined as an urban development model based on the human, technological, and collective capital for developing urban agglomerations [1]. Though there are numerous definitions available in the literature, there is no agreed

M. N. Anwar (✉) · M. Nazir · A. M. Ansari
Department of Computer Science, Jamia Millia Islamia (A Central University),
Jamia Nagar, New Delhi, India
e-mail: mr.maliknadeem@gmail.com

M. Nazir
e-mail: mnazir@jmi.ac.in

A. M. Ansari
e-mail: adeebmansooransari.am@gmail.com

© Springer Nature Singapore Pte Ltd. 2020
S. Ahmed et al. (eds.), *Smart Cities—Opportunities and Challenges*,
Lecture Notes in Civil Engineering 58,
https://doi.org/10.1007/978-981-15-2545-2_33

upon definition for smart cities because of the multidisciplinary dimensions involved and dynamic advancements in technology and in urban development. Smart city uses rapid advances in information and communication technologies (ICT) for efficient management, design, and operation of urban services primarily in the sectors of governance, transportation, energy, health, water, and society [2]. The recent advances in the field of smart cities and its ability to impact the quality of human lives have made it the fastest-growing sector in recent years. The popularity of smart cities has increased further due to the reduction in hardware cost and easy integration with cheap handheld devices and tablets [3].

However, the increase in the number of interconnected services, components, and technologies has exposed the inhabitant of the smart cities to number of security and privacy risks. Due to the highly interconnected nature of the systems, vulnerabilities in any part of the system can have far-reaching effects on the whole system [2]. Secondly, most of the devices and applications used in smart cities are often novel and lack adequate security testing giving rise to a new set of security issues. Although some traditional techniques like encryption, authentication, and various security policies might be helpful to some extent [4], the emerging smart attackers" could easily interfere and exploit the vulnerabilities, thereby compromising the system [5].In view of the above facts, security and privacy are the key issues that have to be tackled at early stages while designing systems for smart cities.

Threat modeling is a useful technique to determine the anticipated threats to the system, its components, and suggest possible mitigation in the form of cost-effective security solutions. Microsoft's STRIDE, a lightweight systematic and comprehensive approach, is widely used for modeling threats of the system at component level. Since smart cities are a complex network of interconnected components, we strongly feel that this method will be useful in modeling threats in the domain. The objective of this paper is to provide a step-by-step demonstration of the STRIDE approach in modeling threats and produce an effective categorization of threats along with the countermeasure in the smart city domain. Rest of the paper is as follows: Sect. 2 describes the STRIDE methodology, Sect. 3 elaborates on the smart city architecture, Sect. 4 discusses the application of the STRIDE methodology in the smart city domain while Sect. 5 concludes the paper along with feature work.

2   STRIDE Methodology

It is a lightweight, practical, and user-friendly technique proposed by Microsoft for modeling threats to a system at the design phase of SDLC [6]. STRIDE classifies different threats of a system into six broad categories [7]:

- **Spoofing (S)**: impersonating as a legitimate user
- **Tampering (T)**: modifying legitimate information
- **Repudiation (R)**: denying a particular event or action performed in the system

Fig. 1 STRIDE
methodology

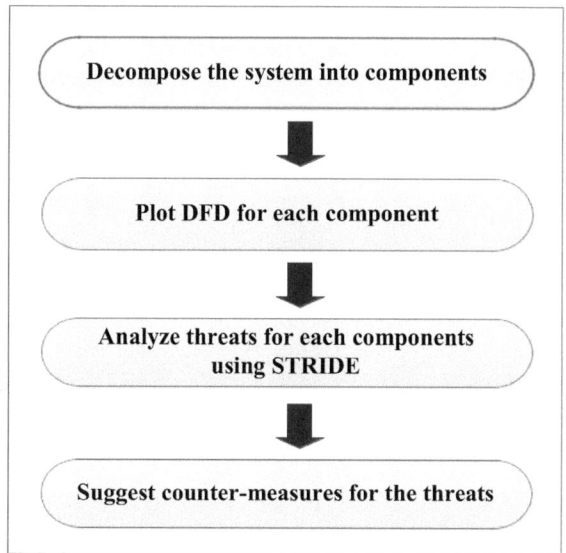

– **Information disclosure (I)**: unauthorized exposure of confidential information
– **Denial of service (D)**: making the system unavailable for legitimate user
– **Elevation of privilege (E)**: getting higher privilege than granted for a particular user.

STRIDE analyzes weakness against each component of the system that can possibly be exploited for possible threats, thereby compromising the whole system. Various methodologies are available in the literature guiding security experts in applying the STRIDE approach. Due to the absence of well-defined methodology, we adopted the high-level steps suggested by [7] for modeling threats using STRIDE. The steps are: (1) decompose the system into components; (2) sketch the data-flow diagram (DFD) for each component of the system; (3) identify threats for each element in the DFD using the STRIDE mnemonics; (4) suggest possible countermeasures to the threat (Fig. 1).

3   Smart City Architecture

Modern-day smart cities are the integration of the advanced wireless technologies and sensor networks along with intelligent computation, providing smart and flexible real-time support to the inhabitants. Any smart city network must ensure high-level customized services for its residents, thereby providing a better living environment and improved utilization of resources. We envisage the smart city architecture as a user-centric model, with important elements like intelligent health care, smart homes, smart transportation, smart grid, and smart governance connected to the user via smart phones. Though smart city is not confined only to these elements, for the sake of simplicity and illustration we have included only these five key components in the smart city setup. An abstract view of the components of the smart home along with their interactions is shown in Fig. 2.

4 Applying STRIDE to the Smart City Domain

This section provides a detailed overview of the STRIDE-based threat modeling approach applied in the smart city domain. The approach consists of four simple steps which are: (1) decomposing the system into components; (2) plotting the DFD data-flow diagram) for each component of the system; (3) identify threats for each element in the DFD using the STRIDE mnemonics; (4) suggest possible counter-measures to the threat. We hereby elaborate on each of these steps in detail.

4.1 Decompose the System into Components

After analyzing the smart home architecture, we have identified five key components, namely smart health care, smart homes, smart transportation, smart grid, and smart governance. The components along with their interactions are described in Fig. 2.

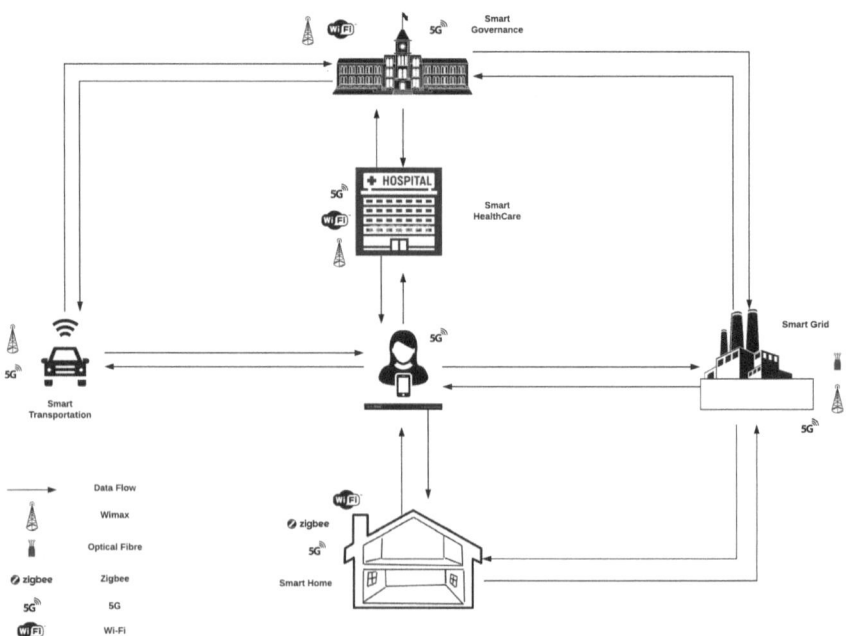

Fig. 2 An abstract view of a smart city

4.2 Plot the DFD for Each Component

DFD is a high-level UML diagram that captures the functioning of individual system components and analyzes the data flow through them [8]. After decomposing the system into manageable components, we have sketched a DFD for each component of the system using Microsoft Visio tool 2010. Due to the limitation of space, the individual DFD is combined together to form one single DFD (Fig. 3). We further annotated the DFD components with four prime threat-prone elements, namely sensing devices(S), communication channel (C), APIS/computation (AC), and database (D). Any DFD component which falls under these four categories is vulnerable. For example, the smart home consists of two components, namely 'devices' (S1) and 'smart meters' (S2) which fall under the sensing devices' category, a communication link component 'sending status and receiving commands' (C5) which fall under the communication channel category (C), the 'home management system' which falls under the application provider interface (API)/computation category (AC) and the 'database' (D2) which falls under the category database. This simple classification in four categories will help the security analyst to identify the threats using the STRIDE mnemonics in the next step (Fig. 3).

4.3 Analyze Threats for Each Component Using STRIDE

After annotating the DFD components in step 2, we started brainstorming sessions for finding the threats in each of these components. After two sessions of brainstorming, we uncovered 36 different threats in the smart city domain, using the STRIDE technique in each of the four threat categories, namely sensing devices(S), communication channel (C), API/computation (AC), and database (D). Table 1 gives a brief description of each of these threat scenarios, along with the STRIDE category to which it belongs.

4.4 Suggest Countermeasure for the Threats

Once the possible threats to the smart city were identified, we deliberated on each of these threats to find possible countermeasure. Each of the threats was analyzed, and possible mitigation for the threats was discussed based on expert knowledge and STRIDE mnemonics. Table 2 gives a list of each of the countermeasure against the previously classified threats.

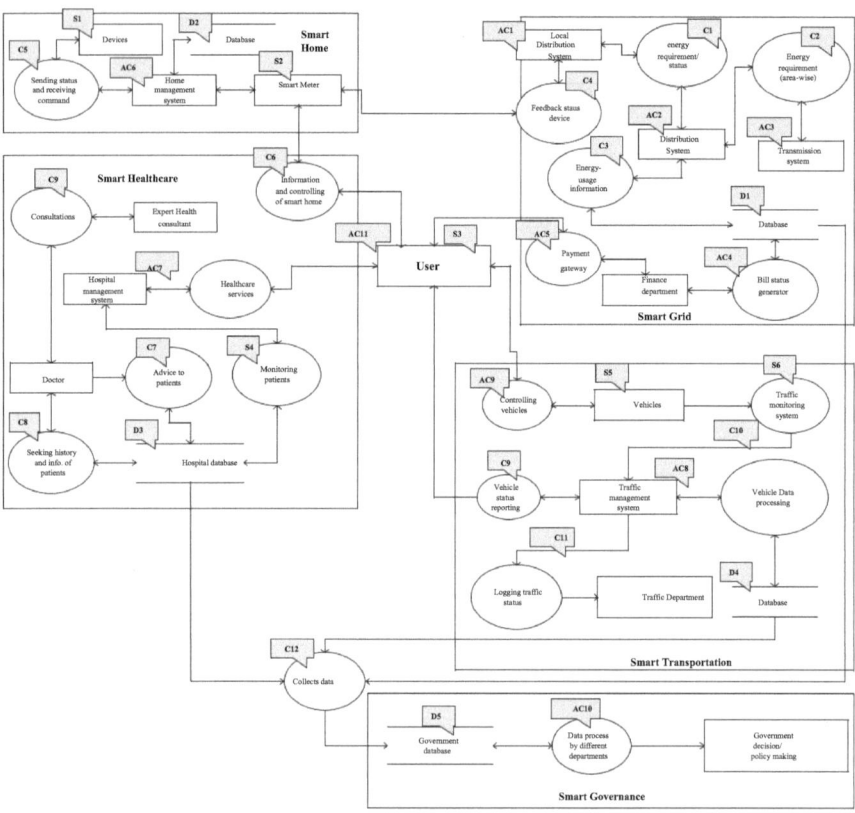

Fig. 3 DFD of smart city architecture

Table 1 Threats and their corresponding STRIDE categorization

Threat components sensing (S)	Threats (T1–T36)	STRIDE category
S1	T1: Attacker can hijack the device from the neighbor home network due to its poor design	S
	T2: Third-party devices can leak information about the configuration of the system	I
S2	T3: Attacker can send malicious code from smart meter to hijack or disrupt functioning of the distribution system in smart grid	T
S3	T4: Wearable or usable things of users can be stolen or lost which can be used by an attacker to extract information of user	S
S4	T5: An eavesdropper can tap the personnel information from the sensors of a patient and may damage its privacy	S
	T6: Attacker can compromise patients device to send private data to different destination	I

(continued)

Table 1 (continued)

Threat components sensing (S)	Threats (T1–T36)	STRIDE category
S5, S6	T7: An attacker can use malwares to disrupt traffic sensors and send wrong information (traffic hijacking) to traffic management system	S
	T8: An attacker can physically disrupt the sensor of vehicle to lower its sensing efficiency and sending large amount of signal from malicious vehicle to drain its battery	D
Communication (C)		
C1, C2, C3, C4	T9: Attacker can tap and manipulate usage information to harm the targeted consumers economically	S, T
	T10: Attacker can tap the communication channel and can jam the radio spectrum to abort the service which can lead to huge economic losses	S, D
C5	T11: An attacker can cause a replay attack on the Zigbee network by recording approved traffic and replay it later	S, I
C6, C7, C8	T12: An attacker can tamper data by altering the packets sent by healthcare devices	T
	T13: Attacker can compromise monitoring devices of the patient and can force them to send data to malicious or wrong destination	S, I
C9, C10, C11	T14: An attacker can use third-party installed applications in vehicles to take control or disrupt communication with traffic controlling system	D
	T15: An attacker can spoof the broadcast messages from the vehicle to disrupt the e-transport ecosystem	S, D
C12	T16: Communication channel can be breached to retrieve classified secret information of government policies and decision	I
	T17: Data encrypted with inefficient encryption policy can be exploited by an attacker to get unauthorized access	I
API's and computation (AC)		
AC1, AC2, AC3	T18: Corrupting the requirement status to make an area blackout or damage distribution system by overloading	T
	T19: Sending energy surplus and requirement status again and again varyingly in order to make the system busy	D
AC4	T20: Attacker may change bill generator code to alter processing leads to wrong outputs	T
AC5	T21: Loose ends of API can cause leakage of user payment information	I
AC6	T22: Less reliable software application outsourced from third party can be exploited by an attacker to take control or corrupt functioning	T

(continued)

Table 1 (continued)

Threat components sensing (S)	Threats (T1–T36)	STRIDE category
AC7	T23: An internal user can change the results of reports and test to effect patient health records	E, T
AC8	T24: Attacker can add malicious code to store inappropriate data in the database to make it unavailable	D
	T25: Compromised system can issue irrelevant alerts to traffic departments to make them unresponsive	D
	T26: Developers of applications can draw information such as location of the owner of vehicle, its routes, and destination which can be sold to adversaries	I
AC9	T27: Compromised system can send wrong directions and locations to vehicles to make traffic blockage and accident	T
AC10	T28: An insider can manipulate processing method to effect policies to harm economy, security, and other factors of governance	E, T
AC11	T29: Insiders can add some hidden features in application which can affect other application processing	E
	T30: Attackers can install malware on third-party applications to steal logged information	S
	T31: Attacker can manipulate or get information from the request data of API's due to weak cryptographic schemes used in it	T
Database (D)		
D1	T32: A skilled attacker can gain access to the smart grid database and tamper with information such as energy usage, electric surcharge etc	T
D2	T33: An attacker can steal the personnel information of the inhabitants of smart home by gaining access to the central home server	I
D3	T34: An internal attacker can misuse its privilege and steal the sensitive information of patients from the hospital database	I, E
D4	T35: Attacker can manipulate the data related to vehicle records, ownership, and movements by gaining unauthorized access using password hijacking	T
D5	T36: An attacker can launch a remote-to-local attack (R2L) to exploit the vulnerabilities of the network for gaining access to the government database and can tamper documents like land record, FIR reports etc	I, T

Table 2 Possible countermeasure for threats (T1–T36)

Threat	Countermeasures
T1	Update the devices regularly and use the latest security patches
T2	Use an advanced network firewall to monitor traffic
T3	Install smart devices that monitor the meter readings and detect anomalies
T4	Use strong passwords in the wearable devices
T5	Use of an optimal sensor scheduling scheme can maximize the secrecy in wireless transmissions
T6	Secure the configuration of the attached devices with regular software update
T7	Install advanced network-based intrusion detection system
T8	Use an advanced network firewall to monitor traffic and detect anomalies in the network
T9	Use strong encryption scheme
T10	Use adaptive broadcasting with partial channel sharing scheme
T11	Monitor the traffic and implement counter-mechanism for every packet
T12	Use lightweight efficient cryptographic solutions for the devices
T13	Use secure routing protocols with special focus on mobility of the devices
T14	Regular software updates with strong authentication for the third-party applications
T15	Detect the attack by monitoring the traffic to observe the sequence of messages and anomalies
T16	Use strong network forensic measures to observe, analyze, and collect evidence of the source of attacks
T17	Use strong encryption scheme
T18	Use secure and efficient communication protocols
T19	Use secure and efficient communication protocols along with attack detection mechanisms
T20	Use devices with codes embedded in them which cannot be modified
T21	Expose only limited data of the API's in the public domain
T22	Regularly update and monitor the third-party applications for security bugs
T23	Strong authorization and role-based access control mechanisms should be used
T24	Use strong security measures for incorrect input and defense in depth mechanisms for applications
T25	Monitor traffics, detect anomalies, and use strong Intrusion detection mechanisms
T26	Strong authorization and role-based access control mechanisms should be used
T27	Use secure routing protocols with a special focus on the mobility of the devices
T28	Use strong authorization and role-based access control mechanisms
T29	Strong authorization and role-based access control mechanisms should be used to manage insiders
T30	Regularly update and monitor the third-party applications for security bugs
T31	Use strong encryption protocols
T32	Use TDE scheme for encrypting the entire database

(continued)

Table 2 (continued)

Threat	Countermeasures
T33	Use strong encryption techniques along with strong authentication schemes
T34	Strong role-based access control mechanisms should be used
T35	Use multilevel authentication systems including OTP
T36	Monitor abnormal traffic and use strong network forensic measures to collect evidence of source of attacks

5   Conclusion and Future Work

The paper models security threats for the smart city domain using STRIDE methodology under four broad threat components. The threats are uncovered using brainstorming techniques, and countermeasure has been suggested for effectively mitigating them. Consequently, 36 different types of threats are outlined and the countermeasure for each threat. These threats are further categorized according to the STRIDE mnemonics. This work may help the system designers in framing security requirements and proposing possible solutions against specific threats related to smart city during the initial phase of software development. In the future, we intend to implement the STRIDE model on a real-world smart city project for analyzing security threats and also framing of pertinent security requirements and match against the set of possible threats outlined in this paper.

References

1. Angelidou M (2014) Smart city policies: a spatial approach. Cities 41:3–11
2. Alibasic A, Al Junaibi R, Aung Z, Woon WL, Omar MA (2017) Cybersecurity for smart cities: a brief review. In: Data analytics for renewable energy integration. Springer International Publishing, pp 22–30
3. Anwar MN, Nazir M, Mustafa K (2017) Security threats taxonomy: smart-home perspective. In: 2017 3rd international conference on advances in computing, communication and automation (ICACCA) (Fall), pp 1–4. IEEE, 2017
4. Martínez-Ballesté A, Pérez-Martínez PA, Solanas A (2013) The pursuit of citizens' privacy: a privacy-aware smart city is possible. IEEE Commun Mag 51(6):136–141
5. Li X, Lu R, Liang X, Shen X, Chen J, Lin X (2011) Smart community: an internet of things application. IEEE Commun Mag 49(11)
6. Shostack A (2008) Experiences threat modeling at microsoft. In: Modeling security workshop. Department of Computing, Lancaster University, UK
7. Howard M, Lipner S (2006) The security development lifecycle, vol. 8. Microsoft Press Redmond
8. Burns SF (2005) Threat modeling: a process to ensure application security. GIAC security essentials certification (GSEC) practical assignment

A Novel Approach in Selection of Municipal Solid Waste Incinerator (MSWI) Ash as an Embankment Material: VIKOR Method

Sonal Saluja, Manju Dominic, Arun Gaur, Kafeel Ahmad and Sadiqa Abbas

Abstract In the present world, municipal solid waste (MSW) generation rates are increasing on a day-to-day basis per year due to the technical advancements and urbanization. This rapid move up in MSW generation causes the problem of waste management in open landfill. Extensive endeavors and very hard attempts are going on in the world to develop sustainable techniques for use of MSW. Incineration of MSW is one of the methods to reduce volume of waste in landfill that contribute toward development of smart cities. Various research programs are being conducted to determine properties of municipal solid waste incinerator (MSWI) ash in order to assess its viability for using as an embankment or fill material. Usually single-criterion decision-making tool is a conservative approach used to evaluate the suitability of material which is not an optimal solution where all the parameters are not looked into it. Keeping in view, the study aims to get the best suitable material by making a compromise solution between various parameters. In the present study, multi-criteria decision-making (MCDM) tool, VIKOR method, is used which incorporates the effect of all the parameters that can enhance the material selection results closer to the ideal condition. Examples are being incorporated to illustrate the suggested method. Alternatives D and E have been ranked 1 in the illustrated examples on the basis of VIKOR ranking and both the alternatives also have an acceptable advantage and suitability over other alternatives.

Keywords MSWI ash · VIKOR method · MCDM tool

S. Saluja (✉) · A. Gaur
Department of Civil Engineering, Malaviya National Institute of Technology, Jaipur, India

S. Saluja · S. Abbas
Department of Civil Engineering, Manav Rachna International Institute of Research and Studies, Faridabad, India

M. Dominic
School of Civil Engineering, Galgotias University, Greater Noida, India

K. Ahmad
Department of Civil Engineering, Jamia Millia Islamia, New Delhi, India

© Springer Nature Singapore Pte Ltd. 2020
S. Ahmed et al. (eds.), *Smart Cities—Opportunities and Challenges*,
Lecture Notes in Civil Engineering 58,
https://doi.org/10.1007/978-981-15-2545-2_34

397

1 Introduction

Due to rapid increase in industrialization and urbanization, MSW generation rate is also increasing hastily throughout the world. MSW consists of organic and inorganic waste material coming from residential, commercial, institutional, and agricultural sectors [1]. The estimated quantity of waste generated in 2016 was 2.01 billion tons and expected to reach 2.59 billion tons by 2030. The composition of MSW is 44% food and green waste, 5% glass, 4% metal, 17% paper and cardboard, 12% plastic, 2% rubber and leather, 2% wood, and 14% others [2]. Municipal solid waste management authorities are now facing problem of unsafe disposal of waste and reducing its storage cost all over the world, particularly in highly populated areas. At present, MSW management has reached an uncontrollable and life-threatening situation. This uncontrolled situation forced solid waste management authorities to work on reduction of waste at source, recycle, reuse, and disposal of waste in order to reduce the adverse effect on public health and environment.

The present study indicates that there is a requirement of smart waste management strategy to achieve sustainable development and to reduce the heap of waste that is open dumped in the cities. In the present days, smart waste management efforts lead to the sustainable development of artificial resources by reusing and recycling of the MSW. It reduces its cost and complexity of handling waste. It also saves the natural material and environment which is associated to keep the city clean and smart. Incineration of MSW is one of the methods that contribute toward smart cities. It reduces waste volume by 90% and mass by 80% in landfill. It produces by-products which mainly consist of bottom ash and fly ash. Although incineration is a sustainable process of recycling of waste and provides energy, it is not an appropriate solution because it generates ashes (fly ash and bottom ash) which must consequently be disposed off [3]. There are more than 2200 incineration plants worldwide using 300 million tons of MSW per annum leaving behind 200–300 kg/tons of bottom ash and 10–50 kg/tons of fly ash which are dumped back to the landfills [4]. To overcome the disposal problem of MSWI ash, it is used as a filler material in road embankment and base layers of pavement. It is also used to produce cement, concrete, hot mix asphalt concrete, ceramics, and as natural aggregate replacement, etc.

Various researchers have investigated chemical, physical, and geotechnical properties of MSWI ash to observe its suitability in various engineering applications [5–7]. Several researchers examined that MSWI fly ash is rich in metal and salts and, therefore, can be used as raw material in the production of cement, artificial aggregates, and ceramics, as a fill or embankment material [3, 8–10]. Xue et al. [11] showed that about 8–16% of MSWI fly ash meets the requirement of stone matrix asphalt mixture. Toraldo et al. [12] investigated the bottom ash engineering properties and concluded that MSWI bottom ash properties are near to that of natural aggregates and meet the environmental laws for reuse of non-hazardous waste in cement bound mixes, asphalt concrete.

While selecting MSWI ash as an embankment or fill material, decision is made based on single criterion that may be strength or permeability of MSWI ash that is not an optimal solution because all the important parameters are not considered in decision making. Selection of suitability of material must be based on all the parameters to achieve better or ideal results. Other evaluation criteria such as optimum moisture content, maximum dry density, permeability, CBR value, unconfined compressive strength, or direct shear strength must be considered in selection of MSWI ash as an embankment or filler material and various other construction materials. It has been realized that to get the optimum combination of waste material, the strength should not only be the establishing criterion. MCDM techniques have become accepted for this purpose. MCDM consists of identification of the alternatives and criteria, establishing of decision matrix, calculating criteria weights, and identification of ranks [13]. VIKOR (*VlseKriterijumskaOptimizacoja I KomopromisnoResenje*) method is an effective MCDM tool that has been proposed in this paper to determine the sustainable solution for the use of MSWI ash.

2 VIKOR Method

Waste material selection for using as an embankment or fill material is a key strategic decision-making process and this has involved the attention of various researchers from last 20–25 years [1–3]. While an incompatible material selection may lead to the failure of the structure, the parameter lowest price might not show the potential of effective waste material; therefore, MCDM technique has gained popularity in establishing the deciding parameter for selecting among different alternatives. In a decision matrix, each criterion has some specific objective, and normalization is used to convert different criteria into a common acceptable feature. The higher values of properties are enviable, called best attribute and those with smaller values are also desirable, called worse attribute. In the present study, maximum dry density and strength is best attribute and permeability is the worse attribute. Various normalization methods are used by various researchers for best and worst attribute of alternative [14]. Jahan et al. [14] proposed comprehensive VIKOR method in which positive, negative, and target values are considered for selection of material with emphasis on feasible solution close to an ideal solution. Tavares et al. [15] used Geographic Information System (GIS) along with Analytical Hierarchy Process (AHP) for selection of waste disposal site location. Sarkar et al. [16] proposed Subjective and Objective Weight Integrated Approach (SOWIA) method for calculating weight of criteria and VIKOR method for final ranking of IIT's. Durga Prasad et al. [17] proposed hybrid decision model (Kano and VIKOR techniques) to solve supplier selection problem. Shemshadi et al. [18] formulated supplier selection as an MCDM problem and solved it by using trapezoidal fuzzy numbers in Shannon entropy to determine objective weight and VIKOR method for final ranking of supplier alternatives. San Cristobal [19] proposed a decision framework for renewable energy project selection in Spain by using Analytical Hierarchy Process and VIKOR method together. Tzimopoulos et al. [20] used VIKOR method in selection of irrigation network in the Thessaloniki Plainth.

3 Applications

To illustrate the appropriateness of VIKOR method in selection of best materials, two examples are presented here. The use of MSWI ash (fly ash and bottom ash) as a filler or embankment material is being studied. VIKOR method has been applied to the selected problem through following steps:

Step I. Formulation of Decision Matrix: Decision matrix (*DM*) has been created as given below:

$$
\mathrm{DM} = \left[x_{ij} \right] = \begin{array}{c} \\ \\ A_1 \\ A_2 \\ .. \\ A_i \\ .. \\ A_m \end{array} \begin{array}{cccccc} C_1 & C_2 & .. & C_j & .. & C_n \\ W_1 & W_2 & .. & W_j & .. & W_n \\ \left[\begin{array}{cccccc} x_{11} & x_{12} & .. & x_{1j} & .. & x_{1n} \\ x_{21} & x_{22} & .. & x_{2j} & .. & x_{2n} \\ .. & .. & .. & .. & .. & .. \\ x_{i1} & x_{i2} & .. & x_{ij} & & x_{in} \\ .. & .. & .. & .. & .. & .. \\ x_{m1} & x_{m2} & .. & x_{m3} & .. & x_{mn} \end{array} \right] \end{array} \tag{1}
$$

where A_i represents the alternatives; C_j is the *j*th performance defining criterion; W_j is the weight of the *j*th criterion; m is the number of alternatives and n is the number of criteria.

Step II. Calculation of projection value: The projection value (ρ_{ij}) for each alternative is calculated by using Eq. (2).

$$
\rho_{ij} = \frac{x_{ij}}{\sum\limits_{i=1}^{m} x_{ij}} \tag{2}
$$

Step III. Calculation of Entropy: After the calculation of projection value, entropy (E_j) of each criterion is calculated by using Eq. (3).

$$E_j = -\zeta \sum_{j=1}^{n} \rho_{ij} \text{ In } \left(\rho_{ij} \right) \tag{3}$$

where ζ is a constant and calculated as:

$$\zeta = \frac{1}{\text{In } (m)} \tag{4}$$

Step IV. Calculation of Dispersion Value: The dispersion value (ψ_j) of the criteria is calculated by using Eq. (5).

$$\psi_j = 1 - E_j \tag{5}$$

Step V. Calculation of Objective Weight: Finally, the objective weight (W_j) of the criteria is calculated by using Eq. (6).

$$W_j = \frac{\psi_j}{\sum_{j=1}^{n} \psi_j} \tag{6}$$

Step VI. Maximum criterion function (f_i^*) and minimum criterion function (f_i^-) value is obtained as below:

$$f_i^* - \begin{bmatrix} \max_i x_{ij}, & \text{for benefit criteria} \\ \min_i x_{ij}, & \text{for cost criteria} \end{bmatrix}, \quad j - 1, 2, \ldots, n$$

Step VII. The maximum utility value (α_i) and individual regret value (β_i) is calculated by using Eqs. (7) and (8).

$$\alpha_i = \sum_{j=1}^{n} \frac{w_j \left[f_i^* - x_{ij} \right]}{\left[f_i^* - f_i^- \right]} \tag{7}$$

$$\beta_i = \text{max of} \left\{ \sum_{j=1}^{n} \frac{w_j \left[f_i^* - x_{ij} \right]}{\left[f_i^* - f_i^- \right]} \right\}, \quad \text{for} \quad j = 1, 2 \ldots n \tag{8}$$

Step VIII. Calculation of the VIKOR ranking: VIKOR ranking index value (Ω_i) is calculated by using the relation given in Eq. (9).

$$\Omega_i = \xi \left(\frac{(\alpha_i - \alpha_i^-)}{(\alpha_i^+ - \alpha_i^-)} \right) + (1 - \xi) \left(\frac{(\beta_i - \beta_i^-)}{(\beta_i^+ - \beta_i^-)} \right) \tag{9}$$

where

$$\alpha_i^+ = \max_i \alpha_i = \max[\alpha_i, i = 1, 2 \ldots m] : \alpha_i^- = \min_i \alpha_i = \min[\alpha_i, i = 1, 2 \ldots m]$$

$$\beta_i^+ = \max_i \beta_i = \max[\beta_i, i = 1, 2 \ldots m] : \beta_i^- = \min_i \beta_i = \min[\beta_i, , i = 1, 2 \ldots m]$$

where ξ is decision-making index, whose value lies between 0 and 1. If $\xi > 0.5$, then decision is in the favor of maximum group utility; if $\xi = 0.5$, then decision has made on the compromise basis; If $\xi < 0.5$, then it reflects the minimum individual regret value. It is generally taken as 0.5 that makes a compromise between maximum utility value and individual regret value of opponent.

Step IX. The ranking of alternative is done by sorting the value of Ω_i in ascending order, (1, 2… m) and the rank 1 is assigned to the alternative that has minimum value of Ω_i. Alternative A' is the first-place ranking alternative; Alternative A'' is the second-place ranking alternative and so on.

Step X. Proposing of a compromise solution: A compromise solution is reached if two conditions C1 and C2 are satisfied. The condition C1 will be satisfied if

$$\Omega(A'') - \Omega(A') \geq D\Omega \tag{10}$$

where

$$D\Omega = 1/(m - 1) \tag{11}$$

If it is satisfied, then $\Omega(A')$ has an acceptable advantage over $\Omega(A'')$. If it is not satisfied with alternative $\Omega(A'')$, then the equation will be revised to

$$\Omega(A''') - \Omega(A') \geq D\Omega \tag{12}$$

and the same step has to be repeated continuously till the equation

$$\Omega(A^m) - \Omega(A') \geq D\,\Omega \tag{13}$$

has been satisfied and the compromise solution lies between alternative A', A'' …A^m.

The condition (C2) is satisfied if ranking based on Ω_i, must also be the same ranked by α_i or/and β_i.

If C2 is satisfied, then there is an acceptable stability between alternatives. If it is not satisfied, then Alternative A' and A'' is the compromise solution only.

Example-1

MSWIF can be combined with other materials to bring out a combination which can be effectively used in other applications. In this example, MSWIF has been combined with different percentages of cement or lime for using as an embankment or fill material. VIKOR method has been used to select the best combination in terms of the alternatives selected so that the same can be used for chosen application. The selected alternatives and their respective symbols used are given in Table 1.

Various criteria that are selected to evaluate the alternatives, their respective symbol and units used are given in Table 2.

Decision matrix for the selected problem is shown in Table 3 [9]. Subjective weight has been identified as maximum (max) for best criteria (best attribute) and minimum (min) for worse criteria (worse attribute).

The entropy (E_j), dispersion value (ψ_j), and objective weight (W_j) of criteria are being calculated by using Eqs. (3), (5) and (6) and is tabulated in Table 4.

The ranking of each alternative is done by using Eq. (9) and is arranged in ascending order as shown in Table 5. Minimum value of Ω_i is assigned as rank 1. It displays the maximum utility value (α_i), individual regret value (β_i) and VIKOR ranking index (Ω_i) of the alternatives.

Table 1 Alternatives selected for material selection

Symbol	Alternatives
A	MSWIF (100%)
B	MSWIF (95%) + Lime (5%)
C	MSWIF (95%) + Cement (5%)
D	MSWIF (90%) + Cement (10%)
E	MSWIF (90%) + Lime (10%)

Table 2 Criteria used to evaluate the alternatives

Criteria	Symbol	Unit
Optimum moisture content	OMC	%
Maximum dry density	MDD	KN/m^3
California bearing ratio	CBR	%
Unconfined compressive strength	UCS	KN/m^2
Permeability (PR)	PR	10^{-6}cm/s

Table 3 Decision matrix for MSWIF ash selection

Alternative	Criteria				
	OMC (min)	MDD (max)	CBR (max)	UCS (max)	PR (min)
A	23	13.5	44	65	10,000
B	23.5	13.5	83	100	93
C	23.8	13.6	105	106	89
D	24.5	14.1	132	173	85
E	24	13.6	122	137	85

Table 4 Entropy (E_j), dispersion value (ψ_j), and objective weight (W_j) of criteria

Criteria	OMC (min)	MDD (max)	CBR (max)	UCS (max)	PR (min)
E_j	1.0008	1.0008	0.965	0.970	0.121
ψ_j	−0.0008	−0.0008	0.035	0.030	0.879
W_j	−0.0008	−0.0008	0.037	0.032	0.933

Table 5 VIKOR index (Ω_i), $(\zeta = 0.5$ taken)

Alternatives	α_i	β_i	Ω_i.	Ranking based on α_i	Ranking based on β_i	Ranking based on Ω_i.
A	1.0012	0.933	1.000	5	5	5
B	0.0417	0.022	0.033	4	4	4
C	0.0293	0.019	0.025	3	3	3
D	−0.0008	0.000	0.000	1	1	1
E	0.0138	0.011	0.013	2	2	2

As per Table 5, alternative D [MSWIF (90%) + Cement (10%)] has been ranked 1. This is followed by alternatives E, C, B, and A. Alternatives B and A are the least favorable ones. Meanwhile, for $\xi = 0.5$, $\Omega(E) - \Omega(D) = 0.013 < 0.25$ which is not satisfied Eq. (10); so the equation is revised with Eq. (12) and the same step has to be repeated continuously till Eq. (13) has been satisfied. By calculating, we get the remarks as shown in Table 6.

Table 6 Remarks for condition C1

Condition	Remarks
$\Omega(E) - \Omega(D) = 0.013 < 0.25$	Not satisfied
$\Omega(C) - \Omega(D) = 0.025 < 0.25$	Not satisfied
$\Omega(B) - \Omega(D) = 0.033 < 0.25$	Not satisfied
$\Omega(A) - \Omega(D) = 1.000 > 0.25$	Satisfied

So, D, E, C, B, and A are the compromise solution.

At the same time, it must have satisfied condition C2; that is the ranking based on Ω_i must also be ranked by α_i or/and β_i, then from Table 5 it is clear that it satisfied condition C2. So, the present example has an acceptable stability.

Example 2

The combination of MSWI fly ash (MSWIF) with different percentages of MSWI bottom ash (MSWIB) used as an embankment or fill material is considered in second example. The problem which consists of six alternatives and five criteria with their respective symbol and unit is shown in Tables 7 and 8.

Decision matrix for the above problem is shown in Table 9 [21]. Subjective weight has been identified as maximum (max) for best criteria and minimum (min) for worst criteria.

The entropy (E_j), dispersion value (ψ_j), and objective weight (W_j) of criteria are being calculated by using Eqs. (3), (5), and (6) and is shown in Table 10.

The ranking of each alternative is done by using Eq. (9) and arranged in ascending order as shown in Table 11. Minimum value of Ω_i is assigned as rank 1. It displays the maximum utility value (α_i), individual regret value (β_i), and VIKOR ranking index (Ω_i) of the alternatives.

Table 7 Alternatives for material selection

Symbol	Alternatives
A	MSWIF (100%)
B	MSWIF (80%) + MSWIB (20%)
C	MSWIF (60%) + MSWIB (40%)
D	MSWIF (40%) + MSWIB (60%)
E	MSWIF (20%) + MSWIB (80%)
F	MSWIB (100%)

Table 8 Criteria used to evaluate the alternatives

Criteria	Symbol	Unit
Optimum moisture content	OMC	%
Maximum dry density	MDD	KN/m^3
Cohesion value	CV	KN/m^2
Friction angle	FA	Degree
Permeability (PR)	PR	10^7 cm/s

Table 9 Decision matrix for MSWI ash selection

Alternative	Criteria				
	OMC (min)	MDD (max)	CV (max)	FA (max)	PR (min)
A	45.7	10.8	34.1	20.8	2.2
B	43	12.2	33.6	15.7	4.9
C	34.2	13	34.4	22.2	5.3
D	28.8	14.2	25.7	23.3	802
E	25.8	15	14.8	37.3	0.19
F	26.8	15.4	7.7	50.7	0.00014

Table 10 Entropy (E_j), dispersion value (ψ_j), and objective weight (W_j) of criteria

Criteria	OMC (min)	MDD (max)	CV (max)	FA (max)	PR (min)
E_j	0.985	0.995	0.944	1.002	0.745
ψ_j	0.015	0.005	0.056	−0.002	0.255
W_j	0.046	0.015	0.170	−0.006	0.775

Table 11 VIKOR index $(\Omega_{i.})$, $(\zeta = 0.5$ taken$)$

Alternative	α_i	β_i	Ω_i	Ranking based on α_i	Ranking based on β_i	Ranking based on Ω_i.
A	0.266	0.208	0.153	3	3	3
B	0.512	0.464	0.526	4	4	4
C	0.524	0.502	0.565	5	5	5
D	0.837	0.776	1.000	6	6	6
E	0.142	0.125	0.000	1	1	1
F	0.172	0.170	0.057	2	2	2

As per Table 11, alternative E [MSWIF (20%) + MSWIB (80%)] has been ranked 1. This is followed by alternatives F, A, B, C, and D. Alternatives C and D are the least favorable ones. Meanwhile, for $\zeta = 0.5$, $\Omega(F) - \Omega(E) = 0.057 < 0.2$ which has not satisfied Eq. (10); so the equation is revised with Eq. (12) and the same step has to be repeated continuously till Eq. (13) has been satisfied. By calculating, we get the remarks as shown in Table 12.

Table 12 Remarks for condition C1

Condition	Remarks
$\Omega(F) - \Omega(E) = 0.057 < 0.2$	Not satisfied
$\Omega(A) - \Omega(E) = 0.153 < 0.2$	Not satisfied
$\Omega(B) - \Omega(E) = 0.526 > 0.2$	Satisfied

So, E, F, A, and B are the compromise solution.

At the same time, it must have satisfied condition C2; that is the ranking based on Ω_i must also be ranked by α_i or/and β_i, then from Table 11 it is clear that it satisfied condition C2. So, the present example has an acceptable stability.

4 Conclusions

Municipal solid waste is being generated and getting deposited in an alarming way these days due to technical advancements and urbanization. Its disposal has become an immediate harsh problem which needs to be dealt with proper care and in an adequate manner. One of the best ways to rectify this problem is to incinerate the waste that can help to reduce the waste volume in landfill. At the same time, incineration produces ashes which ultimately get disposed off in the landfills. Various researchers have been investigated the possible uses of ashes as an embankment or fill material in pavement construction. In a conservative approach, a single criterion is used to evaluate its effectiveness as an embankment or fill material. During literature review, it has been studied that various parameters have been taken during investigation. When the effect of all parameters has been considered, an optimal decision can be attained. Thus, a MCDM tool will help researchers in considering all the parameters to get a better choice. VIKOR method is one of the best MCDM tools which helps to make a better decision. In this paper, VIKOR method has been used as a MCDM tool to incorporate the effect of all parameters. In this method, the objective weight of each criterion is generated by using Shannon's entropy method. Then maximum utility measure and individual regret measure has been calculated and finally based on VIKOR ranking the suitable waste combination is selected effectively. Two examples have been analyzed in the present study by using this method. In the first example, alternative D [MSWIF (90%) + Cement (10%)] has been selected as a best material by ranking it as 1 and followed by alternatives E, C, B, and A. The author Poran and Ahtchi-Ali [9] has also chosen the same alternative by using the conservative approach in general term but could not be able to rank the alternatives. In the second example, alternative E [MSWIF (20%) + MSWIB (80%)] has been selected as a best material by ranking it as 1 and followed by alternatives F, A, B, C, and D. The authors Muhunthan et al. [21] have chosen alternative F [MSWIB (100%)] that is different than VIKOR method ranking but in both approach bottom ash is the best material, which justify our result. It has been further seen in this study that, it is also possible to determine both the acceptable advantage (compromise between the alternatives) and the acceptable stability by using condition C1 and C2 of this method. From Table 6, the alternative D has an acceptable advantage over other alternatives and from Table 12, the alternative E has an acceptable advantage over alternatives F, A, and B. In both the examples, there is an acceptable stability also. Result of applied example shows that VIKOR method is a useful tool to get the most specific combination of waste with their acceptable advantage and stability over other alternatives also. It can be concluded that MCDM tool is a justifiable and better solution as compared to conservative approach which has been used in the past. All these efforts lead to the reduction of MSW in landfill which ultimately gives natural aesthetic appearance to the smart city.

References

1. Hassan HF (2005) Recycling of municipal solid waste incinerator ash in hot-mix asphalt concrete. Constr Build Mater 19(2):91–98
2. Kaza S, Yao L, Bhada-Tata P, Van Woerden F (2018) What a waste 2.0: a global snapshot of solid waste management to 2050, World Bank Publications
3. Ferreira C, Ribeiro A, Ottosen L (2003) Possible applications for municipal solid waste fly ash. J Hazard Mater 96(2–3):201–216
4. Gupta G, Datta M, Ramana GV, Alappat BJ (2017) Feasibility of using MSW incinerator ash in geotechnical applications. In: Indian geotechnical conference GeoNEst, p 119
5. Bagchi A, Sopcich D (1989) Characterization of MSW incinerator ash. J Environ Eng 115 (2):447–452
6. Pera J, Coutaz L, Ambroise J, Chababbet M (1997) Use of incinerator bottom ash in concrete. Cem Concr Res 27(1):1–5
7. Hamernik JD, Frantz GC (1991) Physical and chemical properties of municipal solid waste fly ash. Mater J 88(3):294–301
8. Sherwood PT, Ryley MD (1986) The use of pulverised fuel ash in road construction. RRL Rep 49
9. Poran CJ, Ahtchi-Ali F (1989) Properties of solid waste incinerator fly ash. J Geotech Eng 115 (8):1118–1133
10. Forteza R, Far M, Segui C, Cerdá V (2004) Characterization of bottom ash in municipal solid waste incinerators for its use in road base. Waste Manag 24(9):899–909
11. Xue Y, Hou H, Zhu S, Zha J (2009) Utilization of municipal solid waste incineration ash in stone mastic asphalt mixture: pavement performance and environmental impact. Constr Build Mater 23(2):989–996
12. Toraldo E, Saponaro S, Careghini A, Mariani E (2013) Use of stabilized bottom ash for bound layers of road pavements. J Environ Manag 121: 117–123
13. Vincke P (1992) Multicriteria decision-aid, Wiley
14. Jahan A, Mustapha F, Ismail MY, Sapuan SM, Bahraminasab M (2011) A comprehensive VIKOR method for material selection. Mater Des 32(3):1215–1221
15. Tavares G, Zsigraiová Z, Semiao V (2011) Multi-criteria GIS-based siting of an incineration plant for municipal solid waste. Waste Manag 31(9–10):1960–1972
16. Sarkar S, Sarkar B (2014) A new way to performance evaluation of technical institutions: VIKOR approach. In: Proceeding of 2014 global sustainability transitions: impacts and innovations, pp 209–216
17. Durga Prasad KG, Prasad MV, Sravan Kumar R, Prasad VSD, Shanmukhi KVSJ (2017) Kano-based VIKOR decision model for supplier selection—a case study. SSRG Int J Mech Eng, 227–231
18. Shemshadi A, Shirazi H, Toreihi M, Tarokh MJ (2011) A fuzzy VIKOR method for supplier selection based on entropy measure for objective weighting. Expert Syst Appl 38(10):12160–12167
19. San Cristóbal JR (2011) Multi-criteria decision-making in the selection of a renewable energy project in spain: the Vikor method. Renew Energy 36(2):498–502

20. Tzimopoulos C, Zormpa D, Evangelides C (2013) Multiple criteria decision making using VIKOR method. Application in irrigation networks in the Thessaloniki Plain. In: Proceedings of the 13th international conference on environmental science and technology, CEST
21. Muhunthan B, Taha R, Said J (2004) Geotechnical engineering properties of incinerator ash mixes. J Air Waste Manag Assoc 54(8):985–991

Reducing Disaster Risks in Indian Smart Cities: A Five-Stage Resilience Maturity Model (RMM) Approach

Omar Bashir

Abstract Smart Cities Mission is one of the flagship programmes of the Government of India launched in 2015 with an aim to develop a total of hundred Smart Cities in different states of India. The primary objective of this programme is the urban redevelopment and retrofitting to make existing cities smart, sustainable, and citizen-friendly which will ultimately lead to economic growth and improvement in the quality of life. The Smart City Mission, however, lacks to incorporate resilience as one of its objectives. India is vulnerable to disasters of all types, ranging from earthquakes, floods, droughts to terrorist attacks. The risk of disaster is compounded in an urban area due to densely populated areas, lack of planned development, stress on existing infrastructure, socio-economic imbalance including others. The focus of this study is to develop a holistic resilience maturity model for a smart city in India that can be used to incorporate resilience in planning, development through a stage-wise maturity. This paper used three Indian Smart Cities—Chennai, Surat, and Pune, all of which are a part of the Rockefeller 100 resilient cities, as the basis of the study. A detailed analysis of the proposals of these three cities was carried out and based on the same a resilience maturity model was developed. Though Indian cities are studied, the maturity model can be applicable to other developing countries having similar smart cities.

Keywords Smart city · Disaster · Resilience · Maturity model · Indian smart city

1 Introduction

India, along with other countries, has witnessed large-scale urbanization from the past few decades. It is estimated that more than half of the world population is now living in cities. [1]. The urban population of India has also increased from around 27.81% in 2001 to 31.16% in 2011, it is projected to increase at a steady rate to

O. Bashir (✉)
RICS School of Built Environment, Amity University, Noida, India
e-mail: obashir@ricssbe.edu.in

© Springer Nature Singapore Pte Ltd. 2020   411
S. Ahmed et al. (eds.), *Smart Cities—Opportunities and Challenges*,
Lecture Notes in Civil Engineering 58,
https://doi.org/10.1007/978-981-15-2545-2_35

nearly 40% in 2030 and will touch 50% by 2050 [2]. However, there is no common definition of the terms—"city" or "urban areas", [3] which are sometimes used interchangeably. A city or an urban area is a constricted geographical area identified by the distinct infrastructure which is relatively better than that of the areas surrounding it and is used to serve the people residing in that geographical area [4]. Cities and urban areas are often the drivers of the economy, especially in emerging economies like India. It is estimated that by the year 2025, cities in emerging economies will be producing nearly two-thirds of the global GDP. However, rapid urbanization leads to increase in the density of population, which may further lead to secondary issues like a burden on the existing infrastructure, lack of housing, development of slums, inefficient transportation, pollution, etc. The lack of adequate and poorly managed infrastructure and services have often been one of the reasons for India's inability to attract foreign direct investments [5].

With a dual aim to improve the quality of life of the inhabitants and to drive economic growth, [6], the Government of India launched the Smart City Mission in 2015. In India, most cities had organic growth, developing at different stages of time organically. Lack of strict norms, regulations, and planning have led to decades of unorganized and unplanned growth in most of the cities in India. Only after a city has the provision of all basic infrastructure and amenities can it start its development towards becoming a Smart City, which can only be achieved when a city has become liveable in all respects.

The Smart Cities Mission in India follows a three-pronged strategy which includes:

1. Provision of basic core infrastructure and amenities—roads, highways, water supply, affordable housing, etc.
2. Smart Solutions—Use of ICT to improve infrastructure and services.
3. Area-Based development—retrofitting, redevelopment, greenfield projects, and pan-city initiative to improve the liveability of the city.

The basic objective of developing a smart city is to improve the quality of life of the citizens by providing convenient living solutions. The hundred chosen Smart Cities have started incorporating the three strategies listed above. Chennai is one of the Smart City chosen by the Smart City Mission. Chennai is the capital city of the Indian state of Tamil Nadu and has recently faced a spate of floods, particularly in 2017 and 2015. The floods of 2015 were the most severe, around 470 people lost their lives [7] and an economic loss of around 50,000 crores [8].

Disaster risk reduction and resilience have not been incorporated fully in the Smart Cities of India and only 30% of the Smart Cities out of the hundred selected cities have made attempts to incorporate resilience in their projects [9].

Fig. 1 Reported disasters in India from 1969 to 2019 [11]

2   Disasters in India

Talking about disasters in the Indian context, nearly 59% of the geographical area of India is vulnerable to earthquakes, ranging from moderate to very high intensity. Around 68% of cultivable land or 18% of the total area is drought-prone. Nearly 12% of the land is prone to flood and the long coastline is prone to cyclones and tsunamis. Hilly regions are also vulnerable to snowstorms, landslides, and avalanches. Twenty-two of the thirty-six states in India are multi disaster-prone. India is also vulnerable to man-made disasters, also chemical, biological, radiological, and nuclear emergencies. These disasters are increased manifolds due to the varied nature of the country's varied geographies, demographic divide and socio-economic conditions, unplanned urbanization, changing climate, epidemics and pandemics, geological hazards which pose a risk to India's economy, population, and development [10] (Fig. 1).

3   Disaster in an Urban Context

The definition of an urban area or a city is usually viewed as a rural–urban continuum in research related to the urban disaster, which includes villages, semi-rural areas, suburbs, satellite towns, metropolitan cities, and megapolis. Urban disasters are those type of disasters which happen in the urban context [3]. Urban areas are often characterized by a high density of population, the effects of a disaster can be increased manifolds in an urban area [12]. Researchers agree that there has been an increase in urban disasters, both in terms of numbers of occurrences and the tune of damage both human and economic [3].

Urban areas usually face four types of stressors—(a) Natural, (b) Technological, (c) Economic, and (d) Man-made. A city needs to be resilient against these four stressors. Natural stressors include force majeure or acts of God including earthquakes, floods, tsunamis, etc. These are external in nature; a city has little or no control over these disasters. It is argued, however, that natural stressors like droughts and famines are not truly external in nature. Moreover, out of the four stressors, only natural stressors are purely external in nature while the other three can be internal or a mix of internal and external nature. As cities have developed complex and dependent technical systems which even with the degree of accuracy and precision are prone to errors and failure. Such errors and failures can have cascading effects on a city and can be classed as technological stressors. Technological stressors are mostly internal in nature. Economic stressors include lack of business opportunities, lack of jobs, diversion of investment, failing infrastructure, and housing. Economic stressors can be both internal or external in nature. War, terrorism, crime, riot, etc., constitute man-made stressors. These man-made stressors generally arise from the citizens living within the city. Man-made stressors are usually internal in nature but can be influenced by external factors as well [13].

All the four stressors coupled with unplanned development, lack of resources and the absence of proper planning and the ever-growing population, urban areas are under a lot of stress. The concept of disaster management in India is still reactive in nature, various disaster management activities only take place after a city has been struck by a disaster. There is a need to change this approach to make it proactive. Every city is unique, so are the risks of disasters it is facing. It is therefore important to strengthen these urban cities by increasing a city's resilience to disasters.

4   Resilience and the Built Environment

The concept of resilience is used in several fields and is defined as not only the ability of a system to cope with stresses or threats but also the ability to remain functional under the stress or threat [14]. From the point of view of risk, resilience can be termed as a system that not only considers the risk which a system faces but also how a well a system can recover from a disaster [15]. The risk of disaster can be reduced by incorporating changes in the land-use or zoning policy of a city, design changes, and retrofitting in critical building and structures, develop an integrated emergency management planning with the urban planning and adapting best practices while keeping sustainability in place [16]. Risk and resilience are the key elements for decision making to protect a system from disasters. Resilience not only increases the system's ability to withstand the threats but also develops an ability within the system to function smoothly during a disaster. This proves helpful while dealing with unknown disasters or threats, a resilient system is ready to deal with such unknown threats [17].

Resilience is dependent on the quality of the built environment. A well designed and constructed built environment can make the city more resilient to threats and can be an enabler to its revival after a disaster, on the other hand, a poorly designed and managed built environment may not only be unable to withstand any disaster but may itself worsen the disaster or may give rise to other associated risks [18]. Built environment does play an important role in a city's resilience by providing and facilitating the safety and protection of the physical and social systems of a city [19]. Cities or specific sectors within the city can be resilient or limited to the city's economy or can be interpolated to the national economy as well [14]. The features of resilience are very similar t other urban agendas like sustainability or governance [20] and economic development [21]. From an urban planning sense, if a city can absorb, adapt and responds to changes without any loss of function, the city can be termed as a resilient city [20].

A resilient city should essentially be developed as a network of the sustainable built environment, including both constructed and natural environments and the citizens [22]. Buildings, spaces, and places which are critical for a city's built environment need to function even during any disaster event. Therefore, there is a need to "design, develop and manage resilience" for these elements of the city's built environment [19]. Every city is unique in terms of its built environment; therefore, it is necessary to consider the city-specific scenario while designing resilient solutions for the city [14].

Resilience prepares a city against any expected to unexpected challenges that may arise. This is possible only by developing an understanding of the strengths and weaknesses, interdependencies and risks, stresses, and relievers of a city— looking at a city holistically. Building urban resilience will help a city face future disasters in a better way, minimizing the loss of life, property, economic activity, and environment and an ability to recover in a short span of time, which could have not been possible beforehand. Building urban resilience is also essential in order to increase the safety and wellbeing of the citizens of the urban area [23, 24]. A built environment of the urban area influences the quality of life of the city's inhabitants and facilitates the same. It becomes essentially important to develop a city that is resilient and can withstand and adapt to any threat posed by any disaster in a way that does not change the basic functionality of the built environment of that urban area [25]. Within a city, there can be differences in the quality of infrastructure, housing, safety or protective infrastructure. This can lead to some areas being more vulnerable than others during disasters [14].

The city-specific vulnerabilities can be addressed by strengthening the built environment, this includes reengineering buildings and infrastructure to make them able to withstand damage due to disasters [25]. The resilience must be "built-in" for any city, this "built-in" resilience is the ability of a city with regards to its physical, economic, social, environmental and institutional to resist and overcome disasters. The primary goal of a city should be to proactively incorporate "built-in" resilience which will help a city to cope with various unseen natural and human-induced disasters [18]. From the point of disaster, resilience can be developed at three levels —individual, household, and community [14].

The Resilient city programme of the World Bank is aimed to help the cities to adapt and withstand disaster while maintaining the essential function of the city. The cities face several vulnerabilities which can be categorized into five broad groups—(a) Climate, (b) Environment, (c) Resources, (d) Infrastructure, and (e) Community. The cities need to have five categories of adaptability, which include—(a) Governance, (b) Institution, (c) Technical capacity, (d) Planning systems, and (e) Funding structures [26].

4.1 Critical Infrastructure (CI)

To develop a resilient city, it is necessary to be able to develop the infrastructure that withstands all types of disasters, can function during the disaster and continues to function the same way after the disaster. This is not possible for existing cities, also this is not economically viable for many cities. The concept of Critical Infrastructure has been evolving since the 1980s [27]. Cities have been putting emphasis on protecting critical infrastructure during a disaster [27–29]. This emphasis is also there since no city can be fully resilient while being cost-effective [29]. The critical infrastructure is defined by the National Infrastructure Protection Plan in the United States of America [27] and the National Infrastructure Commission in the United Kingdom [28]. Critical Infrastructure is divided into two groups—(a) Economic Infrastructure and (b) Social Infrastructure [28].

Economic Infrastructure

1. Transportation systems—rails, roads, airports.
2. Energy systems—generation and distribution systems.
3. Communication systems—internet, telephony.
4. Utility systems—water, electricity, gas.
5. Protective systems—Flood Protection systems.
6. Waste removal systems.

Social Infrastructure

1. Education systems—schools, colleges, universities.
2. Health systems—hospital, laboratories, clinics.
3. Housing systems.
4. Judicial system—courts, prisons.

The Critical Infrastructure should be planned, constructed, operated, and maintained in such a way that it is able to absorb and resist the disaster. Also, the location of critical infrastructure is of prime importance.

However, some researchers [29] are of the opinion that overall resilience should be developed for a city. Within a city, there can be differences in the quality of infrastructure, housing, safety or protective infrastructure. This can lead to some areas being more vulnerable than others during disasters [14].

5 Research Methodology

To establish a benchmark for the three smart cities chosen for this study, the smart cities were ranked against the two established frameworks of resilience. The Sendai Framework for Disaster risk reduction (2015–2030) was established by the United Nations. It has seven targets and four priorities. The Sendai Framework was adopted by the UN General Assembly in 2015. The Sendai Framework is a voluntary, non-binding agreement that allows the stakeholders to reduce disaster risk [30].

The 100 Resilient Cities (100RC) programme is pioneered by the Rockefeller Foundation with an aim to develop a hundred resilient cities around the world. A hundred cities were chosen after inviting application in three batches—2013, 2014 and 2016. The 100 Resilient City programme developed a City Resilience Framework, which contains four dimensions and each dimension has three drivers [31].

The parameters of both frameworks were tabulated into a matrix-based analytical framework. This matrix was then used to analyze the Smart City proposals, which serves as the implementation tool for the smart cities. As the Smart City proposals are elementary in nature, the text analysis technique was used.

Further in the Sendai Framework (2015) priority number 4 was broken into two parts—(4a) Enhancing preparedness for effective response, (4b) Effective recovery, rehabilitation, and reconstruction, to better accommodate two parameters of the smart cities (Fig. 2).

5.1 Maturity Model

A maturity model defines key attributes that are characteristic of a level of maturity within the organization. The maturity model is hierarchical in nature showing progress in levels of maturity [32]. A maturity model is used to define maturity levels of progression of the analyzed object using multi-dimensional criteria [33]. One of the distinctive features of a maturity model is that each cell contains a description of the maturity stage. The maturity model is less complex and other tools used. Using the analysis of the maturity model, suggestions for improvements and priority of action can be derived [32, 34] (Fig. 3).

		Chennai Smart City	Pune Smart City	Surat Smart City
Sendai Framework (2015)	Understanding disaster risk.	Yes	No	Yes
	Strengthening disaster risk governance	No	No	No
	Investing in disaster risk reduction for resilience	Yes	No	Yes
	Enhancing preparedness for effective response	No	No	No
	Effective recovery, rehabilitation and reconstruction	No	No	No
100 Resilient Cities (2015)	Provide reliable communication and mobility.	Yes	Yes	Yes
	Provide and enhance natural and manmade assets.	Yes	Yes	Yes
	Foster long term and integrated planning.	No	No	No
	Promote leadership and effective management.	No	No	No
	Meet basic needs.	Yes	No	Yes
	Ensure public health services.	Yes	Yes	Yes
	Ensure social stability, security and justice.	Yes	Yes	Yes
	Support livelihoods and employment.	No	No	No
	Promote cohesive and engaged communities.	No	No	No
	Foster economic prosperity.	Yes	Yes	Yes
	Ensure continuity of critical services.	No	No	No
	Empower a broad range of stakeholders.	Yes	No	No

Fig. 2 Comparative analysis of SCP's and two frameworks

5.2 Resilience Maturity Model (RMM)

The Maturity model was modified to incorporate the elements required for depicting the maturity stages in developing resilience in a smart city. An RMM is helpful in incorporating resilience systematically and incrementally [23].

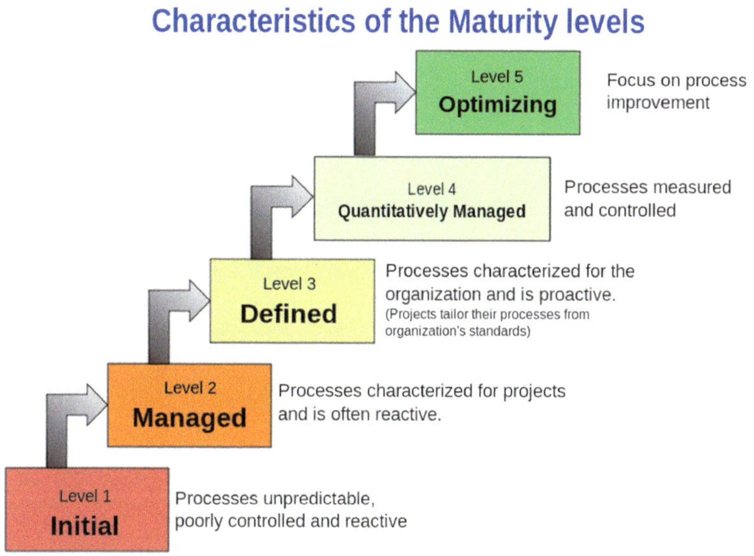

Characteristics of the Maturity levels
Level 5
Optimizing
Focus on process improvement
Level 4
Quantitatively Managed
Processes measured and controlled
Level 3
Defined
Processes characterized for the organization and is proactive.
(Projects tailor their processes from organization's standards)
Level 2
Managed
Processes characterized for projects and is often reactive.
Level 1
Initial
Processes unpredictable, poorly controlled and reactive

Fig. 3 Capability maturity model (CMM) by Godfrey [35]

Recovery Likelehood
Level 1- Adhoc
Level 2 - Reactive
Level 3 - Prepared
Level 4 - Proactive
Level 5 - Resilient
Recovery Readiness

Fig. 4 Five-stage resilience maturity model (RMM)

The Resilience Maturity Model was developed using two scales to measure the level of maturity of each stage. The two scales include—(a) Recovery Likelihood and (b) Recovery readiness.

The incremental stages of the Resilience Maturity Model are mapped against the increment Recovery Likelihood and Recovery Readiness (Fig. 4).

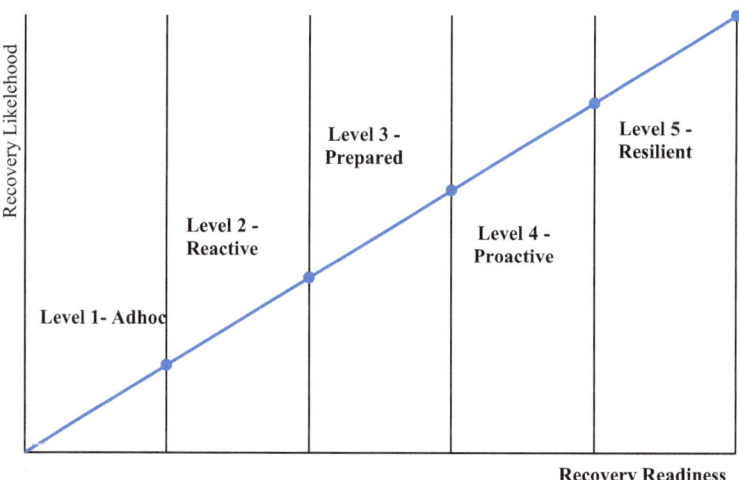

The following sections describe the stages of maturity in the Resilience Maturity Model.

5.2.1 Ad Hoc

It is the first stage in the RMM, the approach towards resilience is not coherent. There may be some element of understanding of disaster risk, but the approach towards dealing with disaster risk is generally fragmented. Disaster management plans may exist in some form but are not integrated into nature. Different emergency agencies, municipalities, and Smart City authorities may have separate sets of plans. There is also a chance that disaster management plans may not work if there are unknown or unexpected or multi-hazard disasters.

5.2.2 Reactive

Some regulations and standard are developed which have elements of disaster management. Disaster management plans are developed to respond to expected and unexpected risks. Education and training programmes are conducted to increase awareness. The importance of Critical Infrastructure is understood.

5.2.3 Prepared

The Institutions become facilitators, laws, regulations, and standards are developed with some elements of disaster management. Involvement of academics and the scientific community to develop methodologies for disaster risk reduction. Enhancement of reliability of critical infrastructure and understanding of the interdependencies of the same. Stakeholders are involved in planning, however, they are not operationalized yet.

5.2.4 Proactive

The stakeholders are fully engaged and have a full understanding of the development of resilience in the city. A strong group of volunteers is established, and joint training is conducted. The institutions are proactive in approach and participate with another city to share knowledge and best practices. The laws and regulations are fully developed, and the disaster management plans are integrated and collaborative.

5.2.5 Resilient

The city has developed fully resilient and can face any expected or unexpected disaster without hampering its functionality. Critical infrastructure is reliable and can function properly. All stakeholders are properly engaged and work together. The city can bounce back from disaster very quickly.

6 Conclusion

The Smart Cities were introduced in India to improve the quality of life and increase economic activity. However, the Smart City Mission does not emphasize on the introduction of resilience into the basic core of these Smart Cities. With rapid urbanization, most cities will be under tremendous stress, including smart cities. Smart Cities of India needs to be resilient to face the impending disaster that India faces. Resilience Maturity Model (RMM) presents a roadmap for the Smart City stakeholders to incorporate resilience in their smart cities and reduce disaster risks.

References

1. UN Population Division (2017) World urbanization prospects 2018. Available: https://population.un.org/wup/. Accessed 9 Feb 2019
2. Chandramouli C (2011) Census of India. Available: http://censusindia.gov.in/2011-prov-results/paper2/data_files/india/Rural_Urban_2011.pdf. Accessed 03 Feb 2019
3. Wamsler C (2014) Cities, disaster risk and adaptation, 1st edn. Routledge, New York
4. Kreimer A, Arnold M, Carlin A (2003) Building safer cities: the future of disaster risk. The World Bank, Washington, D.C.
5. Aijaz R, Hoelscher K (2015) India's smart cities mission: an assessment. ORF Issue Brief, pp 1–12, Dec 2015
6. Gupta K, Hall RP (2017) The Indian perspective of smart cities. Prague
7. PTI (2016) Northeast monsoon claimed 470 lives in Tamil Nadu: Jayalalithaa. Available: https://www.thehindubusinessline.com/news/national/northeast-monsoon-claimed-470-lives-in-tamil-nadu-jayalalithaa/article8064661.ece. Accessed 10 Feb 2019
8. Deccan Chronicle (2019) Tamil Nadu: Chennai floods cause a loss of Rs. 50,000 cr. 2015. Available: https://www.deccanchronicle.com/151206/nation-current-affairs/article/chennai-floods-caused-loss-50-thousand-crore. Accessed 10 Feb 2019
9. Bhatnagar A, Nanada TP, Singh S, Upadhyay K, Sawhney A, Swamy RRD (2018) Analysing the role of India's smart cities mission in achieving sustainable development goal 11 and the new urban agenda. In: Filho WL, Rogers J, Iyer-Raniga U (eds) Sustainable development research in Asia-Pacific region. Springer International Publishing, Switzerland, pp 275–292
10. NDMA (2019) National disaster management authority. Available: https://ndma.gov.in/en/vulnerability-profile.html. Accessed 27 Jan 2019
11. Guha-Sapir D (2019) EM-DAT: the emergency events database - Université Catholique de Louvain (UCL) - CRED. Available at: https://www.emdat.be. Accessed 17 Feb 2019

12. Malalgoda C, Amaratunga D, Haigh R (2013) Creating disaster resilient built environment in urban cities: the role of local governments in Sri Lanka. Int J Disaster Resil Built Environ 4(1):72–94
13. Desouza KC, Flanery TH (2013) Designing, planning, and managing resilient cities: a conceptual framework. Cities 85:89–99
14. Satterthwaite D (2013) The political underpinnings of cities' accumulated resilience to climate change. Environ Urban 25(2):381–391
15. Linkov I, Fox-Lent C, Keisler J, Sala SD, Sieweke J (2014) Risk and resilience lessons from Venice. Environ Syst Decis 34:378–382
16. Bosher L, Dainty A, Carrillo P, Glass J (2007) Built-in resilience to disasters: a pre-emptive approach. Eng Constr Archit Manag 14(5):434–446
17. Baum SD (2015) Risk and resilience for unknown, unquantifiable, systemic, and unlikely/catastrophic threats. Environ Syst Decis 35(2):229–236
18. Bosher L (2008) The need for built in resilience. In: Bosher L (ed) Hazards and the built environment-attaining built-in resilience. Routledge, London, pp 3–19
19. Haigh R, Amaratunga D (2011) Introduction. In: Haigh R, Amaratunga D (eds) Post disaster reconstruction of the built environment: rebuilding for resilience. Willey-Blackwell, Oxford, pp 1–11
20. Tompkins E, Hurlston-McKenzie L-A (2011) Public-private partnerships in the provision of environmental governance: a case of disaster management. In: Boyd E, Folke C (eds) Adapting institutions: governance, complexity and social-ecological resilience. Cambridge University Press, Cambridge GB, pp 171–189
21. Desouza KC, Flanery TH (2013) Designing, planning, and managing resilient cities: a conceptual framework. Cities 35:89–99
22. Godschalk DR (2003) Urban hazard mitigation: creating resilient cities. Nat Hazards Rev 4:136–143
23. Hernantes J, Maraña P, Gimenez R, Sarriegi JM, Labaka L (2019) Towards resilient cities: a maturity model for operationalizing resilience. Cities 84:96–103
24. Spaans M, Waterhout B (2017) Building up resilience in cities worldwide—Rotterdam as participant in the 100 resilient cities programme. Cities 61(1):109–116
25. Malalgoda C, Amaratunga D, Haigh R (2016) Overcoming challenges faced by local governments in creating a resilient built environment in cities. Disaster Prev Manag 25(5): 628–648
26. Global Facility for Disaster Risk Reduction (2015) Investing in urban resilience. The World Bank, Washington
27. O'Rourke TD (2007) Critical infrastructure, interdependencies, and resilience. Available: https://pdfs.semanticscholar.org/6c17/b35ec7555a9f27d5ccb6ca1d357a20b5ce0a.pdf. Accessed 09 Feb 2019
28. Field C, Look R (2018) A value-based approach to infrastructure resilience. Environ Syst Decis 38(3):292–305
29. Pursiainen C (2018) Critical infrastructure resilience: a Nordic model in the making? Int J Disaster Risk Reduct 27:632–641
30. UNISDR (2015) Sendai framework for disaster risk reduction. Available: https://www.unisdr.org/we/coordinate/sendai-framework. Accessed 10 Feb 2019
31. Resilient Cities (2015) City resilience framework. Available: http://www.100resilientcities.org/resources/#section-2. Accessed 10 Feb 2019
32. Wendler R (2012) The maturity of maturity model research: a systematic mapping study. Inf Softw Technol 54(12):1317–1339
33. Fleming M (2001) Safety culture maturity model. Health and Safety Executive, Norwich
34. Becker J, Knackstedt R, Pöppelbuß JB (2009) Developing maturity models for IT management. Inf Syst Eng 1(3):213–222
35. Godfrey S (2008) What is CMMI?. NASA Goddard Space Flight Centre, Maryland

Intelligent Urbanism Guiding the Smart City Region Development: Case Study of Bhopal

Aman Singh Rajput

Abstract City can be accurately measured by its impact as a node of various land use units which diffuses the social, political and economic effects in the region. City region reflects association of multiple municipalities' and scales of government in which responsibility for urban and regional development is distributed formally as well as informally. Smart cities are dominating the current paradigm of planning with a focus on the use of information and communication technology for the development of the society. In India, it is exemplified under the 'Smart Cities Mission' where the emphasis is laid on the promotion of core infrastructure and application of information and communication technology to improve the quality of life in urban areas. Modern development concept of intelligent urbanism as defined by Christopher Benninger through 10 axioms or principles guides the city planning which would help the regions to achieve the results of India's smart cities mission. The intelligent urban region strives to create a balance between economic, environment and social development- three pillars of sustainability. The concept composed of ten principles integrates the planning and management discourse in the preparation of a spatial plan for the region. The paper tries to highlight the impacts of spatial development and planning of the Bhopal city region, in light of these ten principles considering its functionality. The paper tries to understand how the evolution in the planning of Bhopal has impacted the spatial and sectoral development of the region and its achievement of 'smartness'.

Keywords Urbanization · City region · Sub-city · Principles of intelligent urbanism · Smart region

A. S. Rajput (✉)
Department of Regional Planning, School of Planning and Architecture,
New Delhi, India

© Springer Nature Singapore Pte Ltd. 2020
S. Ahmed et al. (eds.), *Smart Cities—Opportunities and Challenges*,
Lecture Notes in Civil Engineering 58,
https://doi.org/10.1007/978-981-15-2545-2_36

1 Introduction

India with an urban population growth rate of 2.76% from 2001 to 2011 supports 31.1% of the urban population [1]. This level of urbanization poses a great challenge to the urban and regional planners and also provides opportunities for guiding the development towards inclusive, equitable and sustainable growth of city and its region [25]. The cities contribute 63% to the India's GDP which is likely to increase to 75% by 2030. In this view to develop cities as engines of growth for the economy, it is required to make it attractive to the people and for investment. With similar regard mission for the development of 'smart cities' was introduced by Prime Minister Mr. Narendra Modi. In the absence of a universally accepted definition of the concept, this mission aims to fulfil the aspirations and needs of its citizens through developing the regional eco-system supported by comprehensive infrastructure development pillars—institutional, physical, social, and economic. The main focus of the mission is laid on the promotion of core infrastructure and application of information and communication technology to improve the quality of life in urban areas [21]. The features of smart development as defined by the mission [12] are as follows:

   (i) promotion of mixed land uses,
  (ii) inclusive development,
 (iii) create walkable localities,
  (iv) open areas development,
   (v) increasing the modal split options,
  (vi) citizen-friendly and effective governance,
 (vii) city identity and
(viii) Smart solutions to services and infrastructure.

   The effects of inevitable urbanization whether positive or negative is experienced not only by the city residents but also by the non-city residents. Therefore, the process of urbanism is a mode of life identified, but not specifically limited to the city boundaries. The city region in this regard reflects the association of multiple municipalities and scales of government in which responsibility for urban and regional development is distributed formally as well as informally. Viewing the future from the perspective of applied smart city policies and approaches is vital to solve the current problems, however, 'smartness' in its nonconcrete form is a construct of a city that adapts over time and cosmos [18]. In similar lines smart city's are envisioned as the 'urban centre of the future' [15]. Thus, besides looking into what's the future of any region it is important to study and analyze the past lessons and examples of how generations handled the challenges of planning and develop the smart region it is today. This paper tries to highlight the same in the case of Bhopal city region, with the discussion around the principles of intelligent urbanism.

   The theory of intelligent urbanism is the most significant modern development concept which has been formulated by the Indian Congresses of Modern

Architecture for the plan preparation process. The theory comprises of ten principles or axioms serving as a guideline for the plan preparation process. These are value-based outlines which promote a participatory planning process, developed by Joseph Lluis Sert and formulated by the Christopher Benninger. It was applied in the development plan preparation of the new Capital of Bhutan—Thimphu. The key characteristic is communication of the development and management phases for the solution of various problems and fulfilling the needs of the citizens [3]. These principles overlap with the 'smart features' (refer Table 1) to be developed as per the Smart Cities mission guidelines by the Ministry of Urban Development, India.

The smart city concept and intelligent urbanism are closely related as both of them strive to create a balance between economic, environment and social development—the three pillars of sustainability under the umbrella of political and cultural setting. The further sections of the paper give a brief about the study region and discuss the ten principles of intelligent urbanism in the context of planning and development of the 'smart region' Bhopal.

2 Bhopal City Region

Bhopal city is the capital of Madhya Pradesh state, located above the upper limit of the Vindhyan range on Malwa Plateau. The city growth started in 1956, with the construction of Bharat Heavy Electricals Limited (BHEL) along with its town in the then eastern municipal limits and shifting of Bairagarh cantonment area within the urban limits. In 1959–1960 Government formed a parastatal authority called *Capital Project Administration (CPA)* to undertake the preparation of *General development plan for the Capital project Township (T.T. Nagar)*, to accommodate government offices and to establish capital function of the region. From 1956 to 91 the city expanded enormously in different directions and presumed the status of cosmopolitan city. The hierarchy of legalized spatial planning of Bhopal started

Table 1 Corresponding principles of Intelligent Urbanism leading to smart city features development

S. No.	Smart city feature	Principle of intelligent urbanism (Nos.)
1	Promotion of mixed uses	7, 8
2	Inclusive development	3, 5, 9
3	Create walkable localities	2, 4, 8
4	Open areas development	1, 3,
5	Increasing the modal split options	7, 8
6	Citizen friendly and effective governance	4, 9
7	City identity	2,
8	Smart solutions to services and infrastructure	6, 9

Source Author

with the Interim Development Plan—1962, followed by Bhopal Master Plan—1991 which focussed on the compact city development, hierarchical city structure and self-contained planning units. In 1994, the second development plan "Development Plan for Bhopal—2005" was prepared whose objective was to develop the region with 'self-contained sub-cities' approach. In 2005, under the JnNURM scheme, the City development plan was prepared for the horizon period of 2021 which was focussing on the development of infrastructure facilities. The major highlight of this plan was the proposal of BRTS project covering major parts of the capital city. Some other plans were made in the direction of sustainable development of the region such as Bhopal city mobility plan, low carbon scenario development plan, Smart cities plan etcetera. The city region is taken as the planning boundary for the preparation of Bhopal Development Plan 2031, notified under section 13 (2) of the Madhya Pradesh Town and Country Planning Act, 1973 (refer Fig. 1). It consists of Bhopal Municipal Corporation (BMC) and 88 villages. The total population of the region is 2,021,107 as per census of India 2011 with urban population share of 95%.

Bhopal city region is a perfect amalgamation of the rich cultural heritage, environment, and economic development. The principles of intelligent urbanism would dovetail the smartness achieved in this direction as discussed below.

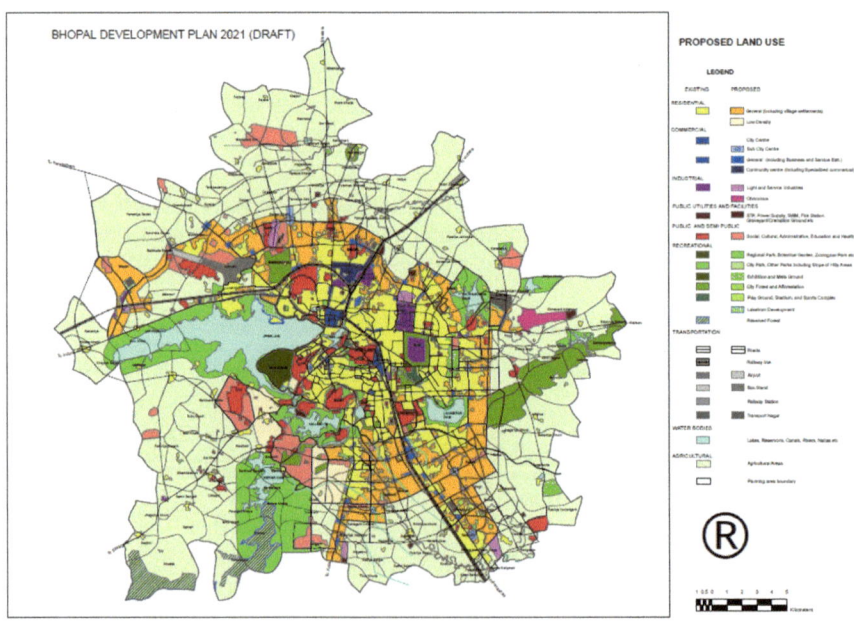

Fig. 1 Bhopal City region. *Source* Bhopal Development Plan—2021 (Draft) available online at http://www.mptownplan.gov.in/bdp2021/Proposed%20landuse.pdf

3   Analysis of Principles of Intelligent Urbanism

This section of the paper discusses in detail the proposals made in the Bhopal Development Plan 1991 and Bhopal Development Plan 2005 in line with the 10 principles of intelligent urbanism and the impacts they had on the urbanism of the city and people.

1. **Balance with Nature**—The principle is closely related to the concept of carrying capacity. During the plan preparation process, reservation of fragile areas for conservation along with natural zone maintenance, intelligent density allocations and integrated land-use planning should be achieved [5]. The use of natural resources in the region should take place up to the regenerative capacity of the nature via its ecological services.

   The Bhopal Development Plan—1991, considers the concept of 'form follows function', i.e., defined functions leads to a more planned form. In the same line, the plan takes into consideration the physical topography of the region as a 'form' for determining the 'functions' for future development (refer Fig. 3a). Example—location of industrial units in the northeast zone considering the wind flow and micro-climatic conditions. In 1984, Bhopal city faced the worst *industrial disaster of gas leak* at Union Carbide Infra limited pesticide plant. As a repercussion, city administrators and planners took the decision to safeguard the ecologically sensitive nature of the region and diverted the overall industrial development towards the south direction in Mandideep. This lead to the major focus for economic development through non polluting (administrative, educational and tourism) sectors.

   The Bhopal Development Plan—2005 considers ecological parameters of the region for development. Nature is a function of geology, geomorphology, climate, physiography, plants, animals and land uses (refer Fig. 2). The development plan proposes the Bhopal urban form through identifying the process involved in nature, values attached to it and examining the externalities of anthropogenic activities. This leads to the delineation of negative areas for future development (refer Fig. 3b) such as:

   (a) **Catchment area of Upper Lake and Van Vihar**: The total catchment area of Upper and Lower lake is 370.6 km^2. Bhoj Wetland Project covers 18% of the city region and is important as it supports more than 700 species [27]. In 1979, *Van Vihar National Park* was declared as a national park that covers an area of 4.45 km^2. The region is proposed for minimum density allocation if needed owing to lower accessibility and geologically faulty region.

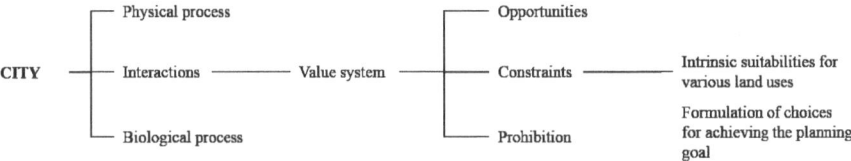

Fig. 2 Ecological concept. *Source* Bhopal Development Plan, 2005, Pg. 10

(b) **Hillocks (Shamla and Fatehgarh)**: Due to the attainment of carrying capacity, the area has potential only for exclusive development.

(c) **Airport area**: partly under the lake catchment area the development is low rise and low density.

Also, housing density allocation to achieve sustainable development had the following criteria for consideration:

1. Existing densities, redevelopment, and regeneration potential,
2. Restrictive nature of development on hilly areas,
3. Changing land values,
4. Carrying capacity of infrastructure and potential for development.

The medium-high to higher density areas are to be developed along the major roads with major work centres and low-density areas in the eccologically sensitive region. These areas are well protected and preserved by the planners through their planning models due to which the lifestyle in Bhopal is less stressful [29]. Apart from this to lower down the carbon emissions in the city seven actions (i) green governance, (ii) holistic development, (iii) low carbon lifestyle, (iv) multi nuclei land use planning, (v) form and flow integration of the city, (vi) nature nurturing, and (vii) rural lifestyle development are proposed [9]. Due to this intelligent approach, the development till now has not taken place in the southwestern direction due to preservation zones of Van Vihar and Bhoj Wetland catchment area. The smart city feature of 'preservation of open spaces' is achieved via this principle in Bhopal.

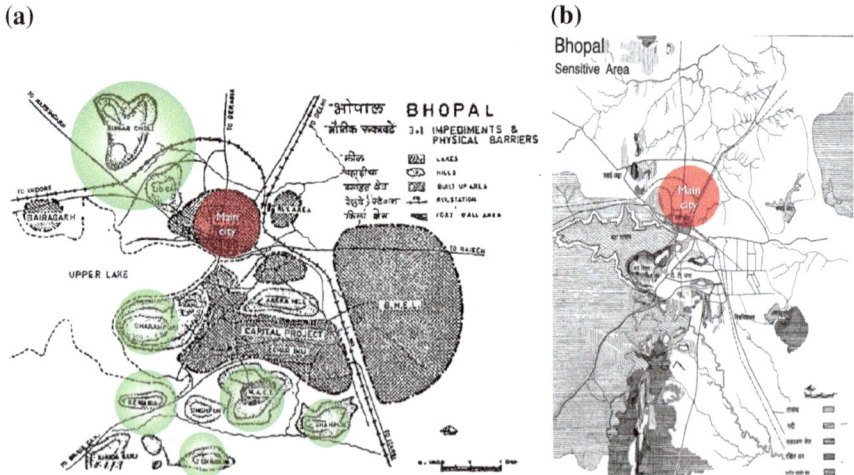

Fig. 3 a Bhopal city region physical barriers; **b** Bhopal city region sensitive regions. *Source* Bhopal Development Plan, 1991, Pg. 40 and Bhopal Development Plan 2005, Pg. 44 with visuals modified

2. **Balance with Tradition**—The second principle emphasizes on safeguarding the unique identity of the city. The major focus should be laid for the maintenance of the views that are developed in history, along with the preservation of the heritage structures. Under the principle, the important natural features in the regional setting such as rivers and water bodies should be given due respect [3]. The city of lakes has an important connection with the water bodies, culturally and spatially. The beauty of the city at the best from the skyline view and lake perspective can be obtained from all around the lakes, especially the VIP road and boat club.

Bhopal Development Plan—1991, in this regard, defines 'special features' which are buildings and monuments having an impression of important historical events in the city's history which differentiate it from other cities. These 'special features' needs to be preserved and improved for which they were divided into seven categories viz, places of archaeological importance, Upper lake, Lower lake, Rani Kamlavati Palace, Religious places and buildings, places of natural landscape and scenic importance and others.

In Bhopal Development Plan—2005, potential natural and built form areas for landscape treatment had been proposed with the objective to develop:

(a) Visual linkage of the lake water.
(b) Skyline of the mosques and minarets.
(c) Parent city's built-up area—hard surfaces and foreground development.
(d) Conspicuous landform—increased accessibility and soft features.

The image of the city is the overall appearance of the man-made and natural elements. The old city area is famous for its cultural and monumental precincts. Thus, the primary exercise is to create an appreciation by the people of the building through height and related horizontal distance (refer Fig. 4). This has led to the evolvement of the future urban form of the region.

Bhopal region has the smart city feature of 'city identity', achieved through the application of this principle. The city region of Bhopal has attained the modern outlook with time but at the same time also holds its root of great cultural significance and protecting the same for its unique identity.

3. **Conviviality**—This principle is related to social development, i.e., creating spaces in the city for the people to interact. The principle proposes to develop hierarchies of open spaces in this regard from a regional level to community level where meetings and rituals can be performed and for public domain wherein different age groups, caste and sexes can socialize [5]. These interactions lead to the development of social relations [36]. The region is a mix of various societies and cultures living together and development should consider the interest of all.

Bhopal Development Plan—1991, the deficiency of socio-cultural and recreational amenities in the region lead to the proposal for the development of a hierarchy of facilities at regional and city level, planning unit level and sectoral level. In order to provide opportunities for economic and social integration, the concept 'Corridor of city-level activities' was applied so that the various planning units are equidistant

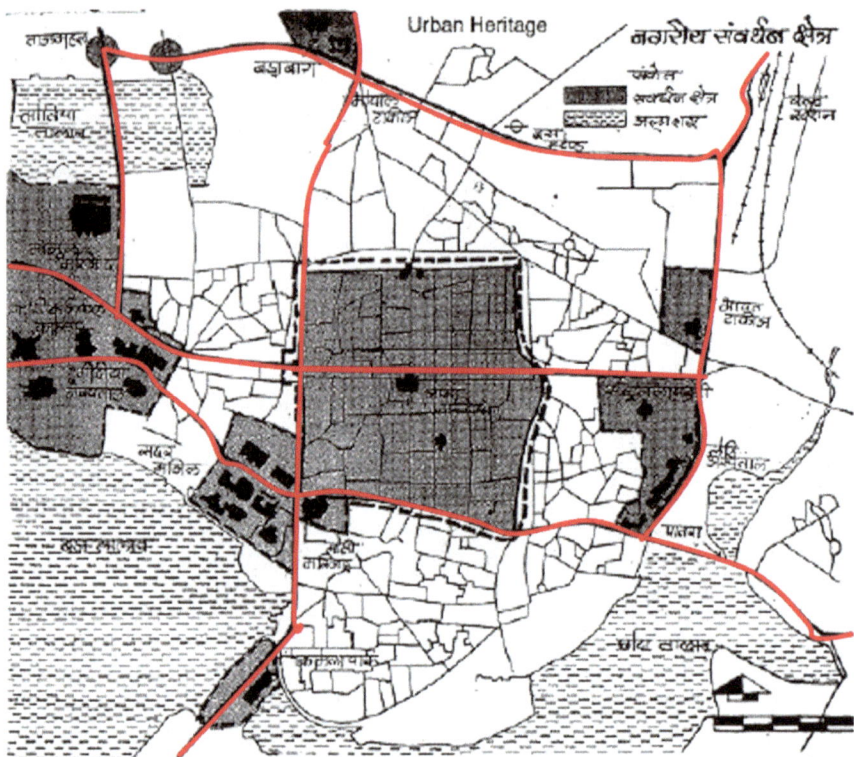

Fig. 4 Bhopal urban conservation region. *Source* Bhopal Development Plan, 2005, Pg. 19 with visuals modified

providing accessibility to all. However, ideally, it was impossible to achieve this due to topographical characteristics of the region thus modified adaptation of such an approach was made (refer Fig. 5). The development of a new center in MP Nagar area was also envisaged owing to growing congestion in the old city area.

Bhopal Development Plan—2005, taking into account the natural features, the plan envisages the development of a system of open spaces. These satisfy the environmental and recreational functions of the residents, in line with the drainage and landscape features of the region. The two levels of open spaces are—(i) regional parks and (ii) local open space system. Apart from this, the plan proposes to develop a botanical garden, zoological park, fish farm, picnic spots, city parks, city forests and central Exhibition ground taking advantage of the site features. Under Chapter-IV 'Development Regulation', the tables 4-T-9 and 4-T-10 specify the minimum area required and population to be served by the facilities. Accessibility to these services is also of paramount importance whose specifications are listed in table 4-T-14.

The availability of recreational green spaces in the region is affected by the demand for housing. In the old city area, very less open space is available and if

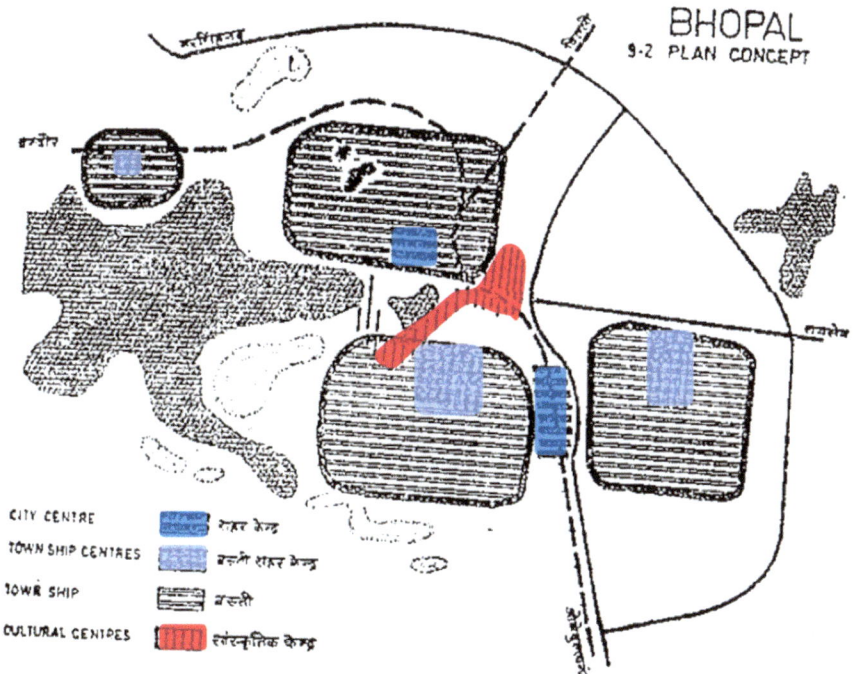

Fig. 5 Bhopal city region plan concept of inclusive development. *Source* Bhopal Development Plan, 1991 with visuals modified

present it is around the architectural monuments and heritage buildings or on the steep slopes with minimum accessibility for recreation purposes. However, the new Bhopal which is developed on the grounds of the Master plan provisions has an ample amount of green spaces as national parks, regional parks, community and city parks etcetera [28]. As per the data released from the National Housing Bank index, Bhopal has shown second-best growth driven by real estate development, as many as 70 new residential projects are being planned in the region [24]. Apart from this, a number of malls have been opened in the city region leading to social cohesion especially in the suburb areas which are new potential areas of residential development. 'Inclusive development' and 'developing open spaces' features of smart cities is achieved from this principle.

4. **Efficiency**—The core of this principle is the effective management of the 'systems' running in the city, viz, transport, infrastructure, and governance. The major problem this principle deals with is related to transport such as accidents, waiting for time loss and impacts on human health [3]. The proposed plan should strive to reduce the dependency on personal vehicles and promote the alternative choices to travel through mixed-use development at the major nodes thereby reducing the travel distance and making public transport affordable.

As per Bhopal Development Plan—1991, for effective management of the transportation 'workplace relationship' was considered. The major employment centers such as BHEL, capital project (T.T. Nagar) and Bairagarh had residential townships in close vicinity reducing inter 'planning units' movement (refer Fig. 6). For effective governance, the plan proposes an integration of various levels of organizations, government, semi-government and private. This would lead to more benefits and quicker returns.

On the other hand, Bhopal Development Plan—2005, envisages the restructuring and re-organizing of the activities in relation to the transport network while considering the options and choices. This is supply-side mechanism different from the previous plan of demand-side mechanism. In order for effective management of the city infrastructure and service strategy, the five key elements that are worked upon—(i) guided development: role of 'facilitator', (ii) creation of infrastructure land bank: land contribution on equitable basis, (iii) Formulating integrated urban development programme: annual action plan and monitoring, (iv) pragmatic approach to development regulation: flexible approach to redefine as per modern outlook and (v) providing framework of time-bound development: reducing the multiplicity of authorities for development.

Due to the effective management of the growing real estate market in the Bhopal city region and the availability of 'virgin land' for development, the buyers have the willingness to pay for a green homes. The government of Madhya Pradesh at present is also developing a Clean Development Agency (CDM) for the preparation

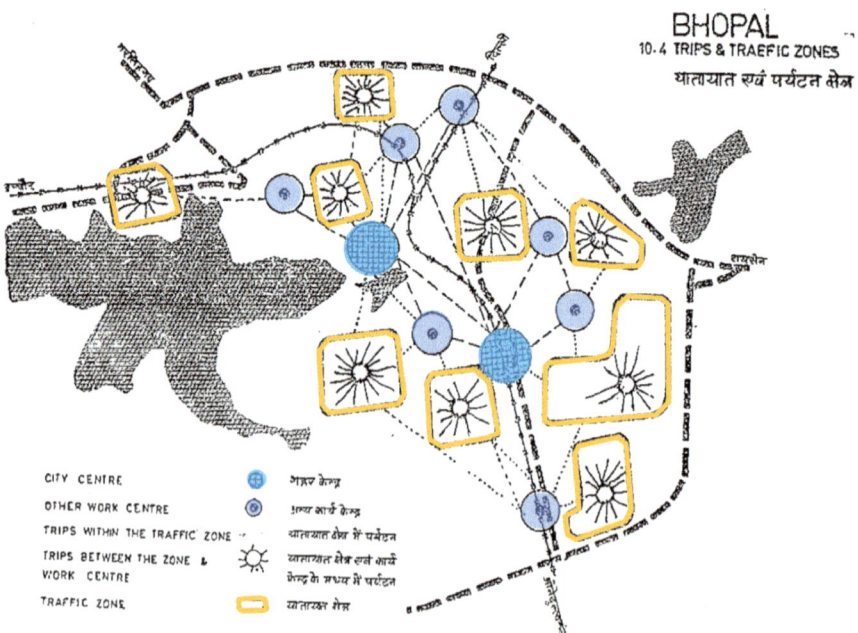

Fig. 6 Bhopal trip and traffic zones. *Source* Bhopal Development Plan, 1991 with visuals modified

of green building principles [13]. Thus, it can be interpreted that the effective management of the transportation system is taken by the government in the direction to reduce GHG emissions and promote public and non-polluting modes of transport. The 'effective governance' smart city feature for the Bhopal city region has been achieved as the city has scored the first position among the 8 urban local bodies of the state. The five parameters of judgement analyzed were presence, urban information, interaction, transaction and e-democracy [17]. Thus, the principle is useful to be incorporated in the planning process discourse to provide the foundation base to the efficient working of the systems.

5. **Human Scale**—This axiom deals with the people friendly spaces and people oriented development of places which are people-friendly and people-oriented development [5]. Taking the considerations of the climatic conditions and locational setting of the region these places should be developed and providing access to all in terms of anthropology [2].

The Bhopal Development plan of 1991 and 2005, accumulate the social services around the major transport stops open spaces and near to the public institutions in order to ensure the 'human dimension'. Example: The Bhoj wetland constructed to provide drinking water to the city also provides recreational and productivity benefits [35]. The government has tried to make the surrounding spaces of this lake people friendly and recreational in nature, and development of aesthetic values around. This leads to 'inclusive development' of the 'smart' city region.

6. **Opportunity Matrix**—Every urban and regional plan provides an opportunity to increase the standard of living and quality of life of the citizens via social and economic development. The city regions are nodes where an individual harness his knowledge and skills. This can be achieved through the provision of physical and social infrastructure. These facilities should not only be physically present but accessible to all [3]. This principle also deals with the major problem of slum and poverty and describes 'these' as prejudices of the region itself via process of gentrification and marginalization.

Bhopal Development Plan—1991, following the attainment of capital city status in 1956, the economic and industrial activity in the region increased at a rapid rate. Major functions of which were administrative but due to its strategic location the city attracted commerce and industries, which lead to the problem of environmental degradation. The wastewater from the manufacturing plants flowing through the natural drainage was polluting the water bodies. Due to topological barriers the water table in some regions was quite low thereby reducing the potential. Owing to the aftermaths of Bhopal Gas Tragedy in 1984 the manufacturing units were shifted to the peripheral areas or the nearby districts with the weak economic base.

For water distribution, a water supply grid has been suggested through the extension of services of treatment and headworks in the periphery. Bhopal has only one source of water supply i.e., Upper Lake due to which development of new sources was proposed. Sewage in the planned area of BHEL is maintained by its own management wherein the effluent is used for vegetable farms. The stormwater drains in the region are pukka and semi pukka flowing the discharge into the Patra Nallah. The city in terms of social infrastructure serves a larger part of the region as a result of which in terms of health infrastructure various specialized hospitals have been proposed. Also owing to the administrative capital and its regional location educational services such as specialized institutions and campus colleges are expanding.

The Bhopal Development Plan—2005, proposes the economic development of the city region to be as a multifunctional capital city through administrative centered, commercial cum industrial centered and educational cum research centred functions. For achieving this institutional zones have been delineated such as capital complex, campus-based institutional complex, universities, and industrial areas. For industrial development, great potential lies around Bhopal being a national priority city. The industrial growth is considered in a regional context which is taking place in outside the regional outlook having the advantage of proximity (i.e., Mandideep and Satlapur (Dist. Raisen) and Pillukheri (Dist. Rajgarh). In the year 2013, the government proposed an IT park to be developed on PPP model by the developer Messers Underhill Technology, New York. This project will provide 50% of non-residents of MP as their employees. In the city region, Govindpura industrial areas have medium and small scale industries along with the largest manufacturing enterprise unit BHEL. Apart from this, the residents are also engaged in retail business and handicrafts. Mandideep is an industrial suburb of the region [19]. From being a traditional industrial base to an education hub in the state, Bhopal has attracted different types of employment centers and is also producing a large number of educated working age group population. The literacy rate of the city region is 83.47% close to that of urban India having 84.11%, while the percentage of working age group (15–59 years) and youth (15–24 years) is 65.22 and 21.30% which is higher than the national average [6]. Many national importance higher secondary schools like DPS, Sanskar Valley, etc., and colleges like SPA Bhopal, NIFT, and IISER, etc., came in suburban areas [11].

Additional water supply sources at Kaliyasot dam and Kerwa dam areas have increased the water supply capacity. However, owing to its population growth the new water supply sources from the Halali river have been proposed. Also, the water from the Narmada river is also proposed to be supplied to the city and thus in the coming years, the development of the city needs to be guided accordingly. Sewerage is the key area that requires innovative solution and use of modern technologies. The plan proposes to examine a way in which the BHEL plan for sewerage can be replicated in various sub-cities, where the waste can be treated around each sub-city and using the same for market gardening and agricultural operations in the green wedge.

The city with a loss of 20% gives 135 lpcd supply but due to lack of efficiency in the production at source and storage capacity. Ironically the city of lakes is also having problems related to water supply and instead of managing its own water sources, the region is managing its own water sources, getting dependent on a distant source of Narmada [16]. Nearly 30% of the city has closed drains and the rest supported by a system of open drains [19]. The solid waste in the region for the last 25 years is dumped at Bhanpur site located within the BMC limits having an area of 57.8 acres. The rural areas use unscientific ways for waste disposal. However, to solve the growing menace due to solid waste a new site at Adampur is developed equipped with engineering landfill facilities like baseliner, composting, collection, and treatment system [7].

The most marginalized sections of the society live either in slums or urban villages. The slum development initiatives in the region are the Patta Act, 1984, through which the leasehold titles either 'permanent' or 'temporary' were granted to the urban poor living in the slum settlement. This patta enabled them to be benefitted by the infrastructure facilities. Other initiative was the Basic Services for the Urban Poor (BSUP), which was a mandatory reform supported under JnNURM, with a goal to provide basic services to all the poor including the security of tenure and improved housing as well as social service delivery [4]. Bhopal Municipal Corporation is also working towards the implementation Slum Free City Plan for Bhopal Metropolitan Area prepared under the Rajiv Awas Yojana. As envisaged in the city development plan under the JnNURM scheme the slum dwellers of Arjun Nagar, Bheem Nagar, Madrasi colony and Rahul Nagar slums are rehabilitated with all the necessary facilities as per Madhya Pradesh Bhumi Vikas Adhiniyam 1986, with a project cost of 0.32 Cr [30]. In the slum areas, though the access to piped water supply and toilets are lacking in 70 and 42% slum HHs in the city along with poor access to sanitation and solid waste management as well as social infrastructure, the perception of the majority is that the situation has improved from previous years [34].

Though the principle deals with the provision of economic, social and physical development, the plans owing to their times do not have any component of the smart solutions attached to it. Thus, it can be recommended that in the next master plan of the city region under this principle 'smart solutions for delivery and maintenance' should be incorporated as they are the needs of the present.

7. **Regional Integration**—Planning for the city and its region is a holistic process as some of the city problems are caused due to regional impacts and vice versa. This principle emphasizes the integration of this factor into plan-making process. Overall development of the city region is obtained when the economic development is not limited to the city itself but dovetailed to the lower order settlements. The ignorance of the hinterland leads to problems of overexploitation of the resources outside the city and reduction in food security [3]. Intelligent urbanism links the city's physical and geographic character as a part of its social and economic region.

As per Bhopal Development Plan—1991, for regional level analysis, the consideration is given to the Capital City Region Plan which consists of Bhopal, Raisen, Sehore, Shajapur, and Rajgarh districts. However, the plan is still in the draft stage of its preparation by the town and country planning organization.

As per the Bhopal Development plan—2005, due consideration is given to the integration of the region and sub-region in the development of Bhopal. The growth of the city is devised both from regional and sub-regional context. The city being a capital has much more regional impact than any other cities which has led to the attainment of its 'multifunctional city' nature. Bhopal being a 'mother city' in the region the plan proposes to strengthen the regional and sub-regional supporting infrastructure and services. The roles of the nearby cities and their growth need to be integrated for the development of Bhopal.

The spatial planning of Bhopal can be identified as self-contained sub-cities mainly attributing to the physical disjuncture caused by the hillocks and water bodies. Taking into consideration the growing importance of the city in its regional outlook, the sub-city concept was proposed to develop the state capital [33]. There are seven sub-cities that are present in Bhopal—Bhopal main city, T.T Nagar, Bairagarh, BHEL, Service township, Misrod, and Neori (refer Fig. 7b). With time each of these sub-cities have gained a special character in the region. Bhopal main city is the old city area famous for its monuments while T.T Nagar sub-city has the major commercial areas viz., new market, Srishti CBD and the DB city mall. Misrod sub-city is located very near to the Mandideep industrial township and thus has a higher percentage of private townships, the Neori sub-city is the newly growing sub-city and the government is also paying attention to its development. BHEL sub-city and its service townships are contained sub-cities and are characterized institutional due to the presence of educational institutions. Bairagarh sub-city is characterized by its wholesale market and environmentally sensitive areas as it lies in the Upper lake watershed region. These sub-cities interact with each other in the region creating an integrated and habitable city region. Bhauri is also one of the areas in the region where new educational institutions such as SPA, IISER are coming up. A network of census towns Berasia, Vidisha, Raisen, Obedullaganj, and Sehore are present around the Bhopal city region (refer Fig. 7a).

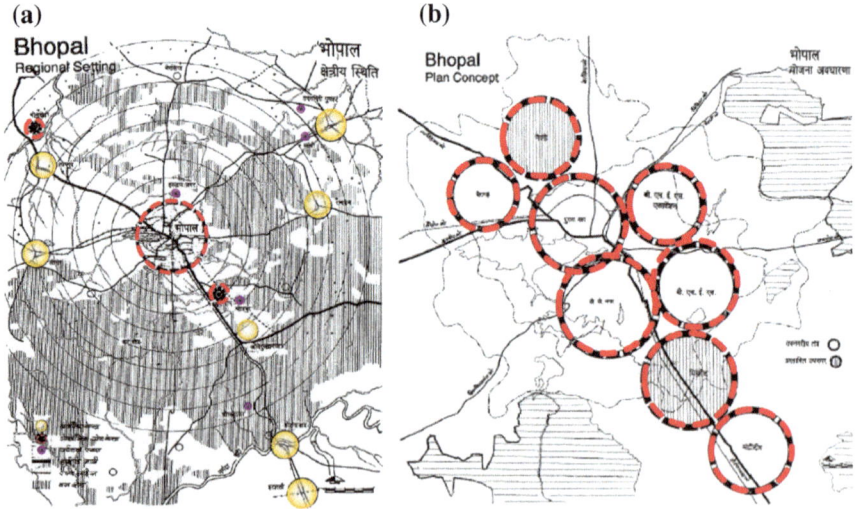

Fig. 7 **a** Bhopal regional setting and **b** Bhopal sub-city concept. *Source* Bhopal Development Plan, 2005 with visuals modified

The city is an administrative hub and the capital acts as a major service center and nodal center of trade and commerce of regional importance [14]. In regard to the regional scale, the city has diverted the industrial development in the south-Mandideep industrial suburb. The shifting has been contained till Mandideep as below it lies the ecologically sensitive zone of Narmada river and Vindhya mountain ranges.

The concept of regional outlook is an important part of the 'smartness' as the city is not a standalone settlement but have linkages forward as well as backward with the hinterland. The sub-city concept is 'smart approach' as the areas are self-sustainable with an economic activity predominance within and thereby reducing longer trips.

8. **Balanced Movement**—This principle deals with an importance of 'land use and transport integration'. Emphasis is laid to balance the different modal split and the public spaces served by cheap transport nodes [5]. In a region, there are multiplicity of modes some connecting within the city and some outside. The theory of intelligent urbanism considers the automobile as an integral part of the city region that cannot be neglected, however, the modes of public transport are considered the most cost-effective solution for the city region. The principle has similar connotations as in principle number 4 'Efficiency' of transport system.

The strategic location along the major transport routes and the presence of BHEL and industrial suburb the region acts as a center of regional trade and commercial center [20]. The region has a radial road pattern with work nodes at the center location (refer Fig. 8) so the people have to travel to these locations [23]. The central location of the Bhopal city region at the state level allows it to act as a major transport and communication center. The NH 12 and NH 86 intersect each other at Bhopal connecting Jaipur, Jabalpur, Raisen, and Dewas. The region is also a transit point to travel between Indore, Hoshangabad, Sehore, Raisen, Kolar, and Vidisha [8]. The city is connected by the broad gauge railway line to Nagpur, Chennai, Delhi and Mumbai. The city is also served by regular air services to Mumbai, Delhi and Indore.

The annual growth rate of vehicular traffic is 10%. As per the modal split, the potential transit users are 74% followed by two-wheelers are 25% and cars are 3%. As the state capital, the city is the center for administration and political institutions and attracts large investment that drives the regional economy. As per the City Mobility plan, the walk is one of the most preferred modes of travel attributing to 45% due to climatic conditions of the region. Further, following the footsteps to more eco-friendly modes the government has launched the nation's first-ever public bicycle sharing system along the BRTS corridor under the smart cities initiatives [8].

Bhopal under the JnNURM scheme has launched Bus Rapid Transit System (BRTS) 'MYBUS' in 2013, and is currently working on the Transit-Oriented Development (TOD) following the opportunities presented by this operational system and in pipeline the metro rail project. The benefits of the TOD plan being

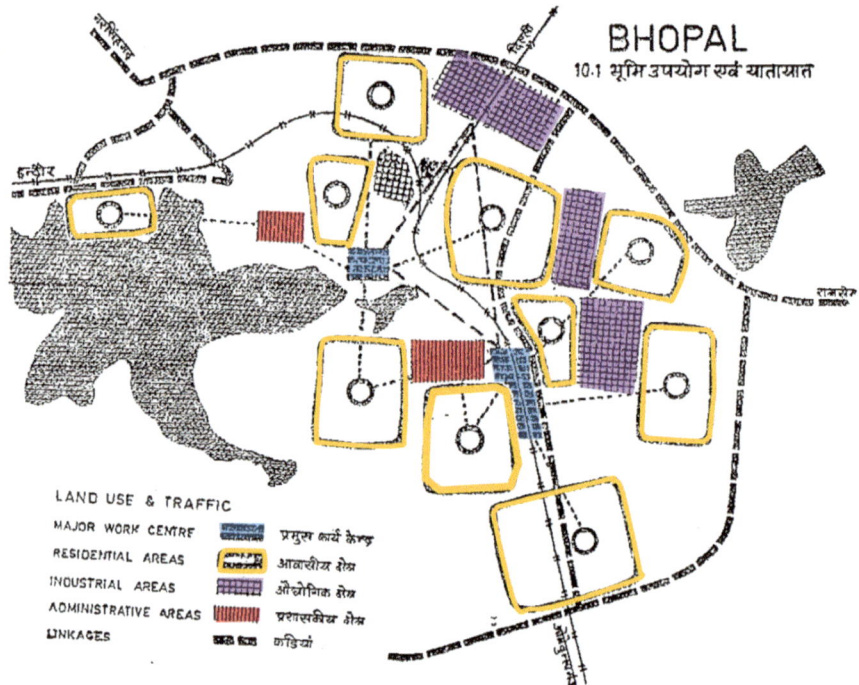

Fig. 8 Bhopal proposed land use and traffic integration. *Source* Bhopal Development Plan 1991 with visuals modified

prepared in consultation with UADD will have the benefits in economic, social and environmental dimensions of smartness. The objectives are having smart approach in terms of multimodal integration, compact growth pattern, optimized densities and land value potential capitalization [22].

The city also has a system of intermediate para transit providing last-mile connectivity. The IPT holds an important position in the city like Bhopal where door to door public transport accessibility can't be provided. As per City Mobility Plan of Bhopal the major IPT modes are auto rickshaws, Tata magic's and private minibuses. There are 7000 auto rickshaws that have an average trip length of 5.47 km. The total minibuses are 339 having a daily ridership of 95,000 and average trip length of 5.01 km [8]. Thus, multiple modal options are available to the residents to select from which in turn increases accessibility to different sections of the society. Smart city features—'walkable localities', 'mixed-use' and 'increasing modal split' are achieved in Bhopal city region via incorporation of this principle in plan preparation process.

9. **Institutional Integrity**—This is the most important principle which guides the implementation of the all other nine principles in the theory. Promoting the concept of institutional framework that can define and legalize city-region development [3]. The theory envisions that the institutions should have clear rules and regulations in order to decrease down the duplicity of the work which finally leads to the creation of governmental web [1]. Public participation should become an integral part of the governance process. There should exist a hierarchy of plans that needs to be made for specific areas such as regional plans for reducing the inequalities, development plan for guiding city development, local areas plans for the development of wards and annual plans as per specific needs and problems of the city. Smart cities are sustainable regions of the citizens, for the citizens, by the citizens. The concept of institution in various disciples refers to the man-made rules that govern the behaviour of men [32]—these had been attributed to 'formal' and 'informal' institutions [37]. Formal institutions are those which acts as political hindrance on the behaviour of the government [26] and the informal institutions are private constraints such as norms and customs.

The Madhya Pradesh Nagar Tatha Gram Nivesh Adhiniyam—1973, lays down the role of town and country planning organization, which is to plan for the area notified in section 13(1) while the Bhopal development authority has the role of implementing it along with other parastatal agencies. In particular, it conceives, promotes and monitors the key projects for developing new growth centers and brings about improvement in sectors like transport, housing, water supply and environment in the region. These assets are then transferred to the Bhopal Municipal Corporation for O&M. The roles at regional level include the provision to make the land available for development and the acquisition of private land, resolving the issues related to tenure and registration. The major issue in the Bhopal Development Authority is the horizontal coordination with the other parastatal agencies leading to delays in approval and construction. The other issues faced by the authority is in the form of fractured regulation in the peri-urban areas of the region [31], i.e., area between the municipal boundary and planning boundary.

10. **Vision**—The most important skill of the planners and the last principle of this theory deals with the accurate predictions of the time horizon. The vision should inculcate all the stakeholders' perspective in a constructive and achievable notion [5]. In terms of SMART vision, it should be specific, measurable, achievable, realistic, and time-bound. The important part is to foresee the impacts of the proposal in the long term [3].

As per Bhopal Development plan—1991, the vision entails avoiding comprehensive change in the city structure simultaneously leaving the potential for future redevelopment. The general approach is to make the city the "living heart" of the city region.

The Bhopal Development Plan—2005, conceives the region as a 'network of sub-cities'. These would be compact and self-sustainable in critical areas such as public transport, physical and social infrastructure. It is also envisioned to have two cycles—city level of sustainable development and sub-city of re-cycling to reduce intra-city networking.

Under the Smart cities mission, in order to make the development proposal citizen-centric a six stage process was proposed. These stages are define, understand, analyze, prioritize, design, and implement. The vision of the Bhopal smart city was to have a clean and green city with smart infrastructure and houses, along with smart governance and health facilities for the next five years. In the next 10–20 years, the residents want Bhopal as a world-class smart city with efficient services round [10].

The vision of Bhopal development plan of the 'multi nucleated compact city development' is an imaginative approach in itself wherein due considerations were given to the ecological setting of the city region and no connections via cutting of hillocks and disrupting the natural resources were made. Also, the vision to safeguard the ecologically sensitive region and shifting the manufacturing base to a larger regional setting also determines the 'smart' thinking approach of the city administrators.

4 Conclusions and Recommendations

The theory of intelligent urbanism and its 10 principles covers the wide array of sustainability pillars in the plan preparation process. The theory can be used to evaluate how intelligently the previous generations have tackled the urban problems and gave the city an image that it holds today. This would entail and describe the process of 'smartness' that has been followed without the applications of information and technology or a digital platform indeed. The concept of smart cities in India has been taken from western influence but the content, context and user is completely different in the nation. The process and features should not only link the development only with the application of ICT but the pillars of sustainable development through these ten principles. Master plans and smart cities need to have a strong linkage and integration since the development plan are the statutory guiding step towards the planning and development of city in India context. In case of Bhopal city region, past decisions were intelligent enough that already made the city region of Bhopal 'smart' in its own terms and the way it is at present. Furthermore, the inclusion of information and communication technology can lead to more efficient use and management of services and infrastructure. In case of India, where most of the statutory towns do not have a master plan in place or are outdated; they should inculcate these principles while formulating, so as to lead towards the creation of 'smart city features'.

References

1. Armor R (2011) Urban India 2011: evidence. Autumn Worldwise, Bangalore
2. Bateson G (1972) Steps to an ecology of mind. Chandler Publications Company, San Francisco
3. Benninger CC (2002) Principles of intelligent urbanism: the case of the new Capital Plan for Bhutan. Athens Center of Ekistics, Athens
4. Bhopal Municipal Corporation (2013) Slum free city plan for Bhopal Metropolitan Area. Mehta and Mehta Associates, Indore
5. Bugadze N (2018) Theory and practice of "intelligent urbanism". Bull Georgian Natl Acad Sci 12(3):145–151
6. Census of India (2013) District census handbook—Bhopal district. Ministry of Home Affairs, Government of India, Delhi
7. Dasgupta T (2015) A case study on Adampur Landfill site at Bhopal in M.P. Int J Adv Eng Technol 8(6):958–964
8. Delhi Integrated Multi Modal Transit System Ltd. (2012) Comprehensive mobility plan for Bhopal (draft report). Bhopal Municipal Corporation, Bhopal
9. Deshpande A, Kapshe M, Mitra S, Puntambekar K (2011) Low carbon society scenario Bhopal—2035. Asia Pacific Integrated Model, Bhopal
10. Frost & Sullivan (2015) Citizen centric model development based on identification of urban issues. Frost and Sullivan, Bhopal
11. Gothi S, Singh DO (2016) Assessing location preferences for housing in Bhopal city. Int J Eng Res Technol 5(08):83–89
12. Government of India (2014) Draft concept note on smart city scheme. Ministry of Urban Development, New Delhi
13. Grover D (2015) Analysing market feasibility of residential green buildings in tier-II cities in India. J Bus Manag 17(3):62–69
14. Gupta A (2013) Measuring urbanization around a regional capital the case of Bhopal district. New Delhi: Centre de Sciences Humaines
15. Hall RE (2000) The vision of a smart city. 2nd International Life Extension Technology Workshop
16. Hans A, Bharat D (2015) Developing a framework for analysing end use water demand dynamics. Int J Eng Res Technol 4(9):42–46
17. Katare J, Malu SK, Banerjee S (2016) An assessment of municipal E-governance service quality using esteves index. Pac Bus Rev Int 1(3):39–47
18. Kong L, Woods O (2018) The ideological alignment of smart urbanism in Singapore: critical reflections on a political paradox. Urban Stud J Ltd 55(4):679–701
19. MANIT (2012) Assessing the impacts of climate change on urban sector in Madhya Pradesh. Development Alternatives, New Delhi
20. Mehta and Mehta Associates (2007) Bhopal—city development plan. Bhopal Municipal Corporation, Bhopal
21. Ministry of urban development (2015) Smart city—mission statement and guidelines. Ministry of Urban Development, New Delhi
22. Ministry of Urban Development, Government of India (2016) Consultancy services for developing guidance documents for transit oriented development, non-motorised transportation plan and public bicycle sharing scheme. Government of Madhya Pradesh, Bhopal
23. Mishra P, Nagpure S, Shukla R (2012) Energy management in Bhopal city Int J Manag Res Dev
24. Naidu K (2015) Among tier II cities, Bhopal is the city you should consider for real estate invetments. PropTiger, Bhopal
25. NITI Aayog (2013) Twelfth five year plan (2012–2017)—faster. More inclusive and sustainable growth. New Delhi: SAGE Publications

26. North DC (1990) Institutions, institutional change and economic performance. Cambridge University Press, Chicago
27. Pani S, Dubey A, Khan MR (2014) Decadal variation in microflora and fauna in 10 water bodies of Bhopal, Madhya Pradesh. Curr World Environ 9(1):137–144
28. Rao P, Puntambekar D (2014) Evaluating the urban green space benefits and functions at macro, meso and micro level: case of Bhopal city. Int J Eng Res Technol 3(6):359–369
29. Rishi P, Khuntia G (2012) Urban environmental stress and behavioral adaptation in Bhopal City of India. Urban Studies Research, Bhopal
30. Saxena PN, Joshi R (2015) Eradication of slums in Bhopal City. Int J Sci Technol Eng 2 (6):104–112
31. Sridharan N (2011) Spatial inequality and the politics of urban expansion. Environ Urban ASIA 2(2):187–204
32. Theurl T, Wicher J (2012) Comparing informal institutions. Institute of Cooperative Reasearch, Munster, Munster
33. Town and Country Planning Organization (1995) Bhopal Development Plan—2005. Directorate of Town and Country Planning, Madhya Pradesh, Bhopal
34. UN-HABITAT (2006) Poverty mapping—a situation analysis of poverty pockets in Bhopal. Bhopal Municipal Corporation, Bhopal
35. Verma M (2001) Economic valuation of Bhoj wetlands for sustainable use. Indian Institute of Forest Management, Bhopal
36. Weber R, Tammi I, Anderson T, Wang S (2016) A spatial analysis of the city-regions: urban form and service accessibility. Nordregio—Nordic Centre for Spatial Development, Stockholm
37. Williamson CR, Kerekes CB (2011) Securing private property: formal versus informal institutions. J Law Econ 54(3):537–572

Environmental Infrastructure for Cardiac Health Care

Md. Shams Tabraiz Alam, Shabana Urooj and A. Q. Ansari

Abstract Coronary Artery Diseases (CAD) is the major cause of demise in many developed countries. In order to convert a city into smart city, the well-being of humanity is of extreme importance. Numerous factors have raised to this horrible condition, some of them are; inadequate exercise, unhealthy diet, lifestyle, smoking and environmental pollution including industrial effluent and vehicle tail emission. Higher concentration of air contamination adversely affects people with the respiratory disorder and other kinds of lungs and cardiac disease. Growth in medicine and health care technology is the backbone for transforming the city to smart city. CAD is one of the majorly observed heart diseases in India and it is distinguished by the settling of greasy ingredients on arteries wall causes swelling, widening and hardening that results in an obstruction of blood flow through artery wall. This paper reviewed the risk and current status, impact of the environment on the heart and its various way to cure. Angioplasty is another technique to unblock the artery vessel using different kinds of stent and is being extensively used in the medication of CAD.

Keywords CAD · Environmental pollutions · Diet · Exercise · Smoking · Angioplasty · Stent

1 Introduction

With the advancement of the industrial revolution, technological advancement and urbanization, environmental pollution are rising exponentially throughout the world and major contributors to environmental pollution are industrial effluent and vehicle

Md. S. T. Alam (✉) · A. Q. Ansari
Department of Electrical Engineering, Faculty of Engineering & Technology,
Jamia Millia Islamia (A Central University), New Delhi, India

S. Urooj
Department of Electrical Engineering, College of Engineering,
Princess Nourah bint Abdulrahman University (On leave from Gautam
Buddha University, INDIA)., Riyadh, Saudi Arabia

© Springer Nature Singapore Pte Ltd. 2020
S. Ahmed et al. (eds.), *Smart Cities—Opportunities and Challenges*,
Lecture Notes in Civil Engineering 58,
https://doi.org/10.1007/978-981-15-2545-2_37

tail emission that leads to asthma and other types of lungs and heart disease [1]. Air contamination is the globally biggest environmental health concern and it is the number one environmental reason for demise in India. It is essential to take care of healthcare and medicinal advancement to make the city smart. The aim of this paper is to bring focus to the statistics of death rate due to the scarcity of aids and amenities in developing and so-called developed countries. It is essential to focus upon the development in diagnosis and therapeutics to make not only the city instead to convert villages to smart. Air pollution is linked to increased cardiovascular disease (CVD), hypertension, diabetes, and pulmonary diseases (e.g., asthma, lung cancer). Coronary artery disease and heart attack are the main cause of premature demise attributable to air contamination, accounting for 80% of cases of premature death in Europe—in other words, more than 440,000 people die prematurely because of air pollution [2]. The Global Burden of Disease study estimate of age-standardized CVD death rate of 272 per 100,000 population in India is higher than the global average of 235 per 100,000 population [3].

India State-Level Disease Burden Initiative, presented a report on Disability-Adjusted Life Year (DALY) due to heart diseases during the year 1990–2016, using all available data resources as part of the Global disease, damage, and Hazardous Factors Study 2016. Overall, CVD heart diseases caused approximately 28.1% of the total demises and 14.1% of the total DALYs in India in the year 2016, compared with 15.2% and 6.9%, respectively, in the year 1990. There was a 9 times difference between states in the DALY rate for CAD, a 6 times difference for heart attack, and a 4 times difference for Rheumatic Heart Disease (RHD). In the year 2016, 238 lakh accepted cases of CAD were predicted in India, and 650 lakh accepted cases of heart attack, and 2.3 times rise in both disease from the year 1990. All information figured in 95% vulnerability interims for the point gauges. In India, less than 70 years aged person death was 56.4% by CVD including diet, 54.6% by the systolic hypertensia, 31.1% by the air pollution, 29.4% by the hyperlipidemia, 29.4% by the tobacco consumption, 16.4% by the high fasting plasma glucose and 14.7% by the high BMI. However, all the data varied by the state to state [4] (Fig. 1).

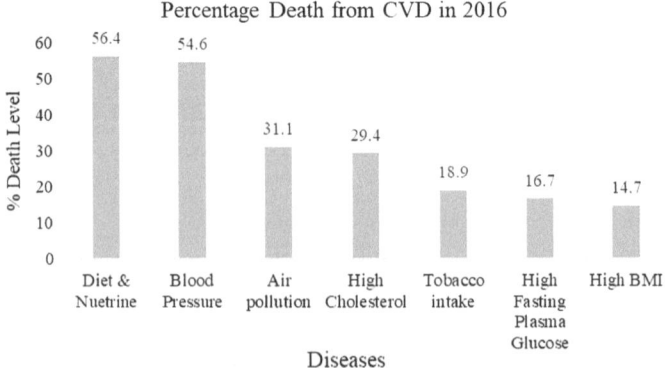

Fig. 1 Percentage death by cardiovascular disease in India of 2016 [4]

A study estimated that India has around 620 lakh CAD sufferers and this is more common in the young generation due to their way of living, unhealthy diet, inadequate physical exercise and somehow bad environmental conditions. For developing a well-conditioned cardiac environment we have to do some physical activity, avoiding smoking, tobacco and liquor intake. And try to motivate every individual not to involve in such a bad activity [5]. Various surveys and belief that the environmental factor and way of life may play a vital role in the rising of CAD and it is assumed that the CAD may be reduced by adopting healthy way of life [6].

According to the data report of RGI (Registrar General of India) for CVD mortality incremental, in India during 1990 the report showing that approx. 20.6% death, it increases 21.4% in 1995, further 24.3% in the year 2000, the rate increase up to 27.5% in the year 2005 and this death rate reached up to 29.5% which is reported in 2013 [7].

2 Developing Environment for Physical Action

In some developed cities due to the hectic lifestyle most of the adult people do not perform the basic physical activities. According to WHO, which suggest that 150 min exercise required for adult people to maintain their CVDs and psychological level in a week. Human activities reducing day by day because of the huge evolution of technological, automation as well as computerization advancement [8]. These advancements directly depend on the physical activities, which leads the several diseases. To develop an environment that is easier for people to be physically active especially walking, cycling, yoga, etc. using hoarding, advertising and motivating people for public awareness. Social networking platforms and mobile handset have to make a barrier to physical action. We should create an environment by using different policies including designing of society in which provision of parks, sports grounds, pathway, cycling paths, transportation path, gym, etc. [9].

3 Diet and Nutrients

CVD is the major cause of demise all over the world. The risk factor of CVD can be reduced by an appropriate nutrients diet [10]. Diet is made up of the combination of the various type of food that effect on human body. To reducing cardiac disease, health fitness is highly connected with diet. In our slight review, we proposed a recent knowledge of diet which directly affects the CVD [11]. Excessive weight, blood pressure, dyslipidemia and inadequate intake of green vegetables and fruits in the diet are the diet-related risk aspects for cardiac disease. A diet containing low fat and sodium, balanced in calories and protein, rich in green vegetables and fruits, Potassium, Fiber, antioxidants, and different vitamins and minerals will reduce the chance of CAD and high blood pressure.

4 Noise and Air Contamination

According to the WHO, unappropriated traffic sound leads the noise pollution and air pollution which directly affects the human body. In 2010, AHA (American Heart Association) gave the connection between PM2.5 espouse with the human heart morbidity and demises. By the use of transportation filter and pollution-free transportation reduces the risk of human demise [12, 13]. Approximately 31.1% of all early demise due to air contamination is because of cardiac disease. Very fine particulate substance called PM2.5 has the capability to enter into lower breathing tract and reach the blood flow in excessive quantity with full of poisonous elements. Industrial and vehicle tail emission produces PM2.5 and Nitrogen Dioxide is the main cause of rising incidence of CAD [14].

5 Tobacco- and Smoking-Free Environment

Tobacco and smoking are a leading cause of several heart diseases which include, severe cardiac arrest, heart stroke, vascular disease, lung problems, etc. Tobacco and smoking increase the risk of atherosclerosis as it leads the thrombosis. However, smoking by the use of either cigar or pipe-supporting cigarettes reduces the risk of rate of heart and lung disease, which is more in conventional cigarettes [15]. A well-organized risk factor of CAD is Tobacco and smoking of cigarettes. In the latest years, many health organizations and NGOs are working with government schemes and policies to prevent smoking in the public area. And providing a slogan "No Smoking, No Tobacco" and increasing taxes on tobacco products [16].

6 Angioplasty and Stent

In recent years, the number of CAD diseases is rising rapidly and it is one of the major cause of demise globally. And the main reason for these issues is caused by atherosclerosis, it is a process of accumulation of fatty material around the artery wall which leads to harden, widen, thicken and even creating a blockage of blood vein which may result in inadequate supply of blood move to the heart. Due to which following issues arise such as cardiac arrest, congestive heart failure and sometime it may lead to demise [17, 18]. Inflatable angioplasty with stenting, bypass medical procedure and atherectomy are a few strategies that can be utilized to recirculate a blocked vein [18]. The transplantation of a stent to revive a vein and reestablish the normal bloodstream is one of the famous strategies [19]. After that, the balloon is inflated that causes gradual stent expanded till target diameter achieved. And inflated balloon gets back to its previous form and removed with catheter equipment [20]. These days, the stents are used as permanent equipment which is manufactured from erosion-resistant elements and may produce issues like long-time endothelial dysfunction, everlasting inflammation in the body, continuing agitation regional reactions [21]. This result has many limitations due to its long duration biological unsuitability [22]. The stent supporting impact should keep for up to a year so as to permit the vein remodeling and curing. At the point decided that the transplant isn't required any longer, so biodegradable substances are an alternative arrangement: after the achievement of their aim, they reduce themselves without a dangerous effect for the human body. Magnesium (Mg) alloys are prominent and famous biodegradable stents and the biocompatibility of Mg alloys is accepted and the Mg is a very important substance for human body [23]. In addition, Mg does not produce any carcinogenic problems [24]. Use of Mg in sufferers is beneficial and secure and it prevents late restenosis and recoil of the stent, due to the enhancement of the characteristics, Mg-based alloys are most likely option to implant in human body [25].

7 Conclusion and Future Work

In India, CVD disease is a leading problem for people health, and this problem arises from environmental pollutants including motor vehicle emissions, smoking, industrial discharges, etc. and way of living of public including lack of exercise, consuming unhealthy food and tobaccos. With raising heart-related problems on public health, we summarized various techniques including angioplasty with stents for the treatment of CVD. Interventional techniques can be followed by doctors for cardiac sufferers. In spite of various benefit from angioplasty, some limitation is also taken into account; which can be reduced by utilizing various non-interventional techniques. Future work will show the advantages and disadvantages of current material which is used in stenting and will address various methods to reduce CVD to save mankind and contribute to transforming the environs into smart.

References

1. Weinhold B (2004) Environmental cardiology: getting to the heart of the matter. Environ Health Perspect 112(15):A880. https://doi.org/10.1289/ehp.112-a880
2. EHN Alliances (2017) Air pollution and cardiovascular diseases—a European heart network paper, 9 Jan 2017
3. Prabhakaran D, Jeemon P, Roy A (2016) Cardiovascular diseases in India current epidemiology and future directions. At CONS CALIFORNIA DIG LIB on Apr 18 (2016). doi:https://doi.org/10.1016/j.jacc.2018.04.042
4. India State-Level Disease Burden Initiative CVD Collaborators (2018) The changing patterns of cardiovascular diseases and their risk factors in the states of India: the Global Burden of Disease Study 1990–2016. Lancet Glob Health 6:e1339–51

5. Parakh N, Karthikeyan G, Bhargava B (2015) Creating healthy heart environment. Indian J Med Res 142(3):235–237. https://doi.org/10.4103/0971-5916.166526
6. Bhatnagar A (2017) Environmental determinants of cardiovascular disease. Environ Heart Dis Circ Res 7. https://doi.org/10.1161/circresaha.117.306458
7. Gupta R, Mohan I, Narula J (2016) Trends in coronary heart disease epidemiology in India. Ann Glob Health 82(2). ISSN 2214 – 9996. http://dx.doi.org/10.1016/j.aogh.2016.04.002
8. Hagströmer M, Troiano RP, Sjöström M, Berrigan D (2010) Levels and patterns of objectively assessed physical activity: a comparison between Sweden and the United States. Am J Epidemiol 171:1055–1064. https://doi.org/10.1093/aje/kwq069
9. Sallis JF, Floyd MF, Rodríguez DA, Saelens BE (2012) Role of built environments in physical activity, obesity, and cardiovascular disease. Circulation 125:729–737. doi:10.1161/CIRCULATIONAHA.110.969022
10. Richter C, Skulas-Ray A, Kris-Etherton P (2017) The role of diet in the prevention and treatment of cardiovascular disease. Nutrit Prev Treat Dis. http://dx.doi.org/10.1016/B978–0-12-802928-2.00027-8
11. Badimon L, Chagas P, Chiva-Blanch G (2017) Diet and cardiovascular disease: effects of foods and nutrients in classical and emerging cardiovascular risk factors. Curr Med Chem. https://doi.org/10.2174/0929867324666170428103206
12. Sørensen M, Pershagen G (2018) Transportation noise linked to cardiovascular disease independent from air pollution. European Heart J Eur Soc Cardiol 1–3. https://doi.org/10.1093/eurheartj/ehy768
13. Lee B-J, Kim B, Lee K (2014) Air pollution exposure and cardiovascular disease. Toxicol Res J Korean Soc Toxicol 30(2):71–75. ISSN: 1976-8257 eISSN: 2234-2753 Thematic Perspectives. http://dx.doi.org/10.5487/TR.2014.30.2.071
14. Mirowsky JE, Peltier RE, Lippmann M, Thurston G, Chen LC, Neas L et al (2015) Repeated measures of inflammation, blood pressure, and heart rate variability associated with traffic exposures in healthy adults. Environ Health 14:66. https://doi.org/10.1186/s12940-015-0049-0
15. Burns DM (2003) Epidemiology of smoking-induced cardiovascular disease. Prog Cardiovasc Dis 46(1):11–29. https://doi.org/10.1016/s0033-0620(03)00079-3
16. Lal PG (ed) (2005) Report of the National Commission on Macroeconomics and Health. Ministry of Health & Family Welfare, New Delhi, India. Byword Editorial Consultants
17. Shankaran K, Karrupaswamy S (2012) Parameterization and optimization of balloon expandable stent. In: Proceedings of the SIMULIA Community Conference
18. Imani M, Goudarzi AM, Ganji DD, Aghili AL (2013) The comprehensive finite element model for stenting: the influence of stent design on the outcome after coronary stent placement. J Theor Appl Mech 51(3):639–648
19. Schuessler A, Bayer U, Siekmeyer G, Steegmueller R, Strobel M, Schuessler A (2007) Manufacturing of stents: optimize the stent with new manufacturing technologies. In: 5th European symposium of vascular biomaterials ESVB
20. Li N, Gu Y (2005) Parametric design analysis and shape optimization of coronary arteries stent structure. In: Proceedings of 6th World Congress of structural and multidisciplinary optimization, Rio de Janeiro, vol. 30
21. Erne P, Schier M, Resink TJ (2006) The road to bioabsorbable stents: reaching clinical reality? Cardiovasc Intervent Radiol 29:11–16. https://doi.org/10.1007/s00270-004-0341-9
22. Li J, Zheng F, Qiu X, Wan P, Tan L, Yang K (2014) Finite element analyses for optimization design of biodegradable magnesium alloy stent. Mater Sci Eng C 42:705–714. https://doi.org/10.1016/j.msec.2014.05.078
23. Moravej M, Mantovani D (2011) Biodegradable metals for cardiovascular stent application: interests and new opportunities. Int J Mol Sci 12:4250–4270. https://doi.org/10.3390/ijms12074250

24. Morajev M, Amira S, Prima F, Rahem A, Fiset M, Montovani D (2011) Effect of electrodeposition current density on the microstructure and the degradation of electroformed iron for degradable stents. Mater Sci Eng B 176:1812–1822. https://doi.org/10.1016/j.mseb.2011.02.031

25. Gomes IV, Puga H, Alves JL, Claro JCP (2017) Finite element analysis of stent expansion: influence of stent geometry on performance parameters. Bioengineering (ENBENG), IEEE 5th Portuguese. https://doi.org/10.1109/enbeng.2017.7889433

Probabilistic Seismic Hazards Maps for District of Pathankot (Punjab)

Shiv Om Puri, Nitish Puri, Sanjeev Naval and Ashwani Jain

Abstract Probabilistic seismic hazard analysis (PSHA) has been carried out for the district of Pathankot. The earthquake data has been collected from different seismological agencies, e.g. NDMA, IMD, ISC-UK and USGS, and a comprehensive earthquake catalogue has been developed. Several ground motion prediction equations have been reviewed to select a GMPE appropriate for carrying out PSHA. Earthquake hazard parameters have been estimated in terms of peak ground acceleration and spectral acceleration. The seismic hazard maps of the district have been prepared at various return periods at different probabilities of exceedance. The results show that the district can experience strong ground motions originating from earthquakes originating in north-west Himalayas. Therefore, structures must be designed using the hazard parameters determined considering the local tectonic setup of the region. The developed hazard maps would help engineers and architects involved in planning and design of earthquake-resistant structures and retrofitting works.

Keywords PSHA · Maximum magnitude potential · Peak ground acceleration · Spectral acceleration · Pathankot

S. O. Puri (✉) · S. Naval
Department of Civil Engineering, DAV Institute of Engineering and Technology,
Jalandhar, Punjab 144008, India

N. Puri
Department of Civil Engineering, Delhi Technical Campus, Greater Noida,
Uttar Pradesh 201306, India

A. Jain
Department of Civil Engineering, National Institute of Technology,
Kurukshetra, Haryana 136119, India

© Springer Nature Singapore Pte Ltd. 2020
S. Ahmed et al. (eds.), *Smart Cities—Opportunities and Challenges*,
Lecture Notes in Civil Engineering 58,
https://doi.org/10.1007/978-981-15-2545-2_38

1 Introduction

Earthquake is a devastating force of nature that can inflict massive damage to structures and is capable of destroying an entire city in few seconds. Hence, it is necessary for the areas located near tectonically active sources to be ready with proper mitigation measures and rescue arrangements. The Himalayas were formed by clashing between Indian plate and Asian plate, which are still converging at a rate of 55 mm/year [1] and the frequent earthquakes occurring in this region are due to the formation and uplift of these mountains. In India, a number of smart cities are being planned in the areas adjacent to Himalayan Thrust System.

Seismic zoning map of India, reported in IS 1893-Part 1 [2], offers broad zoning of the country for earthquake hazard, and it gets updated only when new earthquakes occur. Hence, appropriate steps for seismic hazard assessment are required to obtain reliable estimates of seismic hazard parameters. The aim of a seismic hazard analysis (SHA) is to assess the possible damage that can occur due to earthquakes.

Seismic hazard assessment can be carried out probabilistically (PSHA) and deterministically (DSHA). The DSHA assumes the worst-case scenario earthquake that can occur in an area to estimate strong ground motion parameters for maximum credible earthquakes for the nearest possible distance from the site. The possibility of its occurrence for a specified period throughout the design life of the structure is not considered. It is widely used for the earthquake resistance design of critical structures and as a 'cap' for PSHA [3]. PSHA rectifies several fundamental problems in DSHA, e.g. lack of quantification of uncertainties in location, size of an earthquake with probability of its occurrence. It quantitatively expresses the hazard parameters based on the relationship between existing seismic sources, regional attenuation characteristics and probabilities of occurrence. It determines the chance of exceedance of a specified level of ground motion at a site, which is expressed as function of return period and fault displacement. Due to its capability to accommodate uncertainties, more and more seismic hazard analyses are being carried out using probabilistic approach [4].

The results of the seismic hazard analysis are formulated generally in terms of PGA and Sa. PGA is used to quantify earthquake ground motions, horizontal forces and shear stresses and strains in the earthquake-resistant design procedures. Whereas, Sa is a crude representation of the response of a structure represented by the maximum acceleration experienced by single-degree-of-freedom system undergoing damped vibrations. These are the main inputs to an earthquake-resistant design and hence must be calculated with extreme caution. The present study focusses on seismic hazard analysis of the district of Pathankot in Punjab using a probabilistic approach.

2 Study Area and Tectonic Map

2.1 Study Area

The district of Pathankot is located in north in the state of Punjab in India. It falls in seismic zone IV as per IS 1893-Part 1 [2] and listed among the areas prone to severe earthquake hazards by NDMA. The district is located at the foothills of Kangra, with broad geographical coordinates of 32.32° N and 75.59° E, covering an area of 929 km^2, with two important rivers, Beas and Ravi flowing through it. It shares an international border with the Narowal district of Punjab in Pakistan. It also shares a border with the district Kathua of Jammu and Kashmir, and Chamba and Kangra districts of Himachal Pradesh.

Some disastrous earthquakes that have occurred near the region in the last 200 years are 1934 Bihar–Nepal earthquake (M_w 8.4), 1950 Assam earthquake (M_w 8.7), 1905 Kangra earthquake (M_w 7.8), 1991 Uttarkashi (M_w 6.8), 1993 Killari earthquake (M_w 6.2), 1999 Chamoli earthquake (M_w 6.8), 2005 Kashmir earthquake (M_w 7.6), and 2015 Gorkha earthquake (M_w 7.8). The district of Pathankot being close to the epicenter of the 1905 Kangra earthquake, at a distance of 60 km only, experienced strong ground motions of intensity VII. The havoc produced by these earthquakes has been a wakeup call for the government to take suitable mitigation measures.

An area of 300 km radius around Pathankot district has been selected as the seismic study region (28°–33° N to 73°–77° E).

2.2 Tectonic Map

A tectonic map developed for the seismic study region by Puri and Jain [4] has been adopted for the study (Fig. 1). The tectonic map was developed using Seismotectonic Atlas of India and its Environs (SEISAT) [5].

3 Development of Earthquake Catalogue

A comprehensive earthquake catalogue is a prerequisite for hazard estimation. A reliable seismic hazard assessment of a region strongly depends on the data statistics of the seismic events. Pre-instrumental data are available for damaging earthquakes and have been taken from the catalogue of National Disaster Management Authority of India. The instrumental data have been collected from various national and international earthquake monitoring agencies, e.g. National Disaster Management Authority (NDMA), India Meteorological Department (IMD), International Seismological Centre (ISC-UK) and United States Geological

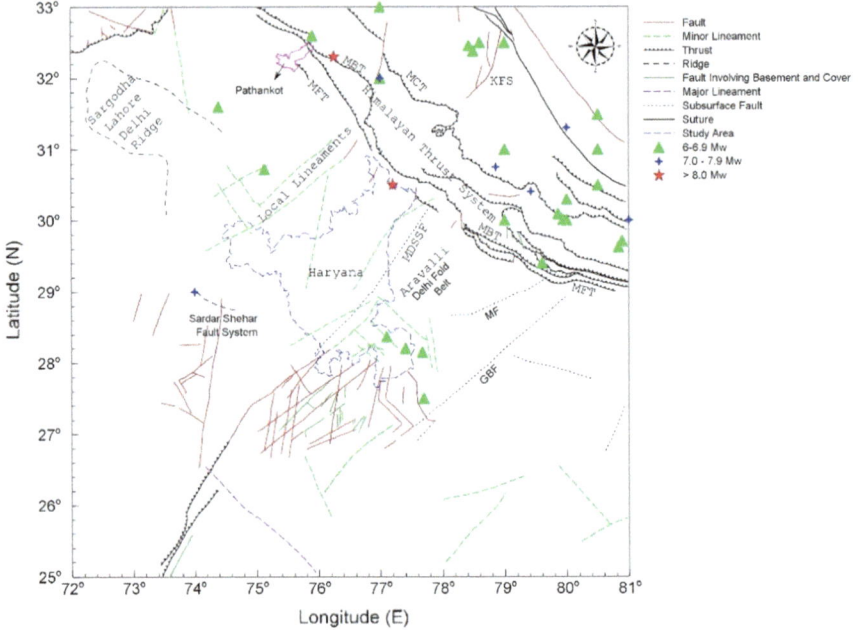

Fig. 1 Tectonic setup of the seismic study region (after Puri and Jain [4])

Survey (USGS). An earthquake catalogue has been developed, for 1800 AD to 2017 AD, by combining the pre-instrumental and instrumental data. The compiled catalogue contains 1256 earthquake events of moment magnitude (M_w) $\geq$ 4 that have occurred up to December 2017 in the seismic study region.

The events with earthquake magnitudes other than M_w have been converted using the available empirical relationships between various earthquake magnitude scales and moment magnitude (M_w). Pre-instrumental data is usually available on MMI scale and has been changed to moment magnitude by Gutenberg–Richter equation, which is as follows:

$$M_w = 2/3 \times \text{MMI} + 1 \tag{1}$$

The following correlations developed by Scordilis [6] between m_b–M_w and M_s–M_w have been used.

$$M_w = 0.85 \times m_b + 1.03, \quad \text{for } 3.5 \leq m_b \leq 6.2 \tag{2}$$

$$M_w = 0.67 \times M_s + 2.07, \quad \text{for } 3.0 \leq M_s \leq 6.1 \tag{3}$$

$$M_\mathrm{w} = 0.99 \times M_\mathrm{s} + 0.08, \quad \text{for } 6.2 \leq M_\mathrm{s} \leq 8.2 \tag{4}$$

For the conversion of local magnitude (M_L) to moment magnitude (M_w), regional correlation is generally preferred. Hence, in the case of India and adjacent areas, the following correlation derived by Kolathayar et al. [7] has been considered.

$$M_\mathrm{w} = 0.815 \times M_\mathrm{L} + 0.767, \quad \text{for } 3.3 \leq M_\mathrm{L} \leq 7.0 \tag{5}$$

The following correlation developed by Yenier et al. [8] between M_d and M_w has been used.

$$M_\mathrm{w} = 0.764 \times M_\mathrm{d} + 1.379, \quad \text{for } 3.7 \leq M_\mathrm{d} \leq 6.0 \tag{6}$$

An epicentral map has been developed using the prepared catalogue (Fig. 2).

To identify the active seismogenic sources, the data from the catalogue have been placed over tectonic map (Fig. 1). Ten tectonic features associated with earthquakes of moment magnitude ≥ 4.0 have been identified as active seismogenic sources and eight sources are potential seismogenic sources (Table 1).

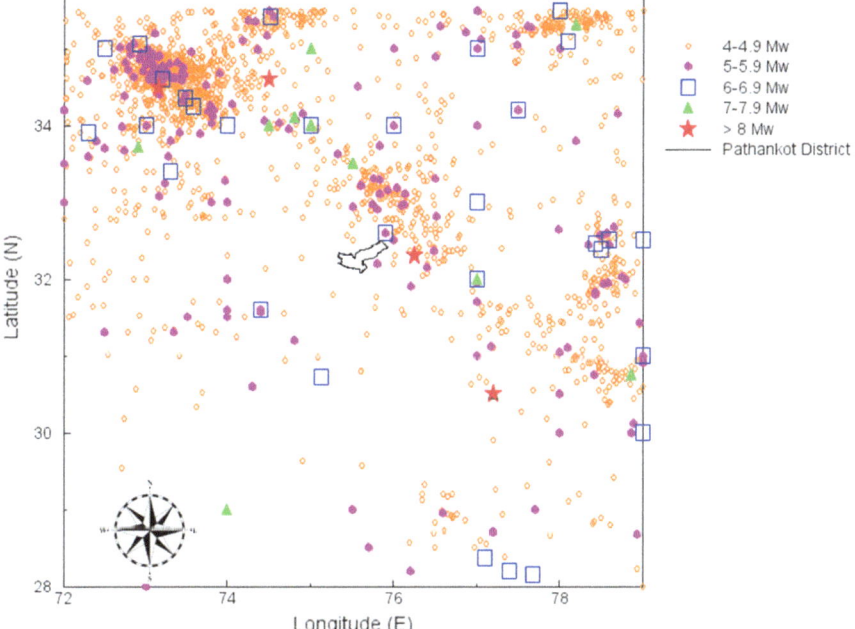

Fig. 2 Epicentral map of study area

Table 1 Maximum observed magnitudes for various sources

S. No.	Seismogenic source	M_{obs}
1	Jwala Mukhi Thrust (JMT)	5.5
2	Kaurik Fault System (KFS)	6.8
3	Lineament System of SLDR (LSLDR)	6
4	Main Boundary Thrust (MBT)	8
5	Main Frontal Thrust (MFT)	5
6	Main Crustal Thrust (MCT)	7.3
7	Sundar Nagar Fault (SNF)	7
8	Sargodha–Lahore–Delhi Ridge (SLDR)	6.5

It has been observed that the district of Pathankot has always been under the threat of big earthquakes due to its proximity with Main Boundary Thrust (MBT) and Main Frontal Thrust (MFT) of Himalayan Thrust System.

4 Estimation of Maximum Magnitude Potential (M_{max})

The earthquakes of $M_w \geq 5$ have been considered for M_{max} estimation. The M_{max} values for the major seismogenic sources in the study area have been estimated by adding an increment 0.5 to the respective values of M_{obs} [9]. The calculated M_{max} values for seismic study area are between 5.5 and 8.5 M_w (Table 2).

Table 2 Maximum magnitude potential for various sources

Seismogenic source	Total fault length (TFL) in kms	M_{max}
SLDR	Area source	7
MBT	825	8.5
MFT	32	5.5
JMT	290	6.0
MCT	769	7.8
SNF	101	7.5
KFS	137	7.3
LSDSR	97	6.5

Table 3 Seismicity parameters for different area sources

Area source	b	a	Range of magnitude (M_w) class	R^2
Himalayan Thrust System	0.75	3.8	4.0–8.0	0.989
Aravalli–Delhi Fold Belt	0.69	2.61	4.0–7.0	0.981
Sargodha–Lahore–Delhi Ridge	0.85	3.27	4.0–6.5	0.959

5 Gutenberg-Richter Seismicity Parameters

The seismicity parameters a and b have been adopted from Puri and Jain [4]. In the study, the study region was divided into three sub-regions considering each sub-region as an area source of earthquakes. Considering the complete part of the catalogue for all the magnitude ranges, the seismicity parameters were calculated for each area source through linear least squares regression method following an exponential distribution of magnitude. The exponential distribution is given in Eq. (7) below:

$$\lambda_m = 10^{a-bM_w} = \exp(\alpha - \beta M_w) \tag{7}$$

where, λ_m = mean yearly rate of exceedance, a = coefficient such that ath power of 10 gives the mean annual number of earthquakes of magnitude ≥ 0, $\alpha = 2.303a$, b = coefficient that describes the possibility of large and small earthquakes and $\beta = 2.303b$.

The reciprocal of λ_m for a given magnitude is called return period (T_R) of an earthquake exceeding that magnitude and is very important for earthquake resistant design. The seismicity parameters estimated for all the area sources have been reported in Table 3. The value of the return period calculated for different area sources demonstrates the capability of tectonic sources in Himalayan Thrust System to generate frequent large earthquakes.

6 Assesement of Seismic Hazard

The global ground motion prediction equation (GMPE) for shallow crustal earthquakes developed by Abrahamson et al. [10] has been used for the estimation of seismic hazard. PSHA has been carried out considering three sub-regions, viz. Himalayan Thrust System, Sargodha–Lahore–Delhi Ridge and Aravalli–Delhi Fold Belt as area sources of earthquakes and based on the developed catalogue, average focal depths have been taken as 15 km, 17 km and 10 km for the selected three sub-regions, respectively.

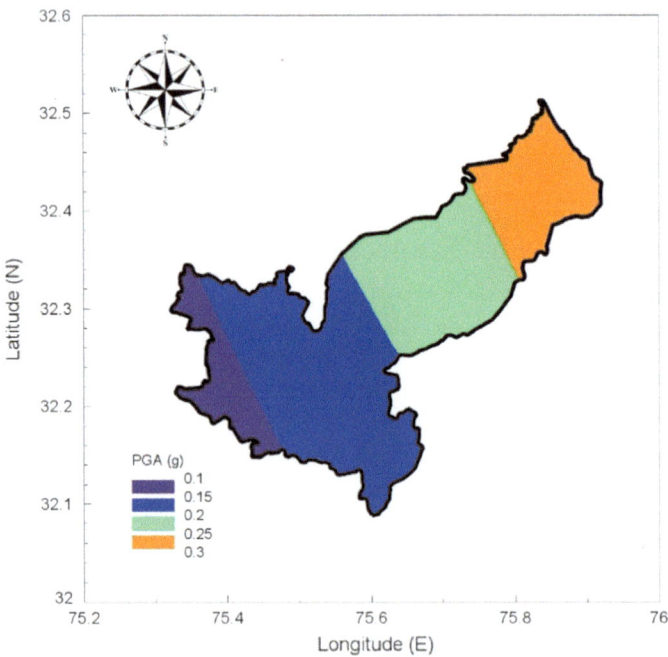

Fig. 3 Seismic hazard map of Pathankot at 10% probability of exceedance in 50 years

The hazard has been calculated for 1, 2 and 10% probability of exceedance in a time frame of 50 years as per the recommendations of Eurocode 8 [11]. For ordinary structures, the seismic hazard map for 10% probability of exceedance in 50 years is recommended. However, for important structures like Nuclear Power Plants and other megastructures, the maps for 2 and 1% probability are used. The PSHA software R-CRISIS v. 18.2 [12] has been used for the purpose. R-CRISIS allows the calculation of results for exceedance probability plots, set of stochastic events, etc. The parameters such as a, b, M_{min}, M_{max}, λ_m, and attenuation models are the input parameters, and PGA and PSA are the outputs.

The estimated PGA values range from 0.126 to 0.268 g, 0.271 to 0.527 g and 0.357 to 0.672 g at 10, 2 and 1% probabilities. The hazard maps have been prepared for return periods of 475, 2475, and 4975 years (Figs. 3, 4 and 5). The response spectra have been developed for various return periods corresponding to maximum observed PGA and shown in Fig. 6 along with those specified in IS 1893 Part-1 for comparison. It has been observed that the IS code underestimates the hazard parameters.

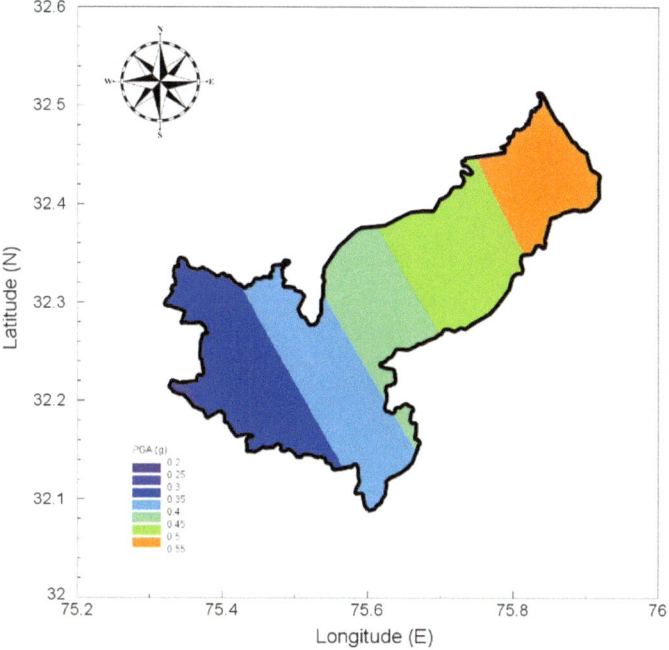

Fig. 4 Seismic hazard map of Pathankot at 2% probability of exceedance in 50 years

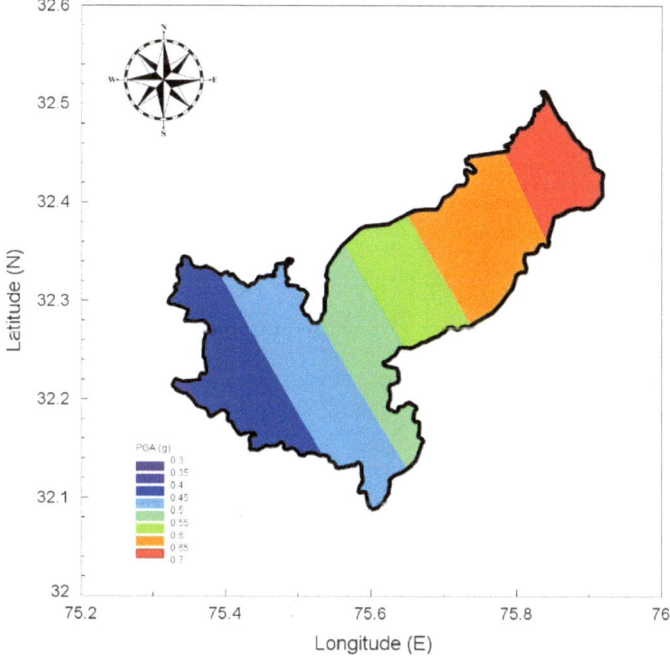

Fig. 5 Seismic hazard map of Pathankot at 1% probability of exceedance in 50 years

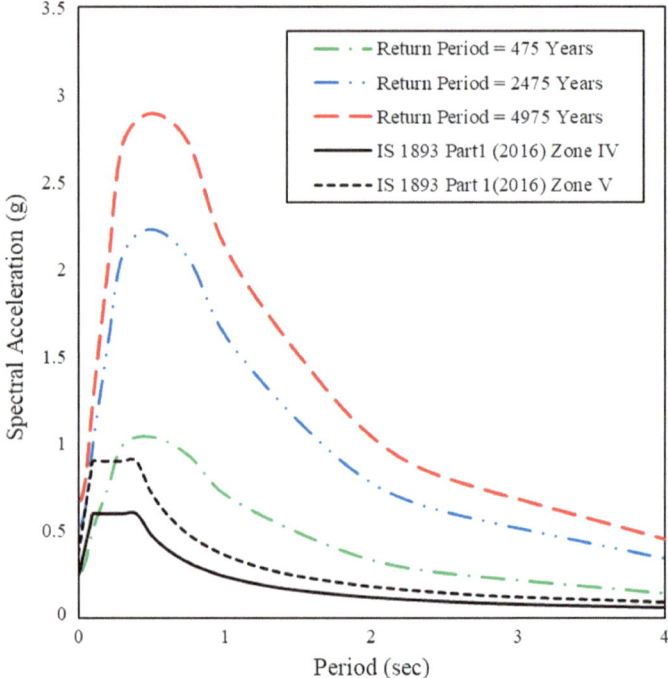

Fig. 6 Comparison of PSHA based response spectra with response spectra specified in IS 1893-Part (1)

7 Conclusions

The seismic hazard analysis has been done using probabilistic approach for estimating hazard parameters for rock sites for district of Pathankot. The district is always under the threat from earthquakes due to its proximity to the Himalayan frontal faults. It has been observed that the district is prone to high to severe earthquake ground motions ranging up to 0.672 g. Hence, a comprehensive estimation of earthquake hazard for the region is required. The results of this study would help and motivate engineers and investigators involved in seismic design of structures and planning mitigation measures.

References

1. Peltzer G, Saucier F (1996) Present day kinematics of Asia derived from geological fault rates. J Geophys Res 101:27943–27956
2. IS 1893—Part 1 (2002) Indian standard criteria for earthquake resistant design of structures, Part 1: General provisions and building. Bureau of Indian Standards, New Delhi
3. Puri N, Jain A (2016) Deterministic seismic hazard analysis for the state of Haryana, India. Indian Geotech J 46(2):164–174
4. Puri N, Jain A (2018) Possible seismic hazards in Chandigarh City of North-western India due to its proximity to Himalayan frontal thrust. J Ind Geophys Union 22(5):485–506
5. Dasgupta S et al (2000) Seismotectonic atlas of India and its environs. Geological Survey of India, Calcutta
6. Scordilis EM (2006) Empirical global relations converting M_s and m_b to moment magnitude. J Seismol 10:225–236
7. Kolathayar S, et al (2012) Spatial variation of seismicity parameters across India and adjoining areas. Nat Hazards 60:1365–1379
8. Yenier E et al (2008) Empirical relationships for magnitude and source-to-site distance conversions using recently compiled Turkish strong-ground motion database. In: Proceedings of 14th world conference on earthquake engineering, Beijing, China
9. Gupta ID (2002) The state of the art in seismic hazard analysis. ISET J Earthq Technol 39 (4):311–346
10. Abrahamson NA et al (2013) Update of the AS08 ground-motion prediction equations based on the NGA-West2 data set. PEER Report 2013/04 Pacific Earthquake Engineering Research Center Headquarters, University of California, Berkeley
11. Eurocode 8: BS-EN 1998-1 (2005) Design of structures for earthquake resistance—part 1: General rules, seismic actions and rules for buildings. European Committee for Standardization, Brussels
12. Ordaz M, Salgado-Gálvez MA (2017) R-CRISIS validation and verification document. Technical Report, Mexico city, Mexico. NDMA (2014) Mw = 8.0 Mandi Earthquake Scenario: Multi-State Exercise and Awareness Campaign. National Disaster Management Authority, Government of India

Analysing Challenges Towards Development of Smart City Using WASPAS

Shahbaz Khan[iD], **Abid Haleem**[iD] and **Mohd. Imran Khan**[iD]

Abstract The concept of 'smart city' is gaining popularity around the globe for a sustainable and liveable urban future in the last two decades, and several developed countries have tried to transform their existing cities into 'smart cities'. Now, a few developing countries are also trying to transform their cities into smart cities or even develop afresh infrastructure. However, these developing countries are facing challenges in the development of smart cities. Therefore, the primary aim of this study is to identify the significant challenges related to the development of smart cities and prioritise in the Indian context. We have identified ten significant challenges through a systematic literature review to accomplish the research objective. Further, these ten challenges are prioritised using 'weighted aggregated sum/product assessment' (WASPAS). The highest ranked challenge is the ICT infrastructure which requires a lot of effort and resource. This study provides a basic understanding of the challenges related to the development of the smart city in the Indian context which can be helpful for the formulation of the policies related to the transformation/development of smart cities. There is some limitation of this study such as the identified challenges are limited and adopted methodology is based on the expert's input which can be biased.

Keywords Smart city challenges · Smart city development · Weighted aggregated sum/product assessment (WASPAS) · Multi-criteria decision-making (MCDM)

1 Introduction

According to the United Nation [1], the cities are '...places where agglomeration economies attain their highest yields, producing cultural, economic and social benefits'. However, increasing urbanisation trends engenders many problems that impact the quality of life in pollution, inequality, insecurity and others. To overcome

S. Khan · A. Haleem (✉) · Mohd. I. Khan
Department of Mechanical Engineering, Jamia Millia Islamia, New Delhi, India
e-mail: ahaleem@jmi.ac.in

© Springer Nature Singapore Pte Ltd. 2020
S. Ahmed et al. (eds.), *Smart Cities—Opportunities and Challenges*,
Lecture Notes in Civil Engineering 58,
https://doi.org/10.1007/978-981-15-2545-2_39

463

these problems, the concept of smart city emerged in the 1990s. Smart cities are using modern technologies (such as IOT and Big data) to deal with the existing urbanisation problems. Currently, smart cities are seen as an instrument to resolve urban challenges in the urbanised world [2, 3]. Therefore, the notion of a smart city has attracted attention and prevalence from the academia, industries and policy-makers worldwide since 1994.

In the last two decades, many cities have started to undertake a more holistic approach to transform/develop smart cities using novel technologies [4]. Smart cities have established to be more than just a buzzword or short-term hype. It is possible to say that 'all cities want to be smart'. For example, India has planned to transform 100 cities into smart cities and China already running more than 500 pilot smart cities [5]. Moreover, the market of smart cities is estimated at 3 trillion USD by 2020 which was 411 billion USD in 2014 [6].

According to the Edwards, [7] 'smart cities are not a panacea for all ills, and they bring their problems'. Although smart cities have a high capability to resolve the most urban problem and emerging global market, the development of smart cities is very challenging. In the context of the emerging economy such as India, the transformation into the smart city is furthermore challenging. It is necessary to identify significant challenges, as to overcome the proper development. Studies have reported issues related to the development of smart cities from a different perspective, but it remains open for the research towards its effective implementation. Therefore, this research has the following objectives:

- Identify the significant challenges towards the smart city development
- Prioritise the identified challenges using the MCDM technique

The remaining section of this paper is as follows: Sect. two deals with the background and literature review of the study and identifies the significant challenges of smart city development. Section 3 provides a brief about the adopted methodology; whereas, Sect. 4 provides analysis and results. Section 5 further provides discussion on the results and provides some strategies to overcome these challenges. Implications of this research are provided in Sect. 6, and finally, Sect. 7 concludes the study and provides scope for future research.

2   Literature Review

The area of smart cities is interdisciplinary. As a result of this, smart cities have several definitions from different perspectives, and they are used inconsistently. In generic terms, 'a smart city is an urban environment that utilises ICT and other related technologies to enhance performance efficiency of regular city operations, and quality of services (QoS) provided to urban citizens' [8].

Primarily, the development of smart cities has been explored in developed countries such as the USA and the European Union. In the context of emerging countries, less attention has been given by academia and policymakers. However, India is gearing up smart city mission projects to provide world-class facilities to its citizens [9]. In June 2015, Ministry of Urban Development (MoUD), India, launched the 'Smart Cities Mission' for 100 cities [10]. Additionally, 'Atal Mission for Rejuvenation and Urban Transformation (AMRUT)' mission has been an announcement for upgradation of infrastructure across 500 cities in India. The private sector also involves in the management of such type of project such as 'Lavasa and Gujarat International Finance Tech City (GIFT city)'.

Several studies have been carried out to identify the significant challenges from a different perspective. Chourabi et al. [11] identified thirteen challenges of the smart city implementation and categorised into three main dimensions, namely IT infrastructure, security and privacy and operational cost. Lam and Ma [12] identified twelve challenges using a hierarchy, and these challenges are categories into three major categories. Braun et al. [13] summarised the significant challenges related to security and privacy in the context of smart cities. Further, Cui et al. [14] also highlighted the security and privacy issue related to smart city developments.

Based on the literature review, twelve challenges of the smart city development are identified and shown in Table 1.

3　Research Methodology

The major challenges related to smart city development are identified through the literature review. Further, these challenges are prioritised using the MCDM method for developing deeper insights. Weighted aggregated sum/product assessment (WASPAS) is utilised for prioritising the challenges. Four academic experts who are working in the area of the smart city provided the input for the WASPAS implementation. The proposed research framework for this study is provided in Fig. 1

3.1　WASPAS

Chakraborty and Zavadskas developed weighted aggregated sum/product assessment (WASPAS) in 2004 [28]. This method comes under the umbrella of MCDM technique [29], and it is the integration of the two decision-making techniques, namely the weighted sum model (WSM) and weighted product model (WPM). This method relics on both additive and multiplicative utility functions. The steps of the WASPAS method are as follows [30]:

Table 1 Challenges of smart city development

S. No.	Challenges	Brief description	References
1	Data sources and characteristics (CH1)	Data come from different sources (such as sensors, experts, models) and having different formats	[15, 16]
2	Data and information sharing (CH2)	Sharing data and information among different city departments is another challenging task	[15, 17]
3	Data quality (CH3)	Data captured and processed by different people under special regimes and tainted with several types of imprecision, imperfection, uncertainty and ambiguity	[18–20]
4	Security and privacy (CH4)	Data may include confidential information and need a high level of security to protect the data	[15, 21]
5	Cost (CH5)	Cost is another major challenge of smart city development because the development of a smart city requires a considerable amount of investment	[22]
6	Smart city population (CH6)	The size of the big data merely depends on the smart city population and management of the big data is a challenge	[23, 24]
7	ICT infrastructure (CH7)	Development of the ICT infrastructure to enhance the performance of the smart cities	[23, 25]
8	Understanding the needs of people (CH8)	The smart city aims to improve people's living experience and quality of life and therefore the information related to the need of peoples is vital	[4]
9	Social polarisation (CH 9)	With fewer labour market opportunities, there is a risk of increasing intolerance and polarisation between those who contribute and those who benefit from social allocations	[26, 27]
10	Green energy (CH10)	The immense number of sensors and IoT devices is operated in the smart cities which required a green energy resource for their sustainability	[22]

Step 1: Construct a decision/evaluation matrix

We have obtained a decision matrix by arranging the linguistic terms of the decision-maker with respect to the evaluation criteria (please see Table 2). The structure of the matrix is expressed as follows:

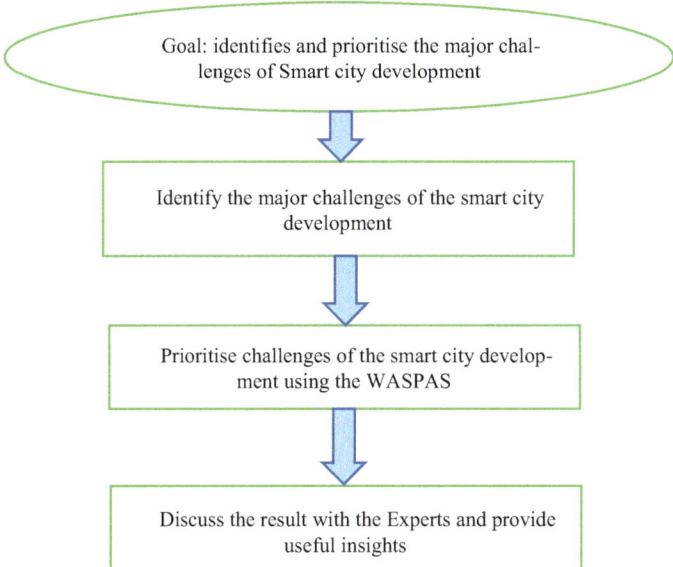

Fig. 1 Proposed research framework for this study

Table 2 Linguistic variables and their corresponding numbers

Linguistic variables	Corresponding Number
Very low (VL)	1
Low (L)	2
Medium (M)	3
High (H)	4
Very high (VH)	5

$$x = \begin{bmatrix} x_{11} & x_{12} & \cdots & x_{1j} & \cdots & x_{1n} \\ x_{21} & x_{22} & \cdots & x_{2j} & \cdots & x_{2n} \\ \cdots & \cdots & \cdots & \cdots & \cdots & \cdots \\ x_{i1} & x_{i2} & \cdots & x_{ij} & \cdots & x_{in} \\ \cdots & \cdots & \cdots & \cdots & \cdots & \cdots \\ x_{m1} & x_{m2} & \cdots & x_{mj} & \cdots & x_{mn} \end{bmatrix} \quad (1)$$

The matrix $[X]_{m \times n}$ represents the decision matrix which contains the m-number of alternative and n-evaluation criteria. The element of the matrix 'x_{ij}' represents the performance of ith alternative with respect to the jth criterion. In our case, the alternatives are the 'challenges', and criteria are the experts. Therefore, in the context of this study, 'x_{ij}' shows the importance of the ith challenge as per the jth expert/decision-maker.

Step 2: Normalisation of decision matrix

Obtained the normalised decision matrix by applying the linear normalisation of performance values as expressed in Eqs. (2) and (3):

For the benefit criterion:

$$x_{ij}^* = \frac{x_{ij}}{\max_i x_{ij}} \tag{2}$$

For the cost/non-benefit criterion:

$$x_{ij}^* = \frac{\min_i x_{ij}}{x_{ij}} \tag{3}$$

where x_{ij}^* is the normalised value of the x_{ij}.

Step 3: Calculate the measures of WSM and WSP

The measures of WSM (S_i) and WPM (P_i) for each alternative using the Eqs. (4) and (5), respectively:

$$S_i = \sum_{j=1}^{n} w_j * x_{ij} \tag{4}$$

$$P_i = \prod_{j=1}^{n} \left(x_{ij}^* \right)^{w_j} \tag{5}$$

where w_j is the weight of the jth criterion and $\sum_j w_j = 1$

Step 4: Calculate the aggregated measure for alternatives

The aggregated measure (Q_i) of the WASPAS method is calculated for each alternative using the Eq. (6):

$$Q_i = \lambda S_i + (1 - \lambda) P_i \tag{6}$$

where λ is the parameter of the WASPAS method that can take value 0–1. Without loss of generality, the value of the parameter (λ) is chosen 0.5, which infers equal importance of the measures of additive (WAS) and multiplicative (WPS).

Step 5: Prioritisation of the alternatives

Prioritised the alternatives according to descending order of Q_i values, i.e. the alternative having the highest value of Q_i is ranked first.

4   Results

The significant challenges of the smart city development in the Indian context are identified through systematic literature review. After finalising the significant challenges, the experts were asked to evaluate these challenges based on their importance on the linguistic scale. The expert's assessment is taken in the decision matrix and shown in Table 3.

These linguistic evolutions are converted into the crisp number as per Table 2 and shown in Table 4.

This initial matrix is transformed into the normalised matrix using Eqs. (3) and (4). The normalised matrix is shown in Table 5.

Further, measures of WSM (S_i) and WSP (P_i) are calculated using Eqs. (5) and (6) and shown in Tables 6 and 7, respectively.

After determining the measure of the S_i and P_i, the aggregated measure (Q_i) is determined using Eq. (6). In our case, we take the value of λ is 0.5 for giving equal importance to the measures of additive (WAS) and multiplicative (WPS) and results are shown in Table 8.

Table 3 Experts evaluation matrix for the challenges of smart city development

Challenges	DM1	DM2	DM3	DM4
CH1	M	H	M	L
CH2	L	VL	VL	VL
CH3	L	L	M	M
CH4	M	M	H	M
CH5	H	M	H	VH
CH6	H	H	VH	H
CH7	VH	H	VH	VH
CH8	M	VL	M	L
CH9	L	L	VL	M
CH10	VL	VL	L	L

Table 4 Initial matrix for the challenges of smart city development

Challenges	DM1	DM2	DM3	DM4
CH1	3	4	3	2
CH2	2	1	1	1
CH3	2	2	3	3
CH4	3	3	4	3
CH5	4	3	4	5
CH6	4	4	5	4
CH7	5	4	5	5
CH8	3	1	3	2
CH9	2	2	1	3
CH10	1	1	2	2

Table 5 Normalised matrix for the challenges of smart city development

Challenges	DM1	DM2	DM3	DM4
CH1	0.6000	1.0000	0.6000	0.4000
CH2	0.4000	0.2500	0.2000	0.2000
CH3	0.4000	0.5000	0.6000	0.6000
CH4	0.6000	0.7500	0.8000	0.6000
CH5	0.8000	0.7500	0.8000	1.0000
CH6	0.8000	1.0000	1.0000	0.8000
CH7	1.0000	1.0000	1.0000	1.0000
CH8	0.6000	0.2500	0.6000	0.4000
CH9	0.4000	0.5000	0.2000	0.6000
CH10	0.2000	0.2500	0.4000	0.4000

Table 6 Weighted normalised matrix for WSM

Challenges	DM1	DM2	DM3	DM4
CH1	0.1500	0.2500	0.1500	0.1000
CH2	0.1000	0.0625	0.0500	0.0500
CH3	0.1000	0.1250	0.1500	0.1500
CH4	0.1500	0.1875	0.2000	0.1500
CH5	0.2000	0.1875	0.2000	0.2500
CH6	0.2000	0.2500	0.2500	0.2000
CH7	0.2500	0.2500	0.2500	0.2500
CH8	0.1500	0.0625	0.1500	0.1000
CH9	0.1000	0.1250	0.0500	0.1500
CH10	0.0500	0.0625	0.1000	0.1000

Table 7 Weighted normalised matrix for WPM

Challenges	DM1	DM2	DM3	DM4
CH1	0.8801	1.0000	0.8801	0.7953
CH2	0.7953	0.7071	0.6687	0.6687
CH3	0.7953	0.8409	0.8801	0.8801
CH4	0.8801	0.9306	0.9457	0.8801
CH5	0.9457	0.9306	0.9457	1.0000
CH6	0.9457	1.0000	1.0000	0.9457
CH7	1.0000	1.0000	1.0000	1.0000
CH8	0.8801	0.7071	0.8801	0.7953
CH9	0.7953	0.8409	0.6687	0.8801
CH10	0.6687	0.7071	0.7953	0.7953

Table 8 Ranking of the challenges of smart city development

Challenges	S_i	P_i	Q_i	Rank
CH1	0.6500	0.6160	0.6330	5
CH2	0.2625	0.2515	0.2570	10
CH3	0.5250	0.5180	0.5215	6
CH4	0.6875	0.6817	0.6846	4
CH5	0.8375	0.8324	0.8349	3
CH6	0.9000	0.8944	0.8972	2
CH7	1.0000	1.0000	1.0000	1
CH8	0.4625	0.4356	0.4490	7
CH9	0.4250	0.3936	0.4093	8
CH10	0.3125	0.2991	0.3058	9

5 Discussion

The significant challenges of smart city development are prioritised using the WASPAS method. Based on the value of the aggregated measure (Q_i) of WASPAS, the ranking of the challenges is ICT infrastructure > smart city population > cost > security and privacy > data sources and characteristics > data quality > understanding the needs of people > social polarisation > green energy > data and information sharing. The highest priority challenge is the 'ICT infrastructure' among the identified challenges. In developing countries like India, ICT infrastructure development is the major issue which requires a lot of resources and government attention. The second major challenge is the 'smart city population' which is a major concern in the context of India. The population directly affect the size of big data, the number of IoT devices and diversity of the data generation. These population-driven issues are somehow complex and become a challenge on the smart city development. Next major challenge is the 'cost' of the implementation of a smart city which is the major limitation of the developing economies. 'Security and privacy' are also a significant challenge of smart city development because a lot of IoT devices are continuously stored and process the confidential data related to the people. These confidential data are required a high level of security. Data characteristics are another major issue related to smart city development. Data are captured from several sensors, experts and models which are in different format. These raw data required different prepossessing which is a chal lenging task with limited resources. Another major challenge is the quality of the data that is measured in terms of credibility and reliability.

The objective of the smart city is to provide the quality of life which determines from the understanding of the peoples need, and it is a complex task. Thus, understanding peoples' need is the major challenge of smart city development. Next challenge in this row is the social polarisation which is the result of the development of the smart cities. Green energy is the next major challenge that is required for the sustainable development of smart cities. The lowest priority challenge among the identified challenge is the 'data and information sharing' among various stakeholders and platforms.

6 Implications of the Study

This study provides an overview of the major challenges related to smart city development in the Indian context. The identified challenges are beneficial in the formulation of the strategic policies for the smart city transformation/development. Further, the ranking of the challenges assists the policymaker to develop the solution for mitigating these challenges. The highest ranked challenge requires high priority and more attention to the policy and decision-makers. From academic perspectives, the study contributes to the literature of smart city research and provides directions for future research.

7 Conclusion

This study identified the major challenges of smart city development in the Indian context. Initially, ten major challenges of smart city development are identified through the literature review. For deeper insights, these challenges are prioritised using the WASPAS method. Some challenges are technological such as 'ICT infrastructure', 'data quality' and 'data sources and characteristics'; some are social challenges such as 'smart city population', 'security and privacy', 'social polarisation', and other are financial challenges such as cost. The highly significant challenge is the 'ICT infrastructure' which requires a lot of effort and resource. This study provides a basic understanding of the smart city challenges in the Indian context which can be helpful for the formation of the policies related to the transformation/development of smart cities.

Similar to other studies, this study also has some limitation which provides the scope for future studies. The first limitation of this study is that the identified challenges are limited which can be extended in future studies. These challenges can be categorised based on the nature of the challenges. Further, the adopted methodology is based on the expert's input which may be biased. This limitation can be eliminated using the fuzzy theory and grey theory with WASPAS. In future studies, these challenges can be evaluated using other MCDM methods such as AHP, ANP, TOPSIS, DEMATAL and many more. Moreover, this study can be replicated in developing countries such as Malaysia.

References

1. United Nations (1996) Report of the United Nations conference on human settlements (Habitat II). 2500, June 1996, pp 3–14. Retrieved from https://documents-dds-ny.un.org/doc/UNDOC/GEN/G96/025/00/PDF/G9602500.pdf?OpenElement
2. Albino V, Berardi U, Dangelico R (2015) Smart cities: definitions, dimensions, performance, and initiatives. J Urban Technol 22(1):3–21. https://doi.org/10.1080/10630732.2014.942092

3. Meijer A, Bolívar M (2015) Governing the smart city: a review of the literature on smart urban governance. Int Rev Admin Sci 82(2):392–408. https://doi.org/10.1177/0020852314564308
4. Chen Y, Ardila-Gomez A, Frame G (2017) Achieving energy savings by intelligent transportation systems investments in the context of smart cities. Transp Res Part D: Transp Environ 54:381–396. https://doi.org/10.1016/j.trd.2017.06.008
5. World Bank (2015) Transport in the smart city—international experience
6. Anthopoulos L, Reddick C (2016) Understanding electronic government research and smart city: a framework and empirical evidence. Inf Polity 21(1):99–117. https://doi.org/10.3233/ip-150371
7. Edwards L (2016) Privacy, security and data protection in smart cities: a critical EU law perspective. SSRN Electron J. https://doi.org/10.2139/ssrn.2711290
8. Silva B, Khan M, Han K (2018) Towards sustainable smart cities: a review of trends, architectures, components, and open challenges in smart cities. Sustain Cities Soc 38:697–713. https://doi.org/10.1016/j.scs.2018.01.053
9. Chatterjee S, Kar A, Gupta M (2017) Alignment of IT authority and citizens of proposed smart cities in India: system security and privacy perspective. Global J Flex Syst Manag 19(1):95–107. https://doi.org/10.1007/s40171-017-0173-5
10. MoUD (2015) Smart cities mission statement and guidelines. Ministry of Urban Development, Govt. of India
11. Chourabi H, Nam T, Walker S, Gil-Garcia JR, Mellouli S, Nahon K, Scholl HJ et al (2012) Understanding smart cities: an integrative framework
12. Lam P, Ma R (2018) Potential pitfalls in the development of smart cities and mitigation measures: an exploratory study. Cities. https://doi.org/10.1016/j.cities.2018.11.014
13. Braun T, Fung B, Iqbal F, Shah B (2018) Security and privacy challenges in smart cities. Sustain Cities Soc 39:499–507. https://doi.org/10.1016/j.scs.2018.02.039
14. Cui L, Xie G, Qu Y, Gao L, Yang Y (2018) Security and privacy in smart cities: challenges and opportunities. IEEE Access 6:46134–46145. https://doi.org/10.1109/access.2018.2853985
15. Lim C, Kim K, Maglio P (2018) Smart cities with big data: reference models, challenges, and considerations. Cities 82:86–99. https://doi.org/10.1016/j.cities.2018.04.011
16. Al Nuaimi E, Al Neyadi H, Mohamed N, Al-Jaroodi J (2015) Applications of big data to smart cities. J Internet Serv Appl 6(1). https://doi.org/10.1186/s13174-015-0041-5
17. Ibrahim M, El-Zaart A, Adams C (2018) Smart sustainable cities roadmap: readiness for transformation towards urban sustainability. Sustain Cities Soc 37:530–540. https://doi.org/10.1016/j.scs.2017.10.008
18. Ben Sta H (2017) Quality and the efficiency of data in "smart cities". Future Gener Comput Syst 74:409–416
19. Bibri S (2018) A foundational framework for smart sustainable city development: theoretical, disciplinary, and discursive dimensions and their synergies. Sustain Cities Soc 38:758–794. https://doi.org/10.1016/j.scs.2017.12.032
20. d'Aquin M, Davies J, Motta E (2015) Smart cities' data: challenges and opportunities for semantic technologies. IEEE Internet Comput 19(6):66–70. https://doi.org/10.1109/mic.2015.130
21. Cao Q, Giyyarpuram M, Farahbakhsh R, Crespi N (2017) Policy-based usage control for a trustworthy data sharing platform in smart cities. Future Gener Comput Syst. https://doi.org/10.1016/j.future.2017.05.039
22. Bibri S, Krogstie J (2017) Smart sustainable cities of the future: an extensive interdisciplinary literature review. Sustain Cities Soc 31:183–212. https://doi.org/10.1016/j.scs.2017.02.016
23. Kummitha R, Crutzen N (2017) How do we understand smart cities? An evolutionary perspective. Cities 67:43–52. https://doi.org/10.1016/j.cities.2017.04.010
24. Fernandez-Anez V, Fernández-Güell J, Giffinger R (2018) Smart city implementation and discourses: an integrated conceptual model. The case of Vienna. Cities 78:4–16. https://doi.org/10.1016/j.cities.2017.12.004

25. Kumar H, Singh M, Gupta M, Madaan J (2018) Moving towards smart cities: solutions that lead to the smart city transformation framework. Technol Forecast Soc Chang. https://doi.org/10.1016/j.techfore.2018.04.024
26. European Commission (2011) Cities of tomorrow—challenges, visions, ways forward. Information Society Policy Link. Publications Office of the European Union. http://dx.doi.org/10.2776/41803
27. Fernández-Güell J, Collado-Lara M, Guzmán-Araña S, Fernández-Añez V (2016) Incorporating a systemic and foresight approach into smart city initiatives: the case of Spanish cities. J Urban Technol 23(3):43–67. https://doi.org/10.1080/10630732.2016.1164441
28. Zavadskas E, Turskis Z, Antucheviciene J (2012) Optimization of weighted aggregated sum-product assessment. Electron Electr Eng 122(6). https://doi.org/10.5755/j01.eee.122.6.1810
29. Haleem A, Khan S, Khan M (2019) Traceability implementation in food supply chain: a grey-DEMATEL approach. Inf Process Agricult. https://doi.org/10.1016/j.inpa.2019.01.003
30. Chakraborty S, Zavadskas E (2014) Applications of WASPAS method in manufacturing decision making. Informatica 25(1):1–20. https://doi.org/10.15388/informatica.2014.01

Challenges of IoT Implementation in Smart City Development

Ibrahim Haleem Khan, Mohd. Imran Khan⬤ and Shahbaz Khan⬤

Abstract The smart city aims to ease the city-related decision by facilitating its citizen with the appropriate information at the right place and at the right time. IoT-based systems provide a foundation for smartification of services by enabling person-to-object and object-to-object communications. However, there is a challenge in integrating the IoT in city services to make the city smart. This paper put forward an objective to identify and prioritise the challenges in the implementation of IoT in the smart city. IoT devices represent emerging decentralised computing era and have capability to communicate with other computing devices over a network. Ten challenges to IoT implementation in making cities smart were identified using literature review and expert's opinion. Further, TOPSIS approach is used to analyse the identified challenges. The findings suggest that the major challenge is IoT interoperability, as companies are developing IoT solution independently by utilising different platforms which result in poor integration in the devices and data security issues. The companies need to develop an open-source platform to promote an interoperable framework. The study will help the practitioner and policy planner in realising the potential challenges in IoT implementation and easing the life of the citizen.

Keywords Smart cities · IoT · TOPSIS · Challenges · Big data

I. H. Khan
Jamia Hamdard, New Delhi 110062, India

Mohd. I. Khan (✉)
Lovely Professional University, Punjab 144411, India
e-mail: imrankhan@st.jmi.ac.in

S. Khan
Jamia Millia Islamia, New Delhi 110025, India

© Springer Nature Singapore Pte Ltd. 2020
S. Ahmed et al. (eds.), *Smart Cities—Opportunities and Challenges*,
Lecture Notes in Civil Engineering 58,
https://doi.org/10.1007/978-981-15-2545-2_40

1   Introduction

Cities as an essential aspect of urban development have always been an attraction to the citizens as the cities provide more opportunities for employment and business, excellent facilities, and availability of resources. It is due to the attraction towards the cities that more and more people dwell in the cities. As the cities become more and more populated, the need to organise resources, transportation, services, infrastructure is also increases. Thus, to provide a sustainable quality life for its citizens, a city needs to become smart. Thus, keeping in view the current scenario, cities need to identify various ways towards managing newer challenges/threats. Cities globally have initiated to see forward for solutions enabling mixed land uses, transportation linkages, and urban services of high-quality with long-lasting economic growth [1].

The concept of the smart city is emerging with an aim to provide a solution to the challenges by rapid uncontrolled urbanisation and increase in population density in the cities. ICT plays a vital role in smartening the services of the cities. The term "smart city" is easy to understand; however, there is no globally recognised definition of the smart city due to its different perspective of the term "smart". In literature several definitions were reported, one of the most popular definition is offered by IBM which defines the smart city as: "the city that makes optimal use of all the interconnected information available today to better understand and control its operations and optimise the use of limited resources".

Smart cities based on the thought of using the Internet of things (IoT) with multiple sensors to meet the end need of modern citizens [2]. IoT systems improve the quality of life by creating an autonomous environment through devices which are able to compute, sense, and network. Future urban landscape equipped with IoT provides a smart solution in the area of transportation, energy management, health, education, city services, surveillance, and technology-related issues [3, 4]. IoT-based smart solutions will impact the life quality of the citizens [5]. Characteristics such as the ability to connect objects and allows to interact with the human pervasively and intelligently, implementation of IoT is crucial for realising the potential of smart cities [6, 7]. The proliferation of IoT-based services is conceptualised to automate, control, and monitor human activities in the smart environment [8, 9]. However, diversity in the domain of application of IoT makes its deployment a challenging task. IoT implementation is a complex and tedious process and needs a lot of planning and resources in terms of finance. Being decentralised makes it adds more to the degree of randomness and raises a security concern. Therefore, this study developed a framework for the deployment of IoT in smart city environment based on the prioritisation of challenges of implementation using the TOPSIS approach.

2 Background and Related Works

The concept of IoT gained interest both in literature and among professionals during the last decade. IoT enables interaction between users and devices. IoT enables individuals to make better decisions about the use of energy and health practices by providing the right amount of information at the right time. These decisions increase the living convenience of the citizen. Several authors have reported the implication of IoT on smartification of city services. Rathore et al. [10] proposed an IoT-based system for smart city development using big data analytics. Luthra et al. [11] identified the challenges of implanting IoT systems in the Indian context. Mehta et al. [12] describe IoT along with its vision, possible application domains, and key challenges faced in making IoT a reality. Grammatikis et al. [13] provide a comprehensive security analysis of the IoT by examining and assessing the potential threats and countermeasures. Bello and Zeadally [2] highlighted the requirement of quality of service in IoT-based services by providing relationship between inter-operation of different communication standards. Tzounis et al. [14] presented the potential use of recent IoT technologies in the agriculture sector and their value for future farmers. This study also explored the challenges in the propagation of IoT in smart farming. As reported above, studies have focused on recognising the challenges IoT implementation, however, no studies have reported challenges faced by IoT implementation in a smart city context. This study put an attempt to identifying and prioritising the challenges for implementation of IoT in the smart city. Next sub-section deals with challenges followed by a brief description and source.

2.1 Challenges for Implementation of IoT in Smart City

The implementation of IoT depends upon various vital factors. An understanding of the challenges of implementation may help management in designing, directing, and controlling services of smart cities. Policymakers can do planning for IoT implementation in order to provide better quality services. After reviewing the relevant literature following (shown in Table 1), challenges were selected for this study.

3 Research Methodology

In this study, the ten significant challenges are identified using the literature review. Further, these challenges are prioritise using the TOPSIS method which is discussed in the upcoming section.

Table 1 Challenges for implementation of IoT in smart cities

S. No.	Challenges	Description	Sources
C1.	Lack of compatibility and interconnection among IoT devices	The inability of IoT devices from different manufacturers to communicate and exchange data	[15, 16]
C2.	Poorly designed IoT devices	Poorly designed and implemented might negatively affect the utilisation of network resources and overall smart city operation	[5, 17]
C3.	Interoperability (homogenous networks)	Interoperability challenge is to make all IoT devices operate in integrated software platforms	[2, 18]
C4.	Lack of standards	Improper regulatory standards pose a challenge in structuring and handling big chunks of unstructured data	[19, 20]
C5.	Data security and privacy concern	IoT devices deal with data containing private/confidential information related to the behaviour of citizens. A weak security protocol or a data breach can lead to profiling of the citizens	[21–23]
C6.	Difficult networking plan implementation	The enormous number of devices connected to networks puts a significant strain on it, and the primary challenge in this area is network implementation	[24, 25]
C7.	Lack of Internet skill (developers and designers)	Internet skills matter for the use of IoT. IoT companies are faced with a shortage of talent to plan, execute, and maintain IoT systems on the market	[9, 26]
C8.	Economic viability	The IoT application employs a vast number of sensing and actuating devices, components and in consequence, its cost and its payback period will be an important factor	[27, 28]
C9.	Mobility of city infrastructure	The mobility of city infrastructure in the smart city requires IoT systems which can deal with a mobile data source	[20, 29, 30]
C10.	Updation and insufficient analysis of data collection	Need for updated hardware and software of an IoT device plays a significant role in senses and security as to enforce data-specific rules and detect anomalies (anomalous data) and traffic pattern	[31]

3.1 TOPSIS

To cater multi-criteria decision making (MCDM), Hwang and Yoon [32] proposed "technique for order performance by similarity to ideal solution (TOPSIS)". This technique became very popular and being used extensively in MCDM situations

[33, 36]. The underlying concept of TOPSIS is that the alternatives which need to be chosen should have the shortest distance from the "positive ideal solution (PIS)"; and the farthest from the "negative ideal solution (NIS)" [34]. The steps for finding solution through utilising TOPSIS are mentioned below [35]:

Step1: *Construct a decision matrix*

Challenges labelled as $B = \{B_1, B_2 \dots B_m\}$ will be evaluated against n criteria, i.e. $C = \{C_1, C_2 \dots C_n\}$. This implies that the decision matrix will have "m" number of rows and "n" number of columns (refer Table 3). Five-point linguistic scale against which importance of the challenges will be rated are shown in Table 2. x_{ij} indicates the elements of the decision matrix, which represent the importance of ith challenges against jth criterion.

Step 2: Obtain normalised decision matrix

Equation (1) shown below is used to calculate the normalised value of x_{ij}

$$r_{ij} = \frac{x_{ij}}{\sqrt{\sum_{j=1}^{n} x_{ij}^2}} \quad i = 1, 2, \dots, m; \quad j = 1, 2, \dots, n \tag{1}$$

Step 3: Determine the weighted normalised decision matrix.

The elements of weighted normalised decision matrix are the product of decision matrix element and its associated weights as represented by Eq. (2).

$$v_{ij} = r_{ij} * w_j \tag{2}$$

where w_j symbolise the weight of the jth criterion, and $\sum_{j=1}^{n} wj = 1$.

Step 4: Find the PIS and NIS by following equations (i.e. Eqs. 3 and 4).

$$A^+ = \{v_1^+, v_2^+, \dots, v_n^+\} \quad i = 1, 2, \dots, m; \quad j = 1, 2, \dots, n \tag{3}$$

$$A^- = \{v_1^-, v_2^-, \dots, v_n^-\} \quad i = 1, 2, \dots, m; \quad j = 1, 2, \dots, n \tag{4}$$

Table 2 Linguistic scale for the importance	Linguistic scale	Importance intensity
	Very low	1
	Low	2
	Medium	3
	High	4
	Very high	5

In this study, as mentioned importance of challenge is prioritised. Thus, PIS is taken as 5, whereas NIS is taken as 0.

Step 5: Estimate the separation measures between the challenges through Euclidean distance Eqs. (5) and (6).

$$D^+ = \sqrt{\sum_{j=1}^{n}(v_{ij} - v_j^+)} \quad i = 1, 2, \ldots, m; \tag{5}$$

$$D^- = \sqrt{\sum_{j=1}^{n}(v_{ij} - v_j^-)} \quad i = 1, 2, \ldots, m; \tag{6}$$

Step 6: Estimate the *relative closeness,*

The relative closeness of the alternative i_{th} is calculated as:

$$CC_i = \frac{D^-}{(D^+ + D^-)} \tag{7}$$

Here, $0 \leq CC_i \leq 1$, $i = 1, 2, \ldots, m.$

Step 7: Descending order of relative closeness will decide the rank of the alternatives.

4 Result

The challenges in the implementation of IoT solutions in smart city are finalised by having a focus group discussion with four members comprising of academician as well as practitioner and same is presented in Sect. 2.1. This focus group discussion helped in gaining deeper insight and developing a consensus. After finalisation, members were asked to evaluate the importance of each challenge using a linguistic scale. Linguistic variables for each challenge are presented in Table 2. Table 3 shows the linguistic assessment of the challenge as obtained through the experts ($E_1, E_2 \ldots E_4$).

Table 4 represents the conversion of linguistic assessments into initial matrix using the corresponding importance intensity as per Table 2.

The elements of weighted decision matrix are obtained using Eq. (2). In this case, all decision-makers have been given equal importance. Hence, the weight given to each expert is $\frac{1}{4} = 0.25$, i.e. $w_j = (0.25, 0.25, 0.25, 0.25)$ for all j. Weighted decision matrix as shown in Table 5.

Table 3 Linguistic assessment by the experts

Challenges	Expert 1	Expert 2	Expert 3	Expert 4
C1.	H	VH	H	VH
C2.	L	M	L	L
C3.	H	VH	VH	VH
C4.	H	H	VH	H
C5.	H	H	VH	VH
C6.	VH	H	M	M
C7.	M	H	H	M
C8.	M	L	M	L
C9.	VH	M	VH	M
C10.	VH	H	H	H

Table 4 Initial matrix

Challenges	Expert 1	Expert 2	Expert 3	Expert 4
C1.	4	5	4	5
C2.	2	3	2	2
C3.	4	5	5	5
C4.	4	4	5	4
C5.	4	4	5	5
C6.	5	4	3	3
C7.	3	4	4	3
C8.	3	2	3	2
C9.	5	3	5	3
C10.	5	4	4	4

Table 5 Weighted decision matrix

Challenges	Expert 1	Expert 2	Expert 3	Expert 4
C1.	0.315244	0.405554	0.306786	0.419591
C2.	0.157622	0.243332	0.153393	0.167836
C3.	0.315244	0.405554	0.383482	0.419591
C4.	0.315244	0.324443	0.383482	0.335673
C5.	0.315244	0.324443	0.383482	0.419591
C6.	0.394055	0.324443	0.230089	0.251754
C7.	0.236433	0.324443	0.306786	0.251754
C8.	0.236433	0.162221	0.230089	0.167836
C9.	0.394055	0.243332	0.383482	0.251754
C10.	0.394055	0.324443	0.306786	0.335673

Table 6 Weighted normalised matrix

Challenges	Expert 1	Expert 2	Expert 3	Expert 4
C1.	0.078811	0.101388	0.076696	0.104898
C2.	0.039406	0.060833	0.038348	0.041959
C3.	0.078811	0.101388	0.095871	0.104898
C4.	0.078811	0.081111	0.095871	0.083918
C5.	0.078811	0.081111	0.095871	0.104898
C6.	0.098514	0.081111	0.057522	0.062939
C7.	0.059108	0.081111	0.076696	0.062939
C8.	0.059108	0.040555	0.057522	0.041959
C9.	0.098514	0.060833	0.095871	0.062939
C10.	0.098514	0.081111	0.076696	0.083918

Table 7 Ranking of the challenges

Challenges	D^+	D^-	CC	Rank
C1.	1.819283	0.182693	0.091256	2
C2.	1.909815	0.092112	0.046012	10
C3.	1.809627	0.191535	0.095712	1
C4.	1.830192	0.170363	0.085158	5
C5.	1.819782	0.181617	0.090745	3
C6.	1.850239	0.153473	0.076594	7
C7.	1.860163	0.141125	0.070517	8
C8.	1.900505	0.101034	0.050478	9
C9.	1.841262	0.162966	0.081311	6
C10.	1.829954	0.170905	0.085416	4

The range of importance intensity belongs to the closed interval (0, 1). Therefore, PIS is 1 and the NIS is 0 (Table 6).

Further, the distance of each challenge from PIS and NIS is calculated using Eqs. (5) and (6). Finally, the relative closeness (CC_i) is calculated using Eq. (7) and shown in Table 7.

Through the analysis, it was found that the "IoT interoperability" is ranked on the top (please refer Fig. 1) followed by "lack of collaboration among IoT devices" and "poorly designed IoT devices" is the least prioritised challenge.

Fig. 1 Web diagram
showing ranking of the
challenges for implementation
of IoT in smart cities

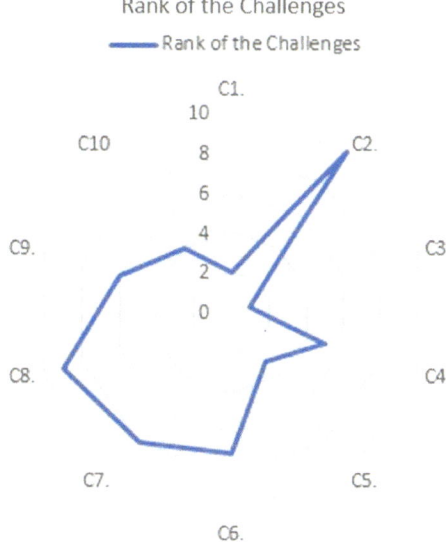

5 Discussion

Rapid growth in the city is causing different issues such as traffic congestion, health issues, environmental degradation, insufficient power, inadequate housing, and increase in crime rates. IoT is considered as the next big prospect to the world of the Internet. IoT as a computing device connects different electronic appliance and devices, interpret and understands human interaction, and also achieving high-quality two-way interaction. However, the implementation of IoT systems to achieve urban sustainability is a multifaceted approach. The findings of this study suggest that interoperability of IoT devices is a significant issue as the interoperability challenge is to make all IoT devices operate in integrated software platforms. The companies are developing IoT solution independently on the base-specific needs utilising different platforms which result in poor integration in the devices which is now only possible through selective pre-programmed API's. The companies and IoT-based organisations need to develop open source platforms to promote an interoperable framework. This will help in the integration of any hardware or software easier across all platform. The next major challenge is the amount of skill required for operating IoT devices as these devices run autonomously. This suggests that users should be able to scrutinise the device hardware and software development. Besides, users should be skilled enough to visualisation, interpret, and analyse advance data. The mobility of city infrastructure is also an important issue as IoT systems in smart city environment must be able to deal with mobile data sources. IoT devices from different manufacturers hinder seamless communication and exchanging of data as these devices capture data in different formats and employ distinct operating systems. The lack of seamless collaboration hinders the smartification of the services by developing island of non-collaborative and non-standardised services. Lack of standards leads to poorly designed IoT devices which negatively impact network utilisation. However, the standardisation of IoT devices will lead to interchangeability and ultimately better utilisation of the resources. Security and privacy in a smart city integrated with IoT devices is an important issue. Data containing private/confidential information related to the behaviour of citizens and policies of government needs to be protected against unauthorised use. Agencies responsible for providing various smart services need to deploy high-security protocol across the network. IoT devices in a smart city environment exchange and communicate data about financial records, health, and medical records of the citizen. High profile hacks and breach of these data will lead to profiling of citizens which defies their private life.

IoT devices can detect anomalies in data and its interconnected system with advanced data visualisation techniques. The traffic inflow and outflow can be monitored which lead to better defence technique against breaches as autonomous systems evolve after implementation. Visualisation is indispensable to depict all devices/sensors that are most indicative of a pending failure points and more interconnected sustainable systems. The consumerisation of IoT devices with advance small but powerful computation devices with miniature sensors is the way to mass adoption as a paradigm shift for brimming device intelligence. IoT devices enable to make tailored data-based decisions satisfying quality of service metrics such as security, cost, service time, energy, autonomy, consumption, accuracy, reliability, and availability.

6 Conclusion Limitation and Future Scope

Smart urbanisation needs efficient services to improve the quality of services. This requires a platform which redefining the traditional urban process parameters. However, this activity requires a collection of large amounts of data for better planning and development by exploiting the merits of IoT systems. This study deals with the challenges faced in the implementation of IoT systems that later prioritised as per their weight using the TOPSIS approach. This study opens up direction and actuates conducting empirical research to quantify the challenges. The identified challenges can further be evaluated other MCDM techniques like AHP, ANP, ISM, and DEMATEL under fuzzy environment and result can be compared. A case-based validation may further improve the findings. Smart cities using IoT do have their challenges and implementation nightmare like data security, AI fuelled autonomy, and mesh interconnection. Implementation of IoT requires a high degree of knowledge of designing and developing IoT and deep consideration of the C points.

References

1. Rathore M, Paul A, Hong W, Seo H, Awan I, Saeed S (2018) Exploiting IoT and big data analytics: defining smart digital city using real-time urban data. Sustain Cities Soc 40:600–610. https://doi.org/10.1016/j.scs.2017.12.022
2. Bello O, Zeadally S (2019) Toward efficient smartification of the Internet of Things (IoT) services. Future Gener Comput Syst 92:663–673. https://doi.org/10.1016/j.future.2017.09.083
3. Kummitha R, Crutzen N (2019) Smart cities and the citizen-driven Internet of Things: a qualitative inquiry into an emerging smart city. Technol Forecast Soc Chang 140:44–53. https://doi.org/10.1016/j.techfore.2018.12.001
4. Li X, Fong P, Dai S, Li Y (2019) Towards sustainable smart cities: an empirical comparative assessment and development pattern optimization in China. J Clean Prod 215:730–743. https://doi.org/10.1016/j.jclepro.2019.01.046
5. Tankard C (2015) The security issues of the Internet of Things. Comput Fraud Secur 2015 (9):11–14. https://doi.org/10.1016/s1361-3723(15)30084-1
6. Kumar H, Singh M, Gupta M, Madaan J (2018) Moving towards smart cities: solutions that lead to the smart city transformation framework. Technol Forecast Soc Chang. https://doi.org/10.1016/j.techfore.2018.04.024
7. Weinberg B, Milne G, Andonova Y, Hajjat F (2015) Internet of Things: convenience vs. privacy and secrecy. Bus Horiz 58(6):615–624. https://doi.org/10.1016/j.bushor.2015.06.005
8. Khajenasiri I, Estebsari A, Verhelst M, Gielen G (2017) A review on Internet of Things solutions for intelligent energy control in buildings for smart city applications. Energy Procedia 111:770–779. https://doi.org/10.1016/j.egypro.2017.03.239
9. Boer P, van Deursen A, van Rompay T (2019) Accepting the internet-of-things in our homes: the role of user skills. Telemat Inf 36:147–156. https://doi.org/10.1016/j.tele.2018.12.004
10. Rathore M, Ahmad A, Paul A, Rho S (2016) Urban planning and building smart cities based on the Internet of Things using big data analytics. Comput Netw 101:63–80. https://doi.org/10.1016/j.comnet.2015.12.023
11. Luthra S, Garg D, Mangla S, Singh Berwal Y (2018) Analyzing challenges to Internet of Things (IoT) adoption and diffusion: an Indian context. Procedia Comput Sci 125:733–739. https://doi.org/10.1016/j.procs.2017.12.094
12. Mehta R, Sahni J, Khanna K (2018) Internet of Things: vision, applications and challenges. Procedia Comput Sci 132:1263–1269. https://doi.org/10.1016/j.procs.2018.05.042
13. Grammatikis PI, Sarigiannidis PG, Moscholios ID (2019) Securing the Internet of Things: challenges, threats and solutions. Internet Things 5:41–70. https://doi.org/10.1016/j.iot.2018.11.003
14. Tzounis A, Katsoulas N, Bartzanas T, Kittas C (2017) Internet of Things in agriculture, recent advances and future challenges. Biosys Eng 164:31–48. https://doi.org/10.1016/j.biosystemseng.2017.09.007
15. Asghari P, Rahmani A, Javadi H (2019) Internet of Things applications: a systematic review. Comput Netw 148:241–261. https://doi.org/10.1016/j.comnet.2018.12.008
16. Atzori L, Iera A, Morabito G (2010) The Internet of Things: a survey. Comput Netw 54 (15):2787–2805. https://doi.org/10.1016/j.comnet.2010.05.010
17. Jin J, Gubbi J, Marusic S, Palaniswami M (2014) An information framework for creating a smart city through Internet of Things. IEEE Internet Things J 1(2):112–121. https://doi.org/10.1109/jiot.2013.2296516
18. Yaqoob I, Hashem I, Ahmed A, Kazmi S, Hong C (2019) Internet of Things forensics: recent advances, taxonomy, requirements, and open challenges. Future Gener Comput Syst 92:265–275. https://doi.org/10.1016/j.future.2018.09.058
19. Roman R, Zhou J, Lopez J (2019) On the features and challenges of security and privacy in distributed Internet of Things

20. Rathore M, Paul A, Ahmad A, Jeon G (2017) IoT-based big data. Int J Semant Web Inf Syst 13(1):28–47. https://doi.org/10.4018/ijswis.2017010103
21. Sagirlar G, Carminati B, Ferrari E (2018) Decentralizing privacy enforcement for Internet of Things smart objects. Comput Netw 143:112–125. https://doi.org/10.1016/j.comnet.2018.07.019
22. Sicari S, Rizzardi A, Grieco L, Coen-Porisini A (2015) Security, privacy and trust in Internet of Things: the road ahead. Comput Netw 76:146–164. https://doi.org/10.1016/j.comnet.2014.11.008
23. Weber R (2015) Internet of Things: privacy issues revisited. Comput Law Secur Rev 31(5):618–627. https://doi.org/10.1016/j.clsr.2015.07.002
24. Serrano W (2018) Digital systems in smart city and infrastructure: digital as a service. Smart Cities 1(1):134–153. https://doi.org/10.3390/smartcities1010008
25. Ahmed E, Yaqoob I, Hashem I, Khan I, Ahmed A, Imran M, Vasilakos A (2017) The role of big data analytics in Internet of Things. Comput Netw 129:459–471. https://doi.org/10.1016/j.comnet.2017.06.013
26. Risteska Stojkoska B, Trivodaliev K (2017) A review of Internet of Things for smart home: challenges and solutions. J Clean Prod 140:1454–1464. https://doi.org/10.1016/j.jclepro.2016.10.006
27. Liu L (2018) IoT and a sustainable city. Energy Procedia 153:342–346. https://doi.org/10.1016/j.egypro.2018.10.080
28. Albishi S, Soh B, Ullah A, Algarni F (2017) Challenges and solutions for applications and technologies in the Internet of Things. Procedia Comput Sci 124:608–614. https://doi.org/10.1016/j.procs.2017.12.196
29. Mital M, Chang V, Choudhary P, Papa A, Pani A (2018) Adoption of Internet of Things in India: a test of competing models using a structured equation modeling approach. Technol Forecast Soc Chang 136:339–346. https://doi.org/10.1016/j.techfore.2017.03.001
30. Marques P, Manfroi D, Deitos E, Cegoni J, Castilhos R, Rochol J et al (2019) An IoT-based smart cities infrastructure architecture applied to a waste management scenario. Ad Hoc Netw 87:200–208. https://doi.org/10.1016/j.adhoc.2018.12.009
31. Ismail N (2019) The impact of the Internet of Things (IoT)—information age. Retrieved from https://www.information-age.com/impact-internet-things-iot-123467503/
32. Hwang CL, Yoon KP (1981) Multiple attribute decision making: methods and applications. Business and economics. Springer, New York
33. Wang T, Lee H (2009) Developing a fuzzy TOPSIS approach based on subjective weights and objective weights. Expert Syst Appl 36(5):8980–8985. https://doi.org/10.1016/j.eswa.2008.11.035
34. Khan M, Khan S, Haleem A (2019) Compensating impact of globalisation through fairtrade practices. Contrib Manag Sci 269–283. https://doi.org/10.1007/978-3-030-11766-5_9
35. Imran Khan M, Khan S, Haleem A, Javaid M (2018) Prioritising barriers towards adoption of sustainable consumption and production practices using TOPSIS. IOP Conf Ser: Mater Sci Eng 404:012011. https://doi.org/10.1088/1757-899x/404/1/012011
36. Khan S, Khan M, Haleem A, Jami A (2019) Prioritising the risks in Halal food supply chain: an MCDM approach. J Islamic Mark (ahead-of-print). https://doi.org/10.1108/jima-10-2018-0206

Challenges of Shallow Tunnelling in Soft Soils—A Review

Sumee Tabassum Amin, S. M. Abbas and Altaf Usmani

Abstract A short but descriptive paper has been presented to understand various researches that have been made recently on shallow tunnelling in soft soils and the challenges faced during the construction. In a smart city, smart means of communication is vital, which should be safe and green mode of transportation. Due to the scarcity of space on the land surface of the urban and developed areas, there has been an increase of demand in the extensive use of underground structures. These constructions are often situated at shallow depths. Hence, there arises the demand for smart tunnels which requires a better understanding of various factors which influence the response of a shallow tunnel in soft soils is highly required. This understanding is enhanced with the help of the previous research papers on this specific issue which are a reliable and trustworthy source of information. Therefore, the sole purpose of this paper is to analyse various research works that have been accomplished in the field of shallow tunnelling in case of soft soils.

Keywords Shallow tunnelling · Tunnel challenges · Soft soils

1 Introduction

In the present generation of rapid transit and smart cities, underground tunnels have become an essential part of urban life and an important means of communication. The transport needs to be safe, green and smart. It must integrate subsystems of vehicles with its surrounding environment and must help in the reduction of traffic accidents as well as less production of fuel consumption and emissions. In the

S. T. Amin (✉)
Department of Civil Engineering, Jamia Millia Islamia, New Delhi 110025, India
e-mail: tabassumamin56@gmail.com

S. M. Abbas
Department of Civil Engineering, Jamia Millia Islamia, New Delhi, India

A. Usmani
Engineering and Technology Division, EIL, New Delhi, India

© Springer Nature Singapore Pte Ltd. 2020
S. Ahmed et al. (eds.), *Smart Cities—Opportunities and Challenges*,
Lecture Notes in Civil Engineering 58,
https://doi.org/10.1007/978-981-15-2545-2_41

upcoming times, the demands on safety will increase for both above ground and underground vehicles. Stringent applications of smart and green technologies will be high in demand for less emission of polluting gases. This will increase the demand of tunnels as they provide safe and environment friendly means of communication. Their analysis and design are of utmost importance as they regulate the performance of the structure under adverse situations. The response of the tunnel is highly dependent on the type of soil surrounding it. The design considerations are different for soft soils and hard soils (basically rocks).

An underground tunnel that is built in soft soil requires advanced techniques in comparison with hard soil. There are a wide range of techniques for tunnelling on soft soils, and the best technique is determined by influential factors such as ground type, age of the soil, duration of the project, monetary provisions and structures around site. To select the most appropriate method for a particular underground tunnel, several factors are taken into consideration including soil conditions, length, depth, diameter, orientation geometry and finance allocated. Another important factor is the risk of nearby structures to ground movement.

An important aspect of soft ground tunnelling is the protection of existing structures, as many shallow tunnels are located in urban environments where settlements caused due to tunnelling are a major issue. Protective and stable measures such as ground improvement should be used to ensure proper tunnelling in soft soils. On top of all, adequate instrumentation and proper monitoring system are essential.

2 Challenges of Tunnelling in Soft Soils:

Due to scarcity of space for the construction of modes of communication in places of heavy traffic and large towns, construction of shallow tunnels is a common solution to the issue. But there are lots of challenges faced in the process. Some of them are noted below:

(i) The stability of the face of a shallow tunnel is of major concern in soft soils, especially in clayey soils and water-bearing soils. Hence, special consideration should be laid on the proper reinforcement of the face of the tunnel, and the ground improvement method should be applied on the surrounding soils.

(ii) Method of excavation is also of utmost importance. In case of soft soils, special earth pressure bearing machines (EPBM) are needed for excavation, with special consideration to the cover depth of the soil above the tunnel.

(iii) Due to the construction of shallow tunnels, there is a significant influence on the buildings and other structures in the close proximity of the tunnel and vice versa. Proper study is carried out to find the maximum probable settlement in the tunnel due to any further above-ground construction. Special attention is paid in case of water-bearing soils.

3   A Review of the Literature

Many researchers have done remarkable work in the field of shallow tunnelling, especially in soft soils, and these findings form an important source of information for further investigation. Let us analyse some of the important works that have been done so far.

3.1   Numerical Analysis of Shallow Tunnels in Soft Ground Using Plaxis2D [1]

In this paper, the authors aim to evaluate horizontal and vertical movements of a shallow tunnel lining in a composition of sand and soft clay under un-drained conditions using PLAXIS2D. The paper also aims to find the effect of different geometrical and geotechnical parameters on the effective stress and tunnel lining movements, along with the influence on a building due to the construction of the tunnel.

A perfectly elasto-plastic soil model comprising sand was designed based on Mohr–Coulomb criteria. The tunnel lining was a reinforced concrete ring with the property of elastic-linear, and the lateral movements of the model were restricted. The pore water pressure was also considered to be effective from the phreatic line.

After analysis, it was found that the diameter of the tunnel has a profound effect on the displacements of the lining. The horizontal and vertical movements increase significantly with larger diameters. The effective stress also increases accordingly.

3.2   Structural Analysis for Shallow Tunnels in Soft Soils [2]

The paper aims to find a new model that has got more exact loads on the lining of the tunnel, with the combination of FEM analysis for shallow tunnels. Further, overburden effects are studied on internal forces and tunnel lining deformations, along with obtaining the optimal C/D ratio for construction of tunnels in soft soils.

With the help of a case study a tunnel in the Netherlands, a 2D analysis was performed on PLAXIS, and a 3D analysis was carried out on ANSYS. The results obtained were compared with the field data.

The obtained results were quite similar to the field data. The maximum bending moments were obtained at the top and bottom of the tunnel lining, and the values were in agreement with the field data. The values of internal forces and lining deformations increase considerably at shallow depth irrespective of the influence of buoyancy. By changing the value of C, the cover depth of the tunnel, one can achieve a minimum value for maximum deformation. A particular C/D ratio can be obtained for a specific d/D ratio to carry out the structural analysis along with uplift analysis.

3.3 Interaction of Twin Circular Shallow Tunnels in Soils—Parametric Study [3]

The authors, in this paper, try to study the effects of twin tunnel construction on the settlement of the ground surface and the resulting contact pressure between the lining and ground. The subsequent change in tunnel diameter is also studied along with the significant influence of compressibility ratio.

A 2D finite element model was constructed, consisting of two tunnels with a spacing of 0.25D, 0.5D and 1D and was analysed, respectively, for different values of compressibility ratio starting from 0.01 to values greater than 0.04. A step analysis was performed in ABAQUS FE software.

It was found that the interaction effect between the two tunnels increases with the increase in 'c', compressibility ratio. The settlements in the tunnel linings increase with the decrease in the spacing between the tunnels. The effect becomes negligible at values higher than the compressibility ratio of 0.04. The construction of a new tunnel has no significant effect on the far side of the old tunnel but has a considerable effect on the near side of the old tunnel.

3.4 Construction Technique of Long-Span Shallow-Buried Tunnel Considering the Optimal Sequence of Pilot-Tunnel Excavation [4]

The most optimum sequence of construction and the method was applied to a long-span shallow tunnel which was then under construction. After applying the chosen method, numerical simulations were carried out to validate the results with the field data.

Since the method of excavation has a profound impact on the ground settlements, it is very essential to find the most optimum method of tunnel excavation. Six pilot tunnels were designed for three different methods of excavation. The first method comprised excavating first three consecutive tunnels and then placing the rest three with a specific gap between them. The second method consisted excavation from both ends towards the centre by placing the central tunnels at the end. The third method comprised placing the inner tunnels first and the outer tunnels and finally the central ones. FEM analysis was carried out on ANSYS to find the most optimal method out of the three. The chosen method was applied to a shallow tunnel under construction, and FEM analysis was carried out similarly to find the values of settlement for comparing with the field data.

The second method came out to be the most optimum method as the settlements were the least for central tunnels out of the six pilot tunnels. It is understood from the study that the method of construction has significant effect on the ground settlements along with the fact that the hydrological and geological conditions of the surrounding rock should be ensured before carrying out the excavation.

3.5 Study of Surface Displacements on Tunnelling Under Buildings Using 3DEC Numerical Modelling [5]

The effects of variation of building storeys are studied on the displacements at the ground surface in the case of granular soils and also due to eccentricities of buildings from the tunnel centre line. Further, the effect of geo-synthetic layer below the footings is also studied.

With the help of three-dimensional distinct element code, an elasto-plastic model with Mohr–Coulomb failure criterion was created. The analysis was carried out for three different layers of soil strata with different values of building eccentricities. Changes were noted at the surface and crown of the tunnel.

The displacement was less in case of inclusion of buildings as compared to without any buildings. In case of dense soils with lesser density, the increment of building load produced lower displacements in the vertical direction. But the deformations increased with the increment in building storeys. In loose soil, the displacements increased horizontally with the increase in building loads. The displacements also increased vertically with the increase in eccentricity of the building from the centre line. Also, the presence of geo-synthetic layer reduces the displacements vertically in loose soils and can be used to enhance the overall quality of soil with loose strata.

3.6 The Settlement of Soft Soil Caused by Tunnelling in Presence of Flow [6]

The authors have attempted the effect of groundwater flow on different soil conditions like sandy and clayey using Mohr–Coulomb criterion due to the construction of shallow tunnels. A 2D plane strain model was designed and analysed in PLAXIS FEM software, considering the soil to be a perfect elasto-plastic soil with Mohr–Coulomb criterion.

From the analysis, it was found that the flow has significant effect on the ground settlements, which is more in case of sand and less in case of clay. This is due to the fact that the permeability of sand is more than clay. Due to construction of tunnel on the hydraulic level, there occurs infiltration of water into the tunnel, and the hydraulic flow increases as the flow approaches the tunnel. Hence, the effects of groundwater flow are quite of significance on a shallow tunnel.

3.7 Settlement Trough Due to Tunnelling in Cohesive Ground [7]

An attempt has been made by the authors to understand the settlement pattern of cohesive soils caused due to tunnelling.

Various analytical and empirical formulas of Peck and Rankin were used to obtain the values of settlements. These values were compared with that of the values of FEM analysis performed with the help of Modf-CRISP, in which the soil medium was modelled as elastic and elasto-plastic.

In case of linear plastic soils, there occurs the formation of heave at different depth of the tunnels such as z/D. Due to the effect of relief in the excavated soil, there is an upward movement of soil in the form of heave. The settlement trough width which is the distance of the tunnel centre line from the point of inflection is overpredicted in comparison with Peck's solution; whereas, it was on agreement with the Rankin's values.

3.8 Interaction of Transverse Surface Settlements for Twin Tunnels in Shallow and Soft Soils: The Case of Istanbul Metro [8]

The author aims to find the surface settlements in transverse direction, along with the effect of interaction between settlements of twin towers in transverse direction and the calculation of surface settlement in transverse direction in parallel tunnels with the help of earth pressure bearing machines (EPBM) shields in sandy and clayey soils.

The research paper is based on Otogar-Kirazli Metro in Istanbul. Two EPBM machines were used to carry out the construction with a delay distance of 100 m. The most important parameter of an EPBM is the face pressure, which was maintained carefully all throughout the construction phase, along with the other important influential parameters like average speed of the machines, excavation time and amount of tail grouting.

The settlement troughs for the twin tunnels were obtained by mapping for different C/D ratio. The graphs obtained were quite asymmetrical as the ground parameters along with the various operational parameters have profound impact on the settlement trough. After the construction of first tunnel, the ground undergoes an overall change in pressure and displacement. This affects the settlement of the second tunnel considerably. It was observed that the settlement above the first tunnel is less in comparison with that of the second tunnel. Also, it was found that the volume loss in case of second tunnel was significantly higher than the first tunnel. This can be attributed to the fact that the second tunnel underwent construction in an already disturbed soil.

Hence, it was understood from the paper that a good quality construction is highly essential to control the stability of a tunnel face and also to minimise the surface settlement. Backfill parameters and face pressure are the most important parameters in carrying out the construction of a tunnel with the help of EPBM.

3.9 A Short Survey on Construction Problems and Numerical Modelling of Shallow Tunnels [9]

The authors, in this paper, studied some of the methods of construction and carried out numerical modelling in case of shallow tunnels.

According to the author, earth pressure balancing machines (EPBM) are used for soft soils. An optimum face pressure has to be maintained to carry out the excavation. In case of water-bearing soils, slurry shield machines are used where the pressure at the face of the tunnel is maintained by pressurised bentonite. Ground treatment before excavation is highly required in case of open excavation and shallow tunnels in soft soils. This can be achieved by jet grouting the 'umbrella arches' at the outer vicinity of the roof of the tunnel. In case of the presence of buildings above the site of excavation, the ground above the roof of the tunnel should be pre-stabilised with the help of jet grouting or by low-pressure grouting.

Finally, the numerical modelling should consist of simulation of face support, support pressures evaluation, surface displacements evaluation and finite element modelling in 2D and 3D.

It is understood that the ground should be treated prior to the excavation, and proper tunnel lining should be used along with jet grouting to ensure the stability of face of the tunnel, especially in case of soft and water-bearing soils. The settlement trough is supposed to follow the Gaussian curve, and the maximum settlement is largely dependent on the amount of volume loss. Ground settlements should be measured at the surface of the tunnel and also in depth and high settlements due to tunnel construction under the foundations of the buildings can be compensated by grouting.

3.10 Influence of Groundwater Effect on Shallow Tunnel: A Case Study [10]

The study is made to find the reason behind the collapse of a roof tunnel and burst of the faces during the second phase of Izmir Metro construction. The region comprised water-bearing soils from groundwater and thereby through the study they try to investigate the failure of injection study in the regime.

Chemical injection method is a very common method to consolidate the soil in a water-bearing soil. But a detailed study was conducted to understand the failure of the method. The soil quality of the area was quite poor, and the groundwater aroused gradually along with the advancement of the tunnel. Back analysis was also performed, but it failed to prohibit the explosion of the tunnel face.

Due to the application of cement injection at the face of the tunnel, the general flow of groundwater was stopped at that interface, and this led to the generation of turbulence in the groundwater flow. This turbulence led to weathering of the soils in the overburden stratum and gradually decreased the yielding potential of the soils. Eventually, this led to the collapse of the face tunnel.

Hence, it is understood that due to the tension failure of the side walls of the tunnel, the tunnel face consequently collapsed. Especially, in case of groundwater regime, jet grouting and chemical injection method should be enhanced in a way to avoid jamming of groundwater flow so that the flow does not convert to turbulent flow.

3.11   Interaction of Longitudinal Surface Settlement Profile in Soft Metro Tunnelling with EPBM [11]

The authors aimed to find the surface settlements in the longitudinal direction, along with the effect of interaction between settlements of twin towers longitudinally and the calculation of surface settlement in the longitudinal direction in parallel tunnels with the help of earth pressure bearing machines (EPBM) shields in sandy and clayey soils.

It is recommended by the author to monitor the surface settlements before and after the excavation for safety and stability of the above structures. The research paper is based on Otogar-Kirazli Metro in Istanbul. The soil in the concerned area comprised sand, clay, gravel and some amount of masonry. Two EPBM machines were used to carry out the construction with a delay distance of 100 m. The most important parameter of an EPBM is the face pressure, which was maintained carefully all throughout the construction phase, along with the other important influential parameters.

It is understood that longitudinal settlements occur mainly due to settlements above and ahead of the tunnel, along the shield, at the shield tail skin and due to lining deformations. Every shield has a considerable effect on the longitudinal settlement of the other tunnel, and the peak settlement is obtained during the excavation of the first tunnel. Due to soil stabilisation techniques, the soil gradually stabilises later. Backfill parameters and face pressures are of utmost importance in calculating the maximum settlement at the face of the tunnel. Surface settlement can be checked by good quality of construction method.

3.12   Deformation Characteristics and Counter Measures of Shallow and Large-Span Tunnel Under-Crossing the Existing Highway in Soft Soil: A Case Study [12]

The authors investigated on a case study of a tunnel in Shenzhen City in China, which was constructed below a highway by double-side-drift method (DSDM). But the method was insufficient to prevent the large ground surface settlements as the tunnel was a shallow tunnel with a large span on soft soil with heavy traffic on the highway, of which the tunnel was an under-crossing of long distance.

The soil in the area mainly comprised silty clay which was weakened by the presence of water. To solve the issue, the double-side-drift method was optimised by controlling the traffic, reinforcing the face of the tunnel, strengthening the water drainage at the bottom of the tunnel and removing the primary support at the adequate time and the temporary support after a considerable delay.

3D finite difference program was designed to carry on the simulation of the optimised method of construction. A soil medium with two different layers with Mohr–Coulomb yield criteria was designed along with the cover depth of the tunnel. Numerical values of the geological parameters were obtained from field investigation report and laboratory tests.

It was found that by adopting various measures of optimisation, like reducing the traffic and reinforcing the tunnel face, the risk of the failure of tunnel can be reduced significantly. By optimising the construction method, the largest settlement on the ground surface was considerably reduced, and therefore, optimal double-side-drift method can be used to ensure the safety of tunnel in such a complicated environment.

3.13   Discussion on the Mechanism of Ground Improvement Method at the Excavation of Shallow Overburden Tunnel in Difficult Ground [13]

In this paper, the authors tried to study the mechanical behaviour of the soil surrounding a shallow tunnel during excavation and also of the stabilised ground by numerical analyses and trapdoor experiments. Application of ground improvement method to enhance the stability of tunnels is also studied.

Initially, the quality of overburden soil of the concerned area was enhanced by ground improvement method. The soil was excavated to the top of the tunnel crown, and the natural soil from the sides of the tunnel was then mixed with cement using the deep or shallow mixing stabilisation technique. The pre-mixed soil is then spread over the tunnel and thereby, compacted by rolling over the crown portion of the tunnel. Backfilling of the excavated soil was carried on and compacted thereupon by rolling till the ground surface. Then, 3D trapdoor experiments were conducted to carry out the simulation of the progress in the process of excavation of tunnel. 3D trapdoor experiments generally consist of lowering of trapdoors (supporting plates) to decrease the effect of confining stress in the local areas. Also, 2D elasto-plastic finite element analyses were carried out to understand the effect of ground improvement method on tunnel excavation process.

The experimental results depicted that the ground improvement method is beneficial in improving the settlement of a shallow overburden tunnel, especially at the cutting face of the tunnel by stabilising the soil. This effect increases with the increase in height and width of the improved area. The area of influence due to excavation of the tunnel becomes narrower with the improvement of the surrounding area around the tunnel by ground stabilisation method. In terms of lower settlements, improved height of the soil was more significant than the improved width of the soil surrounding the vicinity of the tunnel.

3.14 Tunnelling in Soft Ground—A General Report [14]

Some of the undisputed points of importance in the field of tunnelling, especially in soft soils, were studied by the authors.

It has been quoted that the excavation method and construction process are of utmost importance as they should be feasible and should be suitable for the ground stress conditions. The above-ground structures should not be affected by the construction, and the tunnel should be adequate enough not to create any adverse movements or settlements of the overlying structures. The tunnel should be strong and stable enough to tolerate all the possible influences on it during its life.

The deformations in soft ground are quite large and significant which should be controlled and limited, taking into consideration any unwanted settlement on the overlying structures.

Different conditions of collapse have been studied in clay and sand soils. Deformations around the tunnel lining are quite difficult to estimate. Proper in situ stresses should be found out to get the exact displacements. Adequate load factors should be taken into account. Improved ground models should be studied to estimate the ground deformations around the tunnel. Pore pressure changes should be noted during various phases of construction. The tunnel should be fit into the ground, and separation of the bed should not be there above the tunnel crown.

3.15 Ground Movements Due to Shallow Tunnels in Soft Ground: Analytical Solutions [15]

The authors have highlighted simplified analytical closed-form solutions to predict and interpret movements of the ground due to tunnelling in shallow soft ground.

The authors have comprehensively explained different methods of calculations to find the vertical translation of the tunnel through approximate and exact solution. Three-dimensional movements of the ground around the heading of a tunnel can be obtained by integrating line equations in all three directions, with prior assumption of a distribution of volume loss.

It was understood that exact solutions by representing finite dimensions of a shallow tunnel provided a more approximate result as the closed-form solutions from superposition of singular solutions. Three-dimensional effects were restricted to a limited region of the tunnel heading. Hence, proper 3D analysis is required to estimate the relative deformations along any tunnel axis.

3.16 Numerical Analysis of a Tunnel in Residual Soils [16]

In this paper, Azevedo et al. presented the results of a back analysis on a shallow tunnel on residual soils. The tunnel was an extension of the underground transit system at Sao Paulo in Brazil.

Numerical simulation by finite element method was performed by considering the soil as elasto-plastic to represent the behaviour of stress–strain–strength of the concerned soils. Numerical simulation of the process of tunnelling was carried out in four stages: (i) in situ stresses at the initial stage were determined, (ii) stress relief before excavation was simulated, (iii) stress relief during excavation was simulated and (iv) stress relief after installation of liner was simulated. Next, a plane strain analysis of the tunnel in $2D$ direction was carried out on ANLOG, a nonlinear FEM program.

Both loaded and unloaded stress paths were determined to understand the $3D$ characteristics of the excavation method of the tunnel. The results were calibrated with the laboratory test values

It was observed that the stress levels measured around the lining of the tunnel were quite far from a failure. There was just a small area at the bottom of the tunnel which showed failing characteristics but it was quite small in comparison with the overall response of the tunnel. The attempt to simulate the $3D$ characteristics of the tunnel, using a $2D$ simulation method, was quite successful and be further used, provided important parameters like exact stress values are taken into account.

3.17 Probabilistic Analyses of Tunnelling-Induced Ground Movements [17]

Probabilistic and deterministic analyses are used by the authors to estimate the ground movements induced by tunnelling. The deterministic model was simulated by FLAC3D. This deterministic model tried to reproduce some of the major phenomenon that occurs during a slurry shield excavation of the tunnel (e.g. ground displacements due to application of face pressure, due to overcutting, etc.). This model tried to provide useful information regarding the magnitude and nature of the ground movements at the surface. An efficient probabilistic method, CSRSM was used to predict the propagation of uncertainty.

A $3D$ model was designed in FLAC, and various input parameters were chosen arbitrarily. Horizontal and vertical displacements in the y-direction were obtained after several stages of excavation. The results showed negative impact on the above lying structures. Towards the upstream of the tunnel, there was a continuous decompression of the soils due to excavation of soil at the face of the tunnel. This led to the generation of settlements. The probabilistic analysis was carried on to understand the liability of the random variable and the influence of these random input variables on the ground movements. The analysis was also done to correlate the outputs to the random input variables. Six different modes of failure were determined, and the liability of the probabilistic mode of analysis needs to be studied further.

3.18 Some Remarks Concerning EPB and Slurry Shields [18]

In this paper, Anagnostou [18] discussed the issue of tunnelling through shields under sub-optimal geological soil conditions through geotechnical perspective.

Increasing demand of underground modes of communication has increased the requirement of technological advancement in the field of tunnelling. While concerning with underground tunnels, the issue of water-bearing soils is quite important. Shields like slurry shields and EPB shields are quite useful in those hydro-geological and geological conditions. In case of EPBM, the face pressure is equivalent to the pressure of the muck inside the working chamber. Hence, in an EPBM, the response of the tunnel depends upon the mechanical condition of the machine and the geological condition of the surrounding soil. Whereas, in a slurry shield, the response of the tunnel is dependent on the slurry pressure as the only pressure at the support of the face of the tunnel is the slurry pressure. EPBM are more reliable on the ground conditions in comparison with slurry shields, and hence, are susceptible to more deviations. Slurry shields are less prone to wear as compared to EPBM, requiring more maintenance. The damage caused by slurry shields is way more than EPBM as the ground loss is higher. However, selection of the machines is very crucial before carrying out the process of tunnelling. Generally, EPBMs are commonly chosen as it is problematic to use slurry shields in urban areas to have separate plants and also to dispose muck. It is also not wise to choose a machine on the basis of availability in the market. Several other measures of importance should be estimated by people in charge of the tunnelling to get the adequate machine as per the demand of the soil and tunnel design.

3.19 Ground Movements Around Shallow Tunnels in Soft Clays [19]

Mair et al. tried to predict the ground movements due to shallow tunnelling in soft clayey soils by understanding the deformations in un-drained soil conditions and temporary pressures at the support.

A finite element analysis was carried out with the help of the CHRISTINA program which was developed at the University of Cambridge. The program enabled the solution of plain strain and asymmetric boundary problems with the help of Cam-Clay model of soil. A number of centrifugal model tests were conducted to compare with the experimental values required for tunnels undergoing deformation due to self-weight of the surrounding clayey soils.

It was found that the ground loss during tunnelling was more related to the load factors rather than stability ratio only. In situ stresses before and during tunnelling are of quite significance in finite element analyses. Combination of centrifugal testing along with the finite element analysis provides better results in predicting the ground settlements, and the estimation of ground loss is of quite significance in estimating the ground deformations.

3.20   Two Shallow Tunnels in Soft Ground [20]

The author Anagnostou [20] attempted to describe the design considerations and constructional issues during the process of construction of two shallow and motorway tunnels of Greece.

The soil in the area comprised mostly loose and fine grained soils with clay, sand, silt and marl. Due to more permeable soil, seepage issue and water drainage were important matters of concern. Full face method of excavation was selected due to the variable nature of strength and resistance of the surrounding soil. Due to the presence of high permeable soils, drilling was quite difficult, and the boreholes provided for drainage were also ineffective in that area. Due to full face method of excavation and prefabrication of reinforcement for the tunnel, the construction of tunnels in that variable soil condition came out to be successful. But, due to less cohesion in the soil and the instability of the soil along with the support for the lining of the invert of the tunnel, the ground at the crown of the tunnel collapsed. Various recovery and ground improvement measures are thereby required, especially jet grouting is initially required at the tunnel crown area.

4   Summary

Some of the important points understood by analysing the above-mentioned research papers are as follows:

(i)   Optimum diameter should be selected for a shallow tunnel as larger diameters produce correspondingly larger settlements.

(ii)   Ground improvement methods like grouting should be applied to soft soils to avoid the generation of excess pore water pressure on the tunnel in case of water-bearing soils.

(iii)   Adequate spacing should be kept between any two tunnels to counteract any adverse effect on one another, along with other influential factors.

(iv)   The chosen method of excavation should produce least settlements on the tunnel lining and the surrounding soil, to ensure the stability of the tunnel.

(v) Proper cover to depth ratio should be selected for a shallow tunnel such as the further construction of structures above ground nearby has less impact on the tunnel lining settlements.

(vi) Pre-stabilisation of the soil is highly recommended in loose soils and soils with variable strength.

(vii) In situ stresses are quite influential in designing a tunnel in soft ground.

(viii) Reconnaissance of the soil is required in any case of tunnelling as they help in understanding the earlier soil conditions.

(ix) Un-drained soil conditions may lead to the generation of excess pore water pressure in soils. This should be taken into consideration, especially in water-bearing soils as they may produce variable effective stress during the life of the tunnel.

(x) Selection of machines is quite a matter of importance as a wrong selection may lead to collapse of the ground in the tunnelling area.

(xi) Proper investigation of the concerned area will help to implement safe, smooth and stable tunnel construction.

(xii) Environmental Impact Assessment should also be carried out to understand the environmental impacts due to the construction as any adverse effects should be avoided.

References

1. Shabna PS, Sankar N (2016) Numerical analysis of shallow tunnels in soft ground using Plaxis2D. Int J Sci Eng Res 7. ISSN 2229–5518
2. Ngan Vu M, Broere W, Bosch JW (2017) Structural analysis for shallow tunnels in soft soils. Int J Geomech 17(8): 04017038
3. Shalabi FI (2017) Interaction of twin circular shallow tunnels in soils—parametric study. Open J Civil Eng. ISSN Online: 2164-3172
4. Jin B, Liu Y, Yang C, Tan Z, Zhang J (2015) Construction technique of long-span shallow-buried tunnel considering the optimal sequence of pilot-tunnel excavation. http://dx.doi.org/10.1155/2015/491689
5. Rebello N, Sastry VR, Shivashankar R (2014) Study of surface displacements on tunnelling under buildings using 3DEC numerical modelling. http://dx.doi.org/10.1155/2014/828792
6. El Hourari N, Allal MA (2014) The settlement of soft soil caused by tunnelling in presence of flow. https://www.researchgate.net/publication/273775846
7. Fattah MY, Shlash KT, Salim NM (2011) Settlement trough due to tunnelling in cohesive ground. Indian Geotech J 41(2):64–75
8. Ocak I (2012) Interaction of transverse surface settlements for twin tunnels in shallow and soft soils: the case of Istanbul Metro. In: Annual international conference on geological and earth sciences (GEOS 2012)
9. Martins JB (2001) A short survey on construction problems and numerical modelling of shallow tunnels. Department of Civil Engineering, Universidade do Minho, 4800 Guimanaes, Portugal
10. Kucuk K (2011) Influence of ground water effect on shallow tunnel: a case study. J Mining Sci 47(1) (2011)

11. Ocak I (2012) Interaction of longitudinal surface settlement profile in soft metro tunnelling with EPBM. https://www.researchgate.net/publication/259975726
12. Cao C, Shi C, Lei M, Peng L, Bai R (2018) Deformation characteristics and counter measures of shallow and large-span tunnel under-crossing the existing highway in soft soil: a case study. KSCE J Civil Eng 22(8):3170–3181
13. Kishida K, Kimura M (2016) Discussion on the mechanism of ground improvement method at the excavation of shallow overburden tunnel in difficult ground
14. Ward WH, Pender MJ (1981) Tunnelling in soft ground—a general report
15. Pinto F, Whittle AJ (2014) Ground movements due to shallow tunnels in soft ground: analytical solutions. J Geotech Geoenviron Eng 140(4):04013040
16. Azevedo RF, Parreira AB, Zornberg JG (2002) Numerical analysis of a tunnel in residual soils
17. Mollon G, Dias D, Soubra A-H (2013) Probabilistic analyses of tunnelling-induced ground movements
18. Anagnostou G (2008) Some remarks concerning EPB and slurry shields. https://doi.org/10.3929/ethz-a-010819279
19. Mair RJ, Gunn MJ, O'reilly MP (1982) Ground movements around shallow tunnels in soft clays
20. Anagnostou G (2001) Two shallow tunnels in soft ground. AITES-ITA 2001 World Tunnel Congress Progress in Tunnelling after 2000, Milano, 10–13 June 2001

Performance Assessment Indexing of Buildings Through Fuzzy AHP Methodology

Prateek Roshan, Shilpa Pal and Ravindra Kumar

Abstract More than half of the total land area of the Indian subcontinent is prone to earthquakes of moderate to very high intensity. Earthquakes cause damage, and assessment of parameters and factors affecting performance of buildings becomes desirable in order to understand the aspects and phenomenon of the same to make or design better earthquake-resistant buildings. Performance of a building is often expressed in qualitative terms like poor, average, good, better, etc. However, the same can be expressed in quantitative terms too and compared with respect to one another. Analytical hierarchical process (AHP) is a well-known multi-criteria decision-making (MCDM) technique to express qualitative measures in quantitative terms. In order to handle ambiguity of the qualitative assessment by humans, the concept of fuzzy theory was embedded by many researchers to the AHP technique. The current study focuses on the development of performance assessment index (PAI) of buildings using fuzzy AHP technique. The index developed is applied on buildings damaged in 2011 Sikkim earthquake. These buildings are ranked on the basis of the performance score. The advantage of such indexing model is that it can help in anticipating a certain level of performance behaviour, comparing or ranking of the buildings on the basis of performance levels in a the occurrence of a seismic event. In other words, the current study can be used to predict the survivability and performance of a building in case of a likely earthquake. The proposed model for performance evaluation based on fuzzy AHP is simple and hence holds the potential for practical application.

Keywords Damages · Plan irregularity · Lateral load-resisting elements · Analytic hierarchy process (AHP) · Fuzzy logic

P. Roshan (✉)
Department of Civil Engineering, Delhi Technological University,
New Delhi, Delhi, India
e-mail: roshan.prateck@yahoo.in

S. Pal
Delhi Technological University, New Delhi, Delhi, India

R. Kumar
Greater Noida Institute of Technology, Greater Noida, Uttar Pradesh, India

© Springer Nature Singapore Pte Ltd. 2020   503
S. Ahmed et al. (eds.), *Smart Cities—Opportunities and Challenges*,
Lecture Notes in Civil Engineering 58,
https://doi.org/10.1007/978-981-15-2545-2_42

1 Introduction

Every time when an earthquake of substantial magnitude strikes an area, damages of different types and levels are observed in structures. To name a few, configuration of structure, properties of the structural components, ground conditions, quality of construction materials and quality management are some factors that might affect the performance of a structure in a likely earthquake event. Damaged components of the structure can put human lives at risk in various ways. The decision regarding re-habitability of a structure if has been damaged after the seismic event is of a serious concern. The management officials and population become quite concerned and keen to know about occupancy of their buildings.

Observation and knowledge of structural performance of buildings during a seismic activity can undoubtedly help in identifying the strong and weak design aspects, as well as appropriate and desirable material qualities, construction practices and site attributes. Knowledge of such crucial factors, norms and guidelines provides an important step in development of strengthening measures and provisions for various types of buildings. The evaluation is also important in establishing reasonable prevention plans regarding risk assessment, design seismic codes and action plans for risk reduction and for emergency management regarding evacuation plans, repair cost estimate, etc. All these factors would ultimately help in making better and safe earthquake-resistant structures. Ensuring sound, safe and resilient infrastructure and safety of users is indeed a very important milestone in realization of a smart city concept practically achievable.

Performance assessment of a reinforced concrete building is a critical and complex task. So, the decision-making about assessing performance of a building is decomposed into hierarchical models, and subsequently, a fuzzy AHP-based performance assessment index (PAI) model for reinforced concrete buildings has been developed. To handle the ambiguity and uncertainty among the opinions of experts, α-cut method was employed.

A survey questionnaire was developed for the collection of experts' opinion for assigning relative weightage among the different parameters at each level of the proposed model. In total, 21 expert surveys were collected and data was analysed for the generation of PAI. After the development of the model, five sample buildings damaged in Sikkim earthquake 2011 were selected for the application of model.

2 Fuzzy AHP (Analytical Hierarchical Process)

Decision-making regarding how a structure will perform during an earthquake is a complex process [1]. Assessment of the performance of a building in case of a seismic event requires knowledge of its configuration, types and quality of materials used, ground characteristics, efficiency of structural member, and quantitative and qualitative data concerning current state of building and an approach to sum up different types of information into a decision-making process for assessing the likely performance of the entire building.

The analytic hierarchy process (AHP) as presented by Saaty [2] is an effective tool to deal with complex decision-making processes. The human preference approach is uncertain and ambiguous in many situations, and decision-makers might be unable to assign exact numerical values to make comparison judgments [3]. To handle this, the concept of fuzzy theory was embedded by many researchers to the AHP technique. To resolve the imprecision and the ambiguity in assessing the relative importance, fuzzy set theory, introduced by Zadeh [4], has been used and adopted in the current study. This study follows and adopts the concepts of Saaty [2] and Chang et al. [3] to analyse data and reach a consensus among experts. Eigenvector method is used to calculate the (priorities or weights) among elements at the same level of the hierarchical model.

3 Study Area

A M_w 6.9 earthquake struck the adjoining areas of the Nepal–Sikkim border on 18 September 2011 at 18:10 local time, about 68 km north-west of Gangtok and at a focal depth of 19.7 km as reported by United States Geological Survey (USGS) [5]. The earthquake initiated a large number of landslides resulting in significant damage to structures and consequently caused huge infrastructural loss [6]. Sikkim was the most severely affected state of India, followed by West Bengal and Bihar. The maximum seismic intensity was estimated to be around VI+ on the MSK scale. Most multi-storey reinforced concrete buildings were non-engineered and sustained considerable damage due to earthquake shaking, a small number of these collapsed or suffered irreparable structural damages as mentioned in EERI special report [7]. The performance index model is applied on the buildings which were damaged in 2011 Sikkim earthquake [8]. The sample buildings upon which the proposed fuzzy AHP (analytical hierarchical process)-based model is applied are summarized in Table 1.

The information regarding the constructional and architectural features, and the types of seismic damage observed in these buildings during the 2011 Sikkim

Table 1 Buildings for the proposed PAI model application

Seismic event	Studied buildings	Location of buildings in Sikkim
Sikkim 2011	Moonlight School	Chuntang
	Boys' Hostel at SMIT	Gangtok
	House of BDO	Chuntang
	Residential building	Singtam
	Himalchuli Hotel	Gangtok

earthquake are derived from the database of EERI reports [9]. The performance index score for each of the buildings in Table 1 is calculated.

The complex problem of performance assessment of a building can be broken down into a simple, organized and manageable hierarchical format. This hierarchical model follows an analytical and logical sequence and order in which the underlying relationship for each parameter or factor is further subdivided into specific contributing options.

The type of structural force-resisting system used in a building plays a major role in terms of its seismic load-resisting capacity or resiliency. In this study, two types of reinforced concrete buildings are considered, namely shear wall buildings and moment-resisting frames with infill masonry walls.

Shear walls of sufficient rigidity when used in buildings tend to resist seismic forces in a significant manner. It generally behaves as vertical cantilevers and acts as lateral bracing system to the whole structural system while receiving lateral forces from diaphragms and transferring them to the foundation. During seismic excitations of moderate to strong earthquakes, structures with shear wall provisions have tended to perform considerably well [10].

The moment-resistant frames primarily resist lateral forces through the flexural action of columns and beams since they are joined by moment connections. Columns are critical elements since they are responsible for overall strength and stability of a structure. Their strength relative to the connecting beams plays an important role in seismic resistance in controlling sequence of hinge formation among structural members. Ability to deform in-elastically as governed by concrete confinement and shear capacity is critical. The detailing of beam–column connections is also an important factor influencing the seismic performance. Many older frame buildings include masonry infill panels. Unreinforced masonry behaves in a brittle manner and is often regarded as undesirable construction material for seismically active regions; sometimes, they may act as shear walls in controlling deformations, and it may save non-ductile concrete frames until their elastic limit is exceeded. In many cases, non-ductile frames have survived strong earthquakes due to the participation of masonry infill walls, especially when the wall-to-floor area ratio is high.

The seismic-induced inertia load is transferred from the floors to the foundation through the lateral load-resisting system. It advises to avoid discontinuities or changes in this load path so that localized stress concentrations are minimized or avoided. Irregularity in vertical direction also results in abrupt change in strength and stiffness along the height of the building. Vertical and reverse setbacks, variation in column height, soft stories, discontinuity in shear walls, weak columns and strong beams, and any possible modifications introduced to the primary structural system are some of the types of vertical irregularities that needs to be avoided as far as possible to enhance the seismic resiliency [11].

The plan irregularity is an important aspect in determining vulnerability of a building to torsion. It also helps in identifying potential areas of high-stress concentration. A symmetrical plan layout is therefore considered to be a good design practice. Torsion can arise when centre of mass and centre of rigidity do not coincide, and there is asymmetry in strength and stiffness along the periphery of building, presence of re-entrant corners [11].

Seismic design, ductile detailing, quality of construction and materials used determine the resilience of buildings to seismic events as all these are crucial attributes in order to ensure good protection intended in earthquake-resistant design [12]. Inferior material quality and handling errors, lack of proper reinforcement anchorage in beams, columns, joints, improper seismic detailing, etc., are some of poor construction measures.

4 Proposed Performance Assessment Index Model

The flow chart of the current study is shown in Fig. 1.

For the performance assessment, three groups of elements were identified, construction quality, configuration of the building and load-resisting system. Construction quality comprises materials used and execution. Materials used are subdivided into concrete and reinforcing steel, and execution includes proper rebar confinement and proper placement of concrete as its sub-divisions. Building configuration comprises features determining overall plan and elevation of the building. Plan of the building was categorized into symmetric plan, non-symmetric plan and plan of a building with re-entrant corners. Building elevation incorporated the effects of soft storey, in-plane discontinuity, setback and mass distribution to the performance of a building. Superstructure and substructure performances are the two sub-elements of load-resisting system. Further, structural system and components of structural members, namely columns, beams and beam–column joints, contribute to overall efficiency and performance score of superstructure. Type of soil and type of footing provided in building accounted for performance score of the substructure element as proposed in this performance assessment hierarchical model.

4.1 Design of Questionnaire and Data Collection

A set of questionnaire was designed to collect experts' opinion on relative importance of attributes in the form of comparison tables for performance assessment model. In the AHP process, the attributes at each hierarchical level are to be compared with each other. Such sample table is shown from Tables 2, 3, 4 and 5 which has a more convenient rectangular format of comparison tables to collect the experts' opinion.

The comparison between criteria 'a' and 'b' is on scale from 1 to 9. The selection of any values towards left, i.e. 'a', gives more weight to it in comparison with 'b' and vice versa. A value of 9 towards left means criterion 'a' is extremely important with respect to 'b', while a value of 7 means that importance of 'a' is

Fig. 1 Flow chart for
formulation of performance
assessment AHP model

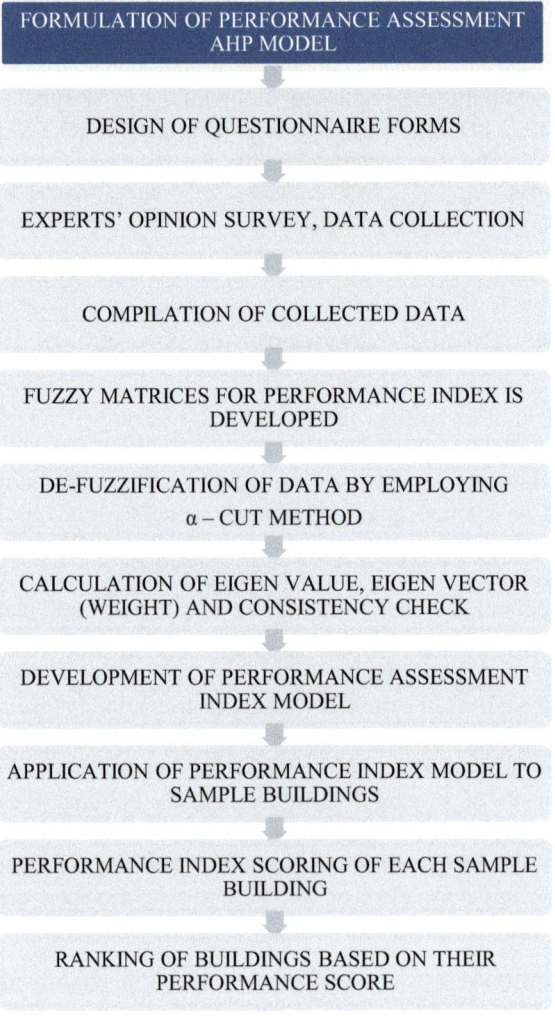

stronger than '*b*' and so on. The selection of the middlebox, i.e. value 1 by an expert, means that both criteria are equally important. These selected values of the relative importance depend on the mindset of the experts.

Similar tabular matrix questionnaire survey scheme was prepared for all levels of hierarchical structure to generate comparison matrices. These questionnaires were sent to various practising structural and earthquake engineers employed in academics and industries. In all responses from 21, such experts were received which have been designated as *E*1, *E*2, *E*3, ..., *E*21, respectively, and the data was compiled.

Table 2 Comparison of the relative preference with respect to: performance assessment index of building

S.No.	Criteria 'a'	Extreme		Very strong		Strong		Moderate		Equal		Moderate		Strong		Very strong		Extreme	Criteria 'b'
		9	8	7	6	5	4	3	2	1	2	3	4	5	6	7	8	9	
								Rating Scale											
1.	Construction quality																		Building configuration
2.	Construction quality																		Load resisting system
3.	Building configuration																		Load resisting system

Table 3 Comparison of the relative preference with respect to: construction quality of building

S.No.	Criteria 'a'	Extreme		Very strong		Strong		Moderate		Equal		Moderate		Strong		Very strong		Extreme	Criteria 'b'
		9	8	7	6	5	4	3	2	1	2	3	4	5	6	7	8	9	
								Rating Scale											
1.	Materials used																		Execution

Table 4 Comparison of the relative preference with respect to: materials used

S.No.	Criteria 'a'	Extreme		Very strong		Strong		Moderate		Equal		Moderate		Strong		Very strong		Extreme	Criteria 'b'
		9	8	7	6	5	4	3	2	1	2	3	4	5	6	7	8	9	
								Rating Scale											
1.	Concrete																		Reinforcing steel

Table 5 Comparison of the relative preference with respect to: reinforcing steel

S.No.	Criteria 'a'	Extreme	Very strong	Strong	Moderate	Equal	Moderate	Strong	Very strong	Extreme	Criteria 'b'
		9	8	7	6	5	4	3	2	1	2	3	4	5	6	7	8	9	
							Rating Scale												
1.	Ribbed bars																		Smooth bars

4.2 Design of Questionnaire Forms

A set of questionnaire was designed to collect experts' opinion on relative importance of attributes in the form of comparison tables for performance assessment model.

4.3 Data Collection and Compilation

The data of comparison received from different experts was arranged and compiled in tabular formats.

The opinions of experts are demonstrated as a triangle between L and U values representing the lower and upper limits of the membership function, respectively, and the M is the geometric mean of experts' opinion representing the major value of the shape function. Graphically the triangular fuzzy number is

$$a_{ij} = \left(L_{ij}, M_{ij}, U_{ij}\right) \tag{1}$$

where $L_{ij} \leq M_{ij} \leq U_{ij}$.
and $L_{ij}, M_{ij}, U_{ij} \in \left[\frac{1}{9}, 1\right] \cup [1, 9]$.

That is, $\tilde{a}_{ij}$ is an element of fuzzy comparison matrix, where L_{ij}, M_{ij}, U_{ij} are lowest, geometric mean and highest values of the experts' opinions, respectively.

$$L_{ij} = \min\left(B_{ijk}\right), \tag{2}$$

Buckley [13] suggested that the geometric mean of experts' opinion for fuzzy comparison values for each criterion may be calculated as given in Eq. (10)

$$M_{ij} = \sqrt[n]{\prod_{k=1}^{n} B_{ijk}} \tag{3}$$

$(k = 1, \ldots, n)$ and

$$U_{ij} = \max\left(B_{ijk}\right) \tag{4}$$

B_{ijk} represents opinions of expert k for the relative comparison of two criteria i and j (Table 6).

4.4 Data Fuzzification

Fuzzy matrices are prepared from the compiled data of the survey opinion tables prepared using L_{ij}, M_{ij} and U_{ij} values in different levels of comparison tables of performance index model (Table 7).

4.5 Data Defuzzification

The fuzzy matrix is defuzzified by transforming these values into a crisp value considering values of α and λ equal to 0.5, using equation (Table 8)

$$\left(a_{ij}^{\alpha}\right)^{\lambda} = \left(\lambda \cdot L_{ij}^{\alpha} + (1 - \lambda)U_{ij}^{\alpha}\right) \quad 0 \leq \alpha \leq 1, 0 \leq \lambda \leq 1 \tag{5}$$

The resultant defuzzified single pairwise comparison matrix for Table 2 is expressed in Table 9.

4.6 Calculation of Eigenvalue, Eigenvector and Consistency Check

Principal eigenvalue and eigenvector of this matrix are calculated. The eigenvector represents the weight vector of the attributes. Eigenvalues and eigenvectors of defuzzified matrix are represented as $\lambda_{\max}$ and W, respectively.

Consistency check is also carried for the comparison matrix (Table 10).

The hierarchical model is based on the aforementioned methodology and calculations, and is reflected in Table 11.

Table 6 Experts' opinion for Table 2 of performance assessment index

Criteria 'A'	Criteria 'B'	E1	E2	E3	E4	E5	E6	E7	E8	E9	E10	E11	E12	E13	E14	E15	E16	E17	E18	E19	E20	E21	L_{ij}	M_{ij}	U_{ij}
Construction quality	Building configuration	3	2	5	1	5	2	7	1	1	7	5	3	1/3	5	4	1/3	1	1/5	1	1/2	3	1/5	1.762	7
Construction quality	Load-resisting system	3	3	1/3	1/5	1	1/3	1	1	1/4	5	1	1/4	3	5	3	1/5	1	1/6	1	5	1/3	1/6	0.915	5
Building configuration	Load-resisting system	1/5	1/5	1/7	1/5	1/6	1/4	1/5	2	2	1/5	1/7	1/7	5	1	1/4	5	1	1/5	1	1/4	1/7	1/7	0.409	5

Table 7 Fuzzified matrix for relative preference with respect to: performance assessment of building

Criteria 'a'/'b'	Construction quality	Building configuration	Load-resisting system
Construction quality	1, 1, 1	1/5, 1.762, 7	1/6, 0.915, 5
Building configuration	1/7, 0.568, 5	1, 1, 1	1/7, 0.409, 5
Load-resisting system	1/5, 1.093, 6	1/5, 2.445, 7	1, 1, 1

Table 8 Calculation of L_{ij}^{α}, U_{ij}^{α} and $(a_{ij}^{\alpha})^{\lambda}$ for Table 2 of performance assessment index

α	λ	L_{ij}	M_{ij}	U_{ij}	L_{ij}^{α}	U_{ij}^{α}	$(a_{ij}^{\alpha})^{\lambda}$
0.5	0.5	1/5	1.762	7	0.9810	4.3810	2.6810
0.5	0.5	1/6	0.915	5	0.5408	2.9575	1.7492
0.5	0.5	1/7	0.409	5	0.2759	2.7045	1.4902

Table 9 Defuzzified matrix for relative preference with respect to: performance assessment of building

Criteria 'a'/'b'	Construction quality	Building configuration	Load-resisting system
Construction quality	1.0000	2.6810	1.7492
Building configuration	0.3730	1.0000	1.4902
Load-resisting system	0.5717	0.6711	1.0000

Table 10 Principal eigenvalue, eigenvector and consistency ratio for matrix of performance index

n	Random index	Eigenvalue, λ_{max}	Eigenvectors, W	Consistency index (CI)	Consistency ratio (CR)
3	0.58	3.0766	0.5143, 0.2577, 0.2280	0.0383	0.0660

The maximum performance value is likely to occur when all the performance criteria meet all the favourable constructional and architectural aspects of a building. The corresponding maximum value of performance index is found to be **0.9164,** and calculation of the same has been presented in detail in Table 11.

The performance score calculation is shown for Moonlight School in Table 11.

Table 11 Performance index and performance score of buildings

| | | MAXIMUM | | MOONLIGHT SCHOOL | |
| | | Weight | Score | Weight | Index Value |
S.No	Particulars				
	TOTAL SCORE		**0.9164**		**0.4009**
A	**Construction Quality**	**0.5143**	**0.9754**	**0.5143**	**0.2354**
A.1	Materials	0.7892	0.9688	0.7892	0.0312
A.1.1	Concrete	0.7259	1.0000	0.7259	0.0000
O1	Good quality	1.0000	1.0000	1.0000	
O2	Poor quality	0.0000		0.0000	1.0000
A.1.2	Reinforcing steel	0.2741	0.8863	0.2741	0.1137
O1	Ribbed bars	0.8863	1.0000	0.8863	
O2	Smooth bars	0.1137		0.1137	1.0000
A.2	Execution	0.2108	1.0000	0.2108	1.0000
A.2.1	Proper rebar confinement	0.7281	1.0000	0.7281	1.0000
O1	Provided	1.0000	1.0000	1.0000	1.0000
O2	Absent	0.0000		0.0000	
A.2.2	Proper placement of concrete	0.2719	1.0000	0.2719	1.0000
O1	Yes	1.0000	1.0000	1.0000	1.0000
O2	No	0.0000		0.0000	
B	**Configuration**	**0.2577**	**0.8251**	**0.2577**	**0.7051**
B.1	Building plan	0.7201	0.7814	0.7201	0.7814
O1	Symmetric	0.7814	1.0000	0.7814	1.0000
O2	Non-symmetric	0.1332		0.1332	
O3	Re-entrant corner	0.0854		0.0854	
B.2	Building elevation	0.2799	0.9374	0.2799	0.5088
B.2.1	Soft-storey	0.0765	1.0000	0.0765	1.0000

Buildings	Performance index score
Moonlight School	0.4009
Boys' Hostel at SMIT	0.7153
House of BDO	0.5178
Residential building	0.4692
Himalchuli Hotel	0.4596

Table 12 Total performance index score of buildings

5 Result and Discussion

Performance assessment index model was decomposed into different units, sub-units, elements and components, a hierarchical model of performance index is developed, and calculations and results are derived based on the aforementioned methodology.

The maximum performance value when the building possesses all the favourable material, design and architectural guidelines anticipated is found to be 0.9164. Performance index score for all the buildings listed in Table 1 was calculated depending upon the performance factors employed in a building. The final performance score of the buildings is given in Table 12.

The obtained weighted score can be normalized and converted to a relative index score at any desired base. The scale has been mapped at the scale of 100 as shown below.

Let WS be the weighted score of any building under consideration and WS_{max} be the maximum possible score in the analysis. The performance index for buildings at the relative scale of 100 can be given a name as 'PI 100' and calculated as

$$PI = \frac{WS \text{ (Weight score of the considered building)}}{WS_{max}\text{ (Maximum score out of all buildings)}} \times 100 \qquad (6)$$

Performance index for buildings at the relative scale of 100 was calculated as shown in Eqs. (7) to (11).

For example, relative performance index (score) of Moonlight School is

$$PI = \frac{WS}{WS_{max}} \times 100 = \frac{0.4009}{0.9164} \times 100 = 43.747 \qquad (7)$$

Relative performance index of Boys' Hostel at SMIT is

$$PI = \frac{WS}{WS_{max}} \times 100 = \frac{0.7153}{0.9164} \times 100 = 78.055 \qquad (8)$$

Table 13 Ranking of buildings based on relative performance score

Buildings	Relative performance score	Rank
Moonlight School	43.747	5
Residential building	51.200	3
House of BDO	56.504	2
Boys' Hostel at SMIT	78.055	1
Himalchuli Hotel	50.153	4

Relative performance index of House of BDO is

$$PI = \frac{WS}{WS_{max}} \times 100 = \frac{0.5178}{0.9164} \times 100 = 56.504 \qquad (9)$$

Relative performance index of Building at Singtam is

$$PI = \frac{WS}{WS_{max}} \times 100 = \frac{0.4692}{0.9164} \times 100 = 51.200 \qquad (10)$$

Relative performance index of Himalchuli Hotel is

$$PI = \frac{WS}{WS_{max}} \times 100 = \frac{0.4596}{0.9164} \times 100 = 50.153 \qquad (11)$$

Table 13 summarizes ranking of buildings in terms of relative PI (or performance score) obtained for each building.

The performance score and ranking of SMIT Boys' Hostel are highest.

Use of good-quality concrete, proper placement, proper rebar confinement, symmetric building plan, uniform mass distribution, strong column–weak beam design and absence of soft storey, setback, floating/hanging columns, short columns, in-plane discontinuity are the factors that accounted for enhanced performance score of Boys' Hostel at SMIT.

In the 2011 Sikkim earthquake out of these five buildings, SMIT Boys' Hostel has performed better than others [9]. So, the result provided by performance index calculations has been found in accordance with what was observed in reality [6].

6 Plotting of Performance Index Values of Studied Buildings

A graph was plotted for performance index for these five buildings, and the obtained graphs are shown in Figs. 2 and 3.

From Figs. 2 and 3, it is clear that as the performance index of a building increases, the better the building has performed in case of the earthquake.

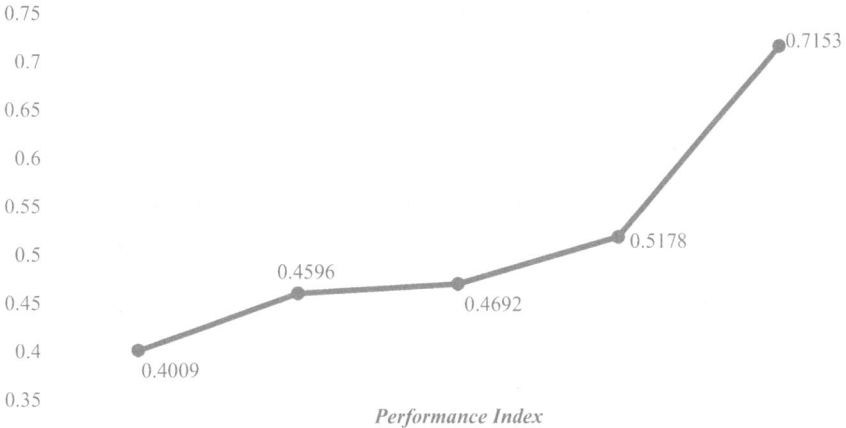

Fig. 2 Performance index score of buildings

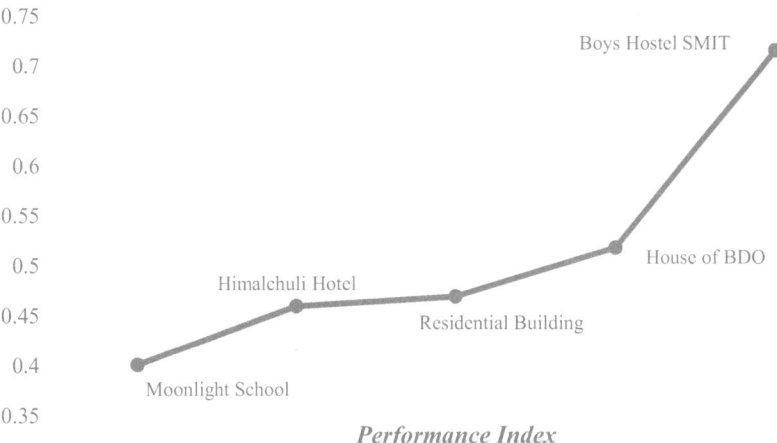

Fig. 3 Buildings in order of their performance index

In other words, it can be concluded that a building with higher performance index has suffered less seismic damage and vice versa.

7 Conclusion

Understanding the structural aspects and parameters affecting the performance of buildings in case of a seismic event is a critical and complex task. So, the decision-making about assessing the same is decomposed into hierarchical models, and subsequently, a performance assessment index (PAI) for reinforced concrete buildings has been developed.

Then, categorizing the factors and parameters contributing to better earthquake-resistant design and their relative importance with respect to each other is determined.

The PAI model is demonstrated through case study of five buildings which were damaged in Sikkim 2011 earthquake to compute their performance score.

These buildings are then ranked on the basis of their performance score. It was found that building with the lowest performance score suffered maximum damage compared to other sample buildings during the earthquake. Also, it is found that building with highest performance score actually survived and performed well compared to other during the seismic event.

The study validates the point that the seismic damage caused to a building is in inverse relation to the performance efficiency and resilience of the building; i.e. the building with high performance score will perform well and suffer less damage in case of a seismic activity.

The presented model is very simple and easy to implement as the value of measurable items or elements for performance assessment index are easily identifiable and quantifiable.

Performance assessment model encapsulates and enlists various provisions to be taken into account to enhance structure's expected behaviour during earthquakes. A proper design and performance-incorporated scheme will ensure risk reduction. As engineers, we strive for greatest authenticity in terms of design, safety and performance of our buildings. In a nutshell, it can be implied that a safe and smart design approach would help make efficient and effective structures, and a subset of smart cities and overall will contribute in better future for the society. The performance index based on fuzzy AHP is simple and hence holds the potential for practical application.

References

1. Carreño ML, Cardona OD, Barbat AH (2007) Neuro-fuzzy assessment of building damage and safety after an earthquake (Chap. VII). In: Intelligent computational paradigms in earthquake engineering. Idea Group Publishing
2. Saaty TL (1980) Analytic hierarchy process. McGraw Hill, N.Y.
3. Chang CW, Wu CR, Lin HL (2009) Applying fuzzy hierarchy multiple attributes to construct an expert decision making process. Exp Syst Appl 36:7363–7368
4. Zadeh L (1965) Fuzzy sets. Inf Control 8:338–353
5. USGS (2011) M 6.9 - Sikkim, India [Online]. Available at: https://earthquake.usgs.gov/earthquakes/eventpage/usp000j88b/executive. United States Geological Survey, Reston, USA. Accessed 18 Sept 2011
6. Rai DC, Mondal G, Singhal V, Parool N, Pradhan T (2012) 2011 Sikkim earthquake: effects on built environment & a perspective on growing seismic risk. 15 WCEE Lisboa 2012

7. EERI (2012) The Mw 6.9 Sikkim-Nepal Border earthquake of September 18, 2011: Learning from Earthquakes, Special Earthquake Report, Earthquake Engineering Research Institute, Oakland, CA
8. Roshan P, Pal S, Kumar R (2018) Seismic damage assessment index of buildings using fuzzy-AHP approach. Int J Tech Innov Mod Eng Sci 4(08)
9. EERI (2013) Online database of concrete buildings damaged in earthquakes [Online]. Available at: http://db.concretecoalition.org/. Earthquake Engineering Research Institute, Oakland, CA. Accessed 4 Nov 2013
10. Saatcioglu M et al (2001) The August 17, 1999, Kocaeli (Turkey) earthquake damage to structures. Can J Civ Eng 30:715–737
11. Arnold C, Reitherman R (1982) Building configuration and seismic design. A Wiley-Interscience Publications, Toronto
12. Mouzzoun M, Moustachi O, Taleb A, Jalal S (2013) Seismic performance assessment of reinforced concrete buildings using pushover analysis, Morocco
13. Buckley JJ (1985) Fuzzy hierarchical analysis. Fuzzy Sets Syst 17:233–247

A Perspective on Migration and Community Engagement in Smart Cities

Pushkar P. Jha and Muhammad A. Iqbal

Abstract This is a conceptual paper interfacing community engagement and migration flows in relation to smart cities' development. The paper notes community engagement as a crucial variable, in general and with reference to the aspired for impact on migration flows. It conceptualizes community engagement as an operationalizable construct for strategic design. The idea of community engagement is there in most multi-stakeholder projects and initiatives. Enhancing design and execution for making it count for superior performance of smart city initiatives is what we seek to develop here. The paper is also oriented to deliver an agenda for field research based on hypotheses it comes forth with.

Keywords Community engagement · Inbound migration · Socio-economic clustering · Willingness · Ability

1 Introduction

In delivering smart city projects, policy aspirations are more than just about developing an effective interface between technology, lifestyle, and regional asset and capabilities. These aim at reducing pressures on urban centres that draw more inbound migration through improvements, and in tandem, enhancing the attractiveness of other urban centres for diffusing migration flows. This could be by diverting more rural-to-urban migration to these secondary urban regions or, inducing migration to them from overburdened urban centres. Frequently, another policy aspiration is to develop effective networking between smart cities and satellite rural regions for socio-economic stability and equitable growth [1].

P. P. Jha (✉)
Faculty of Business and Law, Newcastle Business School, Northumbria University, Newcastle, UK
e-mail: Pushkar.jha@northumbria.ac.uk

M. A. Iqbal
Cambridge Management and Leadership School, Cambridge, UK

© Springer Nature Singapore Pte Ltd. 2020
S. Ahmed et al. (eds.), *Smart Cities—Opportunities and Challenges*,
Lecture Notes in Civil Engineering 58,
https://doi.org/10.1007/978-981-15-2545-2_43

521

The central argument of this paper is to highlight the importance of active engagement by resident communities for achieving such aspirations. This engagement is manifested in the interface between variables of willingness and ability that shape community response. The paper develops this argument and draws lessons from research on community engagement experiences. More importantly, it tries to create an agenda for primary research by presenting hypotheses to help investigate how this can be done effectively. The central questions that the study relates to and for which it shapes a case for further investigation are:

How can the resident community's engagement be purposively interfaced with migration that associates with smart city development?

and

How can rural and urban socio-economic clustering be influenced by design of community engagement strategies?

Research and knowledge domains that relate to these questions include strategies for migration and urban regeneration. Understanding community engagement will be of value to academia in development studies and for policy and practice that deals with community-based projects. Focused insights on the rural and urban interface from the vantage point of urban regeneration will also benefit scholars in development and migration studies.

2 Conceptual Moorings of Community Engagement

Research and practice in the area of urban (and rural) regeneration and associated infrastructure development clearly recognize the importance of community engagement (e.g. [2]). Going deeper this recognition can yield two perspectives; one is that of the *willingness* of the end user community, i.e. the inhabitants of the novel or regenerated urban entity, in this case the smart city. Willingness is a direct consequence of utility perceptions about an intervention, affected by sensitization for buy-in and by past experience of development interventions by the involved agencies. Understanding the level and nature of willingness is crucial to design sensitization in both scope and content. For instance, if willingness, say in a certain segment of the community is already high, investments on sensitization therein would be less useful. On the other if it is low, then also understanding the reasons for this becomes crucial for appropriate sensitization design, whether it is about poor past experiences or simply inability to see benefits and their link through to improving living—the content and delivery will vary. It is important to re-iterate that it is not communication 'but' buy-in for engagement that is strived for under this perspective.

The other perspective is the *ability* of the community which is about their capacity to engage, shaped by variables such as knowledge about the intervention that may relate to access and operations of the new schema, the communities' economic lifestyle and associated contextual rigidities. This is perspective thus less about rigidities that are perceptual and cognitive like 'willingness' discussed above. It is more about behavioural rigidities and resource constraints including knowledge and associated communication (distinct from buy-in and sensitization under 'willingness') about the schema that can relate to engagement [3]. Understanding the precise nature of ability constraints are also critical for directing resources in an optimal fashion for ramping up ability.

Both perspectives can be articulated independently or as in a mutually interacting context. Let us elaborate, for instance, high willingness alone cannot assure engagement. Insufficient knowledge about the operations of an initiative and overbearing constraints can make it difficult to engage. Ability constraints can dilute the impact of high willingness translating into high community engagement. This last aspect can simply make for 'want to but cannot do' scenarios where willingness is high, but ability is low. By extension, a flip side will also exist where ability will be high but willingness low [4].

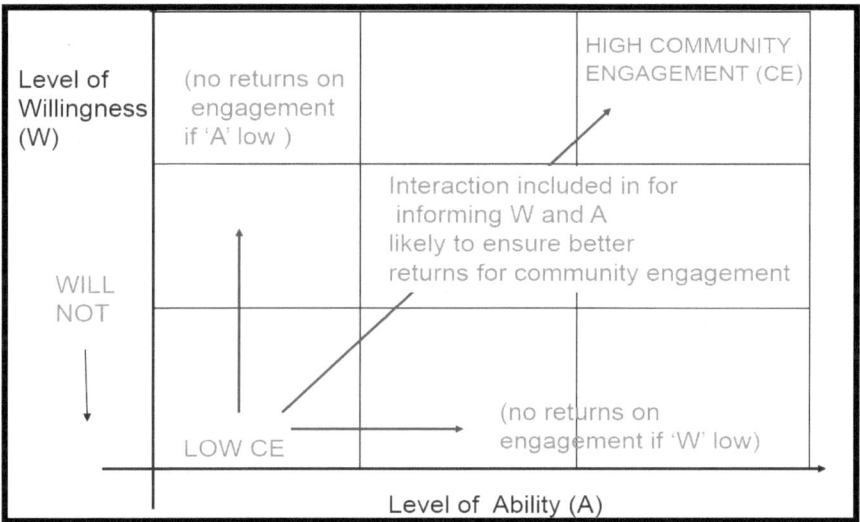

3 Community Engagement and the Migration Context in Smart Cities

Two illustrations in the smart cities' context may be pertinent to flag, just to demonstrate the interface between willingness and ability. Say, for instance, we have an urban region marked by low set housing and plenty of area to get in new housing and also multi-storey buildings. Now while the ability to absorb new housing for migrants may be high, the willingness to disrupt the ambience may be low in the resident community. This needs to be addressed through suitable sensitization. Another case would be that of relative cultural homogeneity in smaller cities, where communities may be willing to absorb inbound migrants (for reasons of services, support and wider economic benefits they get), but the ability to do so because of the rigidities of existing socio-cultural settings would make it difficult. Ability support as focused interventions to provide common grounds and interaction opportunities between migrant sub-communities and existing resident communities could be useful as a mechanism for increasing cohesion over time.

As noted, an active engagement by the resident communities is crucial for achieving such aspirations [5, 6]. Research and practice in development studies clearly recognize the importance of community engagement [6–9]. The two dimensions flagged as contributing to or shaping community engagement thus need to be carefully contextualized, interpreted and fed forward into design. To re-iterate, the first is willingness of the community, in this case the urban centre residents. Willingness is a direct consequence of utility perceptions about an intervention, affected by sensitization for buy-in and by experience of past interventions [10]. The second perspective is the community's ability to engage given resource constraints [11, 12]. The interface between the two is crucial—a high willingness will not advance community engagement if ability is low, and ability will not matter if willingness is low. Community engagement has yet to be examined from a perspective that interfaces willingness and ability as mutually mediating influences. This is becoming of increasing relevance as events and conditions in both developing and developed countries have demonstrated that migration has a crucial determinant in host communities' receptiveness.

Hypothesis 1: The nature of interface between willingness and ability (W&A) of the resident community is crucial to their uptake of the smart city agenda.

Policy and practice informing insights are crucial to understand resident communities' willingness and ability to engage with smart city projects, including in context of inbound migration that follows urban renewal. Satellite rural regions' interface with an urban centre's ecology is important as well to examine how smart cities can be effective as nodal entities—not only seeking superior socio-economic and sustainable outcomes for themselves, but also for the wider regional and national ecology. This goes to the heart of the challenge of reducing regional and rural–urban development disparity facing developing countries in general, and India in particular where urbanization is often argued to have yielded skewed growth and development.

Smart city features would map out differently as well. For instance, willingness and ability intersection (figure provided before) for 'Making governance citizen-friendly and cost-effective—increasingly rely on online services to bring about accountability and transparency' will typically be different than that for 'Applying Smart Solutions to infrastructure and services in area-based development' [13]. The closer these intersection points, less resource intensive and aligned would be the management of enhancement of willingness and ability for the initiative as a whole.

Hypothesis 2: Reducing the difference in how W&A relate to different smart city features will yield superior outcomes.

We have noted the need for effective networking between smart cities and satellite rural regions for socio-economic stability and equitable growth in smart city initiatives. The reasons of inevitable mutual impact satellite rural areas and the urban areas create are oft cited [13]. Co-creation is thus likely to improve design and uptake of intervention's impact on community engagement and outcomes positively. This comes with the caveat of resolving conflicting demands and perception of benefits that can make such co-creation difficult.

Hypothesis 3: Involving satellite rural communities in co-creation of selected interventions within the smart city programme will yield superior outcomes. This provided the nature of willingness and abilities across communities which can be aligned.

4 Conclusions

Examining smart cities from a community engagement, inbound migration and regional development context requires interdisciplinary framing and associated expertise, oft implied in extant research, and something that we propose going forward from this conceptual paper [14–16]. We are looking to further develop this trajectory of hypotheses. This will be for subsequent primary research that examines the very well-situated smart cities initiatives in India, with strong implications for the wider developing countries and Southeast Asian context. These smart cities initiatives are typically, sequentially and parallelly planned over phases; therefore, the opportunity to feed in study findings for impacting design is significant.

References

1. Cosgrave E, Arbuthnot K, Tryfonas T (2013) Living labs, innovation districts and information marketplaces: a systems approach for smart cities. Procedia Comput Sci 16:668–677
2. Saunders T, Baeck P (2015) Rethinking smart cities from the ground up. NESTA, Intel, UNDP. Nesta.org.uk
3. Magis K (2010) Community resilience: an indicator of social sustainability. Soc Nat Resour 23(5):401–416
4. Jha PP, Bhalla A (2018) Life of a PAI: mediation by willingness and ability for beneficiary community engagement. World Dev Perspect 9:27–34
5. Bhattacharya, S., Rathi, S., Patro, S.A. &Tepa, N. (2015). Reconceptualising Smart Cities: Reference Framework- India CSTEP [online]
6. Engasser F, Saunders T (2015) Role of citizens in India's smart cities challenge. Fair observer. Sourced from World Policy Journal podcast [online & broadcasted]

7. Alsop R, Sjoblom D, Namazie C, Patil P (2002) Community-level user groups in 3 world bank-aided projects: do they perform as expected? Social development papers 40. World Bank ESSD, Washington D.C

8. Davis J (2004) Assessing community preference for development projects: are willingness-to-pay robust to mode effects? World Dev 32(4):655–672

9. Schischka J, Dalziel P, Saunders C (2008) Applying Sen's capability approach to poverty alleviation programs: 2 case studies. J Hum Dev 9(2):229–246

10. Hanemann WM (1991) Willingness to Pay and Willingness to Accept: How Much Can They Differ? Am Econ Rev 81(3):635–647

11. Mataria A, Giacaman R, Khatib R, Moatti J-P (2006) Impoverishment and patients' "willingness" and "ability" to pay for improving the quality of health care in Palestine: an assessment using the contingent valuation method. Health Policy 75(3):312–328

12. Sen A (1998) The possibility of social choice. Prize Lecture: Lecture to the memory of Alfred Nobel, Dec 8

13. Smart Cities Mission GoI (2019) Smart city features. Online at http://smartcities.gov.in/content/innerpage/smart-city-features.php. Accessed 01 Feb 2019

14. Brettell CB, Hollifield JF (2014) Migration theory: talking across disciplines. Routledge

15. Handlos LN, Kristiansen M, Norredam M (2015) Wellbeing or welfare benefits—what are the drivers for migration? Scand J Publ Health. https://doi.org/10.1177/1403494815617051

16. Swapan MSH (2014) Realities of community participation in metropolitan planning in Bangladesh: a comparative study of citizens and planning practitioners' perceptions'. Habitat Int 43:191–197

An Investigation on Response of Blast Load on Masonry Structure

Saba Shamim, Shakeel Ahmad and Rehan A. Khan

Abstract The smart cities comprise of low, medium and high rise buildings usually RC (Reinforced Concrete) framed structures having masonry in-filled wall panels both at exterior and interior faces. Any damage caused due to blast/bombing to such structures may cause larger damage to life in vicinity of the event area. Keeping view to this aspect the present endeavour has examined the effect of air blast on masonry in-filled RC framed wall acting from various angle of detonation (0°, 30°, 45°, 60°, and 90°). The deliberation will discuss the findings entertained.

Keywords Masonry · Air blast · Angle of detonation · Macro-modelling

1 Introduction

India in the past few years has suffered many bombing attacks which has not only resulted in the loss of life but, also has caused damage to property and infrastructure. Masonry is the most common type of construction found in India due to its low cost, durability and ease in maintenance. However, masonry is found to be quite weak when subjected to out of plane loading such as earthquake, blast, high and low-velocity impact. Therefore, it has become important for civil engineers to study the effect of such events (blast) on structures so that the 'Smart Cities Mission' of India can be smartly accomplished.

A bomb explosion near a structure may cause catastrophic damage like crumbling of walls, blowing out of windows, and sometimes even ceasing critical life safety structures. Keys and Clubley [1] investigated the breakage patterns and debris distribution of masonry panels subjected to blast loads. Three experimental trials were performed each on ten masonry panels of varying geometries. It was found that blast overpressure, structural geometry and impulse were the main parameters accountable for the initial fragmentation, breakage pattern and debris distribution, respectively. Yuan et al. [2] carried out distinctive finite element

S. Shamim (✉) · S. Ahmad · R. A. Khan
Department of Civil Engineering, Aligarh Muslim University (A.M.U.), Aligarh, India

© Springer Nature Singapore Pte Ltd. 2020
S. Ahmed et al. (eds.), *Smart Cities—Opportunities and Challenges*,
Lecture Notes in Civil Engineering 58,
https://doi.org/10.1007/978-981-15-2545-2_44

527

analysis of masonry infill walls for determining the blast properties using LS-DYNA. The reliability and efficiency of this method was proven to be satisfactory in predicting the dynamic and failure behaviour of masonry walls subjected to blast loads. Alsayed et al. [3] tested half-scale infill masonry walls against C-4 explosives. Both un-strengthened and externally strengthened (using GRFP) were subjected to the action of field blast. Later, the field test results of un-strengthened as well as the strengthened walls were compared and also simulated using ANSYS-AUTODYN. The numerical models were then further studied to see the effect of various parameters such as scaled distance and FRP end anchorage.

The paper has investigated the response of blast load on masonry in-filled RC (Reinforced Concrete) framed wall. The numerical model has been developed in ABAQUS/Explicit software and validated satisfactorily. Thereafter, the effect of the blast from various detonation angles on the numerical model has been studied.

2 Numerical Modelling

Masonry in-filled RC framed wall previously tested in field by Varma et al. [4] against various scaled distances has been taken for the numerical study. The experimental wall having thickness 230 mm has been modelled in Abaqus/explicit tool using macro modelling approach as shown in Fig. 2. Eight-noded SOLID 65 element has been used to model both masonry ($3000 \times 3000 \times 230$ mm) and concrete frame (230×235 mm) whereas, steel reinforcement has been modelled using 3D-TRUSS element. The non- linearity in material properties of masonry, concrete and steel has been considered using Mohr–Coulomb (MC), Concrete Damaged Plasticity (CDP) and elasto–plastic strain hardening criterions, respectively. The mechanical properties of model are tabulated in Table 1.

Mesh size of 100 mm has been used to discretise the model as shown in Fig. 2. The friction coefficient between the RC frame and masonry has been taken as 0.8 [5]. The base of the wall has been considered fixed as shown in Fig. 1. A total of 3490 elements were generated having 5358 nodes.

The developed model has been subjected to blast pressure-time history (Fig. 3) for TNT explosive at various scaled distance to determine the peak displacement at the center of wall. The results of numerical study have been compared with the experimental data in Table 2 and are found to be quite satisfactory thereby, validating the numerical model.

Table 1 Properties of numerical model

	Masonry	Concrete	Steel
Mass density (Υ) (kg/m^3)	2100	2400	8050
Modulus of elasticity (E) (MPa)	2000	22,000	200,000
Poisson's ratio (υ)	0.2	0.15	0.3

Fig. 1 Numerical model for validation

Fig. 2 Numerical model showing mesh of 100 mm size

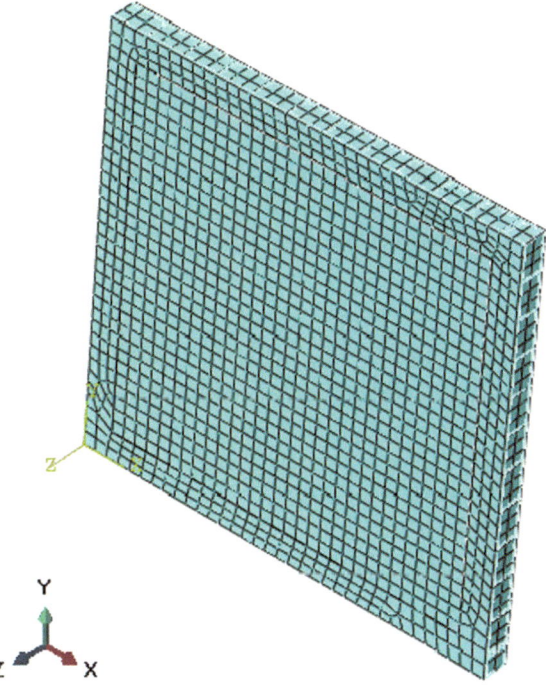

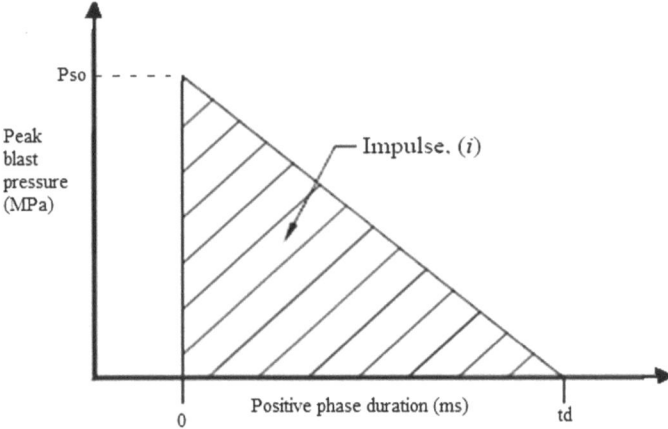

Fig. 3 Idealized pressure-time history for blast load

3   Parametric Study

The numerical model has further been considered to see the effect of air blast occurring at a different angle of detonation at a height of 1.5 m above the ground. The air blast load has been generated using the inbuilt ConWep (Conventional weapon) tool in Abaqus software. In ConWep the amount of charge (in kg of TNT) and stand-off distance (in metre) is specified. Then a Reference Point (RP) is selected in space (x, y, z) where the detonation will occur and a target area is indicated where the responses are to be investigated. Also, for the parametric study the top of the wall has been restrained in Z-direction (perpendicular to the plane of the blast surface) as in real conditions a slab will be there having high stiffness in the in-plane direction (Fig. 4).

3.1   Variation in Angle of Detonation

The numerical model has been subjected to air blast load of 100 kg TNT at stand-off distance 20 m and height 1.5 m above ground. The responses have been investigated to see the effect of variation in angle of detonation 0°, 30°, 45°, 60° and 90°, respectively. The displacement, stresses and pressure time histories obtained after the analysis are shown in Figs. 10, 11and 12. Also, the contour plots for various angles of detonation are shown in Figs. 5, 6, 7, 8 and 9.

Table 2 Comparison of peak displacement at centre of wall for experimental and numerical study

Wall thickness (mm)	Weight of charge (kg of TNT)	Stand-off distance (m)	Scaled distance (m/ kg$^{1/3}$)	Peak blast pressure (MPa)	Positive phase duration $\times 10^{-3}$ (ms)	Peak displacement at centre (mm)		
						Experimental study	Numerical study (present study)	Numerical study [5]
230	21.5	4.0	1.44	1.30	1.73	127.5	134.5	109.5
230	50.6	5.5	1.49	1.84	2.10	Collapse (C)	230.7 (C)	>230 (C)
230	51.4	5.5	1.48	2.01	1.92	C	232.5 (C)	>230 (C)

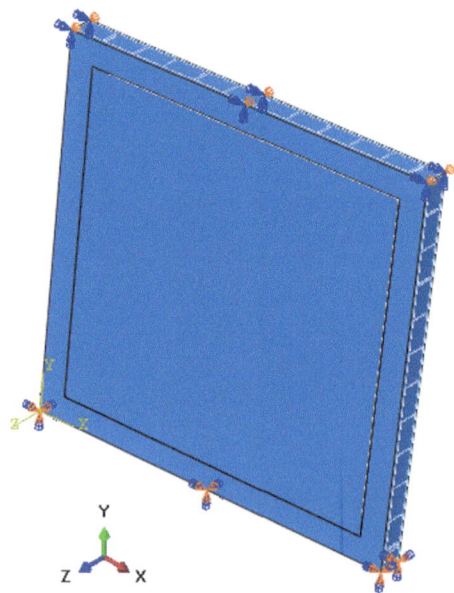

Fig. 4 Support condition of wall model for parametric study

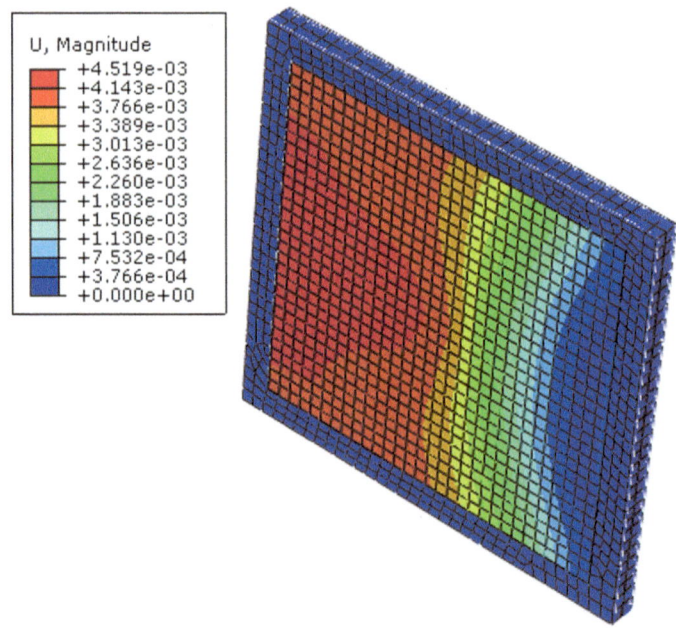

Fig. 5 Contour plot for 100 kg TNT blast @ 20 m stand-off distance at an angle 0°

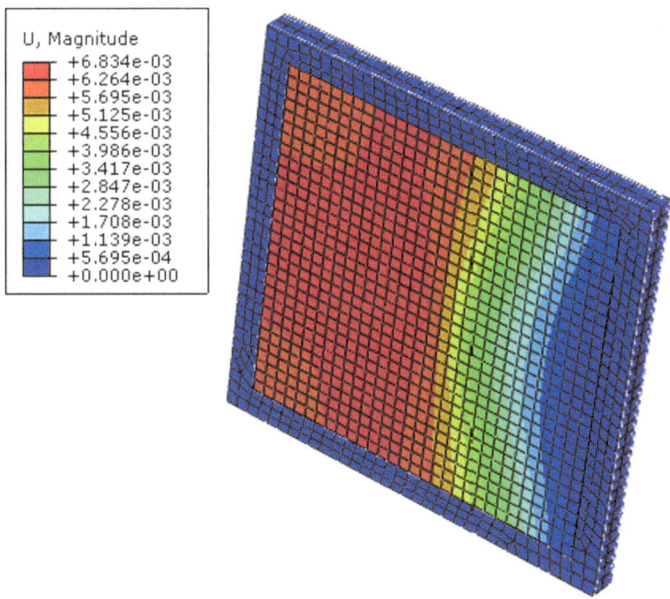

Fig. 6 Contour plot for 100 kg TNT blast @ 20 m stand-off distance at an angle 30°

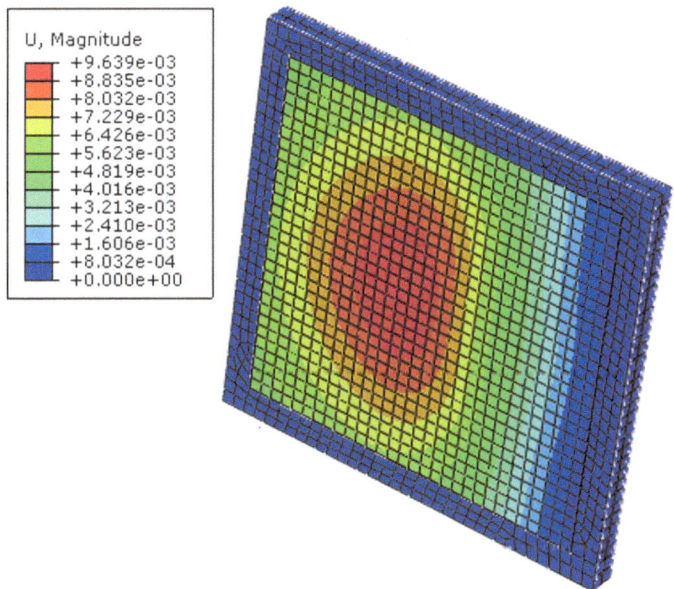

Fig. 7 Contour plot for 100 kg TNT blast @ 20 m stand-off distance at an angle 45°

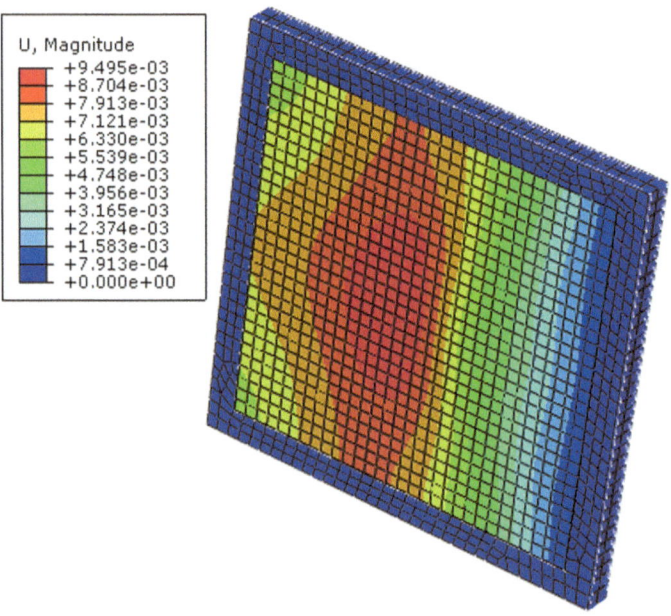

Fig. 8 Contour plot for 100 kg TNT blast @ 20 m stand-off distance at an angle 60°

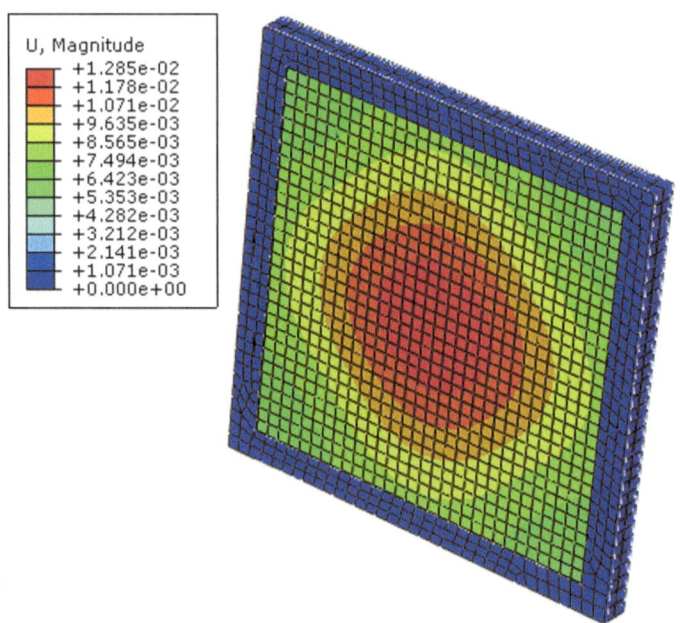

Fig. 9 Contour plot for 100 kg TNT blast @ 20 m stand-off distance at an angle 90°

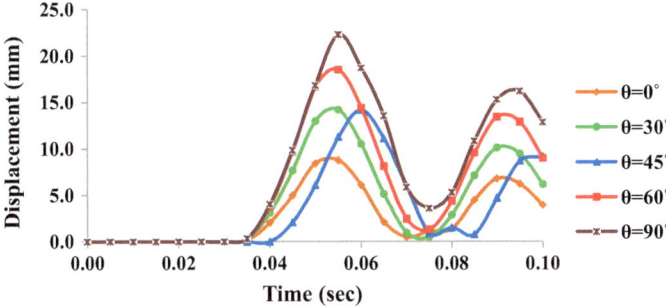

Fig. 10 Displacement versus time variation for 100 kg TNT blast @ 20 m stand-off occurring at various angles

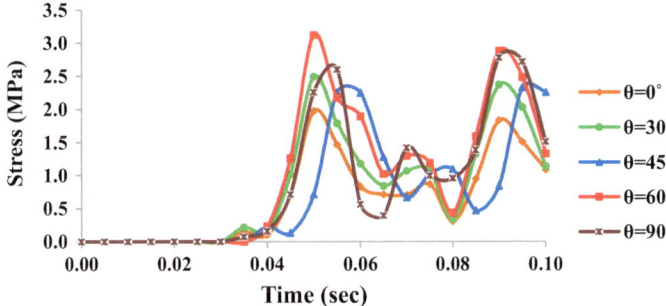

Fig. 11 Stress versus time variation for 100 kg TNT blast @ 20 m stand-off occurring at various angles

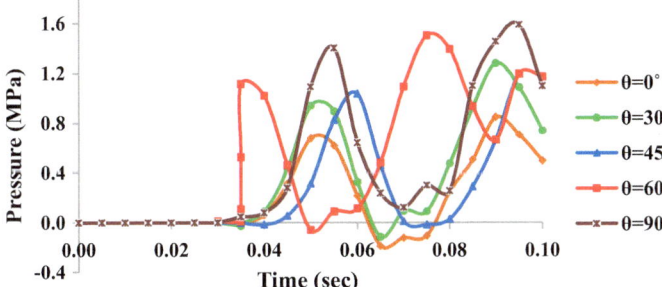

Fig. 12 Pressure versus time variation for 100 kg TNT blast @ 20 m stand-off occurring at various angles

Table 3 Peak values after analysis

Weight of charge (Kg of TNT)	Stand-off distance (m)	Height of detonation above ground (m)	Angle of detonation (°)	Peak displacement (mm)	Peak stress (MPa)	Peak pressure (MPa)
100	20	1.5	0	8.8	1.97	0.85
100	20	1.5	30	14.2	2.49	1.28
100	20	1.5	45	14.2	2.33	1.20
100	20	1.5	60	18.5	3.12	1.39
100	20	1.5	90	22.3	2.78	1.59

4 Results and Discussion

The numerical model subjected to air blast load acting at various angles 0°, 30°, 45°, 60° and 90°, respectively have been studied. The peak values of displacement, stress and pressure developed at the mid/centre of the masonry panel are been tabulated in Table 3. It can be observed that the peak displacement is increasing as the angle of detonation is sweeping towards the front face of the wall. When the blast occurs at 0° (in-plane angle) and stand-off distance 20 m then the peak displacement is minimum compared to when the blast occurs at 90° (out of plane angle).

5 Conclusions

It can be concluded from the results that angle of detonation plays a significant role in the level of damage caused in masonry in-filled RC framed wall. The maximum deflection in-wall has been observed when blast acts at an angle 90° i.e., in out of plane direction. Therefore, this orientation may be considered as critical and can be opted for further research studies.

References

1. Keys RA, Clubley SK (2017) Experimental analysis of debris distribution of masonry panels subjected to long duration blast loading. Eng Struct 130(2017):229–241
2. Yuan X, Chen L, Wu J, Tang J (2012) Numerical simulation of masonry walls subjected to blast loads. Adv Mater Res 461:93–96

3. Alsayed SH, Elsanadedy HM, Al-Zaheri ZM, Salloum YA, Abbas H (2016) Blast response of GFRP-strengthened infill masonry walls. Constr Build Mater 115:438–451
4. Varma RK, Tomar CPS, Parkash S (1997) Damage to brick masonry panel walls under high explosive detonations. Press Vessels Pip Div 351:207–216
5. Pandey AK, Bisht RS (2014) Numerical modelling of infilled clay brick masonry under blast loading. Adv Struct Eng 17(4):2014

Seismic Analysis of Pile Foundation Passing Through Liquefiable Soil

Musabur Rehman and S. M. Abbas

Abstract In soil-pile-structure interaction non-linearity of soil plays very important role. The problem is furthermore complex when the piles are passing through liquefiable soil medium. In practice, for obtaining the internal response of soil-pile subjected to Seismic loading, many researchers and geotechnical engineers have used simple methods. In this paper a simple program based on Finite-element, i.e. "SAP2000" is propounded to examine the influence of seismic loading on deflection and bending moment of laterally loaded piles passing through liquefiable soil. In SAP2000, to simulate soil-pile interaction "Beam on Non-linear Winkler Foundation (BNWF)" model is used. In BNWF model, piles are modelled as a beam element and the surrounding soil is modelled as spring element. 3×3 pile groups in liquefiable and non-liquefiable soil are considered and a parametric study is carried out to analyse their seismic behaviour. The influence of different ratios of depth of pile to depth of liquefiable soil layer and effect of stiffness of the soil on the seismic behaviour of the soil-pile system are examined. Response spectrum and time histories are applied to analyse the response of pile during earthquake.

Keywords Pile foundation · Liquefaction · Seismic analysis

1 Introduction

Earthquakes are natural hazards that can cause damage to life and property. About 500,000 earthquakes are detected in the world each year, out of which 100,000 has felt, and 100 of them cause damage. The main casualties of earthquake are undoubtedly, man-made construction, for example, buildings, bridges, dams, etc., and the loss of life during earthquakes is rather a consequence of the structural collapses than a direct impact. As population grows and more buildings are constructed in seismically active zones, more people are likely to be affected by earthquakes. This observation is justified by the growing number of casualties and

M. Rehman (✉) · S. M. Abbas
Department of Civil Engineering, Jamia Millia Islamia, New Delhi, India

© Springer Nature Singapore Pte Ltd. 2020
S. Ahmed et al. (eds.), *Smart Cities—Opportunities and Challenges*,
Lecture Notes in Civil Engineering 58,
https://doi.org/10.1007/978-981-15-2545-2_45

financial losses in some recent earthquakes, for example, 27 Feb 2010 Chile earthquake, 12 Jan 2010 Haiti earthquake, 12 May 2008 Sichuan earthquake, and. 26 Dec 2004 Sumatra earthquake. Often, ground failures are the direct or indirect causes of the structural failures.

1.1 Literature Survey

Chatterjee and Choudhury [1] studied an analytical procedure to explore the effect of axial load on deflection and bending moment of a laterally loaded pile in liquefiable soil based on finite element technique, subjected to permanent ground displacement, and calculated maximum bending moment at the interface of liquefiable and non-liquefiable soil layers and also have calculated the max displacement and bending moment when thickness of liquefiable soil layer is around 60% of pile length. Shrimal and Maheshwari [2] analyze the design parameters (Deflection and BM) of a single pile (slenderness ratio = 22) in liquefiable soil under seismic loading as well as the effect of increasing PGA studied. For analysis, finite difference based software FLAC3D is employed. Bhuj earthquake (2001), scaled for different earthquake zones of India (viz, 0.10, 0.16 0.24 and 0.36 g as IS 1893:2002 part I) are used for seismic analysis of soli-pile system and response in terms of design forces were compared.

Janalizadeh and Zahmatkesh [3] studied a pseudo-static method of analysis to determine kinematic loads from lateral ground displacements and inertial loads from vibration of the superstructure of piles in liquefiable soil under seismic loading. Three-dimensional (3D) numerical modelling has been conducted by comparing the numerical results with the results of centrifuge tests; it has been found that the use of the p-y curves in liquefiable sand with multiple degradation factors provides acceptable results. Choudhury et al. [4] evaluated the bending moment and lateral deflection of piles at different depths using an analytical process based on finite element approach. Liam Finn [5] studied the various methods and observed that in seismic response analysis of a soil-pile interaction, many methods neglect one or more of the factors that affect the pile's seismic response (i.e. inertial interaction, kinematic interaction, seismic pore water pressures, soil nonlinearity, cross stiffness coupling, and dynamic pile to pile interaction). Phanikanth et al. [6] studied the effects of stiffness degradation for a set of earthquakes with different durations and different amplitudes of earthquakes. Seismic-deformation method is used to evaluate the effects of both inertial and kinematic interactions.

Maheshwari and Sarkar [7] developed a finite-element program to model three-dimensional soil-pile-structure systems in MATLAB. Dashpot and Spring elements were used to model the radiation boundary conditions. By a two-parameter volume change model, pore pressure generation was incorporated for liquefaction. They studied the influence of loading intensity and soil stiffness on seismic response of the soil–pile system.

Haldar and Sivakumar Babu [8] studied two potential pile failure mechanisms, buckling and bending of pile passing through liquefiable soil under seismic excitation. To get the ground displacements, soil stiffness and strength over the soil deposit depth, Liyanapathirana and Poulos [9] studied effective stress-based stress path model. Ultimate lateral soil pressure at the soil–pile structure interface and interaction coefficients were calculated using soil stiffness and strength due to build-up pore pressure.

Kiran et al. [10] a simple procedure to evaluate the lateral capacity of piles in liquefied soil are explained. The lateral capacity of pile is determined using spring model incorporating the effect of excess pore water pressure on soil stiffness. Finn and Fujita [11] studied different design cases and described by case histories, static response of pile displacement and pressure caused due to lateral spreading of ground and studied the dynamic response of pile during strong shaking.

Bhattacharya and Bolton [12] studied, due to rapid and excessive loading, Soil–pile interaction in liquefying soils is very complex phenomenon. The strength and stiffness of soil decrease significantly due to increase in excess pore water pressures due to non-linearity and degradation of shear modulus. Finn et al. [13] Seismic response analyses were conducted on pile groups in liquefiable soils that were subjected to earthquake excitation on the large. There is good agreement between computed and measured free field pore water pressures, pile cap accelerations, time histories of moments and the distributions of maximum moments with depth.

2 Soil–Pile Interaction in Liquefiable Soil

In practice, there are many methods used to model and design the lateral response of the pile foundation and one of the commonly used models is "Beam on Non-linear Winkler Foundation (BNWF)" model in Fig. 1.

In BNWF model, the pile is modelled as a beam element and the surrounding soil is modelled as spring elements. This model is preferred in practice due to its simplicity, mathematical convenience and ability to account for non-linearity. In a BNWF model, LSPI is represented by springs with nonlinear p-y curves, where p is lateral soil pressure/unit length of pile and y is lateral relative pile-soil displacement Fig. 1. The parameters required for a p-y curve definition have been used in practice for 30 years for normal soil conditions (i.e., non-liquefiable soil in this context). However, these parameters change for liquefiable soil because the soil changes its state from solid to fluid during liquefaction and is difficult to estimate.

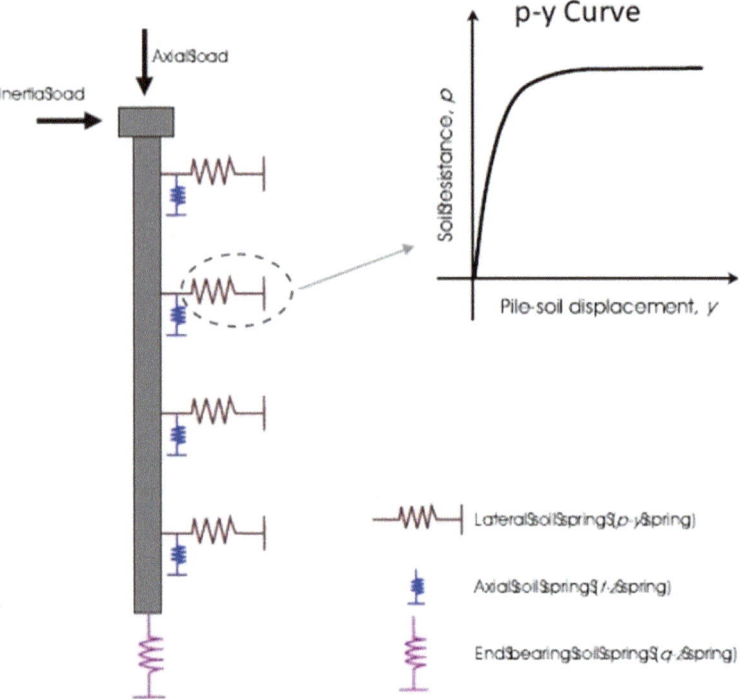

Fig. 1 BNWF model of soil-pile interaction (PSI)

3 Methodology

Soil behaviour is difficult to model, especially when the effects of liquefaction and seismic loading are taken into account. The piles are modelled using frame elements. The Pile Cap is considered as shell element. Material property is assumed to be homogeneous within the material. Dashpot and spring provided in Piles to modelling the soil. Different Stiffness properties were provided at different depth of piles. Response Spectrum IS: 1893:2001 and time history analysis is performed for Northridge Earthquake.

According to the objective of the dissertation, one 3×3 pile group considered in non-liquefiable soil and three pile groups of same size consider in liquefiable soil. Length of the pile is 10 m and the size of the pile cap is 3.8 m $\times$ 3.8 m. Spacing between piles is 1.5 m and side cover is 0.4 m. We take 3 different layers of liquefiable soil (i.e., L_p/L_{ls} = 0.2, 0.4, 0.6, 0.8) Pile and Pile Cap has been assigned the grade of concrete M40 with density equal to 2500 kg/m^3, Young's modulus equal to 3×10^{10} N/m^2 and Poisson's ratio equal to 0.2, and assigned the soil properties in terms of stiffness (spring and dashpot) are given below. The axial load

Table 1 Spring-dashpot properties of soil in lateral direction

Depth (m)	Stiffness of soil (KN/m)			
	Non-liquefiable soil	20% thickness of liquefiable soil	40% thickness of liquefiable soil	60% thickness of liquefiable soil
1	1210	6.1	6.1	6.1
2	1322	5.2	5.2	2
3	2024	2024	5.1	5.1
4	3800	3800	5.2	2
5	4312	4312	4312	6.1
6	6784	6784	6784	6.2
7	9810	9810	9810	10
8	12,100	12,100	12,100	12,100
9	16,700	16,700	16,700	16,700
10	19,200	19,200	19,200	19,200

on the pile cap from the column is expected to be 20,000 kN and Northridge time-history is applied on the pile. All these properties modelled in SAP2000 (Table 1).

3.1 Beam on Non-linear Winkler Foundation (BNWF) Model

The BNWF model for LSPI maintains the subgrade reaction idea but replaces the continuous soil reaction function by discrete point spring reactions. The finer the distribution of these discrete points, the better is the analysis result. By specifying the springs with properties that mimic the reaction behaviour observed in full-scale field tests a far more realistic depiction of LSPI in the analysis is obtained.

In seismic analysis, the BNWF model may either be pseudo-static or dynamic according to the requirements. For pseudo-static BNWF the dynamic seismic forces are computed and applied statically on soil–pile system. The basic differential equation of this model used by JRA for piles subjected to lateral forces is expressed as Eq. 1.

$$E_p I_p \frac{d^4 y}{dz^4} + k_s = F_1 \tag{1}$$

where $E_p I_p$—flexural stiffness of pile, k_s is spring constant of soil, and F_1 is force on pile in y-direction.

The seismic analysis could be improved potentially by using a dynamic BNWF model for soil-pile interaction (Fig. 2b). This model includes the inertia and damping behaviour of soil, which is ignored in the pseudo-static method. The

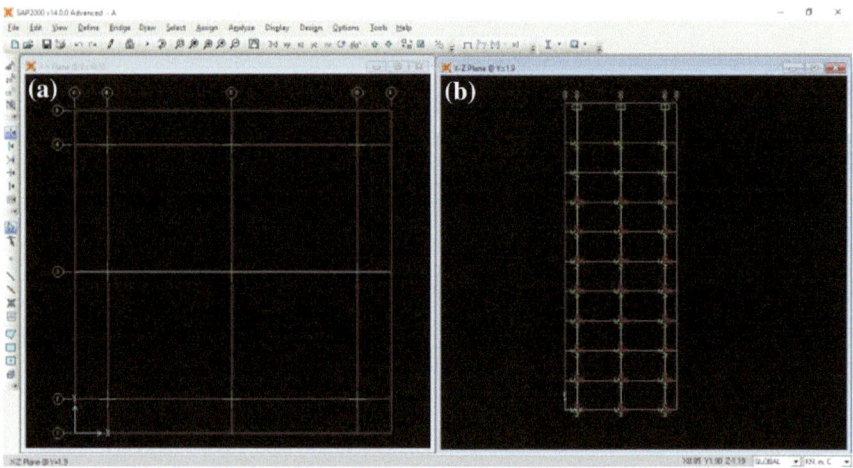

Fig. 2 **a** Plan view of the structure, **b** side view of structure

governing differential equation for dynamic BNWF model can be expressed as Eq. 2.

$$E_p I_P \frac{d^4 y}{dz^4} + m \frac{d^2 y}{dt^2} + c \frac{dy}{dt} + k_s y = F_1 \tag{2}$$

In BNWF model, the stiffness and the damping are two parameters that change significantly during liquefaction and can influence the pile response in liquefiable soil. Hence, their appropriate definition in the BNWF model is of paramount importance. Major discussion in this thesis is given to the stiffness parameter for a pseudo-static model for liquefied soil.

4 Model Description

See Table 2.

Table 2 Model description

Properties	Specification
Depth of pile cap	0.7 m
Size of pile cap	3.8 m × 3.8 m
Diameter of pile	0.5 m
Length of pile	10 m
Pile spacing	1.5 m
Side cove of pile cap	0.4 m
Density of concrete	2500 kg/m^3
Young's modulus of concrete	3×10^{10} N/m^2
Poison's ratio of concrete	0.2

5 Material Properties

The material used for pile modelling in SAP is RCC with M40 grade of concrete and Fe 500 grade of steel. The stress–strain relationship considered is in accordance with IS 456:2000.

T **basic properties of material used are**:

Modulus of elasticity of concrete, E_c = 31,622,776 kN/m^2
Characteristic Compressive strength of concrete, f_{ck} = 40,000 kN/m^2
Poisson's ratio, Concrete (μ_c) = 0.2
Yield stress for steel, f_y = 500,000 kN/m^2
Poisson's ratio, steel (μ_s) = 0.3
Damping of material = 0.05.

6 Seismic Property

See Table 3.

Table 3 Seismic property of site

Zone	IV
Zone factor	0.24
Importance factor	1.5
Soil type	II
Response reduction factor	2.5

7 Structure Modelling on SAP2000

See Fig. 2.

8 Results and Discussion

8.1 Displacement

Table 4 shows various results of Displacement in X-Direction due to Response Spectrum, Time History in X and Y Direction and Time History in X-Direction are given.

8.2 Displacement Graphs

For all results found for different models for displacement p-y curves are plotted for Response Spectrum and Time History between pile length and Displacement for Non-Liquefiable Soil, 20, 40 and 60% Liquefiable Depth of Soil.

8.2.1 Response Spectrum Curves

Figure 3 shows the curves between displacement and pile length for response spectrum for Non-Liquefiable Soil, 20, 40 and 60% Liquefiable Depth of Soil.

Table 4 Displacement due to response spectrum and time history

Soil layer	Displacement (mm)		
	Response spectrum	Time history in X and Y direction	Time history in X direction
Non liquefiable soil	8.9	10.11	12.56
20% liquefiable soil	61.07	71.067	83.076
40% liquefiable soil	94.78	101.78	118.03
60% liquefiable soil	110.207	122.06	131.04

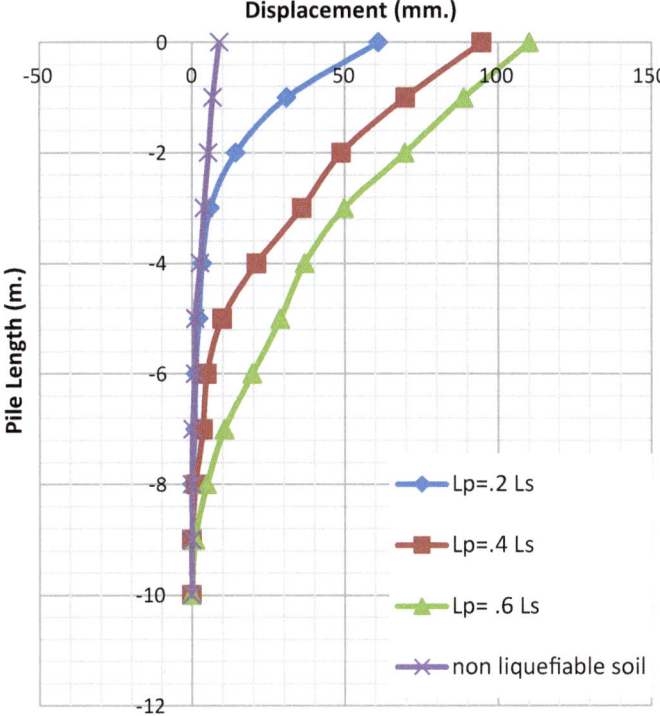

Fig. 3 Plot for response spectrum for different thickness of liquefiable soil layer

8.2.2 Time History in *X* and *Y* Direction

Figure 4 shows the curves between displacement and pile length for Time History in *X* and *Y* Direction for Non-Liquefiable Soil, 20, 40 and 60% Liquefiable Depth of Soil.

8.2.3 Time History in *X* Direction

Figure 5 shows the curves between displacement and pile length for Time History in *X* Direction for Non-Liquefiable Soil, 20, 40 and 60% Liquefiable Depth of Soil.

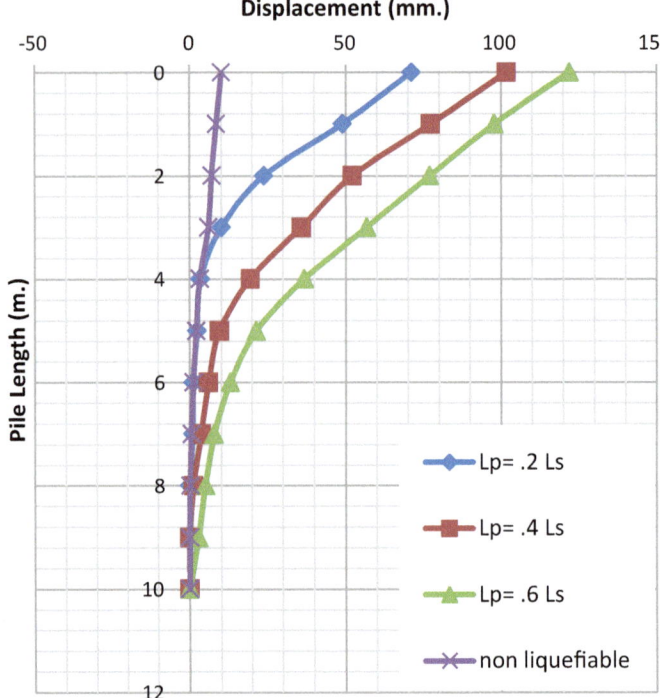

Fig. 4 Plot for time history in *X* and *Y*-Direction for different thickness of liquefiable soil layer

8.3 Bending Moments

Various results of Bending Moment in *X*-Direction due to Response Spectrum, Time History in *X* and *Y* Direction and Time History in *X*-Direction are given Table 5.

8.4 Bending Moment Graphs

For all results found for different models for Bending Moment are plotted for Response Spectrum and Time History between pile length and Bending Moment for Non-Liquefiable Soil, 20, 40 and 60% Liquefiable Depth of Soil.

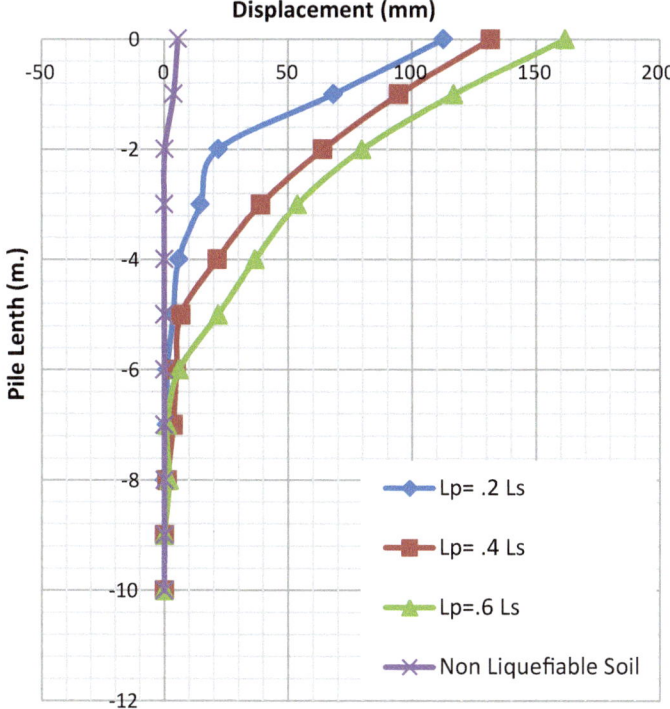

Fig. 5 Plot for time history in *X* and *Y*-Direction for different thickness of liquefiable soil layer

Table 5 Bending moment due to response spectrum and time history

Soil layer	Bending moment (kN m)		
	Response spectrum	Time history in *X* and *Y* direction	Time history in *X* Direction
Non liquefiable soil	68.2304	150.5655	158.1384
20% liquefiable soil	351.915	704.2574	768.4581
40% liquefiable soil	560.3535	1084.85	1107.464
60% liquefiable soil	772.4623	1189.476	1347.815

8.4.1 Response Spectrum Curves

Figure 6 shows the curves between Bending Moment and pile length for response spectrum for Non-Liquefiable Soil, 20, 40 and 60% Liquefiable Depth of Soil.

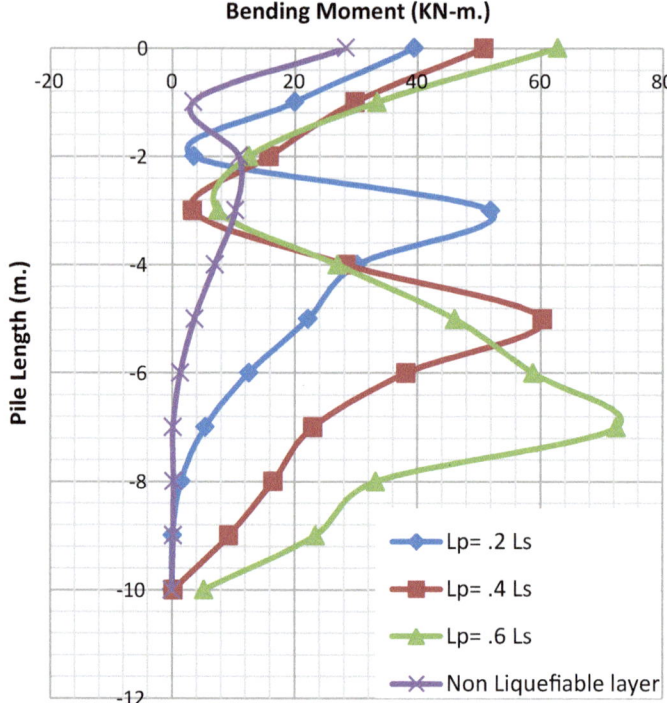

Fig. 6 Plot for response spectrum for different thickness of liquefiable soil layer

8.4.2 Time History in *X* and *Y*-Direction

Figure 7 shows the curves between Bending Moment and pile length for Time History in *X* and *Y*-Direction for Non-Liquefiable Soil, 20, 40 and 60% Liquefiable Depth of Soil.

8.4.3 Time History in *X*-Direction

Figure 8 shows the curves between Bending Moment and pile length for Time History in *X*-Direction for Non-Liquefiable Soil, 20, 40 and 60% Liquefiable Depth of Soil.

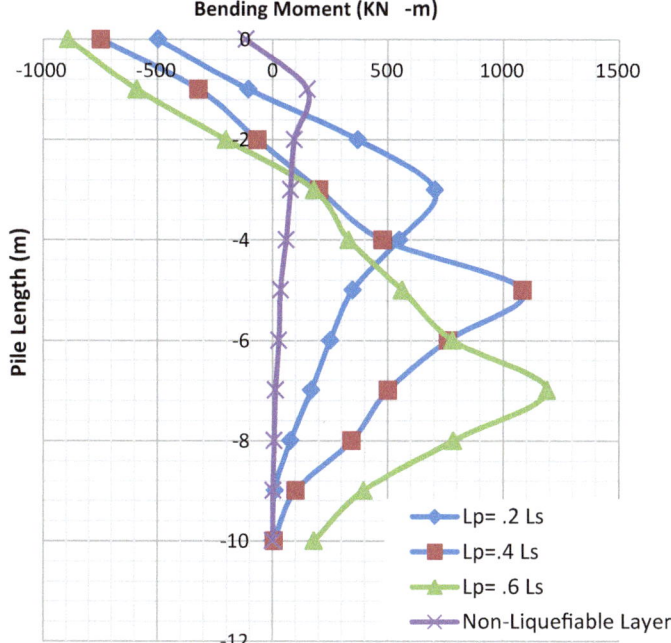

Fig. 7 Plot for time history in *X* and *Y*-Direction for different thickness of liquefiable soil layer

9 Conclusion

Ultimately, it can be seen that the effect of liquefiable soil is a great concern and it affects the interacting parameters as compared to non-liquefiable soils. The analytical results indicate that the results obtained due to liquefiable soil on piles are quite different than that of non-liquefiable soil. The present study provides a brief review of various design approaches to understand the seismic response of piles under liquefiable and non-liquefiable soil.

1. Bending Moment increases about 5 times (for response Spectrum 5.15 times, for time History in *X* and *Y* Direction 4.68 times and for Time History in *X* Direction is 4.85 times) when 20% Layer of Liquefiable soil is considered.
2. Bending Moment increases about 7.5 times (for response Spectrum 8.21 times, for time History in *X* and *Y* Direction 7.2 times and for Time History in *X* Direction is 7 times) when 40% Layer of Liquefiable soil is considered.
3. Bending Moment increases about 9 times (for response Spectrum 11.32 times, for time History in *X* and *Y* Direction 7.9 times and for Time History in *X* Direction is 8.5 times) when 60% Layer of Liquefiable soil is considered.
4. Maximum bending moments on pile occur at the interface of the liquefiable and non-liquefiable soil layers.

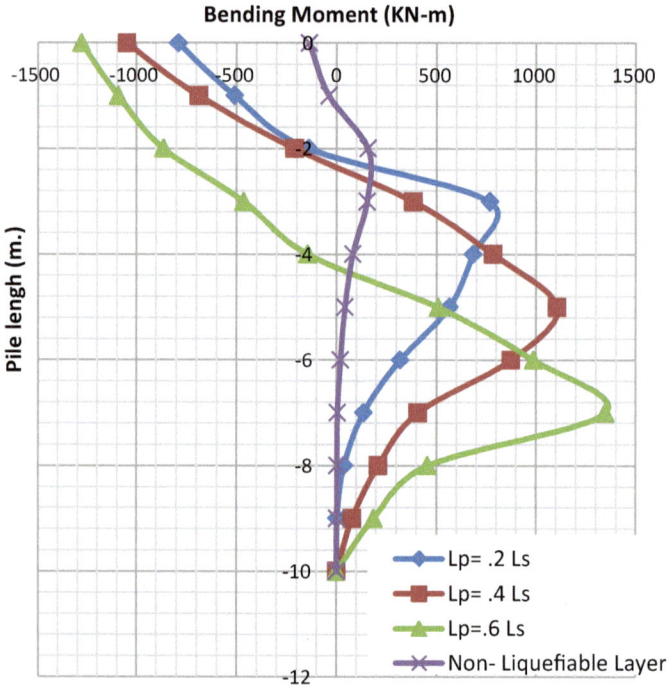

Fig. 8 Plot for time history in *X*-Direction for different thickness of liquefiable soil layer

5. Maximum bending moments on pile have observed when the thickness of the liquefiable soil layer is approximately 60% of pile length. More than this, liquefiable depth of soil, soil fails before pile failure.

6. Displacement increases about 7 times (for response Spectrum 6.86 times, for time History in *X* and *Y* Direction 7.02 times and for Time History in *X* Direction is 6.61 times) when 20% Layer of Liquefiable soil is considered.

7. Displacement increases about 10 times (for response Spectrum 10.65 times, for time History in *X* and *Y* Direction 10.06 times and for Time History in *X* Direction is 9.39 times) when 40% Layer of Liquefiable soil is considered.

8. Displacement increases about 12 times (for response Spectrum 12.38 times, for time History in *X* and *Y* Direction 12.07 times and for Time History in *X* Direction is 10.43 times) when 60% Layer of Liquefiable soil is considered.

10 Recommendation for Further Research

Outcomes from the present research are based on software, more experimental tests are required to verify these results in field condition, it is very difficult to model soil.

1. In this present study, the major focus was to analyse the piles during seismic loading although the soil pile interaction may also influence the results of piles due to axial as well as soil bearing capacity. A very little work on liquefiable soil has been done so far, there is a need to further study about the behaviour of piles in liquefiable soil for lateral, bearing, dynamic as well as axial loading.
2. One can develop p-y curve for different types of soils for different depths of liquefiable soil and to plot curve for different types of soils.
3. For more precise results one can work on PLAXIS3D.

References

1. Chatterjee K, Choudhury D (2017) Influence of seismic motions on behavior of piles in liquefied soils. Int J Numer Anal Methods Geomech 1–26 (2017)
2. Shrimal D, Maheshwari BK (2016) Effect of liquefaction on the response of a single pile under seismic loading. In: Indian geotechnical conference IGC2016, IIT Madras, Chennai, India
3. Janalizadeh A, Zahmatkesh A (2015) Lateral response of pile foundations in liquefiable soils. J Rock Mech Geotech Eng 7:532–539 (2015)
4. Choudhury A, Kumari A, Chatterjeeii K (2014) Coupled behaviour of pile foundations in liquefied and non-liquefied soils during earthquakes including case study. In: International workshop on geotechnics for resilient infrastructure the second Japan-India workshop in geotechnical engineering
5. Liam Finn WD (2014) An overview of the behavior of pile foundations in liquefiable and non-liquefiable soils during earthquake excitation. Soil Dyn Earthq Eng 68(2015):69–77
6. Phanikanth VS, Choudhury D, Reddy GR (2013) Behavior of single pile in liquefied deposits during earthquakes. Int J Geomech (2013)
7. Maheshwari BK, Sarkar R (2011) Seismic behavior of soil-pile-structure interaction in liquefiable soils, parametric study. Int J Geomech 335 (2011)
8. Haldar S, Sivakumar Babu GL (2010) Failure mechanisms of pile foundations in liquefiable soil: parametric study. Int J Geomech (2010)
9. Liyanapathirana DS, Poulos HG (2009) Analysis of pile behaviour in liquefying sloping ground. Comput Geotech 37(2010):115–124
10. Kiran AS, Ramasamy G, Maheshwari BK (2009) Lateral capacity of piles in liquefiable soils. IGC, Guntur, INDIA
11. Finn WDL, Fujita N (2004) Behaviour of piles in liquefable soils during earthquakes: analysis and design issues. In: Fifth international conference on case histories in geotechnical engineering
12. Bhattacharya S, Bolton M (2004) Buckling of piles during earthquake liquefaction. In: 13th world conference on earthquake engineering Vancouver, B.C., Canada 1–6 Aug, pp 95
13. Finn WDL, Thavaraj T, Wilson DW, Boulanger RW, Kutter BL (2007) Seismic response analysis of pile foundations at liquefiable sites. J Civ Eng Environ Technol

Effect of Tunnel Construction on the Settlement of Existing Pile-Supported Superstructure

Rohan Deshmukh and Pravin Patil

Abstract The need for subway transportation systems has been increased in many important cities. The most tunnels are constructed in congested city areas hence the risk of damage to surrounding infrastructure is very high. Any subsurface construction will generate ground vibrations that resulting in ground settlements and lateral movements, which have the potential to cause damage to the existing surface and underground structures. The subway development in computing technology has led to the use of the finite element method (FEM) in the analysis of the surface subsidence has been increased in recent years. In the current study, a 2D analysis of the varying depth of tunnel and effect of existing pile-supported structure on the settlement of ground surface is carried out by using PLAXIS finite element software. The result of the varying depth of the tunnel, existing pile-supported structure above the tunnel and the used of grout anchor on the ground surface settlement shown in terms of the Gaussian distribution curve. Results show that when pile lengths are sufficient to bear superstructure load then stresses are less on tunnel lining so it resulting in a low ground settlement and vice versa.

Keywords Tunnel · Finite element analysis (FEM) · Existing structure · Ground surface settlement

1 Introduction

Worldwide, last nineteenth century the number of people living in urban areas increased by day by day. The lack of available space and high cost in urban space has significantly increased the demand for tunnels in big urban cities. Planning, designing and infrastructure development in urban areas is the basic need for growing nations like India. Garner et al. [1] reported that town planners are now

R. Deshmukh (✉)
Applied Mechanics Department, SVNIT, Surat, Gujarat 395007, India

P. Patil
Department of Civil Engineering, MCOERC, Nashik 422105, India

© Springer Nature Singapore Pte Ltd. 2020
S. Ahmed et al. (eds.), *Smart Cities—Opportunities and Challenges*,
Lecture Notes in Civil Engineering 58,
https://doi.org/10.1007/978-981-15-2545-2_46

555

focusing on underground space for additional infrastructure. Therefore, it necessitates adopting and developing innovative and new technologies for faster growth of construction. India has a long coastal region having complex deep-seated soft clay extended to a depth of 10–30 m. Today, there are more than subways tunneling operating for urban utilities. The surface settlement during the tunneling and optimum tunnel depth could be based on the mathematical, experimental or numerical methods. The present work was aimed to evaluate the location of tunnel depth and effect of the existing structure on the tunnel by observing the deformation of soil mass while tunneling is underway and its control. Goel [2] stated that the rapid growth of population and cities, unavailability and extremely high price of land are the few main reasons, which force town planning authorities to focus on deep underground to accommodate the basic additional infrastructure for the growing population. Tunnelling can be effectively used for public transport, water supply and sewerage disposal. Control over air and noise pollution in cities is possible by using tunnelling for transport, which gives an extra advantage.

1.1 Surface Settlement

The most common empirical method is Gaussian curve to predict ground movements. Peck [7] and Schmidt [9] were the founders to use Gaussian curve (Fig. 1) to show the transverse settlement trough, after the construction of a tunnel and in many cases, it is well explained by Gaussian curve. The ground Loss (GL) and the standard deviation 'i' of the Gaussian curve are the two parameters that need to match with the surface settlement.

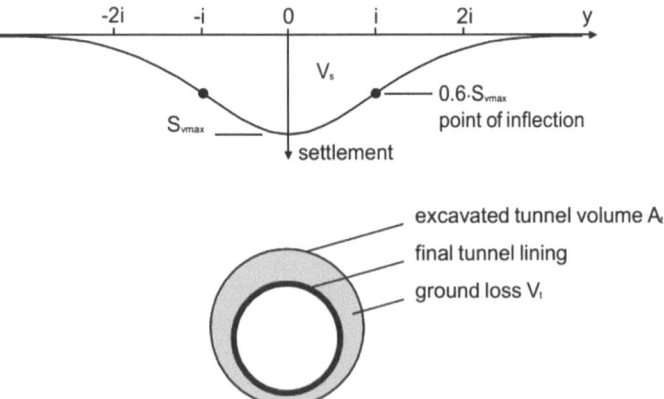

Fig. 1 The Gaussian curve for transverse surface settlement

Cording and Hansmire defined the ground loss as the volume of soil that is displaced across the perimeter of a tunnel. It is often defined in terms of volume lost per unit length of tunnel constructed. The percentage ground loss (GL) is defined as follows:

$$GL = \frac{V_i}{V_o} * 100 \tag{1}$$

where V_i = Trough volume, V_o = Tunnel opening volume ($\pi\ r^2$) and r = Radius of the tunnel. Based on the shape of the Normal distribution curve, Peck [7] showed that the maximum settlement Sv, max can be given by,

$$Sv,max = \frac{0.314 \cdot GL \cdot D^2}{i} \tag{2}$$

where 'D' = diameter of the tunnel and Sv, max = maximum settlement occurring above the tunnel axis. The settlement at various points of the trough is then given by,

$$Sv\ (y) = Sv,max \cdot \exp\left(\frac{-y^2}{2i^2}\right) \tag{3}$$

where 'y' = horizontal distance from the tunnel axis and 'i' = horizontal distance from the tunnel axis to the point of inflection. Peck [7] suggested that the 1–2% of ground loss (GL) is usually found for stiff clay, 2–5% for soft clay and less than 1% for sandy soil. Mair [3] also suggested that Gaussian distribution could be helpful to approximate subsurface settlement profiles.

1.2 The Width of the Settlement

The distance from the tunnel axis to the inflection point 'i', and determining the width of the settlement trough has been the topic of many investigations. Peck [7] developed a relationship between tunnel diameter 'D' and tunnel depth 'Z_o', depending on ground conditions. After Peck [7], many other researchers have come up with similar relationships. O'Reilly and New [6] developed strong correlation between 'i' and tunnel depth from multiple linear regression analysis performed on collected field data. Significant correlation was not observed between 'i' and tunnel diameter or method of construction (except for very shallow tunnels with a ratio of overburden to a diameter less than one). They proposed the following Eq. 4, which is a simplified version of regression lines and useful for many practical purposes,

$$i = K \cdot Z_o \tag{4}$$

where K is a trough width parameter, with $K \approx 0.5$ for clayey ground and $K \approx 0.25$ for sandy soil. The approach of Eq. 4 has also been confirmed by Rankine [8], who presented a variety of tunnel case histories in different types of soils. Mair and Taylor [4] give a large number of tunnelling field data with different linear regressions for tunnels in different types of soil.

2 Finite Element Analysis Using Plaxis-2D

A tunnel has been modeled to demonstrate the effect of tunneling and face pressure on the surface settlement and ground deformation. The water table was also considered in modeled at a 2 m deep from the ground surface. The depth of the tunnel is varying from 10 to 25 m and the diameter of the tunnel is kept constant i.e. 6.05 m (Fig. 2). The 2D model used in the analysis is shown in Fig. 2; model is 60 m wide and 50 m in height. In Table 1, one can observe the various soil material properties used for the finite element analysis using the Mohr–Coulomb material model. Maleki et al. [5] reported that the interface value must be less than 1 for actual soil-structure interaction, it indicates that the interface zone is weaker compared to associated soil layer. Typical range for interface value is set to be between 0.7 and 1.

In Tables 2 and 3, one can see the tunnel lining, beam and anchor material properties that were used in the analysis. The normal stiffness (EA), flexural rigidity (EI), and weight 'w' were chosen based on the properties of the materials of the tunnel lining used. It should be noted that EA and EI are related to the stiffness per unit width and 'w' is the specific weight in units of force per unit area.

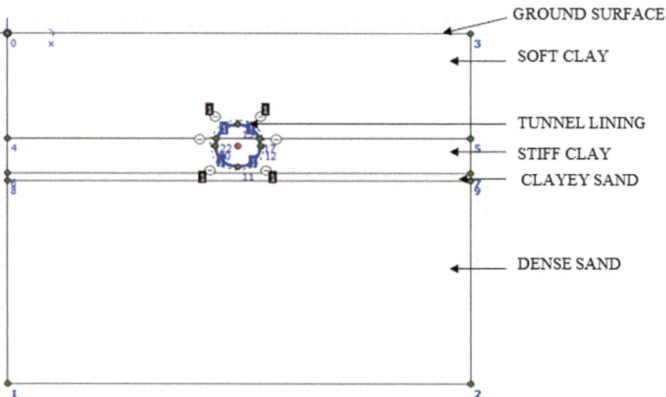

Fig. 2 Installation of soil layers and tunnel

Table 1 Soil material properties

Parameter	Unit	Soft clay	Stiff clay	Clayey sand	Dense sand
Material model Type of material behavior	–	Mohr–Coulomb Drained	Mohr–Coulomb Drained	Mohr–Coulomb Drained	Mohr–Coulomb Drained
Dry unit weight γ_{unsat}	(kN/m^3)	16.00	18.20	19.00	19.50
Saturated unit weight γ_{sat}	(kN/m^3)	18.00	19.50	21.00	21.50
Permeability k_x, k_y	(m/day)	0.000	0.001	0.300	1.000
Young's modulus E_{ref}	(kN/m^2)	6250	47,900	72,000	72,000
Porosity N	(–)	0.330	0.330	0.300	0.300
Shear modulus G_{ref}	(kN/m^2)	2349.6	18,007.5	27,692.3	27,692.3
Cohesion c_{ref}	(kN/m^2)	5.00	5.00	2.00	1.00
Friction angle ϕ	(°)	23.00	26.00	36.00	36.00
Dilatancy angle ψ	(°)	0.00	0.00	0.00	1.00
Reduction factor $R_{inter.}$	(–)	0.70	0.70	1.00	1.00
Interface	–	Neutral	Neutral	Neutral	Neutral

Table 2 Tunnel lining and beam data sets parameters

Parameter	Name	Building raft	Pile toe	Tunnel lining
Type of behavior	Material type	Elastic	Elastic	Elastic
Normal stiffness (kN/m)	EA	1E10	2E6	1.4E7
Flexural rigidity (kN/m^2/m)	EI	1E10	8000	1.43E5
Equivalent thickness (m)	d	3.464	0.219	0.350
Unit weight (kN/m^3)	w	25.00	2.00	8.40
Poisson's ratio	v	0.00	0.20	0.15

Table 3 Anchor material properties

Parameter	Name	Pile	Anchor rod	Grout body
Type of behavior	Material type	Elastic	Elastic	Elastic
Normal stiffness (kN/m)	EA	2,000,000	200,000	100,000
Spacing (m)	L spacing	5.00	2.50	2.50

2.1 Step by Step Analysis

1. In this first step first left and bottom geometry dimensions are given in General setting followed by assigning a proper name to the model file.
2. Preparation of model was conducted by using work plane, the diameter of the tunnel, depth of tunnel, length of varying pile and existing loading.
3. In this first step, properties assigned to the soil and tunnel lining.
4. In this step, the effect of existing structural loading has considered for the analysis.
5. In this step diameter, normal stiffness, flexural rigidity has assigned to the pile.
6. In this step, properties assign for anchor and grout body.
7. After assigning properties the mesh has been generated.
8. In this step generation of phases 1–4 (stage construction).
9. In this first steps the activation of gravity loading and defining the members of the existing structure.
10. In this step activation of tunnel lining and remove soil from tunnel and cluster dry the tunnel.
11. In this first step activation of anchor and grout body.
12. In this step application of tunnel contraction parameter for ground loss coefficient.
13. Selection of point of stress or node point selection.
14. Calculations progress for all three phases.
15. After successful completion of calculation (green tick).
16. The output from calculation the deformed mesh and displacement value.
17. The output from calculation the direction of principal stresses.

3 Results and Discussion

3.1 Effect of Tunnel Depth

For depth of tunnel location varying from 10 to 25 m and corresponding values of vertical settlement (S), were observed to determine the settlement profile for each case. The below-shown graph is the Gaussian distribution curve for tunnel location from 10 to 25 m. The vertical settlement reduces from the depth of 16–25 m for different soil stratification. For 10 m depth of tunnel surface settlement is 15.95 mm but this depth is closed to the ground surface. After applying the existing load, this value will be higher, therefore for tunnel location depth, the optimized value should be selected from 16 to 25 m. At 20 m depth of the tunnel center, it can be seen that there is an optimum reduction in the vertical settlement. After this depth at every level, the vertical settlement is very less, so the 20 m is considered as the optimum depth (Fig. 3).

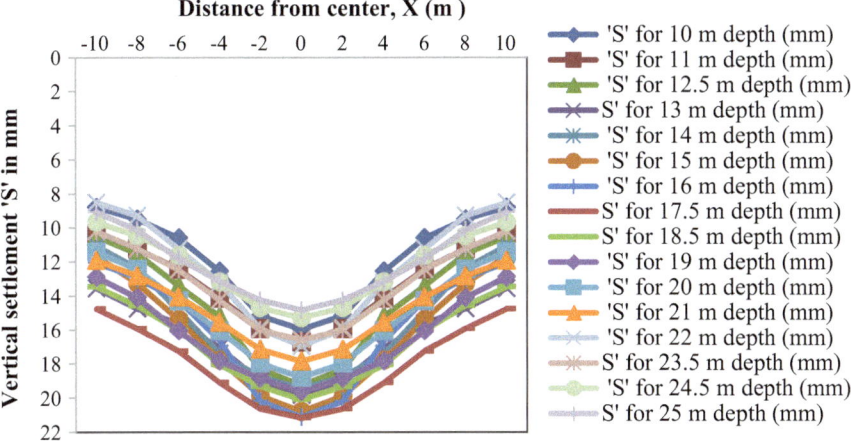

Fig. 3 Ground surface settlement for depth of tunnel location from 10 to 25 m

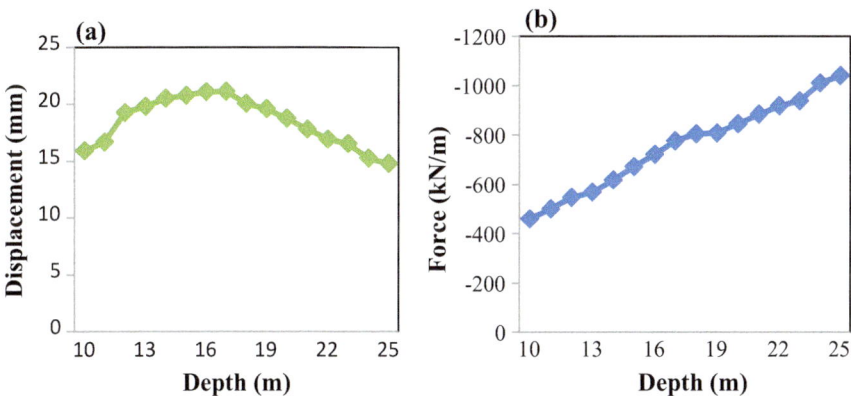

Fig. 4 Result for **a** extreme total displacement and **b** axial force for various depth of the tunnel

Figure 4a shows that extreme total displacement for depth of tunnel location from 10 to 25 m. For 10 m depth of tunnel surface settlement is 15.95 mm but this depth is closed to the ground surface. After applying the existing load this value will be higher, therefore for tunnel location depth, the optimized value should be selected from 16 to 25 m. Figures 4b and 5a, b shows that axial force, shear force and bending moment on tunnel increases due to increasing depth of tunnel from 10 to 25 m. This because the overburden of soil as well as the above loose clayey sand increases, thus relieving tunnel of the stresses (Fig. 6).

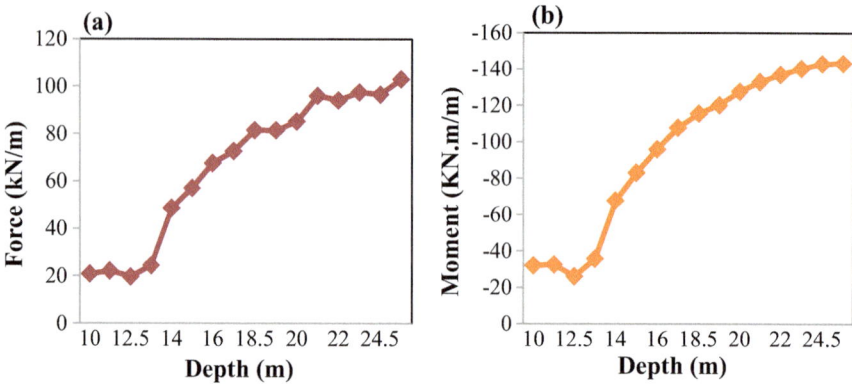

Fig. 5 Result for **a** shear force and **b** bending moment for depth of tunnel location from 10 to 25 m

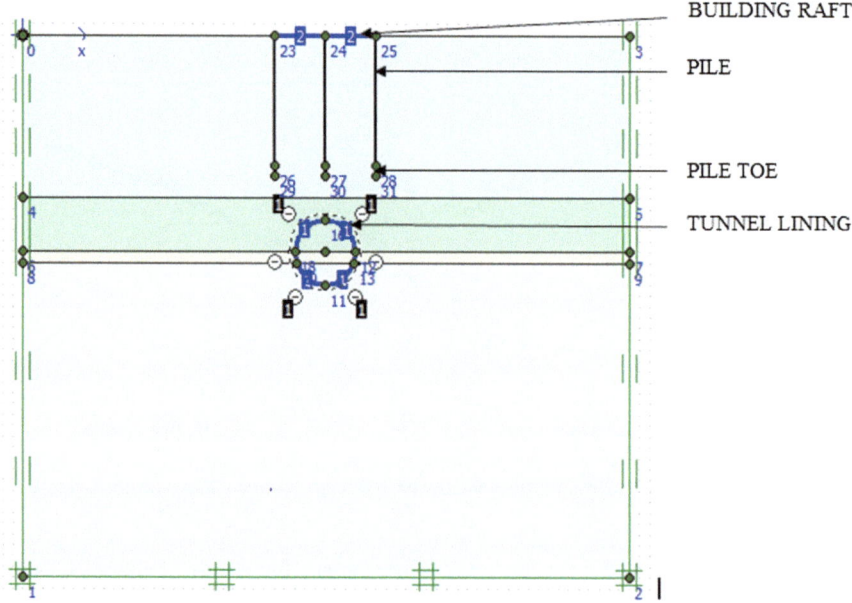

Fig. 6 Model for pile length varying from 5 to 15 m

3.2 Effect of Varying Pile Length

For varying pile length of the tunnel, corresponding to different values of vertical settlement (S) was applied at the ground surface in order to observe the settlement profile for each case. In the same soil stratification model for pile length varying

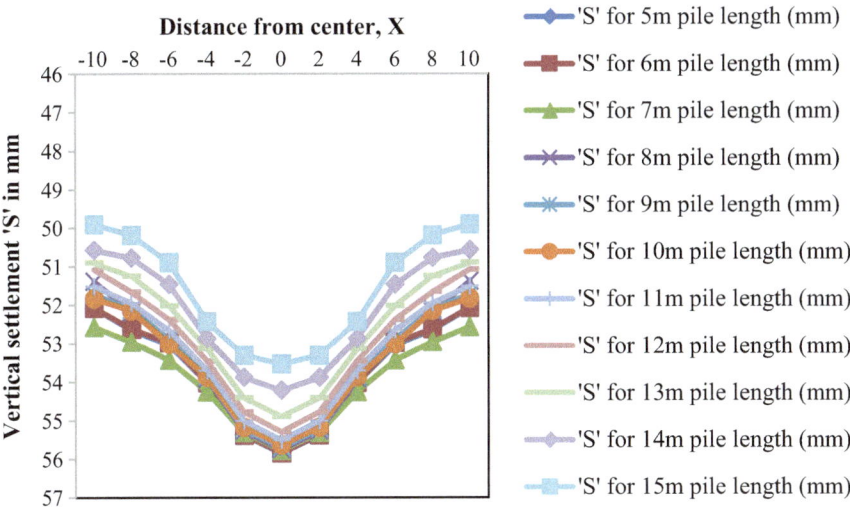

Fig. 7 Ground surface settlement for pile length varying from 5 to 15 m

from 5 to 15 m at pile spacing, 5 m is created. The center of the tunnel is located at 20 m from the ground surface. The material properties for building raft pile and pile toe are used from the above table. The values of the vertical settlement go on reducing order as the depth of the pile gradually increases from 5 m to up to 15 m (Fig. 7).

It is seen that as the length of the pile increases the vertical displacement of the ground surface due to tunnel construction is reducing in nature as seen from the above graph. From 13 m it can be seen that there is a rapid reduction in vertical displacement of the surface, as there is a presence of clay layer from 15 m depth. So as the further increase in depth of piles, the skin friction on pile increases when it is embedded near the clay or in the clay layer. Hence the vertical settlement of surface reduces.

Figure 8a shows that extreme total displacement reduces due to increasing pile length from 5 to 15 m for line spacing between piles is 5 m. Figures 8b, 9a, b) shows that Axial force, Shear Force and Bending moment on tunnel reduces due to increasing pile length from 5 to 15 m for line spacing between pile is 5 m. This because the overburden of structure as well as the above loose clayey sand reduces because of the use of piles, thus relieving tunnel of the stresses.

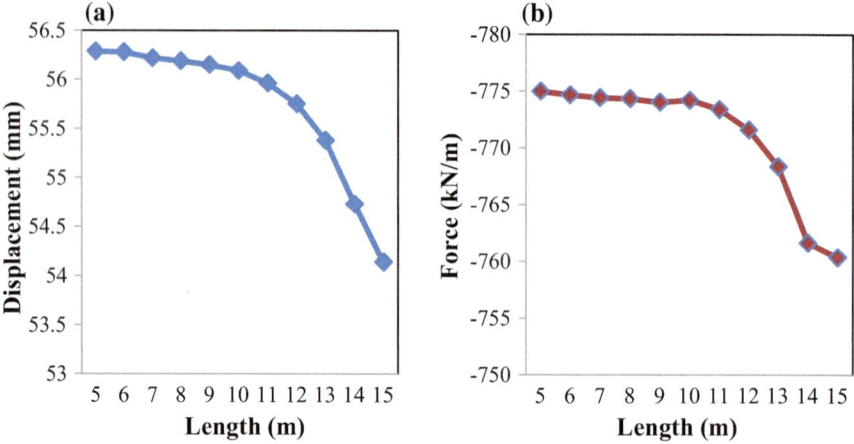

Fig. 8 Result for **a** extreme total displacement and **b** axial force for the various length of the pile

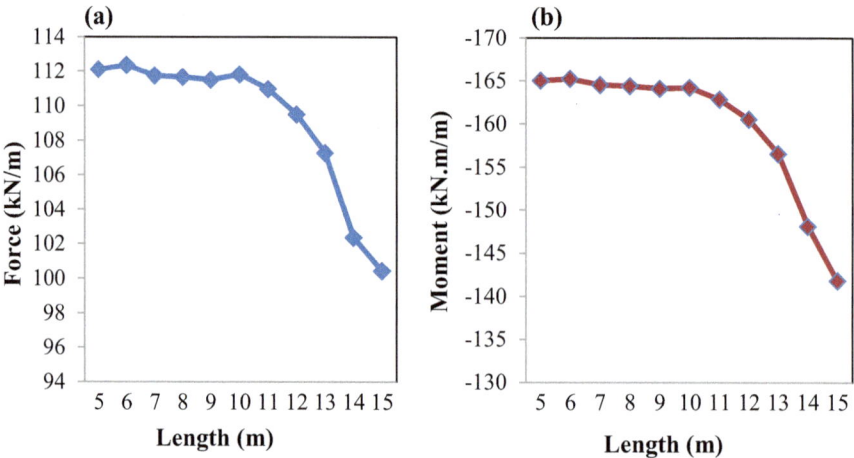

Fig. 9 Result for **a** shear force and **b** bending moment for various length of the pile

3.3 Effect of Existing Building Load

For varying existing building load on the tunnel, corresponding to different values of vertical settlement (S) was observed to determine the settlement profile for each case (Fig. 10).

In the same soil stratification model, an existing building load varying from 50 to 250 kN/m^2 was considered and settlement analysis was carried out. Extreme total deformations increase due to an increase in an existing building load.

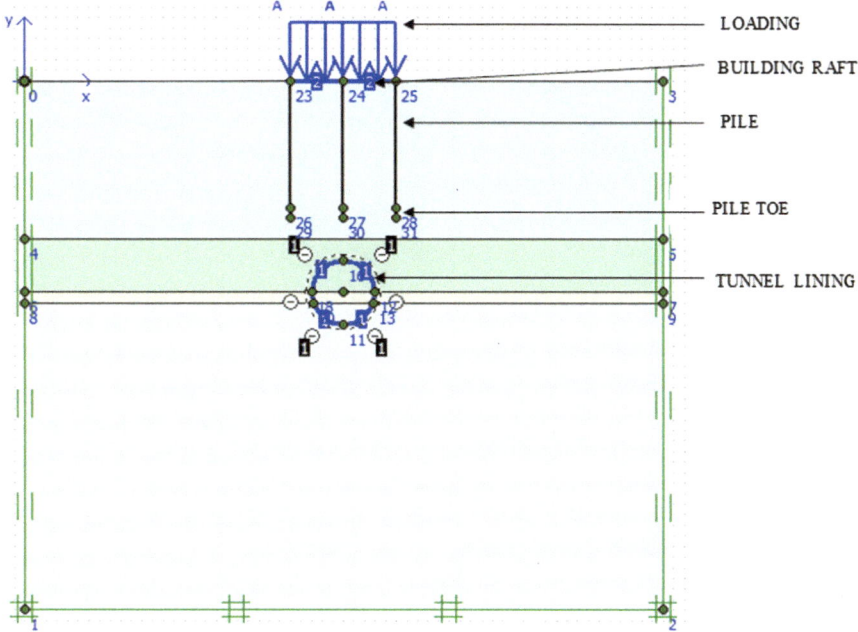

Fig. 10 Model for existing building load from 50 to 250 kN/m²

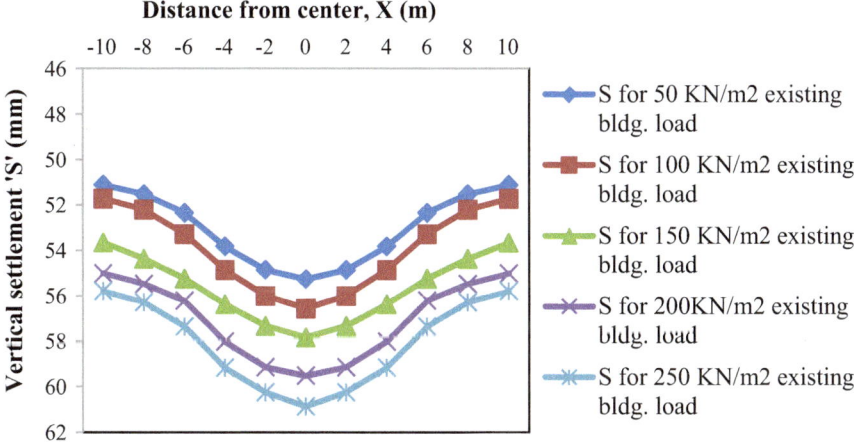

Fig. 11 Ground surface settlement for existing building (bldg.) load from 50 to 250 kN/m²

The settlement analysis found in the case where very small extreme total deformations observed at the surface. The material properties for building raft, pile and pile toe are used from the above table (Fig. 11).

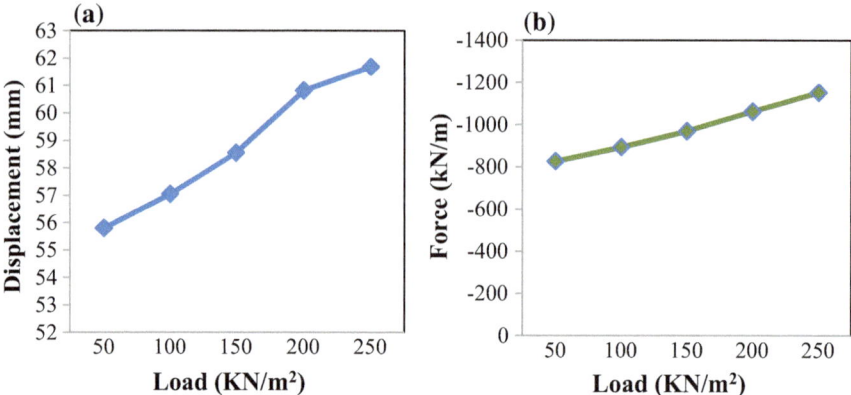

Fig. 12 Result for **a** extreme total displacement and **b** axial force for existing building load

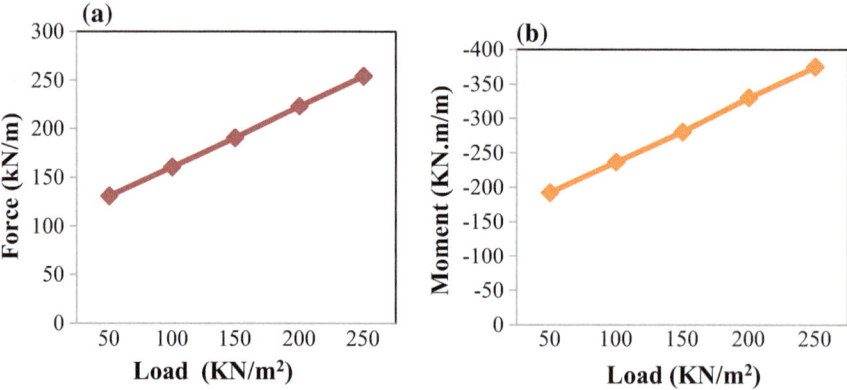

Fig. 13 Result for **a** shear force and **b** bending moment for existing building load

From the above result for existing building load on tunnel shows that vertical settlement increased due to an increase in building load and in the previous case the vertical settlement decreased for different cases. Due to this extreme total displacement on the ground surface, axial force, shear force, and bending moment of the tunnel also increased due to increase in existing building load which is shown in Figs. 12, 13, 14 and 15.

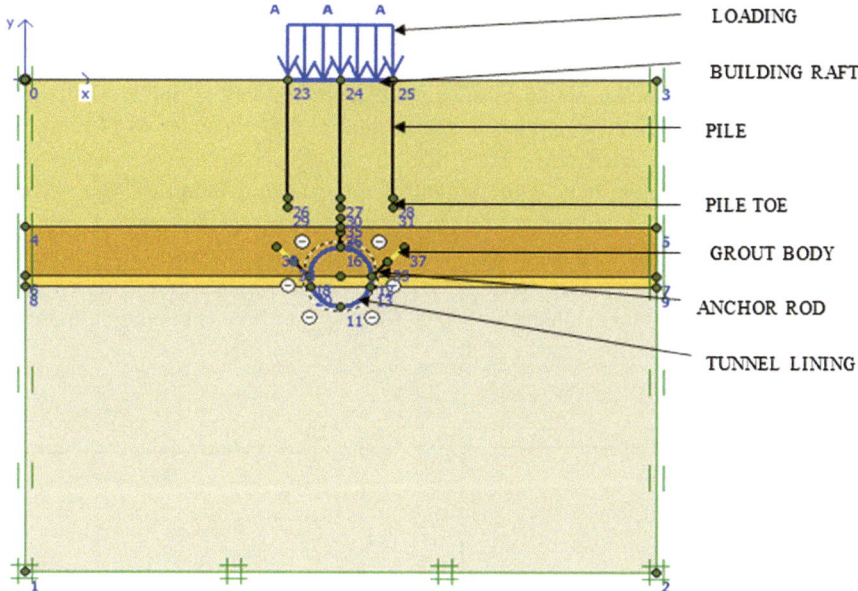

Fig. 14 Model for grout anchor length 1–3 m

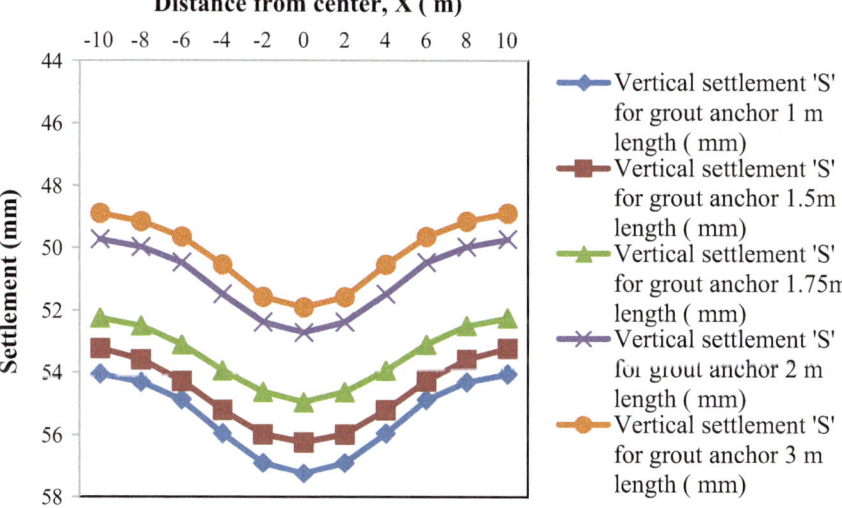

Fig. 15 Ground surface settlement for Grout anchor length from 1 to 3 m

3.4 Effect of Grout Anchor Length

For varying Grout anchor length of the tunnel, corresponding to different values of vertical settlement (S) was applied at the ground surface in order to observe the settlement profile for each case.

Figure 16a it is seen that as the length of grout anchor increases, the vertical displacement of the ground surface due to tunnel construction decreases. From 1 m it can be seen that there is a rapid reduction in vertical displacement of the surface, as there is a presence of clay layer from 15 m depth. Figures 16a, b and 17a, b shows that extreme total displacement, axial force, shear force and bending moment

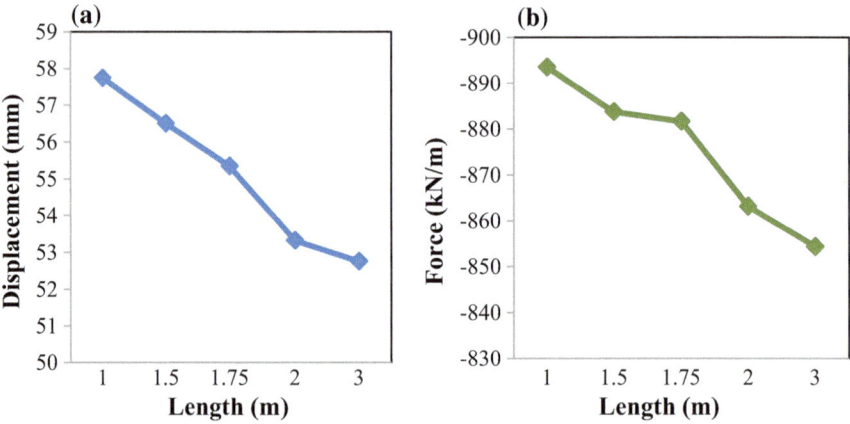

Fig. 16 Result for **a** extreme total displacement and **b** axial force for Grout anchor length from 1 to 3 m

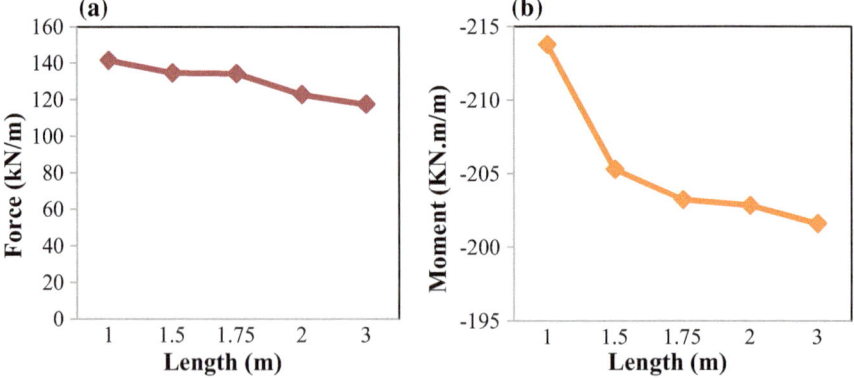

Fig. 17 Result for **a** shear force and **b** bending moment for Grout anchor length from 1 to 3 m

on tunnel reduces due to increasing anchor length from 1 to 3 m for line spacing between pile is 2.5 m. This because the overburden of structure as well as the above loose clayey sand reduces because of the use of grout anchor, thus relieving tunnel of the stresses.

4 Conclusions

(1) The behavior of tunnel depth from 10 to 25 m in given soil conditions is studied with the help of Plaxis 2D software to generate a graph of vertical settlement versus depth of the tunnel. Ground surface settlement showed by Gaussian distribution curve.

(2) Due to an increase in depth of tunnel overburden pressure increases and simultaneously values of axial force, bending moment and shear force on tunnel lining also increase.

(3) As the depth of the center of tunnel increases from 10 to 25 m the surface settlement reduced as per the different soil conditions, it means the influence of tunnel location on ground settlement decreases as the depth of tunnel increases.

(4) Due to the increase in pile length varying from 5 to 15 m, the values of extreme total displacement on the ground surface, axial force, bending moment, and shear force on tunnel also decreases.

(5) Due to increase in existing building load varying from 50 to 250 kN/m^2 the values of extreme total displacement on the ground surface, axial force, bending moment and shear force on tunnel also increases.

(6) As the number of grout anchors increases the values of extreme total displacement on the ground surface, axial force, bending moment and shear force on tunnel also decrease.

(7) As the grout anchor length increases from 1 to 3 m the values of extreme total displacement on the ground surface, axial force, bending moment and shear force on tunnel also decrease.

References

1. Garner CD, Coffman RA (2012) Subway tunnel design using a ground surface settlement profile to characterize an acceptable configuration. Tunn Undergr Space Technol
2. Goel RK (2001) Status of tunnelling and underground construction activities and technologies in India. Tunn Undergr Space Technol
3. Mair RJ, Taylor RN (1997) Theme lecture: bored tunneling in the urban environment. In: Proceedings of the 14th international conference on soil mechanics and foundation engineering, Hamburg, vol 4, pp 2353–2385
4. Mair RJ (1993) Unwin memorial lecture-developments in geotechnical engineering research: application to tunnels and deep excavations. In: Proceedings of the institution of civil engineers. Civ Eng 97(1):27–41

5. Maleki M, Sereshteh H, Mousiv M, Bayat M (2011) An equivalent beam model for the analysis of tunnel-building interaction. Tunn Undergr Space Technol
6. O'Reilly MP, New BM (1982) A settlement above tunnels in the United Kingdom, their magnitude and prediction. Tunn Inst Min Metall 173–181
7. Peck RB (1969) Deep excavations and tunneling in soft ground-state-of-the-art report. In: Proceedings of the seventh international conference on soil mechanics and foundation engineering, Mexico City, pp 225–290
8. Rankine WJ (1988) Ground movements resulting from urban Tunneling. In: Proceedings conferences engineering. Geotechnical underground movements, Nottingham, London Geological Society, pp 79–92
9. Schmidt B (1969) Settlements and ground movements associated with tunneling in soil Ph.D. thesis, University of Illinois

Analysis of Logistical Barriers Faced by MNCs for Business in Indian Smart Cities Using ISM-MICMAC Approach

Nikhil Gandhi, Abid Haleem, Mohd Shuaib and Deepak Kumar

Abstract India is predicted to be one of the world's fastest-growing large economies for this decade, according to projections from the World Bank and the International Monetary Fund. A lot of this growth is attributed to the operations of various multinational companies (MNCs) setting up their businesses in the country. Additionally, the plan set in place by the Indian Government to bring up numerous Smart Cities augurs well for the MNCs from a business point of view. But, most MNCs face logistical problems in connection with transportation of their material and the flow of information. The research objective of this paper is to describe the current state of Indian logistics service and identify the logistics barriers that foreign firms have encountered in India. Identification of the barriers in the system is a good first step towards rectifying the logistical systems. This work lays an affirmation to the observation that an 'Incompatible Supply Chain Model', along with 'Poor Skills of Logistics Professionals' and 'Low Rate of Technology Adoption' collectively act as the primary driving barriers to the Indian logistical system. This paper portrays the interdependence between various factors in the logistics industry that act as barriers for MNCs while carrying out business in India. After listing out the barriers, a hierarchy is formed using the Interpretive Structural Modelling (ISM) and MICMAC techniques to find the individual importance of each barrier and in what capacity it contributes to the problem.

Keywords Logistics service · MNCs · Smart cities · Logistical barriers · ISM · MICMAC

N. Gandhi · M. Shuaib (✉) · D. Kumar
Delhi Technological University, New Delhi 110042, India

A. Haleem
Jamia Millia Islamia University, New Delhi 110025, India

© Springer Nature Singapore Pte Ltd. 2020
S. Ahmed et al. (eds.), *Smart Cities—Opportunities and Challenges*,
Lecture Notes in Civil Engineering 58,
https://doi.org/10.1007/978-981-15-2545-2_47

1 Introduction

Logistics is the management of the flow of things between the points of origin and consumption in order to meet the requirements of customers or corporations [1]. In India, the logistics sector is fast becoming an area of huge focus and also a concern. The arrival of MNCs has further enhanced the need for an efficient logistics structure in the country. The fact that the volume of freight traffic moved has seen a tangible increase over time comes from the recent growth in the Indian economy. The increasing amount of freight movement has spurred the construction of better highways and expansion of railways so as to penetrate deeper into the remote areas of the country. The consistently growing economy invites an even better prospect for trade and operations of the best MNCs of the world.

India's economy has been growing at over 8% for the past few years. The nation spends around 14.4% of GDP on logistics. India's logistics sector is touted to be worth around US$ 307bn by 2020. The sector would witness a CAGR of 15–20% in FY 2016–2020 [2]. Business is booming with the arrival of various home-grown as well as foreign E-commerce firms such as Amazon. It can be safely said that the logistics sector in India is one of the massive opportunities and shows the potential for immense growth.

Despite the positivity that is reflected through statistics, the ground reality continues to stay bleak. Present-day scenario is actually quite negative, taking into account the financial liabilities incurred upon the company because of the delays and logistical shortcomings of the company's ground zero operations. A number of problems occur, both organizational as well as operational, which affect the functioning of MNCs in India. These problems range from very basic shortcomings like selection of an incompatible supply chain model to fundamental issues like poor skills of professionals to everyday operational issues such as transport delays due to traffic. All these problems, when combined, can harm a company's reputation. And in a country like India, where finding a vendor for a lesser price is one of the easiest tasks because of the sheer multitude of service providers, such barriers can act as deal breakers of the highest order.

The entire logistics framework, from warehousing up to last-mile distribution, suffers from a large number of problems, which make these seemingly strong points look like barriers. One would normally assume that with one of the world's most elaborate railway and roadway networks, India would be an efficient transportation market. But going by World Bank's Logistics Performance Index of 2018, India is ranked 44th in the world as a logistics market, dropping 9 places from 35th in 2016. This work should go a long way in listing out the barriers that affect the efficient functioning of MNCs in India so that all the loopholes can be pointed out and acted upon to improve the logistics sector in the country.

2 Literature Review

The importance of an efficient logistics system in a growing country like India is massive. With the aforementioned arrival of MNCs, the scenario has become much more elaborate and sophisticated. The relationship between logistics performance and customer loyalty is generally very close and crucial to a company [3]. For example, online purchases are generally small in quantity but the delivery schedules of these orders are more intricate, so the role of logistics is quite large, and the final customer always has a high expectation from the logistics domain [4]. Many studies show that the average customer considers logistics performance as an important subset of the overall service provided by a company [5, 6]. And a company's logistics capacity has a significant role to play in the logistics performance of an E-commerce firm [7]. Considering fundamental factors like financial stability, operational flexibility, and competency, companies find that outsourcing is the most effective way to fulfil all customers' logistics service requirements [8]. But with an extremely fragmented Indian market functioning on less than sophisticated warehousing and equipment, outsourcing can often become more of a problem than a solution. According to recent successful case studies and relative research works [8–13], the future of logistics should consider classifying logistical services primarily based on the specified company's operational barriers. Comparing this to the viewpoint of MNCs operating in India, and with the expansion of business into newer upcoming smart cities, it comes out as a result of the observation that a lot of bottlenecks are yet to be loosened to take the logistics sector in India to newer heights. And as it is essential for the betterment of any business, the first step, as explored in this work, is to recognize the concerns in each segment.

3 Adopted Approach

This work employs Interpretive Structural Modelling as a tool for creating a hierarchy of the identified barriers. The ISM methodologies' mathematical basis is a structural model used to analyze the complicated relationship between the barriers to logistical performance [14, 15]. The opinions of a selected group of professionals for the study and their practical knowledge decide whether and how the barriers are interactive and thus make it interpretive [16]. On this foundation, relationships between the enlisted barriers are established and an overall structure is portrayed in a graphical model. ISM generally has the following steps:

- Identify and list the barriers affecting the system as shown in Table 1.
- Establish a contextual relationship among the barriers.
- Develop the Structural Self-Interaction Matrix (SSIM) to indicate pairwise relationships among the barriers.
- Construct the reachability Matrix from the SSIM and verify it for transitivity.
- Segregate the Reachability Matrix into different level partitions.

Table 1 Identification of barriers

Barrier	Explanation	References
Inefficiencies in transport (B1)	Inefficiencies in transport are characterized by delays at various points in the supply chain. These delays stem out of excessive use of roadways as a method of transportation, overloading of trucks and unsophisticated trucking systems	[17–22]
Poor condition of storage infrastructure (B2)	All kinds of warehouses in India are still in a developmental curve and require a major facelift. Several bottlenecks exist in all departments that lead to various inconveniences to all parties in the chain	[23–26]
Low rate of technology adoption (B3)	Many new technologies that are already in motion in other countries are yet to be implemented in India. This means less compatibility between the Indian supply chain and logistics systems and the logistics and supply chain systems of other countries	[4, 27]
Poor skills of logistics professionals (B4)	An average Indian logistics professional is much less skilled than an average logistics professional of any country having a better-established logistics framework. This is because of lack of dedicated academic courses solely for logistics and a less sophisticated scenario to apply the logistics principles on	[2, 28–31]
Lack of business process improvement (B5)	Failure to meet deadlines is always a good indication of loopholes in the business process of a firm. The fact that very few deadlines are met on ground level in India, this points to a host of business process issues like need for excessive paperwork, unrealistically large loading/unloading times and so on	[32, 33]
Environmental issues (B6)	Many Governmental and International environmental regulations apply to logistical operations, which are the borderlines in which all companies have to work. Adhering to these regulations can lay some roadblocks that companies may have to meander around time and again	[34, 35]
Poor customer service (B7)	Customer experience is one of the most important barometers to measure a company's success or failure in a region. Last-mile distribution, accurate tracking facilities, feasible prices and on-time performance are some tasks to be taken care of for this	[36, 37]
Disregard of safety regulations (B8)	Be it failure to comply with traffic rules or overloading trucks, disregard of safety regulations always adds an unnecessary element of risk in the supply chain. This irregularity comes from the fact that there are traffic restrictions and other operational shortcomings which service providers look to cover up by ignoring safety rules and laws	[18, 21, 23]

(continued)

Table 1 (continued)

Barrier	Explanation	References
Incompatible supply chain model (B9)	A major factor of any operation is its think tank and the basic blueprint laid down by the company. A supply chain model could work wonders in one region and could completely fail in another region, and these sensitive details should be looked upon very seriously by firms	[38, 39]

- Based on the contextual relationships in the reachability matrix, remove transitive links and draw the ISM Model for the barriers.
- Replace barrier nodes and check model for conceptual irregularities.
- Prepare the MICMAC Analysis chart to check the driving and dependence tendencies of the barriers.

4 Structural Self-Interaction Matrix

Four symbols namely *V A X O* are used to show the direct relation that exists between the two sub-variables or barriers under consideration.

V: for the relation from *i* to *j*, but not in both directions
A: for the relation from *j* to *i*, but not in both directions
X: for both direction relations from *i* to *j* and *j* to *i*
O: if there's no relation from *i* to *j* or *j* to *i*.

For studying the variables, a contextual relationship is chosen such that one variable leads to another. Based on this contextual relationship, a SSIM has been developed and shown in Table 2.

Table 2 Structural self-interaction matrix

	B9	B8	B7	B6	B5	B4	B3	B2	B1
B1	O	A	V	O	O	A	A	O	–
B2	A	X	V	O	O	A	A	–	
B3	O	V	O	V	X	A	–		
B4	V	V	V	O	X	–			
B5	X	A	V	O	–				
B6	O	A	O	–					
B7	A	A	–						
B8	A	–							
B9	–								

5 Reachability Matrix

The SSIM format is transformed into the reachability matrix format by the following method:

- If the (a, b) relation in the SSIM is V, the (a, b) input is 1 and the (b, a) input is 0;
- If the (a, b) relation in the SSIM is A, the (a, b) input is 0 and the (b, a) input is 1;
- If the (a, b) relation in the SSIM is X, the (a, b) input is 1 and the (b, a) input is also 1;
- If the (a, b) relation in the SSIM is O, the (a, b) input is 0 and the (b, a) input is also 0.

Based on the above procedure, the transitivity rule is applied to the initial reachability that is obtained from the SSIM and it is hence converted into the final reachability matrix. The final reachability matrix is as shown in Table 3.

6 Level Partitions

The formulation of the final reachability matrix enables the reachability and antecedent sets for each barrier to be found. The reachability set is found to consist of the barrier itself and the other barriers which it drives, whereas the antecedent set comprises of the barrier itself and the other barriers on which it depends. After that, the intersection of these sets is found. The top-level barrier in the hierarchy, for which the reachability and antecedent sets are the same, would not drive any other barrier above itself and is disassociated from the rest of the barriers. Then, the same process is repeated until the level of each barrier is found. These level partitions help in developing the complete model (Tables 4, 5, 6 and 7) [40, 41].

Table 3 Reachability matrix

	B1	B2	B3	B4	B5	B6	B7	B8	B9	SUM
B1	1	0	0	0	0	0	1	0	0	2
B2	1	1	0	0	1	1	1	1	0	6
B3	1	1	1	1	1	1	1	1	1	9
B4	1	1	1	1	1	1	1	1	1	9
B5	1	1	1	1	1	1	1	1	1	9
B6	0	0	0	0	0	1	0	0	0	1
B7	0	0	0	0	0	0	1	0	0	1
B8	1	1	1	1	1	1	1	1	1	9
B9	1	1	1	1	1	1	1	1	1	9
SUM	7	6	5	5	6	7	8	6	5	

Table 4 First-level iteration

	Reachability	Antecedent	Intersection	Level
*B*1	1,7	1,2,3,4,5,8,9	1	
*B*2	1,2,5,6,7,8	2,3,4,5,8,9	2,5,8	
*B*3	1,2,3,4,5,6,7,8,9	3,4,5,8,9	3,4,5,8,9	
*B*4	1,2,3,4,5,6,7,8,9	3,4,5,8,9	3,4,5,8,9	
*B*5	1,2,3,4,5,6,7,8,9	2,3,4,5,8,9	2,3,4,5,8,9	
*B*6	6	2,3,4,5,6,8,9	6	1
*B*7	7	1,2,3,4,5,7,8,9	7	1
*B*8	1,2,3,4,5,6,7,8,9	2,3,4,5,8,9	2,3,4,5,8,9	
*B*9	1,2,3,4,5,6,7,8,9	3,4,5,8,9	3,4,5,8,9	

Table 5 Second-level iteration

	Reachability	Antecedent	Intersection	Level
*B*1	1	1,2,3,4,5,8,9	1	2
*B*2	1,2,5,8	2,3,4,5,8,9	2,5,8	
*B*3	1,2,3,4,5,8,9	3,4,5,8,9	3,4,5,8,9	
*B*4	1,2,3,4,5,8,9	3,4,5,8,9	3,4,5,8,9	
*B*5	1,2,3,4,5,8,9	2,3,4,5,8,9	2,3,4,5,8,9	
*B*8	1,2,3,4,5,8,9	2,3,4,5,8,9	2,3,4,5,8,9	
*B*9	1,2,3,4,5,8,9	3,4,5,8,9	3,4,5,8,9	

Table 6 Third-level iteration

	Reachability	Antecedent	Intersection	Level
*B*2	2,5,8	2,3,4,5,8,9	2,5,8	3
*B*3	2,3,4,5,8,9	3,4,5,8,9	3,4,5,8,9	
*B*4	2,3,4,5,8,9	3,4,5,8,9	3,4,5,8,9	
*B*5	2,3,4,5,8,9	2,3,4,5,8,9	2,3,4,5,8,9	3
*B*8	2,3,4,5,8,9	2,3,4,5,8,9	2,3,4,5,8,9	3
*B*9	2,3,4,5,8,9	3,4,5,8,9	3,4,5,8,9	

Table 7 Fourth-level iteration

	Reachability	Antecedent	Intersection	Level
*B*3	3,4,9	3,4,9	3,4,9	4
*B*4	3,4,9	3,4,9	3,4,9	4
*B*9	3,4,9	3,4,9	3,4,9	4

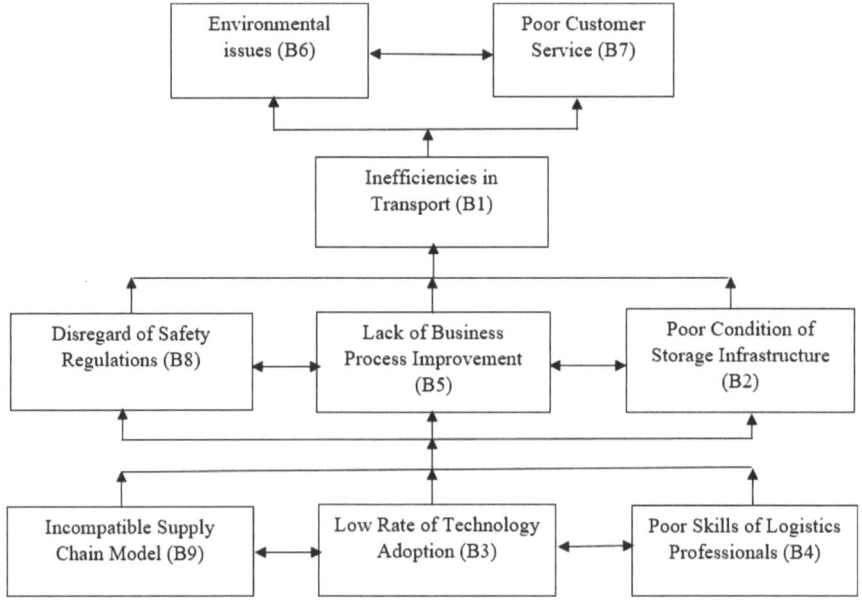

Fig. 1 ISM model

7 Formation of ISM Model

This model represents the direct relationship among different barriers. It states that each barrier has its own importance at its own level as shown in Fig. 1.

The ISM model shows that Incompatible Supply Chain Model (B9), along with barriers Low Rate of Technology Adoption (B3) and Poor Skills of Logistics Professionals (B4), together interdependently acts as the base that enhances all the other barriers. Disregard of Safety Regulations (B8), Lack of Business Process Improvement (B5) and Poor Condition of Storage Infrastructure (B2) further exhibit internally interactive driving capacity towards Inefficiencies in Transport (B1). Environmental Issues (B6) and Poor Customer Service (B7) together show the maximum dependence on other aforementioned barriers. It shows that the root of the problems in the logistical setup in India lies in basic mistakes such as the selection of an incompetent and unfulfilling supply chain model and extends to a company's slow technology adoption pace and grassroots problems like poor skills of logistics professionals in the country.

8 MICMAC Analysis

The objective of the MICMAC analysis is to examine the driver power and the dependence power of the variables. The variables are classified into four clusters as given in Fig. 2.

The first cluster consists of the self-sufficient barriers that have weak driver power and weak dependence. These barriers are somewhat disconnected from the system. The second cluster is made up of the dependent barriers that have good dependence power but less driving power. The third cluster comprises of the linkage barriers that display both the characteristics of driving and dependence in a strong capacity. These barriers are unstable since any action on these barriers will affect all the barriers, including themselves, and cause instability in the system. The fourth cluster is made up of the independent barriers having high driving characteristics but weak dependence power [14, 15]. Variables with a strong driving influence are known as key variables and they are a part of the linkage barriers set [42]. Here, it is seen that Barriers 3, 4 and 9 (5,9) fall in the fourth cluster and show the strongest driving power. Barriers 5 and 8 coincide on (6,9) on the borderline between clusters three and four, exhibiting a stronger driving power than dependence. Barriers 1 (7,2); 6 (7,1) and 7 (8,1) are second cluster barriers that are heavier dependent than drivers whereas Barrier 2 (6,6) coincides with the intersection of the dependence power and driving power mean axes, showing moderate driving and dependence power.

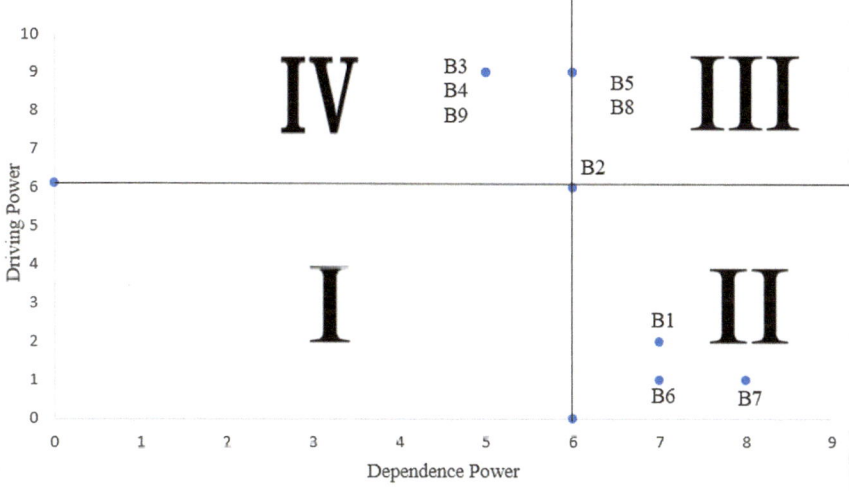

Fig. 2 MICMAC analysis

9 Conclusion

Logistics in India, going strictly by numbers, is on an upward trajectory. However, the ground reality of the same shows a rather negative outcome of the various governmental reforms like revamp of the tax structure with the introduction of GST. This is a cause of concern for MNCs operating in India since a negative logistical performance for especially E-commerce firms could prove to be detrimental to the company's image. The barriers highlighted in this work are, as agreed with industry experts, a broad aggregation of the problems faced by the companies operating in India. The highlighted barriers cover all domains, from grassroots problems like poor skills of professionals to company shortcomings like establishing an unsuitable supply chain model and operational roadblocks like lack of business process improvement. According to the ISM model that comes out of structural hierarchical analysis, Incompatible Supply Chain Model (B9), Low Rate of Technology Adoption (B3) and Poor Skills of Logistics Professionals (B4) are the interdependent base. The base drives the Level 2 barriers that are Disregard of Safety Regulations (B8), Lack of Business Process Improvement (B5), and Poor Condition of Storage Infrastructure (B2), which indicates that the problems faced by the MNCs are present at all levels from planning till execution of any company operation. Naturally, the implied Inefficiencies in Transport (B1) are driven by these core issues. The inefficiencies lead to components of customer service like last-mile distribution to be inadequate. Poor Customer Service (B7) and Environmental Issues (B6) are the barriers that are most dependent on the others, forming the head of the model. Collectively, these issues are the cause of misery for a large fraction of MNCs operating in India and with another wave of urbanization in full swing in the country, the ironing out of these barriers should be on the priority list for companies and the Indian authorities alike, to take business forward in the coming years.

References

1. Chikhalkar R, Khurana S, Khurdelia A (2014) Analyze and design an efficient logistics system: a case study of mid-size FMCG Company in India
2. Agarwal C (2017) Budget 2017: India's logistics sector primary expectation lies with timely implementation of GST, says Chander Agarwal, TCIEXPRESS. Fin Exp, 25 Jan 2017
3. Ramanathan R (2010) The moderating roles of risk and efficiency on the relationship between logistics performance and customer loyalty in E-commerce. Transp Res Part E: Logistics Transp Rev 46(6):950–962
4. Yu Y, Wang X, Zhong RY, Huang GQ (2016) E-commerce logistics in supply chain management: practice perspective. Procedia Cirp 52:179–185
5. Agatz NA, Fleischmann M, Van Nunen JA (2008) Efulfillment and multi-channel distribution–a review. Eur J Oper Res 187(2):339–356
6. Esper TL, Jensen TD, Turnipseed FL, Burton S (2003) The last mile: an examination of effects of online retail delivery strategies on consumers. J Bus Logistics 24(2):177–203

7. Joong-Kun Cho J, Ozment J, Sink H (2008) Logistics capability, logistics outsourcing and firm performance in an E-commerce market. Int J Phys Distrib Logistics Manag 38(5): 336–359

8. Wilding R, Juriado R (2004) Customer perceptions on logistics outsourcing in the European consumer goods industry. Int J Phys Distrib Logistics Manag 34(8):628–644

9. Bolumole YA (2001) The supply chain role of third-party logistics providers. Int J Logistics Manag 12(2):87–102

10. Highfield V (2014) The home depot's Mark Holifield on fulfilling customer needs

11. Knemeyer AM, Corsi TM, Murphy PR (2003) Logistics outsourcing relationships: customer perspectives. J Bus Logistics 24(1):77–109

12. Rabinovich E, Windle R, Dresner M, Corsi T (1999) Outsourcing of integrated logistics functions: an examination of industry practices. Int J Phys Distrib Logistics Manag 29(6): 353–374

13. Rao K, Young RR (1994) Global supply chains: barriers influencing outsourcing of logistics functions. Int J Phys Distrib Logistics Manag 24(6):11–19

14. Kumar D, Jain S, Tyagi M, Kumar P (2018) Quantitative assessment of mutual relationships of issues experienced in greening supply chain using ISM-fuzzy MICMAC approach. Int J Logistics Syst Manag 30(2):162–178

15. Tyagi M, Kumar P, Kumar D (2015) Analysis of interactions among the drivers of green supply chain management. Int J Bus Perform Supply Chain Model 7

16. Shuaib M, Khan U, Haleem A (2016) Modeling knowledge sharing factors and understanding its linkage to competitiveness. Int J Glob Bus Compet 11(1):23–36

17. '3PL: The new sunrise industry', Cargo Connect

18. 'Total transport systems study on traffic flows & modal costs: Report for Planning Commission', RITES

19. Allcargo Logistics-higher dwell time at CFS' and lower freight rates boosts profits', IDFC 'Cold chain market. Research on India

20. Report of the Working Group on Warehouse Receipts & Commodity Futures. Reserve Bank of India

21. Thaller C, Moraitakis N, Rogers H, Sigge D, Clausen U, Pfohl H-C, Hartmann E, Hellingrath B (2011) Analysis of the logistics research in India—white paper

22. Transportation, Logistics, Warehousing and Packaging Sector (2022) National Skill Development Corporation

23. 'Seminar proceedings on Building Warehousing Competitiveness', CII

24. Cushman and Wakefield, Logistics industry real estate's new power house'

25. Deloitte, Logistics and infrastructure: exploring opportunities

26. Ernst & Young LLP, CII Institute of Logistics (2013) The Indian warehousing industry: an overview

27. Neeraja B, Mehta M, Chandani A (2014) Supply chain and logistics for the present day business. Procedia Econ Fin 11:665–675

28. Carter P, Carter JR (2007) The future of supply management—part III. organization + talent. Supply Chain Manag Rev 11(8):37–43

29. Closs DJ (2000) Preface. J Bus Logistics 21(1):1

30. Green A (2010) Building the skills to support a high-performance supply chain. Supply Chain E-Mag

31. Thai V, Cahoon S, Tran H (2011) Skill requirements for logistics professionals: findings and implications. Asia Pac J Mark Logistics 23:553–574

32. By DDC FPO "Top 8 Logistics Challenges Facing the Industry". www.logisticsmgmt.com, 1 Nov 2017

33. van der Aalst WMP, ter Hofstede AHM, Weske M (2003) Business process management: a survey. In: Proceedings of the international conference of business process management, Eindhoven, The Netherlands, 26–27 June 2003

34. Evangelista P, Colicchia C, Creazza A (2017) Is environmental sustainability a strategic priority for logistics service providers? J Environ Manage 198:353–362

35. Aronsson H, Huge Brodin M (2006) The environmental impact of changing logistics structures. Int J Logistics Manag 17(3):394–415
36. Daugherty PJ, Bolumole Y, Grawe SJ (2018) The new age of customer impatience: an agenda for reawakening logistics customer service research. Int J Phys Distrib Logistics Manag
37. Lin Y et al (2016) Exploring the service quality in the e-commerce context: a triadic view. Ind Manag Data Syst 116(3):388–415
38. Bianchi C (2006) Home depot in Chile: case study. J Bus Res 59(3):391–393
39. Bianchi CC, Ostale E (2006) Lessons learned from unsuccessful internationalization attempts: examples of multinational retailers in Chile. J Bus Res 59(1):140–147
40. Babones S (2018) India may be the world's fastest growing economy, but regional disparity is a serious challenge. www.forbes.com 10 Jan 2018
41. Kumar GS, Shirisha P (2014) Transportation the key player in logistics management. J Bus Manag Soc Sci Res (JBM&SSR) 3(1). ISSN No: 2319–5614
42. Ravi V, Shankar R (2004) Analysis of interactions among the barriers of reverse logistics. Technol Forecast Soc Change

Highway Gradient Effects on Hybrid Electric Vehicle Performance

Mohammad Waseem, A. F. Sherwani and Mohd Suhaib

Abstract In the current era, the public is receiving much devotion to hybrid electric vehicles as the subsequent technology pattern in the road transport sector. Three-wheeler 'battery' vehicles are extensively used for intracity transport facility in Indian cities. These vehicles are known as smart e-vehicle and have the potential to mitigate the carbon emissions of conventional vehicles. The driving range of three-wheeler e-vehicle is greatly affected by the gradient of terrain/highway. The projected research emphasis the highway gradient effects on a three-wheeler e-vehicle performance. Next, to examine the effects of road gradient on the three-wheeler e-vehicle, the dynamics motion equations are modelled and derived. Appropriate three-wheeler e-vehicle design parameters and constants are taken from the literature survey. Finally, simulations of the dynamic vehicle model are performed in the MATLAB® simulation tool to compute road gradient effect of zero degrees, three degrees, six degrees, nine degrees, twelve degrees and fifteen degrees, respectively.

Keywords Vehicle dynamics · MATLAB® · Three-wheeler electric vehicle · Road gradient · Hybrid vehicles

1 Introduction

The transportation sector is contributing to several environmental issues such as water, land and air pollutions [1]. Globally, conventional fuel-powered vehicles are the major source of pollution issues and climate change [2–5]. According to the international energy agency report, the road transport sector is responsible to increase global CO_2 emission by 71% during 1990–2014 [6]. In India, 18% of CO_2 emissions are produced by road transportation itself [7, 8]. The most widely used

M. Waseem (✉) · A. F. Sherwani · M. Suhaib
Department of Mechanical Engineering, Faculty of Engineering and Technology,
Jamia Millia Islamia (A Central University), New Delhi 110025, India
e-mail: waseem159088@st.jmi.ac.in

© Springer Nature Singapore Pte Ltd. 2020
S. Ahmed et al. (eds.), *Smart Cities—Opportunities and Challenges*,
Lecture Notes in Civil Engineering 58,
https://doi.org/10.1007/978-981-15-2545-2_48

583

technology in the existing transportation system is the internal combustion engine (ICE). The combustion of fossil fuel in ICE results in harmful gases to the environment. Emissions of carbon contain product from ICE vehicles which are dominating environmental and pollution issues [9, 10].

The automobile industry is undergoing a revolution in designing new electrical platforms for vehicles to counter the sophistication involved with engine and carbon emission issues. Therefore, the alternate engine technology is needed to revamp the ICE vehicles. Electric vehicles are the alternate in place of ICE technology. Electric propulsion system not only diminishes the pollution issue but also conveys precision accuracy of power and easy vehicle handling. Hybrid electric vehicles are investigated by the automobile zone to lessen the application of ignition engine with integrating of electric motor/machine system, i.e. electric propulsion system. The proposed technology by the automobile sector has lessened carbon emissions in the transportation sector as compared to the conventional engine [11, 12].

Bäckryd et al. proposed multidisciplinary optimization technique to improve the design of vehicle structure [13]. Abdullah et al. propose a model updating approach to improve the dynamic properties of go-kart chassis structure [14]. Janarthanan develops a simulation model of a heavy tracked vehicle to demonstrate the transient analysis of longitudinal dynamics [15]. Vibration effect and analysis for passenger electric vehicle with four-wheel drive structure is proposed in [16–18]. Wang et al. propose a particle swarm optimization strategy to design a four-wheeled independently actuated vehicle [19]. Various aspects of parallel, series hybrid architecture e-vehicles are developed in the current literature, but no one has reported highway/road gradient effects on the three-wheeler e-vehicle performance [20, 21]. Work presented above suggests that hybrid electric vehicles are perceived as 'the vehicles of future'. In this research, highway/road gradient effects on the three-wheeler e-vehicle performance are analysed.

2   Three-Wheeler e-Vehicle Active Model

To examine the forces at work of the three-wheeler electric vehicle, the autonomy of the vehicle model is presented as follows: X-direction as the longitudinal, Y-direction as the side, Z-direction as vertical, rolling around the X-direction, pitching around the Y-direction, yawing around the Z-direction. Each wheel of the electric vehicle is a sub-module and has the freedom to turn around the wheel axle [22]. Figure 1 demonstrates the model of the proposed three-wheeler e-vehicle dynamic.

Figure 2 shows the stress acting on each wheel in the *X-direction*, *Y-direction* and Z-direction, respectively. The stress acting on each wheel is assumed by manifestation $F_U (V, W)$, where the subscript U is X, Y and Z, which represents the force acting in the X-direction, Y-direction and Z-direction, respectively. $V = 1$ means front wheel of the vehicle; $V = 2$ means rear wheel of the vehicle; $W = 1$ means the wheel on left side of vehicle; $W = 2$ means the wheel on right side of vehicle; a is the distance from the vehicle centre of gravity to the front wheel axle;

Fig. 1 A dynamic layout
diagram of the three-wheeler
e-vehicle

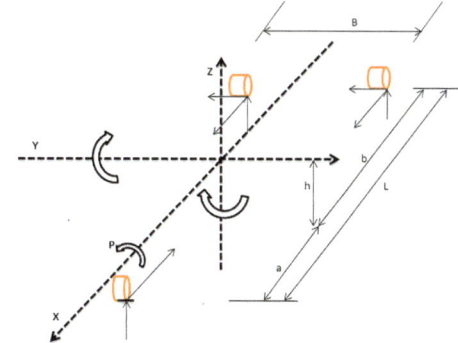

Fig. 2 Forces acting on all
wheels of the three-wheeler
e-vehicle

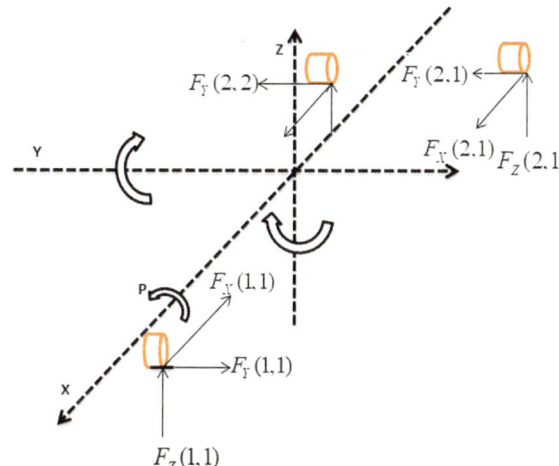

b is the distance from the vehicle centre of gravity to the rear axle; h is the height
between the vehicle centre of gravity and ground level; B is the gap between back
wheels centres of the three-wheeler hybrid e-vehicle; and L is the wheelbase.

3 Longitudinal Motion of the Three-Wheeler e-Vehicle

Considering the motion of the three-wheeled vehicle in the longitudinal direction,
the stress acts on each wheel in X-direction only. The movement deployment of the
three-wheeler e-vehicle in a moving way is estimated by the sum of all forces acting
in that specific course [23]. According to the Newton laws, the dynamic movement
of the three-wheeled vehicle in the longitudinal direction is governed by Eq. (1).

$$\lambda M \dot{V}_X = \sum_{W=1}^{1} F_X(1, W) + \sum_{W=1}^{2} F_X(2, W) - \sum F_{\text{Resistive}} \qquad (1)$$

$$\lambda M \dot{V}_X = F_X(1, 1) + F_X(2, 1) + F_X(2, 2) - \sum F_{\text{Resistive}} \qquad (2)$$

where M is the estimated mass of the e-vehicle, λ is a factor of rotating mass, V_X is velocity in X-direction and $F_{\text{Resistive}}$ is the sum of all resistance force acting the three-wheeled vehicle.

Figure 3 shows all the forces and moments that act on the three-wheeler e-vehicle in the longitudinal direction of the highway with a positive gradient (α). The external resistive agents on the three-wheeler e-vehicle in the longitudinal direction are as follows: the drag force due to air (F_{drag}), the rolling resistance force (F_{roll}) and resistance to the gradient (F_{grad}). F_{TF} is the tractive effort that acts on the front wheel and equal to the stress (F_X (1,1)). Similarly, F_{TR} is the tractive effort that acts on the rear 'left' and 'right' wheels and equivalent to the summation of tensions (F_X (2,1) + (F_X (2,2)). The electrical actuating system provides the necessary total tractive effort (F_{Tractive}). Hence, Eq. (2) of three-wheeled vehicle movement can be expressed as Eq. (3).

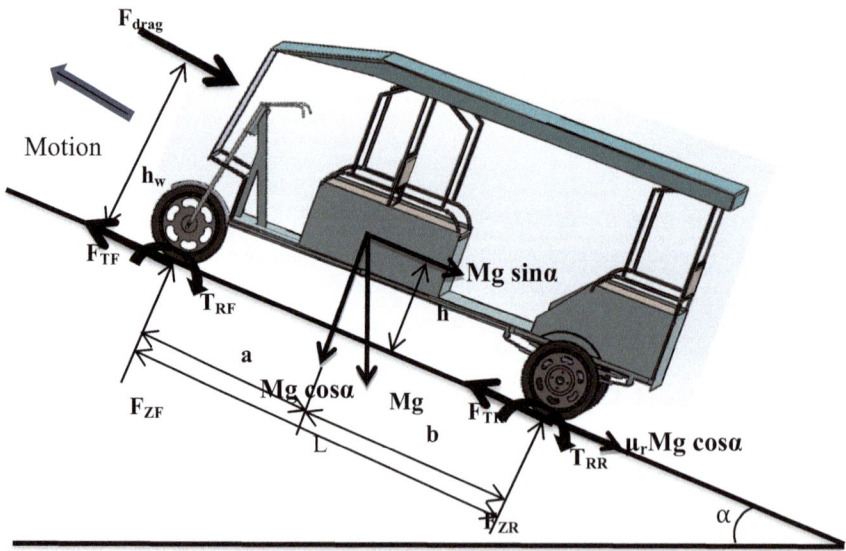

Fig. 3 Free body diagram of three-wheeler e-vehicle on a gradient

$$\lambda M \frac{dV_X}{dt} = (F_{TF} + F_{TR}) - \sum F_{Resistive} \qquad (3)$$

$$\lambda M \frac{dV_X}{dt} = F_{Tractive} - \sum F_{Resistive} \qquad (4)$$

$$F_{Tractive} = \lambda M \frac{dV_X}{dt} + \sum F_{Resistive} \qquad (5)$$

The tractive effort and active opposing pull that act on the three-wheeler e-vehicle in the longitudinal direction is governed by Eq. (6) (please see Eq. (5)).

$$F_{Tractive} = \lambda M \frac{dV_X}{dt} + Mg \sin \alpha + Mg\mu_r \cos \alpha + \frac{1}{2}\rho A_f C_D V_X^2 \qquad (6)$$

where g is gravitation acceleration, α is highway/road gradient, μ_r is coefficient of friction, ρ is the air density, A_f is the front area of the vehicle, C_D is drag coefficient and V_X is the speed of the vehicle.

The vertical normal load acting on the front and rear wheel locations are governed from the Eqs. (7) and (8).

$$F_{ZF} = \frac{Mgb \cos \alpha}{L} - \frac{h}{L}\left(F_{Tractive} - \mu_r Mg \cos \alpha \left(1 - \frac{r_w}{h}\right)\right) \qquad (7)$$

$$F_{ZR} = \frac{Mga \cos \alpha}{L} + \frac{h}{L}\left(F_{Tractive} - \mu_r Mg \cos \alpha \left(1 - \frac{r_w}{h}\right)\right) \qquad (8)$$

4 Selection of Parameters for Three-Wheeler e-Vehicle

Design and development of e-vehicle depend on the mechanical parameters and coefficients associated with the aesthetic and technical look of the vehicle. Hence, parameters determine the overall technical and aesthetic behaviour of a vehicle. Therefore, design parameters associated with the tractive effort ($F_{Tractive}$), the normal loads (F_{ZF}, F_{ZR}) and applied torque ($T_{tractive}$), etc. have been taken from literature and connected work (Table 1).

5 Simulation Result

Three-wheeler hybrid e-vehicle model dynamics performance is simulated and estimated for six different road gradient conditions. To examine the active performance of the e-vehicle model, simulation has been performed in the MATLAB® tool for each highway/road gradient. Thereafter, effort, active load and the applied

Table 1 Three-wheeler hybrid e-vehicle mechanical parameters and coefficient

Name of parameters	Modelling value	SI unit
Friction coefficient (μ_r) [24, 25]	0.80	
Drag coefficient (C_D) [26]	0.30	
Mass factor (λ) [27, 28]	1.05	
Air density (ρ) [27]	1.20	kg/m^3
The front area (A_f)	1.65	m^2
Vehicle speed (V_X)	24.000	km/h
vehicle weight (M)	725.0	kg
Wheel diameter (d_w)	0.508	m
Wheelbase (L)	2.100	m
Vehicle C.G. height (h)	0.275	m

torque of three-wheeler e-vehicle model are computed for zero degrees, three degrees, six degrees, nine degrees, twelve degrees and fifteen degrees gradient of highway/road. Figure 4 shows the variation in the computed torque for the highway gradient of six degrees.

Simulated results are further post-processed by estimating the average value of the applied torque, the active load on the front and rear wheels supporting point, to forecast the road gradient effects on three-wheeler e-vehicle the active performance accurately. Mean values of applied torque on the rear wheels are 108.3317, 173.5099, 238.4582, 302.9987, 366.9543 and 430.1498 (Nm) for highway with gradient order zero degrees, three degrees, six degrees, nine degrees, twelve degrees and fifteen degrees, respectively, to attain 25 km/h of vehicle speed as shown in Fig. 5.

The mean estimated evaluates of dynamic load that acts on the front wheel support are 2.6188, 2.5802, 2.5343, 2.4814, 2.4215 and 2.3548 kN for highway with gradient order of zero degrees, three degrees, six degrees, nine degrees, twelve

Fig. 4 Torque applied to the wheel of three-wheeler e-vehicle

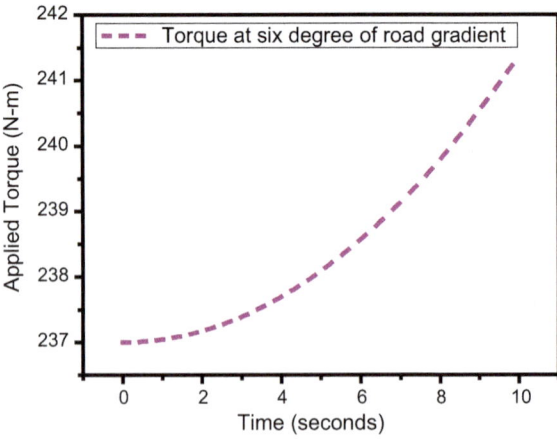

Fig. 5 Effect of gradient on applied torque

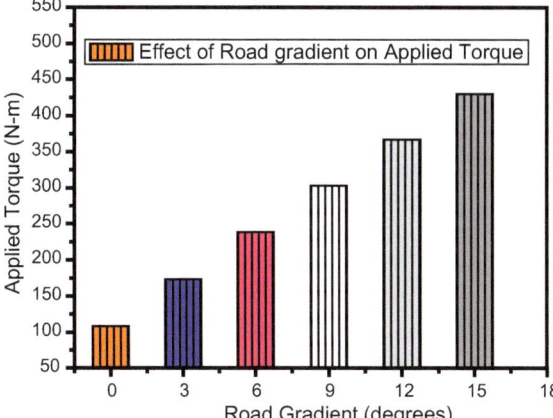

Fig. 6 Effect of the gradient of dynamic load on the front wheel of a three-wheeler e-vehicle

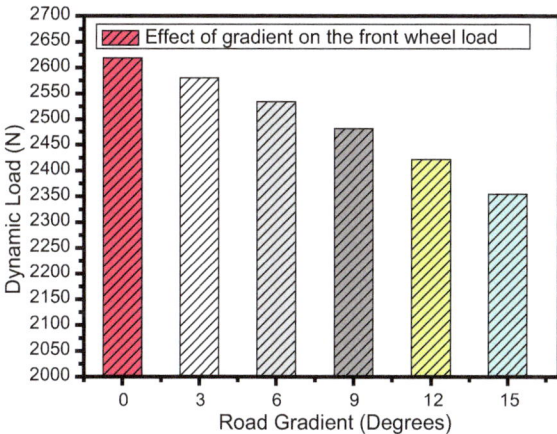

degrees and fifteen degrees, respectively, to reach 25 km/h of vehicle speed as shown in Fig. 6. Mean computed evaluates of the dynamic load that acts on the rear wheel support are 2.2862, 2.3181, 2.3438, 2.3632, 2.3763 and 2.3830 kN for highway with gradient order of zero degrees, three degrees, six degrees, nine degrees, twelve degrees and fifteen degrees to reach 25 km/h of vehicle speed.

6   Discussion and Limitation

The mathematical dynamic modelling of the three-wheeler e-vehicle is presented. After that, the dynamics motion equations of the tractive effort, the torque applied the rear axle, the load that acts on the front and rear wheel are simulated in MATLAB® tool. Thereafter, the appropriate mechanical designing constraints and constants from the literature investigation are determined and assigned in the coded equations. Thereafter, simulations of the dynamic e-vehicle have been performed for 10 s in the MATLAB® programming tool for highway with gradient order of zero degrees, three degrees, six degrees, nine degrees, twelve degrees and fifteen degrees, respectively. Next, for each road gradient condition, simulation output values of applied torque, normal active load that acts on the front and rear wheels support are computed and recorded. Finally, highway/road gradient effects on the three-wheeler e-vehicle active performance are analysed. Time to refuel is the most crucial factor for slow penetration of three-wheeler e-vehicles as slow charging takes 7–8 h. Public awareness is also one of the important factors to adopt e-vehicle in the developing country India. The government of India should start new policy and rules towards sustainable development goals.

7 Conclusion

Active dynamics model of three-wheeler e-vehicle with six degrees of freedom is presented. The active performance of the three-wheeler e-vehicle is simulated for the highway/road gradient order of zero degrees, three degrees, six degrees, nine degrees, twelve degrees and fifteen degrees in MATLAB® tool. Simulation results show that higher torque is required to form the electric propulsion system for the inclined road as compared to the flat road. The simulated results signify that the applied torque on the three-wheeler e-vehicle is 430 Nm for the road gradient of fifteen degrees, while the estimated torque on the e-vehicle is 108 Nm for the gradient of zero degrees. Hence, the power required from the electric motor is more times more for fifteen degrees inclined road as compared to the plane surface road. The percentage increment in the numerical magnitude of the load that acts on the e-vehicle is 4.2% for road gradient of fifteen degrees as compared to zero degrees gradient of the road. Hence, the gradient of the road is a significant factor that affects the dynamic performance of the three-wheeler hybrid vehicle.

Acknowledgements We would like to acknowledge all the faculty members and Laboratories of Mechanical Engineering Department, Jamia Millia Islamia, New Delhi, to support this research.

References

1. Waseem M, Suhaib M, Sherwani AF (2019) Modelling and analysis of gradient effect on the dynamic performance of three-wheeled vehicle system using Simscape. SN Appl Sci 1:225. https://doi.org/10.1007/s42452-019-0235-8
2. Ligen Y, Vrubel H, Girault H (2018) Mobility from renewable electricity: infrastructure comparison for battery and hydrogen fuel cell vehicles. World Electr Veh J 9:3. https://doi.org/10.3390/wevj9010003

3. Stabinsky D, Hoffmaister JP (2015) Establishing institutional arrangements on loss and damage under the UNFCCC: the Warsaw international mechanism for loss and damage. Int J Glob Warm 8:295. https://doi.org/10.1504/IJGW.2015.071967
4. Dinan T (2017) Projected increases in hurricane damage in the United States: the role of climate change and coastal development. Ecol Econ 138:186–198. https://doi.org/10.1016/j. ecolecon.2017.03.034
5. Arnell NW, Gosling SN (2016) The impacts of climate change on river flood risk at the global scale. Clim Change 134:387–401. https://doi.org/10.1007/s10584-014-1084-5
6. International Energy Agency (2016) World energy outlook 2016
7. Digalwar AK, Giridhar G (2015) Interpretive structural modeling approach for development of electric vehicle market in India. Procedia CIRP 26:40–45. https://doi.org/10.1016/j.procir. 2014.07.125
8. Ashrafee F, Morsalin S, Rezwan A (2014) Design and fabrication of a solar powered toy car. In: 1st international conference on electrical engineering and information and communication technology, ICEEICT 2014
9. Alahmad M, Chaaban M, Chaar L (2011) A novel photovoltaic/ battery structure for solar electrical vehicles [PVBS for SEV]. IEEE Veh Power Propuls Conf 1:1–4
10. Simaes MG, Franceschetti NN, Adamowski JC (1998) A solar powered electric vehicle. Appl Power Electron Conf Expo 1:49–55
11. Hannan MA, Azidin FA, Mohamed A (2014) Hybrid electric vehicles and their challenges: a review. Renew Sustain Energy Rev 29:135–150. https://doi.org/10.1016/j.rser.2013.08.097
12. Sarkar T, Sharma M, Gawre SK (2014) A generalized approach to design the electrical power system of a solar electric vehicle. In: 2014 IEEE students' conference on electrical, electronics and computer science. IEEE, pp 1–6
13. Bäckryd RD, Ryberg AB, Nilsson L (2017) Multidisciplinary design optimisation methods for automotive structures. Int J Automot Mech Eng. https://doi.org/10.15282/ijame.14.1.2017. 17.0327
14. Abdullah NAZ, Sani MSM, Husain NA et al (2017) Dynamics properties of a Go-kart chassis structure and its prediction improvement using model updating approach. Int J Automot Mech Eng 14:3887–3897. https://doi.org/10.15282/ijame.14.1.2017.6.0316
15. Janarthanan B, Padmanabhan C, Sujatha C (2012) Longitudinal dynamics of a tracked vehicle: simulation and experiment. J Terramech 49:63–72. https://doi.org/10.1016/j.jterra. 2011.11.001
16. Tang X, Hu X, Yang W, Yu H (2018) Novel Torsional vibration modeling and assessment of a power-split hybrid electric vehicle equipped with a dual-mass flywheel. IEEE Trans Veh Technol 67:1990–2000. https://doi.org/10.1109/TVT.2017.2769084
17. Tarek R, Anjum A, Hoque MA, Azad A (2016) Solar electric ambulance van unfolding medical emergencies of rural Bangladesh. In: GHTC 2016—IEEE global humanitarian technology conference: technology for the benefit of humanity, conference proceedings. pp 514–519
18. Shaha N, Uddin MB (2013) Hybrid energy assisted electric auto rickshaw three-wheeler. In: 2013 international conference on electrical information and communication technology, EICT 2013
19. Zhenpo W, Changhui Q, Xue X, Lei Z (2017) Design and control strategy optimization for four-wheel independently actuated electric vehicles. Energy Procedia 105:2323–2328. https:// doi.org/10.1016/j.egypro.2017.03.667
20. Mulhall P, Lukic SM, Wirasingha SG et al (2010) Solar-assisted electric auto rickshaw three-wheeler. IEEE Trans Veh Technol 59:2298–2307. https://doi.org/10.1109/TVT.2010. 2045138
21. Sameeullah M, Chandel S (2016) Design and analysis of solar electric rickshaw: a green transport model. In: 2016 international conference on energy efficient technologies for sustainability, ICEETS 2016, pp 206–211

22. Peng X, Zhe H, Guifang G et al (2011) Driving and control of torque for direct-wheel-driven electric vehicle with motors in serial. Expert Syst Appl 38:80–86. https://doi.org/10.1016/j. eswa.2010.06.017
23. Eshani M, Gao Y, Gay S, Emadi A (2010) Modern electric, hybrid electric and fuel cell vehicles, 2nd edn
24. Singh KB, Taheri S (2015) Estimation of tire–road friction coefficient and its application in chassis control systems. Syst Sci Control Eng 3:39–61. https://doi.org/10.1080/21642583. 2014.985804
25. Pacejka HB (2012) Tire characteristics and vehicle handling and stability. In: Tire and vehicle dynamics. Elsevier, pp 1–58
26. Hucho W, Sovran G (1993) Aerodynamics of road vehicles. Annu Rev Fluid Mech 25:485–537. https://doi.org/10.1146/annurev.fl.25.010193.002413
27. Maia R, Silva M, Araujo R, Nunes U (2011) Electric vehicle simulator for energy consumption studies in electric mobility systems. IEEE Forum Integr Sustain Transp Syst FISTS 2011:227–232. https://doi.org/10.1109/FISTS.2011.5973655
28. Fang S, Song J, Song H et al (2016) Design and control of a novel two-speed uninterrupted mechanical transmission for electric vehicles. Mech Syst Signal Process 75:473–493. https:// doi.org/10.1016/j.ymssp.2015.07.006

Mechanical Performance Evaluation of Concrete with Waste Coarse Ceramic Aggregate

Hadee Mohammed and Shakeel Ahmed

Abstract This study aims to describe the experimental work to illustrate and evaluate the mechanical behaviour of concrete with waste ceramic (floor tiles) coarse aggregate. Concrete included the following materials natural fine aggregate (NFA), natural coarse aggregate (NCA) and waste ceramic aggregate (WCA). Concrete was prepared with WCA replacing in NCA in absolute volume percentage in of 0, 10, 20, 30, 50 and 100%. All the concretes were prepared with similarity aggregate size gradation and workability. In the hardened case, the concrete performance was evaluated by split tensile strength, compressive strengthen, flexural strengthen and combined flexural and torsion. The results show significant improvement in feasibility the partially replacement of NCA by WCA, in spite of the mechanical behaviour of concrete reductions with the rising replacement of NCA by WCA in case of compressive, tensile and flexure strength, whereas in case of combined flexural and torsion, it is increasing up to 20% replacement.

Keywords Coarse aggregate · Mechanical properties · Recycled aggregate · Concrete · Floor tiles · Ceramic waste aggregate

1 Introduction

Concrete, as construction material, has been applied in the construction industry for many different centuries. Approximately, one ton of concrete is applied per capita per year throughout the world. The following studies have carried out on the concrete strength with waste ceramic and evaluated by compressive strength, split tensile strength, flexural strength and combined flexural and torsion [1].

Lavat et al. [2] studied the recycling of ceramic waste tiles in the manufacture of blended cement. The researcher stated that the use of powdered roof tiles as Pozzolanic product in cement production. The wastes were listed as non glazed,

H. Mohammed (✉) · S. Ahmed (✉)
Department of Civil Engineering, Z. H. College of Engineering and Technology,
Aligarh Muslim University, Aligarh, UP 202002, India

© Springer Nature Singapore Pte Ltd. 2020
S. Ahmed et al. (eds.), *Smart Cities—Opportunities and Challenges*,
Lecture Notes in Civil Engineering 58,
https://doi.org/10.1007/978-981-15-2545-2_49

natural and black-glazed tiles. The compressive strength of resulting models was measured. This result has approved that waste material is used to produce Pozzolanic cement.

The stone powder can be applied as fine aggregate and waste ceramic can be used as a coarse aggregate, this situation was studied by Reddy [3]. It was shown that the compressive strength increased due to use of stone powder as a fine aggregate, while decreased when using ceramic as a coarse aggregate. It was remarked that the strength of concrete increased in excess of 20%, and leads to decrease in excess of 20% below the standard mix. So in concrete, stone powder is more effectively to used as a fine aggregate Hence in concrete, a stone powder can be effectively used as fine aggregate instead of river sand and WAC can be used as a standard coarse aggregate up to 20%.

Medina et al. [4, 5] investigated reuse of sanitary ceramic waste as a coarse aggregate. These wastes were applied as recycling coarse aggregate in partially replacement (15, 20, and 25) % of NCA in the manufacturing of structural concrete. The result conforms that recycled, eco-efficient concrete resent superior mechanical performance compared to standard concrete.

The possibility of using lignite base bottom ash as fine aggregate was researched by Dhavamant et al. [6], and mix ratio consists 100% replacement of NCA by WCA. The result shows that compressive strength with ceramic coarse aggregate concrete was improved. This improvement in compressive strength refers to reduction in w/c ratio, whereas enhancement upto 85% at 28 day and 95% at 56 days for ceramic waste aggregate concrete.

Anderson et al. [7] study tested impact of using clean, waste ceramic floor tiles and wall tiles, on compressive, tensile and flexural strength of hardened concrete. Where test results showed that the clean tiles had little effect on compressive strength. The per cent which used in replacement of course by recycles ceramic aggregate from 20 to 100%. The result showed decrease in compressive strength in comparison to reference concrete as well as split tensile strength has improved over reference concrete while flexural strength has decreased over reference concrete. Gonzalez el al. [8] study aimed to use of industrial clay brick waste from inadmissible pieces in ceramic precast facilities to replace simultaneously NFA and NCA in the production of concrete for precast pre-stressed beams. Replacement percentage of (20, 35, 50, 70,100) %, respectively, were used. The final finding of this research is that decrease in compressive strength reaching to 23% for 100% replacement. None of the reviewed papers used any testing for waste aggregate concrete (WAC) under torsion.

2 Physical and Mechanical Properties

OPC 53 Grades conforming to IS: 8112-1989 [9] is applied in experimental study with physical properties as in Table 1. Setting time and the test of cement crushing value were performed according to IS: 4031-1988 [10], the procedure is shown in

Table 1 Physical properties of material

S. no.	Physical properties	Cement	Coarse aggregate	Fine aggregate	Waste ceramic aggregate
1	Normal consistency (%)	30	–	–	–
2	Specific gravity	3.6	2.55	2.6	2.35
3	Initial setting time	42 min	–	–	–
4	Final setting time	600 min	–	–	–
5	7 days compressive strengthen	21.1	–	–	–
6	Fineness modules	41.4	6.99	2.65	6.98
7	Max size	–	20 mm	–	20 mm
8	Dry compact density (kg/m^3)	–	1618	1319	1325
9	Water absorption (%)	–	0.25	1.6	4.5
10	Crushing value (%)	–	12.86	–	14.33
11	Impact value (%)	–	20.2	–	24.2

(a)   **(b)**   **(c)**

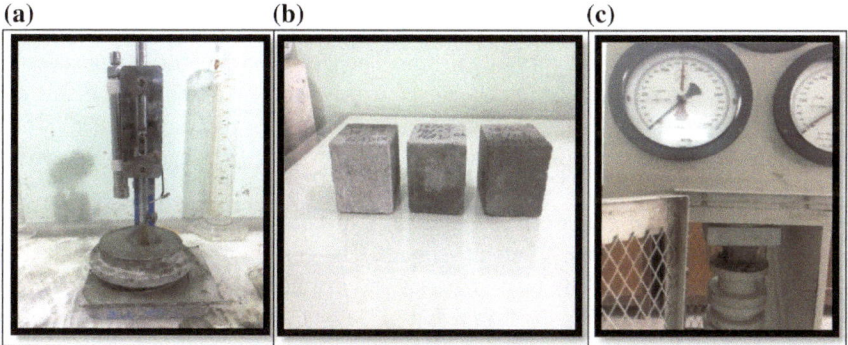

Fig. 1 **a** Setting time test for cement. **b** Cement cube specimens. **c** Testing of cement cube resistance to crushing

Fig. 1. The method which used to determine the relative strength of the cement tested by testing the resistance to crushing under compressive load was cement crushing value [7].

Coarse aggregate is locally prepared as a crushed stones according to a graded aggregate of normal size 10 and 20 mm as per IS: 383-1970 [11], applied in experimental work. Sieve analysis was used per IS 383:1970 [11] because the gradation of aggregate impacted for both hardened and fresh concrete as shown in Fig. 2a. Also some laboratory tests were made in order to find out the physical properties such as dry compact density (kg/m^3), fineness modules, water absorption, specific gravity, crushing value and impact value as per IS 2386-1963 [12] and are shown in Table 1.

(a)   (b)   (c)

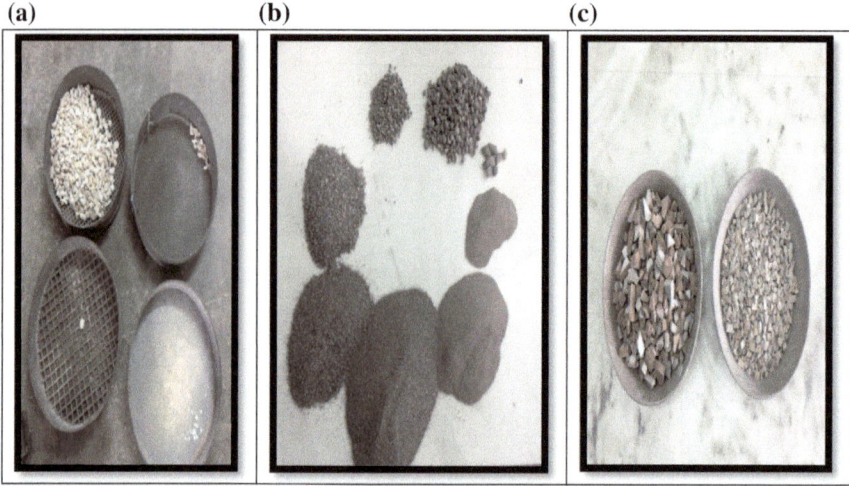

Fig. 2 **a** Coarse aggregate on sieve. **b** Fine aggregate after sieve analysis. **c** Waste ceramic coarse aggregate (20 mm, 10 mm) after sieving

The natural sand was used as a fine aggregate of the requirement of IS: 383:1970 [11], its divisions from 4.75 mm to 150 μm. The sieve analysis has done as per IS 383:1970 [11] as shown in Fig. 2b. Tests were made in order to find out it's physical properties like dry compact density (kg/m^3), specific gravity, water absorption and fineness modules as per IS 2386-1963 [12] and are shown in Table 1.

Waste ceramic coarse aggregates which used in experimental part were collected from many different ceramic stores in Aligarh. The collected waste ceramic coarse aggregates cleaned out of the dirty material and crushed by hammer into different particle size as shown in Fig. 2c. The result of sieving and other physical properties of ceramic waste as determined in the laboratory are shown in Table 1.

3 Mix Proportions

Concrete mix was designed as per IS10262-2009 [13] having a constant W/C ratio of 0.50. The concrete mixes of all specimen were made by definition proportion of (1:1.5:3) (Cement: Fine aggregate: Coarse aggregate), with the constant w/c ratio 0.5. Coarse aggregate was replaced with varying percentage with waste ceramic floor tiles which makes the concrete as most binder. Test increasing replacement ratios starting from 0% replacement (reference concrete) to 100% replacement were designed by aggregate replacement test. The named 25PC refers to reference concrete with (0%) replacement where 25 refers to M25 (grade of concrete) and PC refers to plain concrete. The replacement ratios test for the waste ceramic coarse aggregate series were 10%, 20%, 30%, 50% and 100%, were named 25WAC-10, 25WAC-20, 25WAC-30, 25WAC-50 and 25WAC-100, respectively. The number like 10 in the naming method denotes the replacement ratio, while W refers to waste ceramic, A refers to aggregate replacement and C refer to concrete.

4   Experimental Investigation

The Total of 72 test specimens were divided into six groups as shown below:

(1) **Group 1**: plain concrete

- 3 cubes, 6 beams, 3 cylinders

(2) **Croup 2**:10% replacement of NCA by WCA

- 3cubes, 6 beams, 3 cylinders

(3) **Croup 3**:20% replacement of NCA by WCA

- 3cubes, 6 beams, 3 cylinders

(4) **Croup 4**:30% replacement of NCA by WCA

- 3cubes, 6 beams, 3 cylinders

(5) **Croup 5**:50% replacement of NCA by WCA

- 3cubes, 6 beams, 3 cylinders

(6) **Croup 6**:100% replacement NCA by WCA

- 3cubes, 6 beams, 3 cylinders.

These specimens were casted in cube (150 * 150 * 150 mm), cylinder (length 300 mm * 150 mm diameter) and beam (100 * 100 * 500 mm) moulds and compacted on vibration machine, after 24 h the models were taken out from moulds and kept them in curing tank for 28 days. The concrete models were cured at 27 °C till the age of test. Four tests were conducted to determining compressive strengthen, split tensile strengthen, flexural strengthen and combined flexural and torsion for the optimal concrete mixes as mentioned in Fig. 3a–d for 10% replacement. All the tests were conducted as per IS 516–1959 [14].

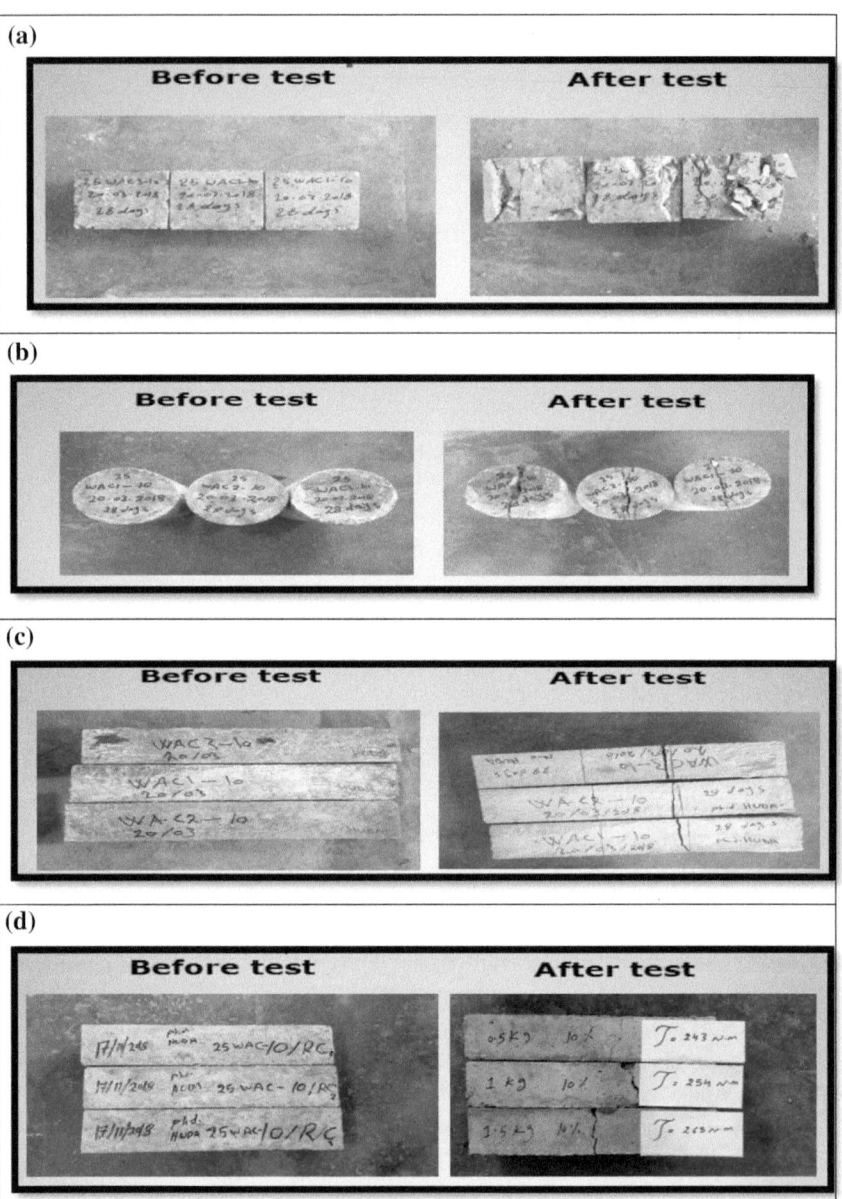

Fig. 3 **a** Compressive strength test: specimen before and after test. **b** Split tensile strength test: specimen before and after test. **c** Flexural strength test: specimen before and after test. **d** Combine flexural and torsion test: specimen before and after test

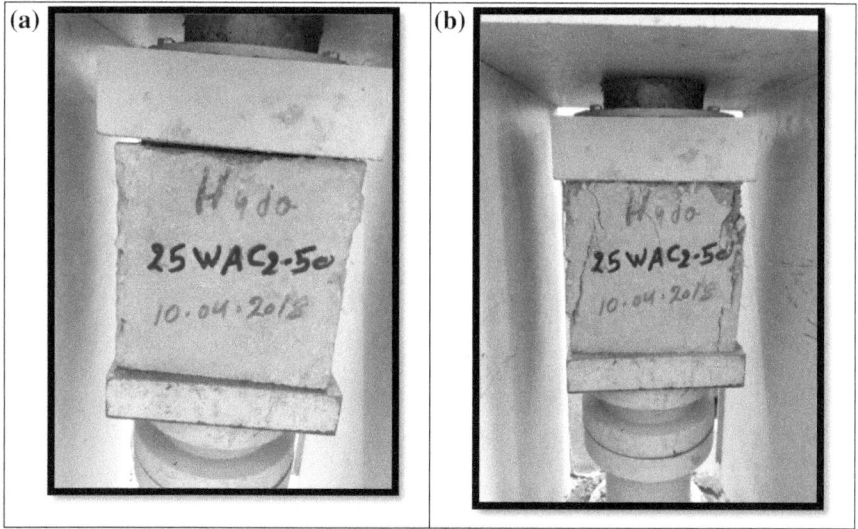

Fig. 4 **a** Sample in compression testing machine before fail **b** Sample in compression testing machine after fail

5 Hardened Concrete Properties

5.1 Compressive Strength

Compression testing machine was used for this test. Load was shedded gradually till the model fail as shown in Fig. 4a, b. The results in Fig. 5 show a typical hardened concrete sample after failure under compression loading test, showing the effect of WCA content into concrete mix on the 28 days compressive strength into the hardened concrete for same W/C ratio. The decrease in compressive strength in all models was noticed, where percentage error for 10% replacement is equal to 7.4% and for 100% replacement is equal to 30.73%. The final results are shown in Table 2. One reason for decreasing compressive strength of samples when increasing the percentage of aggregate replacement because of the increasing in the flaky aggregate, where waste ceramic aggregates have different sizes and shapes. Second reason may be low hardness and abrasion value of waste ceramic aggregate for decrease of strength.

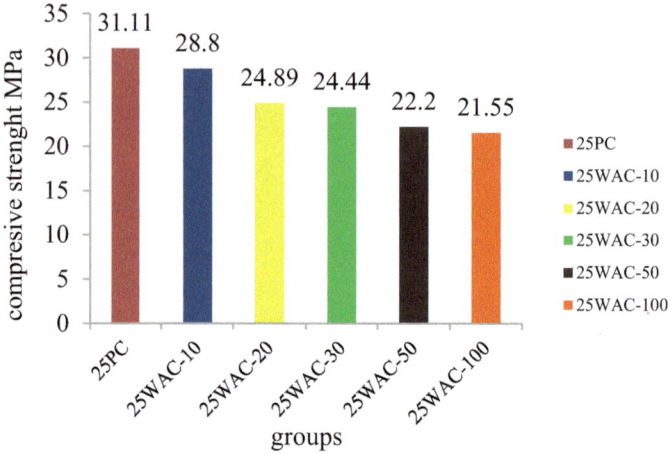

Fig. 5 Compressive strength test result

Table 2 Compressive, tensile, flexural, combined flexural and torsion test result

S. no.	S. name	Compressive strength 28 days (N/mm^2)	Split tensile strength 28 days (N/mm^2)	Flexural strength 28 days (N/mm^2)	Ultimate stress under torsion 243 N m (MPa)	Ultimate stress under torsion 254 N m (MPa)	Ultimate stress under torsion 265 N m (MPa)
1	25PC	31.11	3.69	6.08	3.25	3	2.75
2	25 WAC-10	28.8	3.37	6	4.25	4	3.75
3	25 WAC-20	24.89	3.25	5.6	3.25	3	2.75
4	25 WAC-30	24.44	3.18	5.13	3	2.75	2.5
5	25 WAC-50	22.2	3.14	4.75	2.25	2	1.75
6	25 WAC-100	21.55	2.95	4.67	2	1.75	1.5

5.2 Split Tensile Strength

This test has done by fixing cylindrical model horizontally between the load surface of compression test machine and the load was shed gradually until fail of cylinder along the height which is shown in Fig. 6a, b. The results in Fig. 7 show decrement in split tensile strength in all specimens, where percentage error for 10% replacement is 8.6% and for 100% replacement is 20.05%. The final output result is listed in Table 2. The lower performance of ceramic waste aggregate concrete in compared to plain concrete is possibly due to the intrinsic properties of the adhered mortar and its low adhesiveness to ceramic.

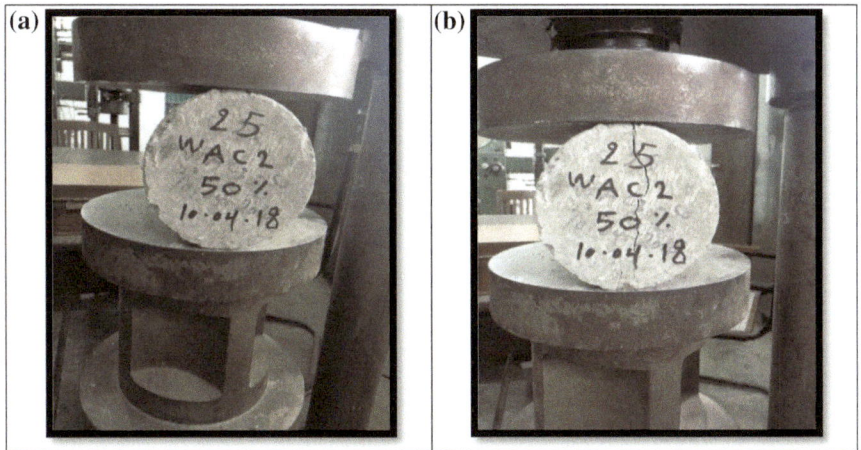

Fig. 6 **a** Sample in split tensile machine before fail. **b** Sample in split tensile machine after fail

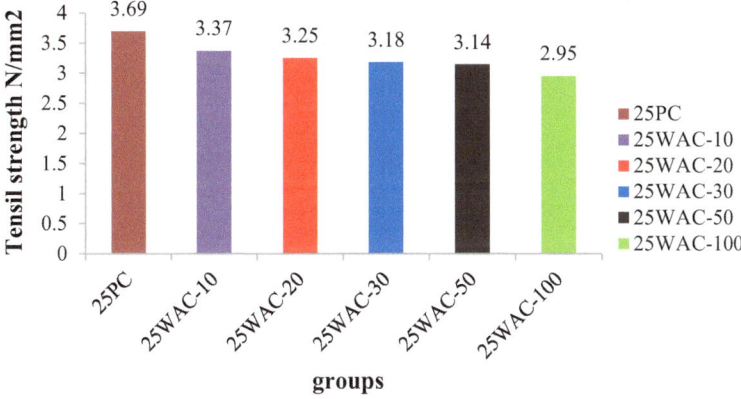

Fig. 7 Split tensile result

5.3 Flexural Strength

This test was used to determine the flexural strength of the models. Two-point load system was chosen as an effective span while testing the specimen. The load is employed until the failure of specimen takes place which is shown in Fig. 8a, b. The results are shown in Fig. 9 noted decrease in flexural strengthen in all models, where percentage of error for 20% replacement is equal to 7.89% and for 100% replacement is equal to 23.19%. The final results are shown in Table 2. The flexural strength decrease may be due to weak-bond formation in waste ceramic coarse aggregate.

Fig. 8 **a** Sample in flexural machine before fail. **b** Sample in split flexural machine after fail

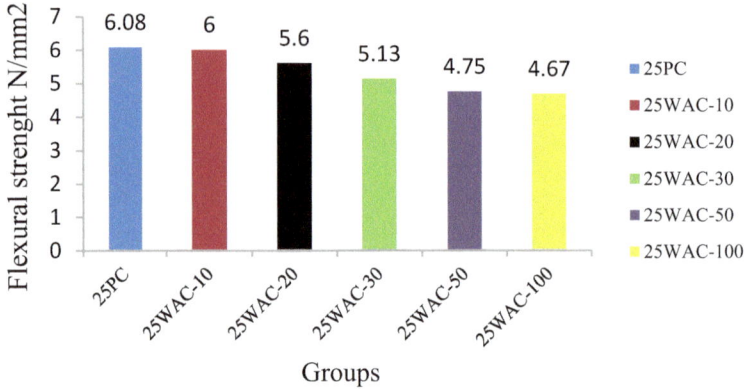

Fig. 9 Flexural strength result

5.4 Combined Flexural and Tensional Strength

This test involves the assessment of the combined effect of flexural and torsion (Fig. 10). Three samples for each group were tested. The load was applied and increased gradually till the failure occurred as shown in Fig. 11a–c. The final results in Fig. 11 noted increase in ultimate bending stress up to 20% replacement. Beyond 20% replacement WCA, ultimate bending stress is decreased as compared to reference concrete, where percentage of error for 20% replacement under torsion 243, 254 and 265 N m is equal to 0.0% and for 100% replacement under torsion 243,

Fig. 10 Experimental set-up for combined test flexural and torsion

254 and 265 N m is equal to 38, 41.66 and 45.45%, respectively. Final results are shown in Table 2. This result increase still may be because of flakiness and Pozzolanic property of ceramic tiles, and it may fill the gap completely, therefore, under torsion it may behave strongly.

6   Results and Discussions

Based on experimental investigations, following results have been found:

1. Compressive strength decreased 20% for 20% replacement where for 100%, it decreased to 30.73%.
2. Splitting tensile strength decreased 12% for 20% replacement where for 100%, it decreased to 20.05%.
3. Flexural strength decreased 7.89% for 20% replacement where for 100%, it decreased to 23.19%.
4. For combined flexural and torsion, ultimate bending stress increasing up to 20% replacement under torsion 243, 254 and 265 N m. For 10% replacement, increment is 30.7%, 30.7%, 30.7%, respectively, whereas for 20% replacement it is 0.0% for all and for 100% replacement under torsion 243, 254 and 265 N m it decreases to 38, 41.66 and 45.45%, respectively.

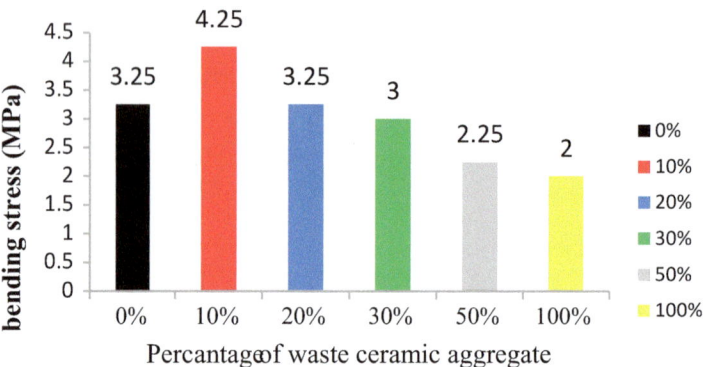

(a) Ultimate bending stress under torsion 243 N.m

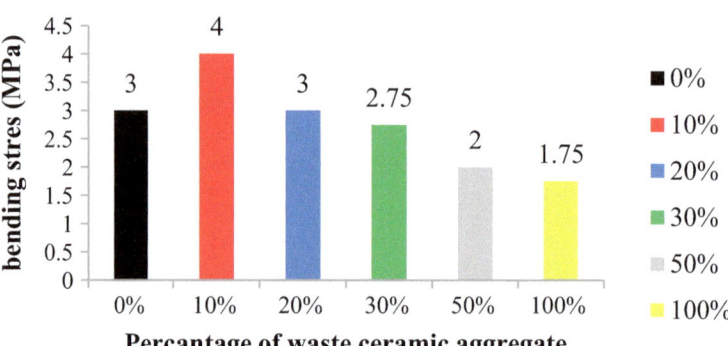

(b) Ultamite bending stress under torsion 254 N.m

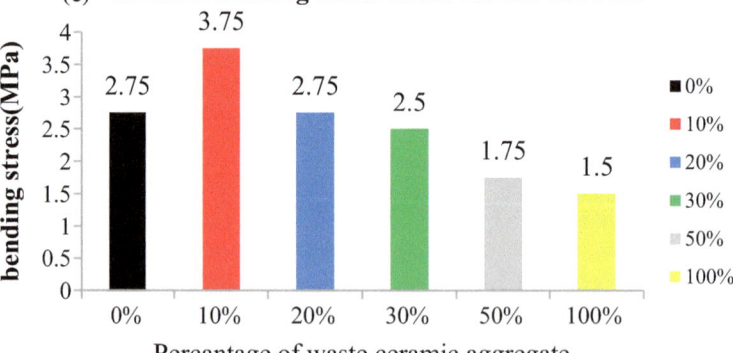

(c) Ultamite bending stress under torsion 265 N.m

Fig. 11 **a** Combined test result under torsion 243 N m. **b** Combined test result under torsion 254 N m. **c** Combined test result under torsion 265 MPa

7 Conclusion

On the basis of the experimental study, it is found that less than 12% strength reduction up to 20% replacement is observed in tensile and flexural strength except compression, whereas for combined state, it is increasing up to 20% replacement. Therefore, it is concluded that the replacement up to 20% is highly appreciable.

References

1. Nepomuceno MC, Isidoro RA, Catarino JP (2018) Mechanical performance evaluation of concrete made with recycled ceramic coarse aggregates from industrial brick waste. Constr Build Mater 165:284–294
2. Lavat AE, Trezza MA, Poggi M (2009) Characterization of ceramic roof tile wastes as Pozzolanic admixture. Waste Manag 29(5):1666–1674
3. Reddy VM (2010) Investigations on stone dust and ceramic scrap as aggregate replacement in concrete. Int J Civ Struct Eng 1(3):661
4. Medina C, De Rojas MS, Frias M (2012) Reuse of sanitary ceramic wastes as coarse aggregate in eco-efficient concretes. Cement Concr Compos 34(1):48–54
5. Medina C, De Rojas MS, Frias M (2013) Properties of recycled ceramic aggregate concretes: water resistance. Cement Concr Compos 40:21–29
6. Dhavamant DS et al (2013) Int J Curr Eng Technol
7. Anderson DJ, Smith ST, Au FT (2016) Mechanical properties of concrete utilising waste ceramic as coarse aggregate. Constr Build Mater 117:20–28
8. González JS, Gayarre FL, Perez CLC, Ros PS, Lopez MAS (2017) Influence of recycled brick aggregates on properties of structural concrete for manufacturing precast pre stressed beams. Constr Build Mater 149:507–514
9. IS: 8112-1989 (1989) Specification for 43 grade ordinary Portland cement. Bureau of Indian Standards, New Delhi
10. IS 4031 (1988) Determination of compressive strength of hydraulic cement. Bureau of Indian Standard, New Delhi
11. IS: 383-1970 (1970) Specification for coarse and fine aggregates from natural sources for concrete. Bureau of Indian Standards, New Delhi
12. IS: 2386 (Part I–VIII)-1963 (1963) Indian standards method of testing for concrete. Bureau of Indian Standards, New Delhi
13. IS: 10262-2009 (2009) Indian standard concrete mix proportioning. Bureau of Indian Standards, New Delhi
14. IS: 516-1959 (1959) Method of tests for strength of concrete. Bureau of Indian Standards, New Delhi

Effective Grid-Connected Solar Home-Based System for Smart Cities in India

Iram Akhtar, Sheeraz Kirmani and Majid Jamil

Abstract The reduction in fossil fuel reserves on a global base has initiated a serious exploration of renewable energy sources to satisfy the present energy demands. Among the different alternative energy sources, the solar energy is a clean and unlimited energy source; hence, it could play a vital role in the solar home-based system in a developing country like India. There is a requirement for a continuous supply of energy, which cannot be satisfied by the alone solar system because of regular variation of the solar irradiance. It has been found that installed capacity is extended 20 GW in February 2018 in India, and it will achieve around 100 GW in 2022. Hence, the grid-connected solar home-based system is now being used to fulfill the energy demand. In this paper, cost of 20 kW grid-connected solar home-based system in India is analyzed. This work presents the finest scheduling of grid-connected solar home-based system for efficient and economical operation of the system. The proposed system is best suitable for a rural area having the lowest price of energy as INR 3.15/kWh.

Keywords Solar home-based system · Grid · Cost · Energy · PVsyst V6.73

1 Introduction

The energy demand is changing day by day; hence, there is a need for a reliable energy source which fulfills the present energy scenario. The human depends on electricity and needs always accessible. The solar home-based system is a clean and reliable way to electrify the rural area. The grid-connected solar system has the advantage of real utilization of energy. When there is surplus power available, then it will be sold out to the utility. When the solar home-based system generates less energy in comparison with load demand, then this extra power can be taken from the grid. But it is a challenging task to connect the solar energy system to the grid.

I. Akhtar (✉) · S. Kirmani · M. Jamil
Department of Electrical Engineering, Faculty of Engineering and Technology, Jamia Millia Islamia, Jamia Nagar, New Delhi 110025, India

© Springer Nature Singapore Pte Ltd. 2020
S. Ahmed et al. (eds.), *Smart Cities—Opportunities and Challenges*,
Lecture Notes in Civil Engineering 58,
https://doi.org/10.1007/978-981-15-2545-2_50

607

Hence, the proper control system is needed to perform this task. Different systems for integration of the solar system to the grid have been developed, but the main concern is stability and the power quality which degrades. Therefore, effective inverter size is chosen according to the power supply to the grid. If the high rating inverter is chosen, then cost would be increased, and when less rating inverter is chosen, then it affects the efficiency of the entire system. The price analysis for 8 kW solar photovoltaic plant is done and established a scheme based on the possible approximations completed for a selected area of 50 m^2. The solar panel efficiency is the main concern, and it is affected by the accuracy of the algorithm used for tracking of maximum power point [1] and the converter power conversion efficiency [2, 3]. Therefore, DC/DC converter efficiency is also important to concern. The frequency modulation switching can be used for controlling the DC/DC converters [4–7]. It has added to the decrease in electromagnetic interference by power spectrum diffusion [8, 9]. The solar system is becoming very famous to electrify the industries/colony because it is an economical solution and it also reduces the carbon footprints hence reduces the effect of global warming. The thermal power plants do not offer a better environment as the efficiency of coal-based power plant reduces, this would increase the fuel intake and carbon emission. So, there is a need for good matching between the renewable energy source and coal-based power plant accommodation [10]. On the other hand, the penetration of solar schemes rises, this is main to a problem of voltage difference and transient voltage variability with the case of less coupling with the grid. The significant penetration of solar units has an influence on the short-range voltage and transient stability of the proposed system, and it is not limited to the distribution network but also affects the entire system. The proposed work provides a design of a solar photovoltaic-based grid-connected system to electrify the different areas in India. This could improve the performance of the power system without impious the norms of system and restrictions and enable an effective contribution of solar power to the grid system. In this paper, solar home-based system design is presented in Sect. 2. The economic operation of the grid-connected solar home-based system is presented in Sect. 3. In Sect. 4, cost analysis of the proposed system is described. In Sect. 4, outcomes and discussion are presented. Finally, concluding declarations are presented in Sect. 5.

2   Grid-Connected Solar Home-Based System

The grid-connected solar home-based system comprises a solar charge controller, electric meter, inverter, and PV modules as presented in Fig. 1. To get the cost of 8 kW solar-based home system in India, there is a need to measure solar radiation over a different period of time. We defined the value from January to December for Delhi site. The input and output of solar radiation are very important to install any solar project. So, we can know about the efficiency of any plant by knowing how much solar irradiance is used for electricity generation. The selected area for the

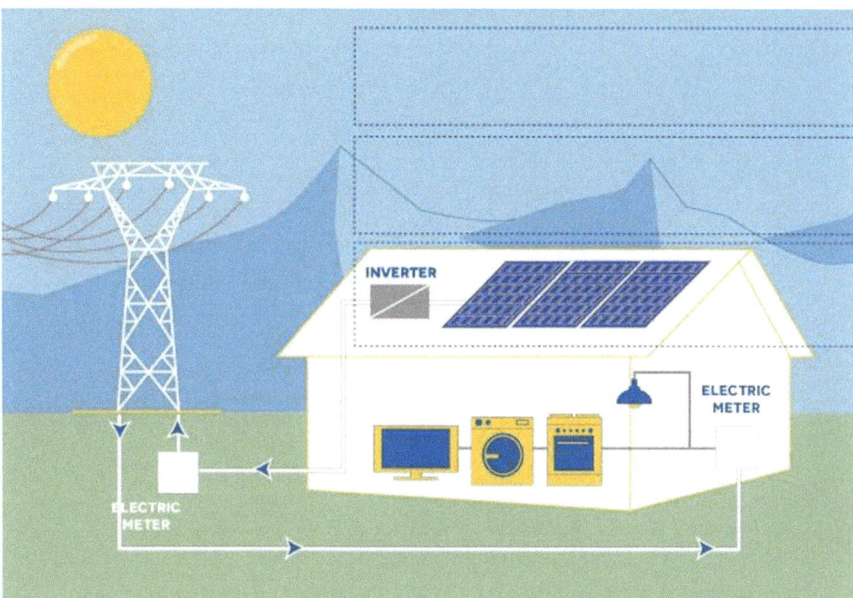

Fig. 1 Grid-connected solar home-based system

expected plant capacity is chosen as 50 m². The 8 KW solar system is linked to the grid via an inverter. The power consumption demand is determined by the connected loads. The 8 KW solar home-based system is considered for electrification of rural area.

The total solar panels energy needed = 8 kWh/day.

The quantity of solar panels needed can be known by considering complete power necessary and the solar panel watt rating.

The number of solar panels required = 8000/270

$$= 29.62.$$

Real requirement = 30 modules.

Therefore, 30 solar modules of 270 W rating are desirable for this grid-connected solar system.

Table 1 shows the solar model specifications. Whereas, the inverter size depends on the appliances' rating so for safety concerns two inverters are taken with 5000 W capacity every one. Table 2 displays the inverter description. Intelligent control is used to control the inverter output, and it should be constant for reliable operation.

The geographical site is very significant and it is essential to collect data from different hubs and Delhi rural area is chosen for this work. In this site, the latitude is 28.58° N and longitude is 77.20° E.

Table 1 Solar model specifications

Parameter	Specifications
The voltage at maximum power (V)	31.01
Current at maximum power (A)	7.93
Module voltage (V)	24
Cell numbers	60
Short-circuit current (A)	8.50
Open-circuit voltage(V)	36.86

Table 2 Inverter description

Parameter	Specifications
Maximum solar array voltage (V)	105
Maximum efficiency (%)	93
Surge power (kVA)	10
Rated power (kVA)	5
Output voltage (V)	230

3 Economical Operation of Grid-Connected Solar System

Economical operation is essential for the effective and reliable operation of the system. When the solar system generates power which is more than the required load, then this power is sold to the utility. Whereas when the solar system in some environmental conditions generates less power than load, then this power is got from the grid.

The economical operation for the grid-connected solar system is defined as below.

Step 1—Observe the solar system output power.
Step 2—Observe the home load.
Step 3—Calculate the extra power by

$$P_{extra} = P_{solar} - P_{load}$$

Step 4—Calculate the grid power by

$$P_{grid} = k * P_{extra}$$

where for lossless system k is equal to 1.
Step 5—Check if, $P_{load} > P_{solar}$, then extra power is got from the grid otherwise power is sold out to utility.

P_{extra} is the extra power, P_{solar} is the solar powers, P_{load} is the load power and P_{grid} is the grid power.

By using these steps, economical operation can be achieved. The electric meter is used to calculate the units taken from the grid or sold out to the grid.

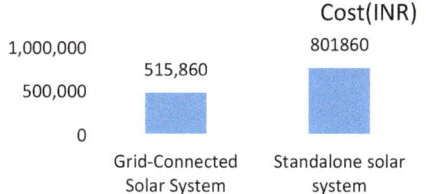

Fig. 2 Cost of the grid-connected solar system and standalone solar system

4   Cost Analysis of Proposed System

Cost of solar panels can be calculated by using the price of a single module. We use the Vikram 270 W Solar Panel Polycrystalline with the cost of 8960 INR. We use 30 modules, and hence, the total cost for the complete panel is equal to 268,800 INR. Whereas the cost of 2 Flin Energy Lite Solar Hybrid Inverter—5 kVA/ 5000 W is equal to 71,000 INR. The system costs (fires, switches, fuses, O&M, etc.) are equal to 128,000 INR approx. The Su-Kam MPPT Solar Charge Controller cost is equal to 48,060 INR. Hence, the total expected grid-connected solar system cost is 515,860 INR.

On the other hand, if the grid-connected system is not used, then an extra battery and battery charge controller is needed. The cost of Exide battery and battery charge controller is 286,000 INR. Hence, the total expected standalone solar system cost is 801,860 INR as shown in Fig. 2. Therefore, we can say that grid-connected solar home-based system gives the saving of 286,000 INR. If solar energy is not accessible, then the system automatically shifts on the main grid and load is supplied by the grid.

5   Results and Discussions

The planned 8 kW grid-connected solar scheme has a peak load of about 8 kW, and therefore, the cost of energy by this system would be 3.15 INR per unit. The expected grid-connected solar system cost is 515,860 INR, and this system is mounted for 25 years. Thus, cost/year would be 20,634.40 INR. The proposed system would provide approximately 20 units per day. The overall bill for these units, when purchased from the grid, is 36,000 INR. By subtracting yearly units' price for the proposed system, we will get the actual saving per year.

$$\text{The total saving} = 36,000 - 20,634.40$$
$$= 15,366 \, \text{INR/year}$$

The performance of 8 kW grid-connected solar system is assessed using PVsyst software. The monthly energy production and maximum global radiation are

Table 3 Different parameters by PVsyst software

Months	GlobHor (kWh/m²)	Coll. plane (kWh/m²)	System output (kWH/day)	System output (kWH)
January	3.81	5.70	38.35	1189
February	4.89	6.59	44.31	1241
March	6.07	7.17	48.21	1494
April	6.83	7.04	47.35	1421
May	7.16	6.61	44.46	1378
June	6.55	5.79	38.96	1169
July	5.37	4.86	32.69	1013
August	5.16	4.96	33.32	1033
September	5.69	6.21	41.77	1253
October	5.31	6.76	45.48	1410
November	4.28	6.27	42.14	1264
December	3.71	5.87	39.45	1223
Year	5.41	6.15	41.34	15,088

calculated, and it varies due to the temperature effect on the solar panel. Table 3 shows the different parameters throughout the year plus the effect of shading. The GlobHor is 3.81 for January and 4.28 for November, and this variation displays that this factor also changes throughout the year. The Coll. plane is 5.70 for January and 6.27 for November. This factor does not alteration fast, and this may be about less vary all over the year.

The solar irradiance changes with time as shown in Fig. 3, and this proves that the solar system does not give same output at all time because the cloudy

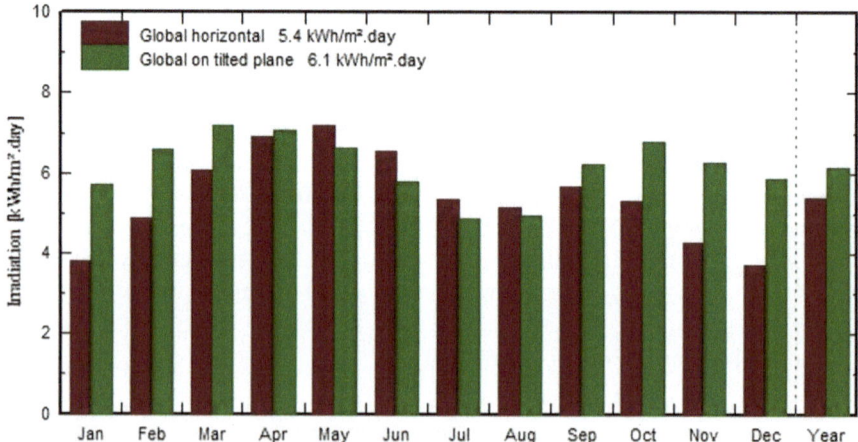

Fig. 3 Solar irradiation variation throughout the year

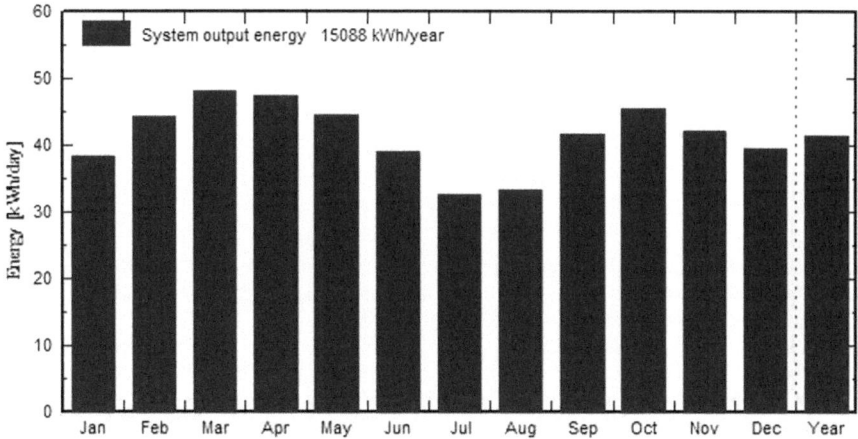

Fig. 4 Output energy distribution during the year

atmosphere varies the irradiance so it is hard to attain the same irradiance throughout the year. Hence, there is a necessity for MPPT manager. MPPT manager provides constant output regardless of the solar irradiance changes.

Output energy distribution is shown in Fig. 4. The output energy varies throughout the year because of the solar irradiance changes with climate conditions. The graph is generated by PVsyst software based on longitudinal and latitude information of the place.

The maximum power of solar modules be determined by the output voltage and current, and it depends on the solar irradiance. Uncertainty solar irradiance is high than maximum power which would be attained easily without using MPPT controller, but due to the restriction of the solar scheme, the irradiance is not constant all the time; hence, MPPT controller is an essential component of any solar-based system.

*1 USD is equal to 68.82 INR as on July 5, 2018.

6　Conclusion

This work presents the economical scheduling and costs analysis of the grid-connected solar home-based system for efficient and reliable operation in the rural area in India. Relating to the grid-connected solar system and standalone solar system that has been used for solar power generation, the outcomes show that the proposed grid-connected solar system is best suitable for the residential purpose in the rural area based on overall cost. The cost/year of the proposed system would be 20,634.40 INR. The overall bill for the same units, when purchased from the grid, is

36,000 INR. Hence, this system gives the total saving 15,366 INR/year. Besides, the performance of 8 kW grid-connected solar system is assessed using PVsyst software.

References

1. de Brito MAG, Galotto L, Sampaio LP, de Azevedo e Melo G, Canesin CA (2013) Evaluation of the main MPPT techniques for photovoltaic applications. IEEE Trans Ind Electron 60 (3):1156–1167. https://doi.org/10.1109/tie.2012.2198036
2. Edwin FF, Xiao W, Khadkikar V (2014) Dynamic modeling and control of interleaved flyback module-integrated converter for PV power applications. IEEE Trans Ind Electron 61 (3):1377–1388. https://doi.org/10.1109/TIE.2013.2258309
3. Andrade AMSS, Schuch L, da Silva Martins ML (2017) High step up PV module integrated converter for PV energy harvest in FREEDM systems. IEEE Trans Ind Appl 53(2):1138–1148. https://doi.org/10.1109/tia.2016.2621110
4. Oederra O, Kortabarria I, de Alegra IM, Andreu J, Grate JI (2017) Three-phase VSI optimal switching loss reduction using variable switching frequency. IEEE Trans Power Electron 32 (8):6570–6576. https://doi.org/10.1109/TPEL.2016.2616583
5. Konjedic T, Koroec L, Trunti M, Restrepo C, Rodi M, Milanovi M (2015) DCM-based zero-voltage switching control of a bidirectional DC-DC converter with variable switching frequency. IEEE Trans Power Electron 31(4):3273–3288. https://doi.org/10.1109/TPEL.2015.2449322
6. Al-Hoor W, Abu-Qahouq JA, Huang L, Mikhael WB, Batarseh I (2009) Adaptive digital controller and design considerations for a variable switching frequency voltage regulator. IEEE Trans Power Electron 24(11):2589–2602. https://doi.org/10.1109/TPEL.2009.2031439
7. Tse KK, Ng RW-M, Chung HS-H, Hui SYR (2003) An evaluation of the spectral characteristics of switching converters with chaotic carrier-frequency modulation. IEEE Trans Ind Electron 50(1):171–182. https://doi.org/10.1109/TIE.2002.807659
8. Yang Z, Li H, Li Z, Halang WA (2016) Spectrum calculation method for a boost converter with chaotic PWM. In: Proceedings of the 18th European conference on power electronics and applications, pp 1–6 (2016). https://doi.org/10.1109/epe.2016.7695481
9. Liou WR, Villaruza HM, Yeh ML, Roblin P (2014) A digitally controlled low-EMI SPWM generation method for inverter applications. IEEE Trans Ind Inf 10(1):73–83. https://doi.org/10.1109/TII.2013.2261078
10. Wang C, Lu Z, Qiao Y (2013) A Consideration of the wind power benefits in day-ahead scheduling of wind-coal intensive power systems. IEEE Trans Power Syst 28(1):236–245. https://doi.org/10.1109/TPWRS.2012.2205280

Exploring the Attributes of Smart City from Organisation's Perspective

A Study Based on Prayagraj, India

Arpita Agrawal

Abstract As, India is resonating its commitment to sustainable development at global fronts, it is essential to consider that the sustainable development of a nation is derived from the well-being of its regions at granular level. However, the dichotomy is that though India is marching toward being a global leader, its cities are finding it difficult to combat with the challenges of migration, urbanization, and disparity. Acknowledging the challenges faced by some of the emerging cities, the Smart Cities Mission was launched in the year 2015 to strengthen the economic, social, and physical infrastructure of 100 cities. However, the achievements of the mission have not been very convincing due to several reasons. One of the major reasons identified is the lack of region-specific study to explore the persisting challenges. Considering the sparsity of such studies, this study is an attempt to explore and highlight the fundamental challenges faced by different types of organizations in Prayagraj (earlier known as Allahabad, one of the 100 cities proposed for smart cities mission). A comprehensive literature review and field survey have been conducted for the study followed by usage of principal component analysis to estimate significance of various attributes of acting as barrier for organizations operating in the city. The findings of the study suggest that factors such as lack of proper power supply, unsupportive government policies and political environment, poor technology adoption rate and unavailability of skilled workforce act as major hurdle. There is a need to focus on removal of major barriers, i.e., energy by using alternate source of energy, ensure supportive investment, and development policy to enhance the attractiveness of the city. The methodology followed could be used to analyze the city and stakeholder specific issues in other proposed smart cities.

Keywords Sustainable development · Regional development planning and policy · Regional economic activity · Factor models

JEL Classification C38 · R11 · R58 · Q01

A. Agrawal (✉)
National Institute of Financial Management, Faridabad, India

© Springer Nature Singapore Pte Ltd. 2020
S. Ahmed et al. (eds.), *Smart Cities—Opportunities and Challenges*,
Lecture Notes in Civil Engineering 58,
https://doi.org/10.1007/978-981-15-2545-2_51

1 Introduction

Literature reviewed highlight that development in India has been uneven across states. Industrial states like Delhi, Gujarat, Maharashtra, etc., have tended to leapfrog in while others like Orissa, UP and Bihar have lagged behind [1, 2]. This has been a major cause of growing regional disparity in India with no sign of convergence [3]. Earlier, economists and development organizations used to undermine regional importance to national growth [4] that has resulted into unbalanced growth. Consequently, some regions are over exploited while still ample scope of development could be seen in certain other regions. This regional disparity has been one of the factors leading to reduced national performance [5]. In the recent years, there is a growing concern about regional disparities in India and this has been endorsed by many researchers [6] too. Further, as per Harter. G. et al., equality is one of the three goals to be achieved in order to ensure overall development [7]. Acharya views such disparities as speed-breakers in the way of India's long-term growth prospects. Moreover, Howes et al. argue that though the economic performance at the country level has been improving yet in order to ensure that reforms pick up momentum, each region should take the initiative [8].

Recently, policymakers have also started showing concern toward the issue of regional disparity and special attention is being paid toward cities which have been left behind. Acknowledging this, Smart Cities Mission was also launched in the year 2015 with a view to promote balanced development and sustainable across 100 cities.

Heterogeneity of cities lead to substantial differences in economic performance and there could not be any single ideal path to attain balanced and sustained development in all regions thus different policies are required for different cities depending upon their specific requirements. Thus, in order to promote development at the city level, it is essential to first identify and understand the pain points of the cities from the view point of vivid stakeholders. Though citizens and other stakeholders do communicate their sentiments and dissatisfaction about different aspects of the city at different forums, however, not much empirical researches have been conducted with a view to understand the issues and solve it systematically. Drèze and Sen suggested that "Uttar Pradesh can be seen as a good case to study the development in the lagging regions regarding a number of important aspects of social progress and well-being" [9]. Thus, to fill this literature gap, the study makes an attempt to identify the major issues faced by the organizations in one of the prospective smart city, "Prayagraj." The reason for selecting Prayagraj is that it situated in Uttar Pradesh, a state with (13) highest number of proposed smart cities. Further, it is an apt case of the city strained with the challenges of urbanization.

The study is based on the variables, i.e., major aspects of a city that impacts organizations, identified from the literature reviewed. These aspects are then analyzed empirically to dig out the challenges faced by the organizations. Further to unveil the potential of the region, a comprehensive study of interrelated and supportive aspects has been undertaken and certain suggestions have been made.

2 Objectives of the Study

- To identify the challenges faced by organizations in Prayagraj
- To explore the strength and weakness of Prayagraj city by using statistical tool
- To recommend strategy for combatting the challenges faced
- To identify the factors to work upon to promote overall development of the city under Smart Cities Mission.

3 Review of Relevant Literature

Harter. G. et al. view smart cities are those that offer a better and more sustainable lifestyle to their constituent in terms of safety, security, transport connectivity, public services, etc. [7]. Different studies focus on different factors of growth like one theory has suggested that urbanization is the crucial factor leading to economic growth of the regions [9]. As per literature reviewed, potential for growth exist in every region across the world [10] and different attributes of a region are viewed as either positive or negative factor the regions own development. In order to draft a strategy for a regions/cities development, it is essential to identify the strengths and weakness of that region/city.

Truly, the role of residents could not be undermined as rightly emphasized that earlier economic growth was tried to be achieved by means of large-scale infrastructure development but today, residents are the motors for development and play superior role in advancement [11]. Further, it is also observed that urbanization has a close nexus with industrialization and overall economic development [12]. However, recently, it has been seen that urbanization has led to challenges such as transport congestion, environmental, health and educational impact of poor living conditions, and increased rate of crime. Further, many cities are not organized in productive ways because of which structural regulatory and institutional constraints hinder productivity and prevent specialization and trade [13].

Demurger et al. view that the variables such as geographical location, proximity to industrial conglomerates, and differential policies of government are crucial to explain difference in economic growth across different regions [14]. Reynolds argues that the administrative and political competences of government are important explanatory variable of development [15].

Furlan et al. emphasize that, success is driven by a combination of various factors such as efficiency, flexibility, innovation, and globalization [16].

Researchers have even mentioned that the successful cities are competitive centers of innovation that has capability to attract young, highly skilled, and talented workers [17]. In order to pace with economies are required to achieve high growth in value-added sectors employing highly qualified human resource and get rid of less profitable sectors, which utilize comparatively less qualified human resources [18]. But the problem is that we cannot ignore these sectors and their workers. Along with the development of city, it is essential to create room for the poorest people also. Lagging regions are required to reorient their economies in order to safeguard jobs and to diversify [19].

As recommended by Commonwealth Association for Public Administration and Management (CAPAM), the solution for the development are found there where public engagement and new technologies converge [20]. Some researchers have emphasized that SME's partnership have critical impact on economic development [21, 22]. While some view that the proximity to renowned universities enhances the competitiveness of the regions [23]. Some researchers have suggested that concentration of a particular economic activity in a region leads to higher growth rates in that region [24, 25]. It has also been said that the well-connected regions are hubs for wealth generation [26].

It has been found that a diversified economic base also supports the growth and attracts human capital [27]. The study conducted by OECD has also mentioned that there are many benefits of having diversified economies [28].

Literature review has brought out that explanation of economic performance of a region goes beyond narrow measures of economic variables to incorporate political and social forces [29–32]. Based on the literature review, the study incorporates 32 major attributes of the city that may affect an organization and uses multivariate analysis to test how these attributes of the city, Prayagraj affects the organizations operating in the city and how the findings of the study could be used to determine the attractiveness of the city. The attributes considered are listed in Table 1.

Table 1 Attributes of the city that affect organizations

Sl. no	Attributes of the city that affect organizations	Sl. no.	Attributes of the city that affect organizations
1	Availability of input	17	Parking spaces
2	Input cost	18	Proximity to the market
3	Proximity to supplier	19	Taxes
4	Unskilled labor availability	20	Training facility
5	Unskilled labor cost	21	Airways
6	Water supply	22	Skilled labor availability
7	Environment and climate	23	Land availability
8	Health and hygiene	24	Land cost
9	Image of city	25	Skilled labor cost
10	Natural resources	26	Political environment
11	Intra-city road transport	27	Government support
12	Inter-city road transport	28	Power supply
13	Telecommunication	29	Railways
14	Technology adoption rate	30	Financial institutions
15	Public authorities in the city	31	Bordering cities
16	Research and development	32	Incentives

4 Methodology

Considering the exploratory nature of the study, a comprehensive literature and field survey have been conducted. Primary as well as secondary data have been used. The secondary sources of data being various government publications, research papers, and other available material, whereas the primary data has been collected from the survey conducted using comprehensive questionnaire and informal interviews.

4.1 Sampling

The research applies non-probability purposive sampling toward organizations located in the city of Prayagraj that was conducted for a month. Samples for the study consisted of organizations in different sectors in the city (Indicated as respondent). The purposive sample of 50 industries in region was taken, out of which 41 industries have responded leading to response rate of 82%.

4.2 Data Collection

Questionnaires were handed over to managing person of the respective organization and were requested to fill them on the spot. It has been ensured that questionnaires are filled by representatives from the organizations having sound knowledge about, its performance and the range of factors affecting them. Moreover, informal face to face interviews have been also conducted in order to get a deeper insight into the problems faced by them. In addition, filled-in questionnaires were thoroughly analyzed to ensure consistency in data provided by the respondents. The data was collected using a well-structured questionnaire comprising of two parts wherein Part I encompassed profile of industry including industry type, location within the city and their performance while Part II included questions related to different attributes of the city which may have impact on performance of organizations, these required the organizations to rank the impact of attribute on a Likert five-point scale [33] ranging from 1—"no influence"—to 5—"very high influence." Higher value means the level of influence of the attribute on performance is very high, while lower value signifies that the given attribute has lesser impact on performance. The attributes considered in the questionnaire have been identified on the basis of the various literature reviewed and are listed in Table 1.

5 Analysis, Results, and Discussions

The results of data analysis with respect to these factors are presented in the following sub-sections.

5.1 Descriptive Analysis

Profile of the Sample: The sample contained 17.1% of education industries, 14.6% of SME's, 12.2% of recreation industries (tourism) followed by 9.8% of hospitality, unorganized retail and heavy industries. The summary of the sample is shown in Fig. 1.

    Average Performance of Industry Sample: The performance has been measured on a given scale of 1–5 where 5 signifies outstanding performance, whereas 4, 3, 2 and 1 signify above average, average, below average, and poor performance, respectively. The summary of sample performance has been shown in Fig. 2.

    Of all the industries considered, organizations from education industry were the best performer with average performance score of 3.7 out of 5, followed by unorganized retail having a score of 3.5, while hospitality, SME, and IT were the worst performers with mean performance score of 1.6, 1.22, and 1.6, respectively. The score of large-scale industry stood at 1.8. Thus, it suggests that the organizations in the above four industries are performing below average and hence require improvement.

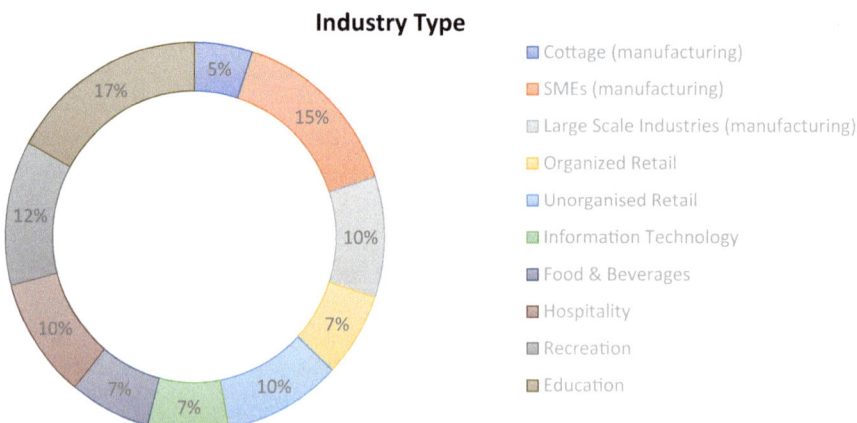

Fig. 1 Profile of sample taken from Prayagraj. *Source* survey conducted

Fig. 2 Average performance
of industry sample. *Source*
survey conducted

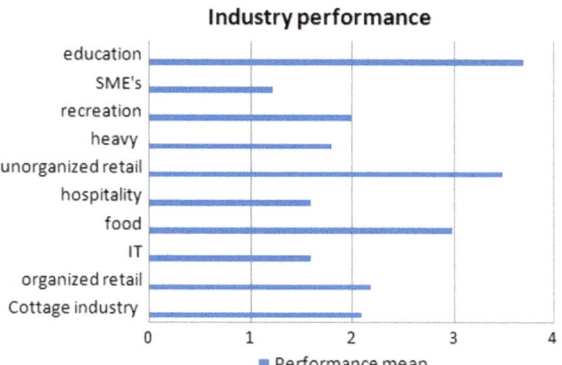

5.2 Inferential Analysis

The data was tabulated and statistically analyzed using the tool SPSS. The analysis has been conducted at two levels, firstly, data was put to "One-Sample Kolmogorov-Smirnov Test" and one-sample t-test, then categorized data was subject to reliability analysis and varimax rotational factor analysis.

First-Level Analysis: As per the finding of One-Sample Kolmogorov-Smirnov Test, the variables proved to be normally distributed. Therefore, parametric test was applied. Further, one-sample t-test has been applied to identify inconsistencies in the dataset and to compare the mean scores of ratings by different organizations. This has assisted in identifying the relative importance of various city attributes for the sample organizations. The results of "one-sample t-test" at $p < 0.01$ reflecting the relative importance of various factors has been shown in Table 2.

Table 2 reveals that the "Means" of most of the evaluation factors are significantly greater than 3 showing above average influence of factors on industry performance. However, the most important ones in descending order can be cited as: power supply (4.1707) followed by technology adoption (3.9024), skill labor availability (3.8049), government support (3.7073), skilled labor cost (3.5123), and intra-city road transport (3.5122). Further the bottlenecks associated with factors like water supply (2.6341), incentives (2.561), natural resources (2.2927), and airways (2.0732) are the least influential for industries present in the region. Thus, the mean scores indicate that power supply is the major obstacle followed by the technology adoption, skill labor availability, etc., for the respondents in the region under study.

Second level analysis: The 32 evaluation attributes have been skimmed down into 8 dimensions using the varimax rotational factor analysis. The main objective of using this is to reduce multiple variables to a lesser number of underlying factors based on correlation among factors [35]. Prior to factor analysis, reliability test for the same was conducted.

Table 2 Relative influence of factors for 41 industries

Sr. no	Evaluation attributes	Mean[a] (avg score on Likert scale)	Std. deviation	p value	Ranking mean
1	Raw material availability	2.9024	1.56213	0	19
2	Raw material cost	2.7805	1.36953	0	22
3	Proximity to supplier	3.0732	1.4898	0	16
4	Unskilled labor availability	3	1.04881	0	18
5	Unskilled labor cost	2.8293	0.99756	0	20
6	Water supply	2.6341	1.31826	0	24
7	Environment and climate	3.439	1.14124	0	8
8	Health and hygiene	3.3171	1.05922	0	11
9	Image of city	3.4146	1.20365	0	9
10	Natural resources	2.2927	1.36462	0	25
11	Intra-city road transport	3.5122	1.12076	0	6
12	Inter-city road transport	3	1.20416	0	18
13	Technology adoption	3.9024	1.26105	0	2
14	Telecommunication	3.4878	1.16452	0	7
15	Public authorities in the city	3	0.86603	0	18
16	R&D	2.7561	1.44535	0	23
17	Parking	3.122	1.63088	0	15
18	Proximity to the market	3.1951	1.43561	0	14
19	Taxes	3.4878	1.00304	0	7
20	Training facility	3.3659	0.99388	0	10
21	Airways	2.0732	1.3673	0	26
22	Skill labor availability	3.8049	0.74898	0	3
23	Land availability	3.2439	1.13535	0	13
24	Land cost	3.1951	1.03004	0	14
25	Skilled labor cost	3.5123	0.9778	0	5
26	Political environment	3.0732	0.93248	0	16
27	Government support	3.7073	0.84392	0	4
28	Power supply	4.1707	0.80319	0	1
29	Railways	3.2927	1.36462	0	12
30	Financial institutions	3.0488	0.77302	0	17
31	Bordering cities	2.8049	1.26924	0	21
32	Incentives	1.9823	1.3465	0	27

[a]Value on Likert scale [34]: 1 = No influence; 2 = Little influence; 3 = Average influence; 4 = Somewhat high influence; 5 = very high influence

Reliability test for Factor Analysis: As the pre-analysis for the suitability of the entire factor analysis, the Kaiser-Meyer-Olkin Measure (KOM) of sampling adequacy and the Bartlett's test of sphericity must be significant at $p < 0.01$, thus indicating the suitability of the sample for factor analytic procedure. So, the data were also subjected to reliability test.

Table 3 KMO and Bartlett's test

Kaiser-Meyer-Olkin measure of sampling adequacy	0.808
Bartlett's test of sphericity (approx. chi-square)	1185.530
Df sig.	496.000

As per the reliability analysis conducted in SPSS, the output is shown in Table 3, The KMO value is 0.808, which is greater than 0.5 and chi-square = 1185.530 at $p < 0.001$. Since, the value of KMO suggests that degree of common variance is middling and the values of Bartlett's test are indicative that sample inter-correlation matrix did not come from a population in which inter-correlation matrix is an identity matrix the sample is suitable for factor analysis.

Initial Analysis: In the initial solution given in Table 4, each variable is standardized to have a mean of 0.0 and a standard deviation of ± 1.0. Thus, the variance of each variable = 1.0. A useful factor must, therefore, account for more than 1.0 unit of variance, or have an Eigen value ≥ 1.

As per the analysis conducted shown in Table 4, factors with Eigen values greater than or equal to 1.0 were retained. The first factor has an Eigen value of 6.758. Since this is greater than 1.0, more than a single variable variance is explained by it, Further, the percent a variance explained is 21.120%

Table 4 Factor analysis—the results of the initial solution

Component	Initial Eigen values			Component	Initial Eigen values		
Sr. no.	Total	% variance	% cumulative	Sr. no.	Total	% variance	% cumulative
1	6.758	21.120	21.120	17	0.275	0.861	95.618
2	4.877	15.242	36.362	18	0.252	0.786	96.404
3	3.746	11.705	48.067	19	0.221	0.690	97.094
4	3.332	10.412	58.479	20	0.195	0.609	97.703
5	2.652	8.289	66.767	21	0.155	0.486	98.189
6	1.841	5.753	72.520	22	0.113	0.355	98.543
7	1.227	3.836	76.356	23	0.104	0.326	98.869
8	1.110	3.469	79.826	24	0.098	0.305	99.175
9	0.911	2.847	82.673	25	0.070	0.220	99.395
10	0.756	2.364	85.037	26	0.066	0.206	99.601
11	0.729	2.277	87.314	27	0.039	0.121	99.721
12	0.685	2.139	89.453	28	0.037	0.114	99.835
13	0.520	1.624	91.077	29	0.028	0.087	99.922
14	0.444	1.387	92.464	30	0.015	0.046	99.968
15	0.378	1.182	93.646	31	0.007	0.023	99.991
16	0.356	1.111	94.757	32	0.003	0.009	100.000

Extraction method Principal component analysis
Source Initial output of factor analysis

$$\% \text{ of variance} = (\text{initial Eigen value/total Eigen value}) * 100$$

$$(6.758/32) * 100 = 21.120\%$$

Similarly, eight dimensions has been extracted till the Eigen value >1, that is 1.110. The rest of the 9–32 factors have Eigen values <1, and, therefore, explain less than a single variable variance. Thus, finally, 8 dimensions were extracted that explained 79.826% per cent of the overall cumulative variance (where, Cumulative variance = Summation of variances having Eigen values greater than 1). The total variation explained by top three components is 21.120%, 15.242%, and 11.705%, respectively.

Factor Loading: The evaluation factors are then classified under 8 dimensions on the basis of their load (correlation) on each dimension, factor loadings greater than or equal to 0.50 have been considered for each dimension; this classification has been done on the basis of varimax rotational analysis executed in SPSS and the output (rotated component matrix) is shown in Table 5.

Table 5 shows that "raw material availability" (0.955), "raw material cost" (0.888), "proximity to supplier" (0.727), "unskilled labor availability" (0.693), "unskilled labor cost," and "water supply" at (0.572) and (0.505), respectively, are highly loaded on dimension 1, while power supply is highly loaded on dimension 8.

Ranking of Components based on Average Mean: On the basis of factor loads, the evaluation factors could be classified into 8 dimensions as given in Table 6.

6 Recommendations Based on Findings

Every region has its own weakness and strength so is the case with region dealt in the study. Considering the city analyzed in the study, a multi-dimensional strategic approach is required to promote the balanced development. Consequently, to come out of the vicious cycle of economic decline, the twofold approach toward development is required that involves eliminating the weaknesses to gain strength and focusing upon the strengths to overcome the weakness.

Elimination of the barriers identified from the study could be long-term strategy for developing the city. This will require ensuring proper power supply (by setting up power plants or depending upon alternative sources of energy like wind mill, biogas, solar power, etc.), supportive government policy (by structural and legislative changes in development policies), and pacing up technological advancement (by funding projects, infrastructure development, conducting training, and awareness programs).

In short–medium term, developing the city by focusing on the strength of the city and setting up organizations that remain unaffected or least affected by the

Table 5 Rotated component matrix

Evaluation factors	Component (factor loads)							
	1	2	3	4	5	6	7	8
Raw material availability	0.955	−0.067	0.013	0.104	−0.071	−0.073	−0.008	0.027
Raw material cost	0.888	0.032	0.005	0.064	−0.081	−0.154	0.015	0.123
Proximity to supplier	0.727	0.298	−0.217	−0.184	−0.191	0.007	0.066	−0.107
Unskilled labor availability	0.693	−0.065	−0.182	−0.019	0.100	0.535	−0.110	−0.025
Unskilled labor cost	0.572	0.239	−0.097	−0.208	0.270	0.528	0.227	−0.124
Water supply	0.505	0.001	0.001	0.485	0.210	−0.176	0.363	0.312
Incentives	−0.467	−0.130	0.463	−0.078	0.292	−0.009	−0.104	0.097
Parking	−0.014	0.858	−0.111	0.158	0.036	0.187	0.200	−0.103
Proximity to the market	0.306	0.847	−0.190	−0.056	0.089	−0.108	0.120	−0.001
R&D	0.044	−0.786	0.395	−0.012	−0.048	0.145	−0.156	0.055
Taxes	0.051	0.635	−0.029	−0.265	0.382	−0.034	−0.232	0.103
Public authorities in city	−0.078	−0.525	−0.148	0.480	−0.194	0.260	−0.403	−0.040
Training facility	0.003	−0.253	0.846	0.064	−0.030	−0.120	−0.038	0.164
Airways	−0.272	−0.185	0.764	0.296	−0.193	0.120	0.163	−0.083
Skilled labor availability	0.197	−0.189	0.710	−0.022	0.213	−0.047	−0.284	0.051
Telecommunication	−0.442	−0.094	0.619	0.188	0.095	0.377	−0.162	0.060
Financial institutions	−0.321	0.054	0.485	−0.318	0.448	0.005	−0.340	−0.092
Environment and climate	0.051	0.043	0.055	0.921	−0.118	−0.073	−0.044	−0.160
Health and hygiene	0.013	−0.170	0.049	0.915	0.149	−0.079	−0.054	−0.046
Image of city	−0.473	0.215	0.336	0.551	0.004	−0.242	0.151	−0.034
Natural resources	0.535	0.131	0.221	0.543	0.116	−0.182	0.345	0.121
Land availability	−0.042	0.267	0.128	0.152	0.809	0.114	0.034	0.132
Land cost	−0.059	0.369	−0.051	0.025	0.752	0.096	0.145	0.358

(continued)

Table 5 (continued)

Evaluation factors	Component (factor loads)							
	1	2	3	4	5	6	7	8
Skilled labor cost	−0.087	−0.149	0.080	−0.015	0.709	−0.390	−0.066	−0.265
Railways	0.433	−0.111	0.133	0.143	−0.493	0.161	0.251	−0.406
Political environment	−0.154	0.014	−0.026	−0.095	−0.038	0.849	0.007	0.015
Government support	0.090	−0.400	0.195	−0.150	−0.126	0.673	−0.086	−0.126
Intra-city road transport	0.074	0.182	−0.201	0.017	0.055	−0.037	0.879	−0.019
Inter-city road transport	0.454	0.057	−0.011	0.087	−0.313	0.241	0.593	−0.283
Technology adaption	0.109	−0.490	0.333	0.150	0.010	0.163	−0.546	0.243
Bordering cities	−0.215	0.248	0.340	−0.097	−0.338	0.372	0.383	−0.112
Power supply	0.056	−0.106	0.170	−0.145	0.117	−0.049	−0.121	0.853

Extraction method Principal component analysis. *Rotation method* Varimax with Kaiser normalization

Table 6 Factors under 8 dimensions with and their factor loadings

Dimensions (extracted from factor analysis)	Factor loading	Variance explained	Average[a] mean	Ranking
Dimension 1 (basic resources)		15.760%	2.869	8
Raw material availability	0.955			
Raw material cost	0.888			
Proximity to supplier	0.727			
Unskilled labor availability	0.693			
Unskilled labor cost	0.572			
Water supply	0.505			
Dimension 2 (market place attractiveness)		12.232%	3.268	5
Parking	0.858			
Proximity to the market	0.847			
Taxes	0.635			
Dimension 3 (avenues of growth)		10.935%	3.327	3
Training facility	0.846			
Airways	0.764			
Skill labor availability	0.710			
Telecommunication	0.619			
Technology adaption	0.533			
Dimension 4 (inherited city resources)		10.347%	3.116	7
Environment and climate	0.921			
Health and hygiene	0.915			
Image of city	0.551			
Natural resources	0.543			
Dimension 5 (land and human)		9.408%	3.317	4
Land availability	0.809			
Land cost	0.752			
Skilled labor cost	0.709			
Dimension 6 (government policies)		8.245%	3.390	2
Political environment	0.849			
Government support	0.673			
Dimension 7 (Connectivity)		8.073%	3.256	6
Intra-city road transport	0.879			
Inter-city road transport	0.593			
Dimension 8 (power supply)		4.826%	4.171	1
Power supply	0.853			

[a]Average mean of dimension = $((\sum$ mean of variables under it)/No. of variables under it). This has been ranked in descending order. These 8 dimensions explain total variance of 79.826% are named as "Basic Resources," "Market Place Attractiveness," "Avenues of Growth," "Inherited City Resources," "Land and Human," "Government Policies," "Connectivity," and "Power Supply." The average mean of these dimensions has been calculated on the basis of mean calculated in Table 2. This average mean signifies the level of influence of each dimensions, the highest influence is of power supply followed by government policies, while the least influence is of basic resources

Table 7 Organizations that could be set up under short–medium term development strategy

| Industry type | Impact of attributes on organizations in respective industries | | | | | | |
	Power supply	Government and political support	Growth opportunities	Land availability and cost	Market place attractiveness	Connectivity	Basic resources
Tourism and hospitality	Low	Medium	Low	Medium–high	High	High	Low
Vocational training	Low	Low	Medium	Low	Low	Low	Low
Cultural industries	Low	Medium	Low	Low	Low	Low	Low
Labor-intensive industries	Low	Medium	Medium	High	High	Medium	High
Warehouses	Low	Low	Low	High	Low	Medium	Low
Transportation and logistic	Low	Low	Medium	High	Medium	High	Low

prevailing problems could be the development strategy. Prayagraj could focus on its strengths such as availability of quality skill base, strong road and rail transportation network, good educational institutes, large consumer market, central location, proximity to industrially developed cities, and cultural heritage to attract organizations.

The given Table 7 shows that organizations under certain type of industries could be set up in Prayagraj. This has been recommended on the basis that these are least affected by the prevailing challenges in the city.

Moreover, the collaborative relation between the educational institutes and industries could be an approach for developing knowledge-based economy for the long-term benefit of the region. Further the simultaneous growth of education and IT industries present a lucrative avenue for the city's growth. Vertical and horizontal cooperations would be helpful in absorption and utilization of local talent within the region by providing a clear roadmap for progression of career and region as well. From the history, it could be seen that, Helsinki and other Institutional cities were the major drivers of growth in Finland during the later years of 1990s [36]. Thus, Building up sustainable relations between education and commercial institutions will be benefitting both the parties. This collaboration could be in terms of resource sharing, promoting Research and development initiatives, training and development of employees and students, testing and simulation, outsourcing, branding, and corporate communication.

7 Practical Implications

The study uses an innovative casestudy-based approach along with survey method to elicit the critical attributes of the city that hinder organizations operating in a particular city. The findings of the study would help the government, investors, analysts, and scholars in formulating policies and business models keeping in view the persisting opportunities and the hurdles. This would enhance the competitiveness of the city and thereby aid the comprehensive and balanced development starting from the granular level.

8 Concluding Remarks

The findings show that the crucial attributes of the city that are viewed as hurdle for different types of organizations are lack of proper power supply, unsupportive government policies and political environment, poor technology adoption rate, and unavailability of skilled workforce. Further, the study also highlights that despite of certain challenges significant growth opportunity resides in various untapped and innovative markets. In order to ensure the development of the city as a truly smart city, it is essential to think locally while acting globally so that we could "make the best use of" the resources that already exist.

Also, for long-term development, there is a need to attract skilled human resource and investment, upgrade and strengthen infrastructure, and offer conducive business environment. For this, government support and good governance are important pre-requisite. Further, a co-operative and collaborative approach could make it possible to leverage the significant active resources available in the city. The promotion of cooperation and technical assistance among industries and institutions can be effective for retaining highly skilled manpower in the city and ensuring balanced development.

9 Limitations of the Study

No research is without limitations; this study is no exception. Due to the paucity of time and resources, the study faces certain limitations. The sample considers 41 organizations for the study. However, in order to generalize the findings of the study, a considerable large number of organizations from vivid industries should be considered. A large number of factors were identified from the literature survey; however, the decision was made to limit the number and content of the questions; this might lead to exclusion of some factors having crucial impact. Further, the data considered in the study are self-reported and so it might be biased.

References

1. Bhattacharya BB, Sakthivel S (2004) Regional growth and disparity in india: a comparison of pre and post-reform decades. Institute of Economic Growth University of Delhi
2. Central Statistical Organization (2017) Statement: per capita net state domestic product at constant (1999–2000) prices
3. De P (2008) Fourth international Russia-India-China annual conference, RIS, New Delhi, 20–21 Nov 2008
4. OECD (2009) Regions matter: economic recovery, innovation and sustainable growth, 12 Dec 2009. ISBN: 9789264076518
5. Wu Y (2006) Regional growth, disparity and convergence in China and India: a comparative study, the ACESA
6. Acharya SN (2002). India's medium-term growth prospects. Economic and Political Weekly, 13 July 2002
7. Harter G, Sinha J, Sharma A, Dave S (2008) The role of ICT in City development, sustainable urbanization. Booz & Company, www.booz.com/media/uploads/Sustainable_Urbanization. pdf
8. Howes S, Lahiri AK, Stern N (2003) State-level reforms in India: towards more effective government. Macmillan India Limited, New Delhi
9. Dreze J, Sen A (1996) India: economic development and social opportunity. Oxford University Press

10. Ahluwalia MS (2002) State level performance under economic reforms in India. Economic Policy Reforms and the Indian Economy. University of Chicago Press, Chicago
11. OECD (2009) How regions grow: trends and analysis, 24 June 2009. ISBN: 9789264039452
12. Lucas J (1988) The mechanics of economic development. J Monet Econ. North Holland, Feb 1988
13. Lall SV, Henderson JV, Venables AJ (2017) Africa's cities: opening doors to the world. Washington
14. Dé Murger S, Sachs JD, Woo WT, Bao S, Chang G (2002) Geography, economic policy, and regional development in China. Asian economic papers. MIT Press
15. Reynolds L (1983) The spread of economic growth to the third world: 1850–1980. J Econ Lit
16. Furlan A, Grandinetti R (2007) Business networks and the internationalization of local cluster suppliers. University of Padova, accessed at https://www.impgroup.org/uploads/papers/6711.pdf
17. Kurt G, Gornig M, Werwatz A (2005) Economic growth of agglomerations and geographic concentration of industries, Evidence for Germany, Berlin
18. Simmie J (2003) Innovation and urban regions and national and international nodes for the transfer and sharing of knowledge, regional studies
19. CAPAM (2010) CAPM 2010 international innovations awards
20. Sherer S (2003) Critical success factors for manufacturing networks as perceived by network coordinators. J Small Bus Manage, SMEs, Prentice Hall, London
21. Brunetto Y, Wharton R (2007) Moderating role of trust in SME owner/managers' decision-making about collaboration, J Small Bus Manage
22. Firoz ABM (2004) Urban growth dynamics of Khulna City: a case study on ward no. 09, 20 &24, URP Khulna, Khulna University
23. Sachs JD (2004) Stages of economic development (transcript). The Chinese Academy of Arts and Sciences, Beijing
24. Henderson VJ (2004) Brown university. Urbanization and city growth
25. Hummon NP, Zemotel L, Bullen AGR, De Angelis JP (1986) Importance of transportation to advanced technology companies, transportation research
26. OECD & United Nations OSAA (2011) Economic diversification in Africa: a review of selected countries
27. OECD (2010) Territorial reviews: competitive cities in the global economy. ISBN: 9264027092
28. United Nations (2010) The Millennium development goals report 2010
29. Simmie J (2003) Trading places: competitive cities in the global economy. European Planning Studies
30. Winden WV, Woets P (2004) European cities in the knowledge economy—the cases of Amsterdam. Dortmund, Eindhoven
31. Yin RK (1994) Case study research: design and methods. Sage, Thousand Oaks, CA
32. Barro RJ (1991) Economic growth in a cross section of countries. Quart J Econom, May 1991
33. Sala-i-Martin RJ, Barro R (1995) Technological diffusion, convergence, and growth, economics. Working papers 116, Department of Economics and Business, Universitat Pompeu Fabra
34. Likert RA (1932) A technique for the measurement of attitudes. Arch Psychol
35. Charles MF (2009) Notes on factor analysis, Criminal Justice Center, Sam Houston State University
36. OECD (2002) Territorial review on Helsinki

ANN-Based Prediction of PM$_{2.5}$ for Delhi

Maninder Kaur⑩, Pratul Arvind⑩ and Anubha Mandal

Abstract Air pollution is one of the prime factors responsible for poor health of mankind. With the advent of technology, industrialization and urbanization, there has been an increase in the pollutants which are hazardous to the mankind. As per the WHO statistics 2016, Delhi, the capital of India is fifth most polluted city. In the present work, an attempt has been made to develop an artificial intelligence-based prediction model. Meteorological parameters such as temperature, vertical wind speed, wind direction, solar radiation, relative humidity and wind speed have been incorporated as input parameters in order to predict PM$_{2.5}$ concentration present in the air. The authors have been successful to develop an algorithm which is able to forecast PM$_{2.5}$ up to one day advance. The efficacy of the algorithm is determined by coefficient of correlation, mean square error and root mean square error, respectively. The results obtained are very promising after an extensive simulation of the neural network for both 70/15/15 and 80/10/10, respectively. The maximum R value obtained is 0.923.

Keywords Air pollution · PM$_{2.5}$ · Artificial neural network (NARX) · Regression · Mean square error

1 Introduction

Pollution can be defined as the debasement of physical and biological components of the environment. Any substance which is present in such a concentration that has an adverse effect on the environment, public and property may be termed as a pollutant. In order to safeguard human health, the World Health Organization (WHO) has laid certain standards. These guidelines describe the maximum concentration of the pollutant present in the environment. Further to cater air pollution,

M. Kaur · A. Mandal
Delhi Technological University, New Delhi, Delhi, India

P. Arvind (✉)
Dr. Akhilesh Das Gupta Institute of Technology and Management, New Delhi, Delhi, India

© Springer Nature Singapore Pte Ltd. 2020
S. Ahmed et al. (eds.), *Smart Cities—Opportunities and Challenges*,
Lecture Notes in Civil Engineering 58,
https://doi.org/10.1007/978-981-15-2545-2_52

633

the United States Environmental Protection Agency (USEPA) has introduced National Ambient Air Quality Standards (NAAQS) [1]. The standard includes permissible limits for particulate matter (PM_{10} and $PM_{2.5}$), oxides of sulphur and nitrogen, carbon monoxide, ozone and lead, respectively. In addition to the above-mentioned pollutants, NAAQS for India also defines permissible limits for benzene, benzo(a)pyrene, arsenic and nickel.

Rapid industrialization and urbanization have led to exacerbation of the air pollution on the global scale. This leads to severe health issues like demurral breathing, cardiovascular diseases and irritation to eyes. Prolonged exposure to poor air quality can cause severe damage to immune system of the body. According to recent WHO statistics, fourteen Indian cities have been identified to be worst affected by air pollution. New Delhi, the capital of the country is ranked fifth amongst all Indian polluted cities as per the statistics of WHO [2]. Its $PM_{2.5}$ concentration was observed three times more than the national safe standard as prescribed by NAAQS. Thereby, making the city inhabitable as it is unable to fulfil the guidelines as laid down in [3, 4] for a smart city.

It is worth mentioning that a smart city is an urban area that combines information and communication technology (ICT), along with several instruments linked with the network (the Internet of Things or IoT) in order to enhance the effectiveness of city operations and services in connection with the public. It further aims to embed the information and communication technologies into the government systems using various technologies such as cloud computing, big data, mobile computing and data vitalization [5]. All the above aspects will hold true if a proper pollution-free environment is provided for the inhabitants and a check is kept to the raising pollutants with growth of industrialization and urbanization.

Since air pollution has become an upcoming challenge in developing countries, it requires regular and effective monitoring. There is an urgent need for pragmatic solution to create awareness and get rid of air pollutants. In order to protect public health, forecasting of pollutants is required. It helps the policy makers to undertake timely solutions for air pollution prevention and its mitigation which is an efficient step towards smart city.

In the present work, an attempt has been made to develop an artificial intelligence-based prediction model. Meteorological parameters such as temperature, vertical wind speed, wind direction, solar radiation, relative humidity and wind speed are fed as input parameters in order to predict $PM_{2.5}$ concentration present in the air. The authors have been successful to develop an algorithm which is able to forecast $PM_{2.5}$ up to one day advance. The research article gives a brief write up about the latest research work carried out for the development of the proposed algorithm. Further, the methodology along with in-depth analyses has been presented in the paper at later stage.

2 Motivation

In order to predict the concentration of an air pollutant, several research works based on linear and nonlinear approach have been carried out. The development of predictive models begins with linear statistical approach. There had been a rigorous application of multiple linear regression, auto regressive moving average (ARMA) and their combinations [6–8]. The metrological parameters and pollutant concentration were the prime inputs used in the model development. Even though the results obtained were satisfactory, but still, it can be improved by examining the correlation amongst variables [9].

With the advent of technology, artificial neural network (ANN) has found its impact for prediction. The modelling has proven to be a reliable air pollution time series prediction tool [10–12]. The nonlinearity amongst the variables can be solved for complex systems such as environmental pollution. Further, multi-layer perceptron (MLP) can be applied in resolving problems of environmental pollution due to its capability of establishing a significant link between predictors and predictands as revealed in various studies. Different ANN algorithms such as forward selection, backward propagation, backward elimination and genetic algorithm technique have found their use in prediction [13]. Also, different techniques such as principal component analyses (PCA), classification and regression trees were used to identify significant variables which were responsible for prediction [14–16]. The predicted result was the best for models based on ANN-MLP using back propagation technique with fewer inputs [17–19].

Further, the researchers gave much better prediction performance than the conventional linear models, but still, there was a need to develop more robust models considering the dynamic nature of the atmospheric pollution. A different approach, nonlinear autoregressive exogenous (NARX) neural network model has been used to predict daily direct solar radiation [20]. NARX is a time series model which can be used to model a variety of nonlinear dynamic models.

3 Study Area and Data Set

Delhi, the national capital of India, is one of the busiest metropolitan cities of the country. It has been ranked sixth amongst the metropolitan cities of the world in terms of economic growth. Delhi is situated in Northern India between the latitudes of 28°–24′ 17″ and 28°–53′–00″ N and longitudes of 76°–50′–24″ and 77°–20′–37″ E. Delhi has a total area of 1483 km^2. As per 2012 census, the population of the city is 1.9 crores. The climate of Delhi is variable with extreme weather conditions. As observed by the statutory bodies, the air quality seems to deteriorate in the winter season.

Table 1 Description of the study area

S. no.	Station name	Category	Coordinates
1	Shadipur	Industrial, Residential	28.6510° N, 77.1562°E
2	R. K. Puram	Residential	28.5660° N, 77.1767° E
3	Mandir Marg	Residential	28.630362° N, 77.197293° E
4	IHBAS, Dilshad Garden	Residential	28.6812° N, 77.3047° E
5	D.T.U.	Commercial	28.7501° N, 77.1177° E

Fig. 1 Location of five air quality monitoring stations, Delhi (*Courtesy* Google Map)

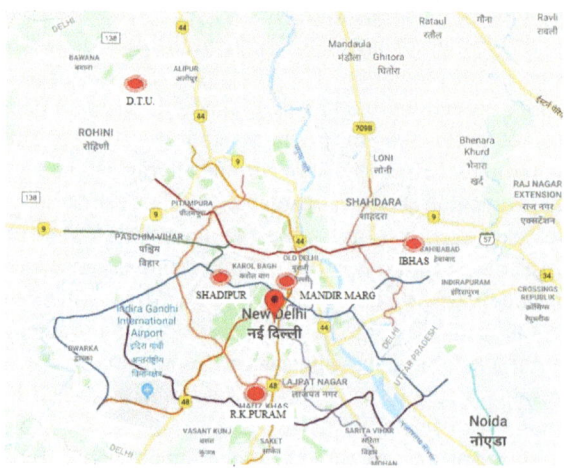

The Central Pollution Control Board (CPCB), a statutory body of India, is responsible for monitoring of ambient air quality across the country. In the present work, 24-h average daily data has been collected for five stations for a period of two years (2017–2018). The data collected includes the concentration of $PM_{2.5}$, meteorological data that includes, wind speed (WS), wind direction (WD), temperature (T), solar radiation (SR), relative humidity (RH), vertical wind speed (VWS), respectively. Table 1 shows the detailed description of the study area selected for the present work. Figure 1 shows the geographical representation of the area considered for the study in the present work.

The above sites were selected on the basis of industrial as well as increased population density in their vicinity. The selected area for the study consists of industries, metro station, administrative and commercial buildings. The precursors of $PM_{2.5}$ can be dust, ash and smog, which lead to premature deaths due to respiratory problems. Hence, $PM_{2.5}$ a prime respiratory pollutant has been predicted for the present work.

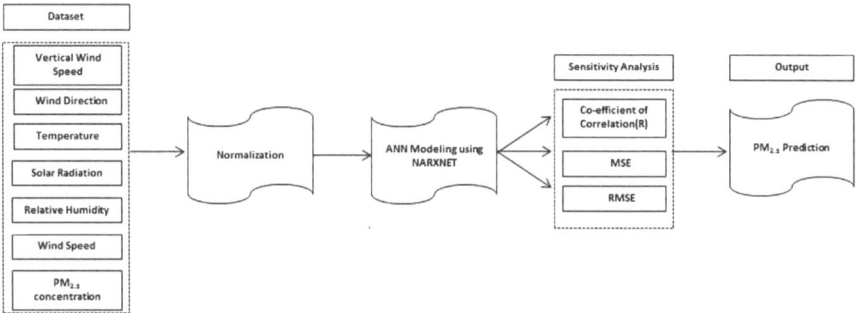

Fig. 2 Block diagram of the proposed technique

4 Methodology

The methodology adopted for prediction of PM$_{2.5}$ has been explained in the block diagram given below in Fig. 2.

4.1 Normalization

The data set including meteorological and pollutant concentration was normalized between 0 and 1, thereby transforming into the input as required by the neural network for prediction. The input data is normalized using Eq. (1).

$$x_{\text{new}} = 0.8 * \frac{(x - x_{\min})}{(x_{\max} - x_{\min})} + 0.1 \tag{1}$$

4.2 NARX Neural Network

An artificial neural network (ANN) is a model that processes information and is encouraged by the biological nervous systems, such as brain, neurons. It has been extensively used in field of classification of images, signals, etc. [21, 22]. In the present work, Levenberg–Marquardt training method is used. It is an iterative technique that locates the minimum of a multivariate function which is given in the form of sum of squares of nonlinear real-valued functions [23].

The nonlinear autoregressive network with exogenous inputs (NARX) is a recurrent dynamic network. It has a weighted feedback connection between layers of neurons that allows lagged values of variables to be considered in model for

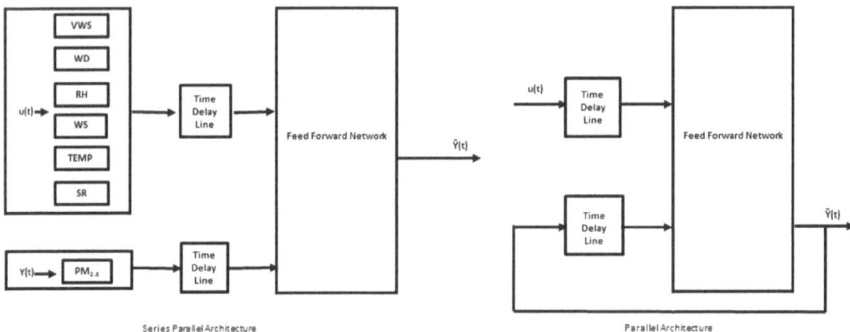

Fig. 3 NARX neural network

efficient time series modelling. Figure 3 shows two distinguish architectures of NARX model, series–parallel (open loop) architecture and parallel architecture (closed loop). Mathematically, the nonlinear time series input–output representation of NARX is given by the following equation.

$$Y(t) = f(y(t-1), y(t-2), \ldots, y(t-Ny), u(t-1), u(t-2), \ldots, u(t-Nu) \quad (2)$$

where $u(t)$ and $y(t)$ represent input and output of the network at time t. Nu and Ny are the input and output order, and the function f is a nonlinear function which is approximated by feedforward neural network. During the training phase, series–parallel architecture is used and is converted into parallel architecture after training. NARX (closed loop) is beneficial for multi-step ahead prediction based on feedforward architecture and back propagation. It also acts as a nonlinear filter, wherein target output is free of noise present in the input.

In the present work, two network architectures, 70% training, 15% validation, 15% testing and 80% training, 10% validation, 10% testing, have been used, respectively. The results have been analysed by extensive simulation by varying the hidden layer neurons from 1 to 100. The maximum result fetched at optimal neuron is presented in the paper. All the computational techniques have been carried out in MATLAB R2018b [24].

4.3 Sensitivity Analysis

The performance of the developed models is evaluated using three statistical indices, namely coefficient of correlation(R), mean square error (MSE) and root mean square error (RMSE).

The coefficient of correlation is defined as

$$R = \frac{\sum (Q_o - M_o)(Q_p - M_p)}{\sqrt{\sum (Q_o - M_o)^2 + \sum (Q_p - M_p)^2}} \tag{3}$$

The mean square error can be described as follows:

$$\text{MSE} = \frac{1}{N} \sum (Q_o - Q_p) \tag{4}$$

The root mean square error is defined as

$$\text{RMSE} = \left[\sum \frac{1}{N} \sqrt{Q_o - Q_p} \right]^{\frac{1}{2}} \tag{5}$$

where Q_0 is observed concentrations, Q_p is estimated concentrations, M_o refers to mean of the observed concentrations, M_p is mean of the estimated concentrations, and N is the total no. of observations of the data set.

5 Results and Discussions

After normalization, the input parameters were fed to NARX neural network. The PM$_{2.5}$ concentration was predicted. The authors were successful in predicting up to one day in advance. An extensive simulation was carried out by ANN from 1 neuron to 100 neurons for getting the best result. Further, the input data was trained, validated and tested in 70/15/15 and 80/10/10, respectively. The results are represented by coefficient of correlation, mean square error and root mean square error, respectively.

The regression value obtained by the proposed method for 70/15/15 and 80/10/10 has been depicted in Figs. 4 and 5, respectively.

The results have been compared with [14]. It is obvious from the above figure that the R value obtained for all the five locations is better than the algorithm developed [14] for the same input. The R value of DTU, Mandir Marg, Shadipur, RK Puram is 0.919, 0.864, 0.890, 0.87 which is improved when compared with the result obtained by algorithm in [14], i.e. 0.683, 0.645, 0.621, 0.716, respectively.

Further, the algorithm has been trained and compared for architecture 80/10/10. Figure 4 shows the results of the R values. The R value of DTU, Mandir Marg, Shadipur, RK Puram is 0.923, 0.886, 0.895, 0.897 which outperforms the result obtained by algorithm in [14], i.e. 0.75, 0.67, 0.66, 0.73, respectively.

The R value of IHBAS, Dilshad Garden for 70/15/15 is 0.408 as compared to 0.356 and for 80/10/10 is 0.65 as compared to 0.49, respectively. The results are not up to the mark due to the fact that there is a wide fluctuation in the input parameters.

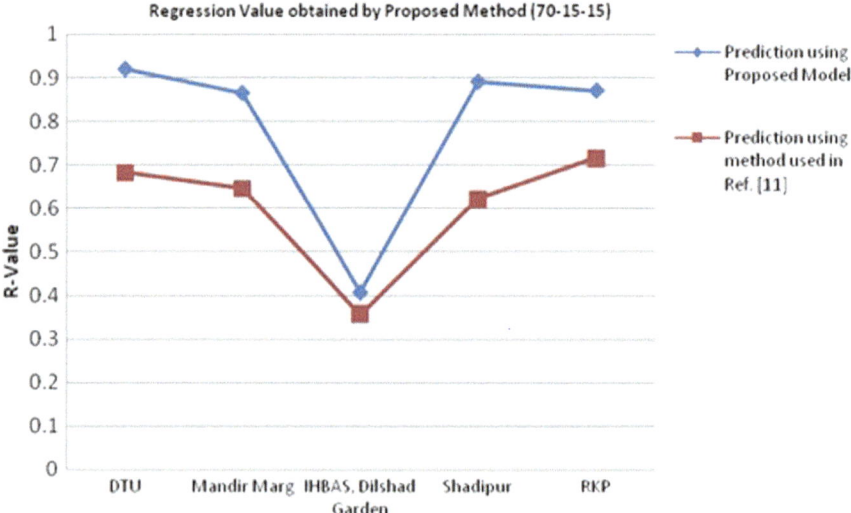

Fig. 4 Regression value obtained by proposed method (70/15/15)

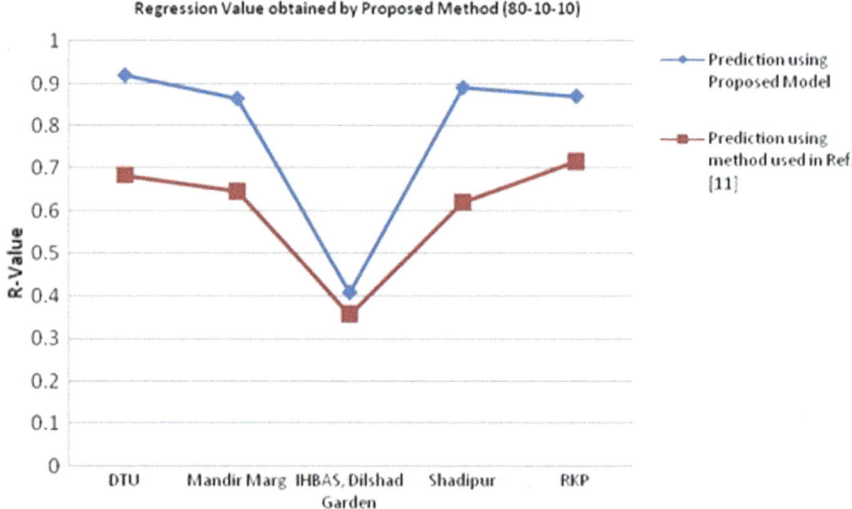

Fig. 5 Regression value obtained by proposed method (80/10/10)

Further, the results are introspected for further analyses for the proposed method as well as the existing methodology [14]. Table 2 gives the value of regression, mean square error and root mean square error for the proposed methodology for both 70/15/15 and 80/10/10, respectively.

Table 2 Proposed method

S. no.	Station name	ANN architecture (70/15/15)				ANN architecture (80/10/10)			
		R	MSE	RMSE	Optimum neuron	R	MSE	RMSE	Optimum neuron
1	DTU	0.919	0.0017	0.041	65	0.923	0.0084	0.091	16
2	Mandi- Marg	0.864	0.00131	0.036	75	0.886	0.0013	0.036	45
3	IHBAS, Dilshad Garden	0.408	5.972×10^{-8}	0.0077	25	0.65	6.38×10^{-8}	0.0007	45
4	Shadipur	0.890	0.00013	0.011	100	0.895	8.66×10^{-5}	0.009	10
5	R. K. Puram	0.87	0.0023	0.047	25	0.897	0.004	0.063	100

Table 3 Results obtained from [14]

S. no.	Station name	ANN architecture (70/15/15)				ANN architecture (80/10/10)			
		R	MSE	RMSE	Optimum neuron	R	MSE	RMSE	Optimum neuron
1	DTU	0.683	0.24	0.48	12	0.75	0.36	0.6	25
2	Mandir Marg	0.645	0.29	0.53	65	0.67	0.41	0.64	45
3	IHBAS, Dilshad Garden	0.356	0.23	0.47	85	0.49	0.37	0.60	75
4	Shadipur	0.621	0.18	0.42	75	0.66	0.33	0.57	15
5	R. K. Puram	0.716	0.31	0.55	16	0.73	0.49	0.70	100

It is quite evident from the above table that the values obtained at 80/10/10 fetch better result as compared to 70/15/15. The value of RMSE is in between the range of 0.007 to 0.04 for 70/15/15 and 0.0007 to 0.091 for 80/10/10, respectively. Here, optimal neurons refer to the neuron where maximum result is obtained.

Table 3 gives the value of regression, mean square error and root mean square error for the proposed methodology in 70/15/15 and 80/10/10, respectively, for [14]. In order to prove the efficacy of the proposed algorithm, the database was fed to [14], and results have been discussed.

It is quite evident from the above table that the values obtained at 80/10/10 fetch better result as compared to 70/15/15. The value of RMSE is in between the range of 0.42 to 0.55 for 70/15/15 and 0.5 to 0.7 for 80/10/10, respectively. Here, optimal neurons refer to the neuron where maximum result is obtained.

From the above discussion, it is evident that the proposed method outperforms the existing method [14]. It is clear that NARX network for 80/10/10 gives better result, and using the database, predictions have been made up to one day in advance.

6 Conclusion

In the present work, an attempt was made to develop an artificial intelligence-based prediction model. Meteorological parameters such as temperature, vertical wind speed, wind direction, solar radiation, relative humidity and wind speed have been used as input parameters in order to predict $PM_{2.5}$ concentration present in the air which would contribute for becoming the national capital as smart city. The authors have been successful to develop an algorithm which is able to forecast $PM_{2.5}$ up to one day advance. Also, it is quite evident that the proposed method outperforms the existing algorithm [14]. The results are represented by coefficient of correlation, mean square error and root mean square error, respectively. The results obtained are very promising. In the future work, the authors would like to improve the acute nonlinearity of the data as obtained in IHBAS, Dilshad Garden and predict more than one day in advance.

References

1. USEPA homepage, https://www.epa.gov/criteria-air-pollutants/naaqs-table
2. World Health Organization https://www.who.int/airpollution/data/cities-2016/en/
3. Chourabi H, Nam T, Walker S, Gil-Garcia J, Mellouli S, Nahon K, Pardo T, Scholl H (2012) Understanding smart cities: an integrative framework. In: 45th Hawaii international conference on system sciences, pp 2289–2297. IEEE Press, Hawaii. https://doi.org/10. 1109/hicss.2012.615
4. Joshi S, Saxena S, Godbole T, Shreya (2016) Developing smart cities: an integrated framework. In: 6th international conference on advances on computing & communications, ICACC 2016. Proc Comput Sci 93:902–909. Elsevier, Cochin, India (2016). https://doi.org/ 10.1016/j.procs.2016.07.258
5. ChuanTao Y, Zhang X, Hui C, Jingyuan W, Daven C, Bertrand D (2015) A literature survey on smart cities. Science China Press, Springer, Berlin. https://doi.org/10.1007/s11432-015-5397-4
6. Goyal P, Chan AT, Jaiswal N (2006) Statistical models for the prediction of respirable suspended particulate matter in urban cities. Atmos Environ 40:2068–2077
7. Sharma P, Chandra A, Kaushik SC (2009) Forecasts using Box-Jenkins models for the ambient air quality data of Delhi City. Environ Monit Assess 157:105–112
8. Banja M, Papanastasiou DK, Poupkou A, Dimitris M (2012) Development of a short–term ozone prediction tool in Tirana area based on meteorological variables. Atmos Pollut Res 3:32–38
9. Patricio P, Reyes J (2002) Prediction of maximum of 24-h average of PM$_{10}$ concentrations 30 h in advance in Santiago, Chile. Atmos Env 36:4555–4561
10. Gardner MW, Dorling SR (1998) Artificial neural networks (the multilayer perceptron)—a review of applications in atmospheric sciences. Atmos Environ 32:2627–2636
11. Perez P, Reyes J (2001) Prediction of particulate air pollution using neural techniques. Neural Comput Appl 10(2):165–171
12. Niska H, Hiltunen T, Karppinen A, Ruu Skanen J, Koleh Mainen T (2004) Evolving the neural network model for forecasting air pollution time series. Eng Appl Artif Intell 17(2): 159–167
13. Elangasinghe MA, Singhal N, Dirks KN, Salmond JA (2014) Development of an ANN-based air pollution forecasting system with explicit knowledge through sensitivity analysis. Atmos Pollut Res 5:696–708
14. Azid A, Juahir H, Toriman ME, Kamarudin MKA, Saudi ASM, Hasnam CNC, Aziz NAA, Azaman F, Latif MT, Zainuddin SFM, Osman MR, Yamin M (2014) Prediction of the level of air pollution using principal component analysis and artificial neural network techniques: a case study in Malaysia. Water Air Soil Pollut 225:2063
15. Mishra D, Goyal P (2015) Development of artificial intelligence based NO$_2$ forecasting models at Taj Mahal, Agra. Atmos Pollut Res 3;99–106
16. Durao RM, Mendes MT, Pereira MJ (2016) Forecasting O$_3$ levels in industrial area surroundings up to 24 h in advance, combining classification trees and MLP models. Atmos Pollut Res 7:961–970
17. Russo A, Lind PG, Raischel F, Trigo R, Mendes M (2015) Neural network forecast of daily pollution concentration using optimal meteorological data at synoptic and local scales. Atmos Pollut Res 6:540–549
18. Rahimi A (2017) Short-term prediction of NO$_2$ and NO$_x$ concentrations using multilayer perceptron neural network: a case study of Tabriz, Iran. Rahimi Ecol Process 6:4
19. Gao M, Yin L, Ning J (2018) Artificial neural network model for ozone concentration estimation and Monte Carlo analysis. Atmos Environ 184:129–139

20. Boussaada Z, Curea O, Remaci A, Camblong H, Bellaaj NM (2018) A nonlinear autoregressive exogenous (NARX) neural network model for the prediction of the daily direct solar radiation. Energies 11:620

21. Arvind P et al (2012) A wavelet packet transform approach for locating faults in distribution system. In: IEEE symposium on computers & informatics (ISCI), Penang, pp 113–118

22. Arvind P et al (2012) Comparison between wavelet and wavelet packet transform features for classification of faults in distribution system. In: American institute of physics conference series

23. Liu H (2010) On the Levenberg-Marquardt training method for feed-forward neural networks. In: Sixth IEEE international conference on natural computation (ICNC), vol 1, pp 456–460

24. MATLAB R2018b, The MathWorks, Inc

Streamflow Modelling for a Peninsular Basin in India

Mohd Izharuddin Ansari, L. N. Thakural and S. Anbu Kumar

Abstract Water is one of the key issues in the development of smart cities. Water availability and water demand are the two key aspects in the management of water resources. The present work deals with the water availability in a river basin as it is a basic source for water supply in most of the urban centres but today's climate change is considered as one of the most grave issues in front of humanity. Thus, it is of utmost importance to analyse the effect of extreme climatic events on water availability in a river basin. Study of river water flow essentially involves mathematical modelling to simulate and forecast future scenario, enabling the water resources planners in making realistic water management strategies to support smart cities. Hence, the present paper intends to study climatic variability in the Chaliyar river basin through rainfall–runoff modelling and describes the application of Nedbor-Afstromnings Model (NAM), to analyse its suitability, efficiency and performance in Chaliyar basin of Kerala, India.

Keywords Climatic change · Smart cities · Rainfall–runoff modelling · NAM · Chaliyar basin

1 Introduction

One of the requirements for smart cities is water availability. For most of the cities, the primary source of water supply is the water availability in river. In present day world, it is very important to study the changing climatic conditions and the resulting rainfall–runoff pattern.

M. I. Ansari (✉)
Jamia Millia Islamia, New Delhi, India

L. N. Thakural
National Institute of Hydrology, Roorkee, India

S. Anbu Kumar
Delhi Technological University, New Delhi, India

© Springer Nature Singapore Pte Ltd. 2020
S. Ahmed et al. (eds.), *Smart Cities—Opportunities and Challenges*,
Lecture Notes in Civil Engineering 58,
https://doi.org/10.1007/978-981-15-2545-2_53

645

The clash among water supply and water use will become more serious in future with economic development and increase in population. For ensuring the appropriate management and utilization, interpreting the probable effects of extreme climatic events on water resources is very necessary. Since runoff information is not available for majority of river basins in India, hence, rainfall–runoff model can be developed to assess runoff from the basin.

The rainfall–runoff process or hydrological modelling is a complex activity [1]. Based on complexity and purpose, many streamflow models are available, Shoemkar et al. [2]. In present study, an appropriate hydrological model, the (Nedbor-Afstromings model (NAM) has been identified and rainfall–runoff modelling has been performed for Chaliyar basin in Kerala state.

2 Area of Study

The area of study is the Chaliyar basin, Kerala, India, which lies between 11° 30′ N and 11° 10′ N latitudes and 75° 50′ E and 76° 30′ E longitudes. In Kerala, Chaliyar is third largest river, which emanates from the Elambalari, Nilgiri District, Tamil Nadu, having an elevation of approximately 2066 m from mean sea level (MSL).

Chaliyar, a perennial stream, joins Lakshadweep Sea south of Kozhikode near Beypore after flowing through distance of around 169 km by the name of "Beypore" River [3].

Total drainage area of the river is 2918 km^2, from which 2530 km^2 falls in Kerala and the remaining one falls in Tamil Nadu. Out of total area of the basin, only an area of 1996.4 km^2 was considered in the study for hydrological modelling using MIKE 11 NAM model (Fig. 1).

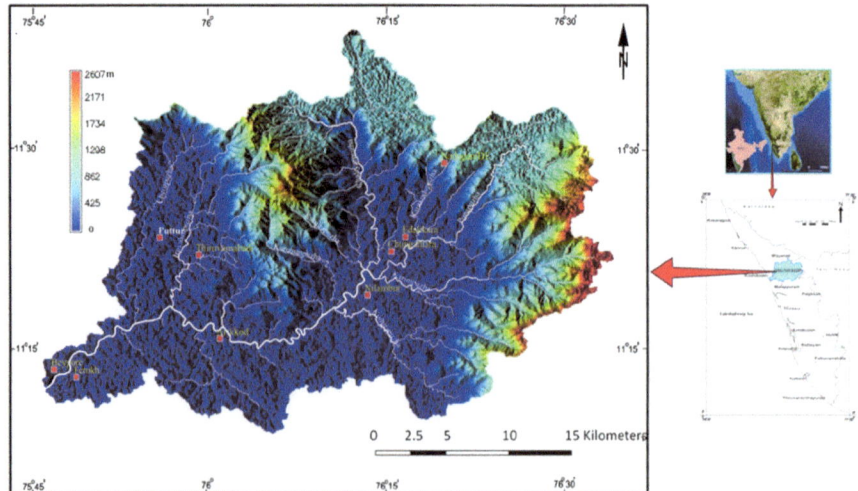

Fig. 1 Digital elevation model (DEM) for the Chaliyar basin. *Source* Google Earth [4]

3 Methods and Materials

3.1 *Digital Elevation Model and Watershed Delineation*

To delineate area of study, for Chaliyar river basin, Kerala, 90-m DEM was used. By partitioning watershed from entire basin, the user is able to reference different areas from one another spatially.

Shuttle Radar Topographic Mission (SRTM) was used to prepare the digital elevation model (DEM) at 90-m resolution. DEM is used to derive slope, aspect, flow direction and accumulation, and stream network information (Fig. 4) also. For our study, digital elevation model was acquired from earthexplorer.usgs.gov site and used to delineate the watershed.

To delineate the watershed, initially, the reprojection of SRTM DEM (90 m) in ARCGIS was done, i.e. the SRTM DEM (90 m) was reprojected to UTM spatial reference. Salient features regarding this are as follows:

Projection system-UTM, Zone 43 N
Spheroid-WGS84
Datum-WGS84.

After this processing of DEM took place in ARCGIS for watershed delineation (Figs. 2, 3, and 4).

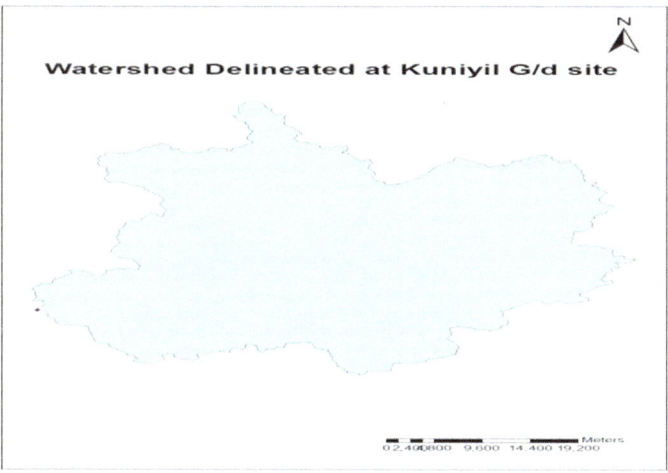

Fig. 2 Our study area for hydrological modelling delineated with the help of DEM and Arc GIS at Kuniyil Gauge discharge site

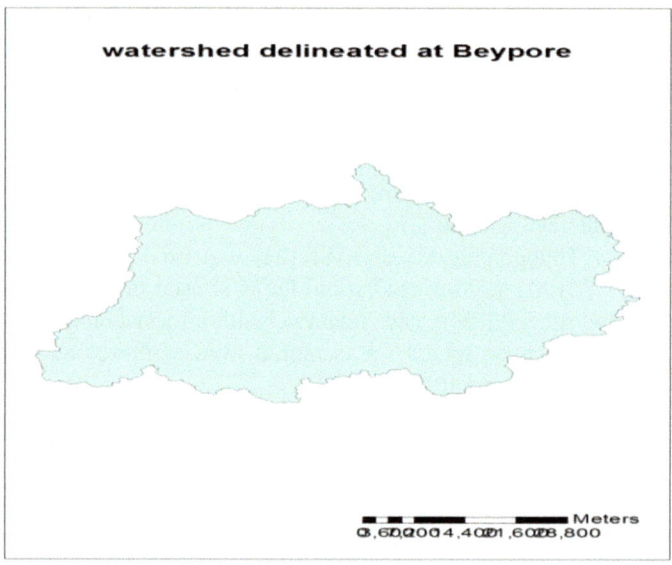

Fig. 3 Map of study area delineated from Chaliyar river basin with the help of DEM and Arc GIS at Beypore

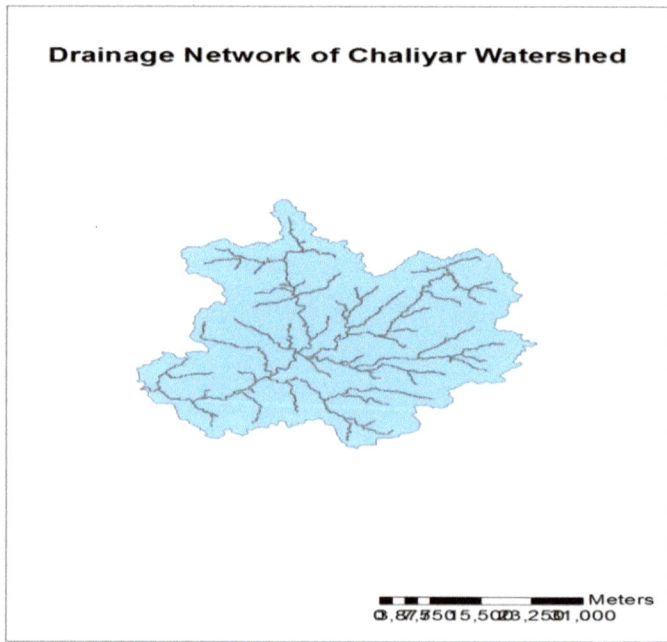

Fig. 4 Drainage of Chaliyar network prepared with the help of DEM

3.2 Mike 11 Nam Model

Many hydrological deterministic models are developed for simulating the rainfall–runoff process for river watersheds, but most have complicated structures and need various observed data for calibration [5]. The Mike 11 NAM model is widely used in so many Asian countries not only because of its simple structure but also because of its fewer data requirements. However, this hydrological model still needs extensive time and effort to calibrate various model parameters.

MIKE11 NAM is a rainfall–runoff model which is part of the MIKE 11 software. MIKE 11 is developed by Danish Hydraulic Institute, Denmark [6]. As default, NAM consists of 9 parameters which represent ground water storage, surface-zone storage and root-zone storage.

By constantly analysing water content for four different and interrelated storages [7], NAM simulates rainfall–runoff process as shown in Fig. 5. These mutually interrelated storage include snow storage, root-zone storage, ground water storage and surface storage. Precipitation time series and potential evapotranspiration time series are basic meteorological data required.

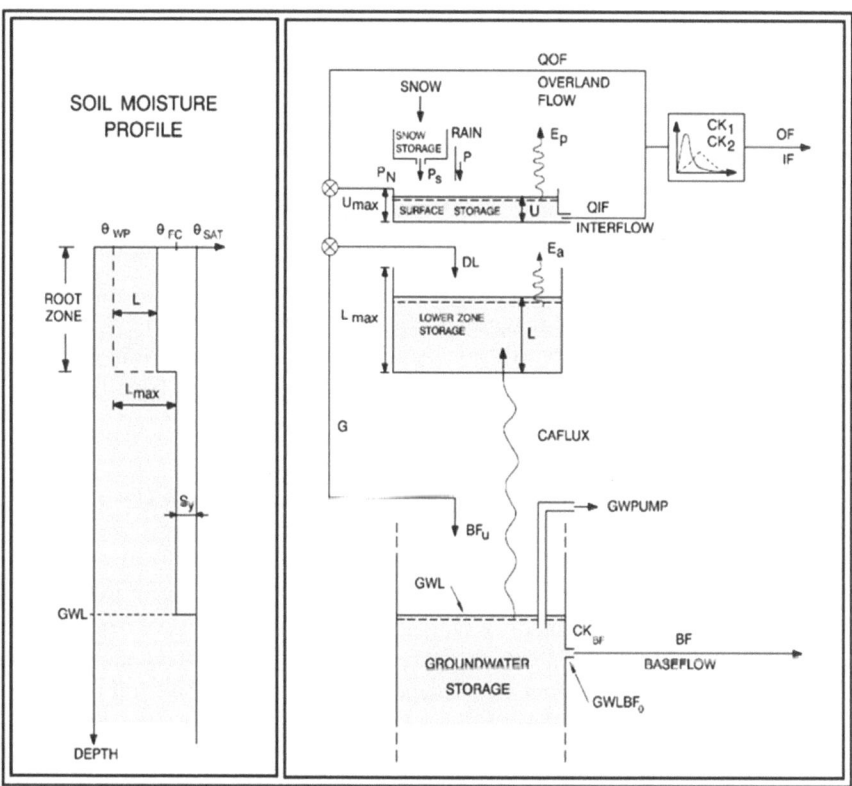

Fig. 5 Structure of NAM model for rainfall–runoff simulation [8]

Data Requirements

- Data required for MIKE 11 NAM are:

  (1) Model parameter
  (2) Initial conditions
  (3) Meteorological data
  (4) Streamflow data for model calibration and validation.

- Meteorological data required are:

  (1) Rainfall
  (2) Potential evapotranspiration.

Data Inputs

Rainfall Data

In present study, the point precipitation at Ambalavayal, Nilambur and Manjeri was obtained from the CWRDM, Kerala. Thiessen polygon method was used to calculate the areal precipitation from point precipitation. The Thiessen polygon was generated from Arc GIS 9.3 software. Among three rain gauge stations, Nilambur was found to be the most influencing station covering maximum area. The weightages denoting the extent of influence of the individual rain gauge station over study area was estimated (Fig. 6).

From Table 1, we can conclude that the weight area for Nilambur is quite larger as compared to other two stations; hence, rainfall data only for the Nilambur station is used in modelling.

The rainfall data was the major input for the rainfall–runoff modelling. The consistency of rainfall was checked through drawing the graph between rainfall and runoff, if straight-line graph occurs, it indicates the consistent data and if the graph is non-straight line, it indicated that the data have been subjected to several change. In the present study, when annual rainfall was plotted against the annual runoff for the period from 2001 to 2005 which is shown in Fig. 7, a straight-line graph was observed. The correlation coefficient of the linear regression has been resulted as 0.8963, showing good correlation between rainfall and runoff.

Discharge Data

In present study, daily discharge data of Chaliyar basin in metre cube per second (m^3/s) obtained from the Kuniyil gauging site for a period ranging from 2001 to 2005 were used and the discharge was treated as instantaneous.

Before using the data for modelling, it is required to check the consistency of data. For checking the consistency of rainfall and runoff, coefficient of runoff should be calculated. Runoff coefficient is the ratio between runoff and rainfall. It is dimensionless term, varies from 0 to 1 depending on land use and soil types.

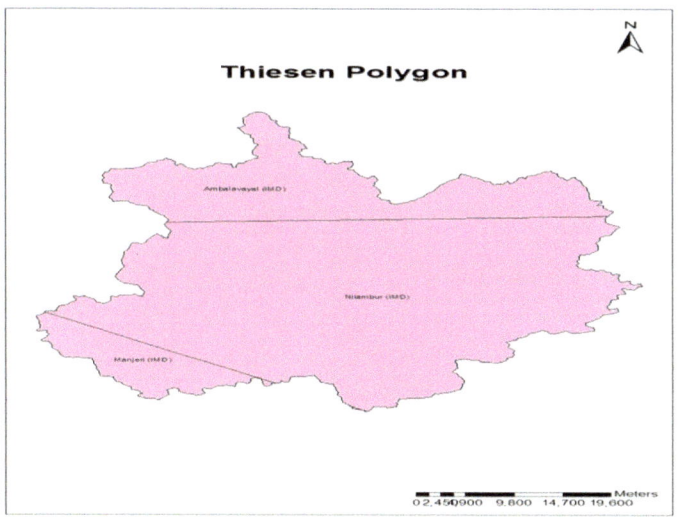

Fig. 6 Thiessen polygon map of the study area

Table 1 Thiessen weightages for raingauge stations	Station	Raingauge station	Weight area (%)
	1	Ambalavayal	23
	2	Nilambur	70
	3	Manjeri	7

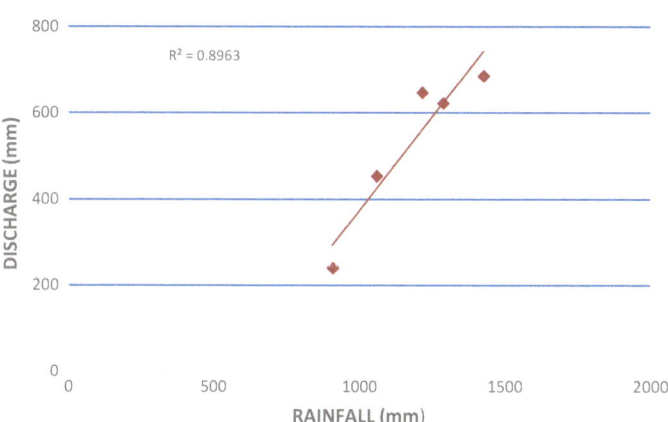

Fig. 7 Graph showing linear relation between rainfall–runoff

Potential Evapotranspiration

Potential evapotranspiration (ET$_0$) is the water that got evaporated and transpired. Calculation of potential evapotranspiration was important due to its severe effect on runoff in the form of surface evaporation. More evaporation causes the additional loss of water from the basin which was the part of runoff.

In the present study, daily ET$_0$ *has* been estimated by employing Hargreaves method. The Hargreaves method [9] is given as

$$\text{PET} = 0.0009384 \times R_{\text{ext}} \times \left(T_{\text{avg} + 17.8} \right) \times \sqrt{\left(T_{\text{max}} - T_{\text{min}} \right)} \tag{1}$$

where

PET   Potential evapotranspiration (mm/day)
R_{ext} Daily extra-terrestrial radiation (watt/m^2)
T_{avg} Daily average temperature (°C)
T_{max} Daily maximum temperature (°C)
T_{min} Daily minimum temperature (°C).

The potential evapotranspiration was calculated for the periods from 2001 to 2005 using the parameters like maximum, minimum temperature and solar radiation.

3.3 Objective Function

Objective of model calibration is: selection of model parameters so that the model simulates the hydrological behaviour of the basin as closely as possible Madsen [10]. MIKE 11 NAM answers these questions using multiobjective approach. It implies that various numerical measures are considered for optimization process.

3.4 Nam Model Set-up

In present study, rainfall–runoff modelling was carried out by setting up MIKE 11 NAM at the river gauging site Kuniyil in Chaliyar river basin.

The NAM model requires input information like daily rainfall, runoff and potential evapotranspiration data in *dfso* format to be assigned in the model. Thus, all the input time series data required was converted into *dfso* time series files using MIKE ZERO software. The best possible input data time series of five years period from 2001 to 2005 were used for the modelling. During the modelling, the calibration was carried out for three years period from 2001 to 2003 and validation carried out for the remaining two years, i.e. 2004 to 2005. After assigning the input information, the NAM model was run by using the auto-calibration mode for the time step of one day.

3.5 Model Calibration

The process of model calibration is normally done either manually or by using computer-based automatic procedures.

In present study, once the NAM model was set-up, calibration was done for three years period from 2001 to 2003. The default parameters of the model were same during calibration and model was run in auto-calibration mode which then resulted in obtaining set of model parameters for the calibration period. Output results from calibration were checked for R^2 (coefficient of determination) and analysed graphically to assess agreement between simulated runoff and observed runoff [11]. Using trial and error method, the parameters were again adjusted one by one to get best set model parameters of the NAM which could mimic runoff having appreciable degree of agreement with observed streamflow. The parameters thus obtained from refinement were then used in validation of the model.

3.6 Model Validation

The ability of model to estimate runoff for the periods other than those used for the calibration process is tested in validation. Thus, validation data must not be the same as those used for calibration but must represent a situation similar to that in which the model is to be applied operationally. In this study, the NAM was then validated for the left over period of two years 2004–2005. Values of parameters obtained amid calibration have been used for validation & model was run without auto-calibration mode. Analysis was done for the statistics obtained as simulated results and model outputs were tested graphically to judge the agreement among the simulated runoff and observed runoff to check the capacity of calibrated model, so as to simulate the runoff for the extended period of time.

4 Results and Discussions

4.1 Model Calibration

Mike 11 NAM was initiated by application of data of time series of daily precipitation for three rainfall station, time series of potential evapotranspiration and observed runoff time series of Kuniyil G/d site for the period from 2001 to 2005 in

Table 2 Model calibration result (all values are in mm)

Period	Q-obs	Q-sim	% Diff	RF	PET	AET	GWR	OF	IF	BF
2001	1613.1	1386.7	0.2	1911	477.7	423.3	524.1	807.3	90.4	20.4
2002	1419.4	1055.4	0.3	1509.7	556.2	477.6	390.3	567.4	84.4	16.8
2003	1107.5	1259.5	−0.1	1781.5	552.7	450.9	485.7	727.2	94.0	18.3
Total	4140	3701.6	0.1	5202.2	1586.5	1351.8	1400.1	2101.8	268.8	55.5

Coefficient of determination = 0.62, WBL = 6.98%

Table 3 Model parameter value and their range during calibration

Parameter	Values of the parameter	Parameter range	
		Lower bound	Upper bound
U_{max}	10	10	20
L_{max}	106	100	300
CQOF	0.665	0.1	1
C_{KIF}	228.6	200	1000
$CK_{1.2}$	35	10	50
T_{OF}	0.131	0	0.99
T_{IF}	0.213	0	0.99
TG	0.148	0	0.99
CKBF	1000	1000	4000

the Chaliyar basin. The area of the basin up to Kuniyil G/d site was 1996.4 km². Initially, the model was run to simulate the runoff using the auto-calibration mode of MIKE 11 NAM and parameters were fixed. After that model was calibrated by conducting trials by modifying, the model parameters to reduce the error between observed streamflow data and simulated streamflow data. The model was refined to obtain the best match between observed and simulated runoff. The statistics of simulation during model calibration are shown in Table 2. The values of all the nine parameters obtained during the model calibration and the range of these parameters are shown in Table 3.

Table 2 shows the statistics of simulated runoff and other parts of hydrologic cycle like inter flow, over land and base flow and ground water recharge. The coefficient of determination, for model calibration, comes out to be 0.62 which shows good correlation among the observed and simulated runoff. In observed and simulated flows, the difference was 0.2%, which was reasonable and shown good match. Also, it was observed calibration period, out of total rainfall of 4140 mm, the simulated discharge was 3701.6 mm out of which 2101.8 mm formed the overland flow, 268.8 and 55.5 mm water becomes contributed as inter flow and base flow, respectively, and remaining amount of 1400.1 mm of water contributed to ground water recharge.

4.2   Comparison Between Observed Runoff and Simulated Runoff for Calibration

From Fig. 8, showing hydrographs of different events of runoff during calibration period, it was observed that the shapes of the hydrograph of both observed and simulated show good match for majority of runoff events [12]. These graphs indicated the good match between the observed and simulated discharge at Kuniyil site in Chaliyar basin.

From the analysis of double mass curve illustrating accumulative runoff during the calibration period of three years, shown in Fig. 9, it was observed that cumulative observed runoff and simulated runoff were matching with high degree of accuracy. Thus, it can be concluded that the model has been calibrated up to its best and it could further be tested by validation (Fig. 10).

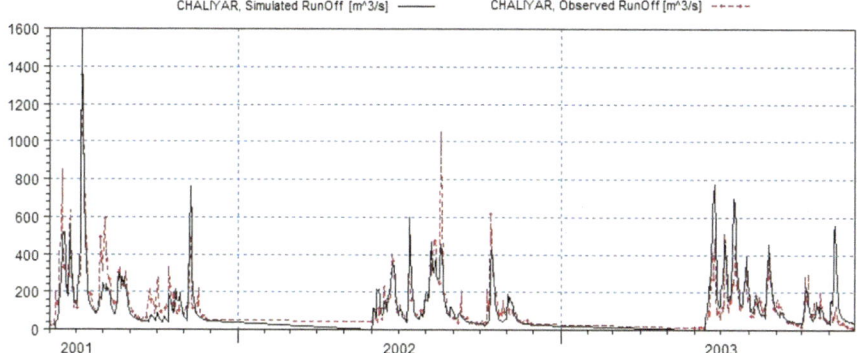

Fig. 8 Comparison of observed and simulated runoff hydrograph during calibration

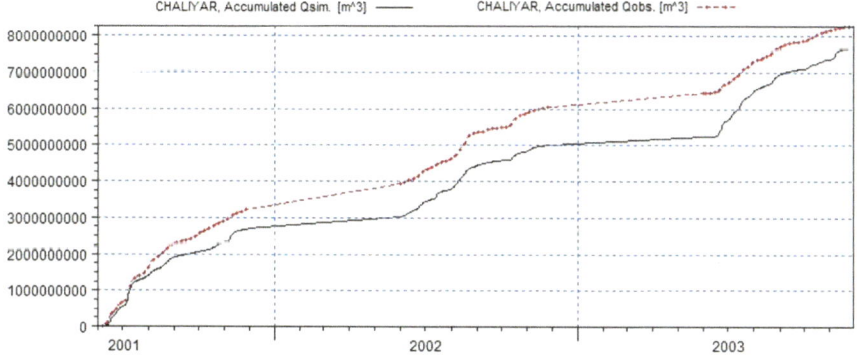

Fig. 9 Double mass curve during calibration period

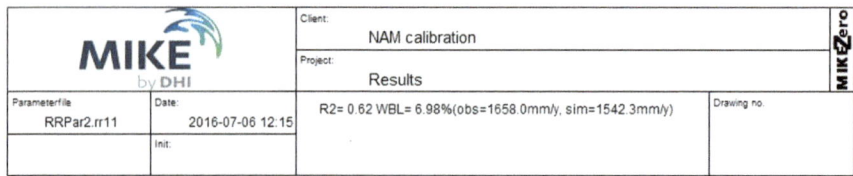

Fig. 10 R^2 & WBL value for calibration period

4.3 Model Validation

For validation, model was run without using auto-calibration mode using calibrated parameters for left over duration from 2004 to 2005 and results were compared with calibration outputs. The coefficient of determination for validation period of the model was seen as 0.65 which shows that peak and shape of the hydrograph were matching well and good agreement with low flows. The difference between total observed and simulated runoff during validation was 0.5%. Figure 11 shows the graphical representation of the results obtained during model validation which shows that model parameter obtained during calibration were almost accurate. The analysis of model validation results indicated that the NAM model was performing well and seems to be capable of generating or predicting runoff time series for extended time period with accuracy in Chaliyar Basin. Double mass curve during the validation was presented in the Fig. 12 which shows that observed cumulative runoff matches well with the simulated runoff. Thus, it can also be concluded that the NAM model thus developed in Chaliyar basin up to Kuniyil can be used to simulate the runoff in other sub-basin of similar characteristics (Fig. 13; Table 4).

From Figs. 11 and 12, we can say that the coefficient of correlation for the validation period of the model was observed as 0.65 which shows that peak and shape of the hydrograph were matching well and good agreement with low flows

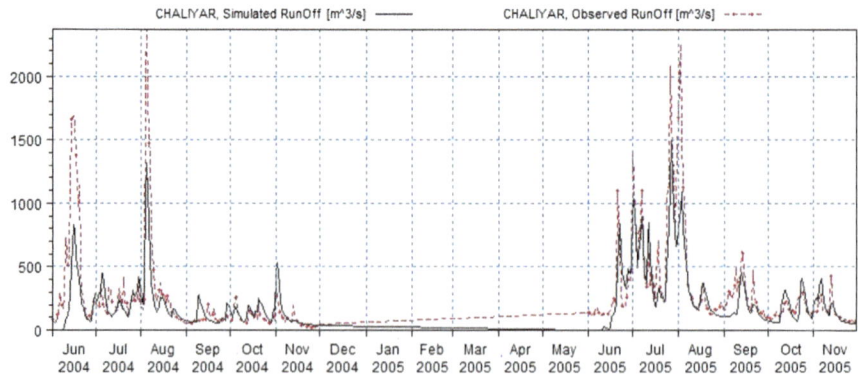

Fig. 11 Comparison of observed and simulated runoff during model validation

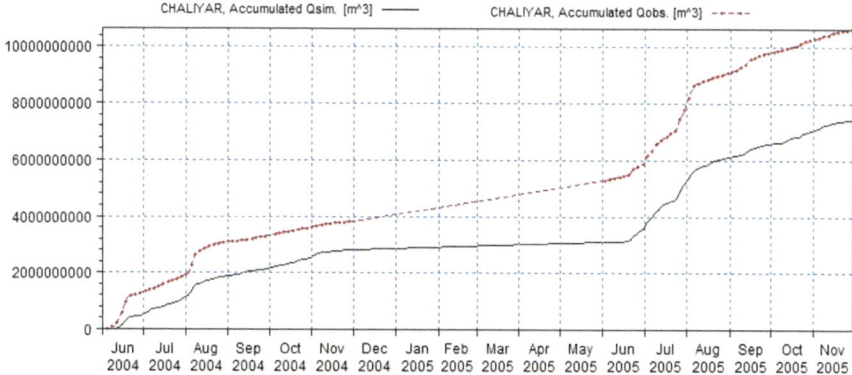

Fig. 12 Double mass curve during model validation

			Client:		
MIKE by DHI			NAM calibration		MIKE Zero
			Project:		
			Results		
Parameterfile RRPar2.rr11	Date: 2016-07-06 12:20		R2= 0.65 WBL= 30.20%(obs=3558.4mm/y, sim=2483.8mm/y)	Drawing no.	
	Init:				

Fig. 13 R^2 and WBL values for validation period

Table 4 Model validation result (all values are in mm)

Period	Q-obs	Q-sim	% Diff	RF	PET	AET	GWR	OF	IF	BF
2004	1904.7	1442.6	0.3	2026	568.3	502.9	554	832.3	87.5	21.8
2005	3424.3	2193.9	0.6	2728	504.0	453.1	797	1353.7	104.5	30.7
Total	5329	3636.6	0.5	4754.6	1072.3	956	1351	2186	192.0	52.4
Coefficient of determination = 0.65, WBL = 30.0										

which indicates that the NAM model was performing well and seems to be capable of generating or predicting runoff time series for extended time period with accuracy in Chaliyar basin. Although a lesser agreement has been reported in water balance, which is due to the dissimilarity in magnitude of rainfall during calibration and validation period. As from Table 5, the values of rainfall vary from 1707.7 to 2169.9 for calibration whereas for validation, it is much higher. Moreover, water balance difference is higher because of small duration of data during validation period.

Table 5 Yearly rainfall

Year	Ambalavayal	Manjeri	Nilambur
2001	1425.1	2377.1	2169.9
2002	1167.4	1911.1	1707.7
2003	1546.3	1969.6	1927.5
2004	1943.6	2257.8	2440.2
2005	2150.4	2358.8	2905.6

5 Conclusion

Hydrological modelling is of foremost importance for appropriate planning, designing and decision-making activities of water resources. A simple, logistic and systematic modelling of rainfall–runoff is an important and challenging issue in recent changing environments to properly manage water resources for socio-economic development of the society in the region.

In the present study, a lumped conceptual model of MIKE 11 NAM has been used. The model has been applied for modelling of streamflow for Chaliyar basin, Kerala, India with an area of 1996.4 km^2.

The hydrological model MIKE 11 NAM has also been successfully applied for modelling hydrological characteristic of the Chaliyar basin. It is a lumped, conceptual model and does not require lot of data to simulate the daily discharges. Thus, it is an useful tool for modelling hydrological characteristic of the Chaliyar basin. The model yielded satisfactory and reliable results with coefficient of determination and water balance error as 0.62 and 6.98%, respectively, for calibration and coefficient of determination as 0.65 for validation. The capability of the model was revealed by a good match of simulated data with the observed data and an overall good agreement in hydrograph shape regarding timing, rate and volume.

The rainfall–runoff model thus developed seems to be capable of predicting runoff for extended time period in Chaliyar river basin. Moreover, the data availability for longer duration could further enhance the modelling result.

Hence, finally, in the present work, rainfall–runoff model has been developed, calibrated and validated using flow data at river gauge station, Kuniyil, in Chaliyar Basin, Kerala, India.

Finally, it can be concluded that water resources development planners, scientists and engineers in the region should design strategies and plans by taking into account spatial and temporal distribution and changing patterns for meteorological and hydrological parameters.

References

1. Shamsudin S, Hashim N (2002) Rainfall-Runoff modelling using MIKE 11 NAM. J Civ Eng 15
2. Shoemaker L, Lahlou M, Bryer M, Kumar D, Kratt K (1997) Compendium of tools for watershed assessment and TMDL development. U.S. Environmental Protection Agency, EPA841-B-97-006
3. Thakural LN, Anbu Kumar S, Ansari MI (2018) Trend analysis of rainfall for the chaliyar basin. S India Int J Res Appl Sci Eng Technol 6(7):91–100
4. Ambili V et al (2014) Tectonic effects on the longitudinal profiles of the Chaliyar River and its tributaries, southwest India. Geomorphol 217:37–47
5. Linsley RL (1982) Rainfall-Runoff models-an overview in Rainfall-Runoff relationship. In: Proceedings of the international symposium on Rainfall-Runoff modelling, 18–21 May 1982, pp 3–22
6. Satish Kumar K et al (2018) River basin modelling for Shipra River using MIKE BASIN. ISH J Hydraul Eng 1–12
7. Danish Hydraulic Institute (2003) MIKE BASIN: Rainfall-Runoff modeling Reference manual, 60p
8. Galkate RV et al (2018) Rainfall runoff modeling using conceptual NAM model. Scientists, National Institute of Hydrology, Regional Centre, Bhopal
9. Regional calibration of Hargreaves equation for estimating reference ET in a semiarid environment. Agric Water Manag 81:257–281
10. Madsen H (2000) Automatic calibration and uncertainty assessment in Rainfall-Runoff modelling. In: 2000 joint conference on water resources engineering and water resources planning & management, Hyatt Regency Minneapolis, USA. American society of civil engineers conference proceeding, pp 1–10
11. Ashutosh et al (2014) Rainfall-Runoff modeling using MIKE 11 NAM model for vinayakpur intercepted catchment, Chhattisgarh. Indian J Dryland Agric Res Dev 29(2):1.33
12. Nash JE (1958) The form of instantaneous unit hydrograph. J Hydrol V 3:114–121

CNTFET-Based Input Buffer
for High-Speed Data Transmission

Hasan Shakir, Yasser Najeeb and M. Nizamuddin

Abstract In this paper, design and simulation of carbon nanotube (CNT)-based input buffer for high-speed digital transmission has been performed. Design and simulation of CNT-based input buffer are compared with that of conventional metal–oxide–semiconductor (MOS)-based input buffer at 45-nm-technology node. The structures, used in this scheme, have a differential amplifier. A differential amplifier is used to amplify the difference of the two input signals which are transmitted on a dissimilar signal path. The proposed structure is designed and simulated using HSPICE. The key characteristics of the proposed device, like response time graph and DC analysis for both conventional MOS and CNT-based input buffer, have been computed $V_{dd} = 1$ V, width of channels $W_P = W_N = W_{CNTFET} = 381.5$ nm, number of CNTs (N) = 20, pitch (S) = 20 nm, diameter of CNT (D_{CNT}) = 1.5 nm. It has been observed that CNT-based input buffer results in high performance in comparison with MOS-based input buffer. It is also observed that the performance of CNT-based input buffer mainly depends upon CNT number (N); inter-CNT pitch (S), and diameter (D_{CNT}) of CNTFET.

Keywords MOS · CNT · CNTFET · Input buffer · Simulation

1 Introduction

Enduring advancements in the reduction of technology node are gaining much attention in the field of nanotechnology. Along with these advancements, it is also necessary to reduce the threshold voltage and power consumption. Reduction in threshold voltage gives rise to increased leakage power and short-circuit power. It has been observed that clock drivers which are made up of CMOS inverters have quite large leakage and short-circuit current. High-speed digital signals pass through several digital circuitries, and these result in various distortions of the

H. Shakir · Y. Najeeb · M. Nizamuddin (✉)
Electronics and Communication Engineering, Jamia Millia Islamia, New Delhi, India
e-mail: mnizamuddin1@jmi.ac.in

© Springer Nature Singapore Pte Ltd. 2020
S. Ahmed et al. (eds.), *Smart Cities—Opportunities and Challenges*,
Lecture Notes in Civil Engineering 58,
https://doi.org/10.1007/978-981-15-2545-2_54

signal due to the addition of noise and the propagation delay caused by the components. Delays are caused because of slow rise and fall. This results in an indecorous instruction given to the on-chip circuitry. Therefore, it is obligatory to use an input buffer at the input of on-chip circuits. It helps to clean the distorted signal. Input buffers are used to slice the data signal at the correct time interlude. This is dependent on the switching point voltage. The term 'switching point voltage' delineates the voltage at which the transitions are from low logic to high logic or vice versa [1–5].

Input buffer circuits are used for numerous digital applications. It is extensively used in memory devices. Usually, input buffers engaging differential amplifier links the input signal with the main memory. High-speed buffers are essential components of communication systems. Complementary metal–oxide–semiconductor (CMOS) integrated circuits are used in applications which possess large capacitive load. To drive these circuits at high speed, tapered buffers are required. Tapered buffers are made up of several inverters with each gate having different sizing factor. To drive large off-chip capacitances in input/output circuits, these buffers are required.

1.1 Input Buffer

These are the circuits which are to covert an error signal into an error-free digital signal. Basically, it provides a clean digital signal to the on-chip circuitry. If the input signal possesses slow rise and fall times, then it would result in an altered waveform having different pulse width. Input buffers are used to slice the data in the correct position, so as to avoid timing errors. If slicing done is not accurate, then it results in error. Henceforth, in high-speed systems, transmission should be accurate [1].

1.2 Carbon Nanotube (CNT) and Carbon Nanotube Field-Effect Transistors (CNTFETs)

Carbon nanotube (CNT) is rolled sheets of graphite. They possess innumerable properties which are valuable for nanotechnology, material science, optics. Carbon nanotubes are created with a length to diameter ratio of 132,000,000:1. Due to the valency of carbon, many different allotropes are formed. The famous one includes graphite, diamond, and fullerene. Nanotubes are member of fullerene structural family. Graphene sheets when rolled at a particular chiral angle form nanotubes. The radius and the rolling angle outline the properties of nanotube. In carbon nanotube field-effect transistors (CNTFETs) one or more semiconducting SWCNTs are used as the channel of the device. These are classified as multi-walled nanotubes (MWNTs) and single-walled nanotubes (SWNTs). MWCNTs have more than one tube. The angle of the arrangement of the carbon atoms along with the nanotube and its indices (n_1, n_2) decides whether a SWCNT is semiconducting or metallic. If $n_1 - n_2 = 3k$ (such that k belongs to z), then SWCNT is said to be conducting otherwise semiconducting [6–8] (Fig. 1).

One of the most important applications of CNTs is the development of a CNTFET. In a CNTFET, the channel between source and drain regions/electrodes is realized using semiconducting single-wall CNTs. The existence of an intrinsic band gap is the foremost purpose to construct CNTFETs. The channel of field-effect transistors (FETs) is made up of CNT in CNTFETs.

Pitch in the CNTFETs is defined as the distance between the mid of two adjoining SWCNTs under the same gate voltage. The width of the gate and the contacts of the device are directly affected. The gate width is determined by the pitch. By setting least gate width W_{min} and the number of tubes N, the gate width can be approximated as:

$$W_g = \text{Max}(W_{min}, N * S) \tag{1}$$

where W_{min} is the minimum gate width, S is pitch, and N is the count of nanotubes under the gate [6, 7].

1.3 Differential Amplifier

Differential amplifier is defined as an electronic amplifier which is used to amplify the difference of the two input voltages and suppress any voltage common to the two inputs.

Output voltage of a differential amplifier is given by:

$$V_{out} = A(V_+ - V_-) \tag{2}$$

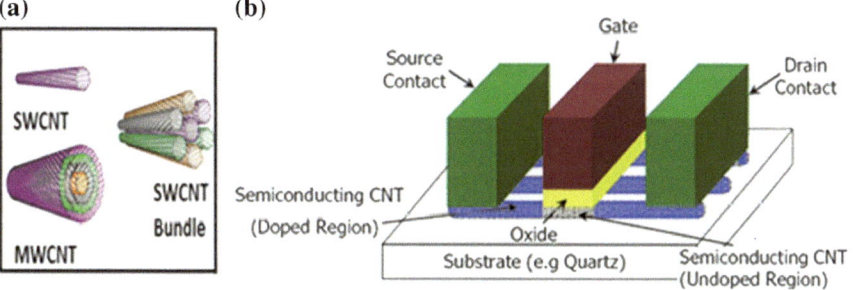

Fig. 1 **a** SWCNT and MWCNTs, **b** schematic of a CNTFET

where V_+, V_- are the two inputs to the amplifier, A is the amplifier gain, and V_{out} is the output voltage.

In order to send an error-free signal (or to precisely slice input data), a reference voltage is to be transmitted on a different signal path, along with the input data. Otherwise, data can be transmitted differentially. Thus, a differential amplifier is required. It amplifies the difference of the two input signals [9–14].

1.4 Conventional MOS Input Buffer

From this plot, it is clearly seen that the output obtained for a given input is not identical, as it contains glitches with no steep rise or fall. Hence, MOSFET-based input buffer device does not act as a perfect input buffer. Design parameters of proposed input buffers are load capacitance, $V_{dd} = 1$ V, width of channels $W_P = W_N = 381.5$ nm, using BSIMv4.6.1 Berkeley predictive technology model at the 45-nm-technology node for MOSFETs (Figs. 2, 3 and 4).

1.5 CNT-Based Input Buffer

From this plot, we can conclude that for the given input pulse, the output pulse is steeper without any glitches and the rise time fall time of the input pulse reduces. Hence, CNT-based input buffer device acts as a perfect input buffer. Proposed CNT-based input buffer is simulated using highly accurate Verilog-A Stanford

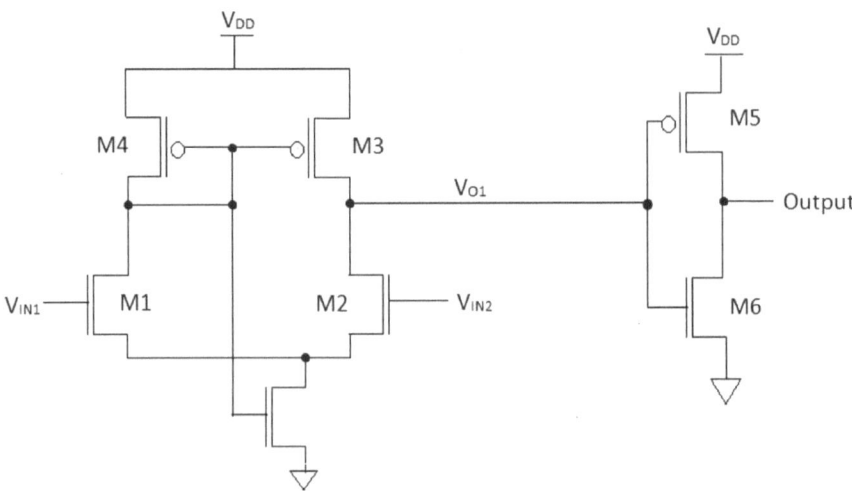

Fig. 2 Input buffer based on conventional MOS [1]

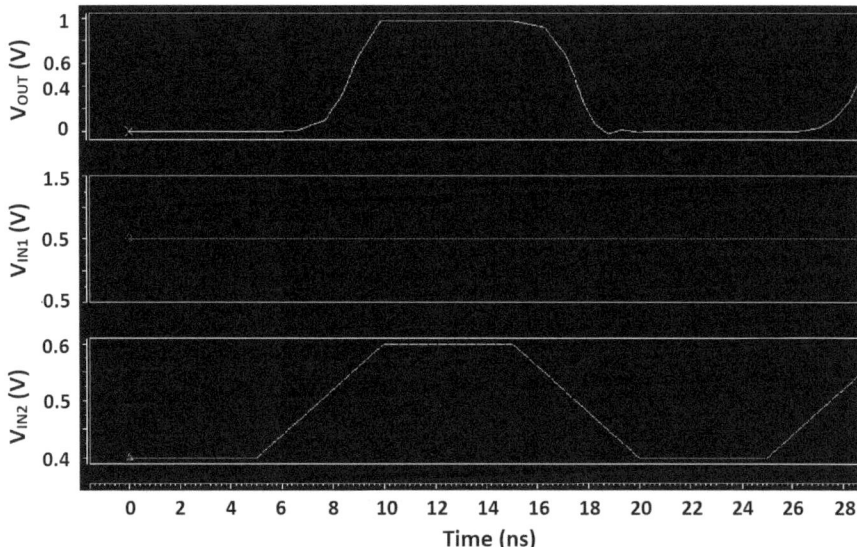

Fig. 3 Output voltage versus time for a given input, using MOS-based input buffer

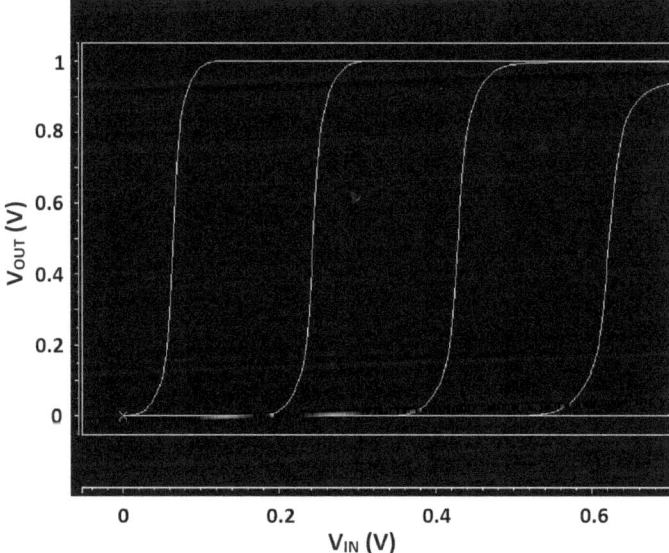

Fig. 4 DC response of MOS-based input buffer

model with V_{dd} = 1 V, Width of channels W_{CNTFET} = 381.5 nm, number of CNTs (N) = 20, pitch (S) = 20 nm, diameter of CNT (D_{CNT}) = 1.5 nm (Figs. 5, 6, and 7; Table 1).

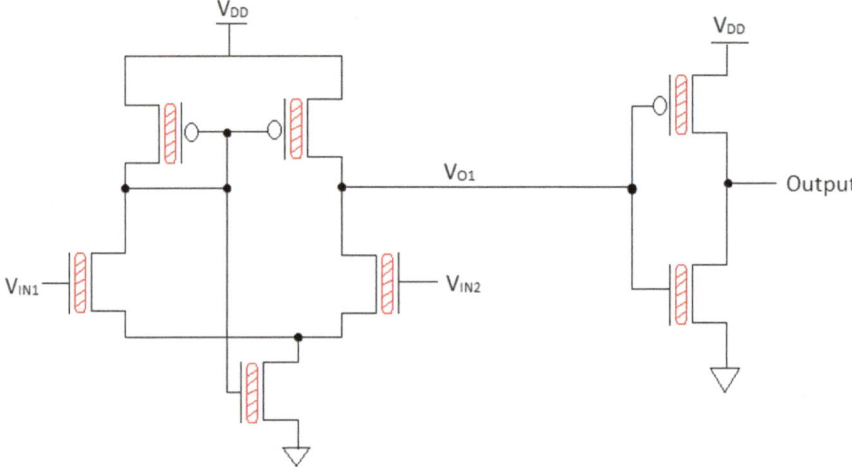

Fig. 5 Proposed CNT-based input buffer for high-speed digital transmission

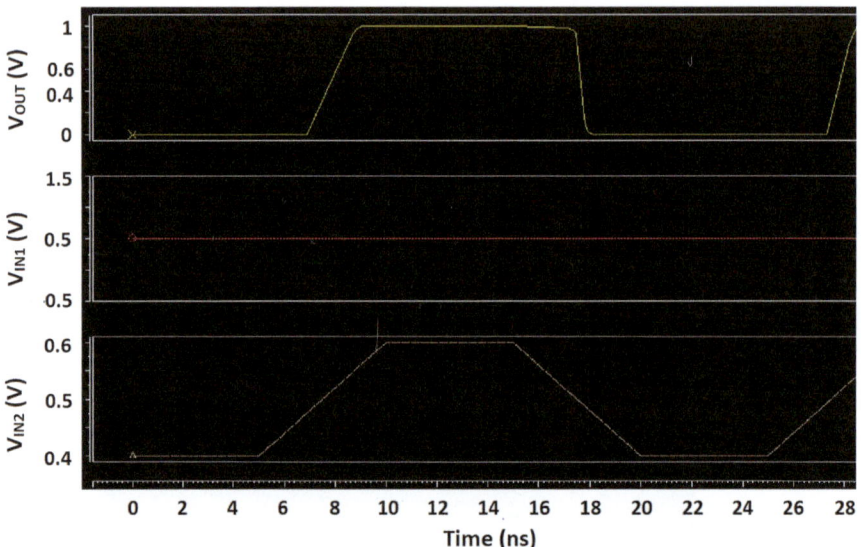

Fig. 6 Output voltage versus time for a given input, using CNT-based input buffer

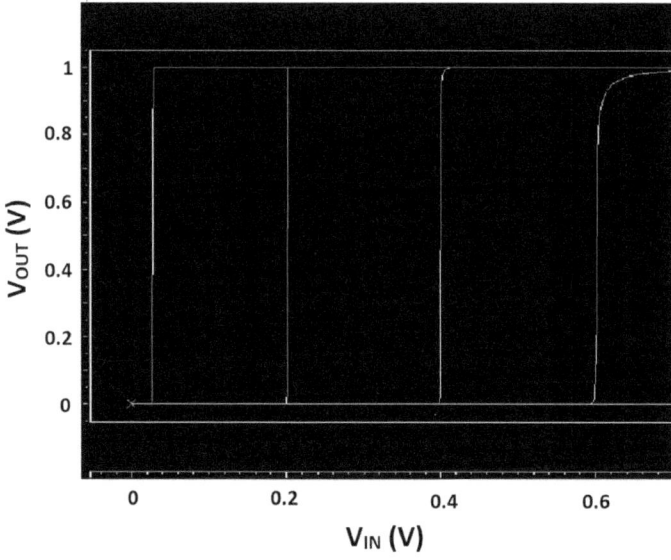

Fig. 7 DC response of proposed CNT-based input buffer

Table 1 CNTFET device parameters used in the proposed CNT-based input buffer

Device parameter	Description	Default value
Lch	Physical channel length	45.0 nm
Lgeff	MFP in the intrinsic CNT channel	200.0 nm
Lss	Doped CNT source-side extension length	32.0 nm
Ldd	Doped CNT drain-side extension length	32.0 nm
Efi	Fermi level of doped S/D tube	0.6 eV
Kgate	Dielectric constant of high-k top gate	16.0
Tox	Thickness of high-k top gate	4.0 nm
Csub	Coupling capacitance b/w channel and the substrate	20.0 pF/m

2 Conclusion

Carbon nanotube (CNT)-based input buffer for high-speed digital transmission is proposed. Design and simulation are performed using Verilog-A Stanford models for CNT-based input buffer and BSIMv4.6.1 Berkeley predictive technology model for MOSFETs at 45 nm technology node using HSPICE. **The key characteristics of the proposed device, like response time and DC analysis, have been computed. CNT-based input buffer results in high performance in comparison with MOS-based input buffer. It has been observed that their performance can be improved further by using optimized values of CNT number, inter-CNT pitch, and diameter.**

Conflict of Interest The authors declare that they have no conflict of interest.

References

1. Baker RJ (2000) CMOS circuit design, layout and simulation, 2nd edn. PHI, New Delhi
2. Rezaei M, Zhian-Tabasy E, Ashtiani SJ (2008) Slew rate enhancement method for folded-cascode amplifiers. Electron Lett 44(21):1226–1228
3. Tsuruya Y et al (2012) A nano-watt power CMOS amplifier with adaptive biasing for power-aware analog LSIs. In: Proceedings of the ESSCIRC, pp 69–72
4. A. L. Coban, P. E. Allen (1995) A low-voltage CMOS op amp with rail-to-rail constant-gm input stage and high-gain output stage, vol 2. In: Circuits and systems ISCAS '95, pp 1548–1551
5. Gulati K et al (1998) A high-swing CMOS telescopic operational amplifier. IEEE J Solid-State Circuits 33(12):2010–2019
6. Loan SA, Nizamuddin M et al (2014) Design of a novel high gain carbon nanotube based operational transconductance amplier. In: Proceedings of the IMECS, Hong Kong, pp 797–800
7. Loan SA et al (2015) Design and comparative analysis of high performance carbon nanotube based operational transconductance amplifiers. NANO 10(3):1550039
8. Sajad A et al (2014) High performance carbon nanotube based cascode operational transconductance amplifiers. In: Proceedings of the WCE, London pp 265–9
9. Far A (2016) Small size class AB amplifier for energy harvesting with ultra low power high gain and high CMRR. In: 2016 IEEE international autumn meeting on power electronics and computing (ROPEC), pp 1–5
10. Far A (2017) Low noise rail-to-rail amplifier runs fast at ultra low currents and targets energy harvesting. In: Submitted to ROPEC
11. Far A (2017) Enhanced gain low voltage rail-to-rail buffer amplifier suitable for energy harvesting. In: Submitted to ROPEC
12. Lu CW, Hsiao CM (2009) 1 V rail-to-rail constant-gm CMOS op amp. Electron Lett 45 (11):529–530
13. De Langen KJ et al (1998) Compact low-voltage power-efficient op. amp. cells for VLSI. IEEE J Solid-State Circuits 33(10):1482–1496
14. A. Far (2016) Amplifier for energy harvesting: low voltage ultra low current rail-to-rail input-output high speed. In: IEEE international autumn meeting on power electronics and computing (ROPEC), pp 1–6

A Study on Modelled Granular Column of Various Diameters in Soils

Ankush Chaudhary, Rahul Siddarth, A. K. Sahu and S. M. Abbas

Abstract In order to develop a smart city, the role of smart transportation infrastructure system is essential, which will improve the ability to connect and improve the quality of human being by saving time and energy. In the present paper, an attempt has been made to improve the soil for the construction of smart structures. In the present paper, the aim is to study the variation of load carrying capacity and shear parameters of soil after introducing granular columns of varying diameter. Results show that the granular columns derive the strength from the soil confining them. The granular columns also help in easy drainage and reduction in pore pressure. A series of CBR and direct shear test were performed after the installation of granular columns at the centre of the specimen by varying the diameter in the soil where it was found that there was an improvement in load carrying capacity and shear strength parameters of the soil. The study also presents swelling behaviour of soil against time.

Keywords Soil · Granular columns · Stone dust · Direct shear test · BCS · CBR test · OMC · MDD

1 Introduction

Smart cities may be considered as a community that uses the advanced technology and information to improve the infrastructures for the betterment of human being which not only increases the safety but a strong and sustainable infrastructure can be made.

A. Chaudhary (✉)
Department of Civil Engineering, KIET, Ghaziabad, India
e-mail: ankush.chaudhary@kiet.edu

R. Siddarth · A. K. Sahu
Department of Civil Engineering, DTU, New Delhi, India

S. M. Abbas
Department of Civil Engineering, JMI, New Delhi, India

© Springer Nature Singapore Pte Ltd. 2020
S. Ahmed et al. (eds.), *Smart Cities—Opportunities and Challenges*,
Lecture Notes in Civil Engineering 58,
https://doi.org/10.1007/978-981-15-2545-2_55

For any infrastructure, the strong and durable substructure is essential, but in case of problematic soils, some ground improvement techniques are required. In order to construct a smart structure on expansive soil, a study on granular sand column was carried out [1].

Other than the strength criteria, one of the other problems encountered by the soil engineer is the expansive nature of soil found in major parts of the world. The central part of India is covered with such soil. These soils are highly expansive when brought in contact of water and thus offer high uplift capacity to the structures built upon it leading to destruction of integrity of the structure; hence, it becomes very important to study such soils and methods to stabilize them.

Reinforcing the soil for modifying the bearing capacity, shear strength parameters, minimizing settlement as well as swelling and so it has been one of the important goals of studies and researches conducted on soil [2, 3]. These studies have proven to be very useful in every small and large construction projects related to soil where soil serves as foundation material or load bearing strata. The previous studies reveal that the granular column offers resistance to shear stress due to its orientation as well as in axial loading condition due to the confining pressure offered by soil surrounding the column when the column bulges under the vertical loading [4, 5].

The present study aims at studying the effect of granular columns on shear strength parameters and swelling characteristics offered by the soil with and without granular columns of varying diameter. The scope of the study conducted can be applied in the following fields:

- Ground improvement
- Analysis of shear parameters of soil
- Stability analysis of a structure.

Granular columns construction was a cost-effective method developed in order to encounter expansive soils. The expansive soils refer to clay and fine silts which posses high water content and are subjected to low undrained strength [6]. In such kind of soils, granular columns are driven space closely to form a grid of granular columns that imparts higher strength and stiffness to the existing soil [1, 7].

2 Materials Used

The materials used for the study are various types of soils such as sand, black cotton soil, and stone dust [7]. The geotechnical properties of various materials are obtained as per various parts of code IS-2720 (Table 1,2,3).

Table 1 Properties of silty sand and poorly graded sand

Description	Silty sand (SS)	Poorly graded sand
Specific gravity	2.66	2.63
Soil type	SM	SP
Maximum dry density	18.83 KN/m^3	15.50 KN/m^3
Optimum moisture content	11.56%	14.24%
Cohesion (c)	8.05 KN/m^2	0 KN/m^2
Angle of internal friction (Ø)	22°	29.6°

Table 2 Properties of stone dust

Description	Value
Specific gravity	2.77
Soil type	SP
Maximum dry density	18.88 KN/m^3
Optimum moisture content	12.22%
Cohesion (c)	0.04 KN/m^2
Angle of internal friction (Ø)	77.2°

Table 3 Properties of black cotton soil (BCS)

Description	Value
Specific gravity	2.73
Soil type	CH
Maximum dry density	17.75 KN/m^3
Optimum moisture content	21.23%
Cohesion (c)	56.57 KN/m^2
Angle of internal friction (Ø)	3.15°
Liquid limit	56.60%
Plastic limit	25.32%
Plasticity index	31.28%
Activity	1.145
UCS value	110.35 KN/m^2

3 Experimental Investigation

This portion covers the study and experiment conducted to observe the effect of granular columns of varying diameter on engineering properties of soil taken for consideration. The motive of the study was to analyse the variation in CBR value, cohesion value, and angle of internal friction and swelling behaviour of soil after the installation of granular columns at the centre of area. As mentioned earlier, soils considered are BCS and silty sand, while the material for column was kept as stone dust and sand.

The tests conducted were:

- California bearing ratio test
- Direct shear test.

3.1 California Bearing Ratio Test

- CBR test is performed according to IS2720 PART 16.
- According to IRC 37: 2012, the CBR results depend on various factors, and wide variation in values can be expected.
- In the current investigation, the improvement in the given soil has been achieved through installation of granular columns of varying diameter.
- The soil was compacted to MDD at OMC.
- All the CBR tests were performed in soaked condition.

3.2 Direct Shear Test

- Test performed as per IS 2720-part 13-1972
- The soil was compacted to MDD at OMC
- The tests were performed with a single objective to see the variation in cohesion and angle of internal friction in the soil after the installation of columns.

3.3 The Experimental Investigation Shall Be Conducted in the Following Stages

- Study the 'Variation in CBR value for black cotton soil with and without installation of granular column'.
- Study the 'variation in CBR value for silty sand with and without installation of granular column'.
- Study the 'Variation in shear parameter for black cotton soil with and without installation of granular column using DST'.
- Study the 'variation in shear parameter for silty sand with and without installation of granular column using DST'.
- Study the 'variation in the swelling behaviour of black cotton soil with and without installation of granular column (Tables 4 and 5; Fig. 1).

Table 4 Results of CBR test on soils with and without granular column

Description	CBR value (%)
Plain BCS	2.13
BCS + 10 mm stone dust column	2.74
BCS + 15 mm stone dust column	3.23
BCS + 20 mm stone dust column	3.52
BCS + 10 mm sand column	2.56
BCS + 15 mm sand column	2.91
BCS + 20 mm sand column	3.06
SM (silty sand)	3.53
SM Soil + 10 mm sand column	3.70
SM Soil + 15 mm sand column	3.83
SM Soil + 20 mm sand column	4.11
SM Soil + 10 mm stone dust column	3.75
SM Soil + 15 mm stone dust column	4.00
SM Soil + 20 mm stone dust column	4.21

Table 5 Results of direct shear test with and without granular column

Description	Cohesion (c) (KN/m^2)	Angle of internal friction (Ø)
Plain BCS	56.57	3.14°
BCS + 10 mm stone dust column	48.34	5.37°
BCS + 15 mm stone dust column	42.77	9.14°
BCS + 20 mm stone dust column	37.95	13.28°
BCS + 10 mm sand column	50.18	4.91°
BCS + 15 mm sand column	46.76	8.19°
BCS + 20 mm sand column	39.16	12.78°
SM soil + 10 mm sand column	2.03	18.00°
SM soil + 15 mm sand column	1.67	19.13°
SM soil + 20 mm sand column	1.10	33.66°
SM soil + 10 mm stone dust column	6.35	16.43°
SM soil + 15 mm stone dust column	4.34	22.80°
SM soil + 20 mm stone dust column	4.06	31.04°

(a)

(b)

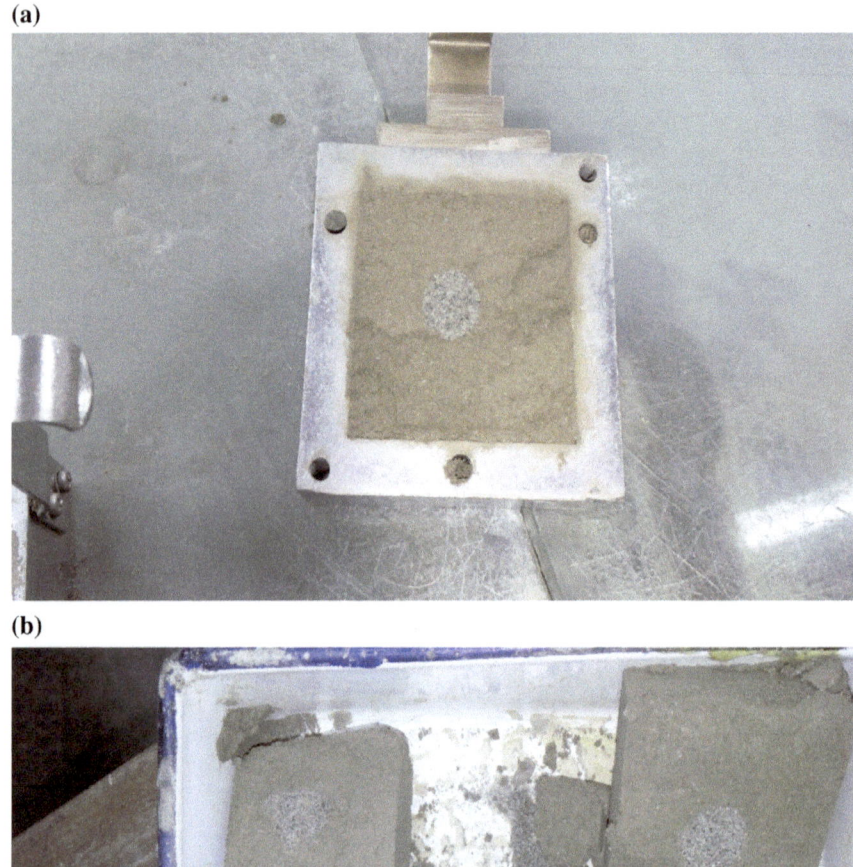

Fig. 1 **a** DST box with silt soil with granular material columns. **b** Failed samples

3.4 Swelling Behaviour of BCS

In order to evaluate the effect of granular column on swelling behaviour of BCS, the soil was kept in soaking condition compacted at OMC in CBR mould. The swelling was recorded every 24 h for a duration of four days. This was adopted for both soils in plain and in reinforced condition. The following results were obtained (Tables 6 and 7; Figs. 2 and 3).

Table 6 Swelling for BCS with and without granular column of stone dust

Time (h)	Swelling (mm)			
	BCS	10 mm column	15 column	20 column
24	6.45	6.55	6.65	6.35
48	11.10	10.75	10.30	9.70
72	12.65	12.05	11.50	10.90
96	13.00	12.50	12.10	11.10

Table 7 Swelling for BCS with and without granular column of sand

Time (h)	Swelling (mm)			
	Plain BCS	10 mm column	15 column	20 column
24	6.45	6.25	6.35	6.25
48	11.10	10.85	10.65	9.50
72	12.65	12.50	11.30	10.70
96	13.00	12.75	11.90	11.0

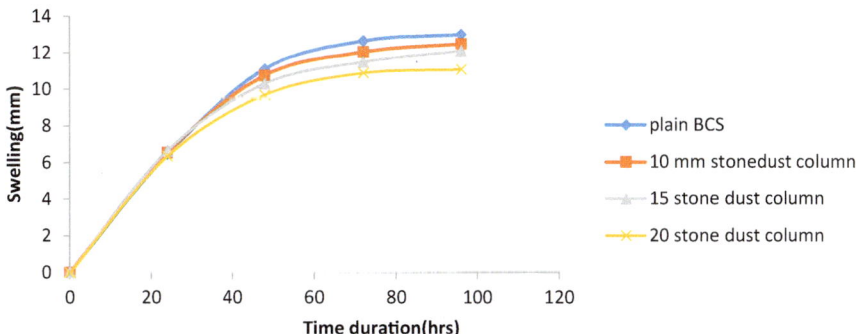

Fig. 2 Swelling behaviour of BCS with and without granular column of stone dust

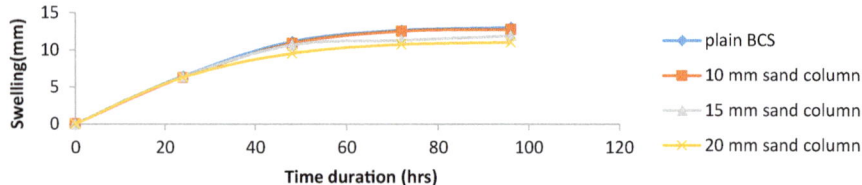

Fig. 3 Swelling for BCS with and without granular column of sand

4   Results and Discussion

The results and discussion with respect to CBR and direct shear tests on black cotton soil (BCS) and silty sand (SS) with and without granular column made up of poorly graded sand and stone dust are presented.

4.1   CBR Test

The results above show that after the installation of granular columns, there has been an increase in the load taking capacity of soils considered, and the possible reason for such behaviour of soil can be deduced from the previous research studies. When the load is applied on the soil in the CBR mould, the load is concentrated on the granular column; under the effect of load, the column tries to fail, but its failure is resisted by the lateral support and resistance provided by the surrounding soil (Table 8).

However, when we compare the BCS to SS soil which is basically comprising of the granular material itself, the granular columns prove more fruitful in BCS; the granular columns in SS soil give low value due to the presence of sand in soil which lowers the cohesion value restricting the lateral support to bulging action of granular column when loaded (Fig. 4; Table 9).

The tabular form of data presented above after conduction of DSTs shows that after the soil was installed with granular columns, there was a decrease in cohesion value of the soil. The decrement of cohesion value increases with the diameter of the column; the reason for happening so is the reduction in the area of plane of failure as the diameter of column is increased.

With the increase in the diameter of column, there is an increase in ϕ value of soil again for the similar reason as stated above; however, the effect of granular column on ϕ value of SS soil is unpredictable which was due to the presence of fine sand particles in the soil.

A better understanding can be developed from the graph plotted Figs. 5 and 6.

After studying the diameter variation of reinforcements on soil's shear parameters and the load carrying capacity [8], it also becomes very important to go through the result which we got from the swelling tests that were performed during

Table 8 Variation of CBR with diameter of granular columns

Soil with granular column of diameter (mm)	BCS				SS soil			
	Stone dust		Sand		Stone dust		Sand	
	CBR (%)	% increase	CBR (%)	% increase	CBR (%)	% increase	CBR (%)	% increase
00	2.130	0	2.130	0	3.53	0	3.530	0
10	2.746	28.92	2.560	20.18	3.76	6.51	3.700	4.81
15	3.235	51.8	2.919	37.04	4.0	13.31	3.832	8.55
20	3.520	65.2	3.063	43.8	4.21	19.26	4.110	16.43

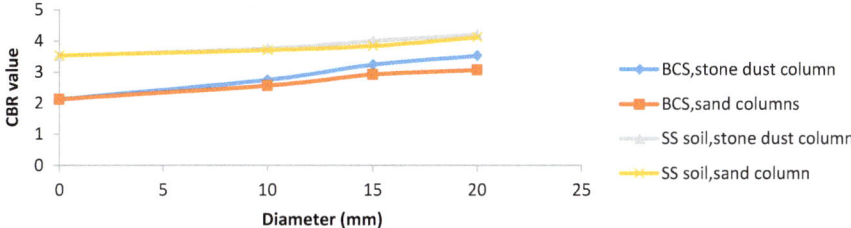

Fig. 4 Variation of CBR for given soil with varying diameter

Table 9 Variation of c−φ with diameter of granular columns

Soil with granular column of diameter	BCS				SS soil			
	Stone dust		Sand		Stone dust		Sand	
	c (KN/m^2)	φ	c (KN/m^2)	φ	c (KN/m^2)	φ	c (KN/m^2)	Φ
Plain	56.57	3.14°	56.57	3.14°	8.05	22.00°	8.05	22.00°
10 mm	48.34	5.37°	50.18	4.91°	6.35	16.43°	2.033	18.00°
15 mm	42.77	9.14°	48.76	8.19°	4.34	22.8°	1.67	19.13°
20 mm	37.95	13.27°	39.16	12.78°	4.066	31.04°	1.103	33.66°

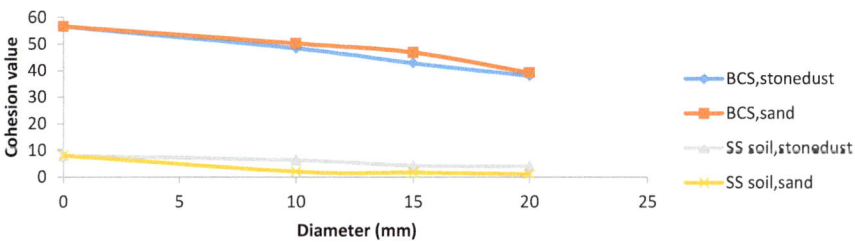

Fig. 5 Variation of cohesion value for given soil with varying diameter

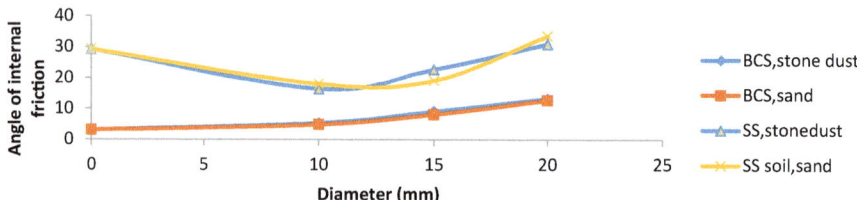

Fig. 6 Variation of φ value for given soil with varying diameter

the investigation program. It was seen that soil like BCS that could swell up to a large extent was cut down short after installation of granular columns. The possible reason for this could be easy drainage facility allowed by the granular columns and minimization of pore pressure caused by them. They also allow expansion in lateral direction. Hence, in this manner, swelling was reduced [9].

However, when it comes to choose between the two granular material, considering the alternations it caused in load bearing capacity, shear strength criteria, and the swelling criteria; it is rational to choose stone dust over sand (Table 10 and 11; Figs. 7, 8 and 9) [10].

Table 10 Variation of CBR with respect to area replacement ratio

Type of cases		BCS				SS (silty sand)			
Diameter of granular column	A_r (%)	Stone dust		Sand		Stone dust		Sand	
		CBR (%)	% increase	CBR (%)	% increase	CBR (%)	% increase	CBR (%)	% increase
0	0	2.130	0	2.13	0	3.53	0	3.53	0
10 mm	0.44	2.746	28.92	2.56	20.18	3.76	6.51	3.70	4.81
15 mm	0.10	3.235	51.8	2.92	37.04	4.0	13.31	3.83	8.55
20 mm	1.77	3.520	65.2	3.06	43.8	4.21	19.26	4.11	16.43

Table 11 Variation of $c-\varphi$ with respect to area replacement ratio

		BCS				SS soil			
Diameter of column	A_r (%)	Stone dust		Sand		Stone dust		Sand	
		c (KN/m²)	φ	c (KN/m²)	φ	c (KN/m²)	φ	c (KN/m²)	φ
Plain	0	56.57	3.14°	56.57	3.14°	8.05	29.4°	8.05	29.4°
10 mm	0.21	48.34	5.37°	50.18	4.91°	6.35	16.43°	2.033	18.00°
15 mm	4.90	42.77	9.14°	48.76	8.19°	4.34	22.8°	1.67	19.13°
20 mm	8.72	37.95	13.28°	39.16	12.78°	4.066	31.04°	1.103	33.66°

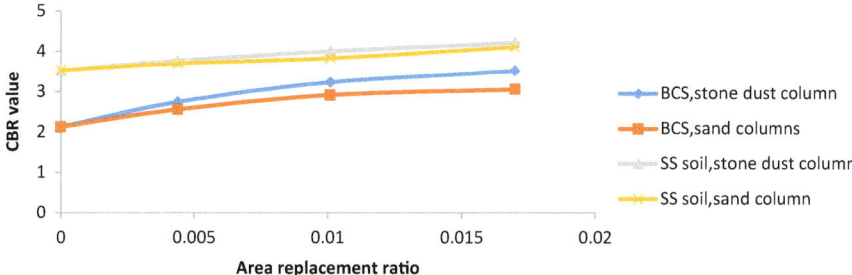

Fig. 7 Variation of CBR with respect to area replacement ratio

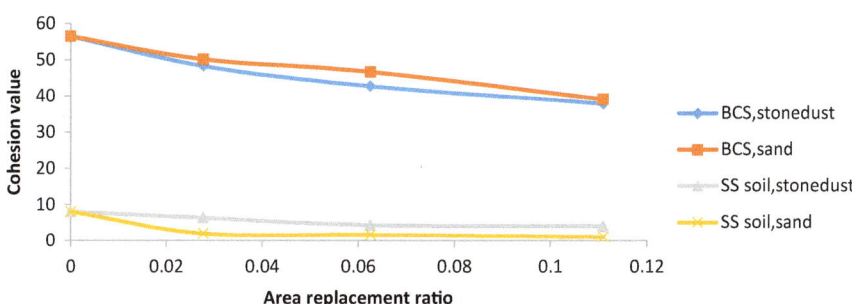

Fig. 8 Variation of c with respect to area replacement ratio

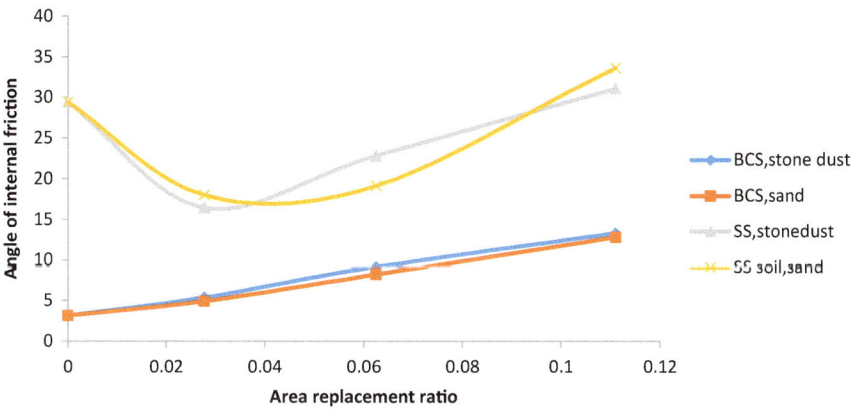

Fig. 9 Variation of φ with respect to area replacement ratio

Let us also discuss the variation of CBR and $c - \phi$ value variation with area replacement ratio where:

Area replacement ratio

= (area of granular column/area of the mould in which sample is placed).

$A_r = A_g/A_m$

Diameter of CBR mould = 150 mm.

5 Conclusions

The smart cities can be developed with the smart infrastructures; however, it is difficult to construct such structures on problematic soils. The experimental results of the present study ensure the improvement of problematic soils by intrusion of granular sand columns. Finally, the following conclusion:

- The results of both CBR and DST reveal that the strength of the silty soil and black cotton soil increases due to installation of granular columns.
- The granular column made up of stone dust gives better results as compared to sand columns in both the soils adopted for the study. Therefore, stone dust obtained from the rock crushing industries often discarded as a waste material should be used wisely.
- With increase in % area of fraction of granular material, the strength of overall composite-reinforced soil increases.
- The swelling behaviour of black cotton soil is also reduced after the installation of granular column. This is an extra advantage that could be induced in expansive soil after the installation of granular columns.

References

1. Mir BA, Juneja A (2012) Strength behavior of composite ground reinforced with sand columns. In: Indian geotechnical conference
2. Juran I, Guermazi A (1988) Settlement response of soft soils reinforced by compacted sand columns. J Geotech Engrg 1988(114):930–943
3. Black JV, Sivakumar V, Bell A (2011) The settlement performance of stone column foundation. Geotech J 61(11):909–922. https://doi.org/10.1680/geot.9.P.014
4. Frikha W, Bouassida M, Canou J (2014) Parametric study of a clayey specimen reinforced by a granular column. https://doi.org/10.1061/(asce)gm.1943-5622.0000419. © 2014 American Society of Civil Engineers
5. Andreou P, Frikha W, Frank R, Canou J, Papadopoulos V, Dupla JC (2008) Experimental study on sand and gravel columns in clay. Ground Improv J 161(4):189–198

6. Najjar SS, Sadek S, Maakaroun T (2010) Effect of sand columns on the undrained load response of soft clays. J Geotech Geoenviron Eng 2010(136):1263–1277
7. Juran I, Riccobono O (1988) Reinforcing soft soils with artificially cemented sand columns. J Geotech Engrg 117:1042–1060 (1991)
8. Greenwood DA (1970) Mechanical improvement of soils below ground surface. Ground Eng J
9. Hughes JMO, Withers NJ (1974) Reinforcing of soft cohesive soils with stone columns. Ground Eng J
10. Najjar SS (2012) A state-of-the-art review of stone/sand-column reinforced clay systems. Geotech Geol Eng 31:355–386. https://doi.org/10.1007/s10706-012-9603-5. (2013)
11. Yoo W, Kim BI, Cho W (2013) Model test study on the behaviour of geotextile-encased sand pile in soft clay ground. KSCE J Civ Eng 19(3):592–601. https://doi.org/10.1007/s12205-012-0473-4
12. Frikha W, Bouassida M, Canou J (2013) Observed behaviour of laterally expanded stone column in soft soil. Geotech Geol, Eng
13. Bergado DT, Lam FL (1987) Full scale load test of granular piles with different densities and different proportions of gravel and sand in the soft Bangkok clay. Soils Found J 27(1):86–93. https://doi.org/10.3208/sandf1972.27.86
14. Sivakumar V, McKelvey D, Graham J, Hughus D (2004) Triaxial tests on model sand columns in clay. Can Geotech J 41(2):299–312
15. McKelvey D, Sivakumar V, Bell. A, Graham J (2004) Modelling vibrated stone columns in soft clay. In: Proceedings of the institute of civil engineers geotechnical engineering
16. Black JV, Sivakumar V, McKinley JD (2007) Performance of clay samples reinforced with vertical granular columns. Can Geotech J 44(1):89–95. https://doi.org/10.1139/t06-08
17. Cimentada A, Da Costa A, Canizal J, Sagaseta C (2011) Laboratory study on radial consolidation and deformation in clay reinforced with stone columns. Can Geotech J 48(1):36–52
18. Gniel J, Bouazza A (2009) Improvement of soft soils using geo-grid encased stone columns. Geotext GeomembrS 27(2009):167–175

Vulnerability Assessment of a Reinforced Concrete Building Frame

Adnan Hussain, Asif Husain and Md. Imteyaz Ansari

Abstract Severe earthquakes are considered as a major hazard with a potential to cause substantial physical losses, economic losses and casualties. At the same time due to population inversion, new settlements are compelled to find their space in higher earthquake-prone zones. The probability of such terrible earthquake events cannot be ruled out for the near future. It is, therefore, quantifying risk and developing strategies which are vital for the disaster mitigation. In this study, a general procedure of seismic vulnerability assessment based on the fragility function is presented. The fragility function corresponding to seismic performance of reinforced concrete frame is obtained and presented as tool for vulnerability assessment. The work presented here aims to direct a quick and simplistic approach towards vulnerability assessment of RC framed building with the help of fragility curves.

Keywords Earthquakes · Vulnerability assessment · Seismic performance evaluation · RC C frame · Fragility curves

1 Introduction

Seismic vulnerability of any structure is a measure of "how vulnerable the structure is with respect to a given state of damage for a given level of ground shaking". How vulnerable a structure is can be quantified on the basis of levels of damage. The seismic vulnerability of a structure refers to its damageability under the given intensity of ground motion. The main reason for doing the vulnerability assessment for any structure is to predict the probability of certain damage levels in the building type under the probable seismic hazard. Assessing the seismic hazard (intensity of

A. Hussain · A. Husain · Md. I. Ansari (✉)
Department of Civil Engineering, Jamia Millia Islamia, New Delhi, India

© Springer Nature Singapore Pte Ltd. 2020
S. Ahmed et al. (eds.), *Smart Cities—Opportunities and Challenges*,
Lecture Notes in Civil Engineering 58,
https://doi.org/10.1007/978-981-15-2545-2_56

683

ground shaking at a site) and the seismic vulnerability of a structure at the site helps evaluate the seismic risk associated with that structure.

RISK = HAZARD * VULNEARBILITY * EXPOSURE

The vulnerability of any structure can be well addressed by one of the notable tools known as fragility curve. Fragility curve is drawn to indicate and present the conditional probability of exceedence for any structural response from its performance limit at provided level of seismic excitation. These curves are considered as viable tool for the evaluation of the structural damage probability under probable level of seismic excitations and may be represented as a function of ground motion indices' parameter or design parameter. Fragility curves show the probability of exceedence for any structural response from its threshold limit versus ground motion parameter like peak ground acceleration or peak ground velocity. Individual point in the fragility curve represents the probability of exceedence of the damage parameter, which may be the base shear, the lateral drift and the storey drift, etc., over and above the threshold value prescribed, at an earthquake intensity parameter.

$$\text{Fragility} = P[D \geq C | \text{IM}] \tag{1}$$

Sabetta et al. [1] derived and presented the empirical fragility curves for three different structural classes using six damage levels according to the Medvedev–Sponheuer–Karnik (MSK) macroseismic scale. Murao and Yamazaki [2] also developed the empirical fragility curves addressing the damage probability for the Japanese building stock using multiple datasets collected after the 1995 Kobe earthquake. The studies involved the estimation of ground motion measures in terms of PGV. Rota et al. [3] derived fragility curves for several building typologies characterizing the Italian building stock using a dataset o f about 150,000 buildings from post-earthquake surveys. Seismic severity was represented as PGA for each municipality evaluated from attenuation relationships. Using an analytical procedure, [4] derived fragility curves for RC structures which were displacement based and suggested the combined use of pushover analysis and the capacity spectrum method to dodge the repetition of analyses due to unavoidable increased number of ground motions thus to reduce the computational effort. Omine et al. [5] developed fragility curves to relate a single ground motion parameter with structural damage. They used PGA and PGV to develop fragility surfaces and found that PGV is better for expressing fragility for severe damage while both PGA and PGV are useful at lower damage levels. Singhal and Kiremidjian [6] developed fragility curves for low-, mid-, and high-rise RC frames based on the results obtained from nonlinear dynamic analysis of the structures. Five damage levels were characterized using Park–Ang's global index. Singhal and Kiremidjian [7] presented a method to obtain fragility curves from the past earthquake damage data through a Bayesian statistical analysis method. The Park and Ang damage index at specified ground motion intervals followed a lognormal probability distribution. Kappos et al. [8, 9] developed a hybrid method for obtaining the fragility curves by combining statistical data and analytical procedures whose data are unavailable.

In this paper, a general methodology for developing the analytical fragility curves for a RC building is described. Such an approach is based on nonlinear static and dynamic analyses of the complete building. The methodology is presented by taking advantage of the capabilities of the software SAP2000, a finite element-based software developed by CSI, which is capable of executing nonlinear time history analyses on the buildings.

2 Seismic Vulnerability

2.1 Analytical Fragility Function

This probability is usually modelled as a cumulative normal distribution function as follows:

$$P_f = \emptyset \left(\frac{\ln(S_d/S_c)}{\sqrt{\beta_c^2 + \beta_d^2}} \right) \qquad (2)$$

where the structural demand mean value in terms of earthquake intensity parameter is denoted by S_d, the median value of structural capacity is represented by S_c, the lognormal standard deviation of structural capacity is represented by β_c, the lognormal standard deviation of structural demand is presented by β_d, and $\emptyset(\cdot)$ is the cumulative lognormal distribution function.

2.2 Intensity Measure

An intensity measure (IM) in the fragility function can be defined as the reference ground motion parameter with respect to which the probability of exceedence of a given damage state is plotted. There are many intensity measures developed till now; among them, the preferred IMs generally used in seismic risk assessment are peak ground acceleration (PGA), peak ground velocity (PGV), spectral acceleration (S_a), spectral displacement (S_d), etc. In the present study, the peak ground acceleration (PGA) is considered as the IM.

2.3 Damage States

In seismic risk assessment of any building, there requires the definition of damage state to obtain the fragility curves. In general sense, the probable performance levels

Table 1 Damage state definition as per Park and Ang, 1985

Building damage state	Damage index range
No damage	Less than equal to 0.14
Slight damage	Greater than 0.14 and less than equal to 0.4
Moderate damage	Greater than 0.4 and less than equal to 0.6
Extensive damage	Greater than 0.6 and less than 1
Collapse	Greater than equal to 1

of any building are defined by the threshold damage levels called limit states. A limit state denotes the edge limits for the performance of the building corresponding to certain damage conditions. However, the damage state denotes the certain damage levels occurred during the excitation between the successive limit states. Alternatively, probable damage data extracted from the fragility curves are represented on the discrete damage scale known as discrete damage probability matrix. In the empirical procedures, post-earthquake damage statistics are used to produce the fragility curves. However, in the current study, analytical fragility functions are obtained for five damage states as per Park and Ang, 1985. These damage states are presented in Table 1.

2.4 Probabilistic Seismic Demand Model

Probabilistic seismic demand model represents a definite mathematical relationship connecting the engineering demand parameter (EDP), i.e. the parameter used to describe seismic demand on the structure and the intensity measure (IM).

$$\ln(\text{EDP}) = \ln(a) + b \ln(\text{IM}) \tag{3}$$

a and b are the coefficients evaluated using regression analysis.

2.5 Dispersion Components

The dispersion component denoted by β, also called the lognormal standard deviation component, describes the total variability of fragility curve damage states [10]. The dispersion component is calculated as root of sum of squares of the variabilities for demand and capacity.

$$\beta_{s,ds} = \frac{\sqrt{\beta_c^2 + \beta_{d|\text{IM}}^2}}{b} \tag{4}$$

The variability in demand for each damage state is calculated as the standard deviation of the demand quantity assuming a lognormal distribution. This factor accounts for variability in both demand and capacity, assuming a lognormal variation for both.

$$\beta_{d|\text{IM}} = \sqrt{\frac{\sum_{i=1}^{N}(\ln(\text{IM}) - \ln(\text{IM}_m))^2}{N-2}} \tag{5}$$

where N is sample size of data. The final form of fragility function may be represented as follows:

$$P[ds|\text{IM}] = \emptyset\left(\frac{\ln(\text{IM}) - \ln(\text{IM}_m)}{\beta_{s,ds}}\right) \tag{6}$$

3 Building Description and Modelling

The modelling of a typical building frame is done using SAP2000. In this study, a RC framed building having ten stories and four bays are considered for the analysis. The frame is designed as per the IS 456: 2000 guidelines. Materials used in the model are M30 grade concrete and Fe415 grade steel. The model is incorporated with each storey height of 3 m and bay widths of 5 and 4 m in X-direction and Y-direction, respectively. The bases of the columns are considered as fixed in the analyses. The self-weights of frames (beams and columns) along with the dead load (due to infill walls and slabs) and live loads assigned for all beams are considered in accordance with IS 875. The structure is designed for soil type-II and zone factor of 0.24. The description of the building model is presented in Table 2, and the three-dimensional model of the building is presented in Fig. 1.

Table 2 Brief details of the model		
Storey height	3000 mm
Bay width (X-direction)	5000 mm
Bay width (Y-direction)	4000 mm
Size of beams	250 mm × 450 mm
Size of columns	450 mm × 450 mm
Thickness of slab	150 mm
Thickness of exterior wall	250 mm
Thickness of interior wall	150 mm
Height of parapet wall	1500 mm

Fig. 1 Three-dimensional
view of the modelled building

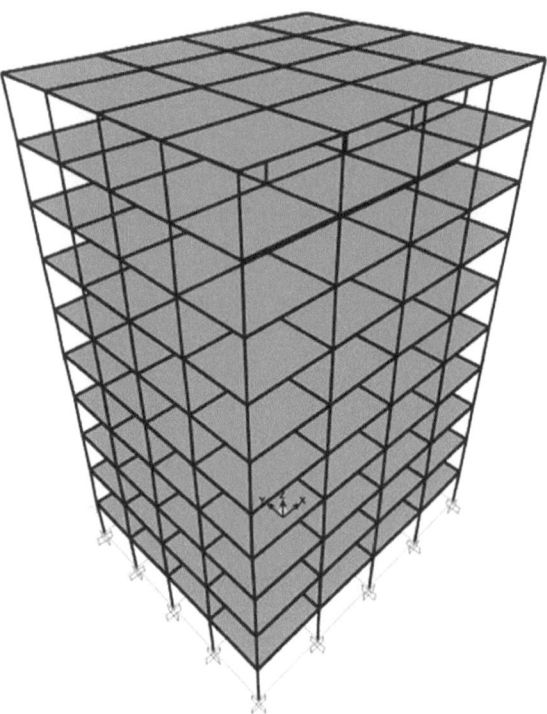

4 Results and Analysis

The damage observed in the building is a linear function of displacement ductility,
energy ductility and cumulative hysteretic energy ductility. For this study, the
damage incurred at the building can be reasonably assessed using the damage index
model presented by Park and Ang, 1985. As per the Park and Ang, 1985, damage
index model structural damage under seismic excitation is expressed as a combination of the damages caused due to repeated cyclic loading and due to excessive
deformation. The model is presented in Eq. (7).

$$\mathrm{DI} = \frac{\delta_m}{\delta_u} + \frac{\beta}{Q_y \delta_u} \int \mathrm{d}E \tag{7}$$

where δ_m, δ_u, Q_y and $\mathrm{d}E$ are the maximum deformation under seismic loading, the
ultimate deformation capacity under monotonic loading, obtained yield strength and
the absorbed hysteretic energy considered as incremental, respectively; β represents
the parameter for the effect of cyclic loading on structural damage taken as 0.15 for
systems with fairly stable hysteretic behaviour and $\delta_u = \mu d_y$ where μ is displacement ductility and d_y as yield displacement.

　　or

$$\mathrm{DI} = \frac{\mu_d}{\mu_u} + \frac{\beta}{\mu_u} \frac{\int \mathrm{d}E}{Q_y \delta_y} \tag{8}$$

where μ_d is the relative displacement ductility and μ_u is the ultimate ductility. Now, for a perfectly elasto-plastic system subjected to a single plastic excursion [11], the total energy is represented as:

$$\frac{\int \mathrm{d}E}{Q_y \delta_y} = \frac{\delta_p Q_y}{Q_y \delta_y} = \frac{\delta_p}{\delta_y} = \mu_p \tag{9}$$

where μ_p is the obtained plastic ductility. Thus, the simplified expression for damage index can be written as

$$\mathrm{DI} = \frac{\mu_d + \beta \mu_p}{\mu_u} \tag{10}$$

The computation of fragility curves requires the execution of certain processes and analysis which are presented as follows:

- A linear response spectrum analysis is the foremost analysis performed on the structure which is followed by designing the structure in accordance with IS 456-2000.
- In the second step, a well-known nonlinear static analysis, i.e. pushover analysis, is carried out on the model to obtain the pushover curve. This curve is used for calculating the performance point (target displacement) and the energy dissipated by the inelastic response of the structure.
- In the third step, nonlinear time history analyses are performed considering scaled El Centro earthquake (scaled for PGA range 0.05–1 g with an interval of 0.05 g). The earthquake time histories are made compatible with the response spectrum corresponding to zone 4.
- In the next step, the damage indices and subsequent damage states are computed for the RCC frame.
- In the next step, a regression analysis is carried out using the simulated response data obtained from the analyses to establish, the probabilistic function of the structural performance. Structural demand, i.e. structural response probability, is then plotted as a function of a ground-shaking parameter, for example, peak ground acceleration or spectral acceleration.
- Then, a probabilistic seismic demand model as well as the dispersion components of the model is calculated.
- Then, the fragility curves are plotted as a function of the selected ground-shaking parameter.

The analysis of the modelled building frame is carried out in two directions, i.e. along X-axis and Y-axis, addressed as Case-I and Case-II, respectively. Figure 2 shows the fragility curves obtained for Case-1.

For a PGA value of 0.24 g, the probabilities of exceeding the various damage states are obtained from the fragility curves and tabulated in Table 3.

From this data, the discreet probabilities of achieving each of these damage states are calculated as 9.24, 26.80, 18.84, 18.26 and 26.86% corresponding to collapse, extensive, moderate, slight and no damage states, respectively. Fragility curves for the case-II analysis are also obtained and presented in Fig. 3.

For a PGA value of 0.24 g, the probabilities of exceeding the various damage states are obtained from the fragility curves and tabulated in Table 4.

Fig. 2 Fragility curve for analysis Case-I

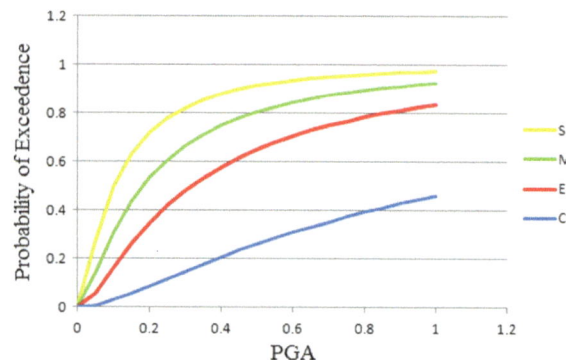

Damage state	Probability of exceedence (%)
Slight damage state	73.14
Moderate damage state	54.88
Extensive damage state	36.04
Collapse damage state	9.24

Table 3 Exceedence probability for different damage states (Case-I)

Fig. 3 Fragility curve for analysis Case-II

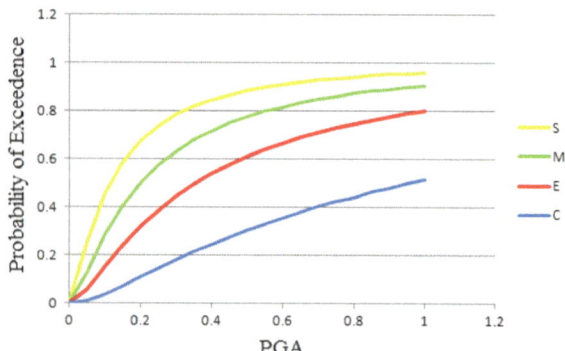

Table 4 Exceedence probability for different damage states (Case-II)

Damage State	Probability of exceedence (%)
Slight damage state	68.75
Moderate damage state	51.58
Extensive damage state	33.31
Collapse damage state	11.60

Table 5 Comparison of discrete probabilities of reaching various damage states

Damage state	Case-1 (%)	Case-2 (%)	Remarks
No damage	26.86	31.25	Increase
Slight damage	18.26	17.17	Decrease
Moderate damage	18.84	18.27	Decrease
Extensive damage	26.80	21.71	Decrease
Collapse	9.24	11.60	Increase

From this data, the discreet probabilities of achieving each of these damage states are calculated as 11.60, 21.71, 18.27, 17.17 and 31.25% corresponding to collapse, extensive, moderate, slight and no damage states, respectively. Discrete probabilities for various damage states obtained in the two cases are tabulated in Table 5 and were compared.

5 Conclusion

This study presents an analytical methodology for developing seismic fragility curves for building frame. The selection of a ground-shaking intensity parameter is another consideration which requires the engineering judgement. Traditionally, peak ground acceleration (PGA) is considered as the earthquake ground motion intensity parameter for obtaining the fragility curves. Based on the fragility curves, developed for multi-storied RC frame building by considering nonlinear time history analyses and pushover analyses, the following conclusions can be projected:

- Talking about pushover analysis, the ultimate displacement comes out to be more in x-direction than in y-direction. Similarly, the energy value also follows the same trend.
- The maximum displacement values obtained from nonlinear time history analyses are found to be more in x-direction as compared to those obtained in y-direction. Moreover, displacement follows an almost linear variation with the increasing values of PGA.
- The discrete probabilities of no damage and collapse increase with the direction changing from x to y. This shows that the building is more vulnerable in y-direction.

- For the PGA values less than 0.5 g, the fragility curves are observed to be distantly placed when comparing the same corresponding to the PGA values greater than 0.5 g.
- By considering the analysis results of irregular buildings, even more general conclusions can be deduced, as the present study was based on the regular building.

References

1. Sabetta F, Goretti A, Lucantoni A (1998) Empirical fragility curves from damage surveys and estimated strong ground motion. In: Proceedings of the 11th European conference on earthquake engineering, Paris, France
2. Murao O, Yamazaki F (2000) Development of fragility curves for buildings based on damage survey data of a local government after the 1995 Hyogoken-Nanbu earthquake. J Struct Constr Eng 527:189–196. (Transactions of AIJ, Architectural Institute of Japan)
3. Rota M, Penna A, Strobbia CL (2008) Processing Italian damage data to derive typological fragility curves. Soil Dyn Earthq Eng 28(10–11):933–947
4. Rossetto T, Elnashai A (2005) A new analytical procedure for the derivation of displacement-based vulnerability curves for populations of RC structures. Eng Struct 7 (3):397–409
5. Omine H, Hayashi T, Yashiro H, Fukushima S (2008) Seismic risk analysis method using both PGA and PGV. In: Proceedings of the 14th world conference on earthquake engineering, international association for earthquake engineering, Beijing, China
6. Singhal A, Kiremidjian AS (1996) Method for probabilistic evaluation of seismic structural damage. J Struct Eng 122:1459–1467
7. Singhal A, Kiremidjian AS (1998) Bayesian updating of fragilities with application to RC frames. J Struct Eng 124(8):922–929
8. Kappos AJ, Panagopoulos G, Panagiotopoulos C, Penelis G (2006) A hybrid method for the vulnerability assessment of R/C and URM buildings. Bull Earthq Eng 4(4):391–413
9. Kappos AJ, Panagopoulos G (2010) Fragility curves for reinforced concrete buildings in Greece. Struct Infrastruct Eng 6(1–2):39–53
10. HAZUS-MH MR (2003) Multi-hazard loss estimation methodology: earthquake model, Department of Homeland Security, FEMA, Washington
11. Terán-Gilmore A, Jirsa JO (2004) The concept of cumulative ductility strength spectra and its use within performance- based seismic design. ISET J Earthq Technol 41(1):183–200

Air Quality Scenario of the World's Most Polluted City Kanpur: A Case Study

Sarah Khan and Quamrul Hassan

Abstract Rapid industrialization and urbanization are paving a way for emerging economies to become more advanced, but these activities also trigger environmental problems. Among many of these problems, the biggest and the most persistent is the air pollution. According to the WHO database, Indian cities are leading the list of world's most polluted cities, with 14 of the 15 cities featuring in the list are Indians, which has been declared badly affected due to air pollution, and the worst among them is Kanpur. Years of studies and research have recognized the industrial sector as mainly responsible for polluting the city. As reported by Indian Institute of Technology, Kanpur, during winter months, the major contributor to air pollution is particulate matter like dust and soot accounting for around 76%, 15% has been contributed by biomass burning and about 8% by emissions from vehicle, whereas, in summer season, the percentage contribution of particulate matter came down to 35% with equal contribution from vehicular emissions. Meteorological data reveals that 20–80% part of the day mostly during winter months, the average wind speed remains between 2 and 4 m/s. This shows that the dispersion of pollutants in the winter season is very less, trapping particulates and toxic metals in the atmosphere to remain persistent for months. The exposure to these pollutants resulted in harmful diseases linked to the cardiovascular system, respiratory systems, nervous system, premature birth, mortality, and illness. Various efforts have been initiated by the authorities to control the increasing level of pollution, like constructing new roads and pavements, mass rapid transit service to cut car pollution, planting more trees and promoting battery-operated transport. The Ministry of Environment, Forest and Climate Change is also making a budgetary allocation of about 7 billion rupees ($104 million) for installing more systems to monitor air quality in cities and installing equipment to settle the dust like water sprinklers. Despite all these efforts, the Air Quality Index of the city has remained much below the national average. In this scenario, this paper focuses on the studies made so far associated with the causes, sources, impacts, and outcomes related to air pollution levels in the city from the available literature. This paper has

S. Khan (✉) · Q. Hassan
Jamia Millia Islamia University, New Delhi, India

© Springer Nature Singapore Pte Ltd. 2020
S. Ahmed et al. (eds.), *Smart Cities—Opportunities and Challenges*,
Lecture Notes in Civil Engineering 58,
https://doi.org/10.1007/978-981-15-2545-2_57

also highlighted the air pollution scenario of neighboring country China and its policies intervention in battling against a similar situation.

Keyword Air pollution · Air quality · Particulate matter · Air pollutant

1 Introduction

Water and air, the two essential elements of life on which all life forms depend, have become the dumping grounds for all kinds of wastes. The basic minimum requirement to live a healthy life is to breathe clean air, which has now become difficult to achieve. Rapid industrialization, technological advancement, and urbanization have provided us better living standards but at the cost of damaging the natural resources which are the building blocks for the sustenance of life. There has been an upsurge in the concentration level of various gaseous pollutants and particulate matter (PM) due to widespread anthropogenic activities of various kinds. The pollutants which release directly from the source are termed as a primary pollutant, and these primary pollutants result in the formation of new pollutants due to the chemical reaction among them, called the secondary pollutants. These are building up in the atmosphere creating environmental hazards and affecting health globally. A report published by the World Health Organization (WHO) specifies that some seven million people have died in 2016 due to the increasing level of pollutants in the atmosphere [1]. Each year WHO releases list of world most polluted cities to bring the attention of global communities to take some actions, but little or no effort was made to improve the air quality. This year, WHO ranked 14 Indian cities in the list of 15 most polluted cities worldwide with Kanpur being the most affected based on data recorded in 2016 for PM2.5 levels [2]. The parameters adopted for selecting the cities were particulate matter PM10 and PM2.5. Particulate matter with size less than 2.5 μm (PM2.5) is highly dangerous, penetrating deep into the lungs or cardiovascular system. The recorded level of PM2.5 in Kanpur was between 120 and 150, as compared to the WHO safe limit of 25. Air pollution is a global concern. Every country must monitor air pollution concentration levels at regular intervals. Quantifying air quality is a composite phenomenon, however, there are some pollutants whose effect on air quality is highly significant, and these include particulate matter (PM), nitrogen dioxide (NO_2) and sulfur dioxide (SO_2) [3]. For measuring the pollutant concentrations, National Ambient Air Quality Monitoring Programme (NAMP) has been initiated by Central Pollution Control Board (CPCB) in 1984–1985. Under this Plan, 342 stations for monitoring air quality were set up in 127 Indian cities [4], and National Ambient Air Quality Standards have been formulated as a measure of comparison, to ascertain that the concentration level of pollutants does not exceed these standards at any location. Based on the National Air Monitoring Programme (NAMP), the air quality is categorized under four broad categories based on the exceedance factor. The categories are critical (*C*), high pollution (*H*), moderate pollution (*M*) and low

Table 1 Air quality categories based on NAMP

Pollution level	Annual mean concentration range ($\mu g/m^3$)					
	Industrial, residential, rural and other areas			Ecologically sensitive area		
	SO_2	NO_2	PM10	SO_2	NO_2	PM10
Low (L)	0–25	0–20	0–30	0–10	0–15	0–30
Moderate (M)	26–50	21–40	31–60	11–20	16–30	31–60
High (H)	51–75	41–60	61–90	21–30	31–45	61–90
Critical (C)	>75	>60	>90	>30	>45	>90

Note Table is taken from importance of national ambient air quality standards, by Anushua, published in Green Clean Guide dated 24 January 2016 [5]

pollution, with exceedance factor for *C* level are greater than 1.5 and *L* level less than 0.5 [5] (Table 1).

2 Study Area

Kanpur is the major industrial centre of North India, situated on the banks of river Ganga, with its own historical, religious and commercial importance. It has been popularized by the sobriquet "Manchester of East" and "Leather city of the world" as it is the hub of the textile and leather industry in the region.

2.1 Geographical Aspects and Sources of Pollution

According to the 2011 census, Kanpur serves 2.7 million people growing at the rate of 2.47% annually with 6800 persons living per square km. The city is divided into urban area spread over 322.84 km^2 and rural area 2832.16 km^2 [6]. The city is surrounded by two rivers, the Ganges encircling the north and the river Pandu flowing south. The city follows rivers and the railway lines linearly. The centre of the city is occupied by various industries such as heavy engineering, tanneries, textile, leather, and fertilizers. The key sources of air pollution are emissions from industries, vehicular emissions, dust from the road and domestic cooking. Air pollution in the industrial and residential areas of the city remains 5–6 times above the National Ambient Air Quality Standards [5].

2.2 Meteorological Aspects

The city is situated at an average height of 128 metres above the mean sea level, between 26.47° N latitude and 80.35° E 137 m longitude. The climate of the city lies in the range of humid subtropical region having a characteristic cool dry winter and comparatively hotter summer. The maximum temperature recorded in summer is about 44 °C and the minimum temperature dropping to nearly 0 °C in winter. The annual precipitation level is about 820 mm received mostly in the monsoon season between June and September. The meteorological data based on the climatic pattern has been prepared using meteoblue climate diagrams show that maximum rainfall occurs during July and August, while the temperature remains high during May and June. The diagrams have been prepared by using data over 30 years also show that wind speed remains between 12 and 28 m/s throughout the year [7].

3 Air Quality Monitoring and Pollution Status

The activities responsible for rampant pollution level in the city are as follows:

1. Transportation sector mainly private diesel tempos and buses contribute to the vehicular emission.
2. Commercial activities comprise diesel/kerosene generators frequently used during power cuts.
3. Industrial activities, the major being thermal power plant, textile industries, leather industry, tanneries among other sources.
4. Domestic activities, institutional and official activities.
5. Miscellaneous sources like dust emission from poorly maintained and partially paved roads.

As per a report by IIT, Kanpur (2010) report, during winter months, the major contributor to air pollution is particulate matter like dust and soot accounting for around 76, 15% has been contributed by biomass burning and about 8% by emissions from vehicle, whereas, in summer season, the percentage contribution of particulate matter came down to 35% with equal contribution from vehicular emissions [8].

Air Pollution Knowledge Assessment City Program (2015) has created a source-wise percentage share of particulate matter with size less than 2.5 μm, as shown in Fig. 1 [9]. Various studies have been conducted on monitoring, emission inventory, seasonal variation, assessment and apportionment study as available from literary sources, and one of the studies includes a project which had conducted regular air quality sampling and analysis of PM10, SO_2 and NO_2 at Vikash Nagar, Kanpur, which had generated air quality data for the period 2005–2007. The study yielded that the PM10 level ranges from 50 to 600 $\mu g/m^3$, and PM2.5 levels are 25–200 $\mu g/m^3$. The data for NO_x and PM10 does not co-relate, and this indicates PM10 is contributed by other sources also, apart from combustion. TSPM and PM10 link to the same source. The winter months recorded the worst level both for PM10 and NO_2, and monsoon season recorded the satisfactory level both for PM10 and NO_2 [10]. The study carried out by New Delhi-based SIM Air (air pollution research and modeling body) considered private vehicles in Kanpur responsible for the high level of CO_2 emissions (84%). Another study has compiled the composite index of air quality. The major contribution (more than 90%) for the index is from particulate matter. Air Quality Index has been prepared for the seven sampling locations based on the land use pattern. This study has determined that during cold months, air quality remains bad. The same has been reported for March, April and early parts of May. During this dry season, winds blow carrying road dust ensuing high SPM. Air quality then advances during the rainy season and after that [11].

One of the studies has been conducted at urban locations during June 2009 to May 2013 on long-term near-surface measurements of atmospheric pollutants reporting that the mean concentrations of sulfur dioxides were 3.0 ppb, oxides of nitrogen were 5.7 ppb, carbon monoxide level was 721 ppb, and ozone level was 27.9 ppb over the period. The winter months were under the cover of SO_2, NO_x and CO, released by various anthropogenic sources like vehicles, domestic cooking, bricks manufacturing, thermal power plants, stable burning which remain persistent due to low dispersion, whereas O_3 concentration touched high during summer season [12].

Air quality assessment, emissions inventory and source apportionment studies for Kanpur City, conducted by the Indian Institute of Technology, Kanpur (2010), have divided air quality monitoring stations based on predominant land use patterns. The predominant pollutants analyzed were suspended particulate matter, sulfur dioxide, oxides of nitrogen, carbon monoxides, organic carbon, ions, elements, volatile organic compounds. The conclusion drawn from this study again pointed particulate matter as the main culprit where the levels were 2.5–3.5 times higher than the standards. There was an abundance of soil particles like Si, Fe, Ca, Na, indicating the cause of pollution may be road dust. For NO_2, 50–70% contribution is from vehicles [10].

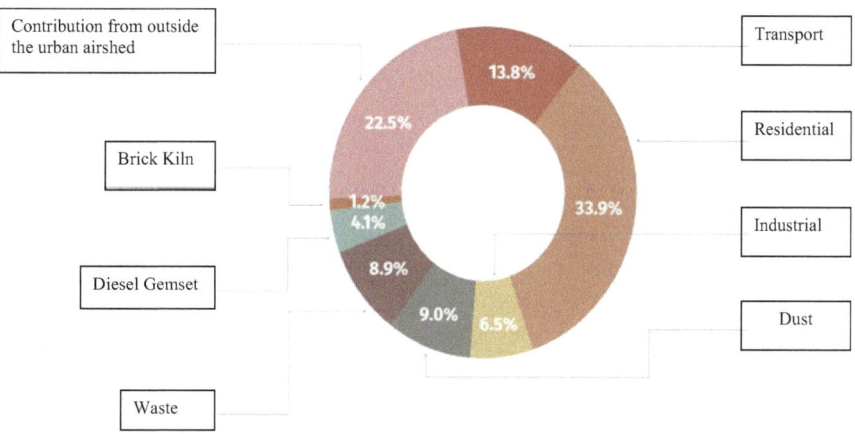

Fig. 1 PM2.5 concentration: source-wise percentage share in 2015. *Source* The Air Pollution Knowledge Assessment (APnA) city program report, 2015

A recent study conducted by urban emission info in 2015 shows that the monitoring at various locations in Kanpur is done under National Ambient Monitoring Program (NAMP), functioned under Central Pollution Control Board (CPCB, New Delhi), with one continuous air monitoring station (CAMS) and eight manual stations. The CMS monitors all the criteria pollutants, while manual stations record data on PM10, SO_2 and NO_2 [2]. PM2.5 was determined using satellite data derived concentration. The resultant concentration for PM10 has been reported for the year 2002–2014, shows that the trend in PM10 level has remained on the higher side of the Annual Indian Standard (60 $\mu g/m^3$). From 2004 to 2008, the range of maximum concentration level was 305–532 $\mu g/m^3$, from 2008 to 2010, the range goes little down from 250 to 208 $\mu g/m^3$, and from 2010 to 2015, the level remains mostly on the higher side with maximum reporting up to 717 $\mu g/m^3$. The data for SO_2 shows that from 2000 to 2010, the level remains below the Annual Indian Standard (50 $\mu g/m^3$), and after 2010, there was a sudden rise in level with maximum touching 85 $\mu g/m^3$. For NO_2 concentration, the trend follows the same pattern as SO_2 with level remains in sync with Annual Indian Standard (40 $\mu g/m^3$), while after 2010, the maximum recorded level is 106 $\mu g/m^3$ after 2014. Concentration based on satellite feeds and global chemical transport models shows Annual PM2.5 concentration varies from 80 $\mu g/m^3$ in 1998 fluctuating in between to reach the highest level in 2004 of 100 $\mu g/m^3$, after 2004, the trend declines to be recorded as 85 $\mu g/m^3$ in 2014, and the National Standard has been given 40 $\mu g/m^3$ and WHO standard as 20 $\mu g/m^3$. The study has also prepared an emissions inventory for SO_2, NO_x, CO, Non-methane Volatile Organic Compounds (NMVOCs), CO_2 and particulate matter, for the year 2015 and also predicted the levels up to 2030. For the year 2015, the total emission in tons/year from all the sources combined was approx. 35,000 and is been estimated for the year 2030 to be 40,000 tons/year approx. [2].

4   Air Quality Trends

India has a vast network of monitoring stations designed to gather data on air quality. The system consists of 731 functioning stations spreading across 312 cities/towns in the country, monitoring four major air pollutants—SO_2, NO_2, respirable suspended particulate matter and fine particulate matter [13]. There are eight manual monitoring stations currently operating under NAMP in Kanpur at the locations as shown in Table 2 [14].

The air quality trend for the three major pollutants, PM10, SO_2 and, NO_2 obtained for the past ten years has been compiled from various literary sources and presented in this study.

Figure 2 shows the previous ten years data of particulate matter taking an average of the concentration recorded over the eight monitoring stations. The trend line shows that the PM10 level remains high over the ten-year period despite taking corrective measure [13].

Table 2 Air quality monitoring station in Kanpur

Sr. No.	Monitoring stations	Description of site
1	Forest & Training Centre, Kidwai Nagar	Residential
2	Chamber of Commerce Darshanpurwa	Commercial
3	Fazalganj, Panki	Industrial
4	Dabauli/Shastri NGR	Residential
5	Jajmau/Awas Vikas	Residential
6	I.I.T. Campus	Residential
7	Dada Nagar	Residential
8	Ramadevi	Commercial

Source ENVIS Centre: Uttar Pradesh status of environment and related issues

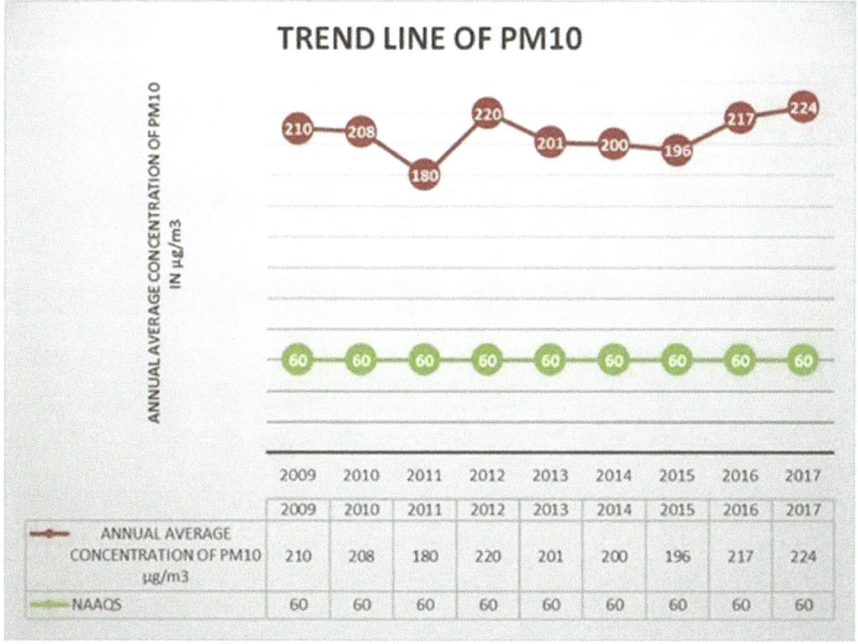

Fig. 2 Annual average concentration of PM10 for the years 2009–2017. *Source* CPCB-ambient air quality data for the year-wise

The trend in SO$_2$ concentration maintains a low profile and remains well below the national average throughout the past ten years [13]. The falling trend has been resulted due to policy interventions like decreasing the percentage of sulfur in diesel, use of CNG in automobiles, use of LPG as domestic fuel, etc. [15].

The trend in NO$_2$ concentration has maintained a steady rate over the years with a slight increase in the last two years [13]. This may be due to the stringent measure adopted in terms of vehicular emission and fuel quality.

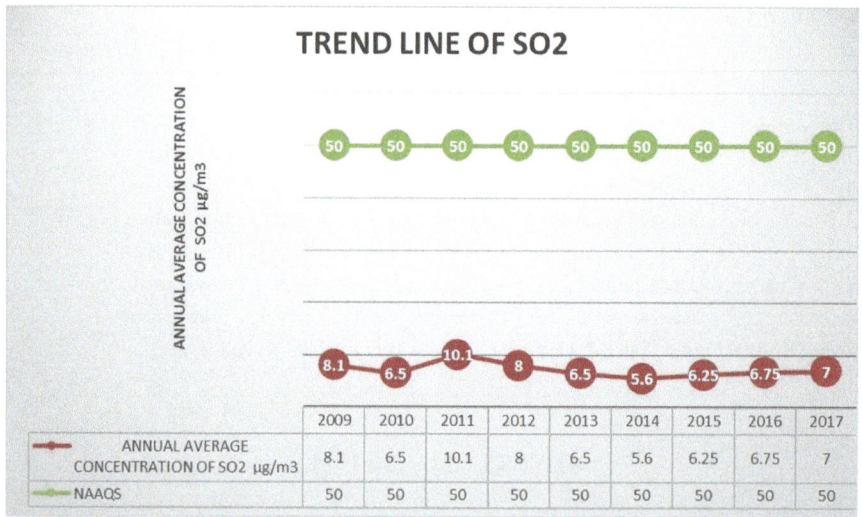

Fig. 3 Annual average concentration of SO$_2$ for the years 2009–2017. *Source* CPCB-ambient air quality data for the year-wise

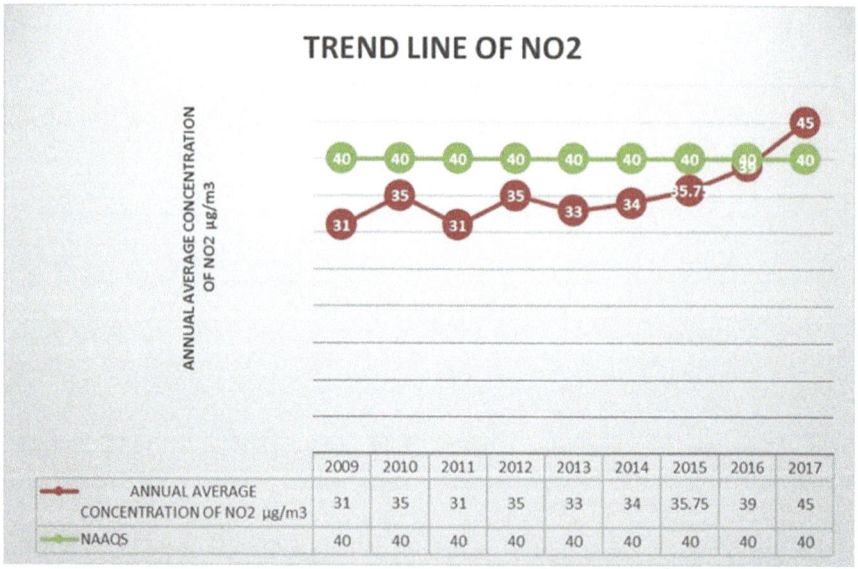

Fig. 4 Annual average concentration of NO$_2$ for the years 2009–2017. *Source* CPCB-ambient air quality data for the year-wise

5　Impacts of Air Pollution

The impact of air pollution ranges from mild to severe especially from the health perspective. These include breathing problems, asthma, physiological changes in heart and lungs, neurological disturbances, preterm birth, infant mortality, and morbidity. The World Health Organization (WHO) evaluates that annually 800,000 deaths occur due to air pollution [16]. There are effects associated with both short-term and long-term exposure to air pollution. According to medical research, long-term exposure to air pollutants is mainly responsible for increasing the chances of respiratory diseases like allergies, asthma, flu, heart diseases, and lung cancer. The victims include young children, the elderly, persons who are already suffering from other diseases and persons with poor background [17]. WHO recognizes air pollution under non-communicable diseases (NCDs) category, responsible for adult deaths reporting from 24% heart disease, 25% from stroke, 43% from chronic obstructive pulmonary disease and 29% from lung cancer [17]. According to a paper published in the Lancet, Uttar Pradesh had the second-highest victims of air pollution among all Indian states as of 2016, predominant cities being Kanpur and Lucknow [18]. According to UPENVIS, 0.4 million disability-adjusted life years are lost every year in Uttar Pradesh due to air pollution. This costs the state about Rs. 2.6 billion. A study by the Usha Gupta Institute of Economic Growth and the Bhimrao Ambedkar College has done the economic study predicting as saving Rs. 213 million if Kanpur meets the air quality standards [18].

Although air pollution had a greater impact on health, other issues due to the increase in pollutant concentration are many. Particulates reduce visibility to a significant level, and it becomes difficult to see beyond 40–60 km (25–35 miles) that is about one-fifth the distance one could normally see without any pollution. NO_x and SO_x are responsible for producing acid rain which is equally harmful to living and material things.

6　Government and Local's Initiatives

Central and state governments, pollution monitoring bodies and other local bodies have undertaken various measures to curb this growing menace of air pollution, but the result is far from achieving the objective either due to lack of proper implementation of plans, poor coordination or target specific strategies. Some of the initiatives taken under the plan are tightening emission standards, strengthening the "pollution under control" system by laying the rules for using new equipment and strictly adhering to the rules for the used vehicles, introduced public vehicles running on CNG and phasing out old vehicles. For combating particulate matter, the concerned authorities have closed down 12 polluting industries and initiated other advance techniques to clean the air, which includes planting more trees along roadsides, removing construction debris from roads, accelerated construction and demolition work, proper check on the use of adulterated petrol, better power supply to cut off the use of generators, green buffer zones have been created, and garbage heaps have been removed among others [19].

Some of the measures as recommended by the studies discussed in the previous section have been presented here. They have suggested that the city needs at least 27 continuous air monitoring stations to record data from all perspective [2]. The city needs to stimulate public transport, better road structure to curb road dust. An energy-efficient plan to check the emissions from the power plant or shift towards using green technology for electricity production is much needed. Some of the studies have projected vehicular growth based on vehicle category as two-wheelers: 8% three-wheelers: 18% four-wheelers: 9% bus and trucks 16% and LCV: 16%. Considering the above projections, some of the action plans recommended by the studies are implementation of BS-V and BS-VI norms, switch to electric and hybrid vehicle, ethanol mixing, bio-diesel, maintenance of vehicles, banning of 15-year-old private vehicles, increasing use of public transport, particulate control system in industry, changing the location of air polluting industries, use of LPG as domestic fuel, sufficient supply of electricity, better construction practices, constructing paved roads, wetting the road to prevent dust suspension, ban on open burning. To achieve the best possible results, these strategies should be used in combination. The study has also analyzed the reduction in pollutant levels in the quantitative term by implementing the above measures as by banning old private vehicles alone, led to the reduction in PM10 level from 26% in 2012 to 21% in 2017. It may be further noted that reducing vehicular load is directly proportional to road dust reduction. This can be done by preventing the entry of vehicles in critical areas, especially during winter months. Another important measure suggested is to construct a network of mass rapid transit systems, elevated railways, flyovers at all railway intersections leading to a 20% reduction in vehicular emission and considerable improvements in air quality [8].

Recently, IIT-K has sent a proposal to the state government to seek approval from the civil aviation ministry for creating artificial rains using cloud cover, which has been under consideration. IIT Kanpur was given a grant of Rs. 15 lakh from the state government to experiment on the option of artificial rains. The scientists have finished their experiments and are ready to execute [20].

7   China's Air Pollution Scenario and Policy Interventions

In China, the early phase of economic development was powered by the manufacturing sector mainly dependent on coal-burning and industrial processes. The era after 1980 is mainly responsible for emitting CO_2, SO_2, NO_x, and PM at a large scale. Recently, vehicular emission has also emerged as another source, contributing to the increase in the concentration level of these pollutants, especially in the urban areas. Global Burden of Disease Study (GBD) Report (2010) has estimated that one year of exposure to pollutants leads to 1.2 million premature deaths in China [21]. On the economic front, as shown by multiple studies, the increasing pollution level is responsible for the losses in the form of gross domestic product (GDP) between 1 and 7% [22]. The current scenario comprises SO_2, PM2.5, all particulate matters (PMs) and ground-level ozone (O_3) as major pollutants [23]. The worsening condition is due to the rising level of PM2.5 and O_3. The scientist has explained this rising as the "air pollution complex", a concept formally proposed and defined by Tang in 1997 [24]. SO_2, NO_2 and VOCs get converted to PM2.5 where O_3 concentration is higher due to the increased rate of oxidation [23]. China's air pollution scenario has been studied in three phases. Phase 1 includes the 1970–1990 time period where dominating sources were coal-burning power plants and households and industries contributing to SO_2, TSP, NO_x, and CO emission [24]. Phase 2 was the transition period from 1990 to 2000, in addition to phase 1 pollutant, increasing vehicular emission has contributed to raising the level of NO_x, and releasing of VOCs [25]. Phase 3 was from 2000 till present where all the factors previously responsible get enhanced spreading to the regional level and worsening the quality of megacities.

China's environmental protection system is based on 5 pillars—(1) environmental protection laws and emission standards; (2) time-specific planning; (3) specific regulatory measures; (4) other specific actions and (5) collaboration with state authorities [23]. China's Environmental Protection plans were developed in the early 1980s and become more stringent after the PM2.5 crises happened in 2012–2013. Pollution control planning in China has been divided into three stages.

Stage I—Until 2005
The measures adopted by China during this stage were (1) local governments control the air pollution within their area; (2) focus on coal utilization sources; (3) emission targets for stack concentration; (4) polluter pay principle had been applied. Special measures were adopted in Beijing due to its worst air quality such as substituting coal, desulphurizing, dust collection and combustion techniques emitting lower NO_x in power plants. Some other measures taken were, Increasing vegetation coverage and minimizing road dust and dust from construction sites, and a set of actions on vehicular pollution including setting a limit on emission for new vehicles, the use of quality rich fuel, phasing out of old vehicles [23, 26]. However, due to the growth in the number of vehicles and contributions from nearby industrial centres [23, 27], these efforts become insignificant in improving air quality. These strategies were adopted to target specific pollutants like NO_x and VOCs.

Stage II—2006–2012
The focus of policies during this phase was getting rid of SO_2 completely and the alternatives for saving energy. For both the key factors, specific targets were set, 10% reduction from the 2005 level for SO_2 and reduction in energy consumption to

20% per unit of GDP from the 2005 level [23]. This has been reported by several studies that although the targets were achieved, there was a trivial improvement in air quality. This was due to the lack of diligence on the part of local government. In the 12th FYP, the focus shifted to SO_2 and NO_x [23]. The failure of policies was due to the focus on two parameters, while the transformation process from primary to secondary pollutants involving nonlinear complexity had been neglected.

During this phase, special action was taken by Beijing and surrounding territories as a part of their preparations for the 2008 Olympics. The measures adopted were very rigorous and were implemented in few identified key areas, which were complete shutdown of powerplants, limiting vehicle usage, stringent regulation of dust from construction activities and the prohibition on stubble burning [23, 26]. These measures resulted in substantial improvements in air quality during and after the Olympics, but the effects were temporary and washed out by the end of October 2009. Some of the measures like plant closure and traffic control were adopted during other events also like World Exposition in Shanghai (2010) and the Asian Games (2010) [23].

Stage III—2012–Till Present

To tackle the PM2.5 crises in 2013, the state council and Ministry of Environment Protection have issued a specific action plan featuring ten points agenda. During this phase, both qualitative and quantitative aspects of air quality were taken into consideration by setting proper time-bound targets [23]. The trend line for PM2.5 from the year 2008 to 2015 shows that the policy actions have yielded results [28].

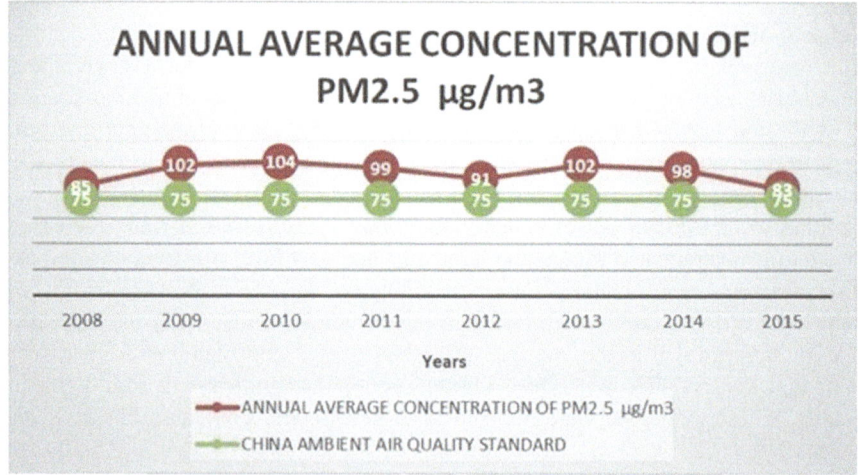

Fig. 5 PM2.5 trend line in Beijing

7.1 China's Handling of Air Pollution Crises

India and China both face the same situation in terms of air quality as both are emerging economies facing similar cost-benefit tradeoffs. But China follows a strict stringent system of environmental protection which works at a regional level, while policy implementation in India has a loose end. Comparing the results of quantitative analysis of pollutants based on reports published in 2015, it has been found that the level of overall particulate matter in India is greater than that in China. The data shows that PM2.5 level was falling in china recording 17% decrement between 2010 and 2015, while in India, it has been increased to 13% during the same period [29]. Before 2008, the pollution level in China was rampant recoding a loss of about 133 million workdays resulted in a loss of 1.34% of GDP. According to a 2015 report by RAND Environment, Energy and Economic Development Program, between 2000 and 2010, air pollution-related health problems in china and lost labour productivity have reached 6.5% per year of China's GDP [30].

Both the countries have geared up to tackle this menace in full-fledged form. According to the World Bank report, 2016, air pollution-related welfare activities account for 8.5% of India's GDP. These activities include measures ranging from a graded response action plan (GRAP) to fixing catalytic converters in vehicles for emission reduction. In comparison with this, China has introduced the "airborne prevention and control action plan" in 2013 and recognized coal as the major force of air pollution and started putting a cap on its use. China has also pursued some innovative solutions like building a 7-m tall smog-free tower in Beijing in 2016, a 328 feet tall air purifier was constructed in Xian in 2017 decreasing the level of PM2.5 by 15% [29]. Apart from this, China has also been investing in clean coal technologies along with the USA, which covers the use of carbon capture and storage technology and improvements in the efficiency of power plants [29]. China is also increasing the efficiency of power generation by using "ultra-supercritical" facilities which decrease coal dependency for energy production. This results in the generation of 286 g of coal equivalent (gce) per kilowatt of power produced, which is less as compared to the USA (375 gce). In its Thirteenth Five-Year Plan released in 2016, the target of China is an 18% reduction in PM2.5 level and 140 million tonnes reduction in coal production by 2020. To achieve the target, 40% of factories in China have been shut down in 2017, resulting in a 50% reduction in PM2.5 levels in the city of Linfen which was designated as the heavily polluted city until then [29].

8 Conclusion

The conclusion of the study presented here identifies particulate matter as the main driver of pollution level in the city has been discussed in various studies previously, contributing nearly 76% to the pollution level. Most of the time particulate fractions

(PM2.5, PM10, SPM) exceeded the standards set by NAMP. Gaseous pollutants like sulfur dioxide concentration have mounted after 2010 and it's still in the rising phase according to the recent studies [2], and nitrogen dioxide level remains at par with PM 10, sufficient in causing potential harm to health. The main reason for the increasing pollution level is the heavy traffic density and unfavorable climatic conditions adversely affecting the surrounding air. The review study suggested that the inhalation of polluted air is responsible for major health problems such as breathing problems, bronchitis, asthma, heart diseases, pneumonia, and lung cancer. Vehicular pollution also affects the vegetation, the roadside plant faces heavy stress due to an increased level of gaseous pollutants, and their biogeochemical cycle is getting disturbed impacting the ecological cycle more broadly.

The paper also presented the air pollution scenario of our neighboring country China as most of the cities in China are also facing the worst air quality situation in the same way as India. Although China has tightened the rope in tackling the situation after 2010 only, their efforts have resulted in refining the air quality in a significant way, while India is still dangling in the middle of the crises with 14 of its cities has been featured in the WHO worst cities list in terms of air quality.

Air pollution is a global phenomenon. Every country is facing the burden of rising pollution levels and is under serious threat. At international forums, talks have been going on among the countries to fight against air pollution on the common ground. Every country has taken responsibility on its part to reduce its emission and set targets according to one's capability. A report published by WHO stated that countries are working deliberately in curbing the pollution menace and appreciated India's efforts in switching to clean household energy use through Pradhan Mantri Ujjwala Yojana.

There is an urgent need to undertake specialized studies targeting specific sources. The heavily polluted area should be marked, and actions pertaining to a particular scenario taking into account climatic and geographical factors should be taken. While initiatives such as closing down coal-fired plants, shifting the location of industries, the odd-even scheme of Delhi government to control traffic move-ment and vehicular emission, ban on crackers during Diwali, ban on stubble burning are welcoming steps, but still, much needs to be done to come down pollution within the standard limits. India needs to streamline its administrative structure in Environment Protection Agencies and follow the example of China where regional administration plays a greater role in handling the issue within its jurisdiction. Decentralizing the efforts by empowering local authorities to take responsibility for the pollution scenarios in their area should help in the long run. Along with the local, state and central government should work towards a common cause with area-specific approaches as air pollution is a complex problem and pollutants just not remain concentrated at a particular place but travel long dis-tances, beyond governmental control.

References

1. WHO Global Urban Ambient Air Pollution Database (2016) https://www.who.int/airpollution/data/cities/en/
2. Times of India, 14 of world's 15 most polluted cities in India, May 2, (2018), http://timesofindia.indiatimes.com/articleshow/63993356.cms?utm_source=contentofinterest&utm_medium=text&utm_campaign=cppst
3. Wu Y (2010) Understanding economic growth in China and India: a comparative study of the selected issue. World Scientific Publishing Co. Ltd. ISBN 9814287784/981-4287- 78-4/9789814287784
4. Central pollution Control Board (CPCB). National Ambient Air Quality Standards (2009). http://www.cpcb.nic.in/NationalAmbientAirQualityStandards.php, 22-5-14
5. Anusha: Importance of national ambient air quality standards (2016) http://greencleanguide.com/importance-of-national-ambient-air-quality-standards/
6. Census 2011-Census of India. www.censusindia.gov.in/2011census/PCA/PCA_Highlights/…/India/Chapter-1.pdf
7. Meteoblue. https://content.meteoblue.com/en/content/view/full/2879
8. Energy Economic times (2018) https://energy.economictimes.indiatimes.com/news/coal/with-worlds-worst-air-kanpurstruggles-to-track-pollution/64183796
9. The Air Pollution Knowledge Assessment (APnA) City Program report (2015) www.urbanemissions.info/india-apna/
10. Sharma M (2010) Air quality assessment, emissions inventory and source apportionment studies for Kanpur City. Indian Institute of Technology, Kanpur
11. Sharma M, Maloo S (2005) Assessment and characterization of ambient air PM10 and PM2.5 in the City of Kanpur, India. Atmos Environ 39:6015–6026
12. Gaur A, Tripathi SN, Kanawade P, Tare V, Shukla SP (2014) Four-year measurements of trace gases (SO_2, NO_x, CO, and O_3) at an urban location, Kanpur, Northern India. J Atmos Chem 71(4):283–301
13. CPCB-National Air quality monitoring programme. www.cpcb.nic.in/monitoring-network-3/
14. ENVIS Centre on control of pollution water, air, and noise, central pollution control board sponsored by Ministry of Environment and Forests, Govt of India. www.cpcbenvis.nic.in/air_quality_data.html
15. Shukla P (2013) Air pollution in select Indian Cities, TERI
16. Arbex MA, de Souza Conceic GM, Cendon SP (2009) Urban air pollution and chronic obstructive pulmonary disease-related emergency department visits. J Epidemiol Community Health 63:777–783
17. World Health Organization (2002) World health report, reducing risk, promoting healthy life, WHO, Geneva, Switzerland, http://www.who.int/whr/2002/en/, 20-03-14
18. Centre for Science and Environment study on Air Pollution on the rise in Kanpur https://www.cseindia.org/air-pollution-on-the-rise-in-kanpur-says-latest-cse-study 558
19. Business Standard, India-Air Pollution (2018) https://www.business-standard.com/article/current-affairs/india-air-pollution-kanpur-delhi-among-14-of-15-most-polluted-cities-in-the-world-shows-who-pollution-index-top-10-highlights
20. Lim SS, Vos T, Flaxman AD, Danaei G, Shibuya K, Adair-Rohani H, AlMazroa MA, Amann M, Anderson HR, Andrews KG (2013) A comparative risk assessment of burden of disease and injury attributable to 67 risk factors and risk factor clusters in 21 regions 1990–2010: a systematic analysis for the global burden of disease study 2010. Lancet 380:2224–2260
21. Zhang SQ, Huang D (2011) Controlling fine particulate pollution and mitigating environmental health damage. Environ Prot 16:25–26
22. Jin Y, Andersson H, Zhang S (2016) Air pollution control policies in China: a retrospective and prospects. Int J Environ Res Public Health 13(12):1219. Published 2016 Dec 9. https://doi.org/10.3390/ijerph13121219

23. Shao M, Tang X, Zhang Y, Li W (2006) City clusters in China: air and surface water pollution. Front Ecol Environ 4:353–361
24. Chan LY, Chu KW, Zou SC, Chan CY, Wang XM, Barbara B, Blake DR, Guo H, Wai Yan T (2006) Characteristics of nonmethane hydrocarbons (NMHCs) in industrial, industrial-urban, and industrial-suburban atmospheres of the Pearl River Delta (PRD) region of south China. J Geophys Res Atmos 111:1937–1952
25. Zhang S, Wu D, Xie X, Hu Q, Zou W, Yi R, An S, Zheng Y, Yue P, Wan W (2008) Research on achieving Beijing's air quality goal: strategies and measures. Peking University, Beijing
26. Wu D, Xu Y, Leung Y, Yung CW (2015) The behavioral impacts of firm-level energy-conservation goals in China's 11th Five-Year Plan. Environ Sci Technol 49:85–92
27. China Power: Is air quality in China a Social Problem (2018)? https://chinapower.csis.org/air-quality/
28. http://www.stateair.net/web/historical/1/1.html
29. Chen S-M, He L-Y (2014) Welfare loss of China's air pollution: how to make personal vehicle transportation policy. China Econ Rev 31 (Beijing: Chinese Economists Society) 106
30. Hindustan Times (2018) https://www.hindustantimes.com/lucknow/kanpur-notices-to-7-governmentgovernment-departments-for-failing-to-check-air-pollution/story

Dynamic Programming-Based Decision-Making Model for Selecting Optimal Air Pollution Control Technologies for an Urban Setting

G. Shiva Kumar, Aparna Sharma, Komal Shukla and Arvind K. Nema

Abstract Increasing population, rapid urbanization and hasty industrialization are assisting gradual shift in human residence from rural to urban. Any city trying to become a smart and sustainable one must provide a better quality of life for its citizens and address the problem of degrading air quality and increasing pollutant emissions. The solution for these problems requires selection and decision-making among a number of candidate emission-control-alternatives that need to satiate a number of technical constraints, policy criteria and regulations. For a given pollutant and emission source, a number of control technologies are available; hence, it is coveted to choose the best combination among them to reduce emissions to a desired standard. Because of its applicability, flexibility and ease of computation in solving large-scale practical problems, dynamic programming-based approach appears to be the appropriate and feasible choice for optimal air pollution control technology selection for an urban metropolitan area. Current study develops a dynamic programming (DP) model that determines the optimal selection strategy, after defining different parameters, at a minimized total cost. The usage and applicability of the proposed model were illustrated with a representative case study of a simulated city with three major sources of pollution. It is inferred that DP is ideal for the 'multiple sources—multiple control technologies—single air pollutant' optimization problem.

Keywords Smart city · Air pollution control · Dynamic programming · Technology selection · Optimization

1 Introduction

United Nations projected an estimate that the proportion of world's population that lives in urban areas would surge from current 55–68% by 2050. This increase in urban population is expected to be 88% in developed countries and 67% in

G. Shiva Kumar (✉) · A. Sharma (✉) · K. Shukla · A. K. Nema
Department of Civil Engineering, Indian Institute of Technology Delhi, New Delhi, India

S. Ahmed et al. (eds.), *Smart Cities—Opportunities and Challenges*,
Lecture Notes in Civil Engineering 58,
https://doi.org/10.1007/978-981-15-2545-2_58

developing countries. Gradual shift in human residence from rural to urban combined with the overall increase in earth's population could add additional 2.5 billion people to world's cities by 2050, with $\sim 90\%$ of this increase taking place in Asia and Africa [57]. Cities and urban areas are growing both in terms of population numbers and in terms of size (or area), thus creating an exceptional demand for resources, services and sanitation. Though world's cities are enjoying more power (technological, political and economic) than ever before in the past, they face augmented number of challenges and threats to efforts of forging their sustainability framework. A city can be termed 'Smart' when it provides improved governance, continual economic growth, efficient management of infrastructure (such as energy, land), better transportation, sustainable housing, maintaining clean unpolluted environment (like water, air) and better quality of life for its citizens through the application of smart solutions.

Air pollution, occurrence of excess quantities of any foreign unwanted particle in the air, is one of the major concerns of this century. Growing population, rapid urbanization and hasty unplanned industrialization along with an increasing demand of energy and an added burden in the growth of number of vehicles are the contributing factors in India [8, 21, 22]. As the level of urbanization in India is predicted to jump from 30.2% (in 2011) to 40.76% (in 2030) [11], the need for developing and maintaining sustainable urban air quality management mechanisms strengthens. Environmental Protection Agency (EPA) of the USA identifies six criteria pollutants—particulate matter, nitrogen dioxide, sulfur dioxide, lead, ground-level ozone and carbon monoxide—which have adverse effects on environment and health. Around 7 million people die every year due to air pollution [60]. Epidemiological studies have reported that PM, SO_2, NO_2, O_3 are responsible for various respiratory problems such as bronchitis, asthma and emphysema along with cardiac morbidity and mortality [6, 16, 40].

One of the main contributors to pollution in metropolitan cities is the emission from vehicles [36]. Though there have been policy interventions like introduction of CNG, construction of metro across the capital, pollution level continues to be as severe as the previous year in New Delhi [23]. Meteorology of the region also strongly influences the concentration of pollutants. Wind speed and direction are the primary parameters which influence the movement of contaminants. Other meteorological parameters such as temperature, relative humidity (RH) also influence the temporal variation of pollutants [23]. Ventilation coefficient is another factor [61]. Ozone levels are reported to be higher during heavy traffic conditions and when temperatures are higher than 30 °C [43, 54]. Studies across multiple sites in Delhi have shown that the concentration of pollutants has been 10–60% lower in the months of summer (May, June and July) and 40–60% higher in winter than the annual average for the year [23]. Cities such as Bangalore are witnessing an increase in population and a massive urbanization program with a city master plan for 2031, the expansion in the use of the current landscape would lead to an increase in travel demands, congestion on roads and henceforth deterioration in the quality of air [24]. A network of stationary and infrequent measurement stations has been conventionally used to monitor air quality across urban areas, but these are expensive to assess the tempo-spatial variation and in the identification of pollution hotspots, required for the real-time control strategies. Optical and chemical analyzers have been used in monitoring air pollutants but these techniques such as gas chromatography are costly. This gives way to developing low-cost micro-scale technologies though the accuracy of the data generated cannot be ascertained [36].

Reducing the pollutant emissions into the atmosphere is indispensable for air pollution control. Many techniques and technologies are applied (or researched) towards this purpose, and the technology for the collection and treatment of pollutants such as dust collectors; electrostatic precipitators; various forms of scrubbers and cyclones; exhaust gas desulfurization and de-nitrification equipments etc. are spreading widely [18]. The main factors required to analyze cost-effectiveness of these pollution control technologies (defined as money spent per ton of pollutant reduced) are (a) the emission rate (pollutant emitted/time), (b) the control effectiveness or efficiency, (c) the annualized capital cost of the control equipment (fixed costs) and (d) the operating and maintenance cost of the control equipment (variable costs). Considering these factors, cost-effectiveness of any control technology is simply "the annualized capital and operating costs divided by the amount of pollution controlled by the technology in 1 year" [50].

Most of the studies that explain the utility of optimization methods in air pollution are focused on developing formal methods like location optimization of air pollution monitoring stations [32], and a derivation of exposure assessment models. Goswami et al. [20], Kukkonen et al. [35] and Lebret et al. [39] have investigated ad hoc fashion of locating monitors in traffic hotspots. Numerous studies in China have come up with multiple inexact optimization techniques, including interval-parameter programming, mixed-integer programming and chance-constrained programming [26]. Development and application of such techniques are extremely limited in India. The cumulative understanding of existing studies still raises uncertainty over optimization of discontinuous and discrete functions.

Air quality management and pollution control planning often pay attention to 'type' and 'measured concentrations' of various pollutants at the monitoring sites relative to the prescribed standards [62]. If variation in distribution of pollutant across the city is considered, then selecting the best set of pollutant control/remediation technologies to use becomes much more complex. Furthermore, it is difficult to use 'cost' as a selection parameter for these technologies. To address these aforementioned complexities, a dynamic programming-based methodology (which is not computer intensive per code run) has been developed that looks at air pollution in a city, caused from a set of identified discrete source emissions followed by optimization of the technology selection for reducing those emissions.

Among all the available alternative potential candidate optimization methods, dynamic programming is used in this current study, with cost as the criteria to be minimized. This paper details the development of an optimization model for air quality management for an urban setting and illustrates its application to a hypothetical city as an example. The developed model allows the decision-makers and the law-making regulatory authorities to run different pollution control scenarios, and to determine whether the reduction in the level of pollutant concentrations corresponding to the 'technology selection-cost estimates' is significant to the planning and decision-making process.

2 Solution Algorithm

Rise of outsized cities and emergence of sizeable groups of towns and city clusters (in proximity to each another) has contributed to rapid urbanization and increasing economic growth in most of world's developing countries [12, 37, 42]. Nevertheless, quality of the environment (air, water and land) within and around these cities (or clusters) has considerably depreciated, with pollution concerns becoming prevalent and ubiquitous [38, 49, 52]. Any city, trying to become a smart and sustainable one, must address the problem of increasing air pollution emissions [31, 52, 58]. In majority of cases, the solution for these problems requires selection and decision-making among a number of candidate emission-control-alternatives that need to satiate a number of technical constraints, policy criteria and regulations [1, 28, 34]. Henceforth, lateral to several other efforts for solving air pollution control obstacles, a voluminous amount of scientific research works have appeared in the literature that approach these issues through the development and advancement of optimization-based decision-making models and their implementation to practical real-life cases [3, 34, 42, 51]. Because of its applicability, flexibility and ease of computation in solving large-scale practical problems, dynamic programming-based approach appears to be the appropriate and feasible choice for optimal air pollution control technology selection for an urban metropolitan area.

2.1 Dynamic Programming

Dynamic programming (DP), developed by Richard Ernest Bellman in the 1950s at the RAND Corporation, is a multistage decision process "designed to handle multi stage decision processes, overcome the shortcomings associated with linear programming, calculus of variations, and other approaches to solving these problems" [53]. Greatest and most imperative reason for creating DP is to be 'computationally feasible' [7, 46]. Though iterative formulations like DP might not be sophisticated alike other mathematical methods, they are and can be faster for formulation and solving, and the solutions may be simplistic to understand. Initially, DP applications in the field of civil engineering have been primarily for the management of water resources, as it can handle complex non-linear constraints, multiple objectives, discontinuous functions and optimize stochastic models [10, 33, 46, 47].

According to Kafkes [30], "Dynamic programming amounts to breaking down an optimization problem into simpler sub-problems and storing the solution to each sub-problem so that each sub-problem is only solved once." The father of dynamic programming, Bellman [4], posits that the chief and powerful principle of DP is *'The principle of optimality.'* This principle has been restated by White [59] as "an optimal policy has the property that all its contractions are optimal." If point 1 (or stage 1) to point M via any intermediate point j is the path of optimality, then any part of the path including 1 to j is also optimal. This allows implementation of piece-wise optimization and iterative methods. To make the optimality principle more concrete, the optimal value function of the decision-making process can be defined in the context of a 'minimum-delay problem' and can be computed using backward induction (or backward dynamic programming framework; starts at right and moves back one stage at a time) or its analog, forward induction (starts at left and moves forward one stage at a time) [5, 44]. Although the spirit of calculations is identical, the elucidation and interpretation are relatively different [9]. Showalter et al. [53] mentioned a powerful feature of dynamic programming, known as 'Invariant Imbedding,' used when a network is solved for optimal trajectory.

The major advantages of DP include: (1) ability to transform a 'single n-dimensional optimization problem' into a series of smaller 'n one-dimensional sub-problems,' that are solved in a sequential and cumulative fashion (which significantly reduces the number of equations that needs solving) [17]; (2) ability to find global maxima or minima, rather than being struck in relatively local optima (which is the case with other methods) [2, 13, 19]; and (3) ability to handle non-linear and discontinuous functions, while also allowing negative decision variables [29]. The key limitation of this sub-optimization procedure is *'The curse of dimensionality,'* where the dimensionality of state space becomes very high and the number necessary calculations might exceed the limits of computation [45, 48, 55]. Finally, DP is not a solution method or algorithm but is an approach (or a way of looking) toward optimization problems.

2.2 Functional Equations in DP

If $[S_k]$ is a set of stages (or points or states) where $k = 1$ to M, and $[C_{ij}]$ is the corresponding cost matrix with C_{ij} being the cost of advancing from any point i to any other intermediary point j. The final objective is to seek the least cost path from 1 to M, and if the provisional objective is to move from any point i to the end, then $f(i)$ is the minimum cost of advancement in that path. If we advance from i to j, the associated cost is C_{ij}; then our residual objective is to optimally move to the end, which upon recalling, requires a minimum cost of $f(j)$. Thus, $C_{ij} + f(j)$ is the cost of forwarding from i to j then optimally till the end (i.e., M).

$$f(i) = \min\left(C_{ij} + f(j)\right) \tag{1}$$

$$f(i) = \min\left(f(j) + C_{ij}\right) \tag{2}$$

According to the aforementioned principle of optimality, the forward functional equation (used to decide the next stage of the path from existing occupied point i) to calculate the optimal cost of moving from i to M is presented in Eq. (1). This equation can be transliterated in the backward induction form as shown in Eq. (2).

3 Model Development

Every developed or developing city, that aims to become a smart one, will have major, identified, point sources of air pollution like coal-burning thermal power plants, kilns of cement manufacturing plants, brickworks, iron and steel smelters, refineries, etc. Consider a city with 'n' major pollutant emission sources, each emitting 'p' number of pollutants. For a given pollutant and emission source, there are 'm' number of pollutant removal/mitigation methods and technologies (including emission control devices, be it new or retrofitted; revised operating procedures, and pollutant capture mechanisms) such as electrostatic precipitators (ESP), various types of cyclones, adsorbents etc. Each of these alternate options has, certain cost and maximum definite reduction capability (efficiency) associated with it. It is coveted to choose the best combination of the pollutant control devices for the sources, in order to reduce, both, the emissions to a desired level and the 'total control cost.' In the subsequent section, a dynamic programming (DP) model that determines the optimal selection strategy will be developed, after defining the different parameters and variables used in the model.

Let δ_i be the production quantity (tonne per year; site-specific average) and β_i be the unit emissions for a particular pollutant (kg of pollutant per tonne of product; adopted from literature for commonly used process equipment) of the ith source (or industry) located inside the city. It is assumed that all 'n' sources currently have no emission control mechanism installed, and no more than one control may be used at any source (i.e., each source must have exactly one control method). It is also assumed that all 'm' controls cannot be used at each source, among 'n' sources. Pollution control alternatives of the 'either/or' variety are not unusual in environmental management problems. The reasons for such restrictions are sometimes technical. It may not be structurally possible to modify existing facilities to provide for more than one waste treatment process. Often, however, the restriction to one control option is based on operational reasons. Pollution control equipment can be difficult to maintain and operate properly, and highly skilled and trained personnel are usually required. By keeping the pollution control system as simple as possible, a factory or municipality many obtain more efficient and reliable waste treatment.

Since every cubic meter of city/community's atmosphere cannot be monitored for a particular pollutant with measuring equipment to conform with the corresponding ambient air quality standard provided by regulatory body, and as air pollution movement is always three dimensional, waste concentrations should be measured (or predicted) at many sites, to arrive at an average concentration in the air. Instinctively, it is best to insure only that average pollutant concentration at certain geographical locations (monitoring or prediction sites) by diminished by 'α%.' The problems that arise are (a) establishing many monitoring stations across the city to measure pollution is a capital-intensive task; (b) predictions of air quality for large areas (like an urban setting) are possible but are very difficult and often produce un-reliable results. These prediction uncertainties are at least partial justification for the assumption of 'Proportional rollback effect,' i.e., "A reduction in pollutant emissions brings about a corresponding reduction in levels of that pollutant in the atmosphere" [14]. With this assumption in place, air quality levels are contemplated indirectly. To illustratively outline the objective for this model using the application of this assumption: The average value of all the measured concentrations of air pollutant under consideration at the various monitoring stations must be reduced by an average of 'α%' to meet a prescribed standard, we infer an objective of reducing total emissions of that pollutant from all its city-sources by 'α%.'

The 'proportional rollback effect' assumption is a beneficial, convenient and useful way of avoiding the difficulties and complexities involved in modeling air pollution transport. Furthermore, there is some empirical evidence that this assumption is valid. According to the eighth annual report of the council on environmental quality of 1977 [14], when hydrocarbon emissions were reduced in San Francisco area by 25% from 1967 to 1976, the daily observed atmospheric oxidant levels associated with those emissions also dropped by 25%. The simple proportional rollback also shows consistence with the improved oxidant air quality trend from 1965 to 1974 in Los Angeles [56].

If environmental regulatory authority (either of the city or the administrative state or the country), after making numerous observations from measuring stations, wishes to reduce concentrations of a specific pollutant in surrounding atmosphere of the city by 'α %' to confirm with the standards, then according to the proportional rollback assumption, the total emissions of that pollutant from the identified sources are also to be reduced by 'α %.' Pollutant emissions are effectively regulated by optimal selection among available alternate treatment options at each source for removing that pollutant from atmospheric discharges. Most of air pollution-related optimization problems require cost-effectiveness approaches. Cost-effectiveness problem in this case is to arrive at the best equipment selection in order to reduce air pollution (reduce pollutant concentration by 'α %') at the least cost possible. The primary objective in the current study is to achieve at minimum cost for reducing the average pollutant concentration by 'α %.'

Let us consider only one pollutant from all 'n' sources. If X_i ($=\delta_i$ β_i) be the pollutant emissions per annum (kg/year) from ith source, then the annual total emissions of the pollutant from all sources will be $\Sigma_{i=1}^{n}(\delta_i\beta_i)$. As the pollutant concentration is targeted to be reduced by 'α %,' the maximum allowable emissions will be $= \gamma = \alpha$ % of $\left[\Sigma_{i=1}^{n}(\delta_i\beta_i)\right]$. In order to reduce the total pollutant emissions to γ, control equipment needs to be selected and installed at all sources, at lowest cost possible. When jth technology/method/equipment, among 'm' technologies, with efficiency η_j is used, the corresponding annual pollutant emissions from ith source will be $= X_{ij}$. It can be seen that the implementation option of 'no control' at any source is infeasible because the emission objective (/standard) of 'γ' kg/year would be violated. Table 1 presents the production levels (δ_i), emission factors (β_i) and annual pollutant discharges (X_i) from all sources, and feasible controls (j) [along with their efficiencies (η_j)] and associated costs (C_{ij}) for various emission sources.

X_i = Pollutant emissions per annum (kg/year) from ith source
γ = Maximum allowable emissions (kg/year) = α % of $\left[\Sigma_{i=1}^n (\delta_i \beta_i) \right]$
When jth technology is used, then corresponding $X_{ij} = (1 - \eta_j) \times \alpha_i \times \beta_i$
C_{ij} = Cost (Rs./t) of feasible emission control method j per unit product run at ith source

where, $i \in \{1, 2, \ldots, n\}$ & $j \in \{1, 2, \ldots, m\}$

Let $C_i(X_{ij})$ be the cost of pollutant emission control at source i when X_{ij} is emitted (Rs./year). These costs are associated as the expenditure required for the emission control method that produces an emission of X_{ij}. Since costs are not direct algebraic functions of X_{ij}, tabular functions must be developed, and the complete set of cost functions is given in Table 2.

The optimization model (objective function) for the problem is given in Eq. (3), and the constraints that the objective function is subjected to are given in Eqs. (4) and (5).

Table 1 Pollutant emission and control alternatives

			Pollutant sources						
			1	2	3	.	.	.	n
Production quantity (t/year)[a]			δ_1	δ_2	δ_3	.	.	.	δ_n
Unit emissions for pollutant (kg/t)[b]			β_1	β_2	β_3	.	.	.	β_n
Pollutant emission/discharge per annum from source (X_i) (kg/year)			$\delta_1\beta_1$	$\delta_2\beta_2$	$\delta_3\beta_3$	.	.	.	$\delta_n\beta_n$
S. No.	Control technologies	Efficiency (%)	Costs for the pollutant sources (Rs./t)						
1	ESP	η_1	C_{11}	C_{12}	C_{13}	.	.	.	C_{1n}
2	Wet scrubber	η_2	C_{21}	C_{22}	×	.	.	.	C_{2n}
3	Multiple cyclone	η_3	×	C_{31}	C_{33}	.	.	.	×
4	Settling chamber	η_4	C_{41}	×	C_{43}	.	.	.	C_{4n}
.	.	.	.	.	.	.	.	.	.
.	.	.	.	.	.	.	.	.	.
.	.	.	.	.	.	.	.	.	.
m	Long-cone cyclone	η_m	C_{m1}	×	C_{m3}	.	.	.	C_{mn}

× = Not feasible
[a]Site and source specific averages
[b]Average values adopted from literature (for predominantly used process equipment)

Table 2 Cost functions

Technology (j)	X_{1j} (kg/ year)	$C_1(X_{1j})$ (Rs./ year)	X_{2j} (kg/ year)	$C_2(X_{2j})$ (Rs./ year)	·	·	·	X_{nj} (kg/ year)	$C_n(X_{nj})$ (Rs./ year)
1	X_{11}	$C_1(X_{11})$	X_{21}	$C_2(X_{21})$	·	·	·	X_{n1}	$C_n(X_{n1})$
2	X_{12}	$C_1(X_{12})$	X_{22}	$C_2(X_{22})$	·	·	·	X_{n2}	$C_n(X_{n2})$
3	X_{13}	$C_1(X_{13})$	X_{23}	$C_2(X_{23})$	·	·	·	X_{n3}	$C_n(X_{n3})$
.	.	.	.	.	.	.	.	.	.
.	.	.	.	.	.	.	.	.	.
.	.	.	.	.	.	.	.	.	.
M	X_{1m}	$C_1(X_{1m})$	X_{2m}	$C_2(X_{2m})$	·	·	·	X_{nm}	$C_n(X_{nm})$

$$\text{Min } Z = \sum_{i,j=1}^{n,m} C_i\left(X_{ij}\right) \tag{3}$$

$$\text{s.t. } \sum_{i=1}^{n} X_{ij} \leq \gamma \tag{4}$$

$$X_{ij} \geq 0 \quad \forall i,j \tag{5}$$

3.1 DP Formulation

All DP problems involving a single resource or state variable (single pollutant, in present case) are one dimensional. To apply DP, the problem needs to be interpreted as a sequential allocation procedure. The resource that can be allocated is the 'γ' (maximum allowable emissions; kg/year) of the pollutant. Portions of the emissions can be apportioned to each source, and the appropriate decision variable is X_{ij}. The model shown in Eq. (3) can also be written in standard DP form by treating the sources as stages (or states) and defining a state variable as S_i. The DP diagram for the problem is shown in Fig. 1.

S_i = Source emissions (kg/year) that can be allotted to remaining $(n - i + 1)$ stages.

The final condition $S_n - X_{nj} \geq 0$ allows solutions that emit less than 'γ' kg of pollutant. Thus, $S_n - X_{nj} = S_{n-1} - X_{(n-1)j} - X_{nj} = S_{n-2} - X_{(n-2)j} - X_{(n-1)j} - X_{nj} = \_ \_ \_ = \gamma - X_{1j} - X_{2j} \cdots - X_{(n-1)j} - X_{nj} \geq 0$, or $X_{1j} + X_{2j} + \cdots + X_{(n-1)j} + X_{nj} \leq \gamma$. If the final condition were specified as $S_n - X_{nj} = 0$, there would be no feasible solution to the problem, as no combination of allowable controls can exactly and accurately produce 'γ' kg of pollutant. The optimal return function for stage i of the model is defined by $f_i(S_i)$.

$f_i(S_i)$ = Minimum cost (Rs./year) of remaining $(n - i + 1)$ stages, when the state variable S_i is emitted from those sources.

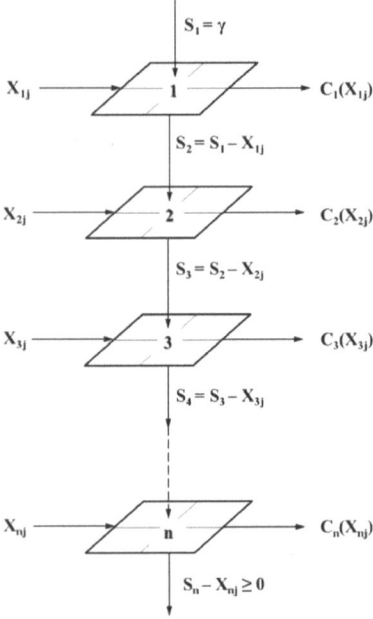

Fig. 1 Methodology model

The general recursive equation of the DP model is

$$f_i(S_i) = \text{Min}\left[C_i(X_{ij}) + f_{i+1}(S_i - X_{ij})\right] \qquad (6)$$

$$\text{s.t.} \quad X_{ij} \leq S_i \qquad (7)$$

Here, $f_{i+1}(\bullet) = 0$ when $i = n$.

4 Implementation of the Model with Illustration

To illustrate the usage and results of the DP-based optimization model, an application of the model (developed in the previous section) to a simulated urban setting was performed. Current section was developed in conjunction with the model development to aid in eliminating the complications and problems that might be encountered in a practical real-life application.

4.1 Site Description

The site used for illustration is oversized, dense, urbanized setting with poor air quality and is purely hypothetical, but the cost data is based on information derived from literature after extensive review. Clearly, this type of site (which sights sustainable and smart development) demands careful analysis and planning prior to implementing any air quality management or action. For model illustration purposes, a single primary air pollutant—total suspended particulate matter (TSP)—is taken. This hypothetical site has a variety of emission sources and the three major sources of TSP within the city are: (a) a coal-burning thermal power station; (b) an iron and steel smelter; and (c) a cement manufacturing plant. Each tonne (1000 kg) of coal burned in the thermal power generation plant produces 95 kg of TSP, while each tonne of cement manufactured leads to emission of 85 kg of TSP [25]. As the major, common and conventional TSP emission source in the iron and steel smelting industry is coal burning [41]; it is considered that each tonne of material processed in that unit corresponds to 95 kg of TSP emissions. The production capacity of cement unit is 2,50,000 t/year, and the coal-burning requirements of the power plant and smelter are 4,00,000 and 3,00,000 t/year of coal, respectively. None of these three emission sources currently have pollution control mechanism in place. Table 3 lists the major sources of TSP in the city along with the available control technologies to curb it.

An environmental control and policy agency, after collecting TSP concentrations across the city's monitoring station, wishes to reduce the average TSP values in the city's atmosphere and surroundings by 80% to meet a standard. According to aforesaid proportional rollback assumption, it is inferred that the objective should be to reduce total TSP emissions from the cement factory, smelter and power plant by 80% at least cost. Emissions are managed by selection of treatment and control options at each source for removing TSP from atmospheric discharges. Several methods are available for TSP removal and are listed in Table 4 along with their efficiencies reported in literature. The table also presents production levels, emission factors and annual pollutant discharges from all sources, and costs associated with feasible controls for various emission sources.

The TSP control objective is to reduce total emissions by 80%. Using the emission factors of 95 kg TSP/t for coal burning and 85 kg TSP/t of cement, the present emissions are

Table 3 TSP sources and its control technologies

I	Source	j	Control
1	Thermal power plant	1	Settling chamber
2	Iron and steel smelter	2	Multiple cyclone
3	Cement factory	3	Long-cone cyclone
		4	Spray scrubber
		5	Electrostatic precipitator

Table 4 TSP emissions, control alternatives and associated unit costs

		Pollutant sources		
		(1) Thermal power plant	(2) Iron and steel smelter	(3) Cement factory
Production quantity (t/year)		4,00,000	3,00,000	2,50,000
Unit emissions for pollutant (kg/t)		95	95	85
Pollutant discharge from source (10^6 kg/year)		38	28.5	21.25
S. No.	Pollution control technologies	Efficiency ($\eta\%$)	Costs for the pollutant sources (Rs./t)[a] (1 US\$ = Rs. 71.18)		
1	Settling chamber (Baffled)	59	71.180	99.652	78.298
2	Multiple cyclone	74	×	×	85.416
3	Lone-cone cyclone	84	×	×	106.770
4	Spray scrubber	94	142.360	156.596	213.540
5	Electrostatic precipitator	97	199.304	213.540	×

× = Not feasible
[a]Average values adopted from literature [25, 27]

Source 1 (Power plant): $4,00,000 \times 95 = 38 \times 10^6$ kg/year
Source 2 (Smelter): $3,00,000 \times 95 = 28.5 \times 10^6$ kg/year
Source 3 (Cement plant): $2,50,000 \times 85 = 21.25 \times 10^6$ kg/year
Total TSP emissions from all 3 sources = 87.75×10^6 kg/year.

This total discharge is to be reduced by 80%, so maximum allowable emissions are 17.6×10^6 kg/year. Therefore, the resource to be allocated, i.e., γ is 17.6×10^6 kg/year.

4.2 Technology Cost Estimates

The option of 'no control' at any source is not feasible because the reduction goal of 17.6×10^6 kg/year could not be reached. When controls are not installed the emissions would be 38×10^6, 28.5×10^6, 21.25×10^6 kg/year for sources 1–3, anyone of which would violate the TSP standard (17.6×10^6 kg/year). From Table 4, there exist 3 feasible control alternatives for sources 1 and 2, and 4 possibilities for source 3. Now, determine the pollutant emissions per annum (kg/year) from ith source when jth control technology is used, i.e., X_{ij} ($=(1 - \eta_j) \times \alpha_i \times \beta_i$). For example: When settling chamber is used at first source, $X_{11} = (1 - 0.59) \times 4,00,000 \times 95 = 15.6 \times 10^6$ kg/year. Similar calculations for other sources and control methods can deliver all the values of $X_{ij} \; \forall \; i, j$. $C_i(X_{ij})$, which is the cost of TSP control at ith source when X_{ij} of TSP is emitted, can be calculated by multiplying unit cost of TSP treatment with production capacity.

Table 5 Complete set of cost functions

Technology (j)	X_{1j} (10^6 kg/year)	$C_1(X_{1j})$ (10^6 Rs./year)	X_{2j} (10^6 kg/year)	$C_2(X_{2j})$ (10^6 Rs./year)	X_{3j} (10^6 kg/year)	$C_3(X_{3j})$ (10^6 Rs./year)
Settling chamber	15.6	28.472	11.7	29.896	8.7	19.575
Multiple cyclone	×	×	×	×	5.5	21.354
Lone-cone cyclone	×	×	×	×	3.4	26.693
Spray scrubber	2.3	56.944	1.7	46.979	1.3	53.385
Electrostatic precipitator	1.1	79.722	0.9	64.062	×	×

For example: When spray scrubbers are installed at thermal power plant (i.e., source 1), $X_{12} = 2.3 \times 10^6$ kg/year and associated cost $C_1(X_{12}) = 142.36 \times 4,00,000 = 56.944 \times 10^6$ Rs./year. Similar evaluations yield the costs corresponding to all $X_{ij} \, \forall \, i, j$. Table 5 gives the complete set of cost functions.

4.3 Results of Implementation

The model shown in Eq. (6) can be applied here by treating the sources as stages and after defining S_i as the state variable.

S_i = TSP emissions (10^6 kg/year) that can be allocated to the remaining ($3 - i + 1$) sources or stages.

The DP stage diagram for the illustration is shown in Fig. 2. The last condition in the image ($S_3 - X_{3j} \geq 0$) allows solutions that emit less than 17.6×10^6 kg of TSP. Hence, $X_{1j} + X_{2j} + X_{3j} \leq 17.6$. The final condition is not written as $S_3 - X_{3j} = 0$, in order to eliminate the 'no feasible solution' scenario. It is desirable to have the initial state variable S_1 equal to a numerical quantity, as shown in Fig. 2.

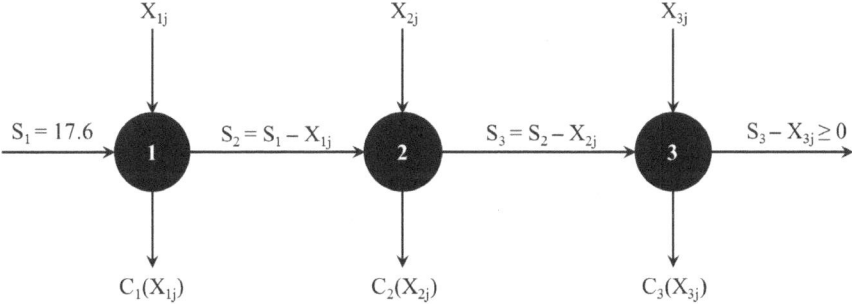

Fig. 2 DP stage diagram for the illustrated example

When the backward DP framework is implemented, the computations for stage 1 are limited to a single value of S_1.

$f_i(S_i)$ = Min cost (Rs./year) of remaining $(3 - i + 1)$ sources when S_i can be emitted by those sources.

Computational Tables. This illustration case study model does not have uniform fixed intervals for decision or state variables, and this requires special consideration in constructing tables of computation for $f_i(S_i)$. The backward induction of the model produced results for stage 3 as shown in Table 6. The maximum value of S_3 is $\gamma = 17.6$. Nevertheless, the decisions, costs and, hence, the value of $f_3(S_3)$ are the same for any S_3 down to 8.7 i.e., $f_3(S_3) = 19.575$, and $X_{3j}^* = 8.7$ for any and all $8.7 \leq S_3 \leq 17.6$. When S_3 drops below 8.7, the largest emission alternative $(X_{3j} = 8.7)$ must also be dropped in order to maintain $S_3 - X_{3j} \geq 0$. In an analogous fashion, all values of S_3 can be categorized into 4 ranges, each of which has distinct decisions and values of $f_3(S_3)$.

Table 7 shows the computations of $f_2(S_2)$, which are lessened substantially by looking in advance into stage-1. At stage 1, $S_1 = 17.6$ and the only possible values for X_{1j} are 15.6, 2.3 and 1.1, hence, it is inferred clearly that the only values of $S_2 = S_1 - X_{1j}$ are 16.5, 15.3 and 2.0. Also, according to Table 7, the case of $S_2 = 2$ is not feasible, as it results in $S_3 (= S_2 - X_{2j})$ values less than 1.3 (its minimum possible value according to Table 6). Further, it is interesting to note that the calculations of $f_2(S_2)$ indicate baffled settling chamber is always the optimal option for source 2 (iron and steel smelter) to implement $(X_{2j} = 11.7)$, irrespective of what controls are implemented for other sources. Table 8 presents the computations of $f_1(S_1)$ i.e., stage 1.

Table 6 Computations for $f_3(S_3)$

S_3 (10^6 kg/year)	X_{3j}	$S_3 - X_{3j}$	$C_3(X_{3j})$ (10^6 Rs./year)	$f_3(S_3)$
17.6	8.7	8.9	19.575	19.575
	5.5	12.1	21.354	
	3.4	14.2	26.693	
	1.3	16.3	53.385	
8.7	8.7	0	19.575	
	5.5	3.2	21.354	
	3.4	5.3	26.693	
	1.3	7.4	53.385	
8.6–5.5	5.5	≥ 0	21.354	21.354
	3.4	≥ 0	26.693	
	1.3	≥ 0	53.385	
5.4–3.4	3.4	≥ 0	26.693	26.693
	1.3	≥ 0	53.385	
3.3–1.3	1.3	≥ 0	53.385	53.385

Table 7 Computations for $f_2(S_2)$

S_2 (10^6 kg/year)	X_{2j}	$S_3 = S_2 - X_{2j}$	$C_2(X_{2j}) + f_3(S_3)$ (10^6 Rs./year)	$f_2(S_2)$
16.5	11.7	4.8	29.896 + 26.693 = 56.589	56.589
	1.7	14.8	46.979 + 19.575 = 66.554	
	0.9	15.6	64.062 + 19.575 = 83.637	
15.3	11.7	3.6	29.896 + 26.693 = 56.589	56.589
	1.7	13.6	46.979 + 19.575 = 66.554	
	0.9	14.4	64.062 + 19.575 = 83.637	
2.0	11.7	−9.7	Infeasible ($S_3 < 1.3$)	
	1.7	0.3		
	0.9	0.1		

Table 8 Computations for $f_1(S_1)$

S_1 (10^6 kg/year)	X_{1j}	$S_2 = S_1 - X_{1j}$	$C_1(X_{1j}) + f_2(S_2)$ (10^6 Rs./year)	$f_1(S_1)$
17.6	15.6	2.0	Infeasible	
	2.3	15.3	56.944 + 56.589 = 113.533	113.533
	1.1	16.5	79.722 + 56.589 = 136.311	

Result. The optimal solution is

Total cost of TSP control = $Z^* = f_1(17.6)$ = Rs. 11,35,33,000 per year;

TSP emissions apportioned to coal-burning thermal power plant and the best suited control technology are: $X_{14}^* = 2.3 \times 10^6$ kg/year and spray scrubber.

TSP emissions apportioned to iron and steel smelter and the best suited control technology are: $X_{21}^* = 11.7 \times 10^6$ kg/year and baffled settling chamber.

TSP emissions apportioned to cement manufacturing factory and the best suited control technology are: $X_{33}^* = 3.4 \times 10^6$ kg/year and long-cone cyclone.

Total emissions from all three sources are 2.3 + 11.7 + 3.4 = 17.4×10^6 kg/year < 17.6×10^6 kg/year.

5 Conclusions

Any smart city employs numerous smart- and adaptive-tools/technologies to increase its economic growth and to improve sustainability while ensuring a better quality of life for its populous. Government of India, in attempt to upgrade quality of living in its current urban landscape, has aimed to develop 100 of its existing cities/towns into smart cities by mid-twenty-first century. A DP-based optimization model has been developed to aid in selecting the best and cost-effective technologies for emission sources in order to regulate the air pollutant concentrations in an urban setting by achieving a given or aspired pollutant reduction level. The model developed in the current study utilizes reported air quality status, which also used for several other purposes including alert services pollutant forecasting and waring. This model analyzes two types of data, the pollution data from emission sources of the city that aspire to become a smart one and cost data reviewed from literature, to arrive at optimal technology selection among various possible candidate alternatives for controlling air pollutants and improve air quality. The objective function of the model was the minimization of the total annual pollutant control cost of all the identified point sources present in the city. The maximum pollutant emissions allowed or apportioned for each source, conforming to the regulatory standards, are also given. The usage of the proposed model was illustrated with a representative case study of a dense city with three major point sources of total suspended particulate matter, which was the pollutant under concern. This developed model, after inclusion of real-word city-specific custom alterations, can help achieve sustainability and other goals of world's governments including India.

6 Recommendations and Inferences

The lack of accurate and reliable cost data for C_{ij} (Rs./t; cost of feasible emission control method j per unit product run at ith source) is one of the major limitations to the application of this research. Further work is required in the area of emission factor (β_i) estimation. Parametric cost estimation of $C_i(X_{ij})$ (i.e., the cost of pollutant control at source i that employs the emission control option j) relies on historical records, which are often rare and changeable in this field. A concentrated effort could have some success in creating cost functions.

Dynamic programming is obviously the optimal method of choice for this 'multiple sources—multiple control technologies—single air pollutant' problem, but this would not be the case if we are considering more than one pollutant. Let us deliberately examine the extension of this above-proposed model for controlling two pollutants.

Y_i = Emissions (kg/year) of second pollutant from ith source
χ_i = Production quantity (t/year) of second pollutant from source i
ψ_i = Unit emissions (kg/t) for second pollutant for source i
Maximum allowable emission for second pollutant = μ % of $\left(\Sigma_{i=1}^{n}\chi_i\psi_i\right) = \theta$
$C_i(X_{ij}, Y_{ij})$ = Annual cost of pollution control associated with source i

Resulting Optimization model:

$$\text{Min } Z = \sum_{i=1}^{n} iC_i\left(X_{ij}, Y_{ij}\right) \tag{8}$$

$$\text{s.t.} \sum_{i,j=1}^{n,m} X_{ij} \leq \gamma \tag{9}$$

$$\sum_{i,j=1}^{n,m} Y_{ij} \leq \theta \tag{10}$$

This 'two pollutant'-based DP model involves allocation of two resources, i.e., allowable levels of both the pollutants (γ and θ). The stage diagram shown in Fig. 3 corresponds to the two-pollutant DP problem.

W_i = State variable for emissions of second pollutant

General Recursive DP Equation:

$$f_i(S_i, W_i) = \text{Min}\left[C_i\left(X_{ij}, Y_{ij}\right) + f_{i+1}\left(S_i - X_{ij}, W_i - Y_{ij}\right)\right] \tag{11}$$

$$X_{ij} \leq S_i \tag{12}$$

$$Y_{ij} \leq W_i \tag{13}$$

The optimal return function of stage i is now function of two variables (S_i and W_i), and the optimization process is with respect to two decision variables (X_i and Y_i) as shown in Eq. (11). Although this change causes no conceptual difficulties, the calculations and computational implications are prodigious.

Fig. 3 DP diagram for the extension of above problem to two pollutants

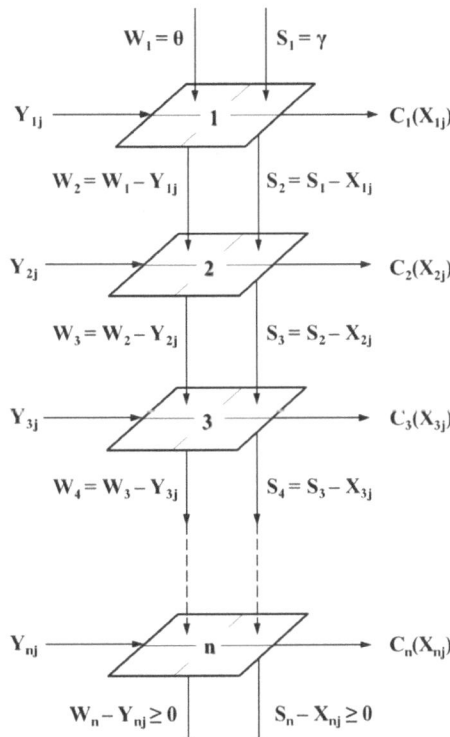

The computations for the single pollutant model involve formation of 'n' tables (each table for each stage of computation), all two dimensional, containing all feasible combinations of S_i and X_i. Compared to this, tables for two-pollutant DP model would be four dimensional, since this would include feasible combinations of S_i, W_i, X_i and Y_i. The sheer number of computations '$curse\ of\ dimensionality$' and the complexity in identification of all feasible computations would probably interdict the use of DP for solving the 'multiple sources—multiple control technologies—single air pollutant' air pollution problem. For the case of multiple pollutants, mixed-integer linear programming (MILP) is suggested for solving air quality modeling problems than DP.

References

1. Abrams R (1975) Optimization models for regional air pollution control. In: Mathematical analysis of decision problems in ecology. Springer, Berlin, Heidelberg, pp 116–6. https://doi.org/10.1007/978-3-642-80924-8_5
2. Amini AA, Weymouth TE, Jain RC (1990) Using dynamic programming for solving variational problems in vision. IEEE Trans Pattern Anal Mach Intell 9:855–867. https://doi.org/10.1109/34.57681
3. Asif Z, Chen Z (2018) Optimization of air pollution control model for mining. Int J Civil Environ Eng 12(4):411–417. https://publications.waset.org/10008805/optimization-of-air-pollution-control-model-for-mining
4. Bellman RE, Dreyfus SE (1962) Applied dynamic programming. Princeton University Press, Princeton
5. Bellman RE, Dreyfus SE (2015) Applied dynamic programming, vol 2050. Princeton University Press, Princeton
6. Bernstein JA, Alexis N, Barnes C, Bernstein IL, Nel A, Peden D, Williams PB et al (2004) Health effects of air pollution. J Allergy Clin Immunol 114(5):1116–1123. https://doi.org/10.1016/j.jaci.2004.08.030
7. Bertsekas DP, Bertsekas DP, Bertsekas DP, Bertsekas DP (1995) Dynamic programming and optimal control, vol 1, No. 2. Athena Scientific, Belmont
8. Bhaskar BV, Mehta VM (2010) Atmospheric particulate pollutants and their relationship with meteorology in Ahmedabad. Aerosol Air Qual Res 10(4):301–315. https://doi.org/10.4209/aaqr.2009.10.0069
9. Bradley SP, Hax AC, Magnanti TL (1977) Dynamic programming. In: Applied mathematical programming [E-reader version], pp 320–362. Retrieved from http://web.mit.edu/15.053/www/AMP-Chapter-11.pdf
10. Caines PE (2018) Linear stochastic systems, vol 77. SIAM
11. Chauhan C (2007) Urbanisation in India faster than rest of the world. Hindustan Times, Retrieved from https://www.hindustantimes.com/india/urbanisation-in-india-faster-than-rest-of-the-world/story-IdmQ4BSqxEZe874AprzfnL.html
12. Cohen AJ, Brauer M, Burnett R, Anderson HR, Frostad J, Estep K, Feigin V et al (2017) Estimates and 25-year trends of the global burden of disease attributable to ambient air pollution: an analysis of data from the global burden of diseases study 2015. Lancet 389 (10082):1907–1918. https://doi.org/10.1016/S0140-6736(17)30505-6
13. Cooper L, Cooper MW (2016) Introduction to dynamic programming: international series in modern applied mathematics and computer science. vol 1. Elsevier

14. Council on Environmental Quality (1977) Environmental quality: the eighth annual report of the council on environmental quality. U.S. Government Printing Office, Washington, D.C, p. 172. Retrieved from https://babel.hathitrust.org/cgi/pt?id=mdp.39015000421662;view=1up;seq=4
15. Craig KJ, De Kock DJ, Snyman JA (1999) Using CFD and mathematical optimization to investigate air pollution due to stacks. Int J Numer Meth Eng 44(4):551–565. https://doi.org/10.1002/(SICI)1097-0207(19990210)44:4<551::AID-NME519>3.0.CO;2-7
16. Curtis L, Rea W, Smith-Willis P, Fenyves E, Pan Y (2006) Adverse health effects of outdoor air pollutants. Environ Int 32(6):815–830. https://doi.org/10.1016/j.envint.2006.03.012
17. Dreyfus S (2002) Richard Bellman on the birth of dynamic programming. Oper Res 50(1):48–51. https://doi.org/10.1287/opre.50.1.48.17791
18. Environmental Cooperation Office, Global Environment Bureau, Ministry of Environment, Japan (1998) Air pollution control technology manual (Copyright 2005). Retrieved from https://www.env.go.jp/earth/coop/coop/document/01-apctme/contents.html
19. Galil Z, Park K (1992) Dynamic programming with convexity, concavity and sparsity. Theoret Comput Sci 92(1):49–76. https://doi.org/10.1016/0304-3975(92)90135-3
20. Goswami E, Larson T, Lumley T, Liu LJS (2002) Spatial characteristics of fine particulate matter: identifying representative monitoring locations in Seattle, Washington. J Air Waste Manag Assoc 52(3):324–333. https://doi.org/10.1080/10473289.2002.10470778
21. Goswami P, Baruah J (2011) Urban air pollution: process identification, impact analysis and evaluation of forecast potential. Meteorol Atmos Phys 110(3–4):103–122. https://doi.org/10.1007/s00703-010-0105-9
22. Goyal P, Sidhartha (2003) Present scenario of air quality in Delhi: a case study of CNG implementation. Atmos Environ 37(38):5423–5431. https://doi.org/10.1016/j.atmosenv.2003.09.005
23. Guttikunda SK, Gurjar BR (2012) Role of meteorology in seasonality of air pollution in megacity Delhi, India. Environ Monit Assess 184(5):3199–3211. https://doi.org/10.1007/s10661-011-2182-8
24. Guttikunda SK, Nishadh KA, Gota S, Singh P, Chanda A, Jawahar P, Asundi J (2019) Air quality, emissions, and source contributions analysis for the Greater Bengaluru region of India. Atmos Pollut Res 10(3):941–953. https://doi.org/10.1016/j.apr.2019.01.002
25. Haith DA (1982) Environmental systems optimization, p 115
26. He YJ, Chen DZ (2008) Hybrid particle swarm optimization algorithm for mixed-integer nonlinear programming. J Zhejiang Univ (Eng Sci) 42(5):747–751. https://doi.org/10.3785/j.issn.1008-973X.2008.05.005
27. Howard DB, Thé J, Soria R, Fann N, Schaeffer R, Saphores JDM (2019) Health benefits and control costs of tightening particulate matter emissions standards for coal power plants-the case of Northeast Brazil. Environ Int 124:420–430. https://doi.org/10.1016/j.envint.2019.01.029
28. Huan L, Kebin H (2012) Traffic optimization: a new way for air pollution control in China's urban areas. https://doi.org/10.1021/es301778b
29. Jiang Y, Jiang ZP (2015) Global adaptive dynamic programming for continuous-time nonlinear systems. IEEE Trans Autom Control 60(11):2917–2929. https://doi.org/10.1109/TAC.2015.2414811
30. Kafkes A (2017) Demystifying dynamic programming. Retrieved from https://medium.freecodecamp.org/demystifying-dynamic-programming-3efafb8d4296
31. Kanada M, Fujita T, Fujii M, Ohnishi S (2013) The long-term impacts of air pollution control policy: historical links between municipal actions and industrial energy efficiency in Kawasaki City, Japan. J Clean Prod 58:92–101. https://doi.org/10.1016/j.jclepro.2013.04.015
32. Kanaroglou PS, Jerrett M, Morrison J, Beckerman B, Arain MA, Gilbert NL, Brook JR (2005) Establishing an air pollution monitoring network for intra-urban population exposure assessment: a location-allocation approach. Atmos Environ 35t, 39(13):2399–2409. https://doi.org/10.1016/j.atmosenv.2004.06.049
33. Kelman J, Stedinger JR, Cooper LA, Hsu E, Yuan SQ (1990) Sampling stochastic dynamic programming applied to reservoir operation. Water Resour Res 26(3):447–454. https://doi.org/10.1029/WR026i003p00447

34. Kondili E (2005) Review of optimization models in the pollution prevention and control. In: Computer aided chemical engineering, vol 20. Elsevier, pp 1627–1632. https://doi.org/10.1016/S1570-7946(05)80113-0

35. Kukkonen J, Härkönen J, Karppinen A, Pohjola M, Pietarila H, Koskentalo T (2001) A semi-empirical model for urban PM10 concentrations, and its evaluation against data from an urban measurement network. Atmos Environ 35(26):4433–4442. https://doi.org/10.1016/S1352-2310(01)00254-0

36. Kumar P, Morawska L, Martani C, Biskos G, Neophytou M, Di Sabatino S, Britter R et al (2015) The rise of low-cost sensing for managing air pollution in cities. Environ Int 75:199–205. https://doi.org/10.1016/j.envint.2014.11.019

37. Kumar R, Joseph AE (2006) Air pollution concentrations of $PM_{2.5}$, PM_{10} and NO_2 at ambient and kerbsite and their correlation in Metro City–Mumbai. Environ Monit Assess 119(1–3): 191–199. https://doi.org/10.1007/s10661-005-9022-7

38. Landrigan PJ (2017) Air pollution and health. Lancet Public Health 2(1):e4–e5. https://doi.org/10.1016/S2468-2667(16)30023-8

39. Lebret E, Briggs D, Van Reeuwijk H, Fischer P, Smallbone K, Harssema H, Elliott P et al (2000). Small area variations in ambient NO_2 concentrations in four European areas. Atmos Environ 34(2):177–185. https://doi.org/10.1016/S1352-2310(99)00292-7

40. Mabahwi NAB, Leh OLH, Omar D (2014) Human health and wellbeing: Human health effect of air pollution. Procedia-Soc Behav Sci 153:221–229. https://doi.org/10.1016/j.sbspro.2014.10.056

41. Owoade KO, Hopke PK, Olise FS, Ogundele LT, Fawole OG, Olaniyi BH, Bashiru MI et al (2015) Chemical compositions and source identification of particulate matter ($PM_{2.5}$ and $PM_{2.5-10}$) from a scrap iron and steel smelting industry along the Ife–Ibadan highway, Nigeria. Atmos Pollut Res 6(1):107–119. https://doi.org/10.5094/APR.2015.013

42. Parvin M, Grammas GW (1976) Optimization models for environmental pollution control: a synthesis. J Environ Econ Manage 3(2):113–128. https://doi.org/10.1016/0095-0696(76)90026-7

43. Plumlee GS, Ziegler TL (1999) Environmental geochemistry. Treatise on Geochemistry

44. Poor HV (1984) Backward forward and backward-forward dynamic programming models under commutativity conditions. In: Proceedings of 23rd IEEE conference decision control, pp 1081–1086. https://doi.org/10.1109/CDC.1984.272179

45. Powell WB (2007) Approximate dynamic programming: solving the curses of dimensionality, vol 703. Wiley

46. Puterman ML (2014) Markov decision processes: discrete stochastic dynamic programming. Wiley, New York

47. Ross SM (2014) Introduction to stochastic dynamic programming. Academic Press

48. Rust J (1997) Using randomization to break the curse of dimensionality. Econometrica: J Econometric Soc, 487–516. https://doi.org/10.2307/2171751

49. Sakawa M, Sawaragi Y (1975) Multiple-criteria optimization of pollution control model. Int J Syst Sci 6(8):741–748. https://doi.org/10.1080/00207727508941858

50. Schnelle KB Jr, Dunn RF, Ternes ME (2015) Air pollution control technology handbook. CRC Press

51. Shaban HI, Elkamel A, Gharbi R (1997) An optimization model for air pollution control decision making. Environ Model Softw 12(1):51–58. https://doi.org/10.1016/S1364-8152(96)00008-4

52. Shao M, Tang X, Zhang Y, Li W (2006) City clusters in China: air and surface water pollution. Front Ecol Environ 4(7):353–361. https://doi.org/10.1890/1540-9295(2006)004[0353:CCICAA]2.0.CO;2

53. Showalter WE, Halpin DW (2008) Dynamic programming approach to optimization of site remediation. J Constr Eng Manage 3(10):820–827. https://doi.org/10.1061/(ASCE)0733-9364(2008)134:10(820)

54. Shukla K, Srivastava PK, Banerjee T, Aneja VP (2017) Trend and variability of atmospheric ozone over middle Indo-Gangetic Plain: impacts of seasonality and precursor gases. Environ Sci Pollut Res 24(1):164–179. https://doi.org/10.1007/s11356-016-7738-2
55. Si J, Barto AG, Powell WB, Wunsch D (eds) (2004) Handbook of learning and approximate dynamic programming, vol 2. Wiley, New York
56. Trijonis JC, Peng TK, McRae GJ, Lees L (1976) Emissions and air quality trends in the South Coast Air Basin. Retrieved from https://authors.library.caltech.edu/25765/
57. United Nations (2018) 2018 Revision of world urbanization prospects. Retrieved from https://www.un.org/development/desa/en/news/population/2018-revision-of-world-urbanization-prospects.html
58. Wang KM (1981) Optimization of an air pollution control model by linear programming. J Chin Inst Eng 4(1):1–11. https://doi.org/10.1080/02533839.1981.9676662
59. White DJ (1969) Dynamic programming, vol 1. Oliver & Boyd, Edinburgh
60. World Health Organization (WHO), Department of Public Health, Environmental and Social Determinants of Health (PHE) (2014) Healthy environments, Healthy People (Press Release). Retrieved from https://www.who.int/phe/eNews_63.pdf
61. Yadav R, Sahu LK, Beig G, Jaaffrey SNA (2016) Role of long-range transport and local meteorology in seasonal variation of surface ozone and its precursors at an urban site in India. Atmos Res 176:96–107. https://doi.org/10.1016/j.atmosres.2016.02.018
62. Zeger SL, Thomas D, Dominici F, Samet JM, Schwartz J, Dockery D, Cohen A (2000) Exposure measurement error in time-series studies of air pollution: concepts and consequences. Environ Health Perspect 108(5):419–426. https://doi.org/10.1289/ehp.00108419

Design of an Energy-Efficient Airport Using TEG on Runways

Shreeja Kacker and Vivek Singh

Abstract This research work aims to design a "green airport" to meet the energy requirements for day-to-day functioning of the airport. Designing a green airport involves re-creating a structure using methods that are environmentally responsible, physically sustainable and resource-efficient. The airport will become a sustainable high-performance structure. **For this purpose, thermoelectric generator is being used to generate and fulfill electricity demand of an airport, using temperature gradient and solar energy. The runway of the airport will be designed to utilize the kinetic and potential energy of the aircraft and convert it into heat/ electrical energy.** The airport taken into consideration for design is the Indira Gandhi International Airport. The electricity cost incurred by an airport using non-renewable sources of energy is calculated and compared to the cost incurred if non-renewable sources were replaced by thermoelectric generator on the runway.

Keywords Green airport · Thermoelectric generator · Runway design · Energy-efficient airport · Green building

1 Introduction

Electricity has become one of the most required amenities in the present day. The various resources from which electricity is being generated in present days include coal, fossil fuels, oil, etc., which are non-renewable in nature. Non-renewable resources are the natural resources which will not replenish themselves once it has been used. Non-renewable resources are the primary sources of the tremendous wealth surpluses required to perpetuate industrialized societies. Over 70% of energy

S. Kacker (✉)
Assistant Professor, Department of Civil Engineering, Delhi Technical Campus, Greater Noida 201306, Uttar Pradesh, India

V. Singh
Assistant Professor, Department of Civil Engineering, JIMS Engineering Management Technical Campus, Greater Noida 201306, Uttar Pradesh, India

© Springer Nature Singapore Pte Ltd. 2020
S. Ahmed et al. (eds.), *Smart Cities—Opportunities and Challenges*,
Lecture Notes in Civil Engineering 58,
https://doi.org/10.1007/978-981-15-2545-2_59

used in industries comes from non-renewable sources. They do substantial harm to the environment, causing water and air pollution, high-intensity damage to public health, depletion of wildlife, habitat loss, over usage of land and water resources, thereby causing global warming emissions. Even the extraction and processing of these energy resources harm and degrade our environment. The major harmful effects are greenhouse gas emissions, air pollution, waste generation, etc.

However, over the years, an increasing gap between the demand and supply of electricity has been observed and has become a matter of concern for us. Airports consume electricity up to 180 M kWh per year with 60% of this being consumed by terminals, while the remaining 40% is allocated to workshops, parking decks, airfield lighting, hangers and other ancillary buildings.

Therefore, keeping in mind, the growing electricity problem, a solution in this regard is needed to meet the growing energy demands without hampering the environment further. This leads to the use of renewable sources of energy which includes solar, geothermal, wind, etc. There are various forms of renewable resources like solar which can be trapped and converted into other forms of energy, i.e., wind power, geothermal powers, etc. Solar energy can be used directly for heating and producing electricity, or it can also be used indirectly via wind, ocean, bio-mass, thermal and hydroelectric power. To meet our increasing demands with the help of renewable source of energy, civil engineers have introduced the concept of green structures. Green structures are known to complement and expand the classical building design concerns of durability, economy, utility and comfort.

In the present study, an attempt has been made to design an energy-efficient airport using TEG on runways generate electricity and meet the energy demands of the airport.

2   Device Description

2.1   Thermoelectric Generator

Thermoelectric generator (TEG) is also called the Seebeck Generator. It can be described as a solid-state device or equipment that directly converts heat energy (heat flux) into electrical energy using the Seebeck effect, i.e., it creates electricity by converting heat energy into electrical energy.

The Seebeck effect is defined as a phenomenon in which a temperature difference between two dissimilar electrical conductors or semiconductors produces a voltage difference between the two substances. When heat is applied to one of the two conductors or semiconductors, heated electrons flow toward the cooler one. If the pair is connected through an electrical circuit, direct current (DC) flows through that circuit (Fig. 1).

The main aim for using these is to use the waste heat generated at the runway either due to friction or by solar energy to generate electricity.

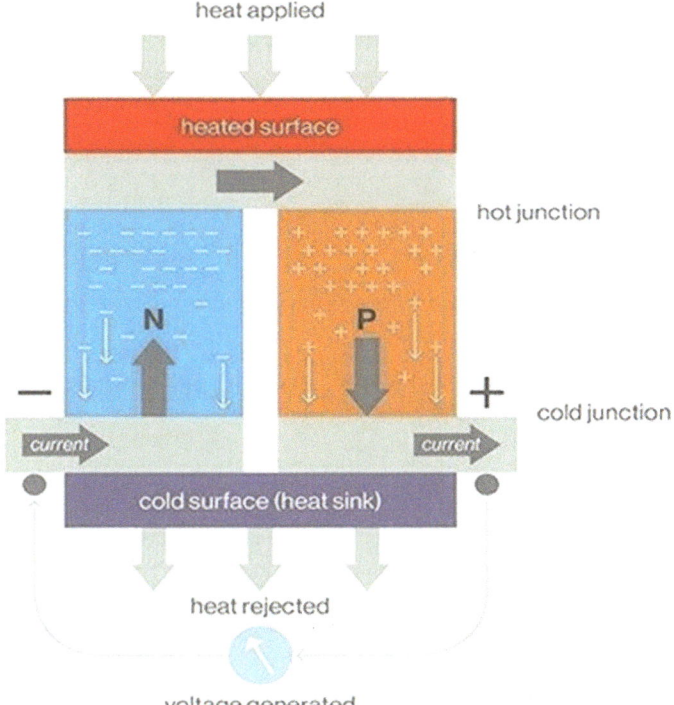

Fig. 1 Seebeck Effect in a Thermoelectric Generator

TEG mainly consists of three components—thermoelectric materials, thermo-electric module and thermoelectric system.

(a) Thermoelectric materials: Thermoelectric materials generate electricity directly from heat energy by converting temperature differences in solids into electric voltage. A good thermoelectric material should possess high electrical conductivity and low thermal conductivity.

(b) Thermoelectric module: A thermoelectric module is a circuit that contains thermoelectric materials. It directly generates electricity from heat energy and consists of two dissimilar thermoelectric materials, namely a p-type (positively charged) semiconductor and an n-type (negatively charged) semiconductor, joined at their ends. As soon as a temperature difference is created between the two materials, electric current flows in the circuit.

(c) Thermoelectric system: A thermoelectric system generates electricity by taking in heat from a source at high temperature. For this, the system needs a large temperature gradient, with the cold side being cooled by air or water. In order to supplement for this heating and cooling of the system on opposite sides, heat exchangers are used on both sides of the modules.

The main advantages of TEG are that it effectively uses the heat that was otherwise wasted, ensuring a convenient power supply. Also, it requires low maintenance. The only drawback with this system is that it requires skilled labor for installation and maintenance.

2.2 Working of Thermoelectric/Seebeck Generator

Thermoelectric power is the generation of electricity due to observed temperature difference. This is known as the Seebeck effect.

Seebeck effect was discovered in 1821 by Thomas J. Seebeck. He pointed out that when two dissimilar materials were heated on one side, an electromagnetic field was generated. Another observation included that this voltage generated is directly proportional to the difference in temperature between both the sides of the materials.

Flow of electrical power in solids can be facilitated by charge carriers. The charge carriers are of two types—the negative electrons or the "n-type" and the positive holes or the "p-type." The charge carriers get distributed on heating one end of the conducting solid, thereby creating a voltage that can be measured (Figs. 2 and 3). Factors on which the output power depends upon—

- Temperature delta
- Quantity of heat that can be moved through module.

The thermoelectric generator is being used to generate the electricity in our paper by using the heat energy generated at the pavement either due to solar or friction.

Fig. 2 Inside view of TEG

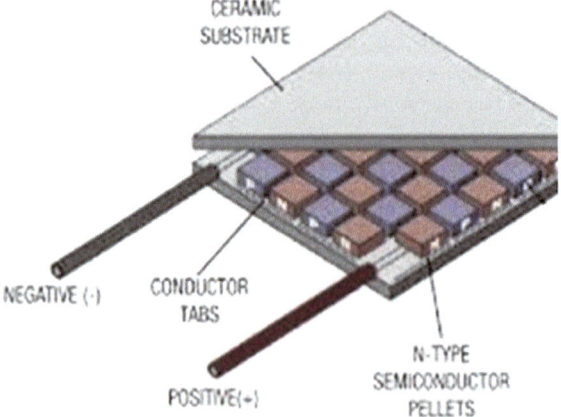

Fig. 3 Circuit diagram of TEG

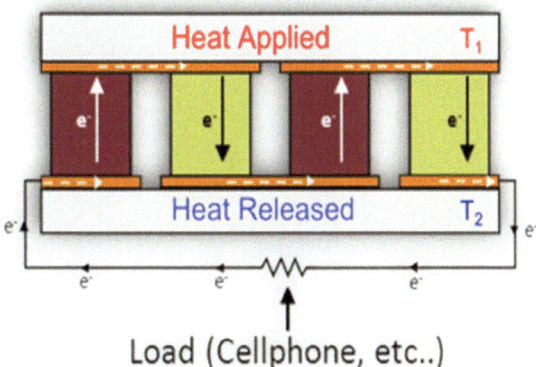

3 Project Setup

To understand the project setup at the site, a small-scale model was developed and readings due to temperature gradient to ensure efficient working of the proposed TEG system.

3.1 Laboratory Setup

The model being proposed included following components:

1. Copper plate—It is used for the heat transfer from the pavement to the TEG. The reason for opting for copper place is its high thermal conductivity which will help in the transfer of the heat energy from the pavement surface to the TEG.
2. TEG—It is used for electricity generation. The model of TEG used is SP 1848-27145.
3. Heat sink—It is used for maintaining a temperature gradient on the either side of the TEG. It is a passive heat exchanging device which transfers the heat energy generated by an electronic or a mechanical device to a fluid medium. This medium may be air or any liquid coolant. From here, the energy is dissipated away from the device, hence ensuring temperature regulation of the device.
4. Multimeter—for determining the current and voltage produced.

The proposed arrangement of all the components was carried out in the laboratory, and images of the same have been attached in Figs. 4 and 5.

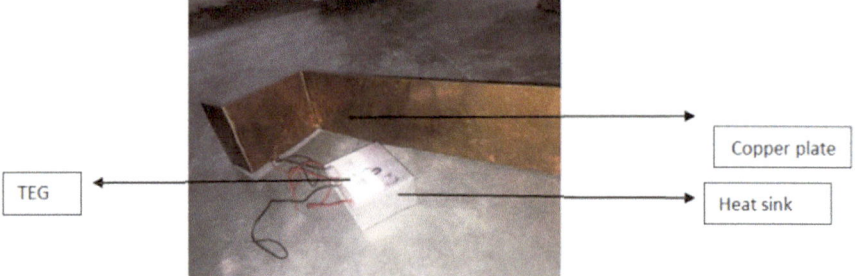

Fig. 4 Components of setup

Fig. 5 Proposed setup

3.2 Checking of Laboratory Setup

The proposed arrangement was checked for its working in the laboratory. For this purpose, a utensil filled with hot water was kept on the heat transfer plate. The TEGs were connected and placed over the aluminum heat sink. Above this heat sink, the heat transfer plate along with hot water container was placed. Readings were taken to ensure that the setup was working efficiently (Fig. 6).

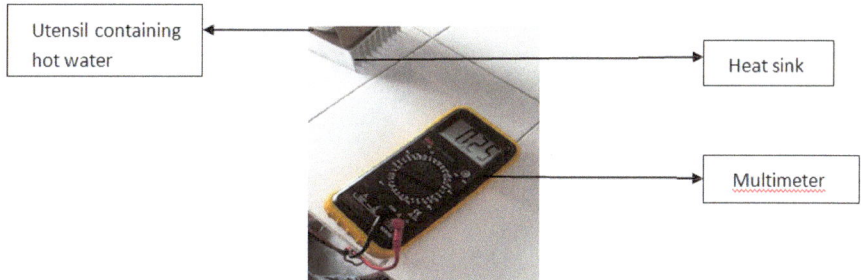

Utensil containing hot water

Heat sink

Multimeter

Fig. 6 Working of proposed setup

3.3 *Experimental Setup*

The experimental setup included the setup that will be placed below the pavement. Hot water was used instead of the solar or frictional heat for checking whether transfer of heat will take place or not and how much electricity will be produced due to temperature variation between two plate ends.

A container for hot water was placed, and the metal plate was inserted in the container for the flow of heat that will reach TEG. A thermometer was used to determine the temperature of the water and the plate. The heat sink was placed below both the TEGs to ensure difference in temperature is being maintained. The multimeter was connected to the TEGs for the voltage and current readings (Figs. 7 and 8).

Fig. 7 Experimental setup

Fig. 8 Multimeter reading
for experimental setup

The readings observed were

Temperature (°C)	Voltage (mV)
60	220
65	225

This shows that some amount of electricity is being produced by the experimental setup.

3.4 Field Setup

A pit of around 20 cm depth was excavated, the setup including heat sink, TEG and copper plate's transferring part was placed, the bricks after applying thermocol sheets were placed so as to ensure the proper touch of plate, and TEG is maintained throughout the reading interval. The heat collector plate was left open to collect

Fig. 9 Field setup

solar heat, and the temperature was measured using thermometer. The readings of voltage and current were noted down at various temperatures of the plate which changes due to change in day temperature (Fig. 9).

4   Observations

4.1   Results Obtained During Field Test

S. No.	Temperature (°C)	Voltage (V)	Current (A)	Watts (W)
1	49	0.44	0.015	0.0066
2	50	0.46	0.016	0.00736
3	48	0.42	0.014	0.00588
4	44	0.34	0.01	0.0034
5	47	0.40	0.013	0.0052
6	46	0.38	0.012	0.00456
7	45	0.36	0.011	0.00396
	Average watt calculated			0.00528

4.2   Graphical Observations During Field Test

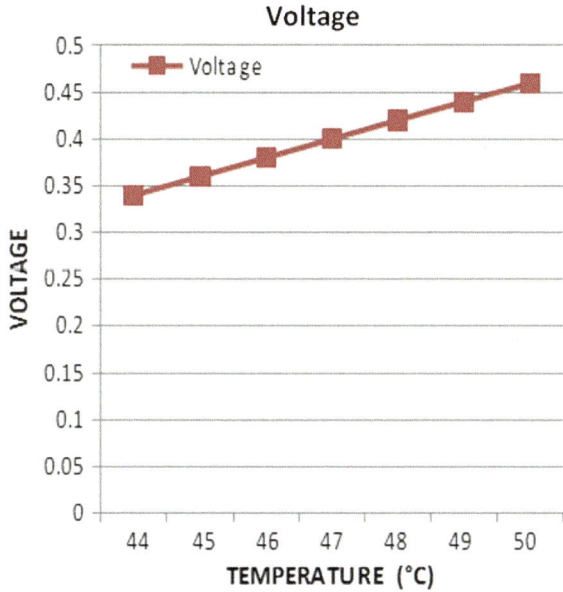

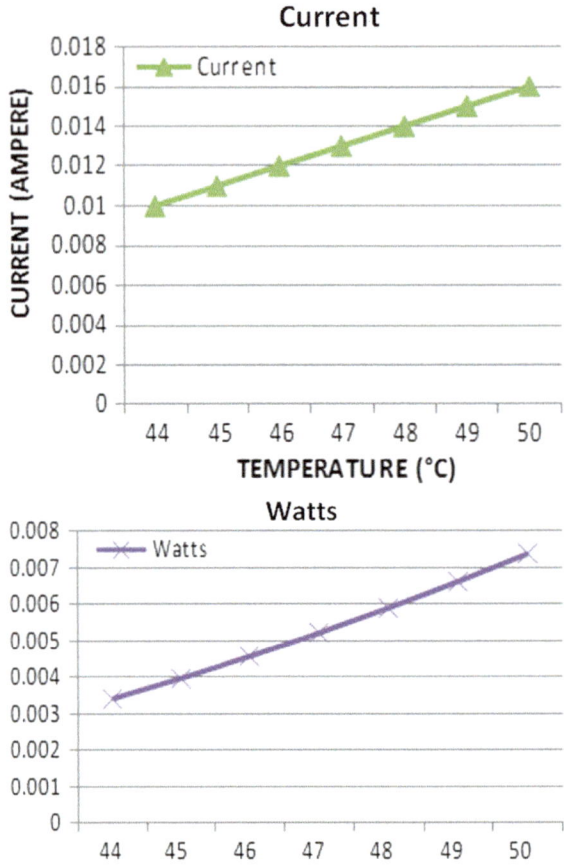

This shows that with the increase in the temperature, the voltage and the current being produced by setup are also increasing. On an average, temperature of runway pavement reaches to around 52 °C so 0.5 V will be produced roughly.

4.3 Calculations

As mentioned earlier, the calculation has been done for Indira Gandhi International (IGI) Airport.

$$\text{Number of runways at IGI airport} = 3$$
$$\text{Length of total 3 runways} = 2813 + 3810 + 4430 = 11,053 \text{ m}$$

Providing setup along the runway alternatively at both the sides at a distance of 10 m for initial trial, therefore, number of setups that can be arranged alternatively at both sides = 2210

$$\text{Average electricity generated} = 0.00528 \text{ W/s}$$

For a single setup,
For solar,

Electricity generated in 8 h = 0.00528 * 60 * 60 * 8 = 152.064 W
Electricity generated in 1 year = 152.064 * 365 = 55503.36 W/year
=0.05550336 MW/year
1 MW = 0.024 mu
For 2210 setup electricity generated = 2.94 Mu.

For friction,

Average rate of heat generated due to friction = 1.35 * 106 W/m^2
Area of the plate provided at surface = (0.45 * 0.12)m^2
Electricity generated due to 1 setup = 0.0729 W
Electricity generated due to 2210 setups = 161.109 MW = 3.866 Mu
Therefore, total amount of electricity generated due to 2210 setups in 1 year
=2.94 + 3.866 = 6.806 Mu

4.4 Energy Requirements

- *For Solar Panels*: Solar energy is a renewable energy which is produced by the sun with the help of solar photovoltaic devices in the form of heat and light. Solar panels are needed to be installed in an open area with no obstruction like trees providing a shade to the panel. Generally, preferred area is the roof of a building. Generally, solar panels are 8–16% efficient and help in saving a good amount of electricity bill. According to the GMR report, IGI airport has two solar plants, one of 2.84 MW and another of 5 MW which can produce a total of 4814 MWh electricity with a total investment of Rs. 32 crores.
- *For TEG*: From the calculations done, it is being noted that our proposed setups could also be used for energy production at the runway pavement. A good amount of energy is being produced per year by our proposed setup with less amount of investment.

5 Cost Analysis

5.1 Comparative Analysis

Investment needed for 5.7 MW solar plant is 32 Crores.
Investment needed for a single proposed setup = 2000 rupees
Therefore, investment needed for 2210 setups = 4,420,000 rupees

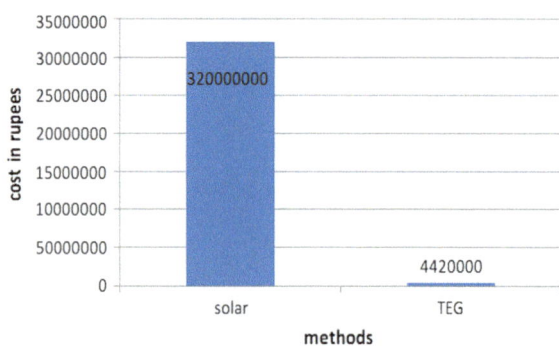

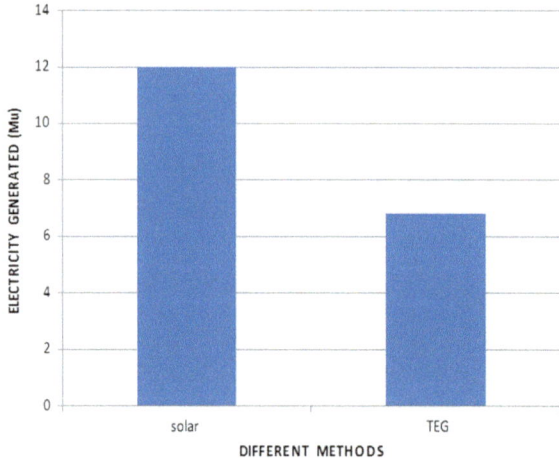

5.2 Site Feasibility

- Temperature difference: A sufficient amount of temperature should be maintained on the either side of thermoelectric generator in order to generate electricity.
- Weather condition: On a rainy day, a less amount of electricity comparing with the calculated one will be produced.
- Flight frequency: Flight frequency is going to affect the performance as during night time, the only source of heat generation on the pavement surface is due to friction, so greater the flight frequency, greater will be electricity generation.
- Time interval: Greater the time interval between two flights, lesser will be the heat production, and thus, a less amount of electricity will be produced.
- Impact factor: To ensure that model is able to take impact on the runway due to vibration generated from aircraft touchdown and its high speed, the width of pavement should be adequate such that no impact of aircraft load is handled by TEGs.

6 Conclusions

In the present study, a TEG runway design model has been proposed to generate electricity using frictional energy of the aircrafts as well as the solar energy.

The proposed model uses the temperature gradient within asphalt pavements of the runway and the taxiway. It has the potential for sustainable generation of electricity. The energy harvesting process is green and environment friendly as well as economical. It does not interfere with the airport movements and aircraft takeoff and landing.

The experimental setup of the proposed TEG model shows that around 6.86 Mu electricity could be generated considering that the setup will receive solar energy during 8 h of the day, and friction will be provided by the aircraft movement on the runway.

Furthermore, this electricity generated has a very low investment cost as compared to that of solar panels. An investment of 50 lakhs can produce an electricity of 6.86 Mu, which covers about 10% of the electricity demands of the airport.

The proposed method is not only reasonable but also is very feasible at site and can be easily installed.

References

1. Thakur A, Dewangan M, Lalwani P (2017) Piezo electric roads. Int J Adv Res Sci Eng. ISSN 2319-8354
2. Kumar P (2013) Piezo-smart roads. Int J Enhanced Res Sci Technol Eng. ISSN 2319-7463
3. Kulkarni A (2013) Solar roadways. Int J Eng Res Appl ISSN 2248-9622
4. Fatima N, Mustafa J (2010) Production of electricity by the method of road power generation. Int J Adv Electr Electron Eng. ISSN 2319-1112
5. Thongsan S, Prasit B, Suriwrong T, Mensin Y, Wansungnern W (2017) Development of solar collector combined with thermoelectric module for solar drying technology. Energy Procedia
6. NASA Technical Note, "NASA TN D-8094"
7. Green Co Best Practices at Delhi Airport (June 2016)

MmWave Networks for Smart Cities

Garima Shukla, M. T. Beg and Brejesh Lal

Abstract In this paper, we have observed that the smart cities leverage information and communication technologies (ICTs) for improved citizen services such as utilities, including energy consumption, transport management, etc., in order to reduce consumption of resources, curb wastage and bring down overall costs. A smart city can be fashioned out of an existing city as well as constructed as a greenfield project. The number of devices increases manifold in smart cities. A significant component of the challenge in communications in smart cities is the challenge of ensuring 5G communication in cellular and HetNets. Ad hoc networks can also be considered for deployment on need basis in smart cities. If the requirements of these formulations merge, then the solutions of one can be deployed for the betterment of the other. The first and foremost requirement of large data transfer in smart cities is bandwidth; fortunately, mmWaves have emerged as promising candidates for 5G networks. These can be deployed for setting up the communications in smart cities as well. The paper examines the protocols, simulators and initial feasibility analysis of introducing mmWave communications in smart cities. The scenario in a smart city has been modelled through wireless mobile network comprising i.i.d. Poisson distributed nodes for ad hoc networks and an infrastructure-based wireless mobile network at varying mobility of nodes.

Keywords Smart city · 5G · Millimetre wave · PHY layer · MAC · Simulator

G. Shukla (✉)
Faculty of Engineering and Technology, Jamia Millia Islamia University,
New Delhi 110025, India

M. T. Beg
Jamia Millia Islamia University, Jamia Nagar, New Delhi 110025, India

B. Lal
Indian Institute of Technology, New Delhi, India

© Springer Nature Singapore Pte Ltd. 2020
S. Ahmed et al. (eds.), *Smart Cities—Opportunities and Challenges*,
Lecture Notes in Civil Engineering 58,
https://doi.org/10.1007/978-981-15-2545-2_60

1 Introduction: Need for Smart Cities to Be 5G Compliant

Smart cities have been conceptualized and initiated. A smart city uses information and communication technologies (ICTs) for improving efficiencies and promoting transparency and overall improvement in delivery of government services [1]. Nearly 31% of India's current population lives in urban areas and contributes 63% of India's GDP. With increasing urbanization, urban areas are expected to house 40% of India's population and contribute 75% of India's GDP by 2030 [2]. This requires comprehensive development of physical, institutional, social and economic infrastructure. "Smartness" through extensive use of communication technology is one response for putting in place an enabling environment for persons to enjoy a comfortable life in a urban setting.

There are several aspects of a smart city. The foremost among them is smart supply of utilities (electricity, water, garbage disposal), then smart regulation of traffic, health services, security and in a way every aspect of governance, commerce, manufacturing, supply chain, etc. Essentially, the service provider (government, private player) ceases to have a direct face-to-face contact, for most of the services, with the residents in smart areas. The work is done through automation and information collection, discreetly, as per the requirements.

These requirements entail real-time data collection, and processing and issuing required information/instructions, unobtrusively. A large number of sensors and devices networked together can deliver the required solutions. While these demands from communications to deliver are ever expanding, especially from wireless communication technology due to its scalability, quickness and ease of deployment and being mobility-friendly, the technology still continues to remain in the 4G stage of development.

New technologies are needed to cope with the quantum-enhanced performance requirements. These technologies will be a quantum jump over Long-Term Evolution (LTE) which gave a new system architecture, all-IP connectivity and a new air interface through orthogonal frequency-division multiplexing (OFDM). The spectrum for all these technologies, however, remained in the microwave band.

Additional efforts were made to harness the potential in the existing communication band by enhancing spectral efficiency through massive-input massive-output (MIMO) (utilizing spatial diversity), carrier aggregation and coordinated interference management (CoMp). These efforts however could only deliver incremental improvements, and a quantum jump over the existing scenario was required [3].

Many believe that 5G will be the key to smart cities. 5G could provide coverage density 100 times greater than current standards. Able to support up to 1 million devices per square kilometre, 5G will deliver the massive improvements in speed, throughput, device deployment, traffic capacity, latency and spectrum efficiency required by the smart city ecosystem.

2 Millimetre Waves

It is increasingly being seen that the frequency spectrum of operation of wireless communications has to be shifted to the millimetre zone from the existing micro-wave zone. Millimetre waves occupy the frequency band from 30 to 300 GHz (some definitions consider frequencies of the order of 20 GHz or more as millimetre waves). Millimetre waves have higher frequency, bandwidth and therefore higher data transmission capacity than microwaves. Data speeds up to 7 Gbps through use of mmWaves have been acknowledged [IEEE 802.11 ad]. However, mmWaves have high atmospheric attenuation and therefore a short range ($\ll$kilometre). However, this allows for better frequency reuse. The short wavelength implies smaller-sized antennas and therefore reduced beam width, which can further increase frequency reuse [3, 4].

Since the existing deployment of wireless communication technology is largely microwave-based, both microwave and mmWave bands are expected to be used by mmWave cellular devices. The microwave band can be used for control channel which is not easily implementable in high frequency due to directionality, blockage and short range of mmWave communication. The higher frequencies of mmWave band will provide the massive amount of bandwidth for data communications. Further, 5G networks are being theorized as combination of several sub-networks. This introduces an element of network heterogeneity which has to be taken into account.

3 Wireless Communication Parameters and Considerations

Wireless communication channel depicts (a) large-scale path loss as well as (b) fading. The fading (both slow fading and fast fading) between two nodes i and j is denoted by h_{ij} and is assumed to be independent and identically distributed (i.i. d.) between any two nodes. The interference at the receiver is given by

$$I(y) = \sum_{x \in \Phi} h_{xy} l(x, y) \tag{1}$$

In the case of wireless networks, the interference due to other transmitters also comes into play. The signal-to-interference and noise ratio between node i and its receiver j is defined as:

$$\text{SINR}(i, j) = \frac{P_L h_{ij} l(i, j)}{\eta + I} \tag{2}$$

The distance loss function $l(i,j) = d_{i,j}^{-\alpha}$ and the path loss function $= h_{ij}l(i-j)$, I is the interference observed at the receiver due to all nodes other than its paired node, and η is the observed noise which is taken as a constant figure.

The first and foremost requirement of wireless communications is ensuring proper SINR. Depending on the magnitude of noise and interference, the system can become noise limited or interference limited. Due to increase in the number of transmitters, the system may progress from being noise limited to interference limited. However, this also depends on the available bandwidth, and in case of mmWaves, the transition from noise limited to interference limited can be experienced in indoor and highly dense deployment of transmitters [5].

Often, the standardization of a wireless protocol involves standardization of MAC. In case of interference-limited systems, resource sharing through TDMA is a preferred option as it is contention free, and in the case of noise-limited systems contention-based access (e.g. through pure ALOHA based on random access) is a preferred option. These considerations have bearings in standardization of protocol for mobile networks. In the case of mmWave personal area networks, a hybrid approach is adopted wherein both contention-free access and contention-based access can be adjusted as per requirements.

Another consideration that is of significance is the channel model to be adopted for mmWave communication [6, 7]. The large-scale channel model, more critically the path loss model, traditionally prescribes, for outdoor or indoor channels, that the average received power decreases exponentially with increasing distance between the transmitter and receiver. This is generally expressed in logarithmic scale as:

$$P_L(d) = P_L(d_0) + 10\alpha\log_{10}(d/d_0) + (X) \tag{2}$$

where $P_L(d)$ is the average power loss at a distance of $d(m)$ relative to a reference distance $d_0(m)$, α is the path loss exponent and X is the attenuation (in dB) caused by fast fading. X is a statistical value and has a Rayleigh distribution for line of sight (LOS) propagation and Nakagami/Rician distribution for non-line of sight (NLOS) distribution. These distributions have to be taken into account while analysing the effects of channel fading. The channel no longer remains a static entity but becomes dynamic and needs to be monitored. Since new challenges are posed, new solutions are required through a new paradigm development. Ultimately, the complexity of the transmitter–receiver, energy and cost considerations have also to be taken into account. Standardization of mmWave protocols is available through IEEE 802.11 ad, only [8].

4 Standardization of 5G Protocol

So far, IEEE 802.11 ad (also known as WiGig) is the only full suite of protocol utilizing mmWaves and having some features of 5G. It is however designed for small distance and operates between 57 and 71 GHz.

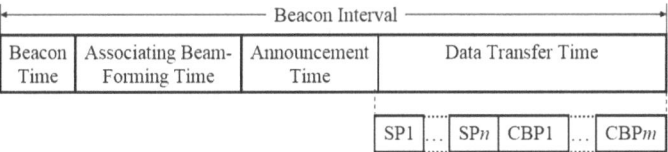

Fig. 1 IEEE 802.11 ad beacon interval for medium access

The IEEE 802.11 ad protocol relies on beamforming and uses beacon intervals for channel access. The standard provides a directional communication scheme that utilizes beamforming antenna gain to overcome increased attenuation in the 60 GHz band (Fig. 1).

The PHY layer utilizes four different categories of modulation and coding schemes (MCSs). The orthogonal frequency-division multiplexing (OFDM) PHY is utilized for highest data rates and for downlink and when there are no stringent requirements of power and complexity. One of the issues with OFDM modulation has been the presence of a large peak-to-average power ratio (PAPR). An optimum choice of MCS can be considered while maintaining communication with a multitude of devices that can be present in a smart city. This scheme is known as adaptive modulation and coding (AMC) and can be implemented through continuous tracking of channel quality indicators (CQIs) [9, 10].

In a smart city setting, with a multitude of users, beamforming can significantly reduce the interference from other transmitters located nearby. However, beamforming entails costs in terms of time for establishing the required beam and complexity of equipment. The gains can, in certain conditions, far outweigh the advantages. Beamforming depends on LOS transmission possibility between the transmitter and the receiver [6, 7, 11, 12]. Since beamforming can be disrupted by occurrence of obstacles, adaptive beamforming has been envisaged where the NLOS transmission can take care of the link while the beam is again established. This will be a fairly common communication scenario in a smart city.

5 Flexibility in Wireless Network Set-Up—Ad hoc Network

The wireless communication infrastructure in a smart city may also require flexibility, and thus the need for consideration of mobile ad hoc networks also comes in. Fortunately, the IEEE 802.11 ad also has a functionality for working without access points and therefore without beamforming in mobile ad hoc networks [13, 14]. This was tried out through open-source simulator ns3 [3, 8, 14–16]. The simulation had the following objectives: (1) To mimic a Poisson independent and identically distributed (i.i.d.) process of mobile ad hoc nodes; (2) to analyse the effect of varying the speed of arrivals of nodes; and (3) to analyse the throughput of such a network and its potential in carrying large data with low delay.

Table 1 Simulation parameters for the ad hoc smart city network

S. no.	Simulation parameter	Simulation value
1	Antenna	Omnidirectional 60 GHz antenna Tx 0dBm = 1 W, Tx gain, Rx gain = 0 Energy detection threshold = $-79 + 3$ dBm
2	Socket type and transmission rate	UDP "On–Off", 10 Mbps data rate
3	Channel	Constant speed propagation delay model, Friis propagation loss model
4	MAC	Beam training interval set to zero
5	MCS	OFDM 6756.75 Mbps
6	Mobility models	Random waypoint

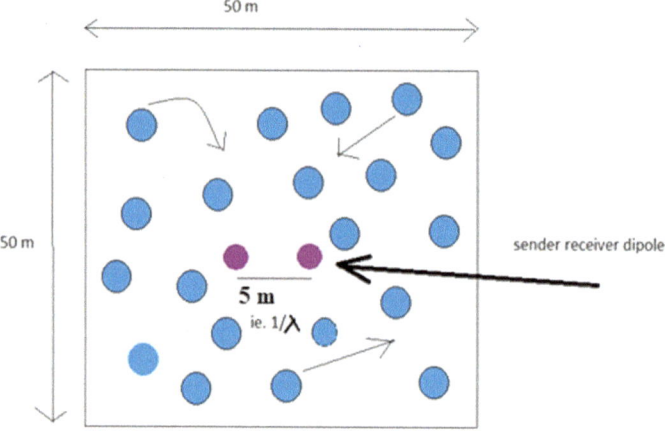

Fig. 2 Simulation scenario

The number of nodes was kept at 22 in a rectangular 2D region of size 50 m 50 m. The last two nodes constitute the sender–receiver dipole. The first 20 nodes are paired up with each other in sender–receiver fashion. The first 10 mobile source nodes broadcast to each of 10 number sinks. This creates a scenario of constantly changing interference pattern around the sender–receiver dipole under study. The simulation parameters (Table 1) for the simulation scenario (Fig. 2) are the following.

The simulation indicated that the link throughput (Fig. 3) for wireless ad hoc network in IEEE 802.11 ad protocol can be sustained at small distances ~ 5 m or so with node mobility.

However, if the IEEE 802.11 ad network is operated in the infrastructure mode, i.e. with an access point, the transmission can be sustained over a distance of up to 50 m also. The equipment in ad hoc network mode can be of utility in personal area networks.

Fig. 3 Link throughput recorded by the simulator

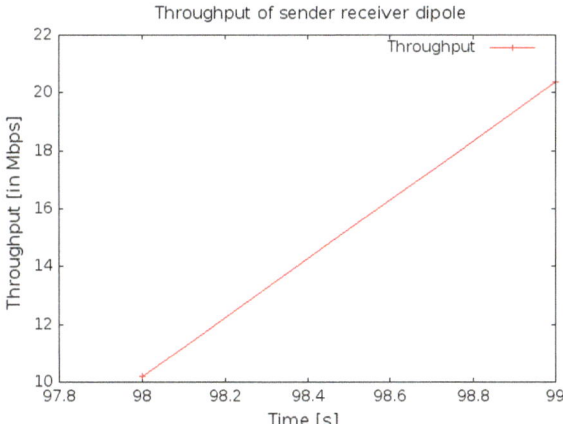

This leads us to the search for a more reliable and robust solution for mmWaves. Other existing solutions for LANs in microwaves in IEEE 802.11 suite of protocols still continue to be the only viable solutions for data transfer in WLANs in ad hoc networks at medium distances (i.e. > 10–20 m) [3, 5].

6 Wireless Network Set-Up in Complex Outdoor Scenario

Various requirements (and use cases) for wireless network set-up in a smart city have been considered in the simulation. Some of these are:

(i) requirement of backhaul for interconnecting access points (this relies on the utilization of bandwidth 800 MHz),
(ii) improving latency (latencies up to 50 ms are easily achievable),
(iii) enabling integration of different radio access technologies (both microwave and mmWave can coexist; the microwaves can be used for link establishment and control overheads),
(iv) data transfer to high-speed users,
(v) adapting to network densification (there will be an explosion in a number of devices that would be connected to the 5G networks).

This simulation was also undertaken through ns3 [15, 16]. The simulation framework involved the evolved packet core set-up, remote host connected through a packet gateway for generating the traffic and backhaul through X2 interfaces.

The existing mmWave module in ns3 employs the traditional contention-based access only, but with the added features of beamforming and MIMO.

A typical scenario of a city with several blocks of buildings interspersed with streets was considered. Within this setting, there are several groups of users, i.e. pedestrians, vehicular traffic, indoor users, etc (Fig. 4).

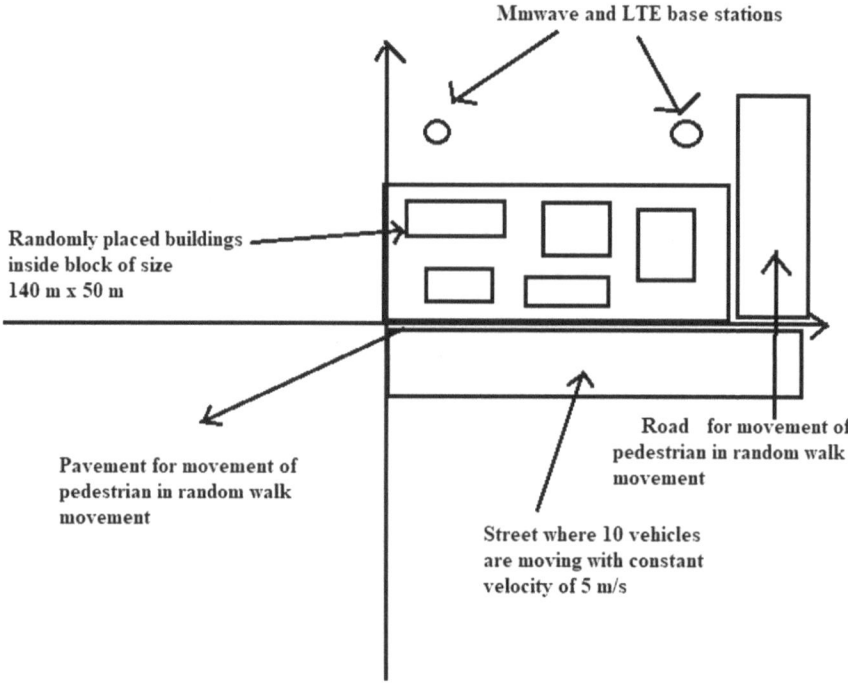

Fig. 4 Mimicking some smart city conditions

Table 2 Parameters considered for the simulation

S. no.	Simulation parameter	Simulation value
1	Antenna	Directional 60 GHz antenna with 12 sectors Tx 10 dBm = 10 W, Tx gain, Rx gain =+10 Energy detection threshold = −79 + 3 dBm
2	Socket type	TCP bulk sender, 10 Mbps data rate
3	Channel model	3GPP propagation loss model 3GPP building propagation loss model
4	Channel condition	(a) Urban microcell (UMi) condition which arises in a narrow street condition ("street canyon") (b) Urban macrocell (UMa). This is expected in widely spaced roads and moving/stationary UEs (c) Indoor office (In) which mimics offices with cubicles and open areas
5	MAC	Contention-based
6	MCS	OFDM 6756.75 Mbps
7	Mobility models	Random waypoint for pedestrians Constant velocity for vehicles

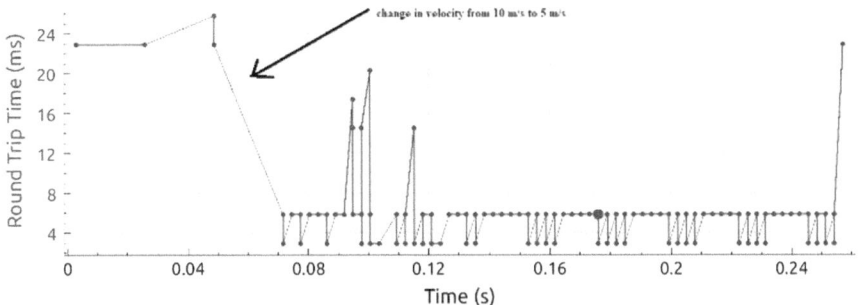

Fig. 5 Change in latency

The performance of the network (Table 2) was analysed by observing the captured packets and simulator output in the form of text files.

The latency (round-trip time delay) is observed to change from a high of 30 to 4 ms after drop in vehicular speeds (Fig. 5). The congestion window, which records the congestion at transmitter/receiver, was observed to be well within the limit at all times. The higher latency in the beginning of the graph in Fig. 5 is on account of delay due to beamforming and beamtracking.

The performance of the network was not observed to be degraded due to increase in the number of users (network densification). This is possible as the large amount of bandwidth available could satisfy all the bandwidth requirements of the users. However, the effect of network densification at several orders of magnitude higher was not tested as the simulator takes inordinately long time and produces inordinately large data for the validation.

7 Conclusion and Future Work

The mmWave scenario was evaluated in a typical smart city setting. Several parameters were introduced. The mmWave network was observed to work as per requirements in both LOS and NLOS conditions. However, at high vehicular speeds degradation in latency was observed. The limitations in the scenario were the use of contention-based MAC and not hybrid MAC. The performance can be improved if the MAC protocols incorporate both contention-free and contention-based accesses to account for latency limitations for critical applications such as traffic regulation and healthcare delivery.

Further research will be undertaken in the realm of improvement of MAC parameters for performance improvement.

References

1. Martins JSB (2018) Towards smart city innovation, arXiv:1810.11665v1 [cs.NI] 27 Oct 2018
2. Smart City Guidelines, Ministry of Urban Development, Government of India, June 2015
3. Andrews JG, Buzzi S, Choi W, Hanly SV, Lozano A, Soong ACK, Charlie Zhang J (2014) What will 5G be? IEEE J Sel Areas Commun 32(6)
4. Rappaport TS, Sun S, Mayzus R, Zhao H, Azar Y, Wang K, Wong GN, Schulz JK, Samimi M, Gutierrez F (2013) 1NYU WIRELESS, Polytechnic Institute of New York University, New York, NY 11201, USA. Millimetre wave mobile communications for 5G cellular: it will work! IEEE Access, 10 May 2013
5. Shokri-Ghadikolaei H, Fischione C (2017) Millimeter wave ad hoc networks: noise-limited or interference-limited?, arXiv:1509.04172v2 [cs.IT] 1 Oct 2015
6. Rappaport TS, Heath RW, Daniels RC, Murdock JN (2015) Millimetre wave wireless communications. Prentice Hall, Pearson
7. Samimi MK, MacCartney GR Jr, Sun S, Rappaport TS (2016) 28 GHz millimetre-wave ultra wideband small-scale fading models in wireless channels. In: 2016 IEEE vehicular technology conference (VTC2016- Spring), 15–18 May 2016
8. Assasa H (2016) IMDEA Networks Institute and Universidad Carlos III de Madrid, Spain, Joerg Widmer, IMDEA Networks Institute, "implementation and evaluation of a WLAN IEEE 802.11 ad Model in ns-3". In: WNS3 '16 proceedings of the workshop on ns-3, 2016, ACM Digital Library
9. Thornburg A, Bai T, Heath RW Jr Performance analysis of mmWave Ad Hoc networks
10. Nitsche T, Cordeiro C, Flores A, Knightly EW, Perahia E, Widmer JC, IEEE 802.11ad: directional 60 GHz communication for Multi-Gbps Wi-Fi
11. Akdeniz MR, Liu Y, Samimi MK, Sun S, Rangan S, Rappaport TS, Erkip E, Millimetre wave channel modelling and cellular capacity evaluation
12. IEEE 802.11ad. Part 11: wireless LAN medium access control (MAC) and physical layer (PHY) specifications—amendment 3: enhancements for very high throughput in the 60 GHz band," 2 "WirelessHD: WirelessHD specification overview," 2009
13. Roh W, Seol J, Park J, Lee B, Lee J, Kim Y, Cho J, Cheun K, Aryanfar F (2014) Millimetre-wave beamforming as an enabling technology for 5G cellular communications: theoretical feasibility and prototype results. IEEE Commun Mag 52(2):106–113
14. Ramanathan R, Redi J, Santivanez C, Wiggins D, Polit S (2005) Adhoc networking with directional antennas: a complete system solution. IEEE J Sel Areas Commun 23(3):496–506
15. Mezzavilla M, Zhang M, Polese M, Ford R, Dutta S, Rangan S, Zorzi M, End-to-end simulation of 5G mmWave Networks arXiv:1705.02882.pdf
16. Rappaport TS, Sun S, Shafi M (2017) 5G channel model with improved accuracy and efficiency in mmWave bands. 5G Tech Focus 1(1)

Aquifer Modelling in Greater Noida Region (U.P) Using MODFLOW

Mohd Saleem, Shobha Ram, Gauhar Mahmood, Mohd Abul Hasan
and Mohd Waseem

Abstract Groundwater is a very useful resource of water for domestic, industrial and agricultural necessities in Greater Noida. Groundwater withdrawal has increased in the last two to three decades due to the requirement of water in domestic, agricultural and industrial field. A part of Yamuna–Hindon River is incorporated for the study area. A careful study of water balance of a basin is essential for the effectiveness of groundwater management. The behaviour of the flow system and water balance was assessed by groundwater flow modelling. MODFLOW was established to quantify groundwater using steady-state finite difference model with 26 groundwater data from tube wells in Greater Noida region. Location map and well inventory are used to evaluate the groundwater level, direction and surface features. Different boundary conditions are used to recharge the ground by rainfall and irrigation returns, seepage losses from unlined canal and horizontal flows using diverse boundary packages available in Visual MODFLOW. In this study, hydraulic conductivity and recharge parameters are analysed by MODFLOW and are shown the sensitiveness of the model. For more optimistic urban groundwater intercessions to be executed in urban cities, we have to energize increasingly incorporated, foundational and quantitative assessment of the possible pathways, mediations and communications. This is progressively conceivable with the utilization of connected and coordinated demonstrating where groundwater procedure models, physical

M. Saleem (✉)
Faculty of Engineering and Technology, University Polytechnic—Jamia Millia Islamia,
New Delhi 110025, India

S. Ram
School of Engineering, Gautam Buddha University, Greater Noida,
U.P. 201308, India

G. Mahmood · M. Waseem
Department of Civil Engineering, Faculty of Engineering and Technology,
Jamia Millia Islamia, New Delhi 110025, India

M. A. Hasan
Civil Engineering Department, King Khalid University, Abha 61421, Saudi Arabia

© Springer Nature Singapore Pte Ltd. 2020
S. Ahmed et al. (eds.), *Smart Cities—Opportunities and Challenges*,
Lecture Notes in Civil Engineering 58,
https://doi.org/10.1007/978-981-15-2545-2_61

models and water industry models can be coupled and illuminate financial appraisal, where the estimation of water, in the entirety of its appearances, is progressively being perceived.

Keywords MODFLOW · Recharge · Rainfall · Groundwater model · Greater Noida

1 Introduction

Freshwater resources are shrinking due to the continuously increasing the water demand for domestic, agricultural and industrial uses [1]. Declining groundwater in the study region has prompted harmful impact to the groundwater assets and domestic, agriculture and industrial clients. A normal decrease of 0.88 m/year demonstrates the pattern of long term groundwater level [2]. In many perspectives, groundwater is favoured over surface water system such as operational costs and pumping are low, it is effectively assessable and siphoned due to shallow profundity. Central Ground Water Board (CGWB) and Groundwater Department of Uttar Pradesh (U.P) government completed different examinations of hydrogeological parameters and were discovered the principal gathering of aquifer [3, 4]. Conjunctive use of groundwater and surface water along with stream aquifer interactions was assessed in Daha region [5, 6]. The river-aquifer association was evaluated in Ganga–Mahaba sub-basin [7].

Groundwater modelling has turned into a standard device for viable groundwater executives with fast increments in computational capacity and wide accessibility of PCs and model programming software. The present work researches the impacts of groundwater advancement and is additionally useful to consider the groundwater flow framework in Yamuna–Hindon inter-stream area utilizing steady-state groundwater models.

The three-dimensional partial differential equation of groundwater flow is expressed as:

$$\frac{d}{dx}\left[Kxx\frac{dh}{dx}\right] + \frac{d}{dy}\left[Kyy\frac{dh}{dy}\right] + \frac{d}{dz}\left[Kzz\frac{dh}{dz}\right] - W = Ss\frac{dh}{dt}$$

where Kxx, Kyy, Kzz are components of the hydraulic conductivity in x, y, z directions, respectively.

Specific storage is Ss, time is t, source or sink term is W and potentiometric head is h.

To simulate the groundwater flow in the study area, the above equation was used in finite difference computer-based MODFLOW software [8].

2 Study Area

Greater Noida is a standout amongst the most essential district for residential and business reason nearby Delhi (Fig. 1). It is the part of the district Gautam Buddha Nagar which lies in the state of Uttar Pradesh (India). Latitude and longitude of the study area are 28.47 44° N and 77.50 40° E, respectively. The population of 124 towns comprises of 107,676 (till March 2014) and the area constitutes around 40,000 ha. River Hindon flows in the western side of the city along with National capital Territory (NCR) area of Delhi. Amongst the latest decade, the quantity of various industries in Greater Noida has established more than 10 times (Greater Noida Master Plan, 2001). Normal temperature lasts between 23 and 45 °C during summer season and a normal precipitation of 93.2 cm (36.7 in.) in monsoon season. The sources of water supply in the region collected through tube wells, trunks, overhead tanks and other supply lines. To monitor the nature of water level, groundwater monitoring wells were developed and the water table was checked four times in a year [9]. On an average, the study area gets 600 mm of yearly precipitation. Past 15 years precipitation information demonstrated that

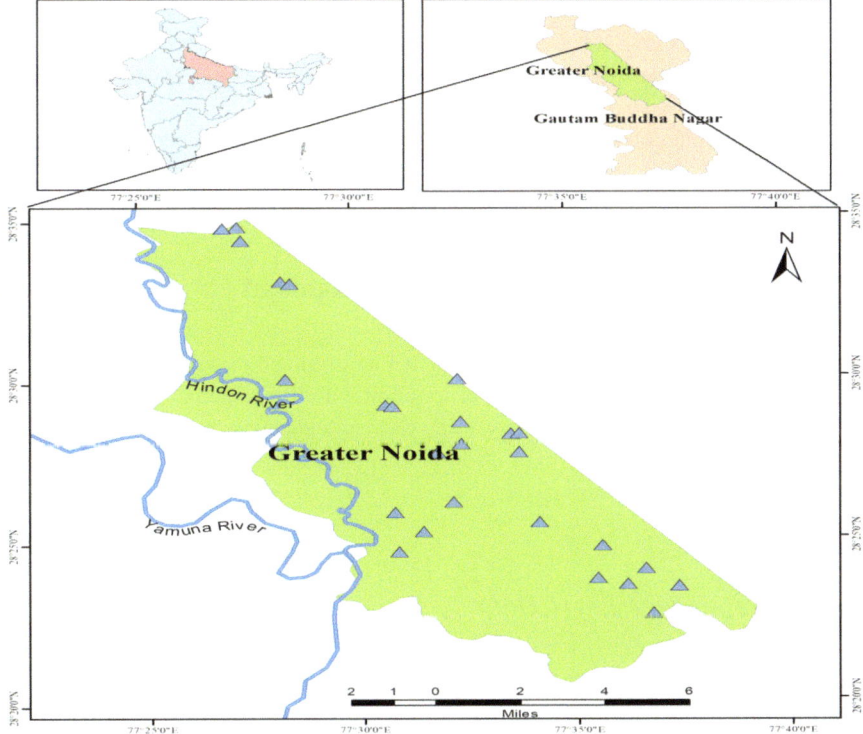

Fig. 1 Study area

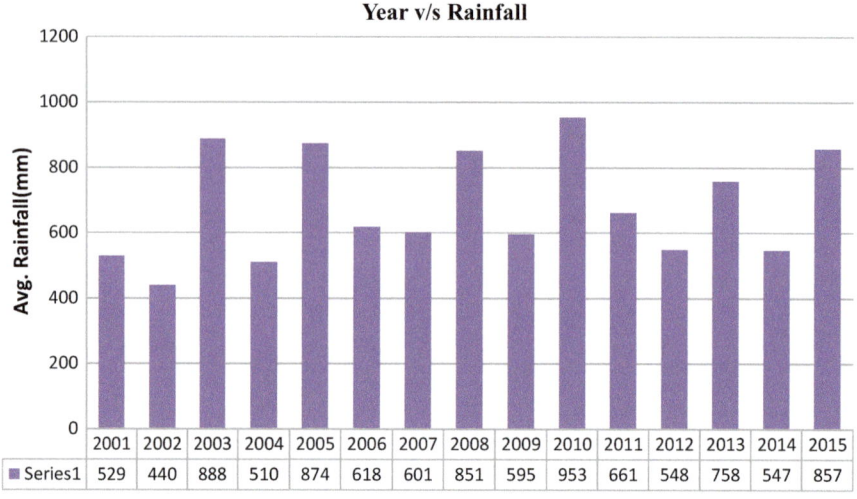

Fig. 2 Rainfall chart

precipitation is inadequate and variable in nature (Fig. 2). Water level fluctuation happens because of the measure of precipitation received by the territory. By and large, water level rises in the course of post-monsoon period. Besides numerous different components, viz. drainage from channel, base succession of rivers, evaporation losses and so forth and furthermore control the outflow and inflow of groundwater. It demonstrates that such regions have moderate to low revive over the groundwater draft amid the period.

3   Model Conceptualization and Development

A model is said to be conceptual model when it is utilized to comprehend the field issue and connected the related field information with the goal that the arrangement can be broke down more anxiously [10]. Conceptualization is a procedure of to amalgamation and encircling up of information applicable to hydrology, geography, hydrogeology and meteorology.

3.1   Groundwater Level Data

To monitoring water level of the study area, 26 tube wells were selected. In December 2013, the water level monitoring system was started and expected to screen every year four times. During the month of May and November, water level data of the pre-monsoon and post-monsoon season were collected. In this period, the whole zone partitioned into various zones based on profundity to water ranges. It was discovered that an enormous zone has shallow to direct profundity to water conditions. In heretic aquifer, the scope of water level was discovered 3.75–15.14 mbgl along with pre-monsoon period while these extents shift from 2.30 to 12.87 mbgl during the post-monsoon season. It was also found that the water level of non command areas was greater than 10 mbgl.

3.2 Aquifer Geometry

The units for the conceptual model with information on hydrogeological properties including cross sections, well logs and geological maps were combined together. Four different types of aquifer groups were observed in the area on the basis of lithological and electrical logs. The range of these groups lies at a depth of 82–132, 95–213, 238–375 and 375–450 m below ground level mbgl [11]. First group of aquifers was restricted in this present study. The alluvial dregs are commonly made out of a fast variation of sand and clay layers. A granular zone broadens the top clay bed downwards up to a most extreme depth of 132 mbgl. The granular zone is divided into two to three subgroups due to the presence of clay beds. Neighbourhood clay focal points are discovered throughout the territory. The water table meets on the top sandy layer with a range from 10 to 80 mbgl.

3.3 Aquifer Parameters

Specific yield and hydraulic conductivity aquifer parameters were anticipated and allocated to various layers utilizing information significant from past investigations. The hydraulic conductivity esteems were chosen to run the model between 0.83 and 5.31 m/day (Avg. 2.63 m/day). The estimation of specific yield for the whole territory was taken 0.18. Hydraulic conductivity esteems were accomplished from twenty-six tube well tests and were relegated to seven definite zones utilizing the Thiessen Polygon technique. The estimation of conductivity for the first and third layers continued as before as 4 m/day for both the layers are comparable in nature. The estimation of conductivity for second aquitard layer was taken 1 m/day while comparative conductivity esteems was taken for clay layers which were intermittent.

3.4 Recharge

The Groundwater Estimation Committee (1997) [12] stated that the groundwater will be recharge through many sources such as rainfall, canal seepage and irrigation return water. The following formula was used to recharge the groundwater by rainfall,

$$R_{rf} = h * S_y * A + D_G - R_C - R_{SW} - R_{gw} - R_{wc},$$

where R_{rf} is the gross recharge due to precipitation and different sources including reused water, h is the ascent in groundwater level, A is the area of recharge consideration, S_y is specific yield, D_G is gross groundwater draft, R_C is recharge due to seepage from canal, R_{SW} is recharge from surface water irrigation, R_{gw} is recharge from groundwater irrigation and R_{wc} is recharge from water conservation structure.

Surface water system and groundwater water system were determined independently to recharge the groundwater and it was acquired from the accompanying equation,

$$R_F = R_C * Q$$

The drainage factor was used as $0.15 < R_C < 0.45$ (GEC 1997).

Recharge through canal seepage was evaluated with the following expression.

Canal seepage = length ×wetted perimeter ×total running days ×specific loss.

Recharge and discharge parameters were evaluated for monsoon and non-monsoon periods. The whole recharge assessed was according to water table variance methodology [13] and Groundwater Estimation Committee (GEC 1997) logic for groundwater resource estimation. Recharge through water system framework returns and drainage through unlined channels was assessed using models endorsed by GEC-97. During model calibration, site unequivocal recharge data are used essentially as fitting parameters [14, 15]. These assumptions are palatable for the whole deal spread of common groundwater flow system and were used in the course of present research [16].

3.5 Groundwater Draft Through Pumping

During the period of December 2013–November 2018, several fields visited to the study area and collected the existing tube well data. For a similar reason, registration from the Statistical Department was additionally utilized. It is found that the discharge rate of tube well changes from 1500 to 3000 L/min.

3.6 Boundary Conditions

Boundary conditions are required for each model to represent the framework's association with the surrounding area. Northern, southern and eastern parts of study area were assigned general head boundary conditions (Fig. 3). Previous water level data of heads were assigned as general head boundary. The western part representing the Hindon River having the river head and bed bottom elevations at the

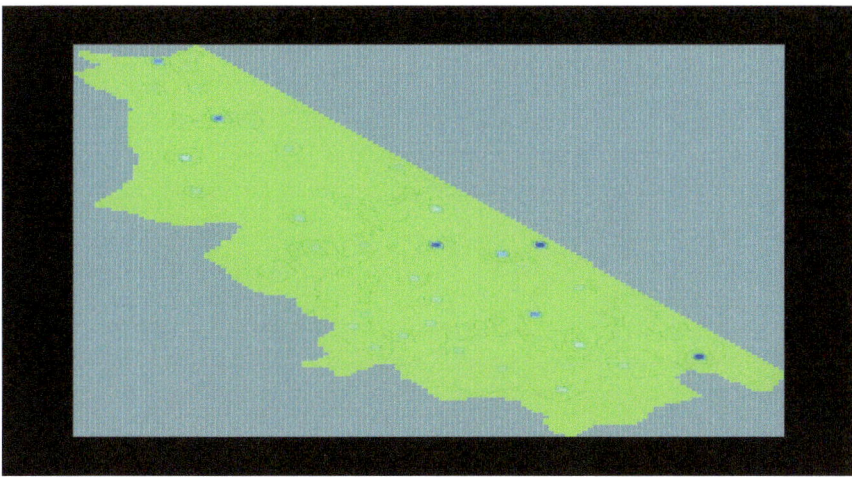

Fig. 3 Heads

initial and end points are 199 m and 198 m amsl and 194 and 193 m amsl, respectively. It has river bed conductance between 200 and 260 m^2/day. The purpose behind using this boundary condition is to keep up a key separation from senselessly extending the model region outward to meet the segment affecting the head in the model [17].

4 Model Calibration

Model calibration includes changing estimations of model info parameters attempting to organize field conditions inside a satisfactory establishment. Calibration is finished by experimentation modification of parameters. After various preliminary runs, computed water levels, heads and drawdown were coordinated decently sensibly with observed values by changing hydraulic conductivity, specific yield and revive values (Figs. 4 and 5).

4.1 Steady-State Calibration

A condition that is existed in the aquifer is said to be steady-state condition when any headway had occurred. Steady-state calibration coordinates the initial heads observed for the aquifer with the hydraulic heads reproduced by MODFLOW [18]. The calibration was mentioned utilizing 26 tube wells monitored during December 2013.

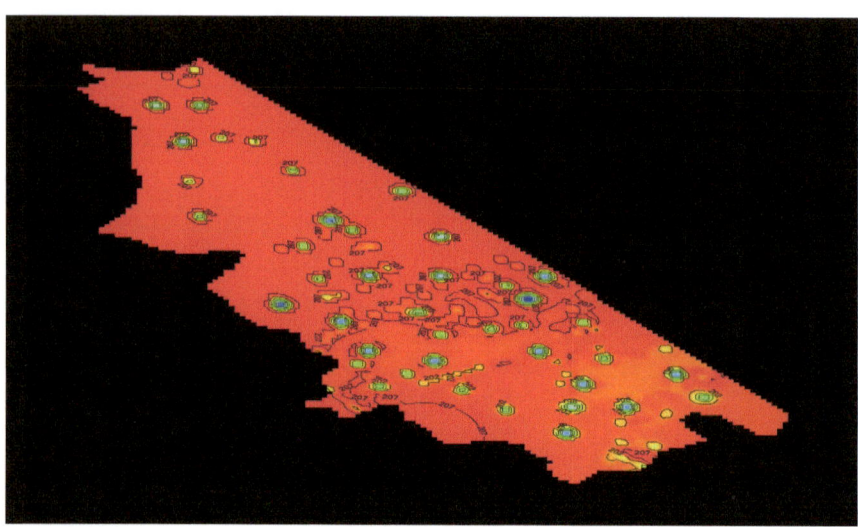

Fig. 4 Water table

Pumping test was used as initial values of hydraulic conductivities for steady-state simulation. The conductivity values were increased by trial and error calibration during various sequential runs till the match between the simulated and observed water level contours were obtained.

5 Results and Discussion

From the mass balance chart of MODFLOW, it was observed that the inflow is greater than outflow before the land use master plan, whereas in the case of after land use master plan, outflow is greater than inflow. Therefore, to overcome this deficit, it is suggested to use rainwater harvesting to recharge the groundwater (Fig. 6a, b). Spontaneous urbanization has definitely changed the seepage attributes of normal catchments by expanding the volume and rate of surface overflow. Over utilization of groundwater sources like wells for drinking water and industrial use has additionally brought about exhausted water levels and drying up of bore wells due to the shortcoming of inflow and outflow for sub-surface water. The first porous ground surface has diminished due to urbanization. Because of the open space transformation into the roads, pavements and development of tempest water channels to deplete the downpour water as fast as conceivable to regular stream, waterway to abstain from flooding of grounds. These surfaces and fast kept running off give no degree for permeation of downpour water to renew the sub-surface

Fig. 5 Drawdown

aquifer causing the dropping of water levels or evaporating of wells. Also, land use and land cover changes have added to the provincial and worldwide atmosphere changes, bringing about sporadic, decreased, inconsistent and questionable rainfalls. Protection and reasonable administration depict helps in the reclamation of the regular affinity. In these circumstances, rainwater harvesting is viewed as a reasonable choice to recharge the groundwater.

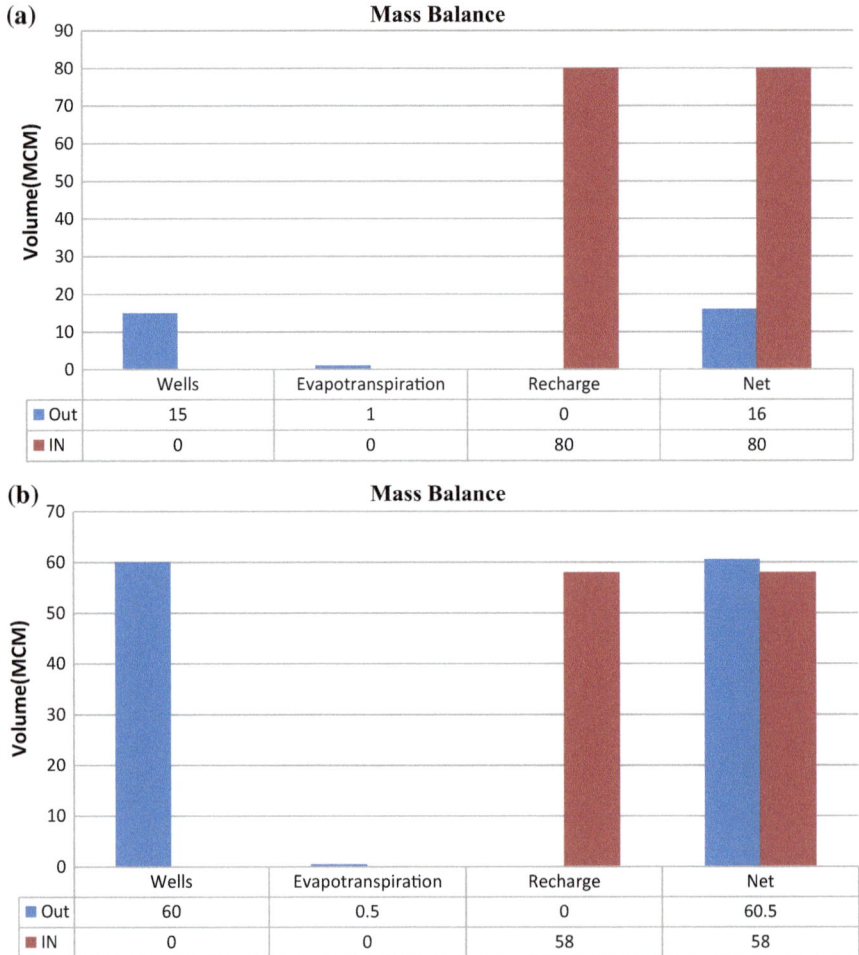

Fig. 6 **a** Mass balance before land use master plan, **b** mass balance after land use master plan

6 Conclusions

Model-based water management has developed throughout the years to improve the quality, quantity and operation costs of the worldwide water supply through complete demonstrating applications. A groundwater model is a modified depiction of a continuously unpredictable reality. They have ended up being useful instruments more than many years for keeping an eye on an extent of groundwater issues and supporting the fundamental authority process. In the present study, a MODFLOW model is made to survey water spending plan of a bit of Greater Noida with the acknowledged boundary conditions and field recognitions. The field inspecting is joined to check model desires. The model calibration has been performed subject to the available data. The model results exhibit that the figured characteristics are in extraordinary well-being of the measure information, which demonstrate the model is steadfast. Incorporated use of surface water and groundwater resources is the perfect technique for getting most extraordinary likely water progression. The familiarity with agriculturists about the groundwater revive can demonstrate indispensable in dealing with the groundwater assets. There is an earnest need to uphold some control on groundwater pumpage to stay away from exploitation of groundwater asset. Modelling is an erratic exercise, parcel of dialogue with specialists, and interview is required for the development of model result. The expanding accessibility of water resources provided by information and communication technologies for the monitoring and conservation of the water assets of the planet has led to the development of smart water management. The SWM approach advances the practical utilization of water assets through coordinated water executives, by incorporating ICT items, arrangements and frameworks went for boosting the financial welfare of a general public without trading off nature. SWM can be connected to numerous areas and urban situations.

References

1. Tamma Rao G G, Gurunadha Rao VVS, Mahesh SL, Padalu G (2011) Application of numerical modeling for groundwater flow and contaminant transport analysis in the basaltic terrain, Bagalkot, India. Saudi Soc Geosci. https://doi.org/10.1007/s12517-011-0461-x
2. Ahmed I, Umar R (2008) "Hydrogeological framework and water balance studies in parts of Krishni-Yamuna interstream area", Western Uttar Pradesh, India. J Environ Geol 53(8):1723–1730
3. Khan AM (1992) "Report on systematic hydrogeological surveys in parts of Muzaffarnagar District", U.P., Central Groundwater Board, Northern Region, Lucknow, India
4. Kumar S (1994) Report on hydrogeological and groundwater resources potential. Muzaffarnagar Districts, U.P., Central Ground Water Board, Northern Region, Lucknow
5. Gupta CP, Thangarajan M, Rao VVSG (1979) Electrical analog model study of a aquifer in Krishni-Hindon interstream region, Uttar Pradesh, India. Groundwater 17(3):284–292
6. Gupta CP, Ahmad S, Rao VVSG (1985) Conjunctive utilization of surface water and groundwater to arrest the water level decline in an alluvial aquifer. J Hydrol 76(3/4):351–361
7. Ala Eldin MEH, Ahmed MS, Gurunadha Rao VVS, Dhar RL (2000) Aquifer modeling of the Ganga-Mahawa sub-basin, a part of the Central Ganga Plain, Uttar Pradesh, India. Hydrol Process 14:297–315
8. McDonald MG, Harbaugh AW (1988) A modular three dimensional finite-difference groundwater flow model. USGS Open File Report 83-875. USGS, Washington D.C
9. https://en.wikipedia.org/wiki/Greater_Noida
10. Anderson MP, Woessner WW (2002) Applied groundwater modeling-Simulation of flow and advective transport. Academic Press Inc., San Diego
11. Bhatnagar NC, Agashe RM, Mishra AK (1982) Subsurface mapping of aquifer system water balance study of Upper Yamuna Basin. Section-hydrogeology, technical report No. 2, Upper Yamuna Project, CGWB, NW region, Chandigarh
12. Ministry of Water Resources (1997) Report of the groundwater resource estimation committee-groundwater resource estimation methodology. Government of India, New Delhi

13. Hearly RW, Cook PG (2002) Using groundwater levels to estimate recharge. J Hydrogeol 10 (1):91–109
14. Vami MR, Usunoff EJ (1999) "Simulation of regional scale groundwater flow in the Azul River Basin", Buenos Aires Province, Argentina. J Hydrogeol 7(2):180–187
15. Kennett-Smith A, Narayan K, Walker G (1996) Calibration of a groundwater model for the upper south east of South Australia. Div. report 96-2. CSIRO Division of Water Resources, Canberra, Australia
16. Jyrkama MI, Skyes JF, Nomani SD (2002) Recharge estimation of transient groundwater modeling. Groundwater 40(6):638–648
17. Waterloo Hydrogeologic Inc. (2005) Visual MODFLOW v.4.1, user's manual. Waterloo, Ontario, Canada
18. Abdullah F, Al-Assa'd T (2006) "Modelling of groundwater flow for Mujib Aquifer", Jordan. J Earth Syst Sci 115(3):289–297

Optimizing Fluoride Removal and Energy Consumption in a Batch Reactor Using Electrocoagulation: A Smart Treatment Technology

Saif Ullah Khan, Mohammad Asif, Faizan Alam, Nadeem Ahmad Khan and Izharul Haq Farooqi

Abstract Electrochemical-based approaches have gained much attention as sustainable, eco-friendly and cleaner methods of treatment technologies as they are less sludge producing. The presence of excess fluoride in drinking water supplies is responsible for dental, skeleton and other forms of fluorosis. Among various defluoridation techniques available, electrocoagulation (EC) process was experimentally applied and optimized aiming higher removal efficiency along with minimum energy consumption. Electrocoagulation process was employed at batch scale using both aluminium and iron electrodes, and a comparative assessment was carried out. The effects of initial pH (4–10), applied current (0.2–1.0 A), initial F^- concentration (5–20 ppm) and reaction time (5–30 min) were explored. The EC reactor was optimized for initial F^- concentration of 20 ppm, applied current of 0.5 A, pH 6 and reaction time of 20 min using aluminium electrodes giving 97.6% removal efficiency and energy consumption of 0.0195 W hour per gm of fluoride. Operational cost was also analysed, and it was found that among the two, aluminium electrodes outclassed iron electrodes in terms of higher removal efficiency proving cost effective as well.

Keywords Fluorosis · Electrocoagulation · Optimization · Removal, energy

1 Introduction

Smarter and sustainable solutions for environmental protection have paved the way for extensive research in the field of water purification technologies. A Clean and reasonable supply of water to meet the needs of growing population has become a

S. U. Khan (✉) · M. Asif · F. Alam · I. H. Farooqi
Department of Civil Engineering, Zakir Husain College of Engineering and Technology, Aligarh Muslim University, Aligarh, India

N. A. Khan
Department of Civil Engineering, Jamia Millia Islamia, New Delhi, India

© Springer Nature Singapore Pte Ltd. 2020
S. Ahmed et al. (eds.), *Smart Cities—Opportunities and Challenges*,
Lecture Notes in Civil Engineering 58,
https://doi.org/10.1007/978-981-15-2545-2_62

great challenge of this century [1]. Among several pollutants, inorganic components are the most concerning as they are responsible for serious health problems. The existence of fluoride, as an inorganic ion, beyond permissible limits in drinking water is a major health concern. Significant groundwater contamination due to fluoride is reported in various countries including China, Southern Africa, Iran, Iraq, Uganda, Sudan, Kenya, India, Ethiopia and Argentina [2–5]. The excessive fluoride content in groundwater occurs naturally due to geological or physico-chemical characteristics of the aquifer or may be attributed to industrial operations in glass manufacturing and semiconductor fabrication. Various regulatory bodies have issued warnings due to adverse health effects such as skeletal fluorosis diseases, other fluorosis deformities, neurological damage, thyroid disorder, mottling of teeth, all of which are caused by excessive and prolonged intake of fluoride [6–8], that too in drinking water. Thus, the World Health Organization has lowered the permissible fluoride value in drinking water to 1.5 mg/l [9]. Subsequently, fluoride removal methods from groundwater or containing it within drinkable limits have engrossed attention all over the world.

Fluoride removal from drinking water has been achieved through several technologies like precipitation [7], ion exchange [10], adsorption [11], electrochemical technologies [12], etc. Each of these approaches has some inadequacies like high costs or forming of secondary waste product due to chemical additions. Thus, it is important to develop such a method of treatment which is efficient as well as economical. Among these adopted technologies, adsorption has been used widely applied. However, studies carried out without optimization have several drawbacks: it produces a lot of sludge and is also poorly selective at high concentrations, proves costly.

Recently, electrochemical treatment techniques have attracted researchers as ecological and environmentally friendly processes. Amid these electrochemical methods, electrocoagulation (EC) is a widely accepted water and wastewater treatment process. Small quantity of chemicals is needed, reducing the volume of sludge generation [13]. Unlike other processes, it does not necessitate large capacities and sophisticated amenities [14]. Extensive investigations on electrocoagulation for various types of wastewater have been already carried out in past [15–19]. Fluoride removal using EC is a quite complex method. A number of electrochemical and physiochemical reactions undergo concurrently. These include oxidation of metal at anode. The use of aluminium and iron electrodes has revealed better results in removing fluoride than their counterparts.

Oxidation: (at anode)

$$Al \rightarrow Al3^+ + 3e^- \tag{1}$$

$$Fe \rightarrow Fe2^+ + 2e^- \tag{2}$$

$$Fe \rightarrow Fe3^{+} 3e^{-} \tag{3}$$

$$2\,H_2O_{(1)} \rightarrow 4H^+_{(aq.)} + O2_{(g)} + 4e^- \tag{4}$$

Reduction: (at cathode)

$$3/2\,H_2O_{(1)} + 2e^- \rightarrow 3/2H_{2(g)} + 3/2OH^- \tag{5}$$

Hydrolysis reaction:

$$Fe^{3+} + 3OH^- \rightarrow Fe(OH)_3 \tag{6}$$

$$Al^{3+} + 3OH^- \rightarrow Al(OH)_3 \tag{7}$$

In the EC cell, dissolution of the anode produces different metal hydroxide complexes of Al^{3+} and Fe^{3+} in the form $Al(OH)^{2+}$, $Al_2(OH)_2^{4+}$, $Al(OH)_3$, $Fe(OH)_3$, $Fe(OH)_4$, $Fe(H_2O)_3(OH)_3$, $Fe(H_2O)_6^{3+}$, $Fe(H_2O)_5(OH)^{2+}$, $Fe(H_2O)_4(OH)^{2+}$, etc., along with charged hydroxo-cationic complexes. The proposed removal mechanism may be:

$$nAl^{3+}\,(aq.) + (3n-m)OH^-(aq.) + mF^-(aq.) \rightarrow Al_nF_m(OH)_{3n-m} \tag{8}$$

$$Al_nF_m(OH)_{3n-m} + OH^-(aq.) \rightarrow Al_aF_{m-1}(OH)_{3n-m+1} + F^-(aq.) \tag{9}$$

$$Fe(OH)_3 + 3F^- \rightarrow FeF3 + 3OH^- \tag{10}$$

Exchange of F^- ions occurs easily with hydroxide ions in the precipitate due to strong affinity between the two; F^- ions gets adsorbed by exchanging the OH^- on $Al_n(OH)_{3n}$ and $Fe_n(OH)_{3n}$ flocs [20]. Similarly, fluoride interacts with ferric oxy-hydroxides as shown in Eq. (8). Moreover, $Fe_n/Al_n)F_m(OH)_{3n-m}$ is formed by co-precipitates of F^- hydroxide and Al/Fe ions [21]. However, very few studies on fluoride removal mechanism by iron compounds are available in the literature. However, one pathway reported involving fluoride interaction with ferric oxyhy-droxides (Eq. 10) [22]. Other reactions such as adsorption onto aluminium/ferric hydroxides or substitution of Fe (III) in ferrous hydroxide also undergo. Besides, oxygen is formed at the anode (Eq. 4) while hydrogen is formed at cathode (Eq. 5) which assists in floc formation. These bubbles attach themselves with the flocs drifting these to the top surface.

This study aims to maximize removal efficiency and minimize energy consumption. Laboratory-scale EC experiments were performed along with varying process parameters such as distance between electrodes, applied current, reaction time, initial F^- concentration and initial pH. Comparison of Al and Fe electrodes has been carried out as well. Several studies optimizing the removal efficiency had been published in the past, but limited works are available in the literature in which energy consumption has been optimized as well.

2 Materials and Techniques

Standard solutions of fluoride were prepared using sodium fluoride (Merck, AR grade). Conductivity was maintained using NaCl to 1000 μS/cm with 5% accuracy. Spectrophotometer (DR 5000 UVVis) for analysing fluoride by SPADNS method was used. pH conductivity was measured by appropriate probes (Hach). Current was supplied using a DC power supply (Scientech 4079).

2.1 Experimental Procedure

Electrocoagulation set-up was comprised of a square reactor 15×15 cm dimension made by plexiglas of 2-litre capacity. Rectangular perforated plates of 2 mm thickness were provided at centre as anode and cathode made up of Al and Fe symmetrically with a distance of 2 cm between them. The surface area of anode and cathode was 100 cm^2. The electrodes were perforated aiming sufficient mixing of dispersed metal coagulant. A diagram presenting electrocoagulation mechanism and laboratory set-up is drawn schematically in Fig. 1a, b. A port was made at middle side wall of reactor for collecting samples post treatment at fixed time after which it was filtered and sent for analysis. The experimental runs were performed in triplicate, and the mean values were reported as results with deviations less than 5%. After every experimental run, rubbing of electrodes using sandpaper and rinsing with Millipore water were done appropriately.

Efficiency of the EC process was computed in terms of % fluoride removal by the formula:

$$\% \text{ Fluoride removal efficiency} = \frac{C_i - C_f}{C_i} \times 100 \tag{11}$$

where C_i is the initial fluoride concentrations while C_f denotes the final fluoride concentrations (in mg L^{-1}), respectively.

3 Results

3.1 Effect of Initial pH

pH is a deterministic factor in defluoridation process as it influences the speciation of dissipating electrodes [7]. Variable pH in the range 4–10 was investigated for its influence on fluoride removal. pH variation is quite sensitive for the formation of Al (OH)$_3$ and Fe(OH)$_3$ flocs. Control over pH of the solution remains a challenge in electrocoagulation process due to defluoridation reaction which alters the pH.

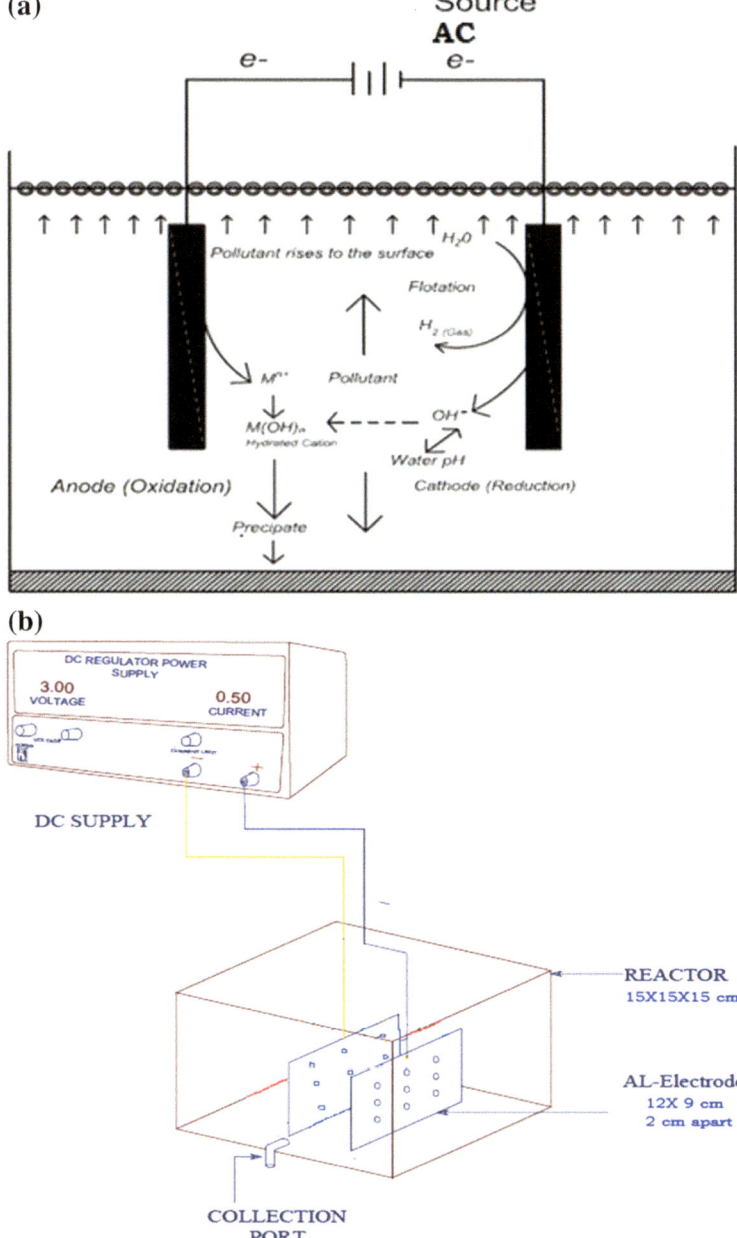

Fig. 1 Representation of **a** electrocoagulation mechanism, **b** laboratory set-up

Fig. 2 Effect of varying pH for both electrode materials on fluoride removal efficiency at optimized conditions of 0.5 A current, initial F conc. of 20 ppm and pH 6

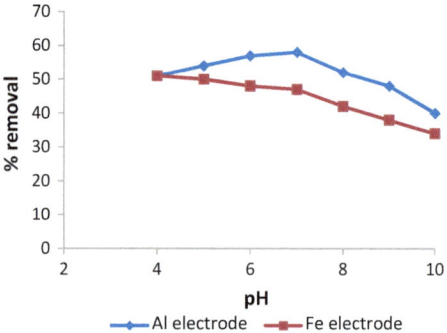

Due to the buffer capacity of aluminium hydroxide, no major shift occurs in the final pH with respect to the initial pH, that is, 4–8 in the electrocoagulation process. The change in pH values was measured during EC treatment and it was observed that on varying the initial pH from 4 to 10, there was a little effect on removal efficiency, which was found maximum in the acidic range. This may be understood by the fact that in exchange of F^- with the OH^- in $Al(OH)_3$, the concentration of OH^- decreases, and equilibrium shifts accordingly. In this way, low pH value assists fluoride removal which is also in accordance with the past studies [12].

Effect of varying initial pH on fluoride removal efficiency is shown in Fig. 2. The aluminium electrodes were found to show better results in terms of removal efficiency. The pH in acidic range favours fluoride removal using Fe electrodes, while in case of aluminium electrodes, the removal efficiency remains almost same, although higher as compared to using iron electrodes.

3.2 Effect on Initial Concentration

Experiments were performed by varying the initial F^- concentration in the range 5 to 20 mg/L at varying current intensity from 0.2 to 0.1 A, optimum pH of 6. In both types of electrodes, the effects of initial concentration on defluoridation are shown in Fig. 3. It can be seen that at the same current intensity (0.5 A) by varying the initial fluoride concentration from 5 to 20 ppm, the removal efficiency decreased from 76 to 54% in case of aluminium electrodes, while it reduced from 67 to 32% in case of iron electrodes. It can be inferred from the plot in Fig. 3 that higher removal efficiency was achieved within short time durations for low initial F^- concentrations. Moreover, for longer durations of time and high initial F^- concentration, the residual F^- concentration reduces to within permissible limits even at low current intensity of 0.2 A, but this limit was reached in a much shorter duration only at higher current intensity of 1 A, which may be attributed due to insufficient aluminium hydroxide complexes formed in the initial reaction time at lower current intensity.

Fig. 3 Effect of varying initial F conc. for both electrode materials on fluoride removal efficiency at optimized conditions of 0.5 and 0.1 A current, initial F^- concentration of 20 ppm, reaction time 10 min and pH 6

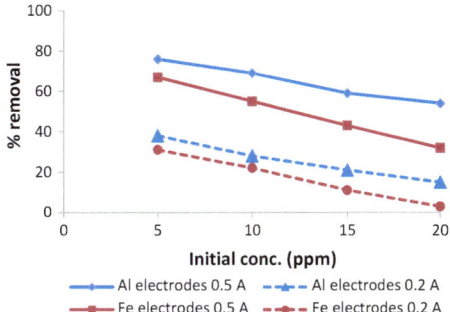

Fig. 4 Effect of varying applied current for both electrode materials on fluoride removal efficiency at initial F concentration of 10 and 20 ppm and pH 6

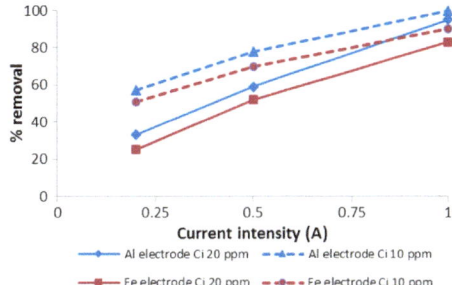

3.3 Effect of Applied Current

The effect of applied current in EC on fluoride removal efficiency was investigated by varying it from 0.2 to 1 A at varying initial concentration of 5, 10, 15 and 20 ppm, keeping optimum pH of 6. In both types of electrodes, the effects of applied current were found significant which are shown in Fig. 4. It can be seen that at the max current intensity (1 A) and initial fluoride concentration of 20 ppm, 100% removal was achieved in just 12 min using aluminium electrodes, while it took around 21 min for the same in case of iron electrodes. It is clear that at higher currents, the coagulant dissipated is sufficient to form the precipitates even in small time durations. In addition, it was observed that for higher current intensity, the bubbles' density increased while their size decreased, ensuring greater upwards flux and rapid removal of pollutant [23, 24]. Furthermore, for longer durations of time and at low current intensities, the residual fluoride concentration reduces within permissible limits, with less sludge generation.

3.4 Effect of Reaction Time

Duration of electrolysis and current applied affects the mass of metal (here aluminium and metal) dispersed into system based on Faraday's law. Thus, these two

Fig. 5 Effect of varying
reaction time for both
electrode materials on fluoride
removal efficiency at initial F
concentration of 10 ppm,
applied current 0.2 A and
pH 6

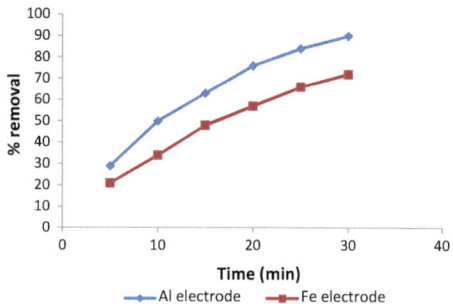

factors in combinations will lead to an increase in metal released inside the reactor. The experiments were carried out accordingly by varying the residence time from 5 to 30 min employing both Al and Fe electrodes one by one to explore its effect on the removal efficiency. The experimental results as shown in Fig. 5 revealed that at higher currents, more aluminium and iron were dissipated to be available per unit time, due to which fluoride removal was spontaneous, especially in case of aluminium. However, to achieve spontaneous removal of fluoride, energy consumption was found quite high.

3.5 Effect of Electrode Distance

In EC, spacing between electrodes plays a vital role for the removal of pollutant from effluent. The electrostatic field depends on the distance between the electrodes. The experiments were conducted by varying the distance between electrodes from 2 to 6 cm as shown in Fig. 6. The maximum pollutant removal efficiency in case of iron electrode was obtained by maintaining a minimum optimum distance of 2 cm between the electrodes, which decreased on increasing this distance. However, in case of aluminium electrodes, the efficiency remains almost same when distance between electrodes was increased to 4 cm; however, it was greatly reduced on further increasing the distance. This may be attributed to the fact that on further

Fig. 6 Effect of varying
electrode distance for both
electrode materials on fluoride
removal efficiency at initial F
concentration of 10 and
20 ppm and pH 6

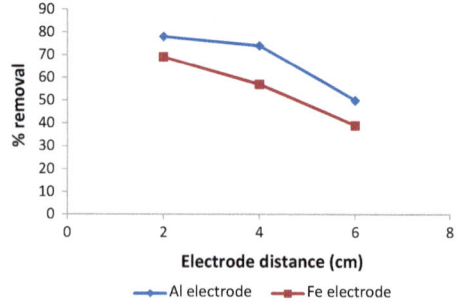

increasing the distance between electrodes more than optimum, the travel time of the ions increases causing less formation of flocs required to coagulate the pollutant [25] resulting in reduced removal efficiency.

3.6 Energy and Electrode Dissociation

Energy consumption in EC may be written as watt-hour per gram removal using the formulae:

$$\text{Energy consumed per gram removal}$$
$$= \frac{EIt \times 60}{(C_0 - C) \times V} \ (\text{J g}^{-1}) \tag{12}$$

$$\text{Energy consumed per gram removal}$$
$$= \frac{EIt \times 60}{(C_0 - C) \times V} \times 2.778 \times 10^{-4} \ (\text{W h g}^{-1}) \tag{13}$$

where "E" is voltage in volts, "I" is current in ampere and "t" is time in minutes. V is the quantity of treatable water in litres. C_0 is the initial concentration while C is the final concentration, respectively.

Electrode dissociation is computed using the Faraday's Law.

$$m = \frac{I.M.T}{Z \times F} \tag{14}$$

where m is mass of material dispersed at an electrode in gram, I is change in coulomb, M is molar mass of metal in grams per mole, T is time, F is faraday's constant equal to 96,500 C mol^{-1} and Z is valency.

The energy consumption with respect to the applied current and fluoride concentration employing both Al and Fe electrodes has been shown in Table 1. From the results, it may be inferred that at low concentration and higher values of current, complete removal of fluoride is attained within the few minutes; the energy supplied

Table 1 Consumption of electrical energy during EC using aluminium and iron electrodes

Applied current (A)	Electrical energy consumption (W h gm^{-1}) aluminium electrodes. Iron electrodes			
	Initial F concentration		Initial F concentration	
	(10.0 mg/l)	(20.0 mg/l)	(10.0 mg/l)	(20.0 mg/l)
0.2	0.0017	0.0026	0.0019	0.0029
0.5	0.0149	0.0195	0.0168	0.0214
1.0	0.0321	0.0380	0.0352	0.0456

thereafter is unusable. Energy disbursed per gram F^- removal increased, near about linearly, on increasing the current applied. An easy justification of this is that for smaller values of current, energy consumed was low, thereby keeping the removal efficiency low as well. Energy consumption increased up till the completion of removal process for higher applied current values, resulting a total increase in the energy consumption. This drift remains only till the removal was incomplete, after that the higher current values assisted in attaining greater removal efficiencies.

4 Cost Estimation

The foremost factors involved in estimating cost incurred in EC is due to electrode and energy consumption while minor factors including disposal of sludge, maintenance, etc., are overlooked in general. So, operating cost,

$$C_{total} = C_{electrode} + C_{enery}.$$

where $C_{electrode}$ and C_{enegy} are the amount incurred on electrode and energy consumption per gram removal of pollutant, respectively.

The hypothetical electrode mass dissipated for the optimized environment was computed using Faraday's law (Eq. 14), which was found to be 0.1016 grams per run. Cost incurred on electricity consumption per run was computed using (Eq. 12). Considering the present market price of aluminium/iron and local residential electricity rates, the typical operational cost per gram removal was found to be 0.956 and [0.102(for Al)/0.096(for Fe)] Indian rupee for electricity and electrode material, respectively.

5 Conclusion

Electrocoagulation process was successfully employed for achieving complete defluoridation. The removal mechanism was explored using both aluminium and iron electrodes. The role of initial pH was also assessed along with coagulation chemistry and formation of different metal complexes. The process was optimized for maximum removal efficiency with minimum energy consumption. The applied current and distance between electrodes were the most significant factors affecting the efficiency of EC process. The reaction time and applied current were the main factors governed as per Faraday's law which were responsible for the amount of production of metal dissipating from electrodes, for which aluminium-based electrodes proved better in achieving higher removal efficiency, at reduced cost. Moreover, the volume of sludge generated in EC is also quite low as compared to coagulant-based approaches. The results proved EC to be a simpler, cleaner, safer, reliable and smarter technology among several defluoridation techniques available.

References

1. Khan SU, Zaidi R, Hassan SZ, Farooqi IH, Azam A (2016) Application of Fe–Cu binary oxide nanoparticles for the removal of hexavalent chromium from aqueous solution. Water Sci Technol 741(1):165–175

2. Nayak B et al (2009) Health effects of groundwater fluoride contamination. Clin Toxicol (Phila) 47(4):292–295

3. World Health Organization (2006) Fluoride in drinking water

4. Khan SU, Noor A, Farooqi IH (2015) GIS application for groundwater management and quality mapping in rural areas of District Agra. India. Int J Water Res Arid Env 4(1):89–96

5. Kumar E, Bhatnagar A, Ji M, Jung W, Lee SH, Kim SJ, Lee G, Song H, Choi JY, Yang JS, Jeon BH (2009) Defluoridation from aqueous solutions by granular ferric hydroxide (GFH). Water Res 43:490–498

6. Hu CY, Lo SL, Kuan WH, Lee YD (2005) Removal of fluoride from semiconductor wastewater by electrocoagulation–flotation. Water Res 39:895

7. Mameri N, Yeddou AR, Lounici H, Belhocine D, Grib H, Bariou B (1998) Defluoridation of septentrional Sahara water of north Africa by electrocoagulation process using bipolar aluminium electrodes. Water Res 32:1604–1612

8. Viswanathan G, Jaswanth A, Gopalakrishnan S, Sivailango S, Aditya G (2009) Determining the optimal fluoride concentration in drinking water for fluoride endemic regions in South India. Sci Total Environ 407(20):5298

9. Brouwer ID, De Bruin A, Dirks OB, Hautvast JGAJ (1988) Unsuitability of World Health Organisation guidelines for fluoride concentrations in drinking water in Senegal. Lancet 331:223–225

10. Tor A (2007) Removal of fluoride from water using anion-exchange membrane under Donnan dialysis condition. J Hazard Mater 141:814–818

11. Wu X, Zhang Y, Dou X, Yang M (2007) Fluoride removal performance of a novel Fe–Al–Ce trimetal oxide adsorbent. Chemosphere 69:1758–1764

12. Zuo Q, Chen X, Li W, Chen G (2008) Combined electrocoagulation and electroflotation for removal of fluoride from drinking water. J Hazard Mater 159:452–457

13. Adhoum N, Monser L, Bellakhal N, Belgaied JE (2004) Treatment of electroplating wastewater containing Cu^{2+}, Zn^{2+} and Cr(VI) by electrocoagulation. J Hazard Mater 112:207–213

14. Mollah MYA, Morkovsky P, Gomes JAG, Kesmez M, Parga J, Cocke DL (2004) Fundamentals, present and future perspectives of electrocoagulation. J Hazard Mater 114:199–210

15. Balla W, Essdki AH, Geourich B, Dassaa A, Chenik H, Azzi M (2010) Electrocoagulation/electro-flotation of reactive, disperse and mixture dyes in an external-loop airlift reactor. J Hazard Mater 184:710–716

16. Drouiche N, Ghaffour N, Lounici H, Mameri M (2007) Electrocoagulation of chemical mechanical polishing wastewater. Desalination 214:31–37

17. Feng J, Sun Y, Zheng Z, Zhang J, Li S, Tian Y (2007) Treatment of tannery wastewater by electrocoagulation. J Environ Sci 19:1409–1415

18. Khan SU, Islam DT, Farooqi IH, Ayub S, Basheer F (2019) Hexavalent chromium removal in an electrocoagulation column reactor: process optimisation using CCD, adsorption kinetics and pH modulated sludge formation. Proc Saf Env Prot 122:118–130

19. Khansorthong S, Hunsom M (2009) Remediation of wastewater from pulp and paper mill industry by the electrochemical technique. Chem Eng J 151:228–234

20. Shen F, Chen X, Gao P, Chen G (2003) Electrochemical removal of fluoride ions from industrial wastewater. Chem Eng Sci 58:987–993

21. Hu CY, Lo SL, Kuan WH (2005) Effects of the molar ratio of hydroxide and fluoride to Al (III) on fluoride removal by coagulation and electrocoagulation. J Colloid Interface Sci 283:472–476

22. Martínez-Miranda V, García-Sánchez JJ, Solache-Ríos M (2011) Fluoride ions behavior in the presence of corrosion products of iron: effects of other anions. Sep Sci Technol 46:1443–1449

23. Bazrafshan E, Ownagh K, Mahvi AH (2012) Application of electrocoagulation process using Iron and Aluminum electrodes for fluoride removal from aqueous environment. E-J Chem 9(4):2297–2308

24. Malakootian M, Yousefi N (2009) Efficiency of electrocoagulation process using aluminum electrodes in removal of hardness from water. Iran J Environ Health Sci Eng 6(2):131–136

25. Aoudj S, Khelifa A, Drouiche N, Belkada R, Miroud D (2015) Simultaneous removal of chromium (VI) and fluoride by electrocoagulation- electroflotation: application of a hybrid Fe–Al anode. Chem Eng J. 267:153–162

Towards a Unifying Approach to City Sustainability on the Changing Expression of Urbanisation in Bengaluru

Juhi Priyanka Horo and Milap Punia

Abstract On the path of accelerated urbanisation, India contains four of the world's ten largest cities Mumbai, Delhi, Bengaluru, and Kolkata. Bengaluru is prominent among the world's fastest growing cities that have attracted abound of the population, capital, and investments with robust connectivity. As a result, the pressure of the metropolis has swept over its carrying capacity, thereby affecting the resources, infrastructure, ecological diversity and land therein. Various 'cities' related multifaceted categories such as 'sustainable cities', 'smart cities', 'digital cities', 'resilient cities', 'low carbon cities', 'eco cities', and 'liveable cities' are into discourse which are still framing the distinctive character of cities which would have impression on the theoretical understanding and on the policymaking. This raises an issue of sustainability and quality of life for its citizens now and for future as well, under which, Bengaluru has to make key decisions on land use, sprawling peripheries, infrastructure, transport, and energy. This demands for a careful and rational provision of planning keeping in mind the issue of imbalance without forsaking the pre-requisite of the regions. The following research makes an attempt to analyse the urban environment; an emphasis on the associations of distributed aspects of urban growth, their relation and the resultant influences taking the study area of Bengaluru of Karnataka state. The outgrowth of centre towards periphery is gaining importance wherein the spatial as well as temporal changes in urbanisation helps in witnessing the prevalent land use changes provide insight of rate, extent, and direction of sprawl.

Keywords Shannon's entropy · Bengaluru · Urbanisation · Sustainability · Urban expansion

J. P. Horo (✉) · M. Punia
Centre for the Study of Regional Development (CSRD), School of Social Sciences (SSS), Jawaharlal Nehru University (JNU), New Delhi 110067, India

© Springer Nature Singapore Pte Ltd. 2020
S. Ahmed et al. (eds.), *Smart Cities—Opportunities and Challenges*,
Lecture Notes in Civil Engineering 58,
https://doi.org/10.1007/978-981-15-2545-2_63

1   Introduction

Human beings in their process of interactions with the environment and endowments have been persistent in altering the process of recycling of services by deliberate actions as responsive recipients. In the process of interactions, human faces impediments on its way of obtaining desired outcomes, hence struggle its own way to combat the situations, that either be constructive or destructive. Considering urban component as one of the dynamic domains of human interference, the outgrowth from the centre towards periphery is gaining importance and is discovered in rapidly urbanising metropolitan cities. Urbanisation unfolds the challenges of increased mobility and heterogeneity that has implications over the dire need of urban social and territorial sustainability rather than operationalize concrete methods of achieving sustainability [1]. Under the consideration of the complexity of larger cities, they also have the capability to overcome the challenges of cities, but it depends completely on the cities where such capacities need to be developed.

Bengaluru presents a perfect example of rapidly increasing urban landscape at the cost of natural ecosystems. As per Nagendra, Bengaluru had 19,800 lakes in 1830 which provided 50% of water supply in Bengaluru. Bellandur Lake in Bengaluru has been in news flash for the upsurge of foam and froth along with smoke and flames as a result of unusually heavy rain in the city [2, 3]. An estimate of 400–600 million litres of untreated sewage is diverted to the lake every day, thus creating an unhealthy environment and life miserable for the citizens [4–8]. Despite series of intervention on the rejuvenation of the lake, the implementation is still in process [9]. The Respirable Suspended Particulate Matter (RSPM) has exceeded the national permissible limit in Bengaluru [10–13]. The increased concentration of the mentioned chemicals, loss of wetlands and vegetation has altered the natural heat budget of the area and is widely hampering the natural setting of the environment by converting it into urban heat island and causing reasons for future flooding within the urban limits [14]. The increased concentration of the mentioned chemicals, loss of wetlands and vegetation has altered the natural heat budget of the area and is widely hampering the natural setting of the environment by converting it into urban heat island and causing reasons for future flooding within the urban limits [14]. The vehicular stress imposed by the improved connectivity and distressed urbanisation cause problems for surrounding areas, especially the times of high traffic congestion and high speeds [15]. These provision of urban services and thereby the management of resources have gained popularity to study the nexus of the urban systems; what Broto and Sudhira call 'Engineering modernity' [16].

The objectives of the research explored the importance of city sustainability with respect to urban expansion of Bengaluru to become a smart city. The effect of urban transition is such that it is proportional to the scale of population shift [17]. It is also because the cities, especially Indian cities replicate the growth model of the international urbanized economies in order to battle in the pace of world competency. The growth model of Bengaluru needs to explore through the research; the changes in urban dynamics in the spatio-temporal context in association with the spatially disseminated constituents of urban growth. Such development happens to be biased as it leads only the affordable class but it creates social and spatial fragmentation that subjugates the destitute class. As far as the directional change is concerned, the usual trend of sprawl is either radial around the city centre, on the urban fringe, the edge of an urban area or linear extension along highways [14]. The further research focuses on the zonal and directional changes of urban growth in Bengaluru to validate the trend. The urban expansion of Bengaluru is indicative of the fact that much of the urban development has been a concentration in and around. Vaddiraju [18] predicates this pattern of urbanisation of Bengaluru similar to the exclusionary urbanisation evoked by Kundu [19].

2 Study Area

Geographically, Bengaluru is located to the South-east in Karnataka extending from 12°49′5″N to 13°8′32″N and 77°27′29″E to 77°47′2″E. The erstwhile Bangalore Mahanagara Area (BMA) constituted a population of 5.8 million in 2001 to 8.4 million in 2011 now renamed as Bruhat Bengaluru Mahanagara Palike (BBMP) [20, 21]. The areal extent of Bengaluru Mahanagara Palike (BMP) of 226 km² has now increased to 741 km² of BBMP forming the largest municipal corporation in India (Fig. 1). The population density has increased from 7,880 persons per sq. km to 11,330 persons per sq. km during the mentioned decade. In 1961, with a population of 1.2 million, Bengaluru stood the sixth-largest city in India and the fastest city of Mysore State and its administrative capital. Coinciding with the uncontrolled rise in its population, the city underwent a change in the economic structure and, therefore, emerged an industrial centre. According to Census 1961, Bengaluru constituted of two Class II towns, four Class IV towns, five Class V towns and one Class VI town [22]. Bengaluru became one of the 'Million-plus cities' of India in 1961 and was notified as Greater Bangalore in December 2006.

3 Data Used and Methods

The research is based on the quantitative exploration of the urban growth in terms of the spatially disseminated constituents. In order to deal with the perspective, the base maps were used from the District Census Handbook (DCH), BBMP (2001 and 2011), Census of India. The boundaries of BBMP, wards, etc. were retained from the above-mentioned source. Furthermore, the spatial aspects need to quantify the research area to grasp the knowledge about the area. For this, Primary Census Abstract (PCA) of BBMP wards (2001 and 2011) was obtained from Census of India website. The large-scale changes in land use and land cover are efficiently monitored by the remote sensing datasets and are available in profusion. For the purpose of the research on spatial constituents, Landsat-4 and Landsat-8 images

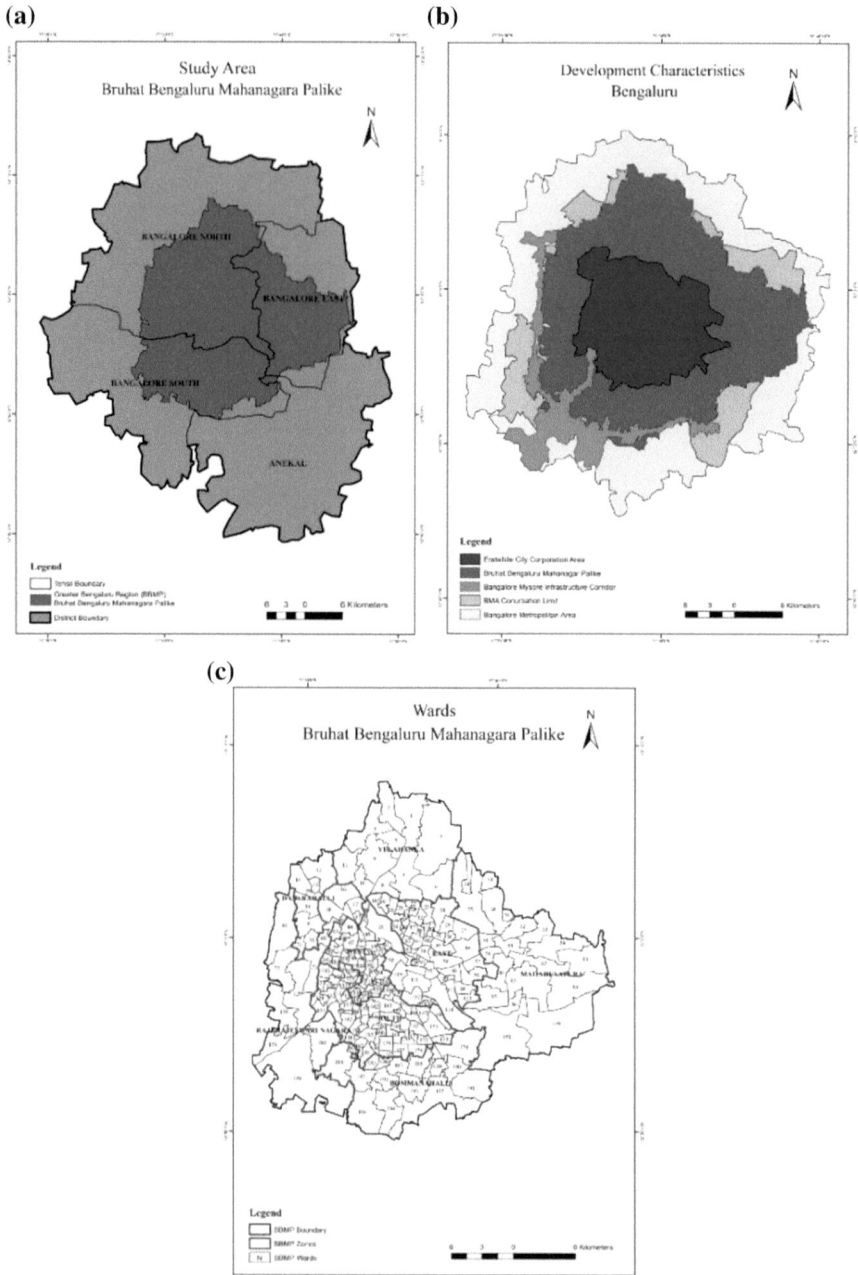

Fig. 1 **a** Study area (BBMP); **b** development characteristics of Bengaluru; **c** BBMP zones and wards. *Source* Administrative Atlas of Karnataka 2011 and BBMP

were taken from United States Geological Survey (USGS) Earth Explorer for the years of 2001 (March 27), 2005 (February 18), 2011 (March 7), and 2005 (January 13). The software used in the processing of large datasets of the research includes Arc Map 10.2.2, Erdas Imagine 2014, Microsoft Excel, and Microsoft Word.

As per the objectives mentioned framed, following methodology has been put forward for the research:

3.1 Shannon's Entropy Index (Form and Type of Urban Growth)

The force of energy; one of the fundamental driving forces of nature which is omnipresent is what is known as entropy which drives all creation and destruction. The concept of entropy is much related to the second law of thermodynamics which infers that entropy of an isolated system never decreases. The isolated system here means that neither the mass nor the energy is bound to follow in or out. If there are 'n' numbers of ways, a system can function then the entropy will increase with an increase in 'n'. According to Tiwary [23], 'Entropy is a way to quantify the overall disorder or chaos of a system'. Quantitatively, it can be better expressed in terms of the research on urban growth in the article.

Shannon's entropy [24] can be computed for different areas to investigate the form and type of urban growth and the sprawl pattern. The formula for Shannon's entropy (H_n) is:

$$H_n = -\sum P_i \log_e(P_i)$$

where

P_i = Proportion of the variable in the ith zone (e.g. proportion of built-up area in each ward)
n = Total number of zones (e.g. total number of wards)

The range of entropy values extends between 0 and log n, wherein, homogeneity of the urban from will posses values towards 0 and values closer to log n reveal that the distribution is much dispersed and of varied types. Larger values of entropy correspond to the delineation of sprawl.

3.2 Shannon's Entropy Values Classification

As already mentioned, the value of Shannon's entropy ranges from zero (0) to infinity, and so there is no upbound limit. In this study, the values of the index for the regions of BBMP ranges between 0.400 and 1.800, so the number of a range is

Table 1 Range and palette for Shannon's entropy values

Range	Palette	Name
<0.400		Tzavorite green
0.401–0.500		Lemongrass
0.501–0.600		Light apple
0.601–0.700		Medium key lime
0.701–0.800		Fern green
0.801–0.900		Yucca yellow
0.901–1.000		Topaz sand
1.001–1.100		Medium yellow
1.101–1.200		Autunite yellow
1.201–1.300		Solar yellow
1.301–1.400		Rose quartz
1.401–1.500		Medium coral light
1.501–1.600		Medium coral
1.601–1.700		Rose dust
1.701–1.800		Tulip pink

prepared accordingly [25]. Arc Map 10.2.2 provides several palettes that makes the visualisation easy and identifiable (Table 1).

3.3 Classification of Land Use and Land Cover Change

The conventional techniques of mapping the land use changes imply heavy expenses and time. However, with improvement in technological assistance, the quantification of impervious surfaces has authenticated the research proficiency along with the spatial data and analysis through the advanced methods of Remote Sensing and Geographical Information System (GIS). Comparisons of land use and land cover over a span of years determine crisp transformations taking over time and space [26, 27]. Considering the above statement, similar applications have been made for study taking into account the satellite images of BBMP for the years of 2001, 2005, 2011, and 2015. The sub-categorical typologies enable a better understanding of urban diversity. These categories are: 'Built-up', 'Agriculture', 'Fallow land', 'Vegetation/Forest', 'Waterbodies', and 'Rocky Highlands'.

4 Spatial and Temporal Assessment of Urban Growth of Bengaluru in (2001–2015)

4.1 Analysis and Discussion

BBMP is subdivided into 8 administrative zones: West, South, Rajarajeshwari Nagar, Mahadevapura, East, Dasarahalli, Yelahanka, and Bommanahalli. East administrative zone comprises of an abound of wards, constituting BBMP with a total of 198 wards.

Since the focus area of the study is BBMP, an investigation from broader to narrow level gives a clear and precise picture of the scenario. The satellite images have been obtained from Landsat-4 and Landsat-8 of USGS Earth Explorer and the study area of BBMP has been extracted from each image. The land use and land cover changes are prominent when looked at the satellite images but classification gives a much clearer image of the scenario across time.

Figure 2 provides a classified image of land use and land cover. The built-up area of BBMP (Table 2; Fig. 3) has increased from 290.81 km^2 (42.11%) in 2001 to 327.89 km^2 (47.48%) in 2005 to 407.38 km^2 (58.97%) in 2011 to 500.13 km^2 (72.43%) in 2015. The agricultural fields initially experienced an increase by 2005 (6.62 km^2 in 2001 and 7.62 km^2 in 2005) but the years of 2011 and 2015 have shown a decrease in them (5.34 km^2 in 2011 and 5.21 km^2 in 2015). After the built-up, fallow land occupies a larger share of land use and land cover. The fallow land possessed 242.45 km^2 in 2001 but with the alternate use of land, it has suffered a reduction in its share as well as in area. It has now reduced to 9.95 km^2 in 2015. The major effect has also been on barren/waste land as it shot up to 2005 but it has also reduced to a considerable extent due to the alternative use of the land. The situation of vegetation/forest cover has been quite fluctuating, being 26.39 km^2 in 2001, peaked to 36.80 km^2 in 2005, reduced in 2011 with 30.01 km^2 and again picked up with 32.24 km^2 in 2015. The patches of waterbodies are decreasing from 9.09 km^2 (2001) to 7.39 km^2 (2015) in its area and share. The fluctuations in rocky highland are because of the forest areas over it. These regions with limited forest have shown a high percentage than when covered with more greenery. According to Indian Institute of Science (IISc), the resultant effects of urbanisation and expansion of Bengaluru between 1973 and 2016 have been such that it has led to 1005% increase in paved surfaces and a loss of its vegetation by 88%, whereas water bodies have declined by 85% between 2000 and 2014 [28].

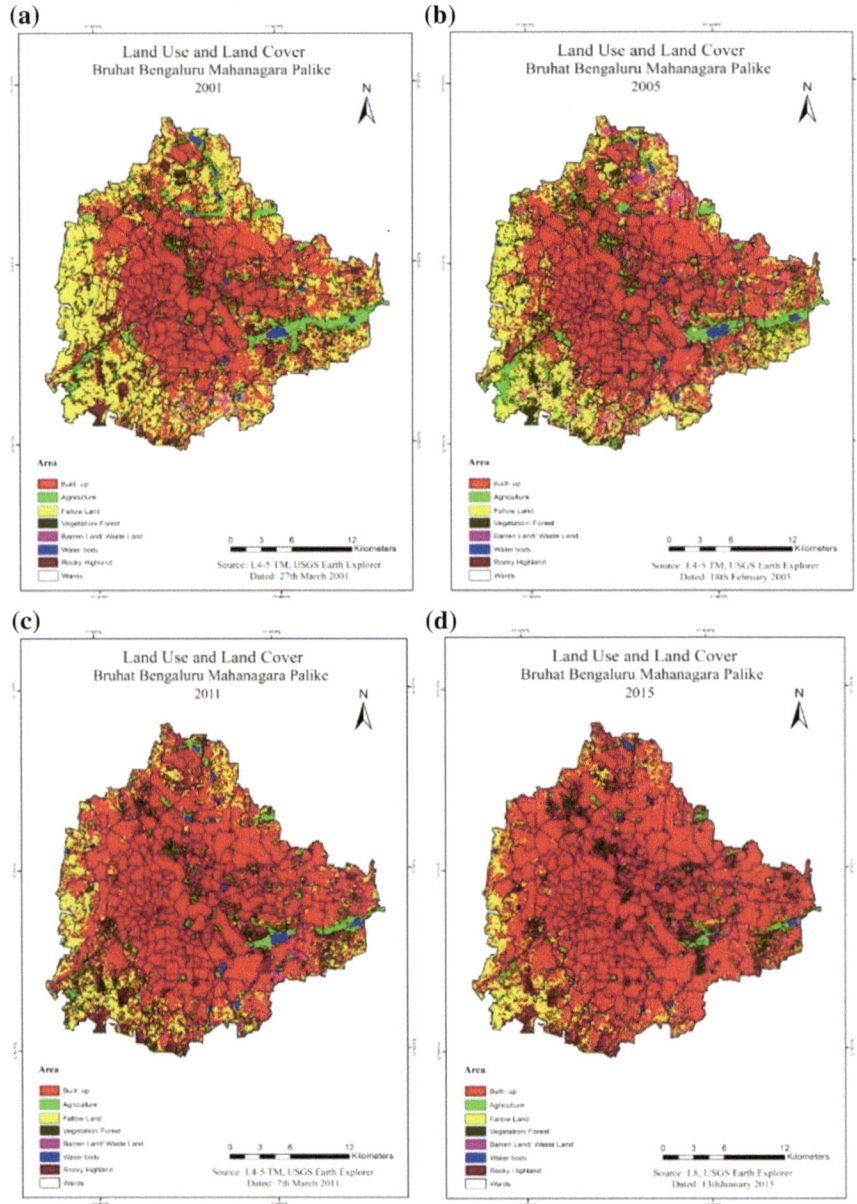

Fig. 2 Land use and land cover (BBMP); **a** 2001, **b** 2005 (BBMP, ward-wise), **c** 2011, and **d** 2015

Table 2 Change (absolute and percentage) in a built-up area, 2001–2015 (BBMP)

Year	Area (km²)	Absolute change	Percentage of built-up	Percentage change in built-up area
2001	290.81	–	42.11	–
2005	327.89	37.08	47.48	5.37
2011	407.38	79.49	58.97	11.49
2015	500.13	92.75	72.43	13.46

Fig. 3 Change in built-up area, 2001–2015 (BBMP, ward-wise)

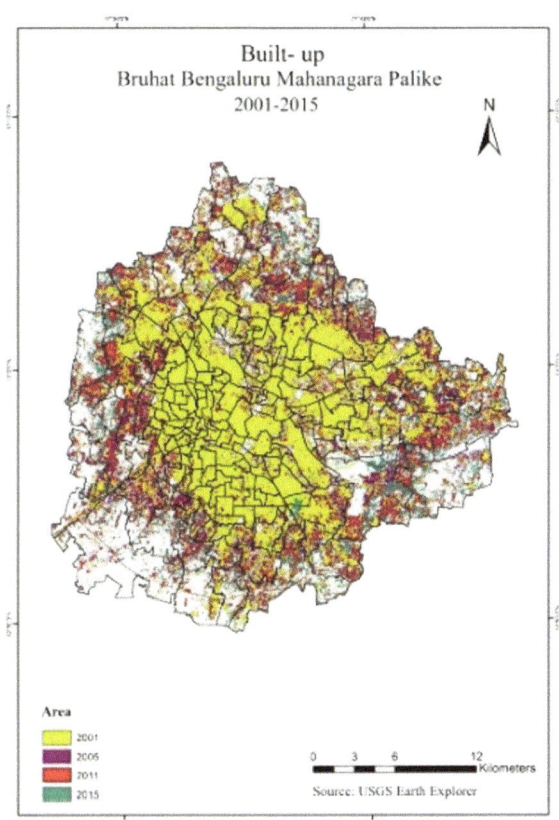

4.2 Structural and Directional Transformations of Urban Growth of Bengaluru (2001–2015)

The study of patterns of urban growth of Bengaluru has been divided into zones and directions that will help us to know not only the sprawling pattern but also the direction and rate of change. In order to delineate the form and direction of urban growth, Shannon's entropy index has been applied to the study. The areal extent of

Fig. 4 Multiple buffer zones
and directions of BBMP
(North, East, South, and
West)

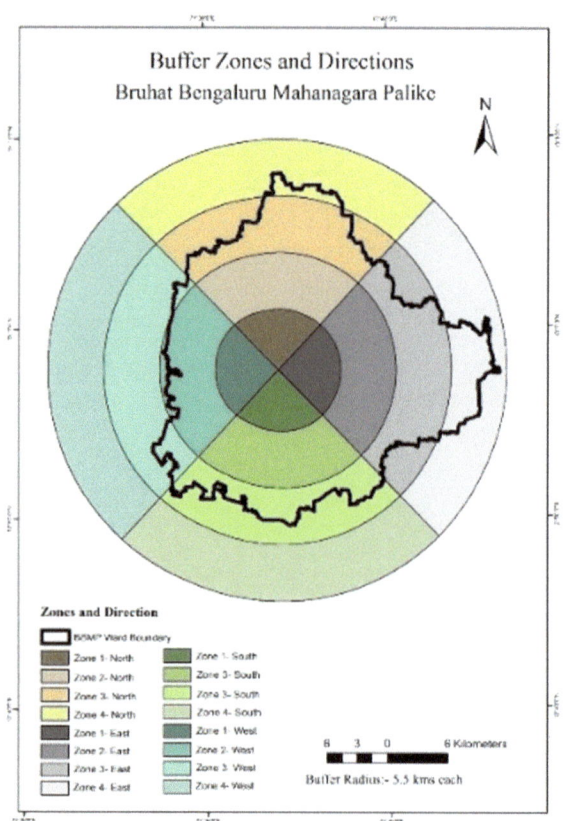

BBMP is the boundary limit of the index and the analysis has been made in terms of zones. The formation of the zone has been prepared in Arc Map 10.2.2 with the method of multiple buffers, resultant 4 concentric zones (circles) of 5.5 km radius each. The zones have been named as Zone 1, Zone 2, Zone 3, and Zone 4 radiating from centre to the periphery and further they were subdivided into different direction: North, South, East, and West (Fig. 4).

Figures 5 and 6 show the form of urban growth along with the direction of influence from the year 2001–2015. Comparison of all the zones, Zone 1 of 2001 is compact, whereas as the movement is towards the periphery (Zone 4), the diversification is increasing. Considering the Western direction, this section shows remarkably highest diversification than the others direction. This implies that the West direction is diversified in its urban form. Next higher order is East. This region is less diversified than West but is more when compared with East and South. Highest diversification is seen in West of Zone 1, South of Zone 2, East of Zone 3, and Zone 4. East and South have higher diversification than North and West in all the Zones. In the year 2005 of the Zone 1, South (0.507) is less diversified than the other directions. However, all zones show increasing diversification. Zone 1 shows

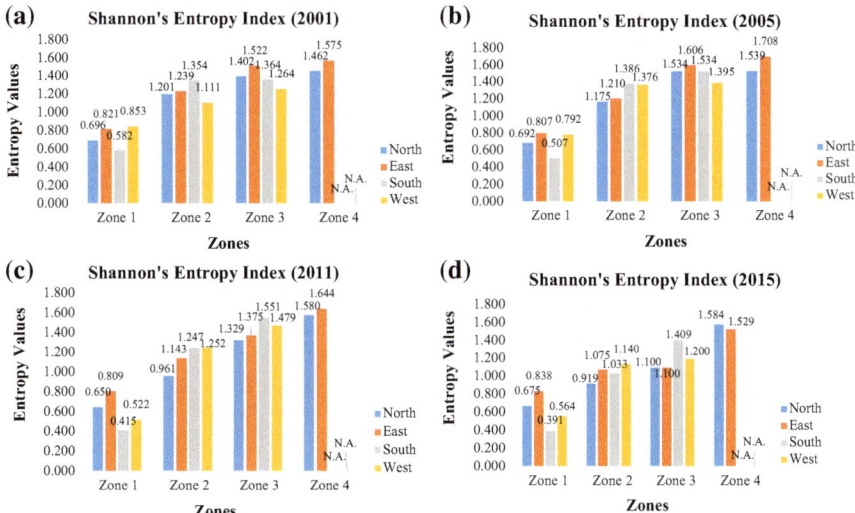

Fig. 5 Shannon's entropy index of BBMP; zone- and direction-wise: **a** 2001, **b** 2005, **c** 2011, and **d** 2015

fluctuating trend among direction, whereas the other zones show a gradual trend. Zones 2, 3, and 4 show equal level of urban growth which projects a higher diversification and lowing in South and West direction. The figure of 2011 represents the values of only three directions of Zone 4 because South and West direction does not fall under the limits of the study area. The North of Zone 1 shows a higher level of compactness but slight change has been observed from 0.650 to 0.809 in East causing slight diversity in land use. South and West are compact in nature. Zone 4 has higher diversification of urban growth than Zone 1, Zone 2, and Zone 3, but the rate of change is highest among South and West directions of all the Zones. In 2015, in Zone 1, there are differences in values causing fluctuations in directions. However, the rate at which these differ is nearly similar in Zone 2 and remains significantly fluctuating in Zone 3. South in all zones is attaining a disproportionate positive change. Another noticeable fact here is found in Zone 2. Zone 2 in previous years shown increasing diversification but the year of 2015 have recorded increasing compactness in all the direction. The North of all zones has experienced increasing diversification. However, the rate of diversification is decreasing. Same is the case with East, but this region is more diversified than North. The South of all zones presents a different picture. Though this region is more compact in nature in comparison to North and East but South was more diversified in Zone 2, exceeding North and East. The West direction is still increasing its diversity showing no signs of reduction.

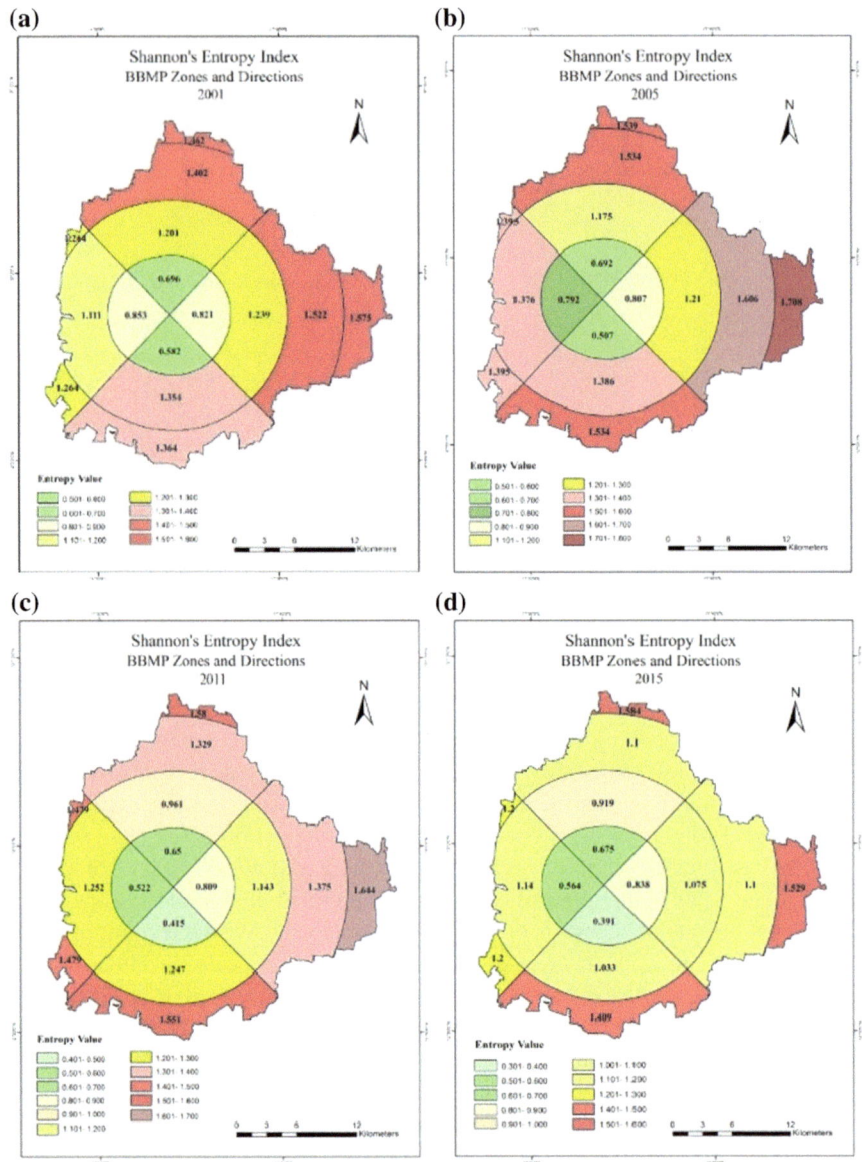

Fig. 6 Shannon's entropy index (BBMP zones): Direction-wise: **a** 2001, **b** 2006, **c** 2011, and **d** 2015

5 Inference around Bengaluru's Sustainability and Smart Cities

After the third attempt made, Bengaluru was shortlisted for Smart Cities Mission in 2017 and with the consequent pace of smart city project, the objective remains to make life of its citizens more remarkable [29–31]. This research dealt with Shannon's entropy index which detects the degree of compactness and the degree of diversification in any region. Due to restricted limits of a territorial region, the urban growth happens within its limits and cause densification in terms of its form (Fig. 7). Similar results have been procured through this study. Bengaluru Shannon's entropy values reveal that the development of settlements is taking place from centre to the periphery. In the initial years of study, there happens to be an increasing entropy value mostly in all the direction that suggests that there is a high diversity among the land use and land cover, all experiencing a change from centre to the periphery. But with the passing years, the values are decreasing that suggest that with due course of time the land use and land cover pattern have been changed to the more unified character. This suggests that Bengaluru has been expanding from the centre to periphery by engulfing the peripheral rural areas into its urban fold. This builds an expectation of growth of areas neighbouring Bengaluru [32]. Although the city-level planning and management for the restoration of the ecosystem services are dynamically active, yet it becomes essential to have data on the distribution of different components of a city and their geographical setting that can imply over better planning and management [14]. Bengaluru is seen as the

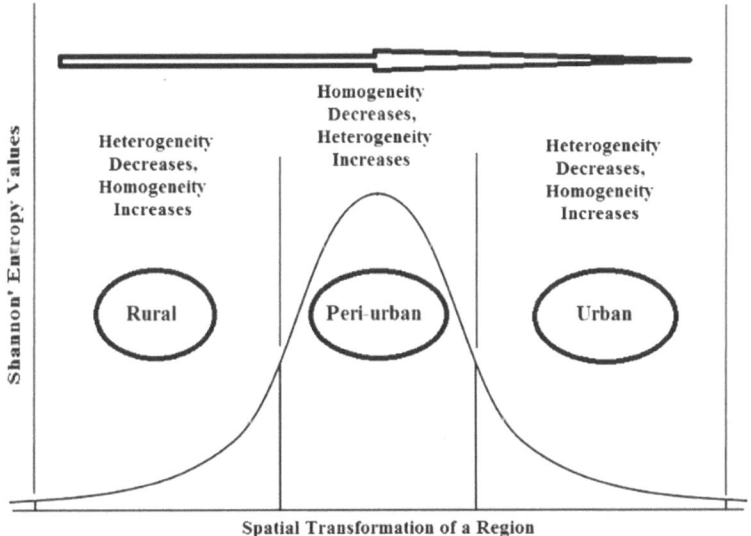

Fig. 7 Schematic representation of Shannon's entropy index. *Source* [25]

important artery of the global city, the integral sustainability of itself is into mystery thereby leading to a mismatch between the city in the frame and the city in its sustainability. In order to accommodate and direct the population fairly to adjacent perpheries, the city needs to explore spatially and we temporally so as to project itself as an inspiration city of Smart Cities Mission. Not only this, but this research will also help to exploit the resources efficiently, unstimulating clustering and avoiding adherence to irrational and irrepressible mode of consuming them.

The reflection of Shannon's entropy value proceeds with lower values at the beginning that suggest that the region has high homogeneity. Due to restricted limits of a territorial region, the urban growth happens within its limits and cause densification in terms of its form. The pre-existed rural region has low entropy values due to its homogenous agricultural nature with less of mixed uses. When the city reached a point where further sprawl within the boundary gets discontinued, the territorial extent is increased by the inclusion of peripheral areas which possess of less developed character. The entropy value increases, and this is caused by mixed urban and rural use. This is the transitional phase (Peri-urban) from rural to being urban. Once the rural region takes the shape of an urban region, the value again decreases due to the region attains a homogenous character.

Further research from this work lies on the demarcation of the large-scale changes that have caused much of the undesirable changes and to explore the possible solution for the same within the specified geographical setting. Individually, assessment can be performed depending on the urban component in consideration. However, this research incorporates large tracts of territorial limits; minute differences cannot be inferred from the above research. The form and pattern of growth as analysed through Shannon's entropy index provides the directional change of urban land uses over the years that will help to figure out the dominant direction of a rural–urban interface and underlies the causes of such growth at the location level where interventions can be made through proper planning. Furthermore, the locational change has an advantage over the transport sector because stronger the connectivity, the propensity of urban expansion will be more. This will also help in focusing on the other channels of connectivity that can decongest the city considerably and proper exploration can channelize the implementation process better.

6 Conclusion

Cities present an example of a socio-ecological system where there always to have an opportunity if interactions with the environment and ecologies but at present, the way, these interactions are modified; the interactions have produced negative impacts and environmental damage. Therefore, it becomes essential to draw the attention of the authorities to make them realise the substandard levels of living at the cost of inadequate modes of sustenance and non-resistance towards the unethical spurt in modernisation and urbanisation. We are living in a system where every individual is craving just for sustenance; compromising with the sustainability whether in a city or in a village; ignoring the others' aspects of living that shapes the internal composition of the human body. The villagers apparently lead to the nourishment of the environment as their sustenance itself depends on the environment so it becomes one-to-one relationship with the same. But as far as city dwellers are in question, the inhabitants draw their needs from the environment but are discourteous towards repayment towards the same. Since the management authority is seated for the planning and implementation of the whole of the city, the driving force is to alert in order to make its citizens aware of the surroundings and environment. The matter of sustainability is not just limited to sustainable development wherein the resources are utilized in a way to keep it for future but the time has approached where the resources are not only to be saved but also to be rejuvenated at a rate higher than the pace of degradation; keeping in consideration the constant hike in population, the never-ending human needs from the environment and the rush of achieving high standards of living. The technologies can resort to faster rates of achieving a purpose but it can never repay the damages it makes to the environment. Of course, the pace at which technologies have advanced can never rejuvenate the environment at the same pace as nature has its own pace of nourishing itself but sooner the matter is brought into priority, faster will be the awareness among the human beings and also the implementation. The time has already reached where we need to build sustainability upon the built features of the city. It is not just about the city of Bengaluru but is a tale of every Indian city being remembered of the cool, mild, breeze climate throughout the year-the reality of which; at present speaks of the contrast experiences. The increased impervious surfaces along with the need for improved intercity links especially though roadways and railways have led to the widespread clearing of the natural-positioned constituents. All the more, in order to decongest the irreversible population of the city, the territorial limits are getting expanded and have incorporated the healthy peripheries into the overburdening limits of the city thereby adding to the degradation of the rural–urban interface. This becomes all the more brainstorming effort as the planning and implementation now require proportionately more solutions of sustainability over the already built spaces that can re-orient the material ecologies of cities for a better future; a better future not for human but a better future for nature.

References

1. Pattaroni L, Pedrazzini Y (2010) Insecurity and segregation: rejecting an urbanism of fear. In: Jacquet P, Pachauri RK, Tubiana L (eds) Cities: steering towards sustainability. The Energy and Resources Institute (TERI), pp 177–186
2. Nagendra H (2010) Maps, lakes and citizens. Retrieved from http://www.india-seminar.com/2010/613/613_harini_nagendra.htm/ 20 Apr 2017
3. Wirth K (2017) Communities take command of lake clean-ups in Bengaluru. Citizen Matters. Retrieved from http://bengaluru.citizenmatters.in/bangalore-lake-revival-solutions-part-2-21106/. Accessed 17 Mar 2018

4. Mundoli S, Manjunath B, Nagendra H (2014) Effects of urbanisation on the use of lakes as commons in the peri-urban interface of Bengaluru, India. Int J Urban Sustain Dev 7(1):89–108

5. Bhasthi D (2017) City of burning lakes: experts fear Bangalore will be uninhabitable by 2025. The Guardian. Retrieved from https://www.theguardian.com/cities/2017/mar/01/burning-lakes-experts-fear-bangalore-uninhabitable-2025/. Accessed 11 Feb 2018

6. Thakur A (2017a) Despite ban, plastics bags continue to choke Bengaluru. Deccan Chronicle. Retrieved from http://www.deccanchronicle.com/nation/current-affairs/030917/despite-ban-plastic-bags-continue-to-choke-bengaluru.html/. Accessed 1 Nov 2017

7. Thakur A (2017b) Killing pollution in Bengaluru: just metro not enough. Deccan Chronicle. Retrieved from http://www.deccanchronicle.com/nation/current-affairs/190917/killing-pollution-in-bengaluru-just-metro-not-enough.html/. Accessed 15 Oct 2017

8. Katoch M (2017) How this Bengaluru quarry went from garbage dump to beautiful park. The Better India. Retrieved from https://www.thebetterindia.com/116053/bagalur-landfill-park-bbmp/. Accessed 7 Jan 2018

9. Wirth K (2017) Clear and present danger: what's the future of Bengaluru's rejuvenated lakes. Citizen Matters. Retrieved from http://bengaluru.citizenmatters.in/problems-for-rejuvenated-lakes-bangalore-21169/. Accessed 8 Nov 2017

10. Rukmini S (2015) Bengaluru fares worse than Delhi air quality. The Hindu. Retrieved from http://www.thehindu.com/news/national/air-quality-levels-bengaluru-fares-worse-than-delhi/article7074817.ece/ Accessed 28 Feb 2017

11. Saha D (2015) A Bengaluru neighbourhood's toxic air portends India's future: as a national debate grows over toxic air, some national-level solutions that appear expensive are far cheaper than the costs of doing nothing. Business Standard. Retrieved from http://www.business-standard.com/article/current-affairs/a-bengaluru-neighbourhood-s-toxic-air-portends-india-s-future-115041100200_1.html/. Accessed 8 June 2017

12. Air pollution: Bengaluru's (2015) The Hindu. Retrieved from http://respromasks.com/2015/07/21/air-pollution-bengalurus-rspm-levels-exceed-the-allowed-limit-by-12-283/. Accessed 27 Sept 2017

13. IANS (2017) Could this reduce air pollution in Bengaluru? K'taka offers sops for electric vehicles. The News Minute. Retrieved from https://www.thenewsminute.com/article/could-reduce-air-pollution-bengaluru-ktaka-offers-sops-electric-vehicles-68366/. Accessed 20 Dec 2017

14. Sudhira HS, Ramachandra TV, Subrahmanya MB (2007) Bangalore. Cities 24(5):379–390

15. Mathews S (2015) A review of urban transport scenario in Bengaluru city. Int J Manag Soc Sci Res Rev 1(11):82–89

16. Castán Broto V, Sudhira HS (2019) Engineering modernity: water, electricity, and the infrastructure landscapes of Bangalore, India. Urban Studies, pp 1–19

17. Barbier C, Jozan R, Renard V, Sundar S (2010) Cities: steering towards sustainability. In: Jacquet P, Pachauri RK, Tubiana L (eds) Cities: steering towards sustainability. The Energy and Resources Institute (TERI), pp 17–25

18. Vaddiraju AK (2013) A tale of many cities. Econ Polit Wkly 48(2):67

19. Kundu A (2003) Urbanisation and urban governance: Search for a perspective beyond Neo-Liberalism. Econ Polit Wkly, 3079–3087

20. Census of India (2001) Primary census abstract: Karnataka. Office of Registrar General and Census Commissioner, New Delhi

21. Census of India (2011) Primary census abstract: Karnataka. Office of Registrar General and Census Commissioner, New Delhi

22. Suri KB (1970) Impact of an expanding metropolis. Bangalore: a study. Econ Polit Wkly, 16–20

23. Tiwary P (2015) An ice cube on a summer day. Teacher Plus, pp 28–30

24. Yeh AGO, Xia L (2001) Measurement and monitoring of urban sprawl in a rapidly growing region using entropy. Photogramm Eng Remote Sens 67(1):83–90

25. Horo JP, Punia M (2018) Urban dynamics assessment of Ghaziabad as a suburb of National Capital Region, India. GeoJournal. Springer, pp 1–17
26. Sudhira HS, Ramachandra TV, Karthik SR, Jagadish KS (2003) Urban growth analysis using spatial and temporal data. J Indian Soc Remote Sens 31(4):299–311
27. Sudhira HS, Ramachandra TV, Jagadish KS (2004) Urban sprawl: metrics, dynamics and modelling using GIS. Int J Appl Earth Obs Geoinf 5:29–39
28. Halbe A (n.d.) Satellite images show green cover of Bangalore reducing alarmingly fast. Research Matters. Retrieved from https://researchmatters.in/article/satellite-images-show-green-cover-bangalore-reducing-alarmingly-fast/. Accessed 4 Mar 2018
29. MoUD (2017) Smart city proposal: Bengaluru. Ministry of Urban Development, Government of Karnataka, Government of India
30. Aranha J (2017) Bengaluru gears up for smart city project: free Wi-Fi, E-Toilets, water ATMs and more. The Better India. Retrieved from https://www.thebetterindia.com/108856/bengaluru-smart-city-project-wifi-toilets/. Accessed 6 Mar 2018
31. Akshatha M (2018) Bengaluru's smart city project is yet to take off. The Economic Times. Retrieved from https://economictimes.indiatimes.com/news/politics-and-nation/bengalurus-smart-city-project-is-yet-to-take-off/articleshow/66159314.cms. Accessed 15 Jan 2019
32. Kumar RS (2018) What does Rs. 1,700 crore 'Smart City' plan for Bengaluru involve? Citizen Matters. Retrieved from http://bengaluru.citizenmatters.in/smart-city-bengaluru-plans-1700-croes-25659. Accessed 18 July 2018

Urban Flooding—A Case Study of Chennai Floods of 2015

Mohammad Sharif, M. S. Dhillon, Sharad Chandra, Manoj Kumar and Vasanthakumar V

Abstract Urban areas are the centres of economic activities having vital infrastructure. Damage to these infrastructures can have a bearing not only on the country but also globally. Urban flooding is generally caused due to inadequate drainage system in an urban area. Owing to rapid urbanization, the extent of open spaces for percolation of surface water in the ground has considerably reduced. Inadequate drainage system coupled with large intensity rainfall has often been the major cause of urban flooding. Water may even enter the sewage system in one place and then get deposited somewhere else in the city on the streets. Major cities such as Mumbai in 2005 and Chennai in 2015 have witnessed the loss of life and property, disruption in transport and power and incidence of epidemics. This demands the need for management of urban flooding. The present paper analyses the reasons for Chennai flood of 2015 that affected several sub-urban areas of the city, especially those in the basin of River Adayar. The factors analysed in this paper are the effect of the series of rainfall events in the month of November–December 2015, operation of Chembarambakkam Lake, effect of increased urbanization and infrastructure and the local storages which play a major role in flood moderation in catchment. This paper also discusses the how well the flood events have been

M. Sharif
Department of Civil Engineering, Jamia Millia Islamia, New Delhi, India
e-mail: msharif@jmi.ac.in

M. S. Dhillon · S. Chandra · M. Kumar · Vasanthakumar V (✉)
Central Water Commission, New Delhi, India
e-mail: vasanthakumar-cwc@nic.in

M. S. Dhillon
e-mail: cefmgmt@nic.in

S. Chandra
e-mail: sharadchandra-cwc@nic.in

M. Kumar
e-mail: manojkumar2-cwc@nic.in

© Springer Nature Singapore Pte Ltd. 2020
S. Ahmed et al. (eds.), *Smart Cities—Opportunities and Challenges*,
Lecture Notes in Civil Engineering 58,
https://doi.org/10.1007/978-981-15-2545-2_64

managed in terms of emergency preparedness. The paper finally tries to comprehend the key areas where the improvement is required in terms of urban flood
preparedness.

Keywords Reservoir operation · Urbanization · Lakes

1 Introduction

Urban flooding has generally been attributed to a large increase in concrete/
impervious surface, unplanned usage of urban land, lack of proper drainage, loss of
wetlands, less groundwater recharge, tidal effects in coastal cities and very heavy
storms—cloud burst.

The present paper attempts to study the devastating floods that hit Chennai city
during November–December 2015 and have claimed more than 400 lives (including other parts of Tamil Nadu) and caused enormous economic damages. The
focus mainly is on flood routing through channels and reservoir, flood control and
reservoir operation. Every smart city needs not only its drainage to be planned in
smart way, but there are other factors which are connected to the city's sustainability, when neglected may lead to catastrophe. This paper highlights the aspects
which need to address while planning a smart city with lesson learnt from Chennai
floods of year 2015.

2 Study Area

The southern part of Chennai city lies in the Adayar Basin in Tamil Nadu. During
the flooding event 2015, the Adayar Basin was heavily flooded. The encompassed
area consisting of three polygons in yellow outline is the study area for analysis of
rainfall pattern, urbanization, reservoir releases and development of inundation
model (Fig. 1).

3 Rainfall Event

Chennai city received 539 mm of rain in December 2015 as against the monthly
average of 191 mm. Chennai rain in December was almost 3 times more than the
normal rainfall in December. There was very heavy rainfall on November 8, 9, 12,
13, 15, 23 and December 1. During the 24 h ending 8:30 a.m. on 2 December 2015,
"extremely heavy rainfall" was reported in Chennai: 49 cm at Tambaram in
Kancheepuram district, 47 cm at Chembarambakkam in Tiruvallur district, 42 cm
at Marakkanam in Tiruvallur district, 39 cm each in Chengalpattu (Kancheepuram

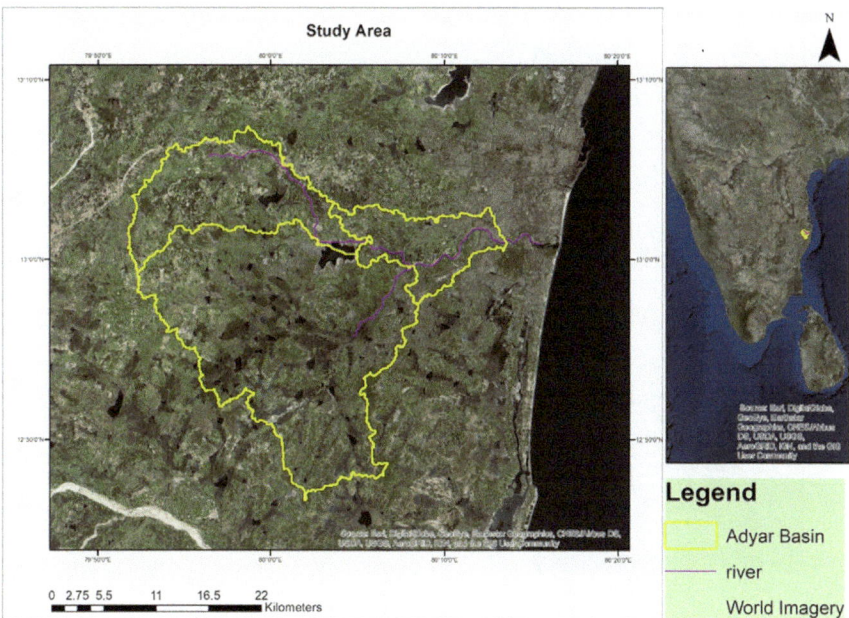

Fig. 1 Catchement map of Adayar River

district) and Ponneri (Tiruvallur district) and 38 cm each in Sriperumbudur and Cheyyur (both in Kancheepuram district), 35 cm at Chennai airport and 34 cm in Mamallapuram, Poonamallee, Red Hills and Chennai city [1]. The following images show the rainfall estimates from the TRMM based 3B43 product which gives the monthly accumulated rainfall data. These images clearly shows that the catchments of rivers such as Kosasthalai River and Adayar River draining in the Chennai city have received high amount of rainfall in the month of November itself, which made soil in the catchment highly saturated. As high amount of rainfall happened in 2–3 December, the catchments generated huge amount of run-off creating massive floods (Figs. 2 and 3).

4 Reservoir Operation

Chennai city's drinking water supply is managed by small reservoirs such as Poondi, Chembarambakkam and Puzhal situated in the rivers Kosasthalai, Adayar and water from River Krishna. The capacities of the reservoirs are 3231, 3645 and 3300 Mcft [Chennai Metropolitan Water Supply and Sewerage Board (CMWSSB)] [2]. The capacity of these lakes are very small when compared to the normal reservoirs but the flooding created by release of one of the reservoir

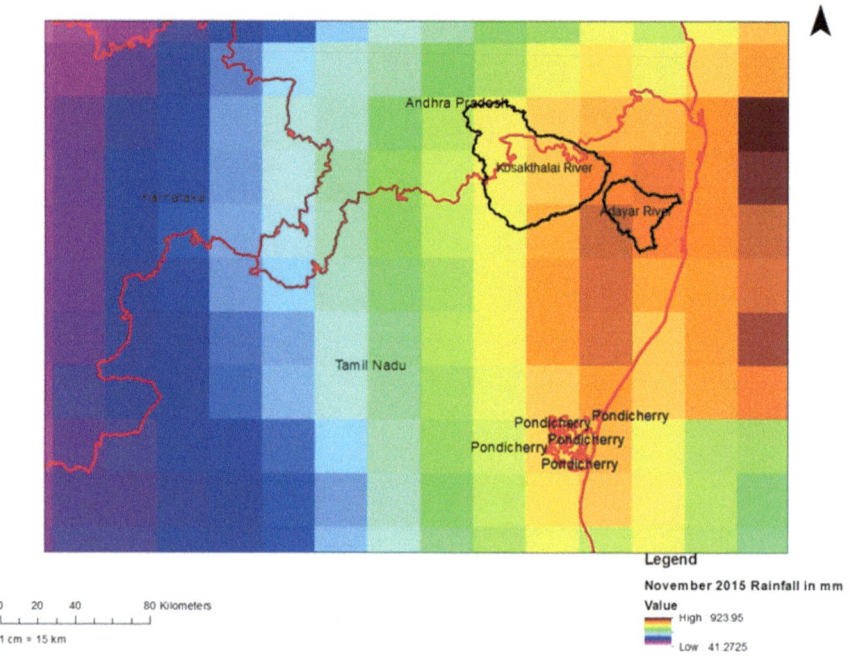

Fig. 2 Monthly accumulated rainfall (TRMM-3B43) November 2015

(Chembarambakkam reservoir) in Adayar) due to the operation discussed in this chapter.

The reservoir inflow–outflow graphs shown in Figs. 4 and 5 shows that the during the flood event of Nov–Dec 2015 the inflow received by the reservoir has been completely released from the reservoir, in fact excess volume of water has been released from the reservoir clearly showing the issue in the operation of the reservoir during floods. These reservoirs function as source of drinking water supply and the reservoir managers are not trained for the operation during the floods.

5 Increased Urbanization and Infrastructure Development

Sub-urban areas of Chennai is rapidly urbanizing region. The Adayar River which is situated in the southern side of Chennai city is characterized by swamps, marsh lands, low-lying areas, streams and interconnected regional tanks which are converted into special economic zones of big corporations, middle class housing. This rapid development has taken place in the past three decades [3]. The rapid increase

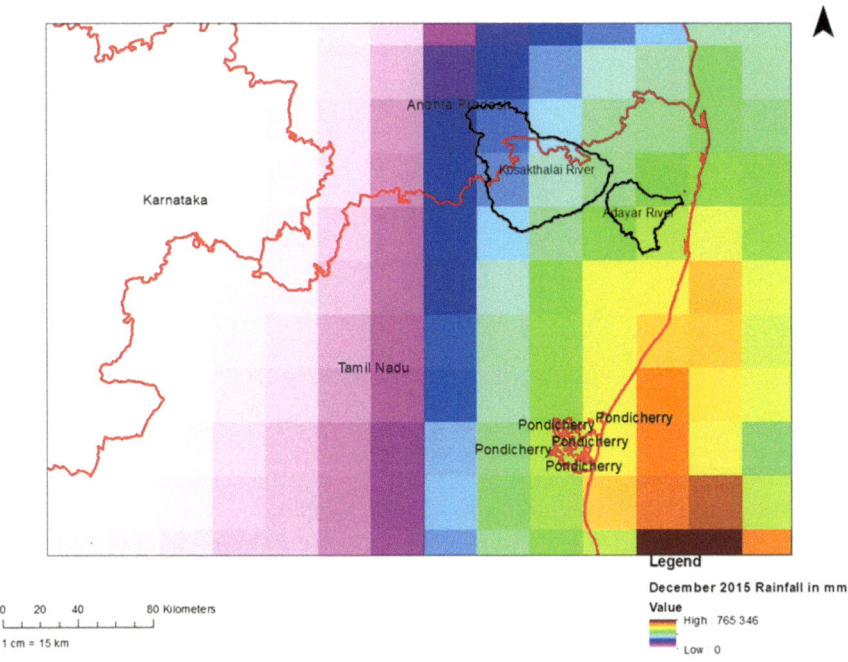

Fig. 3 Monthly accumulated rainfall (TRMM-3B43) December 2015

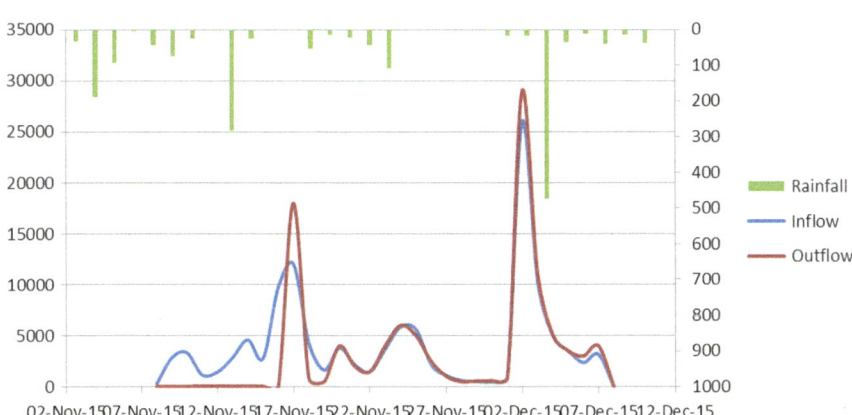

Fig. 4 Inflow–outflow and rainfall at Chembarambakkam reservoir

in the paved area due to creation of industries and housing complex in the catchment of Adayar leads to the increase in the run-off from the catchment generating
more run-off volume.

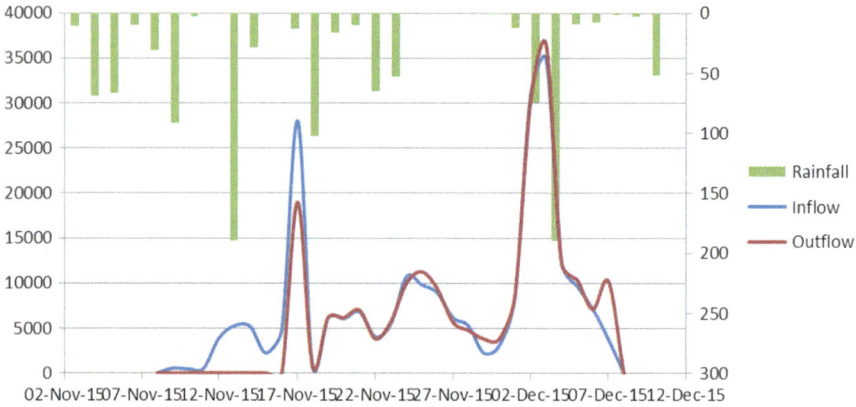

Fig. 5 Inflow-outflow and rainfall at Poondi reservoir

In this study, Normalized difference Built-up Index has been used to calculate the variation in the urbanized area between the years 2000 and 2015 using Google Earth engine platform. NDBI has been used in finding out the urbanized are in Ankara using Google Earth engine [4].

$$NDBI = (Red - NIR)/(Red + NIR)$$

Sentinel 5 MSS image has been used for year 2015, and Landsat 7 image has been used for the year 2000. NDBI has been calculated for the years 2000 and 2015, and the difference has been calculated to increase in urbanized area. The threshold to classify the new built-up area has been carried out using trial and error method with validation with latest and old Google Earth imageries. The calculation shows that there has been 23% increase in the built-up area (Fig. 6).

The Chennai International Airport has been renovated in the year 2011. River Adayar flows near the airport. Due to the renovation activity, the runway was extended, and it has to be laid over the River Adayar. During the floods of 2015, the capacity of the culvert below the runway was not sufficient enough to discharge the floodwater causing major inundation in the airport disturbing the normal operation of the airport for a week (Figs. 7 and 8).

6 Effect of Local Storages

Catchment of the rivers surrounding Chennai city has big and small ponds connected by a working overflow system. The water was allowed to spread into fields and thousands of smaller ponds, with the entire region acting as a "sponge" to absorb the excess water. This interconnected system of ponds' overflow finally finds their way to lakes and drain through the river. Due to excessive urbanization

Increased Built-up Area Calculated using NDBI index in Adayar River Basin

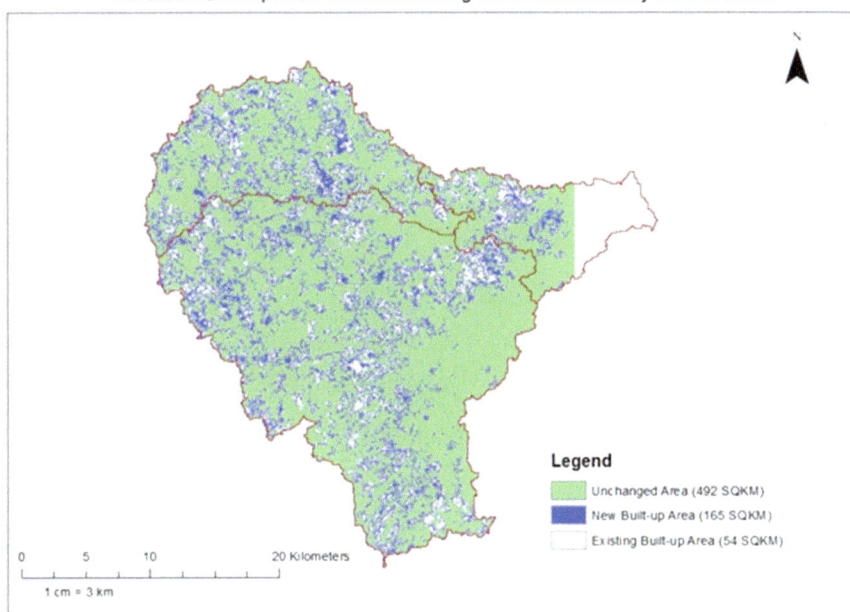

Fig. 6 Increased built-up area in Adayar Basin classified using NDBI

Fig. 7 Chennai Airport before (2004) and after (2017) renovation

and encroachment of these big and small ponds, the capacity and the connectivity of these ponds have been reduced and lost. Due to the reduced capacity, the storage of floodwater in the ponds, which moderate floodwater did not happen effectively and the missing connectivity of the lakes also hindered the process of efficient transfer of floodwater from one pond to other. This was one of the major reasons for Chennai flood 2015. The below images shows the water bodies and their inundated area before (July) and after the Chennai floods (December). The satellite images used are from Sentinel SAR (Synthetic Aperture Radar) data (Fig. 9).

Fig. 8 Chennai Airport during December 2015 floods

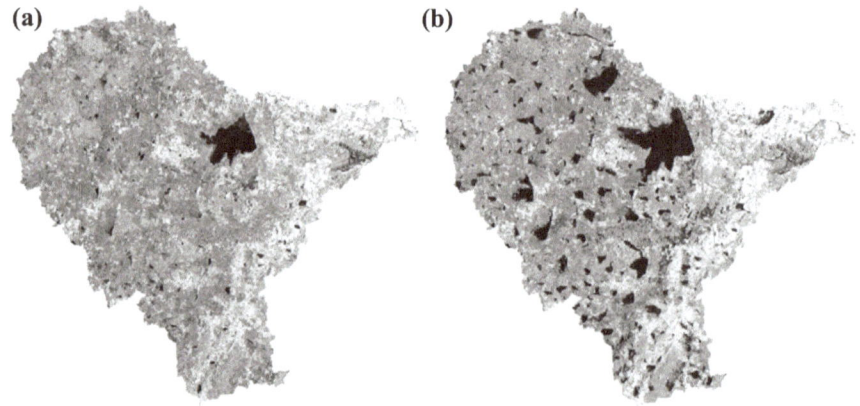

Fig. 9 Water bodies in Adayar Basin. **a** July 2015. **b** December 2015

7 Emergency Management and Rehabilitation Activity During the Flood Event

Graphs showing the release of the Chembarambakkam and Poondi reservoir clearly show that the after 16th of November the inflow received due to the rainfall has been completely released by the dam managers. Extraordinary emergency should be issued by the dam owners to the Chennai corporation, and public should have warned about the impending flood situation, since the monsoon season was still continuing. The reservoir was maintained in the same level of 83.7 since 17

November to 02 December 2015. Since the information related to the impending flood in the Adayar River was issued to Chennai corporation or public. Both public and disaster managers were took by surprise when the almost 450 mm rainfall occurred in on 2 December, and a discharge of 29,000 cusecs was released from Chembarambakkam reservoir causing major flooding densely populated areas along the banks of River Adayar.

Short-term efforts of rescue and relief have been achieved effectively because of the media attention and general sense of distress in the community as well as external support. There has been a lot of activity during the immediate flood response situation in Chennai, but activity related to rehabilitation of people who have lost their homes and belongings and who have been forced to migrate could not be done effectively. There was no proper data about the migrant workers. It was not clear how many of the migrant workers survived the floods and even among the survived it was not clear whether they were eligible for the relief support provided by the government.

The long-term rehabilitation requires meticulous planning and coordination. The major disasters that happened in India during the last decade suggest that we still have much to learn in terms of integrating resources in order to achieve effectiveness in long-term disaster management. Indeed, strengthening community capacities that can reduce vulnerabilities drastically and providing information to the government about the vulnerable communities especially to women and children can help the system to prepare effective disaster plans. Similarly, strengthening policy decisions, integration of rehabilitation services and sharing of resources can be a crucial element in development. There is a need for systems to be in place to ensure sustained efforts in the long term [5].

8 Analysis of Flood Events Using Mathematical Modelling Approach

The 2D mathematical modelling has also been attempted in this study for simulating the flood downstream of the Chembarambakkam Reservoir in Adayar River and its tributaries using MIKE 21. The run-off has been calculated at two upstream boundaries of the catchment using IMD observed rainfall data of that period. For Bathometry preparation JAXA 30 m DEM is used and resampled it to 90 m resolution. The model run for the period 28/11/2015–02/12/2015 and the maximum simulated inundation have been calculated for Adayar Basin. During the flood events, the Chennai Airport was fully submerged and the model's result also depicting the same. Water depth near Airport was between 2 and 4 m as per model simulation result (Figs. 10 and 11).

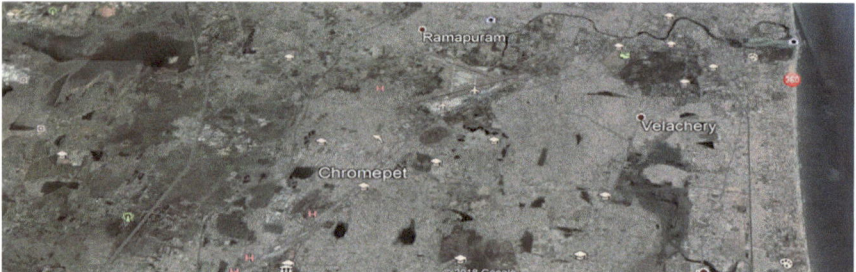

Fig. 10 Google Earth image of Chennai city

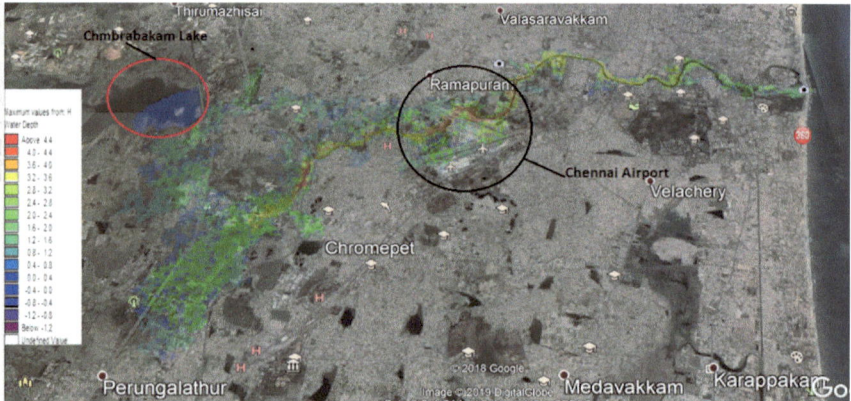

Fig. 11 Simulated inundation in Chennai city

9 Results and Conclusion

Any planning for effective urban flood management has to take into the consideration of entire watershed, identifying problems, causes and remediation, preparedness and mitigation, early warning and communication, response, awareness generation, community capacity development, vulnerability and risk assessment—reduce vulnerability, hazard mapping–flood level mapping, identify damages, insurance and risk transfer, flood inundation mapping based on different return period of storm, Inflow and level forecast, public awareness programme, etc.

The analysis of Chennai floods highlights the issues prevailing in effective management of the flood situation in Chennai. The preparedness of the dam authorities in managing flood situation needs to be improved in the areas of providing prior emergency warning to municipal authorities and public, using the inflow forecasting techniques based on meteorological warning. The dam authorities need to keep updating their compendium of rules and regulations for issuing emergency warning at regular interval. Since these catchments are very small in nature, the lead time in forecast in very less, so the dependency on the meteorological forecast needs to be improved and models for simulating the inflow needs to be well calibrated and updated based on rapidly urbanizing catchment.

Rivers draining in the Chennai city has its catchment with huge number of small and large ponds, which are interconnected in nature. These systems of ponds play a major role in moderation of the floodwater. As a result of rapid urbanization, the area of the ponds is decreasing as well the connectivity between these ponds are lost leading to decrease in efficiency of these ponds in moderation of the floods. This aspect needs to be seriously taken into consideration and plan needs to be devised to revive these systems of ponds and its connectivity. Public awareness has to be created about how important these tanks are in reducing the impact of the floods.

Effective management of disaster rescue and relief activities needs to be taken up seriously. Chennai Corporation has already created area wise disaster management plans, focusing on how the corporation and public have to react during the disaster such as floods. The true success of this disaster management plan lies in creating awareness among people about the disaster management plans through disaster education centres and regular awareness programs.

Creation of Hazard maps based on hypothetical flooding scenarios will educate the people about the impending danger and how they can be well informed about the magnitude of the disaster they might be facing in future. This will improve the preparedness for the disaster managers and public in responding to the disaster.

References

1. Chennai Floods (2015) A rapid assessment, 2016. Interdisciplinary Centre for Water Research Indian Institute of Science, Bangalore
2. Reservoir storage details, Chennai Metropolitan Water Supply and Sewerage Board (CMWSSB) http://www.chennaimetrowater.tn.nic.in/reserve.asp
3. Mukherjee D (2015) Effect of urbanization on flood—a review with recent flood in Chennai (India). Int J Eng Sci Res Technol. http://doi.org/10.5281/zenodo.57002
4. Celik N (2018) Change detection of urban areas in Ankara through Google Earth engine. 978-1-5386-4695-3/18/$31.00 ©2018, IEEE
5. Mariaselvam S, Gopichandran V (2016) The Chennai floods of 2015: urgent need for ethical disaster management guidelines. Indian J Med Ethics 2 April–June 2016. https://doi.org/10.20529/ijme.2016.025

MPP Technique for Solar PV Module Through Modified PSO Using Cuk Converter Under Varying Insolation Conditions

Parvaiz Ahmad, Anwar Shahzad Siddiqui and Uzair Malik

Abstract Photo electricity is treated as the most essential natural energy source and has been used extensively for photovoltaic (PV) power production. But the insignificant efficiency and nonlinear behaviour with numerous local peak power points under varying insolation conditions of PV systems halt the progress of solar electricity power production. We know output behaviours of PV system are affected by surrounding elements like solar irradiance, temperature and shading conditions, which have a direct impact on peak power point tracking of solar electricity. Therefore, an excellent (MPPT) technique is of the utmost need for efficiency advancement of PGS and functional under dynamic environmental conditions. The same issue is overcome by an algorithm, namely particle swarm optimization (PSO) along with Cuk converter.

Keywords Particle swarm optimization (PSO) · MPPT · Solar electricity production system

1 Introduction

With the expansion of human society, the need for electricity is growing at an alarming rate. Merely due to lack of sufficient energy resources to carry out this growing electricity demand, the incessant use of conventional resources has affected the environment poorly and ultimately activated global warming. Conventional

P. Ahmad (✉)
Department of Electrical and Electronics Engineering, Al-Falah University,
Faridabad, Haryana, India

A. S. Siddiqui
Department of Electrical Engineering, Jamia Millia Islamia New Delhi,
New Delhi, India

U. Malik
Department of Electrical Engineering, Indian Institute of Technology Roorkee,
Roorkee, U.P, India

© Springer Nature Singapore Pte Ltd. 2020   809
S. Ahmed et al. (eds.), *Smart Cities—Opportunities and Challenges*,
Lecture Notes in Civil Engineering 58,
https://doi.org/10.1007/978-981-15-2545-2_65

energy resources fail to fulfil this growing demand, and perturbation in environmental issues has generated extensive engaging in utilization of renewable energy resources. Photo generation is one of the major energy resources and has been used extensively to mitigate growing electricity demand [1–5]. Solar energy is universally available and can be harnessed easily. Energy of sun is universally available and can be harnessed easily. The solar energy when trapped can be fed to the local grid after its conversion to AC using suitable DC–AC converters. Therefore, it can be used to facilitate power to backward areas, where dream of electricity is yet to be fulfilled.

To overcome this speedy energy crisis and concern about environmental issues, we have to devise an efficient algorithm which can extract power from incoming solar radiation under varying insolation conditions. Because MPP of PV panel is affected by panel temperature, moisture, solar irradiance and due to stochastic nature of light, PV graph under the same condition comes with multiple peaks. In PGS, each photovoltaic array consists of multiple solar plates interconnected in series and shunt configuration to generate desired voltage and loading capacity. PSC can have an appreciable influence on panel output of photo generation which in turn depends on system configuration and shading pattern in PV module. Due to apprehension of multiple peaks, it curtails the effectiveness of traditional MPPT algorithm like INC, Perturb and Observe, neural network [6, 7] have demonstrated to be much boring with its need of regular training and computational difficulty. Essentially these practices are working on 'hill climbing' rule of leading to next operating point in guidance towards which generation enhances. Therefore, there is a speedy demand to design an efficient MPPT model that can well identify GMPP under same juncture. As it is obvious that current gradually decreases, voltage increases when insolation decreases. We know there is a peak power generated, for a specified current voltage combination. This peak is called maximum power point of photovoltaic array. It is evident that allowing photovoltaic array to function in or around this peak power always will be more futile for fluctuating values of irradiance and temperature. Bio-inspired algorithm improved PSO is extensively used and turned out to be profitable. In PSO with boost converter as interface, we encounter more losses and need of filter at both input and output side, which adds further burden. This improved PSO with Cuk converter is presented here, and its duty cycle is adjusted. Moreover, power exchange in Cuk converter is via passive element, namely capacitor therefore losses are negligible. Pulse generator generates pulse for DC–DC Cuk converter.

2 Basic Model of Photovoltaic Cell

Figure 1 shows PV cell with its basic equivalent circuit. Eventually, the nonlinear behaviour of V and I under varying insolation conditions is shown with the help of single diode model. Only one diode representation is least complicated and comparatively accurate. The actual one diode behaviour is depicted in Fig. 1 which comprises a current source, (photovoltaic current) connected across diode and a

Fig. 1 Circuit diagram of a
PV cell

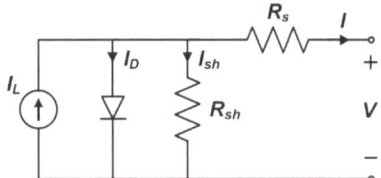

Fig. 2 *I-V* and *P-V* characteristics of PV cell

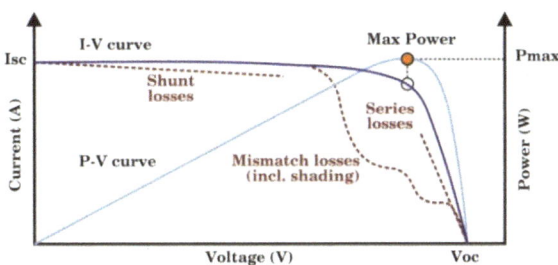

series resistance (Rs). The magnitude of the shunt resistance (Rsh) is very small in one diode behaviour. The *I-V* characteristics of PGS are given by the following equation.

$$I = I_{\text{pvcell}} - I_{\text{ocell}}\left[\exp\left\{\frac{qv}{aKT_c}\right\} - 1\right]$$

where I_{pvcell} is the photo current, whose level will be based on the cell area and will depend upon irradiance. V is the voltage of the diode, Boltzmann constant is K, where q is charge of electron, T_c is operating temperature usually 25 degree, and $I_{\text{ocell}}$is reverse saturation current of diode (Fig. 2).

3   Need of MPPT

We know reduction in power is proportional to shaded area. To increase power from a given solar module under varying insolation, various algorithms can be put forth for extracting peak power from solar PV module under *b* varying irradiance. Power converters act as interface, as we cannot couple PV module directly to load, essentially it is placed amidst PV array and load. DC converter has cycle of operation which can be perturbed so that peak power could be available at load side even under fluctuating circumstances. Usually, Cuk converter is used as interface because of its numerous advantages.

Fig. 3 Block diagram of
MPPT with Cuk converter

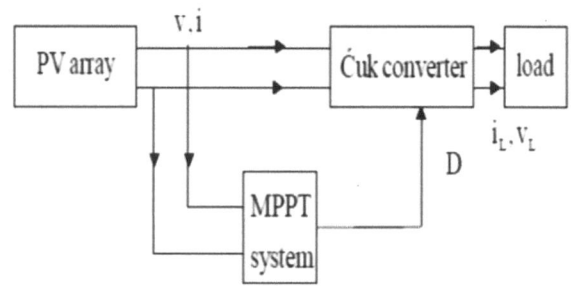

3.1 Modified PSO Technique and Its Application Through Cuk Converter

The PSO was proposed by Russell in the year 1995. Same technique has substantial future due to its simple implementation, fast computation ability and accuracy and having potential to track the MPP under semi shading environment. It does also a search that is much unspecified than searches done by traditional techniques. With PSO, updating of duty cycle is not constant. Here, we vary duty cycle until MPPT is attained. During operation of traditional methods, there the duty cycle is disturbed by a permanent value (Fig. 3).

Output equation for Cuk converter for maximizing the output in accordance with perturbation in input under varying insolation conditions is depicted as

$E = -(D/1 - D)*V$ Volts

where V = Supply Voltage

D = duty ratio.

4 PSO with Inertia Algorithm

PSO is an advanced technique. PSO technique is similar to social behaviour of bird flocking and fish schooling. It comprises two parts: first one is learning and next is communication. The algorithm finds the best value while comparing each iteration. Best value or top value is called as extreme point. For same technique, particles are used which could share information with one another. Agent is called as particle. Every particle studies its highest and then checks it with another for identifying extreme peak. The latest PSO method can be studied following below-mentioned equation, and particles having more inertia or weight facilitate GMPP while particles without inertia facilitate local extreme points (Fig. 4).

$$V_{i(K+1)} = W(K)V_i(K) + C_1 r_1 (P_{\text{best}} - X_i(K)) + C_2 r_2 (g_{\text{best}} - X_i(K))$$

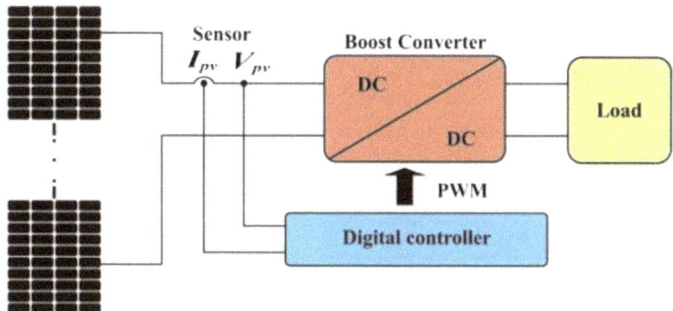

Fig. 4 Block diagram of MPPT

position of particle is $X_i(K)$, speed of particle is given by $V_i(K)$, and iteration is represented by k. Inertia, r_1, r_2 are variables which are diffused haphazardly. P_{best} represents personal best place as obtained by particles. g_{best} is computed in order to get extreme best position.

5 Implementation of PSO with Inertia to MPPT

1. MPPT path is defined by converter's duty cycle. Duty cycle being a vital parameter and estimation part then comes PGS module energy which is produced under half shading part. Surplus particles are required for getting an accurate point of convergence, but at same juncture it may need tremendous mathematical task consequently tracking speed and efficiency remains good while identifying peak point on PV graph.
2. Agents are kept in a haphazard fashion if there is no credible data about MPP. Particles can also be kept at permanent path.
3. PWM is used which may instruct Cuk converter as soon as it receives V and I from solar PV array. Even command from PWM may deviate as the values of V and I fluctuate under varying insolation.
4. In this step, personal best of all particles is identified during all iterations. GMPP is located and therefore desired operating point is identified.
5. Position and velocity of particles are checked. By going through this, peak power is traced in a miniature duration. Performance will depend upon global best and personal best position. If the particles are not verified, then it would take much time to identify GMPP.

6 Modelling of PV Array

It is obvious through the figure that arrays are producing unequal voltage as the irradiance is not same for different arrays. This pushes the PV nature towards worst nonlinear condition which consists of many peaks. That is why improved PSO is used, which can work under semi shading conditions followed by better results.

A. Unshaded Condition

In uniformly shaded PV array, all the modules get same solar insolation, and hence, every module produces same voltage as they are subjected to same solar insolation PV behaviour during uniformly shaded condition comprising only single point. So, there would be no minimal extreme and top extreme, it would be having only single extreme point.

It is apparent from the figure that all modules are having same sunlight and thereby produce equal voltage.

B. Partial shading Conditions

In PSC, voltage produced by each module is not uniform, as various cells accept varying insolation. Consequently, voltage produced might be non-uniform. PV nature would comprise countless peaks called as minimal extreme and best extreme, and maximum point is called as GMPP. Essentially under semi shading condition, PV behaviour turns out to be much nonlinear, with countless peak points. Multiple points get erected due to varying insolation.

1. Simulation, Results and Discussion

To go through semi shading issue, a number of PV arrays are selected. The same task is accomplished by connecting number of arrays in shunt and series (Fig. 5).

Making apparent from above model that all panels are getting dissimilar insolation, this slight variation may occur due to clouds, settling of dust elements on panels, clubbing of photo PV arrays and moisture. This makes PV behaviour to comprise surplus peaks termed as local extreme, and global extreme peak is called GMPP as available in Fig. 6 PV characteristics become significantly nonlinear. *I-V* behaviour during half shading conditions is shown in Fig. 7; optimization technique PSO is brought into front for identifying GMPP. PSO with inertia is important for identifying global extreme power point fast as compared to traditional ways (Fig. 8) (Table 1).

7 Conclusion

Under unshaded conditions, the modules are producing uniform voltage thereby making the PV behaviour of uniformly shaded to comprise only a single point but under varying conditions, and modules are not able to give equal voltage. Photo behaviour during fluctuating circumstances consists of surplus points. PSO without

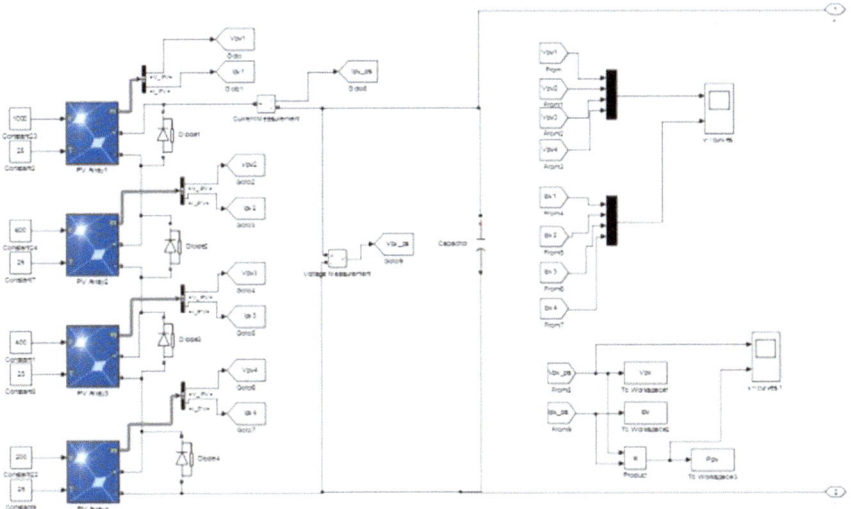

Fig. 5 Simulink representation of semi shaded PGS

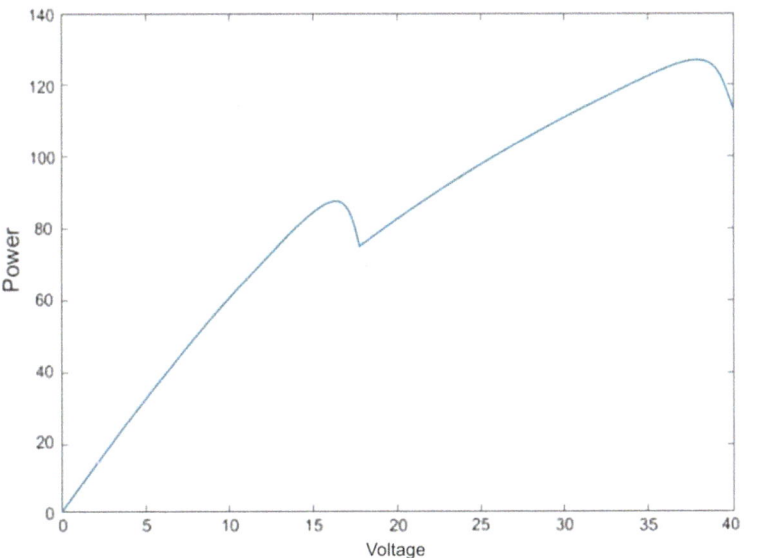

Fig. 6 PV behaviour under varying insolation condition

inertia facilitates towards local extreme, and points take very long time to merge towards the highest power point. PSO with inertia facilitates GMPP and is enough for identifying global extreme peak point in a nap interval of time, and therefore, can locate exactly maximum points compared to traditionally improved PSO.

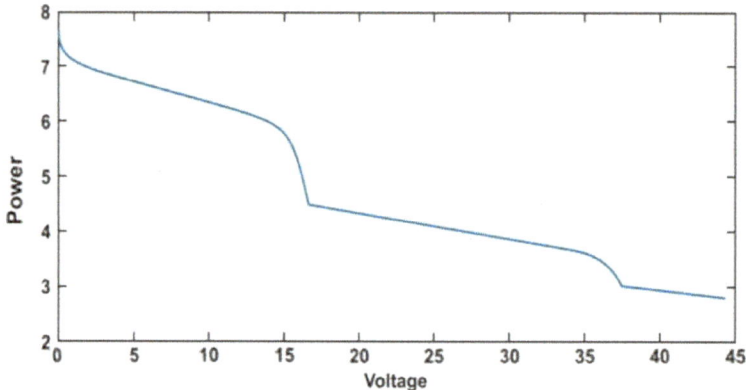

Fig. 7 *I-V* characteristics under partial shading condition

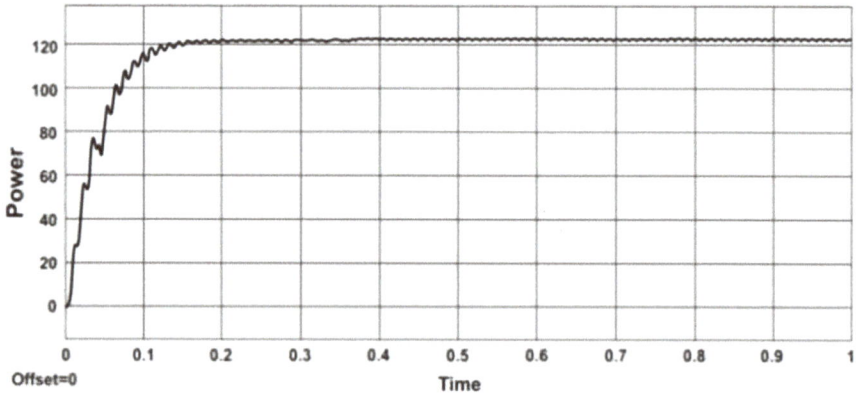

Fig. 8 MPPT through PSO

Table 1 Parameters of PSO using Cuk converter

Count of particles	5	Vmax	1
Duty cycle	0.2	Vmin	0.1
Highest duty cycle	0.95	D1max	3
Sampling time	0.1 s	D1min	2
Highest iteration	31	D2max	2.5
Switching frequency	$f_s = 50$ kHz		

References

1. Kumar N, Hussain I, Singh B, Panigrahi BK (2017) Rapid MPPT for uniformly and partial shaded PV system by using JayaDE algorithm in highly fluctuating atmospheric conditions. IEEE Trans Ind Inform 13(5):2406–2416
2. Kumar N, Hussain I, Singh B, Panigrahi BK (2016) Maximum power peak detection of partially shaded PV panel by using intelligent monkey king evolution algorithm. In: 2016 IEEE international conference on power electronics, drives and energy systems (PEDES), Trivandrum, pp 1–6
3. Sundareswaran K, Vigneshkumar V, Sankar P, Simon SP, Nayak PS, Palani S (2016) Development of an improved P & O algorithm 12(1): 187–200, Feb 2016
4. Kumar N, Hussain I, Singh B, Panigrahi BK (2017) Single sensor-based MPPT of partially shade. 983–992, Sept 2017
5. Hassan H, Geliel MA, Abu-Zeid M (2014) A proposed fuzzy controller for MPPT of a photovoltaic system. In: 2014 IEEE conference on energy conversion (CENCON), Johor Bahru, pp 164–169
6. Sundareswaran K, Sankar P, Nayak PSR, Simon SP, Palani S (2015) Enhanced energy output from a PV system under partial shaded conditions through artificial Bee colony. IEEE Trans Sustain Energy 6(1):198–209
7. www. Mnre.gov.in, ministry of non-conventional renewable energy govt. of India

A Need for City-Specific Water Policies—A Case Study of Kabul City: Afghanistan

Masoom Khalil, Sirajuddin Ahmed, Mukesh Kumar, Mansoorul Haqe Khan and R. K. Joshi

Abstract Kabul, the capital city of Afghanistan is facing the serious problem of scarcity of water. Over-exploitation of the groundwater and the contamination of water are adding to the problem. There are poor management and misuse of water due to instability and the long war in the country. A chain of a large number of shallow wells is created to meet the requirement of rising population. It led to the over-exploitation of groundwater and the water table is declining very fast. Some of the existing wells are already dried. The grave situation can be imagined by the fact that around 85% of the population of the city is dependent on groundwater sources particularly hand pumps installed by individual households. Another problem of the city is the contamination of groundwater resources due to lack of a sewage system and treatment before its disposal. This paper elaborates the current status of water resources in terms of availability and its quality. Basin boundaries of water systems do not follow the administrative boundaries. Hence, cooperation and commitment at local, national and international levels, among various institutions, and organizations are needed to manage the looming water crisis in the capital of Afghanistan. This paper stresses the need for city-specific water policy for effective management of water resources. Population growth regulation in water-scarce areas is also required to contain the impact of the water crisis. The water policy will ensure environmental sustainability, enhanced productivity, groundwater recharge and biodiversity.

Keywords Surface water · Groundwater · Water quality · Contamination

M. Khalil (✉) · S. Ahmed · M. H. Khan
Department of Civil Engineering, Jamia Millia Islamia University, New Delhi, India

M. Kumar
Flood and Irrigation Department, New Delhi, Delhi, India

R. K. Joshi
Department of Science and Technology, Govt. of India, New Delhi, Delhi, India

© Springer Nature Singapore Pte Ltd. 2020
S. Ahmed et al. (eds.), *Smart Cities—Opportunities and Challenges*,
Lecture Notes in Civil Engineering 58,
https://doi.org/10.1007/978-981-15-2545-2_66

1 Introduction

Everyone knows about the war in Afghanistan, but few know about the crisis of water in Afghanistan, Afghanistan was once a booming country with beautiful sightseeing cities with abundant food and water supply. Now most of the cities are facing serious water crises. The major causes of the crisis are geographical constraints, global climate change phenomenon, and lack of reservoirs, canals, public awareness on potable water and infrastructure and sanitation. Around 30–35% of water is obtained from rivers coming out of the mountains of Afghanistan. Amu, Darya, Northern, Harirod-Murghab, Hilmand and Kabul are the main rivers of Afghanistan. Out of these Kabul is a seasonal river, which passes through the Kabul city. The river basin map of Afghanistan is shown as Fig. 1. The scarcity of water has led to conflict among the communities. Oxfam found that 43% of local conflicts in Afghan communities are over the water. Around 25% of the deaths among the children are due to the contaminated and bad sanitation [1].

Population shifting from rural to urban areas, led to increase in the proportion of people living in cities. This process is termed as urbanization. It results in an increase in size and numbers in the cities and towns, with vertical and horizontal expansions. It puts tremendous pressure on the diminishing natural resources [2]. In most of the developing countries, the pace of infrastructure development is unable to match the pace of urbanization [3]. Increase in water demand, inadequate sewage collection and treatment infrastructure, absence of wastewater treatment plants and paved surface, severely impacts the water resources and hydrology of the urbanized areas. Crooks et al. [4] analysed the impact of urbanization of the Thames basin and

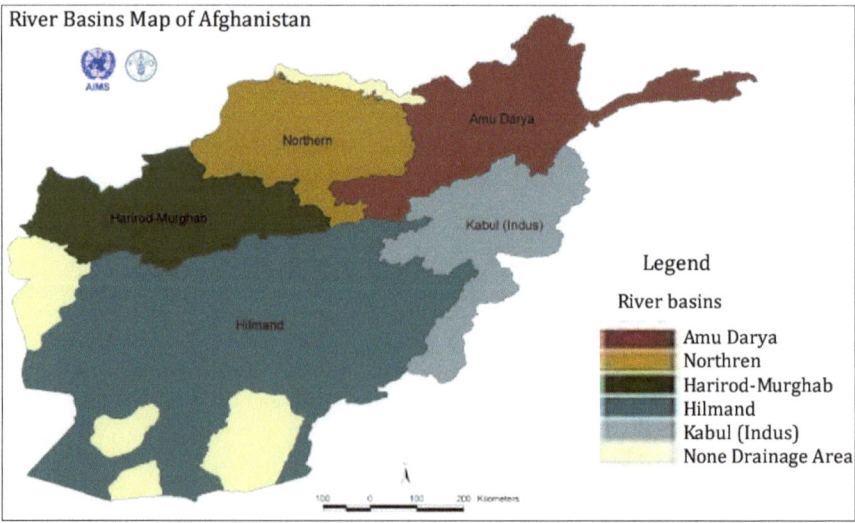

Fig. 1 River basin map of Afghanistan. *Source* Habib [16]

stated that urbanization resulted in storages reduction [4]. In a particular situation, an existing infrastructure may be satisfactorily fulfilling demand, but with the altered scenario and new developments can make the existing system inadequate [5]. A sustainable drainage and sewerage network is essential for preventing the contamination of water resources [6].

A team of BBC Persian journalists tested the Kabul water quality in the Ministry of Rural Rehabilitation and Development (MRRD) laboratories. They have found at least 20 harmful bacteria in 100 mi of a water sample and these samples were collected from the western and northern parts of the Kabul City. Afghanistan National Environmental Protection Agency (NEPA) found that 70% of Kabul groundwater is contaminated with pernicious chemicals and bacteria. According to the Afghanistan Investment Support Agency (AISA) and Environmental Health Directorate of Ministry of Public Health (MoPH), there were 87 registered private companies for potable water supply in 2016, and 58 for these companies are operating in Kabul city [7].

The groundwater resource of Kabul city is clearly faced with long-term sustainability concerns, such as inoperable or drying of 30% of the shallow water-supply wells. The current extraction of groundwater is more than 120,000 m^3/day. The population of Kabul city in 2015 was 4 million with the water demand of 123.4 million m^3/year [8]. The estimated groundwater quantity was 44 million m^3/year with the consumption of 50 L/day per capita [9, 10]. However, only 27.5% population of Kabul city has access to the piped water-supply system [11]. The supply of water to Kabul city is an important and a risky issue due to the groundwater over-abstraction, and scarcity of the source. The main sources of contamination of groundwater are pit latrines, ditches, sewage, septic tanks, solid disposal, lack of sewerage treatment plants, and lack of water-distribution networks. Besides these issues, the Kabul city experienced rapid and uncontrolled population growth, as a result of the high growth rate, security and return of refugees to the country from neighbouring countries. And the demand for water is increasing day by day; therefore, the public awareness and finding the alternative solution for the potable water is a concerning issue in Kabul city [10]. The water quality in terms of physical, chemical and biological parameters when compared with the permissible limit of the World Health Organization (WHO) shows the deteriorated condition of water resources. It stresses the need of water treatment plants in different community centres to provide the drinking water for the inhabitants of Kabul City.

2 Surface Water Quantity

The average annual precipitating of Kabul basin between 1957 and 1977 was 330 mm as per the data of Ministry of Water and Power in 1981. And the average streamflow was 8×10^{-4} $(m^3/s)/km^2$ or rate over the basin 7×10^{-5} m per day. The highest average streamflow was in the month of May with a flow rate of 0.55 m^3/s, while the lowest flow rate occurs in the month of January, with a flow

Table 1 Agricultural land by Kabul river basin

Type of land	Kabul basin (ha)
Surface water	450
Groundwater	99
Rain-feed agriculture land	9

Source Habib [16]

Table 2 Forecasted development of irrigated farming and assumed water consumption of the Kabul basin is as follows

Basin	Forecasted development (1000, ha)			Assumed water consumption (km^3)		
Years	1990	2000	2020	1990	2000	2020
Kabul	110	146	146	1.1	1.5	1.5

Source Habib [16]

rate of 0.05 m^3/s. As the Kabul basin is 80 km-long valley from Paghman to Kohi Safi, which contains the Kabul city and surrounding area with their communities [12]. And the mean annual evaporation rate of the Kabul basin is approximately 1600 mm [13, 14]. Peak flow values of discharge can be determined by application of annual maximum or peak over threshold data series [15].

The Kabul basin water for agricultural land with respect to the type of land use as follows [16] (Tables 1 and 2).

As per the hydrographic system of Afghanistan, the Kabul basin is the eastern basin, which is the only basin that has an outlet to the sea and contains 12% of the total surface water of the country [16]. To use and maintain the potential of surface water of the upper Kabul river basin for future, the feasibility study has been done by the Japan International Cooperation Agency (JICA) in 2011. The study included the Shahtoot dam on the Maidan River and Gat dam on the Logar River (Figs. 2 and 3). As the study shows that the development of the Gat dam will have an impact on the social environment. The Shahtoot dam is expected to provide a mean annual quantity of 87.2 million m^3/year water to supply the Kabul City. As the Shahtoot dam operates the mean flow of the Maidan River will be dramatically reduced and the groundwater average amount will be reduced to 33 million m^3/year up to 2025 [8]. Therefore, the goal of the Ministry of Urban Development Affairs is to provide the 120.4 million m^3/year water from the upper Kabul basin by operating the Shahtoot dam with the availability of 87.2 million m^3/year and local groundwater with the availability of 33.2 million m^3/year for 6.6 million population, with per capita demand of 50 L/day [11, 17].

3 Groundwater Quantity

The Afghanistan Geological Survey (AGS) and the USGS have established a network of 70 wells to monitor and determine the groundwater levels of Kabul Basin. The range of seasonal fluctuation of water level was from 1 m to more than 7 m the seasonal water level. As the hydrography for the central Kabul shows that

Fig. 2 Location of Kabul basin [19]. *Source* USGS [19]

the water level decreases to 25 m. The reason that the groundwater-level declines, shows that demand for consumption of water increases proportionally to the population [18]. The Kabul aquifer has mainly four quaternary interconnected aquifers. The upper Kabul aquifer, which is the primary aquifer with 80 m thickness in the valleys and the upper aquifer is underlain by secondary aquifer with a thickness of 1000 m in some parts valley more than 1000 m [19]. And the upper aquifer has two aquifers (Darlaman-Paghaman sub basin), the lower Kabul aquifers (Kabul-Logar sub-basin). For the purpose of water supply, the quaternary aquifers are mostly used. The sources that are used to recharge the groundwater are recharge from the river bed, recharge from precipitation, foothill recharge, snow melting, irrigation channels, sewage percolation and leakage of septic tanks [11, 13, 14, 18]. The current extraction of groundwater is more than 120,000 m^3/day. The population of Kabul city in 2015 was 4 million and with a groundwater quantity of 44 million m^3/year with the consumption of 50 L/day per capita [9, 10]. However, only 27.5% of the total population of Kabul city have access to the piped water-supply network [11]. Therefore, the study recommends developing the surface water. Though, it is desirable to the upper Logar water aquifer, which is estimated with a potential of

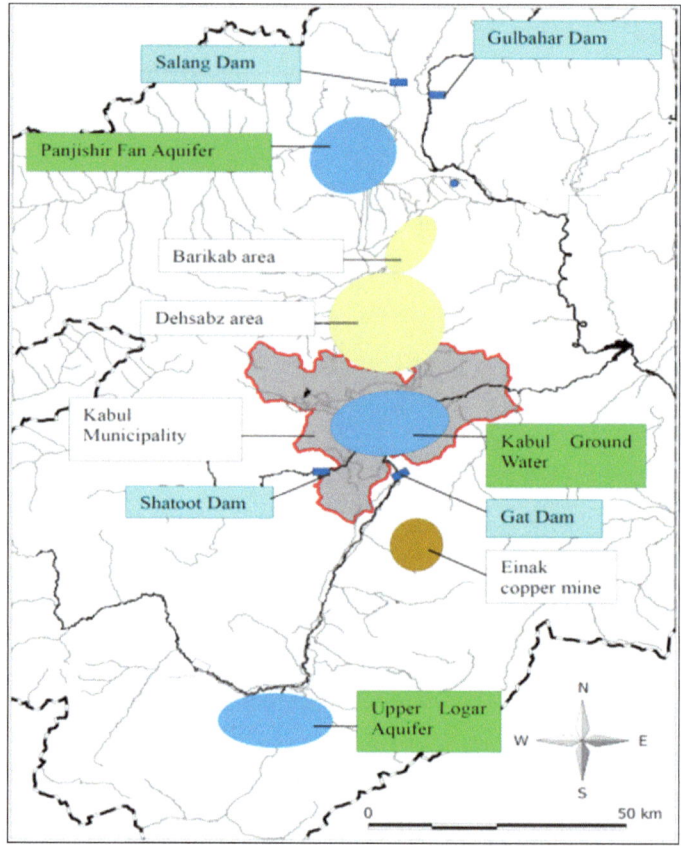

Fig. 3 Location of potential water resource [11]. *Source* Zaryab [14]

63 million m^3/year. It is also possible to be done the artificial recharge (rivers and snowmelt water) of the groundwater for storing the water under the ground and use it for future. It also requires further investigation [14].

4 Surface Water Quality

The Kabul river basin is not a continuum river, which is a periodic and seasonal river with average streamflow 8×10^{-4} (m^3/s)/km^2 or rate over the basin 7×10^{-5} m per day. The highest average streamflow in the month of May with a flow rate of 0.55 m^3/s, while the lowest flow rate month with the mean streamflow is January, with a flow rate of 0.05 m^3/s (Fig. 4). Kabul basin is 80 km-long valley from Paghman to Kohi Safi, which contains the Kabul city and surrounding area with their communities, as the groundwater quality was analysed in the last

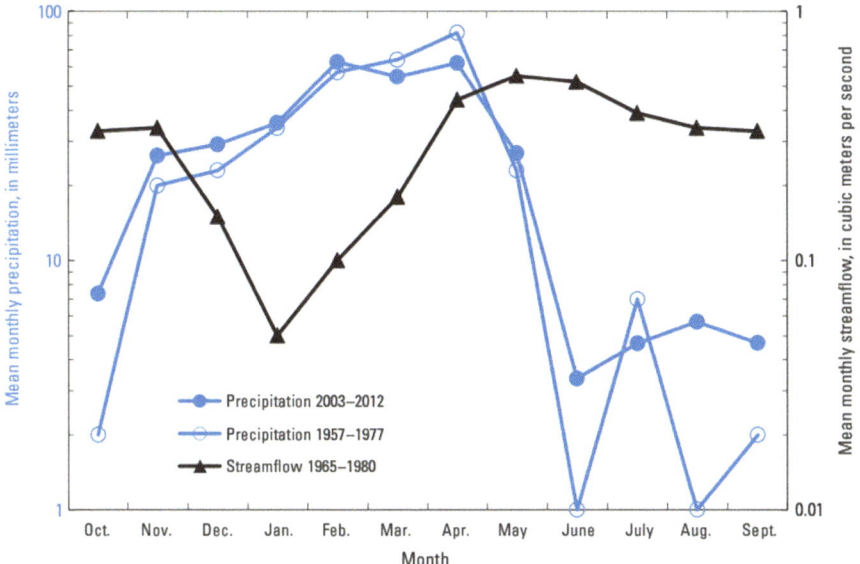

Fig. 4 Mean monthly precipitation in Kabul, from 1957 to 1977 and 2003 to 2012 [12]. *Source* Thomas, Micheal 2014

18 years by the different organization such as DACAAR, USGS, BGR and JICA. In respect to the groundwater, the surface water also required physical, chemical and biological tests. It required further investigation on Kabul River and its tributaries and dams. One sample of surface water collected from Chakari River, which indicates the pH value 7.9 and dissolved solids 250 mg/L [12].

5 Groundwater Quality

The groundwater quality was analysed in the last 18 years by the various organizations such as DACAAR, USGS, BGR and JICA. And they have compared it with the WHO standards and the consequence of their analyzation for the Kabul city and overall Afghanistan is tabulated (Tables 3, 4 and 5).

5.1 Biological Parameters

Coliform bacteria groundwater has existed approximately from 58 to 70% of the wells and urban area of Kabul city [20]. As the DACAAR study shows in 2007, 26% of water contaminated with coliform bacteria, and the WHO limit is 0 coli/100 mg [20]. But the latest information has shown that about 20 harmful bacteria in

Table 3 Physical parameters

Parameters	Results	Permissible limits of (WHO)	Reference range
Water temperature	19–25 Co	20 Co	–
Electrical conductance	(306-13,899) µs/cm	1500 µs/cm	(19% $\leq$ 1500 $\leq$ 81%) mg/L
Transparency or turbidity	>5 NTU	$\leq$ 5 NTU	–
Total dissolved solids (TDS)	(833–1328) mg/L	600 mg/L	–

Source Saffi and Paikob-e-Naswar [20]—[8, 9, 18, 20, 21]

Table 4 Chemical parameters

Parameters	Results	Permissible limits of (WHO)	Reference range
pH	7–7.9	6.5–8.5	–
Chemical oxygen demand (COD) (mg/L)	(4–23)	250	
Total hardness (TH) (mg/L)	(50–1850)	180	>80% very hard
Nitrate (NO$_2$) (mg/L)	(0.08–90)	50	(90% $\leq$ 50 $\leq$ 10%)
Nitrite (NO$_3$) (mg/L)	(0.001–1.01)	0.2	(4% $\leq$ 0.2 $\leq$ 96%)
Sulphate (mg/L)	(1–2900)	250	(81% $\leq$ 250 $\leq$ 19%)
Iron (mg/L)	<0.3	0.3	
Manganese (mg/L)	<0.4	0.4	
Sodium (mg/L)	(7–1010)	200	(82% $\leq$ 200 $\leq$ 18%)
Chloride (mg/L)	(1.3–512)	250	(7% $\leq$ 250 $\leq$ 93%)
Potassium (mg/L)	<10	10	(6% $\leq$ 10 $\leq$ 90%)
Silica (mg/L)	(0.04–74)	10	(8% $\leq$ 10 $\leq$ 92%)
Fluoride (mg/L)	(0.5–10.7)	1.5	(17% $\leq$ 1.5 $\leq$ 83%)
Aluminium (mg/L)	<0.2	0.2	–
Boron (mg/L)	(0.05–3.9)	0.5	(15% $\leq$ 0.5 $\leq$ 85%)
Bromide (mg/L)	(0.01–0.8)	0.1	(20% $\leq$ 0.1 $\leq$ 80%)
Arsenic (mg/L)	(0.0005–0.01)	0.01	–
Copper (mg/L)	<1	1	–

Source Saffi and Paikob-e-Naswar [20]—[8, 9, 18, 20, 21]

100 mm of water detected in the western and northern parts of the Kabul City. And also the Afghanistan National Environmental Protection Agency (NEPA) warned that 70% of Kabul groundwater is contaminated with pernicious chemicals and bacteria [7].

Table 5 Concentration of 10 toxic elements in the Kabul river basin [22]

Elements	Permissible limit (µg/L)	Average (µg/L)
Cr	50	6.2
Ni	70	4.1
Cu	2000	6.5
As	10	3.9
U	30	25.1
Cd	3	0.0
Sb	20	0.1
Ba	700	79.5
Hg	6	0.0
Pb	10	0.2

Source Kato el al. [22]

6 Discussion and Recommendation

The existing surface water resources are getting contaminated due to the dumping of garbage and other solid waste materials in water bodies. The majority of the population is using septic tanks and soak pits. Most of the households do not have piped water supply. In the absence of the sewerage system and its treatment, groundwater is contaminated. To improve the groundwater quality and access to potable water, the following points should be considered.

- Collection of garbage and waste, to the disposal site to prevent its dumping into water bodies and their subsequent contamination.
- Providing a sewage treatment system, with the sewer network system.
- Removal sewage from septic tanks of houses and latrines, or conversion of septic tank of the courtyard to mini-treatment plants.
- Wait for some wet years to flush out the pollutions by groundwater flow.
- Creating public awareness about groundwater pollution through schools, universities and social media.
- Improving the surface water systems to satisfy the required demand of water.
- Preparing water treatment plants to facilitate safe and clean water for the citizens
- Developing the water-distribution network.

7 Conclusion

The availability of water is not only vital for the survival of life, but for the economic growth of the country. The sustainability of water supply to Kabul city is at great risk due to over-exploitation of groundwater resources and the contamination of groundwater resources due to untreated sewage in the city. The analysis of groundwater quality shows that the quality of groundwater has deteriorated seriously, particularly in the urban areas of the Kabul city. The majority of households are dependent on groundwater in the absence of piped water supply. Around 85% of the population uses water from shallow wells or hand pumps. The existing groundwater sources are capable to provide water to around 2 million populations, while the existing population of Kabul city is around 6 million. It stressed the need for bringing water from outside. Besides these issues, the Kabul city is experiencing rapid and uncontrolled population growth. The high growth of the population in Kabul city is due to the high birth rate, the improved security situation in the city led to migration from rural areas and the return of refugees to the country from neighbouring countries. The recent droughts have stressed the need for water policy for the Kabul city. The source of water should be augmented. Over-exploitation of groundwater has already made many wells inoperable and dry. Groundwater is contaminated at many places, making the groundwater unfit for use as potable water. The climate change phenomenon and diminishing glaciers in Afghanistan are the issues of the highest concern for future water availability in the Kabul Basin. The time is for recharge and augments the groundwater resources. Dams are to be made to store water for its diversion and use in Kabul city. In addition to the optimal use of water and bringing water from other sources, the treatment of sewage and the use of recycled water should be part of the future water policy of the Kabul city.

References

1. Jazeera Al (2012) Afghanistan Water Crisis. www.hydratelife.org/afghanistan-water-crisis/ DOA: 25/02/2019
2. Kumar M, Sharif M, Ahmed S (2019) Impact of urbanization on the River Yamuna Basin. Int J River Basin Manag, 1–27 (just-accepted)
3. Kumar M, Sharif M, Jain VK, Ahmed S (2016) Delhi urban flooding—an overview. In: Indian building congress, seminar on towards building smart and sustainable infrastructure in urban development, vol 23, number 2, October 2016, pp 44–49, ISSN 2349-7475
4. Crooks S, Cheetham R, Davies H, Goodsell G (2000) EUROTAS (European river flood occurrence and total risk assessment system) Final Report
5. Sharif M, Kumar M, Ahamd S (2016) Need of flood proof housing in flood plains of river Yamuna in Delhi. In: International conference n on building utilities 2016, organized by department of mechanical engineering, faculty of engineering and technology, Jamia Millia Islamia, Enriched Publication, New Delhi, India, pp 435–441 (978-163535676-2)
6. Kumar M, Sharif M, Ahmed S (2017) Flood risk management strategies for national capital territory of Delhi, India. ISH J Hydaul Eng. http://doi.org/10.1080/09715010.2017.1408434
7. Blue Gold, the quest for the household water in Kabul city (2018). https://reliefweb.int/report/Afghanistan/blue-gold-quest-household-water-Kabul-city. DOA: 25/02/2019
8. JICA (2011) Japan international cooperation agency study team, draft Kabul City master plan. Product of technical cooperation project for promotion of Kabul metropolitan area development sub project for revise the Kabul city master plan
9. Beller consults (2004) Feasibility study for the extension of the Kabul water supply system, Interim Report, (Funded by Kreditanstalt fur Wiederaufbau (KfW))

10. Eqrar N (2015) Groundwater quantity and quality problems in Kabul city-Kabul-University
11. Mack TJ, Akbari MA, Ashoor MH, Chornack MP, Coplen TB, Emerson DG, Hubbard BE, Litke DW, Michel RL, Plummer LN, Rezai MT, Senay GB, Verdin JP, Verstraeten IM (2010) Conceptual model of water resources in the Kabul Basin, Afghanistan. U.S. Geological survey scientific investigations report 2009–5262, p 240
12. U.S Geological Survey (USGS) Hydrogeology and water quality of the Chakari basin, Afghanistan (2014–5113) Project No. 263
13. Taher MR, Chornack MP, Mack TJ (2013) Groundwater-level monitoring and sustainability in the Kabul Basin, Afghanistan, 2004 to 2013. In: 4th international Hindu kush geosciences conference, Kabul Polytechnic University, Kabul Afghanistan
14. Zaryab A, Noori AR, Wegerich K, Kleve B (2017) Assessment of water quality and quantity trends in Kabul aquifers with an outline for future drinking water supplies. Kabul Polytechnic University, Kabul, Afghanistan
15. Kumar M, Sharif M, Ahmed S (2018) Flood estimation at Hathnikund Barrage, river Yamuna, India using the peak-over-threshold method. ISH J Hydraul Eng, 1–10
16. Habib H (2014) Water related problems in Afghanistan. Int J Educ Stud 1(3):137–144
17. Myslil V, Eqrar NM, Hafisi M (1982) Hydrogeology of Kabul Basin–report for United Nations children found und das ministry of water and power democratic republic of Afghanistan [unpublished]
18. DACAAR (2011) Danish Committee for Aid to Afghan refugees, update on "National groundwater monitoring wells network activities in Afghanistan" from July 2007 to December 2010. DACAAR, Kabul, p 23
19. U.S Geological Survey (USGS) (2010) Availability of water in the Kabul basin, Afghanistan
20. Saffi MH, Paikob-e-Naswar W (2011) Groundwater natural resources and quality concern in Kabul Basin, Afghanistan
21. Houben G, Tunnermeier T (2005) Hydrogeology of the Kabul Basin-part groundwater geochemistry and microbiology. Foreign Office of the Federal Republic of Germany, BGR
22. Kato M, Azimi MD, Fayaz SH, Shah MD, Hoque MZ, Hamajima N, Yoshinaga M (2016) Uranium in well drinking water of Kabul, Afghanistan and its effective, low-cost depuration using Mg-Fe based hydrotalcite-like compounds. Chemosphere 165: 27–32

Durability of Soil Blended with Flyash

Aasia Mukhtar, Suruchi Sneha, Sadiqa Abbas and S. M. Abbas

Abstract Coal-based thermal power plants are the main source for electricity generation in India which leads to the production of huge quantities of flyash and later its disposal problems. To solve this problem many researchers have used soil–flyash mix as a smart material for the construction of roads and embankments. The researchers have proposed different quantities of flyash in the soil to get the engineering behaviour in terms of better stability but a few of them have taken into account the durability consideration. The present project deals with the durability of various mixes of soil and flyash. Locally available soil and flyash obtained from CRRI, New Delhi was mixed at different proportions. Index properties of these materials were determined initially. Heavy Proctor Compaction Tests were performed on these mixes to get the OMCs to be considered for the determination of CBR values. The durability of these mixes was determined in the form of CBR values passing through five weathering cycles to simulate five years in actual. To simulate the Delhi climate, the CBR moulds were kept for 4 days at 50 °C (representing summer season) → 1 day at room temperature → 4 days submerged in water (representing monsoon season) → 3–4 days at room temperature → 4 days at 0 °C (representing winter season) → 1 day at room temperature as one weathering cycle. After five weathering cycles, CBR values were determined for the mixes. It is found that the CBR values before weathering cycles decrease with respect to that of pure soil up to 10% flyash. The soil + 15% flyash mix got more CBR value. For the soil, the CBR value decreases after passing through the five weathering cycles. The CBR values after five weathering cycles increase with increasing percentage of flyash. This shows that the use of flyash in the soil not only increases the CBR but increases the durability of the mix.

Keywords Flyash · CBR · Durability

A. Mukhtar · S. Sneha (✉) · S. M. Abbas
Department of Civil Engineering, Jamia Millia Islamia, Jamia Nagar, New Delhi, India

S. Abbas
Department of Civil Engineering, Manav Rachna International Institute of Research and Studies, Faridabad, Haryana, India

© Springer Nature Singapore Pte Ltd. 2020
S. Ahmed et al. (eds.), *Smart Cities—Opportunities and Challenges*,
Lecture Notes in Civil Engineering 58,
https://doi.org/10.1007/978-981-15-2545-2_67

831

1 Introduction

Rapid industrialization and urbanization over past couple of decades have led to serious environmental problems. In particular, environmental pollution due to the generation and management of waste materials has become one of the most emergent problems. The disposal issue of such waste products has opened a new window of challenges for engineers. A wide spectrum of waste materials is being discarded in large quantities across the globe, a part of which could be used for geotechnical purposes. Disposal issue of the waste products is a challenge. Some of those materials are not biodegradable and often leads to waste disposal crisis and environmental pollution. The present article seeks the possibilities of whether some of these waste products can be utilized as a durable construction material.

Different waste materials such as recycled crushed aggregates, recycled glass, flyash/pond ash, crumb rubber tires, incinerator ash and municipal solid waste (MSW) are used for geotechnical purpose. Researches have been carried out to mix such wastes with soil and understand their behaviour. In other words, to determine their properties, these materials were mixed with soils in different compositions and tested.

Recycled concrete aggregate (RCA) or crushed concrete is produced from construction and demolition debris. RCA used as a base and sub-base is more prevalent in the construction of roadways that typically have lower traffic volumes.

Glass aggregate is a relatively new material for construction. Glass aggregate is generally durable, robust and easy to place and compact. The percentage of glass content, gradation and compaction level and the type or source of glass greatly affect the geotechnical parameters of cullet aggregate.

Flyash is beneficial in roadway constructions, which can be included in embankments and pavements' structural layers such as base/sub-base layers and shoulders to create long lasting and sustainable structure. This utilization of waste material such as flyash for the stabilization of soil proves to be a great advantage in city improvement and city renewal which are the strategic components of development in Smart Cities Mission. In the approach of the Smart Cities Mission, the objective is to provide a clean and sustainable environment which can be achieved by cleaning up the waste flyash dumped from various thermal power plants and utilizing it as smart material in construction of roads and embankments [1].

Recent experiences using of flyash in embankments as fill materials have shown that flyash is an excellent fill material for engineering purposes, the performance is as good as naturally available borrow materials. The geotechnical characteristics of flyash such as moisture-density relationship, particle size distribution, permeability and strength are of major importance in embankment design consideration.

According to Moulton [2], natural soils have 1–5% air voids at maximum dry density whereas the same for flyash is 5–15%. The higher void content tends to limit the build-up of pore pressures during compaction and thus allowing flyash to be compacted over a large range of water contents. Yudhbir and Honjo [3] have reported that flyashes with high carbon content gave lower dry density values and higher OMC values. Ferguson [4] studied the feasibility of using class C flyash from Kansas Power Plant and Light Jeffrey Energy Centre for the stabilization of subgrade materials. The addition of flyash altered the compaction characteristics of both granular and cohesive materials. When the compaction tests were carried out, immediately after mixing, the compaction characteristics changed due to change in material gradation. When the compaction tests were delayed, hydration products bonded the particles in the loose state and thus disruption of these aggregations was required to densify the material.

A good number of researchers have studied different materials mixed with soil to understand soil stability, however not much work has been done in the area of soil durability, when mixed with different materials. This article presents the durability study conducted on locally available soil blended with different percentages of flyash in form of CBR values. The CBR values have been determined before and after five weather cycles as per Delhi's climate conditions as Delhi experiences a humid subtropical climate, the temperature varies from 2 to 47 °C in a cycle, it receives about 712 mm average annual rainfall spread across 27 rainy days.

2 Literature Review

Literature review was conducted in reference to the properties of the soils blended with various percentages of flyash.

Ansary et al. [5] studied the effect of flyash stabilization on geotechnical properties of coastal soil to develop a general approach for soil improvement in coastal area using flyash. The study was limited to chemical stabilization by flyash; the author concluded that flyash stabilization of the coastal soils under study rendered it suitable for road construction.

A study was conducted by Meng et al. [6] for solidification/stabilization of soil using flyash. In the study carried out, chromium and lead present in the contaminated clayey sand soil were immobilized using flyash and quicklime (CaO). He also performed unconfined compressive strength and swell tests to investigate the reusability of stabilized waste for various construction activity. It was observed that heavy metal leachability of the contaminated sandy soil was considerably reduced by adding flyash and quicklime and observed values were below the regulatory non-hazardous metal leachability limits. However, addition of flyash and quicklime leads to the formation of excessive ettringite which is a pozzolanic product formed due to sulphate presence. The pozzolanic material (ettringite) thus formed may lead to swelling and deterioration of stabilized soil matrix when it is brought in water contact. It was also observed during the study that addition of flyash causes a drastic increase in the strength of soil mix which is completely untreated, coarse grained and having with low content of clay.

Misra [7] blended C class flyash with several different clay soils of varying plasticity to study moisture-density relationships and strength behaviour of stabilized soils. It was observed that the characteristics of stabilized soil are dependent on the soil mineral type and plasticity. Class C flyash is advantageous because of its self-cementing property and hence can be used in soil stabilization. Since flyash is a by-product; uniformity of its physical and chemical properties is significant for quality control. It was found that the flyash used in these experiments has a rapid hydration characteristic. Addition of moisture to clay-flyash blends with very less or no delay helped to achieve a higher density and strength. Conversely, delay in compaction gave low densities and strength.

Edil et al. [8] studied the efficiency of self-cementing flyashes produced from combustion of sub-bituminous coal at electric power plants for stabilization of fine-grained soils. A laboratory study was conducted in which soil–flyash mixtures were prepared for different flyash contents ranging from 10 to 30% to assess how adding flyash can improve the CBR and resilient modulus (M_r) of wet fine-grained subgrade soils. The California bearing ratio (CBR) of soil–flyash mixes increased with an increase in flyash content and decreased with an increase in compaction water content. CBR value increased considerably when flyash was mixed with wet or highly plastic, i.e. poor fine-grained soil.

Pandian et al. [9] conducted experimental study to understand the variation in California bearing ratio (CBR) values of black cotton soil due to addition of flyash for both unsoaked and soaked conditions. The study indicates that CBR values of black cotton soil mixed with flyash increased up to an optimum flyash content for flyash acting as a coarser material. When black cotton soil acts as a binder, the CBR values decrease beyond the optimum flyash content and then increased to an optimum value. Author concluded that the proper proportion of black cotton soil and flyash improves the CBR values. The effective utilization of flyash, on the one hand, proves to be an effective admixture for improving the soil quality and, on the other hand, affords a means of disposal of the industrial by-product without adversely affecting the environment.

Very limited studies have been carried out to assess durability of soils blended with flyash under various load conditions over time. To investigate durability of flyash stabilized soil, specimens were exposed to the repeated climatic condition in the laboratory simulating the effect of weather cycle of Delhi seeking the three seasons in a year viz. summer season, rainy season and winter season. The weather cycle of 5 years is taken into consideration for testing the durability of soil mixed with flyash. The flyash content in the analysed samples was 5, 10 and 15%. CBR values of various samples of soil mixed with different proportions of flyash before and after undergoing the effect of weather representing the Delhi climate have been used to assess soil durability.

3 Materials and Methodology

3.1 Collection of Materials

(i) The soil sample was collected locally from a park near the building of Faculty of Engineering and Technology, Jamia Millia Islamia, New Delhi, India.

(ii) The flyash was collected from transportation department of Central Road Research Institute (CRRI), New Delhi, India.

3.2 Laboratory Testing

A series of laboratory tests were conducted on the soil and flyash to find out their physical properties as mentioned below:

(i) Gradation test was carried out on oven-dried soil following the standard procedure of sieve analysis (IS: 2720 Part 4) [10].

(ii) Specific gravity of the soil and flyash sample was determined using the Pycnometer as per IS: 2720 Part 3 (1980) [11].

(iii) The optimum moisture content (OMC) and the maximum dry density (MDD) as per IS: 2720 Part 8 [12] were determined for the following mixtures:

- Soil sample
- Soil + 5% flyash (by weight)
- Soil + 10% flyash (by weight)
- Soil + 15% flyash (by weight).

(iv) The California bearing ratio (CBR) as per IS 2720 (Part 16), 1979 [13] of the above-mentioned combinations was determined.

(v) To investigate durability of flyash stabilized soils, specimens were exposed to the repeated temperature variation in the laboratory. For this purpose, the CBR tests were conducted after the CBR mould passed through the 5 years' weather cycles of Delhi's climate.

3.3 Weather Cycle Simulation

The climate cycle of 5 years is taken into consideration for testing the durability of soil mixed with flyash. Delhi experiences summer, rainy and winter seasons annually. To simulate the summer season, well-compacted specimens are kept in the oven at 50 °C for 4 days. Then, the specimens were kept at room temperature for 1 day. This was followed by simulation of the monsoon season. The specimen was soaked in water for next 4 days then kept at room temperature for 2–3 days to

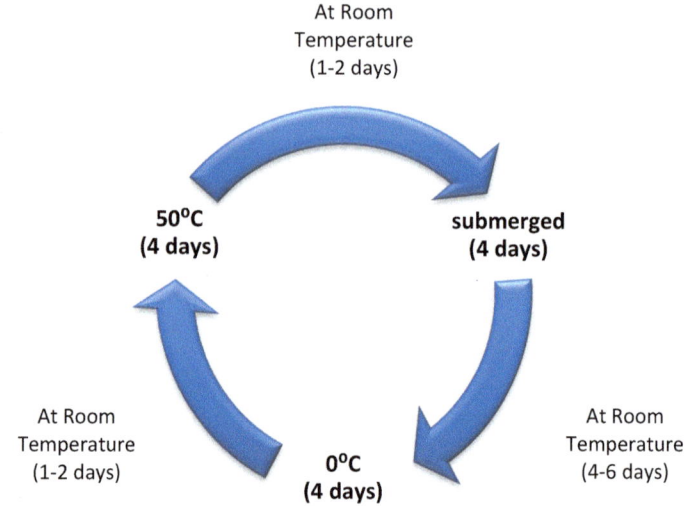

At Room
Temperature
(1-2 days)

50°C
(4 days)

submerged
(4 days)

At Room
Temperature
(1-2 days)

0°C
(4 days)

At Room
Temperature
(4-6 days)

Fig. 1 One weather cycle simulated to Delhi's climate

drain out excess water. Then, the simulation for winter season was done, and the specimen was kept in fridge at 0 °C for 4 days and then left at room temperature for a day. The weather cycle simulation is depicted in Fig. 1. This cycle was repeated five times for each specimen before conducting the CBR test on it again.

3.4 Durability Study

Stabilized soil has to be durable so that it can maintain strength under all adverse conditions. A good stabilizer should also retain its bonding with soil and not only help in gaining strength during the cyclic seasonal changes. The durability is judged by comparing the CBR values of various samples of soil–flyash mixes with different proportions before and after undergoing the effect of temperature variation representing the climate change behaviour.

4 Result and Discussion

Result obtained from the laboratory tests on soil, flyash and soil–flyash mixes is presented in this section.

Table 1 Specific gravity of soil and flyash

Sr. no.	Sample	Specific gravity
1.	Soil	2.55
2.	Flyash	1.97

4.1 Specific Gravity

The values of specific gravity of soil and flyash are shown in Table 1. It is observed that flyash sample is lighter than the soil sample.

4.2 Sieve Analysis

Sieve analysis of soil sample is shown in Fig. 2. The soil type is sandy silt which is well graded containing particle sizes varying between coarse and fine particles.

Uniformity coefficient of soil, $C_u = D_{60}/D_{10} = 26.92$
Coefficient of curvature, $C_C = (D_{30})^2/(D_{10} \times D_{60}) = 0.33$
Fineness Modulus, F.M $= \sum$Cumulative Per cent Retained$/100 = 5.1$.

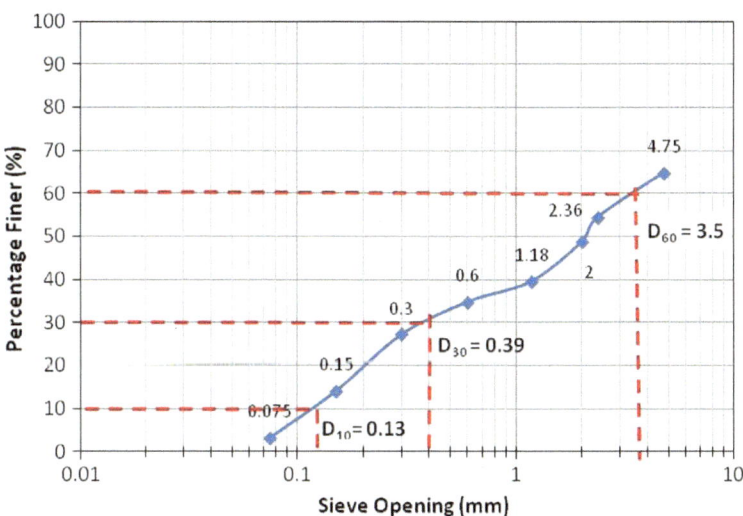

Fig. 2 Sieve analysis of soil sample

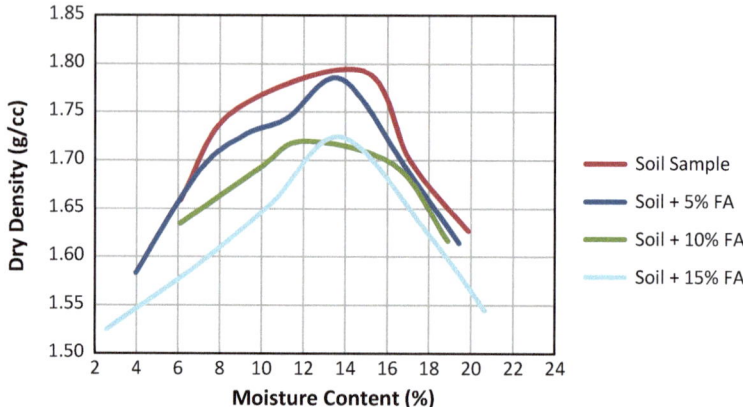

Fig. 3 Comparison of dry density versus moisture content of different soil and flyash mixes

4.3 Modified Proctor Compaction Test

Comparison of dry density versus moisture content of various samples is shown in graph in Fig. 3.

Variations of MDD and OMC for different proportions of soil and flyash mixes are shown in Figs. 4 and 5. Figure 4 shows a downward curve, representing a decrease in maximum dry density with increase in percentage of flyash added to the

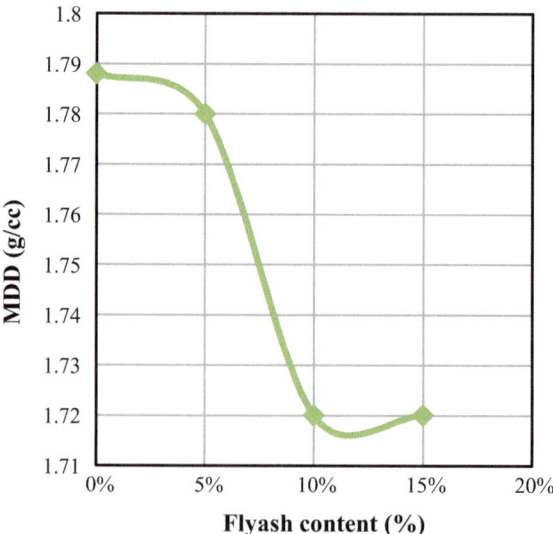

Fig. 4 Variation of MDD of soil with flyash content

Fig. 5 Variation of OMC of
soil with flyash content

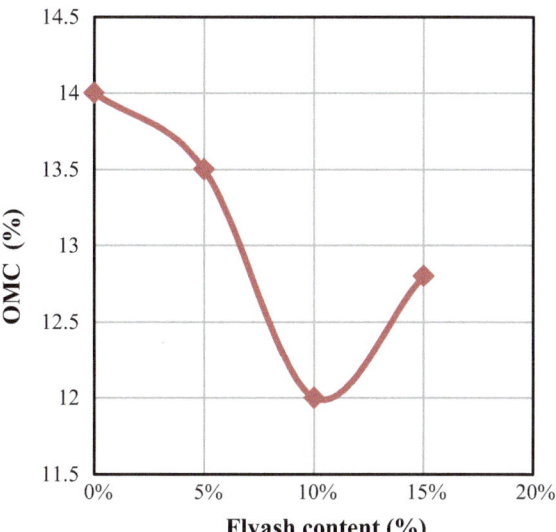

soil. Although at soil + 10% and 15% flyash sample, MDD is constant. Figure 5
shows a decrement in optimum moisture content (OMC) for soil mixed with 5%
and 10% flyash compared to the plain soil. For soil + 15% flyash sample, there is
an increase in OMC value compared to soil + 10% flyash.

4.4 CBR Test

Load versus penetration for different samples for before and after weather cycle are
shown in Figs. 6 and 7. Comparison of CBR values before and after weather cycle
for various mixtures of soil and flyash mixes is shown in graph in Fig. 8 corresponding to Table 2.

Figure 8 shows the comparison of the CBR values of the various soil and flyash
mixes before and after weather cycle. The CBR value for soil + 5% flyash mix and
10% Flyash mix decreases 2.92% and 0.97% before weather cycle. The CBR value
for soil + 15% flyash increases 8.52% before weather cycle. The CBR value for soil
mixed with 5% flyash decreases 1.46% after weather cycle. The CBR value for soil
mixed with 10% and 15% flyash increases 3.43% and 15.06% after weather cycle.

The decrease in the CBR values may be caused due to lack of bonding between
soil and flyash particles due to small amount of flyash or limited pozzolanic reactions of flyash caused by varying temperature. Also Fig. 6 shows a large increment
of CBR value between soil + 10% FA and soil +15% FA. This behaviour may be
due to development of a sufficient bond between particles and due to enhanced
pozzolanic reaction.

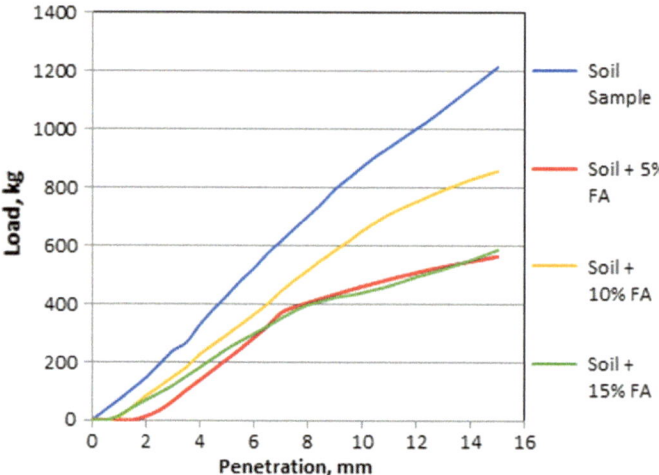

Fig. 6 Load versus penetration for different soil and flyash mixes before weather cycle

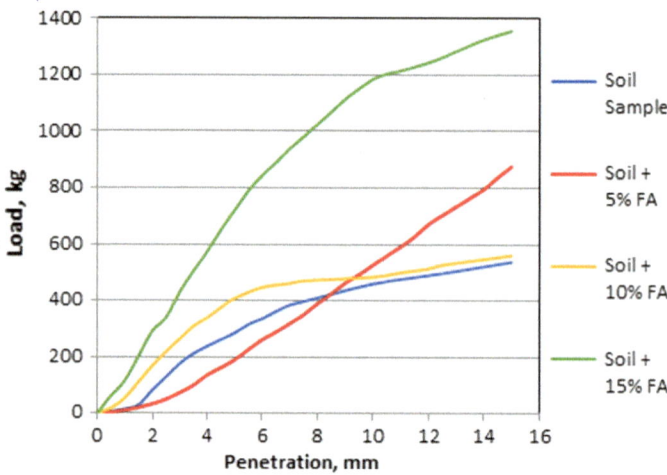

Fig. 7 Load versus penetration for different soil and flyash mixes after weather cycle

Percentage changes in CBR values of different samples before and after weather cycle are shown in Table 3 and Fig. 9. It can be seen that percentage increase in CBR value before and after weather cycle increases with increase in percentage of flyash mixed with the soil. For original soil sample and soil mixed with 5% flyash

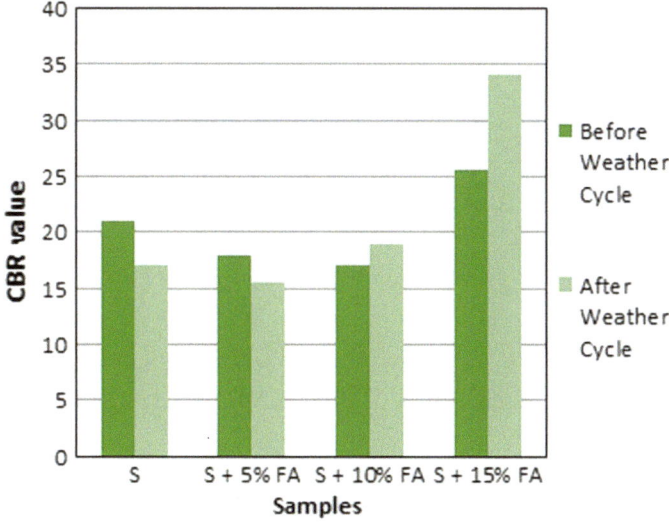

Fig. 8 CBR at 5.0 mm penetration for different soil and flyash mixes

Table 2 CBR values of different soil and flyash mixes

Sample	CBR at 5.0 mm	
	Before weather cycle	After weather cycle
S	20.92	17.03
S + 5% FA	18	15.57
S + 10% FA	17.03	19
S + 15% FA	25.55	34.06

Table 3 Percentage change in CBR values after weather cycle

Sample	Per cent increase after weather cycle for 5.0 mm penetration
S	−3.89
S + 5% FA	−2.43
S + 10% FA	1.97
S + 15% FA	8.51

by its weight, the CBR value decreases by 3.89% and 2.43% after weather cycle, respectively. Whereas for soil mixed with 10% and 15% flyash, the CBR values increase by 1.97% and 8.51% after weather cycle.

The maximum increase in CBR value before and after weather cycle can be seen for soil mixed with 15% flyash.

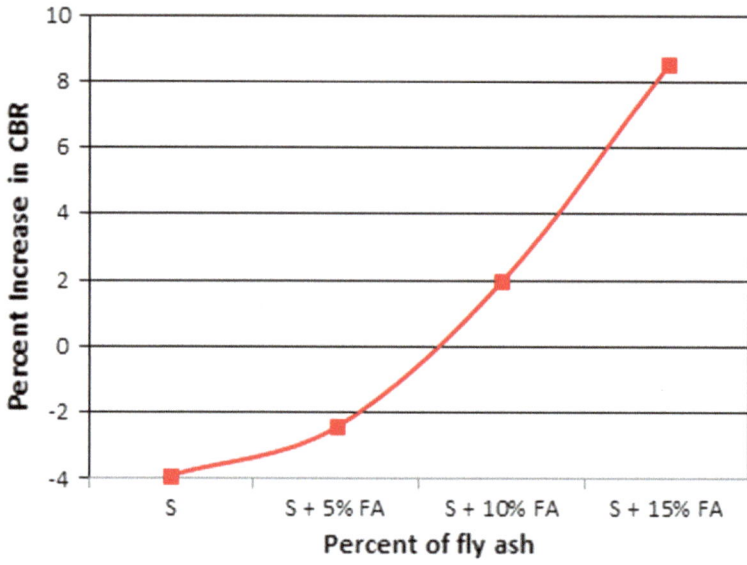

Fig. 9 Per cent variation in CBR for different soil and flyash mixes

5 Conclusion

The objective of this study was to determine the durability of various mixes of soil and flyash in the form of CBR values before and after five weathering cycles. Locally available soil and flyash obtained from CRRI, New Delhi was mixed at different proportions and tests were conducted. Flyash was mixed with soil at three different percentages (5, 10 and 15%).

The following conclusions are drawn:

i. MDD of the mixes decreases with increasing percentage of flyash while OMC decreases most for soil + 10% flyash, thereafter it increases.
ii. For all the cases, the CBR values were obtained at 5.0 mm as it was more than that of 2.5 mm penetration.
iii. The CBR values, before weather cycles, decrease with increase in flyash content up to 10%, thereafter the CBR values increase.
iv. The CBR values, after weather cycles, decrease with addition of 5% flyash then increase with increase in flyash content.
v. Comparing the CBR values before and after weather cycles following conclusions is drawn:

- For soil and soil + 5% flyash the CBR values decrease, whereas,
- For rich mixes, the CBR values increase,
- Therefore, it can be inferred that the durability of pure soil and soil + 5% flyash decreases, whereas, on addition of more flyash, the durability increases.

vi. The soil is becoming more durable with increase in flyash content beyond 5%.

The optimum dose of flyash would be obtained if more quantity of flyash is mixed with the soil.

On the basis of this experimental study, it can be concluded that soil blended with more than 5% flyash is more durable considering the effects of weather changes according to Delhi's climate cycle. Thus, use of higher percentage of flyash for soil stabilization in the roads and embankment construction will increase the durability of roads and embankments. This productive use of flyash will lead to considerable environmental as well as health benefits, reducing land air and water pollution thereby decreasing the health hazards caused by them. Also, this smart construction technique will lead to economic growth and improve the quality of life of people which is the main purpose of Smart Cities Mission [1].

References

1. Smart Cities Mission (2017) Ministry of Housing and Urban Affairs, Government of India. http://smartcities.gov.in/content/innerpage/what-is-smart-city.php
2. Moulton KL (1978) Technology and utilization of power plant ash in structural fills and embankments. West Virginia University, pp 791–806
3. Yudhbir, Honjo Y (1991) Applications of geotechnical engineering of environmental control. Theme lecture 5, 9 ARC, Bangkok, Thailand, vol 2, pp 431–469
4. Ferguson G (1993) Use of self-cementing flyashes as a soil-stabilizing agent. In: Proceedings of session on flyash for soil improvement. ASCE geotechnical special publication, vol 36, pp 1–14
5. Ansary MA, Noor MA, Islam M (2006) Effect of flyash stabilization on geotechnical properties of Chittagong coastal soil. In: A collection of papers of the geotechnical symposium in Rome, 16–17 Mar
6. Meng X, Hua Z, Dermatas D, Wang W, Kuo HY (1998) Immobilization of mercury (II) in contaminated soil with used tire rubber. J Hazard Mater 57.231–241
7. Misra A (1998) Stabilization characteristics of clays using class C flyash. Transp Res Rec 1161:46–54
8. Edil TB, Acosta HA, Benson CH (2006) Stabilizing soft fine-grained soils with flyash. J Mater Civ Eng 18(2):283–294
9. Pandian NS, Krishna KC, Sridharan A (2001) California bearing ratio behavior of soil/flyash mixtures. J Test Eval 29(2):220–226
10. IS 2720 (Part 4) 1979, Method of tests for soils: Part 4, Grain size analysis
11. IS: 2720 (Part 1) 1979, Method of tests for soils: Part 3, Determination of specific gravity
12. IS 2720 (Part 8) 1979, Method of tests for soils: Part 7, Determination of water content-dry density relation using heavy compaction
13. IS 2720 (Part 16) 1979, Method of tests for soils: Part 16, Laboratory determination of CBR by Indian Standards

Impact of Nanotechnology in the Development of Smart Cities

Akanksha Gupta, Vinod Kumar, Sirajuddin Ahmed and Siddhartha Gautam

Abstract Due to rapid urbanization and limited availability of resources, it is need of the hour to make cities sustainable, greener and smarter. The idea to make cities smart is essential in point of view of increasing population in urban areas and to meet the demand smartly. Nanotechnology has applications in various fields such as in industries, agriculture, biomedical and military equipment. Nanotechnology has the potential to make cities smart by using different nanomaterials for energy storage, smart building construction, infrastructure, smart textiles, environmental remediation and nanoscale photonics technologies to mention a few. In the present article, authors have attempted to provide insights into the ways by which nanotechnology can be utilized for developing smart cities.

Keywords Nanomaterials · Energy · Infrastructure · Transportation · Environmental remediation

A. Gupta
Department of Chemistry, Sri Venkateswara College, University of Delhi, New Delhi, India

V. Kumar (✉)
Special Centre for Nano Sciences, Jawaharlal Nehru University, Delhi, India
e-mail: vinod7674@gmail.com; vinodkumar@kmc.du.ac.in

V. Kumar
Department of Chemistry, Kirori Mal College, University of Delhi, New Delhi, India

S. Ahmed
Department of Civil Engineering, Jamia Millia Islamia (Central University), Jamia Nagar, New Delhi 110025, India

S. Gautam
Delhi Pollution Control Committee, Government of NCT of Delhi, New Delhi, India

© Springer Nature Singapore Pte Ltd. 2020
S. Ahmed et al. (eds.), *Smart Cities—Opportunities and Challenges*,
Lecture Notes in Civil Engineering 58,
https://doi.org/10.1007/978-981-15-2545-2_68

1 Introduction

Exponential growth has been experienced in the urban regions from the past few decades. Currently, around 54% of the population in the world exists in an urban area, and it is estimated that by 2050, around 66% will live in cities [1, 2]. This speedy growth will challenge cities to meet the huge demand for limited resources such as food, land, clothes, energy throughout the world. Therefore, it is indispensable to provide sustainable access to basic, smarter and better facilities. Further, if the extraction of materials from limited sources continues, which is also a major cause of global climate change, then it will be difficult to meet all the demands of the huge population. The answer to these problems lies in the development of smart cities across all sectors of the society with "smart" solutions [3]. The physical, social, institutional and economic infrastructure has to be focused comprehensively. The idea to make cities smarter can be done by utilizing technologies to improve in terms of infrastructure and city operations [4–8]. This includes smart electricity distribution, smart buildings, smart health care, smart materials to name a few (Fig. 1).

Nanotechnology has emerged to be one of the important fields of science that can be utilized to make cities smart [9, 10]. Nanomaterials can be 0D (quantum dots), 1D (nanoplates), 2D (nanofibres, nanotubes, nanorods, nanowires) or 3D (nanoparticles) [11]. With the aid of nanotechnology, it is possible to tune the structure and properties of materials to obtain desired materials. Nanomaterials are highly flexible, lightweight, conductive and durable and possess high surface area. Nanoparticles permit us to attain unbelievable properties of materials, which are

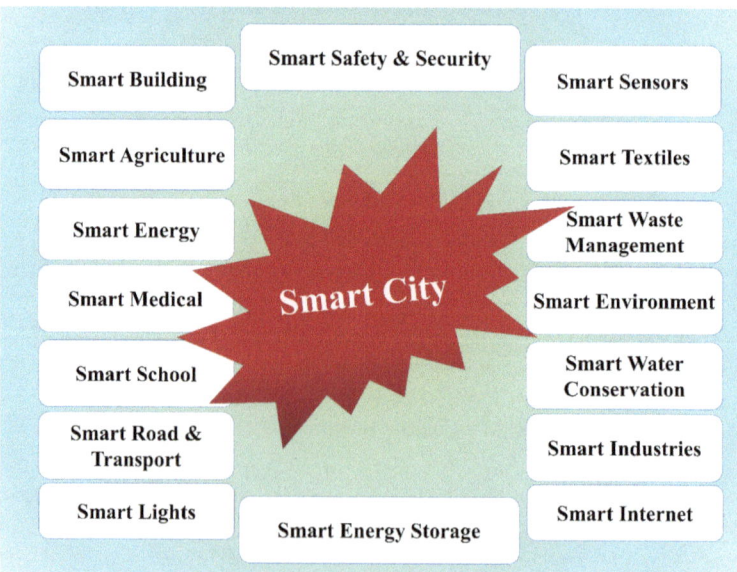

Fig. 1 Emphasis on different fields for smart city

Fig. 2 Nanotechnology for the development of smart city

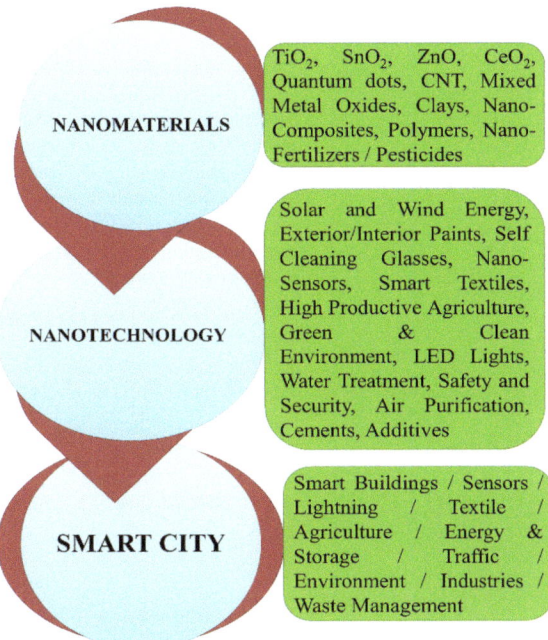

important in every area of our daily lives. For example, incorporation of nanoscale additives into polymer composites can generate automobile parts, baseball bats, tennis rackets which are lightweight, stiff, durable and resilient [12]. Research on carbon nanotubes has been increased drastically from the past few decades to develop sensors that may find application in smart cities to detect harmful gases. By the application of nanotechnology on the existing technology, materials can be made smaller, faster or lighter (Fig. 2). Nanotechnology has an important role in all of everyday routine work, in safety, health, Internet, infrastructure, etc., in urban environment (Table 1). The World Nano Foundation has well-defined it as the science of the ultra-small for the profit of all. It is also imperative that the methodology adopted must be economic and environment-friendly.

2 Applications of Nanotechnology/Nanomaterials

2.1 Nanomaterials in Energy

Energy is one of the crucial factors, which absolutely needs to be addressed for the development of smart cities. Nanotechnology has applications in traditional energy sources as well as in developing alternate energy sources to meet the demand of huge energy. Scientists are more interested to generate clean, efficient, affordable energy from renewable sources of energy.

Table 1 List of nanomaterials along with their uses and benefits for building smart cities

S. no.	Nanomaterials	Uses	Benefit	References
1	TiO_2	As a pigment in interior and exterior paints, as a thin film on glass windows, photocatalyst, adsorbent, as a membrane, hydrophilicity, climate control, sensors	Antimicrobial properties, self-cleaning effect, no need to clean high storey building, degrade and adsorb pollutants, water purification and wastewater treatment	[13–26]
2	ZnO	Bactericidal, UV protection, photocatalyst, nanogenerator	Antimicrobial properties, UV protection, water purification, energy generation	[24, 27–32]
3	SnO_2	Sensors, adsorbent, transparent conducting oxides, photocatalyst	Sense various gases, adsorb harmful gases, component of solar panels, degrade pollutants, water purification	[24, 33–35]
4	Ag/Ag_2O	Antimicrobial, textiles, food packaging, photocatalyst	Water purification, inhibit and kill bacteria	[25, 29, 36, 37]
5	SiO_2	Rheological behaviour under uniaxial extensional flow, improved mechanical properties, fire retardant and insulation	Fibre dispersion and robust properties, durability properties of self-compacting concrete, improve segregation resistance	[38–42]
6	CeO_2	Sensors, agriculture	Remediation of environmental pollution, change the thermogenic metabolism	[24, 43]
7	CNT	Nanosensors, grids	Fine insulation performance, not attacked by chemical and flame retardancy, water intake, uptake of essential nutrients from the soil, enhances stiffness and decreases the porosity, possess high Young's modulus, and durable	[44–52]
8	Nanoclays	Protects wood surfaces by using multilayer UV coatings, filling of the cracks in concrete if formed	Rheological properties and corrosion protection, fire resistant	[42, 53, 54]

(continued)

Table 1 (continued)

S. no.	Nanomaterials	Uses	Benefit	References
9	Nanopolymer	Fills the cracks in concrete if formed	Evaluation of damage induced by moisture	[55, 56]
10	Nanoporous materials	Nanogenerator, smart electronics	Self-powered without wire, green energy source	[56]
12	Nano- and micro-electrical–mechanical systems	Sensors, high-performance flexible printed antenna structures for the emerging Internet of Skins (IoS) and 5G applications	To detect and control corrosion, stresses, temperature, smoke, moisture, noise, vibrations and cracks	[50]
14	Ceramo-polymeric nanocomposite	Biomimetics, biomechanics and tissue engineering	Health sector, provide equivalence between artificial and natural construction materials	[57–59]
15	Nanofertilizers (NPK, zeolites)	Agriculture, the slow release will increase fertility	Highly porous and allow the retention of the soil	[48, 49, 60]
16	Nanosensors	Agriculture	Used in nanorange and mixed with soils	[24, 48, 49, 60]
17	Nanopesticides	Agriculture	Less effect on environment	[48, 49, 60]
18	Metal oxides-CNT	Micro-sized electrochemical sensors, environment	Detect dust, pollutants, heavy metals	[24, 61]
19	Li-containing mixed metal oxides	Energy, sensors	Li-ion batteries, Wi-Fi	[62–65]
20	Mixed metal oxides	Energy, transistor, photocatalyst, electrochemical sensors, solar panels, magnetic properties	Generate H_2 fuel through water splitting, methane reforming, used in equipment	[66–69]

- Nanotechnology helps in improving the efficient fuel production from raw petroleum with the help of better catalysis so that there will be less fuel consumption in vehicles and power plants.
- Carbon nanotubes (CNT) wires have lower resistance than high tension wires can be utilized in electric grid in smart cities to reduce the transmission power loss [51, 52].
- Nanomaterials can be utilized to develop efficient solar panels to generate electricity from sunlight. Additionally, epoxy-containing CNT can be used on the blades of the windmill to make it stronger, longer and lightweight [70]. Nanostructured materials will be affordable and can solve the problem to meet the huge energy demand.

- Nanotechnology can be utilized for the storage of electrical energy in batteries and super-capacitors. High cell voltage, energy and power density make lithium-ion batteries as one of the crucial electrical energy storage material. Due to novel ceramic, high-performance electrode and heat-resistant materials, nanotechnology is promising to improve the capacity and safety of lithium-ion batteries [62].
- Nanomaterials (mixed metal oxides, doped TiO_2, etc.) can be used to generate clean fuel, i.e. H_2 by water-splitting reaction in the presence of sunlight [66].
- Piezoelectric nanogenerator (ZnO, $BaTiO_3$, etc.) can be used as energy harvesters [32, 71].
- Nanoporous metal organic compounds are also very useful in terms of operation in fuel cells for portable electronic devices [72].

2.2 Nanomaterials in Environmental Remediation

With the development of urban areas with rapid growth of the population, it will have a direct impact on the environment. Hence, it is essential to monitor and clean up environmental contaminants.

- Nanomaterials can be used to detect water impurities and in the treatment of water pollution by using several photocatalysts such as TiO_2, SnO_2, ZnO [14, 33].
- Molybdenum disulphide (MoS_2) membrane with nanopores can be utilized for efficient filtering of water [73].
- CNT and other nanocomposites tend to adsorb pollutants to clean environmental atmosphere.
- Smart sensing—Nanotechnology-based sensors can be used to sense, recognize and monitor chemical and biological agents in the soil, water and in the air [74]. It is crucial to improve and monitor the quality of the inhabitants. Smart sensors must be selective, sensitive, high response time, reversible and cost-effective. Materials such as metal oxides, semiconductors and polymers can be utilized. Spectroscopy is an important field for optical sensors.

2.3 Nanomaterials in Transportation

Nanotechnology aids in developing various innovative and improved infrastructure for urban transportation [75].

- It has been investigated that nanoparticles of oxides uniformly dispersed on stainless steel improve mechanical and thermal properties of it [76].
- Performance, toughness and durability of highway and transportation infrastructure gears can be improved by nanoengineering of cementitious materials such as steel, asphalt, concrete, and it will reduce their life cycle cost as well.

- With the help of nanoscale devices and sensors, performance and structural integrity of tunnels, pavements, rail tracks can be monitored.
- Nanoelectronics and nanoscale-based communication devices can afford assistance to improve transportation infrastructure to adjust travel routes, avoid congestion, collisions and supportive for emergency situations [77].
- Further, smart alloys can be developed for their application in the foot over bridges, flyovers, etc., which helps in load-bearing capability [78].
- Applications of graphene on highways have already been on trial basis on ∼1 km stretch of road in Rome, Italy, by Directa Plus and Iterchimica. Graphene can be used as an additive into the asphalt which can improve the road's quality. It is expected that it helps in improving the overall lifetime of roads by 6–12 years and additionally helps in resisting few of the climate issues.

2.4 Smart Materials

Smart materials are indispensable for the sustainable development of smart cities. These materials respond immediately to any alteration in magnetic, electric or heat waves. They have the ability to modify their properties and shape in the presence of applied electric, magnetic and thermal fields. Smart materials are advantageous due to their high strength, toughness, increased durability and resistant to corrosion. Few reported smart materials are ceramics, piezoelectric polymers, optical fibres, shape memory alloy. Piezoelectric materials can generate charge on surface in response to an applied mechanical stress and can undergo a material distortion in response to an applied electric field. This novel ability enables the material to be used as a sensor and as an actuator. The SMAs have tendency to change its shape at low temperature and can resume its shape at higher temperature. To develop actuators and sensors for smart buildings, magnetostrictive materials are important due to its tendency to convert magnetic energy into kinetic energy and vice versa [79, 80].

2.5 Nanoscale Photonics Technologies

Photonics can have a great impact on developing smart cities in concern with smart lighting, sensing (discussed in Sect. 2.2) and communication technologies, even for smart water supplies. This technology aids in information and communication such as in the progress of social networks, Internet and maintaining infrastructure that can make cities smarter [81–83].

- Smart lighting–Light-emitting diodes (LEDs) are essentially more energy efficient for traffic control instead of incandescent bulbs for urban lighting, which is one of the current applications of photonics. Further, their emission wavelength can be tailored from the UV to infrared using suitable semiconducting compounds. LEDs also have applications in vertical farming. LEDs can be used as lighting source by which wavelength of emission can be tuned for different plants to minimize loss of heat and attraction of pests.

- Smart communication and signal processing.

2.6 Smart Internet of Things (IoT)

Internet technology has helped immensely which makes the world connected and has changed the lives completely since its use. However, it should be remembered that IoT consumes energy and creates toxic pollution and e-waste which is not good for environment. Therefore, scientists' interest has been shifted towards green IoT which consumes power efficiently and is environmentally friendly. Green IoT and green nanotechnology give the idea as the suitable solutions to yield smart and sustainable agriculture and food industry [82, 84].

2.7 Smart Textiles

Smart textiles are materials that are designed and equipped to comprise technologies which offer the wearer with improved functionality. These textiles possess several applications, such as the capability to connect with other devices, transfer energy, convert into other useful materials and shield the wearer from environmental risks. Research on the development of these materials has been increased widely to generate wearable textile that can monitor health, safety and healthy lifestyle. Coating of nanoparticles on the surface of textiles enhances the performance and functionality of textiles by varying different properties such as water-repellence, self-cleaning, antibacterial, UV protection. With the application of nanotechnology, durability of the fabrics increases which is very helpful in the development of smart cities [85–87].

2.8 Smart Agriculture

Green revolution has led to the excessive usage of chemical pesticides and fertilizers, which has caused harm to soil variety and the development of resistance against pests. Excessive usage of fertilizers, which contain ~ 40–70% nitrogen, ~ 80–90% phosphorous, ~ 50–70% potassium, get lost to the environment that results in pollution and recently, and it has been reported to be one of the factors for global climate change. Nanotechnology provides various solutions for the problems in the field of agriculture. Nanoparticles or nanochip materials and advanced biosensors are very helpful for farming. Nano-mediated fertilizers, herbicides and pesticides aid in slow and controlled release of agrochemicals and nutrients to plants. Nanoparticles such as Ag, ZnO, TiO_2 are very helpful for the treatment of plant diseases. Nanotechnology facilitated plant disease detection kits are widespread and are beneficial in immediate and initial recognition of viral diseases. The current nanotechnology techniques have made possible to identify and rectify the numerous complications of agriculture. Smart agriculture by nanotechnology will be strongly required for the smart cities [48, 49, 88].

2.9 Smart Health Care

Nanotechnology helps in the diagnosis and treatment of diseases. Nanoparticles are brilliant host media for labelling appropriate drugs for various types of tumours [89]. These novelties help to target specific affected cancer tissues/organs whereas healthy tissue remains unaffected and thus provides the healthier treatment.

- Nanomaterials are helpful in biomedical imaging for medical diagnostics. Nanocrystals can be used to locate and recognize biological activities and specific cells. These have the optical detection tendency 1000 times better than the traditional biological tests such as MRIs [89, 90].
- Magnetic nanoparticles such as Fe_3O_4 are discovered for hyperthermia treatment to abolish target specific cancer cells in a well-organized manner.
- Numerous noble metal-based nanoparticles are being investigated at various clinical trial levels to cure organ-specific cancers.
- Gold nanoparticles can be used for the early detection of Alzheimer's disease.
- Biosensors based on nanowires, nanochannels are being developed for molecular imaging purpose which helps to recognize genetic disorders and detect rare illness [91].

2.10 Smart Building

Nanotechnology provides innovative methods for the development of smart buildings for smart cities. Nanotechnology and advanced nanomaterials can improve energy efficiency and can help in environmental performance for building materials and designing. Various novel cement composites are developed by adding TiO_2, SiO_2, clay and Al_2O_3 particles of nanosizes into it. Also, nanopaints have been investigated that increase the capability of the paints to withstand the environment risks. Paints containing nanosized TiO_2 and CNTs are very operative against the industrial toxins. Nanotechnology has wonderful possibilities in the construction of smart buildings [92].

3 Conclusion

The application of nanotechnology will have a great impact on the development of smart cities. Nanomaterials make the performance of materials smarter, faster and cheaper. Several nanomaterials can be utilized to make the infrastructure and transportation smarter. Nanomaterials are also helpful in environmental remediation, water purification and energy storage. Smart materials will provide sustainable development for smart cities. Still, there are several challenges that need to be addressed for the fully development of smart cities to reality.

Acknowledgements The author thanks SERB (YSS/2015/001120), Government of India for financial support.

References

1. Goldstone JA (2010) The new population bomb: the four megatrends that will change the world. Foreign Aff 89:31
2. Bloom DE (2011) 7 billion and counting. Science 333(6042):562–569
3. Anttiroiko A-V, Valkama P, Bailey SJ (2014) Smart cities in the new service economy: building platforms for smart services. AI Soc 29(3):323–334
4. Gautam S et al (2017) Cost-effective treatment technology for small size sewage treatment plants in India
5. Dhoble YN, Ahmed S (2018) Sustainability of wastewater treatment in subtropical region: aerobic vs anaerobic process. Int J Eng Res Dev 14(1):51–66
6. Parihar RS et al (2017) Characterisation and management of municipal solid waste in Bhopal, Madhya Pradesh, India. In: Proceedings of the institution of civil engineers—waste and resource management. Thomas Telford Ltd
7. Parihar RS et al (2018) MSWM in Bhopal city: a critical analysis and a roadmap for its sustainable management. In: Proceedings of the institution of civil engineers—municipal engineer. Thomas Telford Ltd
8. Kumar M, Sharif M, Ahmed S (2017) Flood risk management strategies for national capital territory of Delhi, India. ISH J Hydraul Eng 1–12
9. Barrionuevo JM, Berrone P, Ricart JE (2012) Smart cities, sustainable progress. IESE Insight 14(14):50–57
10. Casini M (2016) Smart buildings: advanced materials and nanotechnology to improve energy-efficiency and environmental performance. Woodhead Publishing
11. Zeng HC (2007) Oriented attachment: a versatile approach for construction of nanomaterials. Int J Nanotechnol 4(4):329–346
12. Mallick PK (2007) Fiber-reinforced composites: materials, manufacturing, and design. CRC Press
13. Konstantinou IK, Albanis TA (2004) TiO_2-assisted photocatalytic degradation of azo dyes in aqueous solution: kinetic and mechanistic investigations: a review. Appl Catal B 49(1):1–14
14. Akpan UG, Hameed BH (2009) Parameters affecting the photocatalytic degradation of dyes using TiO_2-based photocatalysts: a review. J Hazard Mater 170(2–3):520–529
15. Leong S et al (2014) TiO_2 based photocatalytic membranes: a review. J Membr Sci 472: 167–184
16. Braun JH, Baidins A, Marganski RE (1992) TiO_2 pigment technology: a review. Prog Org Coat 20(2):105–138

17. Macwan D, Dave PN, Chaturvedi S (2011) A review on nano-TiO$_2$ sol–gel type syntheses and its applications. J Mater Sci 46(11):3669–3686
18. Allahverdiyev AM et al (2011) Antimicrobial effects of TiO$_2$ and Ag$_2$O nanoparticles against drug-resistant bacteria and leishmania parasites. Future Microbiol 6(8):933–940
19. Foster HA et al (2011) Photocatalytic disinfection using titanium dioxide: spectrum and mechanism of antimicrobial activity. Appl Microbiol Biotechnol 90(6):1847–1868
20. Fu G, Vary PS, Lin C-T (2005) Anatase TiO$_2$ nanocomposites for antimicrobial coatings. J Phys Chem B 109(18):8889–8898
21. Muranyi P, Schraml C, Wunderlich J (2010) Antimicrobial efficiency of titanium dioxide-coated surfaces. J Appl Microbiol 108(6):1966–1973
22. Banerjee S, Dionysiou DD, Pillai SC (2015) Self-cleaning applications of TiO$_2$ by photo-induced hydrophilicity and photocatalysis. Appl Catal B 176:396–428
23. Ganesh VA et al (2011) A review on self-cleaning coatings. J Mater Chem 21(41):16304–16322
24. Eranna G (2016) Metal oxide nanostructures as gas sensing devices. CRC Press
25. Page K et al (2007) Titania and silver–titania composite films on glass—potent antimicrobial coatings. J Mater Chem 17(1):95–104
26. Nagarajan R, Kumar V, Ahmad S (2012) Anion doped binary oxides, SnO$_2$, TiO$_2$ and ZnO: fabrication procedures, fascinating properties and future prospects
27. Li Y-F, Wang B-H, Huang W-X (2003) Study on bactericidal performance of nano-scale ZnO modified interior wall paint. Paint Coat Ind China 33(8):3–5
28. Hochmannova L, Vytrasova J (2010) Photocatalytic and antimicrobial effects of interior paints. Prog Org Coat 67(1):1–5
29. Simoncic B, Tomsic B (2010) Structures of novel antimicrobial agents for textiles—a review. Text Res J 80(16):1721–1737
30. Iacovangelo CD (2001) Infrared reflecting coatings. Google patents
31. Austin RR (1992) Multilayer anti-reflection coating using zinc oxide to provide ultraviolet blocking. Google patents
32. Briscoe J, Dunn S (2015) Piezoelectric nanogenerators—a review of nanostructured piezoelectric energy harvesters. Nano Energy 14:15–29
33. Kumar V, Govind A, Nagarajan R (2011) Optical and photocatalytic properties of heavily F$^-$-doped SnO$_2$ nanocrystals by a novel single-source precursor approach. Inorg Chem 50 (12):5637–5645
34. Kumar V et al (2019) Facile synthesis of Ce–doped SnO$_2$ nanoparticles: a promising photocatalyst for hydrogen evolution and dyes degradation. ChemistrySelect 4(13):3722–3729
35. Kumar V, Uma S, Nagarajan R (2014) Optical and magnetic properties of (Er, F) co-doped SnO$_2$ nanocrystals. Turk J Phys 38(3):450–462
36. Appendini P, Hotchkiss JH (2002) Review of antimicrobial food packaging. Innov Food Sci Emerg Technol 3(2):113–126
37. Aziz N et al (2015) Facile algae-derived route to biogenic silver nanoparticles: synthesis, antibacterial, and photocatalytic properties. Langmuir 31(42):11605–11612
38. Park JU et al (2006) Rheological behavior of polymer/layered silicate nanocomposites under uniaxial extensional flow. Macromol Res 14(3):318–323
39. Beigi MH et al (2013) An experimental survey on combined effects of fibers and nanosilica on the mechanical, rheological, and durability properties of self-compacting concrete. Mater Des 50:1019–1029
40. Rana AK et al (2009) Significance of nanotechnology in construction engineering. Int J Recent Trends Eng 1(4):46
41. Ganesh VK (2012) Nanotechnology in civil engineering. Eur Sci J 8(27)
42. Nkeuwa WN, Riedl B, Landry V (2014) Wood surfaces protected with transparent multilayer UV-cured coatings reinforced with nanosilica and nanoclay. Part I: morphological study and effect of relative humidity on adhesion strength. J Coat Technol Res 11(3):283–301

43. Gupta H (2018) Role of nanocomposites in agriculture. In: Nano hybrids and composites. Trans Tech Publications
44. Gammampila R et al (2013) Application of nanomaterials in the sustainable built environment
45. Lee J, Mahendra S, Alvarez PJ (2010) Nanomaterials in the construction industry: a review of their applications and environmental health and safety considerations. ACS Nano 4(7):3580–3590
46. Feldman D (2014) Polymer nanocomposites in building, construction. J Macromol Sci Part A 51(3):203–209
47. Gopalakrishnan K et al (2011) Nanotechnology in civil infrastructure: a paradigm shift. Springer
48. Antonacci A et al (2018) Nanostructured (bio) sensors for smart agriculture. TrAC Trends Anal Chem 98:95–103
49. Rameshaiah G, Pallavi J, Shabnam S (2015) Nano fertilizers and nano sensors—an attempt for developing smart agriculture. Int J Eng Res Gen Sci 3(1):314–320
50. Eid A et al (2018) Nanotechnology-enabled additively-manufactured RF and millimeter-wave electronics. In: 2018 IEEE 13th nanotechnology materials and devices conference (NMDC). IEEE
51. Vajtai R et al (2002) Building carbon nanotubes and their smart architectures. Smart Mater Struct 11(5):691
52. Schlögl R (2011) Chemistry's role in regenerative energy. Angew Chem Int Ed 50(29):6424–6426
53. Devi RR et al (2013) Synergistic effect of nanoTiO$_2$ and nanoclay on mechanical, flame retardancy, UV stability, and antibacterial properties of wood polymer composites. Polym Bull 70(4):1397–1413
54. Olivier M-G et al (2011) Study of the effect of nanoclay incorporation on the rheological properties and corrosion protection by a silane layer. Prog Org Coat 72(1–2):15–20
55. Shaffie E et al (2016) Moisture-induced damage evaluation of nanopolymer-modified binder in stone mastic asphalt (SMA) mixtures. In: InCIEC 2015. Springer, pp 947–957
56. Ghosh SK et al (2016) Porous polymer composite membrane based nanogenerator: a realization of self-powered wireless green energy source for smart electronics applications. J Appl Phys 120(17):174501
57. Hwang J et al (2015) Biomimetics: forecasting the future of science, engineering, and medicine. Int J Nanomed 10:5701
58. Bhushan B (2017) Springer handbook of nanotechnology. Springer
59. Aversa R et al (2016) Hybrid ceramo-polymeric nanocomposite for biomimetic scaffolds design and preparation. Am J Eng Appl Sci 9(4)
60. Duhan JS et al (2017) Nanotechnology: the new perspective in precision agriculture. Biotechnol Rep 15:11–23
61. Cui L, Wu J, Ju H (2015) Electrochemical sensing of heavy metal ions with inorganic, organic and bio-materials. Biosens Bioelectron 63:276–286
62. Nihtianov S, Tan Z, George B (2017) New trends in smart sensors for industrial applications —part I. IEEE Trans Ind Electron 64(9):7281–7283
63. Kumar V et al (2012) Novel lithium-containing honeycomb structures. Inorg Chem 51(20):10471–10473
64. Gupta A, Kumar V, Uma S (2015) Interesting cationic (Li$^+$/Fe^{3+}/Te^{6+}) variations in new rocksalt ordered structures. J Chem Sci 127(2):225–233
65. Kumar V, Gupta A, Uma S (2013) Formation of honeycomb ordered monoclinic Li$_2$M$_2$TeO$_6$ (M = Cu, Ni) and disordered orthorhombic Li$_2$Ni$_2$TeO$_6$ oxides. Dalton Trans 42(42):14992–14998
66. Yuan C et al (2014) Mixed transition-metal oxides: design, synthesis, and energy-related applications. Angew Chem Int Ed 53(6):1488–1504
67. Tejuca LG, Fierro JL (1992) Properties and applications of perovskite-type oxides. CRC Press
68. Fierro JLG (2005) Metal oxides: chemistry and applications. CRC Press

69. Kumar V, Uma S (2011) Investigation of cation (Sn^{2+}) and anion (N^{3-}) substitution in favor of visible light photocatalytic activity in the layered perovskite $K_2La_2Ti_3O_{10}$. J Hazard Mater 189(1–2):502–508

70. Mishnaevsky L et al (2017) Materials for wind turbine blades: an overview. Materials 10 (11):1285

71. Fan FR, Tang W, Wang ZL (2016) Flexible nanogenerators for energy harvesting and self-powered electronics. Adv Mater 28(22):4283–4305

72. Salunkhe RR et al (2016) Nanoarchitectures for metal–organic framework-derived nanoporous carbons toward supercapacitor applications. Acc Chem Res 49(12):2796–2806

73. Li W et al (2016) Tunable, strain-controlled nanoporous MoS_2 filter for water desalination. ACS Nano 10(2):1829–1835

74. Abegaz BW, Datta T, Mahajan SM (2018) Sensor technologies for the energy-water nexus—a review. Appl Energy 210:451–466

75. Aithal P, Aithal S (2016) Nanotechnological innovations & business environment for Indian automobile sector: a futuristic approach

76. Qin W et al (2017) Incorporation of silicon dioxide nanoparticles at the carbon fiber-epoxy matrix interphase and its effect on composite mechanical properties. Polym Compos 38 (7):1474–1482

77. Roco MC (2017) Overview: affirmation of nanotechnology between 2000 and 2030. In: Nanotechnology commercialization: manufacturing processes and products, pp 1–23

78. Rahim MRU et al (2018) Axial crushing comparison of sinusoidal thin-walled corrugated tubes. Mater Today Proc 5(9):19431–19440

79. Ramaswami A et al (2016) Meta-principles for developing smart, sustainable, and healthy cities. Science 352(6288):940–943

80. Höjer M, Wangel J (2015) Smart sustainable cities: definition and challenges. In: ICT innovations for sustainability. Springer, pp 333–349

81. Smalley JS et al (2016) Photonics for smart cities. In: Smart cities technologies. IntechOpen

82. Szymanski TH (2016) Securing the industrial-tactile Internet of Things with deterministic silicon photonics switches. IEEE Access 4:8236–8249

83. Arnon S (2015) Visible light communication. Cambridge University Press

84. Alsamhi S et al (2018) Greening internet of things for smart everythings with a green-environment life: a survey and future prospects. arXiv preprint arXiv:1805.00844

85. Syduzzaman M et al (2015) Smart textiles and nano-technology: a general overview. J Text Sci Eng 5:1000181

86. Pakdel E et al (2018) Nanocoatings for smart textiles. In: Smart textiles: wearable nanotechnology

87. Yilmaz ND (2018) Nanocomposites for smart textiles. In: Smart textiles: wearable nanotechnology, pp 211–245

88. Manjunatha S, Biradar D, Aladakatti YR (2016) Nanotechnology and its applications in agriculture: a review. J Farm Sci 29(1):1–13

89. Kumar V et al (2019) Nanotechnology: nanomedicine, nanotoxicity and future challenges. Nanosci Nanotechnol Asia 9(1):64–78

90. Gupta A, Kumar S, Kumar V (2019) Challenges for assessing toxicity of nanomaterials. In: Biochemical toxicology-heavy metals and nanomaterials, IntechOpen, https://doi.org/10. 5772/intechopen.89601

91. Wiek A et al (2013) Nanotechnology in the city: sustainability challenges and anticipatory governance. J Urban Technol 20(2):45–62

92. Khitab A, Tausif Arshad M (2014) Nano construction materials. Rev Adv Mater Sci 38(2)

Growth with Optimization: Can Smart Cities Assist in Environmental Limpidness in India?

Shirin Rais

Abstract Recently, the 'Smart Cities' project in India is a topic of discussion at most of the academic forums. The need for smart cities was felt in order to accelerate the pace and size of economic growth and hence economic development in Indian economy. The present paper tries to investigate the environmental aspect of Smart Cities Mission in India. The study argues that smart cities will in future help in protecting our environment from deterioration if job opportunities are also created simultaneously in the periphery regions of future smart cities which will minimize the migration rates in the smart cities. The study finds that there are 23 future smart cities in India which will experience problem of severe environmental degradation due to intra-state and inter-state migration if satellite regions are not developed all together in terms of employment opportunities specially in the primary and secondary sectors. This is specially required for the cities which are already in the list of world's top most polluted cities and are also named in the list of future smart cities of India. The study asserts that the development of periphery regions will trim down the movement of people from less developed regions to future smart cities, thereby, reducing the burden on natural as well as man-made resources in the future smart cities.

Keywords Urbanization · Migration · Environmental degradation · Smart cities · Sustainability

1 Introduction

Cities are considered as an engine of 'economic growth'. Urbanization and industrial growth are prerequisite for achieving sufficient economic growth rates [22]. In the recent years, rapid urbanization has become the most common feature all over the world. By 2050, the global population is expected to increase to 9.2 billion, of which 6.4 billion will be living in the urban areas. Digital and communication technologies

S. Rais (✉)
Department of Economics, Aligarh Muslim University, Aligarh 202001, India

© Springer Nature Singapore Pte Ltd. 2020   859
S. Ahmed et al. (eds.), *Smart Cities—Opportunities and Challenges*,
Lecture Notes in Civil Engineering 58,
https://doi.org/10.1007/978-981-15-2545-2_69

are the key features of present-day smart cities as it helps—to reduce costs and resource consumption, to enhance quality of urban services, and to interact more effectively and actively with its citizens. In addition to the term 'smart city', various other terms based on the technology usage are also used as a substitute to the term 'smart city', such as intelligent, digital, virtual and ubiquitous [1].

The escalating tendency of urbanization in India is one of the reasons that led to the development of the smart cities' concept. Urbanization in India has led to an increase in the pressure on natural and man-made resources. It is assumed that smart cities in the future will help in the optimization of resources and will promote environmental sustainability since urbanization in India is expected to increase manifolds in near future, and hence, the issue of sustainability will arise. One of the reasons for this is that in India, Gross Domestic Product (GDP) is mainly generated from economic activities in the urban areas. Therefore, these urban areas have witnessed large-scale migration from rural to urban areas mainly in search of jobs, which led to the establishment of slums followed by filthiness, unhygienic environment causing major environmental issues in these urban areas. Another problem posed by the urbanization is the increased pressure on resources both natural and man-made which in long run creates an issue of sustainability and will lead to pollutions of various kinds, i.e., air, water and land pollutions.

In this background, the present study will investigate the sway of smart cities' mission on environmental sustainability in India. The paper argues that if periphery areas are not developed, then the formation of smart cities will lead to an increase in the burden on natural and man-made resources in these cities which may lead to severe environmental degradation in many future smart cities. The study finds that in order to have a sustainable working of the smart economy, we need to develop the periphery regions also so that the problem of migration could be avoided. The components or the ingredients of smart cities are very important in determining the impact of smart cities on environmental sustainability. Undoubtedly, the anticipations attached with smart city mission in solving environmental issues are a baffling and enthralling issue.

2 Review of Literature

Glaeser and Kahn [10] in their study have tried to quantify carbon dioxide emission that took place during new constructions in different parts of America. The study found that during the construction, California has experienced less pollution, whereas Texas and Oklahoma were found to be the highest emitters of carbon dioxide. The paper asserts that spatial distribution of population in a city is an important determinant of greenhouse gas production. Social and environmental sustainability has been considered as a major component of smart cities by Caragliu et al. [7]. The concept that the compact city is environmental friendly has been challenged by Gaigne et al. [9] in general equilibrium framework. They argued that higher population density generates negative externalities.

Lazaroiu and Roscia [14] have developed six indicators to assess the performance of the smart city. These are smart economy, smart mobility, smart environment, smart people, smart living and smart governance. The components of 'smart environment' are attractiveness of natural conditions, pollution, environmental protection and sustainable resource management. Tranos and Gertner [21] said that urban interdependencies played an important role in determining urban development and called for the need to tackle the issue of environmental sustainability at a larger level. The paper asserts that the concept of smart city lacks proper global perspective. Zhang et al. [26] have focused on the positive role played by intelligent transportation system (ITS) which will be stimulated in smart cities' regime.

Camagani et al. [6] viewed that urban sustainability of a city depends on the three elements: physical, social and economic. He said that positive and negative externalities results from the interaction which takes place between these three elements. The paper provides possible intervention policies which should be taken to achieve 'sustainable city'. Komninos et al. [12] said that many cities of the world are facing problems related to environmental indicators and laid stress on the development of smart environment. Lee et al. [15] have considered urban sustainability as a crucial aspect for both old cities and smart cities and less smart cities. Their study said that the smart cities will motivate and encourage the citizens to use environment-friendly energy which will ultimately reduce carbon footprints.

Smart cities have been considered as a solution for the present-day pollution problem by Anttiroiko et al. [4]. The study argues that the problem of environmental deterioration will intensify with a surge in urbanisation. The authors call for a bottom-up approach at planning level to solve many issues including environmental sustainability in cities. Kramer et al. [13] viewed that the use of information and communication technology will help in reduction in the sage of energy in smart cities. Trade openness has been blamed to be one of the main causes of carbon emissions in India by Yang and Zhao [25]. They have suggested the use of alternative cleaner energy resources to mitigate environmental crisis.

Alok and Vashisht [2] in their study have considered environment as an important factor in determining the selection of smart cities in India. Balasubramaniam et al. [5] have performed an exhaustive analysis of possessing sustainable transportation planning to reduce environmental degradation in smart cities. The authors have argued for the development of autonomous smart car with the suggested technologies in the study. Crowdsensing has been considered as a powerful tool for environmental monitoring by using small mobile sensors and smartphones. The study by Alvear et al. [3] uses the word smart environment in context of smart cities and dealt with the issue air pollution. The study recommends the use of 'Arduino Nano platform' in this direction.

In a very interesting study by Dwevedi et al. [8], six environmental indicators have been cited by the author which should be incorporated while developing smart cities. These are landscape and geography, climate, atmospheric pollution, water resources, energy resources and urban green space. The study said that the success of smart cities depends on big data because of the extensive coverage of data sets for environmental indicators. Therefore, the study also called for the development of skills for the assessment of big data. Owing to the rapid surge in the pollution-related health issues nowadays, Miles et al. [16] in his paper have laid emphasis on the role of smart cities in acting as an instrument for timely detection of atmospheric pollution and focused on the use of decision support system (DSS) for the finding pollution levels. Ramos et al. [18] have taken into consideration a very important aspect of the rising pollution levels in most of the cities. The study had provided a solution for giving pollution-free routes in smart cities which will be observed by air-quality sensor network. Smith et al. [20], in their study, have raised an interesting question regarding the cities selected for the smart cities' mission. The study says that most of the cities in the selected list have an organised urban area low slum population and better infrastructure. Their study says that there are many other cities which should be included in the smart cities project so that their condition could be improved. The study calls for an exact definition of term 'smart city' so that policies and funds could be organised accordingly. They have considered rapid urbanization in India as a main reason for the development of smart cities' projects in India. The application of least mean square adaptive algorithm has been emphasised for distributed environmental parameters estimations in smart city by Wu et al. [24]. Goddard and Tett [11], in their study, have held 'urbanization' responsible for a significant increase in the temperature of UK.

3 Essence of Smart Cities

Smart city definition is again a mystifying issue. The regular definition of smart cities focuses mostly on smart economy mainly based on digitalization, intensive use of information, communication and technology (ICT) and related items. It is not possible to give a specific definition of smart city. To understand the possible sway of smart cities, we need to understand the essence of smart cities. It does not necessarily mean the development of specific areas (or boosting economic indicators) which help in generating gigantic gross domestic product (GDP) but also social sectors along with improvement of environmental indicators. The smart city concept should be observed by horizontal expansion rather than vertical expansion. If we are going to develop only a particular city in the name of smart city and overlook the surrounding underdeveloped areas, then this will create huge problem of migration, wherein people living in the surrounding underdeveloped areas will think that smart city will provide them better choices of jobs, and hence, huge migration will take place which will create slums in smart cities; density of population will increase, thereby leading to the emergence of various types problems such as hygiene, environmental degradation and so on. In addition to this, the demand for both natural and man-made resources will increase, and hence apart from other problems, environmental sustainability will emerge as a major issue for these future smart cities. The future smart cities will face similar situation which most of the present-day cities are facing such as Delhi and Mumbai. Incidentally, many of these migration infested cities such as Delhi, Mumbai and Faridabad are also in the list of smart cities.

In the future smart cities, investment in economic indicators must be accompanied by investments in job-oriented economic activities and in the social sectors also such as health and education. There is a need to enhance the formation of human capital in smart cities as well as in the periphery areas. This can play a decisive role in adding the component of sustainability in the development process. Human capital formation should be considered as an essential component of smart cities project. India is lagging behind in this sector due to lack of expenditure in the health and education sector which in turn reduces overall per capita productivity.

The pattern of formation of smart cities plays an important role in determining the effects on the welfare of the society. Smart cities' mission will be worth only when holistic development takes place.

4 Role of Smart Cities in Environmental Protection

There is no specific reason for the growth of smart cities. A popular hypothesis is that people with high levels of human capital move toward cities in search of high salaries since cities provide more opportunities. Many researches have proved that salaries in highly educated cities are higher than in less educated cities, even after controlling for individual worker characteristics [17, 19]. This is a major factor behind the movement of people from rural to urban regions.

Smart cities will help in bringing environmental sustainability by optimizing the use of both natural and man-made resources. After the formation of smart cities, economic activities will be well organized, the problem of mainly air, water and land pollution will be solved to a large extend with the help of smart cities. Roads will be organised, thereby reducing pollutants released during traffic, congested road networks, long distances, etc. A major issue faced by cities whether they are rich or poor is the problem of garbage dumping which creates environmental issue such as problem of landfills. Smart cities are supposed to solve this problem also. Landfills lead to land degradation in the form of groundwater pollution, poor fertility and adverse health effects from the consumption of fruits and vegetables which are grown on polluted land. These and many other related environmental issues are expected to be answered by the formation of smart cities. An organised pattern of urbanization is expected to solve most of the environmental issues.

However, if the focus of smart cities remains on the acceleration of the frequency of Net State Domestic Products (NSDPs), then environmental sustainability will be challenged and things may go in an unfavorable direction. If the smart cities experience the construction of industrial units, then results will be perplex if we add the 'migration', aspect also in it. The combination of industrialization and migration will prove to be perilous for many future smart cities. The study finds that the cities which are chosen for smart city mission e.g. Aligarh and which are nearer to metropolitan cities and large cities like Delhi, will not face the problem of migration and environmental issues associated with migration and hence in the future smart cities which are closer to rich or developed cities, environmental indicators will be more healthy in comparisons to other cities.

Table 1 shows the list of future smart cities which are prone to environmental degradation associated with migration. In this case, cities are adversely affected both due to inter-state and intra-state migration of people in search of employment. The environmental impacts on smart cities depend on degree of migration, population, intensity of industrial works and pollution mitigation policies.

Table 1 Cities prone to environmental degradation due to migration after they become smart city

State	City	Distance in kilometers	Population
1. Severely affected			
Andhra Pradesh	Kakinada, Tirupati, Visakhapatnam	Kakinada to Tirupati = 634 km Visakhapatnam to Tirupati = 760 km Kakinada to Visakhapatnam = 153 km	Kakinada = 3.42 lakh Tirupati = 3.74 lakh Visakhapatnam = 20.4 lakh
Kerala	Kochi		Kochi = 6.77 lakh
Telangana	Warangal, Hyderabad	145 km	Warangal = 6.15 lakh Hyderabad = 68.1 lakh
Orissa	Rourkela, Bhubaneswar	355 km	Rourkela = 4.83 lakh Bhubaneswar = 8.38 lakh
Chhattisgarh	Bilaspur, Raipur	115 km	Bilaspur = 13,654 Raipur = 10.1 lakh
Maharashtra	Solapur		Solapur = 9.51 lakh
Bihar	Bihar Sharif, Muzaffarpur, Bhagalpur	Bihar Sharif to Muzaffarpur = 137 km Bihar Sharif to Bhagalpur = 180 km Muzaffarpur to Bhagalpur = 241 km	Bihar Sharif = 3.7 lakh Muzaffarpur = 3.54 lakh Bhagalpur = 50.6 lakh
Jharkhand	Ranchi		Ranchi = 21.9 lakh
Haryana	Dharamshala		Dharamshala = 53,543
2. Moderately affected			
Rajasthan	Udaipur, Kota, Ajmer, Jaipur	Udaipur to Kota = 283 km Kota to Ajmer = 208 km Ajmer to Jaipur = 262 km Jaipur to Udaipur = 395 km	Udaipur = 4.51 lakh Kota = 10 lakh Ajmer = 5.42 lakh Jaipur = 30.7 lakh
Goa	Panaji		Panaji = 40,017
Assam	Guwahati		Guwahati = 9.57 lakh
Arunachal Pradesh	Pasighat		Pasighat = 24,656
3. Less affected			
Remaining cities in the list			

Source Distance data taken from map data 2019 Google population data taken from census 2011

As a result of migration, future smart cities will experience the problem of environmental degradation due to increase in pressure on the resources. The severe affect will be felt in 23 cities where air, water and land pollution are likely to affect the region causing detrimental irreversible long-term impairment. Among them, 16 future smart cities will be severely affected, whereas 07 future smart cities will be moderately affected (Table 1). In addition to this, there are 20 cities in the neighboring states which will experience the problem of environmental degradation due to migration. These are the cities which are nearer to areas devoid of smart cities. Warangal, Hyderabad, Visakhapatnam, Rourkela, Nagpur, Raipur, Ranchi, Varanasi, Kakinada and Tirupati will be the gravely affected cities which will experience large-scale inter-state migration also leading to increasing demand for resources (natural as well as man-made resources) and hence will pose threat to environmental degradation.

These smart cities, due to lucrative job opportunities, will cater to population from every corner of the non-smart cities in the state, and hence, density of population will increase, thereby reducing the per capita availability of required resources along with the polluted environment leading to an overall messy situation in these smart cities.

The main issue of migrants which Uttar Pradesh, Delhi, Punjab, J&K, Rajasthan is facing is migration mainly from Bihar and Jharkhand. With so many numbers of smart cities coming up in Uttar Pradesh, further migration from Bihar will be directed toward Uttar Pradesh and Delhi specially.

Table 2 shows the number of Indian cities in the list of top 100 polluted cities of the world. Among the top 10 most polluted cities of the world, 9 cities are in India. Also, many of the cities listed in the topmost polluted cities in the world are also selected for the future smart city project. Therefore, these cities are certainly on the edge of environmental disaster if migration from neighbouring regions takes place.

For example, Patna which is one of the most polluted cities of the world is already dealing with the problem of migration and this along with the designation of Patna as smart city will further increase migration and hence will create pollution of various kinds. Therefore, in order to have a sustainable growth of smart city like Patna we need to develop the periphery regions also so that the problem of migration could be solved.

The environmental consequences of smart cities depend on the way the program is instigated. Converting already rich cities into smart cities in many cases will further increase the migration of population in these regions, and hence, the pressure of population on man-made and natural resources will increase. Delhi which is already facing lot of issues related to air quality, water scarcity, land degradation, per capita availability of land, etc., will become a perilous place to live. Similar will be the case with Faridabad, Pune, Kanpur, Ahmedabad and other listed cities in Table 1.

Table 2 WHO list of most polluted cities in India (2018)

Rank of cities in the world	Cities	P.M 2.5 level (micrograms per cubic meters)
1	Kanpur	173
2	Faridabad	172
3	Varanasi	151
4	Gaya	149
5	Patna	144
6	Delhi	143
7	Lucknow	138
9	Agra	131
10	Gurgaon	120
10	Muzaffarpur	120
14	Jaipur	105
15	Patiala	101
17	Jodhpur	98
33	Nagpur	84
55	Kolkata	74
87	Ahmedabad	65
94	Mumbai	64

Source WHO [23]

The pressure on future smart cities because of migration can be categorized in India in following ways:

1. **Pressure due to migration:**

   a. Those cities which are already hub of migration and at the same time are also on the list of one of the most polluted cities in the world will experience severe environmental crisis post-formation of smart cities. Converting them into smart cities will further increase the pressure on air, water and land quality in these cities. The worst-affected cities will be Kanpur, Faridabad, Varanasi, Patna, Delhi, Lucknow, Agra, Jaipur, Kolkata, Ahmedabad and Mumbai. These are the cities which already had a migration-related issue mainly for employment purpose. Therefore, along with the development of these cities into smart cities there is a need to develop the periphery regions or satellite areas also so that people's migration could be halted or minimized.

   b. Migration will take place due to sparsely located smart cities. Because there is lack of employment opportunities available in their cities in comparison to the nearest smart city, therefore, this will lead to huge migration to nearest smart cities which will be intra-state as well as inter-state in nature. Example of such smart cities is Tirupati, Vellore, Chennai, Kakinada, Visakhapatnam, Hyderabad, Warangal, Tumakuru, Bhubaneswar, Ranchi, Rourkela, Haldia and Kolkata.

2. **Pressure because they are already most polluted cities in the world:**

According to the WHO classification, 9 out of 10 cities in the list of world's most polluted cities are in India and most of them are listed in the future smart cities list (Table 2). Therefore, smart cities' mission should follow inclusive development approach and must focus on creating employment opportunities in the periphery regions of future smart cities. Otherwise, the development of smart cities will again increase the pressure of migration on certain future smart cities, thereby enhancing pollution of various kinds. These future smart cities can experience an irreparable environmental distress if further migration in these cities takes place. The only way to minimize migration rates is to create job opportunities in the satellite regions of future smart cities.

5   Reverie of Smart City in India

There is a need to provide a desirable definition of smart city which could also take into consideration the aspect of environmental sustainability also. The provision begins with the development of periphery regions or non-smart cities which could reduce the intensity of migration in the smart cities and investment in the health and education sectors which will help in enhancing human capital. This should be followed by developing traffic management, sanitation, roads and bridges, electricity, increasing use of green energy, improving governance, encouraging vocational education and creation of job opportunities.

Efforts should be made to promote the concept of green or environment-friendly city. Green technology should be promoted; i.e., less pollution releasing technology should be used in the production process; use of renewable energy and car sharing should be encouraged specially in crowded cities, encouraging the use of renewable sources of energy, provision for the recycling of waste, introducing carbon tax and carbon trading and tax concessions for using environment-friendly technologies of production. Increase in citizen partnership, improved internet, encouraging use of solar energy and Wi-Fi facilities with enhanced coverage in rural remote areas is also required. For example, the steps taken by Seoul are commendable as they have even provided tablets to people to keep record of their health status. Therefore, the definition of smart cities should be considered in a wider spectrum. We need to have a holistic development.

6   Conclusion

The study has taken into consideration the spillover environmental effects of the future smart cities in India. The smart cities will undeniably lead to an increase in the national income of the country by generating more production possibilities, creating more jobs, increasing per capita income etc., and hence a well-organized urbanization pattern. However, this will be sustainable if environmental aspects are also taken into consideration. The study finds that there are 23 future smart cities in India which will be prone to environmental degradation caused by migration of people from neighboring regions; this is a worrisome issue. Since most of the future smart cities are sparsely located which means that they have to act as a sponge for the job desiring population of many nearby regions. The study opined that there are many cities which are already heavily polluted and they are also included in the list of future smart cities which will inturn accelerate the migration rates in these cities such as Delhi, Kanpur, Patna etc., and in turn will lead to environmental degradation. Future smart cities will certainly help in achieving environmental sustainability through its well-organized structure of economic activities if migration towards these future smart cities could be curtailed by developing the neighbouring areas or periphery regions in terms of employment opportunities specially in the primary and secondary sectors.

References

1. Albino V, Berardi U, Dangelico RM (2015) Smart cities: definitions, dimensions, and performance. J Urban Technol 22(1):3–21
2. Alok VN, Vashisht A (2016) Financing smart cities in India. Indian J Public Adm 63(4): 791–804
3. Alvear O, Calafate CT, Cano JC, Manzoni P (2018) Crowdsensing in smart cities: overview, platforms and environment sensing issues. Sensors 18(2). MDPI. https://doi.org/10.3390/s18020460
4. Anttiroiko AV, Valkama P, Bailey SJ (2014) Smart cities in the new service economy: building platforms for smart services. AI Soc Knowl Cult Commun 29(3):323–334
5. Balasubramaniam A, Paul A, Hong WH, Seo HC, Kim JH (2017) Comparative analysis of intelligent transportation systems for sustainable environment in smart cities. Sustainability 1–12. MDPI
6. Camagani R, Capello R, Nijkamp P (1998) Towards sustainable city policy: an economy-environment technology nexus. Ecol Econ 24(1):103–118
7. Caragliu A, Del Bo C, Nijkamp P (2011) Smart cities in Europe. J Urban Technol 18(2): 65–82
8. Dwevedi R, Krishna V, Kumar A (2018) Environment and big data: role in smart cities of India. Resources 7(4). MDPI. https://doi.org/10.3390/resources7040064
9. Gaigne C, Riou S, Thisse JF (2010) Are compact cities environmentally friendly? J Urban Econ 72(2–3):123–136
10. Glaeser EL, Kahn ME (2010) The greenness of cities: carbon dioxide emissions and urban development. J Urban Econ 67(3):404–418
11. Goddard ILM, Tett SFB (2019) How much has urbanisation affected United Kingdom temperatures? Atmos Sci Lett 20(5):1–6
12. Komninos N, Pallot M, Schaffers H (2013) Special issue on smart cities and the future internet in Europe. J Knowl Econ 13(2):119–134
13. Kramer A, Hojer M, Lovehagen N, Wangel J (2014) Smart sustainable cities exploring ICT solutions for reduced energy use in cities. Environ Model Softw 56:52–62
14. Lazaroiu GC, Roscia M (2012) Definition methodology for the smart cities model. Energy 47:326–332

15. Lee JH, Hancock MG, Hu MC (2013) Towards an effective framework for building smart cities: lessons from Seoul and San Francisco. Technol Forecast Soc Change 89:80–99
16. Miles A, Zaslavsky A, Browne C (2018) IoT-based decision support system for monitoring and mitigating atmospheric pollution in smart cities. J Decis Syst. Published online 9 May. https://doi.org/10.1080/12460125.2018.1468696
17. Moretti E (2004) Estimating the social return to higher education: evidence from longitudinal and repeated cross-sectional data. J Econom 121:175–212
18. Ramos F, Trilles S, Munoz A, Huerta J (2018) Promoting pollution-free routes in smart cities using air quality sensor networks. Sensors 18(8). https://doi.org/10.3390/s18082507
19. Rauch JE (1993) Productivity gains from geographic concentration of human capital: evidence from the cities. J Urban Econ 34:380–400
20. Smith R, Pathak PA, Agarwal G (2018) India's "smart" cities mission: a preliminary examination into India's newest urban development policy. J Urban Aff. Published on 17 May. https://doi.org/10.1080/07352166.2018.1468221
21. Tranos E, Gertner D (2013) Smart networked cities? Innov Eur J Soc Sci Res 25(2):175–190
22. UNFPA (2007) State of the world population 2007: unleashing the potential of urban growth. United Nations Population Fund, New York
23. WHO (2018) WHO global ambient air quality database. Geneva
24. Wu M, Xiong NN, Tan L (2018) An intelligent adaptive algorithm for environment parameter estimation in smart cities. IEEE Access 6:23325–23337. IEEE (Institute of Electrical and Electronics Engineers). Publication 05 Mar 2018. https://doi.org/10.1109/access.2018.2810891
25. Yang Z, Zhao Y (2014) Energy consumption, carbon emissions, and economic growth in India: evidence from directed acyclic graphs. Econ Model 38:533–540
26. Zhang X, Hao S, Rong WG, Cooper DE (2012) Intelligent transportation systems for smart cities: a progress review. Sci China Inform Sci 55(12):2908–2914

Landslide Susceptibility Assessment Due to Construction of Buildings in Garhwal Area Using GIS

Ruchi Saraswat, Surya Parkash and Shilpa Pal

Abstract This paper presents the effect of building construction in hilly areas of Garhwal Himalaya (Dunda to Bhatwari) and its susceptibility with respect to the landslides. For susceptibility, the first step is to prepare the landslide hazard zonation map in terms of various thematic layers like slope, drainage, land use and land cover, geology, structure and roads using GIS and remote sensing. Susceptibility of the element like the construction of buildings and roads has been calculated with respect to LHZ using the weighted linear combination method. This will help in planning future developmental activities and according to which corrective measures can be taken to mitigate the disaster.

Keywords GIS · Garhwal · Landslides · Remote sensing · Susceptibility

1 Introduction

Due to the increase in construction activities, the probability of natural disasters is increasing nowadays. Among various natural disasters, landslides are common natural disasters which mostly occur in the hilly terrain. In Uttarkashi district, many lives have been lost due to rainfall-based landslides, and if proper actions will not be taken then this will keep on rising. So, the need of landslide zonation and its assessment is must. In the study area, the susceptibility assessment with respect to landslide hazard zonation has been calculated with reference to geo environmental conditions to anticipate the occurrence of landslides, and this analysis can support various local authorities to work towards disaster risk reduction plan like stopping

R. Saraswat (✉)
JSS Academy of Technical Education, Noida, India
e-mail: rsaraswat@jssaten.ac.in

S. Parkash
National Institute of Disaster Management, New Delhi, India

S. Pal
Delhi Technological University, New Delhi, India

© Springer Nature Singapore Pte Ltd. 2020
S. Ahmed et al. (eds.), *Smart Cities—Opportunities and Challenges*,
Lecture Notes in Civil Engineering 58,
https://doi.org/10.1007/978-981-15-2545-2_70

the construction work in the susceptible areas and identifying the safety zones for carrying out the developmental activities like urbanisation and industrialisation [20]. Numerous works had been done by different researchers in the field of landslide. The present study uses Geographical Information System-based system to outline the study area into different categories in terms of its susceptibility to landslides. Landslide studies using GIS have been done by many researchers and scientists in the past all across the world [2, 3, 5, 6, 8, 13–15, 19, 21–25]. Several statistical methods have also been developed and used for performing such susceptibility studies. Landslide mapping has been carried out using logistic regression by Atkinson and Massari [1], Lee [7] and Ramakrishnan et al. [18] along with frequency ratio method [11] and information value method [18]. For landslide susceptibility modelling, fuzzy algebraic function was used by Lee [8], Ercanoglu and Gokceoglu [4] and Pistocchi et al. [16]. Artificial neural network was used for landslide susceptibility studies and modelling by Lee et al. [9, 10, 12], Pradhan and Lee [17] and many others. Knowledge-driven approach raster analysis for landslide study has been conducted by Gupta et al. [6]. In this paper, the main emphasis is on landslides susceptibility assessment due to various development activities like construction of buildings and roads which will help in further planning to mitigate the effect of disaster.

2 Study Area

The study area lies in the north-western part of Uttarakhand state falls in Survey of India sheet no. H44G06 and H44G05. The geographical area of my study is 119 km^2. The district is important from religious point of view as the two holy rivers, namely Ganga and Yamuna have their emerging points in this district. Uttarkashi district, the largest district of Uttarakhand, is also important from strategic point of view as it shares its NE boundary with China.

The location of the study area is from Dunda with latitude 30°42″20′ N and longitude 78°21″00′ to Bhatwari village with latitude 30°48″30′ N and longitude 78°37″15′ E, an important town on the Uttarkashi district as shown in Fig. 1. The village is situated on both the banks of Bhagirathi River. Figure 2 shows the topography of study area which contains different types of geographic features, roads, boundary and villages lying in upper part of Himalaya.

3 Landslide Inventory Map

Landslide inventory map gives the details of the known landslide occurrences in that area, including steady, passive, reactivated and most recent landslides. In the mentioned study area, 55 active landslides and 28 inactive landslides have been found out by the field survey. Landslide inventory map for the study area has been

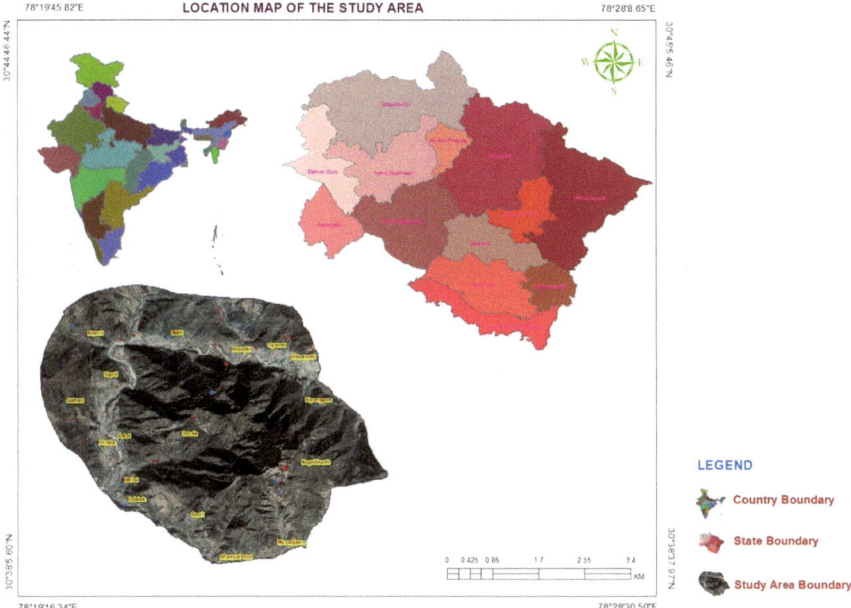

Fig. 1 Study area

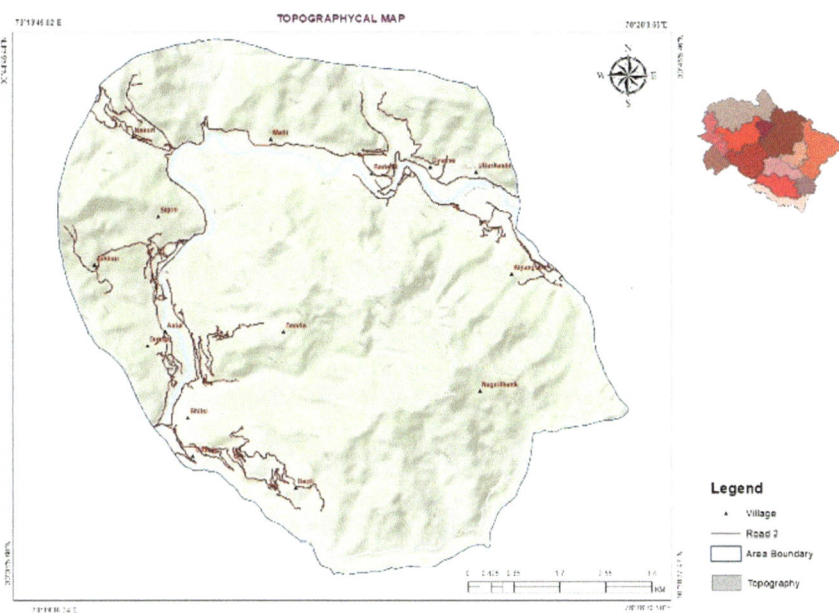

Fig. 2 Topographical map

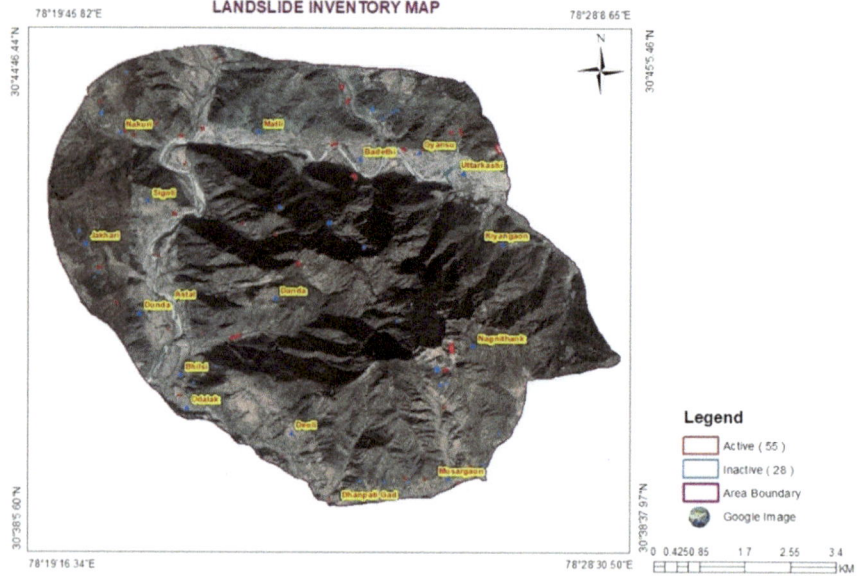

Fig. 3 Landslide inventory map of study area

prepared using the satellite imagery (Landsat and LISS III) as well as using the field survey. The study area consists of total 83 landslides, out of which 55 are active and 28 inactive landslides occurred in the area. Majority of landslides are active landslides in this study area (Fig. 3).

4   Landslide Hazard Zonation

The main aim of LHZ is to find out which area is more prone to landslides and to what extent at that particular location. The LHZ map divides the landslide-prone hilly terrain into different zones according to their degree of susceptibility to landslides. It requires identification of those areas that are affected by landslides and assessing the probability of such landslides occurring within a specific period of time. To achieve the main objective of landslide hazard zonation map, firstly thematic layers of slope, geology, drainage, land use and road network have been prepared using GIS and remote sensing. Further overlaying of the layers is done by using weighting overlay method in which weights have been assigned to each thematic layer according to the relative significance of each factors towards landslides.

For landslide hazard zonation map, the weight of evidence method is used to calculate the statics related to landslides controlling factors affecting landslides. Various thematic layers, like slope, slope aspect, precipitation and geology, were prepared and analysed using GIS-based approach. The positive weight (Wi+) shows the presence of landslides and negative weight (Wi−) shows the absence of landslides that were calculated for each factor affecting landslides. Contrast value (C_w) is the difference between two weights, and it indicates the spatial relationship between predictor variables and landslides. Test of conditional independence is used to generate a landslide susceptibility map. Results show that 90% of the landslides occur in an area having high risk. Table 1 shows the contrast values for every class.

By using landslide hazard zonation map as shown in Fig. 4, the study area has been categorised into different zones like very low, low, low moderate, moderate, high hazard and very high hazard zones. Table 2 shows the percentage of the areas falling in different zones which concludes that high and very high hazard zones are not safe for developmental activity. The obtained landslide hazard zonation map is validated using the landslide inventory map of the study area.

Table 1 Summary of weights of evidence (WOE) and contrast factors for classes

Factor	Class	Wi+	Wi−	C_w
Precipitation (mm/Year)	2550–2650	0.55	−6.33	4.03
	2651–2680	0.00	0.09	−0.09
	2681–2710	0.00	0.02	−0.02
	2711–2750	0.00	0.48	−0.48
	2751–2800	0.00	0.01	−0.01
Aspect	North	0.65	−0.04	0.75
	Northeast	0.03	0.01	0.02
	East	−0.66	0.00	−0.80
	Southeast	−0.52	0.04	−0.58
	South	0.23	−0.05	0.26
	Southwest	0.22	−0.05	0.33
	West	0.17	−0.03	0.22
	Northwest	0.00	0.00	0.00
Slope (degree)	0°–10°	−0.55	1.66	−2.44
	10°–20°	1.59	−0.39	2.00
	20°–30°	2.25	−0.13	2.44
	30°–40°	−9.33	0.00	−9.23
	40°–60°	−6.35	0.00	−6.29
	60°–90°	−0.22	0.00	−0.22
Distance to hill cut (m)	0–100	3.11	−0.52	3.11
	100–200	2.44	−0.21	2.09
	200–300	2.22	−0.17	−0.55
	300–400	0.71	−0.02	−0.78
	>400	−2.11	2.49	−2.66

(continued)

Table 1 (continued)

Factor	Class	Wi+	Wi−	C_w
Distance to road cut (m)	0–100	2.71	−0.54	3.17
	100–200	1.98	−0.13	2.11
	200–300	−0.55	0.01	−0.55
	300–400	−0.80	0.01	−0.83
	>400	−0.88	1.82	−2.77
Soil permeability	Rapid	0.00	0.02	−0.02
	Mixture moderate	0.00	0.02	−0.02
	Moderate	0.00	0.55	−0.55
	Slow	0.00	0.44	−0.44
	Very slow	1.69	−6.66	8.22
Distance to stream (m)	0–100	0.00	0.14	−0.14
	100–200	0.00	0.13	−0.13
	200–300	−1.26	0.09	−1.22
	300–400	0.76	−0.14	0.91
	>400	0.26	−0.66	0.88
Distance to drain (m)	0–100	1.55	−0.70	2.23
	100–200	1.05	−0.65	1.22
	200–300	−1.07	−0.15	−1.09
	300–400	0.01	0.05	−0.48
	>400	−1.22	0.06	−2.22
Geological unit	Dunda quartzite	−2.66	0.91	−3.05
	Bareti quartzite	0.00	0.34	−0.08
	Dharasu formation	0.00	0.07	−0.03
	Gamiri quartzite	−1.05	0.02	2.10
	Khattukhal limestone	2.66	−0.34	3.66
	Epidiorites	−1.19	−0.93	−1.71
Landcover	Built-up area	0.61	−0.61	1.26
	Dense and medium dense forest	2.10	2.40	−0.01
	Low dense forest	1.10	1.21	−2.43
	Agriculture land	0.71	1.10	−0.66
	Open land	3.12	3.61	−0.06

5   Landslide Susceptibility Assessment

Landslide susceptibility assessment with respect to landslide hazard zonation has been done by calculating the susceptibility of various factors including developmental activities like buildings and roads, with respect to landslides.

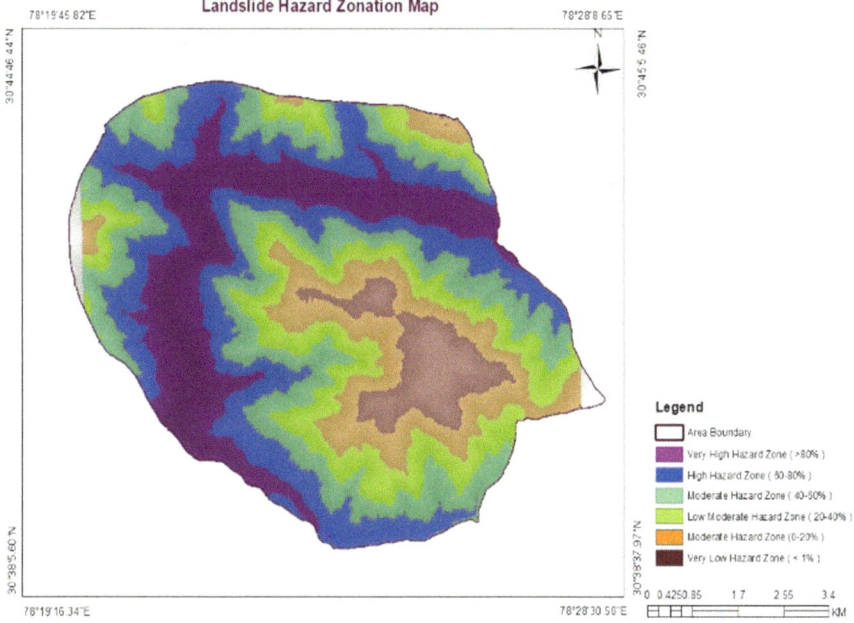

Fig. 4 LHZ map

Table 2 Details of landslide hazard zone

S. no.	Landslide hazard zone	Area (km^2)	Area (%)	Remarks
1	Very low hazard	13.4	0.28	Safe for zone development activity
2	Low hazard	11.29	37.32	
3	Low moderate	15.88	8.94	Local vulnerable zone instability
4	Moderate	25.86	33.13	
5	High hazard	42.25	14.68	Unsafe zone for development activity
6	Very high hazard	11.25	5.66	

5.1 Susceptibility of Geology with Respect to LHZ

Table 3 gives the susceptibility of landslides with geology in km^2. Maximum susceptibility of Gamiri quartzite occurs in the very high hazard zone.

Table 3 Susceptibility of geology with respect to LHZ (km²)

Susceptibility area with geology

Sl. no.		Very low hazards	Low hazards zones	Low moderate hazards zone	Moderate hazards zone	High hazards zone	Very high hazards zone
1	Dunda quartzite	0.6	0.81	0.12	0.62	1.002	0.16
2	Bareti quartzite	1.1	1.2	0.06	0.83	0.61	0.43
3	Dharasu formation	0	0.71	0	0.012	0	0
4	Gamiri quartzite	0.032	0.62	0.62	0.54	0.012	1.02
5	Khattukhal limestone	1.12	0.72	0.87	0.54	0.6	0.28
6	Epidiorites	0.04	0.86	0.43	1.01	0	0.045

Table 4 Susceptibility of slope with respect to LHZ

Susceptibility with slope

Sl. no.	Type of hazards	Slope (in degree)	Susceptibility (in %)
1	Very low hazards	0–15	34.59
2	Low hazards zone	15–30	13.11
3	Low moderate hazards zone	30–45	24.01
4	Moderate hazards zone	45–60	13.15
5	High hazards zone	60–75	8.23
6	Very high hazards zone	>75	6.91

5.2   Susceptibility of Slope with Respect to LHZ

In this study area, there are more chances of landslides in case of flat slope due to increase in the construction of roads. Also, there is decrease in the density of dense forests which increases the chances of landslides. Table 4 gives the susceptibility of slope with respect to categorised hazard zone as the value of slope increases and susceptibility decreases.

5.3   Susceptibility of Drainage with Respect to LHZ

Drainage is an important factor which is taken into consideration while preparing hazard zonation map as the drainage density helps in finding out the main

Table 5 Susceptibility of drainage length with respect to LHZ

Sl. no.	Type of hazards	Drainage length (in km)	Susceptibility (in %)
	Susceptibility with drainage length		
1	Very low hazards	65.7	25.97
2	Low hazards zone	22.4	8.85
3	Low moderate hazards zone	21	8.30
4	Moderate hazards zone	23.2	9.17
5	High hazards zone	43.2	17.08
6	Very high hazards zone	77.5	30.63

characteristics of the soil. Various drainage patterns like first order of length 185.31 km, second order of length 28.92 km and third-order pattern of length 7.07 km have been categorised in this area on the basis of stream order classification. Table 5 illustrates that the susceptibility of the drainage length is maximum in the very high hazard zone.

5.4 Susceptibility of Structure with Respect to LHZ

Various structural features like anticline, syncline, folds and faults have been noticed in this study area, and other structural features have a great impact on the stability of the slopes. Table 6 shows the susceptibility of various structures with respect to LHZ in km^2.

5.5 Susceptibility of Landslides with Respect to Land Use

The land use pattern of the area indicates the percentage of area occupied by agriculture, habitation, snow cover, water bodies, human settlements and open areas

Table 6 Susceptibility of structures with respect to LHZ (all values in km^2)

Sl. no.	Type of structure	Very low hazard	Low hazards zone	Low moderate hazards zone	Moderate hazards zone	High hazards zone	Very high hazards zone
1	Faults	0.3	0.09	0	0.04	1.002	0.12
2	Anticline	1.1	1.2	0.06	0	0.61	0.43
3	Syncline	0	0.71	0	0.089	0.54	0.0082
4	Ridge	0.003	0.66	0.05	0.54	0.004	1.02
5	Other structures	0.12	0.0063	0.0084	0.38	0.06	0.002

Table 7 Susceptibility of land use with respect to LHZ

Susceptibility against land use			
Sl. no.	Type of hazards	Land use	Susceptibility (in %)
1	Very low hazards	Agriculture	5.44
2	Low hazards zone	Open land	9.92
3	Low moderate hazards zone	Other stony land	17.50
4	Moderate hazards zone	Water and river	12.60
5	High hazards zone	Building development	21.12
6	Very high hazards zone	Road development	33.42

Table 8 Susceptibility of roads with respect to LHZ

Susceptibility against road			
Sl. no.	Type of hazards	Road (in km)	Susceptibility (in %)
1	Very low hazards	10	6.02
2	Low hazards zone	23	13.86
3	Low moderate hazards zone	12	7.23
4	Moderate hazards zone	17	10.24
5	High hazards zone	39	23.49
6	Very high hazards zone	65	39.16

is depicted by land cover mapping of the area. Table 7 gives the susceptibility of land use with respect to landslides, and maximum susceptibility of land use occurs in the very high hazard zone.

5.6 Susceptibility of Landslides with Respect to Roads

Road construction is also one of the main factors affecting the landslides in the hilly terrain. Table 8 shows the maximum susceptibility of roads of 39.16% in the very high hazard zone.

5.7 Susceptibility of Buildings with Respect to LHZ

Building map of the study area has been digitised using geographical information system as shown in Fig. 5, and susceptibility of buildings with respect to landslides has been calculated by overlaying the building map over the LHZ of the area as shown in Fig. 6.

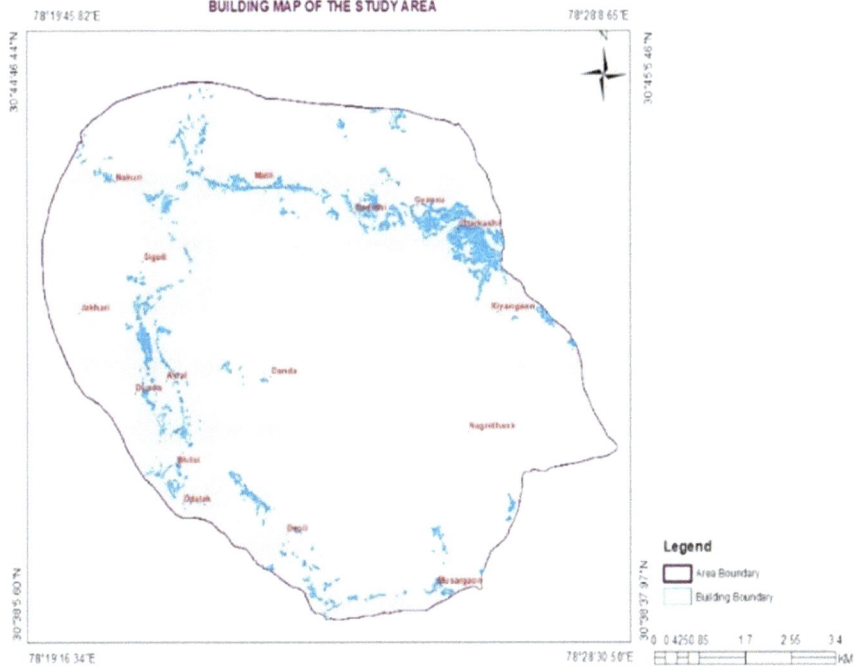

Fig. 5 Building map (2017)

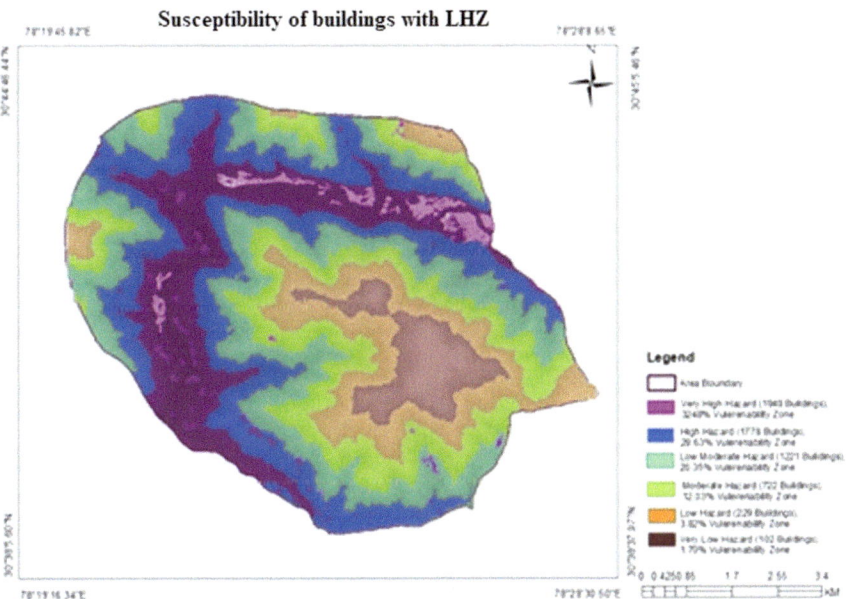

Fig. 6 Susceptibility of buildings with LHZ

6   Landslide Susceptibility Assessment Map

Using the weighted overlay method, landslide susceptibility assessment zonation map was created in GIS environment which helps in finding out the susceptible zones in the area selected for study. Landslide susceptibility assessment map is created by combining the susceptibilities of all the parameters like slope, geology, lithology, drainage, land use and land cover, roads, buildings affecting landslides. From the judgement done, 1% of area is in very low susceptible zone, 17% is in low susceptible zone, 21% is in moderate zone, 26% is in highly susceptible zone, and 35% is in very highly susceptibility zone (Fig. 7). Maximum area comes in the highly susceptible zone, and the factors responsible are buildings, roads and hydel power and bridges as being depicted from Fig. 6. Also, it has been shown from Fig. 7 that the maximum landslide density is in the northern and western part of the study area. Table 7 shows the susceptibility of buildings with landslides, and as the number of buildings is increasing susceptibility also increases.

While finding out the susceptibility assessment due to various construction of buildings using GIS, it can be observed from Table 9 that there are in total 7001 buildings, out of which maximum number of building lies in very high hazard zone, and their susceptibility is around 32.48% which indicates that as the construction activities are increasing, chances of landslides are also increasing; whereas, in very low hazard zone there are 102 buildings, and its susceptibility is around 1.7%. Table 10 gives the percentages of the factors affecting landslides.

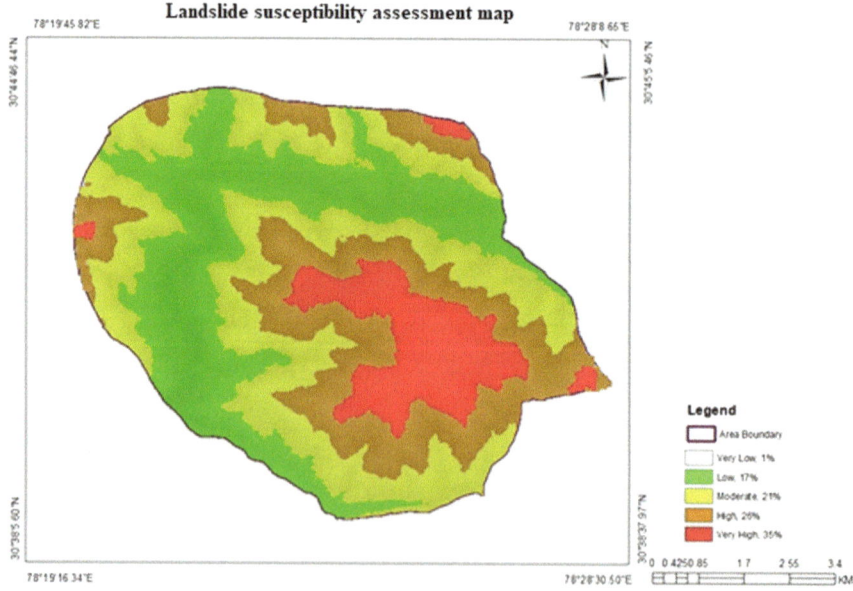

Fig. 7 Landslide susceptibility assessment map

Table 9 Susceptibility of buildings with respect to LHZ

Sl. no.	Type of hazards	Number of buildings	Susceptibility (in %)
1	Very low hazards	102	1.70
2	Low hazards zone	229	3.82
3	Low moderate hazards zone	722	12.03
4	Moderate hazards zone	1221	20.35
5	High hazards zone	1778	29.63
6	Very high hazards zone	1949	32.48

Table 10 Percentage of various contributing factors on landslides

Sl. no.	Controlling factor's	Effect on landslide (in %)
1	Precipitation	1.3
2	Aspect	10.4
3	Slope	14.7
5	Human development (building)	22.4
6	Soil	14.5
7	Geology	6.3
8	Rainfall	5.2
9	Distance to hill cut	6.3
10	Land use	2.4
11	Distance to road cut	15.3
12	Distance to stream	1.2

7 Conclusions

It can be concluded from the present study that there are more chances of landslides where the construction activities like building and road construction are more as finding out while calculating the susceptibility of buildings with respect to landslide hazard zonation. Also, it has been found out that the maximum number of buildings lies in the very high hazard zone which gives rise to the conclusion that developmental activities are one of the major factors responsible for increase in landslides. So, it has been suggested not to plan further developmental activities in a very high hazard zone. For more valid results, history of landslide events, extent of damages and losses in terms of infrastructure should be incorporated which can help various government agencies, local authorities in planning mitigation measures for landslides.

References

1. Atkinson PM, Massari R (1998) Generalized linear modelling of susceptibility to land sliding in the central Apennines, Italy. Comput Geosci 24:373–385
2. Carro M, Amicis MD, Luzi L, Marzorati S (2003) The application of predictive modeling techniques to landslides induced by earthquakes: the case study of the 26 September 1997 Umbria–Marche earthquake (Italy). Eng Geol 69(1–2):139–159
3. Dhakal AS, Siddle RC (2002) Physically based landslide hazard model 'U' method and issues. In: EGS XXVII general assembly, Nice, pp 21–26
4. Ercanoglu M, Gokceoglu C (2004) Use of fuzzy relations to produce landslide susceptibility slope failure risk maps of the Altindag (settlement) region in Turkey. Eng Geol 55:277–296
5. George YL, Long SC, David WW (2007) Vulnerability assessment of rainfall-induced debris flows in Taiwan. Department of Earth Systems and Geo Information Sciences, College of Science, George Mason University, Fairfax, VA 22030, ETATS-UNIS. Available at http://www.springerlink.com/content/uj26871v2831nx.44
6. Gupta M, Ghose MK, Sharma LP (2009) Application of remote sensing and GIS for landslides hazard and assessment of their probabilistic occurrence—a case study of NH31A between Rangpo and Singtam. J Geomat 3(1):13–17
7. Lee S (2005) Application of logistic regression model and its validation for landslide susceptibility mapping using GIS and remote sensing data. Int J Remote Sens 26:1477–1491
8. Lee S (2007) Application and verification of fuzzy algebraic operators to landslide susceptibility mapping. Environ Geol 52:615–623
9. Lee S, Ryu JH, Won JS, Park HJ (2004) Determination and application of the weights for landslide susceptibility mapping using an artificial neural network. Eng Geol 71:289–302
10. Lee S, Ryu JH, Lee MJ, Won JS (2003b) Landslide susceptibility analysis using artificial neural network at Boun, Korea. Env Geol 44:820–833
11. Lee S, Sambath T (2006) Landslide susceptibility mapping in the Damrei Romel area, Cambodia using frequency ratio and logistic regression models. Environ Geol 50:847–855
12. Lee S, Ryu JH, Lee MJ, Won JS (2003a) Landslide susceptibility analysis using GIS and artificial neural network. Earth Surf Process Landf 28:1361–1376
13. Pachauri AK, Gupta PV, Chander R (1998) Landslide zoning in a part of the Garhwal Himalayas. Environ Geol 36(3–4):325–334
14. Pachauri AK, Pant M (1992) Landslide hazard mapping based on geological attributes. Eng Geol 32:81–100
15. Patanakanog B (2001) Landslide hazard potential area in 3 dimension by remote sensing and GIS technique. Land Development Department, Thailand. Available on www.ecy.wa.gov/programs/sea/landslides/help/drainage.html. Accessed on 02.03.2011
16. Pistocchi A, Luzi L, Napolitano P (2002) The use of predictive modeling techniques for optimal exploitation of spatial databases: a case study in landslide hazard mapping with expert system-like methods. Environ Geol 58:251–270
17. Pradhan B, Lee S (2009) Landslide risk analysis using artificial neural network model focusing on different training sites. Int J Phys Sci 4(1):1–15
18. Ramakrishnan D, Ghose MK, Chandran RV, Jeyaram A (2005) Probabilistic techniques, GIS and remote sensing in landslide hazard mitigation: a case study from Sikkim Himalayas, India. Geocarto Int 20(4):53–58
19. Sakellariou MG, Ferentino MD (2001) GIS based estimation of slope stability. Nat Hazards Rev 2(1):12–21
20. Sharma LP, Patel N, Ghose MK, Debnath P (2009) Geographical information system based landslide probabilistic model with trivariate approach—a case study in Sikkim Himalayas. In: 18th United Nation's regional cartographic conference Asia and the Pacific, Bangkok, 26–29 Oct 2009

21. Sivakumar GL, Mukesh MD (2002) Landslide analysis in geographic information systems. Department of Civil Engineering, Indian Institute of Science, Bangalore, India. www.gisdevelopment.net/application/natural_hazards/landslides/nhls009pf.htm. Accessed on 03.03.2011
22. Uromeihy A, Mahdavifar MR (2000) Landslide hazard zonation of Khorshrostam area, Iran. Bull Eng Geol Environ 58:207–213
23. Van Westen CJ, Rengers N, Terlien MTJ, Soeters R (1997) Prediction of the occurrence of slope instability phenomena through GIS-based hazard zonation. Geol Rundsch 86:404–414
24. Varnes DJ (1984) Landslide hazard zonation: a review of principles and practice. UNESCO, Paris, pp 1–55
25. Wadge G (1988) The potential of GIS modelling of gravity flows and slope instabilities. Int J Geogr Inf Syst 2:143–152
26. Wangshu-Quiang, unwin DJ (2007, February) Modelling landslide distribution on loess soils in china: an investigation

Agro Residual Biomass Conversion: A Step Towards Pollution Control and Sustainable Waste Management

M. Khursheed Akram, Iqbal Ahmad Talukdar
and Mohammed Sharib Khan

Abstract Greenhouse gases emissions and exponential growth in pollutants due to excessive consumption of fossil fuels have become a major problem worldwide. At the same time, smog problems due to agricultural residue waste burning are making habitats of Delhi's lives difficult. In the year 2018, we have already witnessed a week "smog out" situation in Delhi which is primarily caused due to crop residues burning in Punjab, Haryana and other neighbouring states. Thus, it is essential methods to curb and control pollution via efficient disposal of agricultural waste. Interestingly, agricultural waste materials such as sugarcane bagasse, rice husk, rice straw and plants leaf have the potential to serve as a substitute of fossil fuels. In general, agricultural waste such as sugarcane bagasse falls under the category of lignocellulosic biomass having lignin, cellulose and hemicellulose as major constituents. These wastes lignocellulosic waste biomass can be converted to fuels and chemicals via different methods such as chemical conversion, biological conversion, catalytic conversion and pyrolysis. However, pyrolysis of lignocellulosic waste is most appropriate and feasible method which does not essentially require the involvement of severe chemicals. In the present manuscript, a detailed overview of various pyrolysis methods will be discussed for waste biomass conversion to produce fuels and chemicals. Post this, the application of pyrolytic processes will be extended to other solid waste materials such as plastic waste, e-waste and municipal solid waste to produce energy and chemicals. Eventually, potential impact of these technologies on pollution control and sustainable waste management will be presented.

M. K. Akram (✉)
Applied Sciences and Humanities Section, University Polytechnic, Faculty of Engineering and Technology, Jamia Millia Islamia, New Delhi 110025, India
e-mail: makram@jmi.ac.in

I. A. Talukdar
Department of Chemistry, Faculty of Natural Sciences, Jamia Millia Islamia,
New Delhi 110025, India

M. S. Khan
Department of Civil Engineering, Faculty of Engineering and Technology,
Jamia Millia Islamia, New Delhi 110025, India

© Springer Nature Singapore Pte Ltd. 2020
S. Ahmed et al. (eds.), *Smart Cities—Opportunities and Challenges*,
Lecture Notes in Civil Engineering 58,
https://doi.org/10.1007/978-981-15-2545-2_71

Keywords Agro biomass · Pollution control · Agriculture · Sustainable waste

1 Introduction

Emission of greenhouse gases and smog is a major problem for the metro cities in India. The greenhouse gases emission is primarily caused due to excess utilization of gasoline, diesel and other higher hydrocarbons which on combustion releases carbon dioxide, thereby causing damage to the atmosphere. In brief, there is a threat to the environment due to excess combustion of transportation fuel all over the world [1]. At the same time, the Indian subcontinent is facing a different kind of pollution problem in the form of smog which is very critical and needs to be solved immediately. The major reason for smog in India is the burning of agro residue which in general is considered a waste material. It is well known that majority of Indian population (approximately 70%) lives in rural areas and relies on agriculture for their livelihood. The major crops which are produced in India in bulk are wheat, rice and sugarcane which on utilization discard wheat straw, rice husk and sugarcane bagasse as major leftover agro residue material. In fact, India is one of the largest producers of sugarcane which makes it one of the largest producers of sugarcane bagasse as well [2]. Due to no economic value of these agro residues and demand to clear the fields for next season crop, most of these materials are burned down; whereas, a small quantity is used for household cooking. In both the conditions, direct burning/combustion of these agro residues causes the emission of carbon dioxide and particulate matters along with other pollutants. Therefore, burning or direct combustion of agro residues is not a sustainable solution albeit we need to get rid of them.

On the other hand, the structural composition of these agro residues indicates that these waste materials can be a great source of energy and chemicals as well as create a closed carbon cycle as shown in Fig. 1 [3]. It is well known that plants consume carbon dioxide for their growth, thus the amount of carbon dioxide released by using agro residue fuel and chemicals will be utilized again by the plants and therefore, no additional carbon dioxide will be emitted into the environment. Scientists have reported several chemicals which can be produced and replace petroleum product especially the US department of energy report is important to evaluate the full potential of biomass including agro residue [4–10]. There are three major constituents of agro residues which are commonly named as cellulose, hemicellulose and lignin. The cellulose and hemicellulose majorly contain six carbon and five carbon sugars, whereas as lignin acts as a binder and holds the cellulose and hemicellulose intact in the biomass. Based on the composition of these agro residues, different chemicals and fuels can be prepared. For example, lignin is used to produce aromatics; whereas, cellulose is used to produce glucose, fructose, and hemicellulose is used to produce furfural.

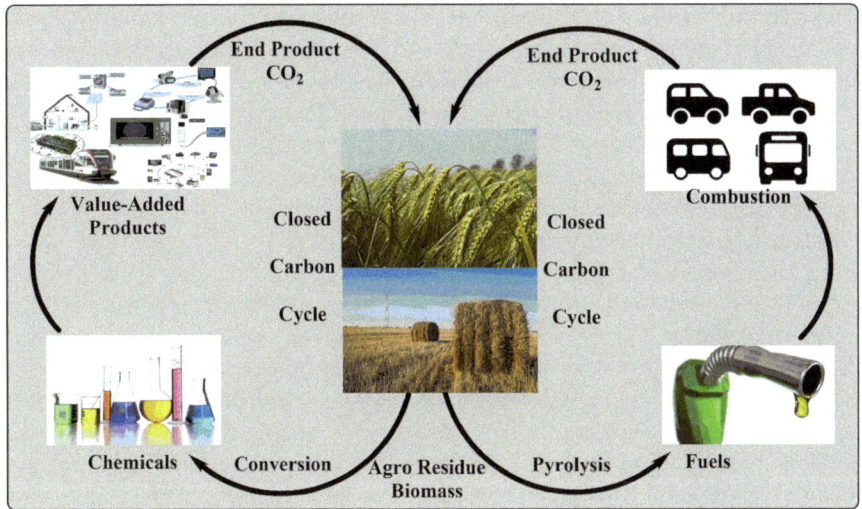

Fig. 1 Biomass as a carbon neutral source for fuels and chemicals production

2 Biomass Conversion Processes

As mentioned earlier, different products can be made from agro residues based on their structural composition. Similarly, different products can be produced using different methods as well. There are several processes reported for agro residue conversion out of which four broad categories are (i) biological process, (ii) chemical process, (iii) pyrolysis and (iv) thermo-chemical process.

The biological processes for lignocellulosic biomass or agro residue conversion primarily proceed via fermentation route in which the catalyst can be either microbes, enzymes, yeast or bacteria, etc. [11]. In these processes, agro residues are first pre-treated to remove lignin and other impurities and then charged into a fermenter. After that, the enzyme or any biological catalyst converts the sugar content into a useful product. The major advantage of biological process is that they operate at room temperature; whereas, slow processing, time needing and large space requirement are the major disadvantages of these processes. Ethanol is one such important chemical produced from agro residues via biological routes.

Chemical conversion is another important route for high value products synthesis from the agro residues. The chemical processes can further be divided into several sub categories; nevertheless hydrolysis and hydrogenation remain the most discussed route. In hydrolysis route, a suitable acid catalyst is used to depolymerize the agro residue via lignin removal into cellulose and hemicellulose on which further reaction yields glucose and xylose. The glucose is converted to fructose and 5-hydroxymethylfurfural; whereas, xylose converts to furfural in the presence of an acid catalyst via dehydration reaction. After that, 5-hydroxymethylfurfural can further be reacted to produce levulinic acid which is an important class of building block platform chemical [12]. On the other hand, hydrogenation is employed for the conversion of either lignin or to produce very high value chemicals. Xylitol is one such example which is produced via hydrogenation of xylulose or hemicellulose [13]. It should be noted that hydrogenation requires the presence of a metal catalyst rather than an acid catalyst. However, if the raw material is raw biomass, then a combination of metal and acid catalyst may be required. Also, majority of the important class of chemical produced from biomass either uses an acid catalyst, or a metal catalyst or combination of both [14].

Another important process for biomass conversion is the pyrolysis which is generally carried out at high temperature as compared to biological and chemical routes which operates at low temperature. In general, any process which operates at high temperature and in the absence of oxygen is termed as pyrolysis [15, 16]. However, the pyrolysis process also is further subdivided into several categories such as torrefaction, slow pyrolysis (carbonization), intermediate pyrolysis and fast pyrolysis process. Torrefaction is simply the removal of moisture and very light volatile components to prepare solid fuel with high energy density and ease of transportation. The torrefaction is generally carried out between 200 and 300 °C for 20–30 min which reduces the weight of the agro residues by 15–20%. The fuel grade torrified yield is nearly 80 wt% or above. Similarly, slow pyrolysis is another important process which is carried out to nearly 350–400 °C temperature in the absence of oxygen to produce liquid and gaseous products in higher amount. This process takes several hours to complete and also known as the carbonization process. However, when residence time is reduced to less than half hour, the process is known as intermediate pyrolysis. The major advantage of this process is that gaseous products' formation can be avoided. Nevertheless, char formation in this process is higher as compared to char formation via slow pyrolysis process. Similarly, the gaseous products can further be reduced, and liquid product yield can be enhanced more than 70 wt% by raising the pyrolysis temperature to 500 °C and reducing the residence time to few seconds. This process with higher operating limit and very short residence time is known as fast pyrolysis. The char quantity is also very low (approximately less than 12 wt%) for the fast pyrolysis process. When there is a need of higher number of gaseous products, then operating temperatures can further be increased to nearly 800 °C with very short residence time of agro residues in the reactor. This process in general is known as gasification process which is generally employed to produce synthesis gas from the biomass.

Also, it is possible to develop several other processes via certain changes and modifications such as putting a catalyst unit for bio oil upgrading or high-quality char formation or synthesis gas conversion to higher hydrocarbons. It is also possible to improve the liquid yield quality via in situ catalysis such as pyrolysis plus simultaneous reforming without using any external catalyst. Ahmad et al. have reported one such process wherein they have utilized produced biochar as a catalyst to upgrade the liquid product obtained from the agro residues [17]. It is also possible to integrate the process for other solid waste such as plastic, e-waste and municipal solid waste as a higher temperature may decompose polymers easily. It should be noted that municipal solid waste, e-waste and plastic waste are another critical area which needs immediate solution besides conversion of agro residues. For this purpose, pyrolysis process in general is preferable as it does not require external chemicals and can be operated by even semi-skilled workers which makes this process user friendly.

3   Conclusions

Although it is difficult to discuss several aspects of agro residues conversion in details in this perspective review article. It is observed that the processes for biomass conversion are either operating at low temperatures or high temperatures. Based on the desired, a suitable process of them can be chosen. However, considering ease of operating and unskilled farming workers in India, the pyrolysis may have a better scope as compared to chemical processes. Also, several types of pyrolysis processes can be employed depending upon the quantity of gaseous, liquid and solid products required. Also, pyrolytic processes can further be extended to tackle other kinds of solid wastes such as plastic waste, municipal solid waste and other types of waste.

References

1. Quereshi S, Ahmad E, Pant KK, Dutta S (2017) Insights into the metal salt catalyzed ethyl levulinate synthesis from biorenewable feedstocks. Catal Today 291:187–194
2. Ahmad E, Pant KK (2018) Chapter 14—Lignin conversion: a key to the concept of lignocellulosic biomass-based integrated biorefinery. Elsevier B.V.
3. Ahmad E, Vani A, Pant KK (2019) An overview of fossil fuel and biomass-based integrated energy systems: co-firing, co-combustion, co-pyrolysis, co-liquefaction and co-gasification. In: Nanda S, Sarangi PK, Vo DVN (eds) Fuel processing and energy utilization, 1st edn. CRC Press
4. Werpy T, Petersen G (2004) Top value added chemicals from biomass
5. Holladay J, White J (PNNL), Manheim A (DOE-HQ) (2004) Top value added chemicals from biomass: volume I—results of screening for potential candidates from sugars and synthesis gas
6. Holladay JE, White JF, Bozell JJ, Johnson D (2007) Top value-added chemicals from biomass: volume II—results of screening for potential candidates from biorefinery lignin
7. Irmak S, Canisag H, Vokoun C, Meryemoglu B (2017) Xylitol production from lignocellulosics: are corn biomass residues good candidates? Biocatal Agric Biotechnol 11(July):220–223
8. Braden DJ, Henao CA, Heltzel J, Maravelias CC, Dumesic JA (2011) Production of liquid hydrocarbon fuels by catalytic conversion of biomass-derived levulinic acid. Green Chem 13(7):1755
9. Stepan E et al (2018) A versatile method for obtaining new oxygenated fuel components from biomass. Ind Crops Prod 113(January):288–297
10. Zakzeski J, Bruijnincx PCA, Jongerius AL, Weckhuysen BM (2010) The catalytic valorization of lignin for the production of renewable chemicals. Chem Rev 110:3552–3599

11. Salvachúa D, Karp EM, Nimlos CT, Vardon DR, Beckham GT (2015) Towards lignin consolidated bioprocessing: simultaneous lignin depolymerization and product generation by bacteria. Green Chem 17(11):4951–4967
12. Kang S, Fu J, Zhang G (2018) From lignocellulosic biomass to levulinic acid: a review on acid-catalyzed hydrolysis. Renew Sustain Energy Rev 94:340–362
13. Carvalho WA, Arcaño YD, Pontes LAM, García ODV, Mandelli D (2018) Xylitol: a review on the progress and challenges of its production by chemical route Catal Today 1
14. Bohre A, Dutta S, Saha B, Abu-Omar MM (2015) Upgrading furfurals to drop-in biofuels: an overview. ACS Sustain Chem Eng 3(7):1263–1277
15. Bridgwater T (2018) Challenges and opportunities in fast pyrolysis of biomass: part II. Johnson Matthey Technol Rev 62(2):150–160
16. Bridgwater AV, Meier D, Radlein D (1999) An overview of fast pyrolysis of biomass. Org Geochem 30(12):1479–1493
17. Ahmad E, Jäger N, Apfelbacher A, Daschner R, Hornung A, Pant KK (2018) Integrated thermo-catalytic reforming of residual sugarcane bagasse in a laboratory scale reactor. Fuel Process Technol 171:277–286

Trend Analysis of Temperature Using CRU Data for Satluj River Basin

Asha Devi Singh, Mukesh Kumar Gupta and Mohammed Sharif

Abstract Water is a precious natural gift to mankind. Smart cities can be build only if uninterrupted supply of water is available. Climate change causes variation in the availability of water. These variation are required to be reassessed in order to construct fully functional smart cities. Statistics reveal that consumption of water has increased threefolds in comparison to rate of increase in world population. If population is doubled, water demand will increase sixfold. Due to climate change, careful planning of water resources is necessary to meet the water demand. Hydrological cycle of particular region depends upon atmospheric temperature. Analysis of temperature trend using CRU data over basin will reflect the increase in temperature over past 113 years and change in precipitation pattern. Precipitation and runoff are affected by increase in temperature. It is therefore important for effective operation of water resources to predict stream flow considering climate change. Forecasting of stream flow helps decision makers to decide the scheduled release from reservoirs and hence improves the operational strategies. CRU data analysis is done to predict temperature trend and to assess the response of the basin to global warming. Using Mann-Kendall nonparametric test, trend of annual surface air temperature of 56 stations is estimated. For annual maximum temperature, thirteen stations exhibited increasing trend, at twenty-six stations, decreasing trend is predicted and seventeen stations indicated zero trend. In case of annual minimum temperature, forty-two stations indicated increasing trend, nine stations reflected negative trend and five stations exhibited no trend. Analysis of last 113 years data for 56 stations predicted enhanced warming and change in precipitation pattern.

Keywords Temperature · Mann-Kendall · CRU data · Climate change · Trends

A. D. Singh (✉) · M. K. Gupta
GB Pant Institute of Technology, New Delhi, India
e-mail: adsingh@gbpit.ac.in

M. K. Gupta
e-mail: mkgupta@gbpit.ac.in

M. Sharif
Jamia Millia Islamia (Central University), New Delhi, India
e-mail: msharif@jmi.ac.in

© Springer Nature Singapore Pte Ltd. 2020
S. Ahmed et al. (eds.), *Smart Cities—Opportunities and Challenges*,
Lecture Notes in Civil Engineering 58,
https://doi.org/10.1007/978-981-15-2545-2_72

1 Introduction

On river Satluj in Bilaspur district of Himachal Pradesh concrete gravity Bhakra Dam is located. Height of Bhakra Dam is 226 m. Its construction was taken up after independence for the welfare of the people of Northern Region of India. In 1960 Indus Water Treaty between India and Pakistan was signed. It was decided in the treaty that water of the river Ravi, Beas and Satluj will be used by India. The nation has gone ahead to fully harness the potential of these rivers. Bhakra Nangal Project is a huge project which comprises Bhakra Dam, Nangal Dam, Nangal Hydel Channel, Ganguwal and Kotla Powerhouses. Bhakra Nangal Project construction work was started in the year 1948 and was completed in 1963. State of Punjab, Haryana, Rajasthan, Himachal Pradesh, Delhi and Chandigarh are deriving benefits from this project in the form of irrigation, power supply and water supply.

Survival of living species depends on water. It is an essential natural resource. Since inception of civilization, human beings have shown inclination to settle near rivers due to guaranteed availability of water. Ocean's temperature has a substantial influence on the climate of a place. Due to global warming, there is rise in Earth's atmosphere and oceans temperature; thus, the world is getting warmer. Due to rise in temperature, the hydrology of a place changes. Preponderance of evidence state that the anthropogenic activities are responsible for rise in thermometer readings all around the world since the beginning of the Industrial Revolution. Hulme [1] IPCC special report (2018) indicates an increase in global temperature which will likely reach 1.5 °C between 2030 and 2052 compared to pre-industrial temperature due to anthropogenic activities responsible for generation of greenhouse gas emission. The global temperature record represents an average increase in temperature over the entire surface of the planet.

Bhakra Dam is located on river Satluj which originates from Himalayan region. The river Satluj originates from Himalayan region which is critical source of water and also considered sensitive to global warming. Due to change in temperature, precipitation pattern changes and causes variation in stream flow. If stream flows change irrespective of magnitude, it poses a serious implication for water distribution, especially for the large reservoir such as Bhakra. Due to rapid urbanization and high rate of population growth, water resource system are stressed and climate change creates an additional stress. Design of hydrological system including reservoirs has traditionally been carried out on the assumption of stationarity of hydrometeorological data, but under the impacts of changing climatic conditions, the assumption of stationarity of hydrometeorological data is not valid. If effects of climate change are considered in design of hydrological systems, water resource system will likely be more reliable. To exploit the potential of the Satluj River, it is mandatory to estimate the inflow of river taking into consideration climate change in order to decide the releases in advance.

2   Literature Review

Due to rapid increase of greenhouse gases concentration in atmosphere, climate models predict abnormal changes in global climate. The increase in greenhouse gas emission is likely to continue throughout the world. Stocker [2] Fifth Assessment Report 5 (AR5) (IPCC) for last three decades has declared that each decade, earth surface is becoming warmer than the previous decade. The earth atmosphere and ocean temperature have increased due to concentration of greenhouse gases causing retreating of glaciers and rising of sea levels. The National Climatic Data Center (NCDC) conducted an analysis of global annual and monthly record of ocean and land surface for 130 years duration; temperature indicates rise of 1 °C above pre-industrial levels of 2015. Surface air temperature is an important parameter that has impact on dynamics of atmospheric process. CO_2 level in atmosphere is increasing, resulting in increase in surface air temperature causing drift in hydrological cycle [3].

Increase in concentration of aerosols and greenhouse gases in atmosphere resulted in perceptible increase of temperature. It is evident that with the increase in atmospheric temperature, pattern of precipitation and runoff will vary considerably, and these changes are considered important for water resource planning. Hydrological systems anticipate not only changes in magnitude of water availability but also timings. Chances of extreme events also increase [4].

The temperature trend of eight stations of region using Mann-Kendall is analyzed by Hamid et al. [5]. Analysis of trends in temperature indicates predominantly increasing trend. Due to this increase in temperature, difficulties arise in water resource management as snow and glacier melts contribute to stream flow at Bhakra reservoir. Using soil and water assessment tool (SWAT), stream flows at Bhakra is predicted by Hamid et al. [6]. This study provides an understanding of the effect of climate change in basin.

Radhakrishnan et al. [7] estimated annual and seasonal mean rainfall and temperature trend of India for duration 1901–2014 using linear regression and Mann-Kendall test. Inferences drawn indicate a negative trend in rainfall, whereas the temperature shows a positive trend.

Kothawale and Singh [8] investigated linear trends of surface and tropospheric temperature of five selected isobaric levels for the period 1971–2015 across India. Study indicates an increase in surface temperature of 0.74 °C for the region with latitude greater than 22 °N and an increase of 0.80 °C for region with latitude less than 22 °N.

Sharif [9] using Climate Wizard tool, three emissions scenarios (SRES A2, A1B, and B1) surface air temperature for Saudi Arabia region is studied. The projection of temperature anomalies was obtained using a set of four GCMs. Among the four models considered, the CCSM4 model projected the maximum increase in average temperature, whereas CSIRO-Mk3.0 model projected the minimum increase.

Using Hadley Centre's high-resolution model, Kulkarni et al. [10] predicted possible impacts of a warmer climate on the Hindukush Himalayan region. Gocic and

Trajkovic [11] used nonparametric Mann-Kendall and Sen's slope methods to analyze annual and seasonal trends for various meteorological variables for Serbia region. Researcher selected twelve weather stations for duration 1980–2010. Study indicated increasing trend for temperature. Singh and Jain [12] have estimated stream flow in Satluj which is generated from rainfall, snow and ice. He estimated the snow-covered area which is contributing 59% of stream flow using satellite data and concluded it is retreating with time due to rise in temperature.

3 Trend Analysis

Trend analysis is generally conducted on a sequence of observations over a period of time on a random variable such as temperature and precipitation to evaluate decrease or increase of value in statistical terms Helsel and Hirsch [13]. In terms of statistics, trend represents probability distribution of observations over a period of time and also describes the changes they have undergone in the central value of the distribution such as mean or median. Trend analysis results are effective in prediction with longer periods of data and help to identify the trends and quantify the rate of change. It will not describe much insight into a particular cause/factor in attributing a trend. The result obtained from trend analysis is more reliable when the period of record is significantly long.

4 Nonparametric Tests

Two tests are generally used to determine trend, but it depends upon the distribution of data. A parametric test is when variability can be evaluated in statistical term. In parametric testing, distribution of data is normal. Helsel and Hirsch [13] recommended that data distribution has to be transformed to make it normal and analysis can be restricted. Parametric test is more reliable than the nonparametric test as data distribution is not normal. The most widely used nonparametric test for the analysis of trend is the Mann-Kendall test Mann [11, 14, 15]. This test is used to determine trend in hydrometeorological variables due to climate change when the data is not normally distributed. The null hypothesis H0 for these data is that there is no trend in the series and three alternative hypotheses are there is a negative trend, null and positive trend. The Mann-Kendall tests are based on the calculation of Kendall's tau measure of association between two samples. The S statistic is the sum of all the counting: A positive value of S indicates an increasing trend, and negative value indicates a decreasing trend.

$$S = \sum_{i=1}^{n-1} \sum_{j=i+1}^{n} \mathrm{Sgn}(y_j - y_i) \tag{1}$$

$$\text{Var}(S) = \frac{n(n-1)(2n+1)}{18} \qquad (2)$$

where n is the number of observations and y_i $(i = 1 - n)$ are the independent observations

$$Z_{mk} = \begin{cases} \frac{s-1}{\sigma_S} & \text{if } S < 0 \\ \frac{s+1}{\sigma_S} & \text{if } S > 0 \\ 0 & \text{if } S = 0 \end{cases}$$

$$\text{sgn}(y_j - y_i) = \begin{cases} 1 & \text{if } (y_j - y_i) > 0 \\ 0 & \text{if } (y_j - y_i) = 0 \\ -1 & \text{if } (y_j - y_i) < 0 \end{cases}$$

Mann-Kendall test is used to determine trend where statistical parameters P value and slope are determined. Lower the value of P, stronger is the trend. Sen's slope method [16] projects the rate of change in the variable during the analysis period and then using the median of these slopes estimates an overall slope. A positive slope indicates an increase in hydrometeorological parameter. Negative value indicates decrease in hydrometeorological parameter.

5 Climatic Research Unit

Natural anthropogenic climate change is researched by Climatic Research Unit (CRU). This institute is leading institute concerned with the study of environment. It is part of the School of Environmental Sciences at the University of East Anglia in Norwich. The aim of this institute is to improve scientific understanding in below-mentioned areas:

- Study of historical data and its impact
- The path and reasons of climate change
- Probable future forecast.

Historical data regarding temperature and precipitation CRU3.21 datasets spanning 1900 to 2013 was downloaded from http://badc.nerc.ac.uk/browse/badc/cru. The CRU time series data is available in ASCII formats. The ASCII data is presented at resolution of $0.5° \times 0.5°$. The data is having a grid of 360 lat $\times$ 720 long. Code to extract data is attached at Annexure 1. The geographical limits of the Satluj basin up to Bhakra dam are latitude 30–33°N and longitude 76–83°E. The data is retrieved at $0.5°$ Latitude $\times 1°$ Longitude. The whole region is divided into 56 nodes. Computed coordinates of 56 nodes in the Satluj river basin are presented in Table 1. The precipitation, maximum temperatures and minimum temperature data at each node are obtained in ASCII format for duration 1901–2013. Developed R script for extracting the data using NetCDF package in R (Annexure 1). Study of annual variation of

Table 1 List of latitude, longitude and coordinates used in Satluj Basin

Node No.	Latitude	Coordinate	Longitude	Coordinate	Node	Coordinate	Longitude	Coordinate
1	30 N	240	76 E	512	2	240	77 E	514
3		240	78 E	516	4	240	79 E	518
5		240	80 E	520	6	240	81 E	522
7		240	82 E	524	8	240	83 E	526
9	30 30'N	241	76 E	512	10	241	77 E	514
11		241	78 E	516	12	241	79 E	518
13		241	80 E	520	14	241	81 E	522
15		241	82 E	524	16	241	83 E	526
17	31 N	242	76 E	512	18	242	77 E	514
19		242	78 E	516	20	242	79 E	518
21		242	80 E	520	22	242	81 E	522
23		242	82 E	524	24	242	83 E	526
25	31 30'N	243	76 E	512	26	243	77 E	514
27		243	78 E	516	28	243	79 E	518
29		243	80 E	520	30	243	81 E	522
31		243	82 E	524	32	243	83 E	526
33	32 N	244	76 E	512	34	244	77 E	514
35		244	78 E	516	36	244	79 E	518
37		244	80 E	520	38	244	81 E	522
39		244	82 E	524	40	244	83 E	526
41	32 30'N	245	76 E	512	42	245	77 E	514
43		245	78 E	516	44	245	79 E	518
45		245	80 E	520	46	245	81 E	522
47		245	82 E	524	48	245	83 E	526
49	33 N	246	76 E	512	50	246	77 E	514
51		246	78 E	516	52	246	79 E	518
53		246	80 E	520	54	246	81 E	522
55		246	82 E	524	56	246	83 E	526

precipitation in mm and variation of temperature of each month for duration 1901–2013 is done using MS Excel add-in program called XLSTAT [17, 18] developed by ADDINSOFT which is used to analyze trends. The statistical significance of the trend using Mann-Kendall test is evaluated at 5% significance level. Results obtained in trend analysis for maximum temperature indicated −ve value of Sen's slope at 26 nodes, +ve value of Sen's slope at 13 nodes and zero value of Sen's slope at 17 nodes, meaning thereby no significant change in temperature at these nodes. For annual minimum temperature, forty-two stations indicated increasing trend, nine stations predicted negative trend and five station exhibited no trend. Mann-Kendall test at 5% significance level for change in annual precipitation, Sen's slope reflected −ve value at 51 nodes indicating decrease in precipitation in this region.

Table 2 Trends at 56 nodes in Satluj Basin for duration (1901–2013)

Statistical parameters	Statistics	Maximum temperature trend	Minimum temperature trend	Precipitation trend
P value at 5% significance	Number of nodes	13	28	47
P value at 10% significance	Number of nodes	16	39	49
Sen's slope	+ve *slope*	13 nodes	42 nodes	5 nodes
	−ve *slope*	26 nodes	9 nodes	51 nodes
	No slope	17 nodes	5 nodes	

6 Observations

Statistical parameters P value and Sen's slope for precipitation and temperature of Satluj river basin for period 1901–2013 at 56 nodes are estimated and reflected in Table 2 using Mann-Kendall trend analysis. The P value is an indicator of the trend. The lower the P value, stronger is the trend. Sen [16] Sen's slope method estimates the rate of change in the variable. If trend of data is +ve, slope indicates rising trend and −ve slope indicates falling trend and no slope exhibit no change in hydrometeorological parameter.

6.1 Trends

6.1.1 Trends for Maximum Temperature

Mann-Kendall nonparametric test statistics is estimated at all 56 nodes. P value and Sen's slope for node 26 which is approximately 65 km toward East of Bhakra having latitude 31.5°N and longitude 77°E is represented in the fig below. To estimate value of Sen's slope, autocorrelation is done using the Hamed and Rao method. Sen's slope for maximum temperature is −0.006 a negative trend indicating decrease in maximum temperature at node 26 reflected through Fig. 1. Sen's slope estimated for all the 56 nodes and trends at various nodes are represented in Fig. 2. From Fig. 2, it is observed that Sen's slope is +ve at 13 nodes and −ve at 26 nodes and exhibits no slope at 17 nodes in river basin. P value at 95% confidence at 13 stations indicates significant trend. At 90% confidence, 16 stations in basin showed significant trend.

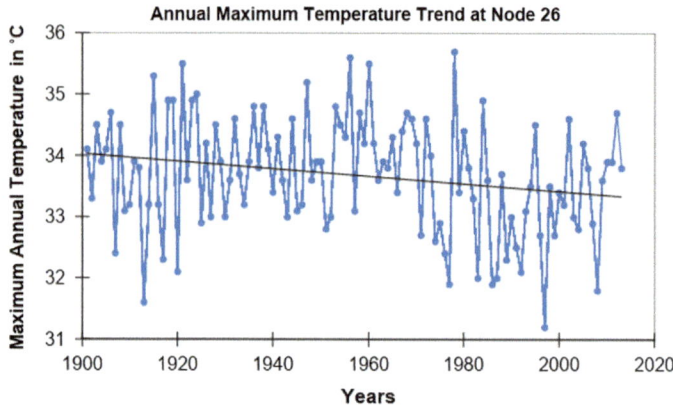

Fig. 1 Annual maximum temperature trend at node 26

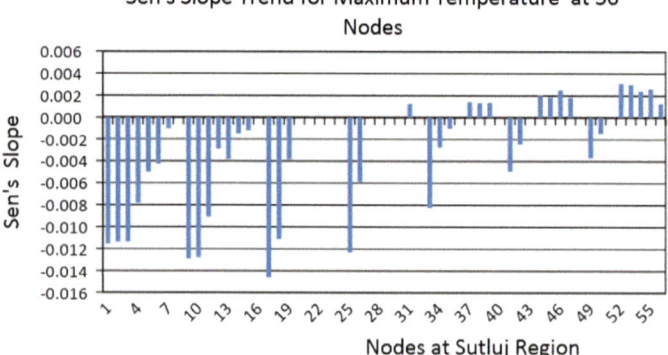

Fig. 2 Sen's slope for maximum temperature for 56 nodes

6.1.2 Trends for Minimum Annual Temperature

Minimum annual temperature trend at node 26 is represented in Fig. 3. P value at node 26 is 76.86% which indicates no trend is observed at this node. The Sen's slope at this node is zero and infers no change in minimum temperature for duration 1901–2013. Figure 4 represents value of Sen's slope at 56 nodes for minimum temperature which predicts increase in minimum temperature. Sen's slope is +ve at 42 nodes, −ve at 9 nodes and at 5 nodes exhibit no change . At 28 stations, P value is less than 5% hence predicts significant trend.

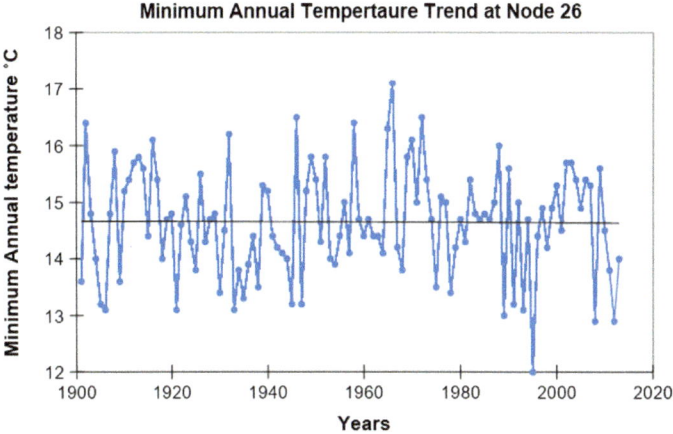

Fig. 3 Minimum annual temperature trend at node 26

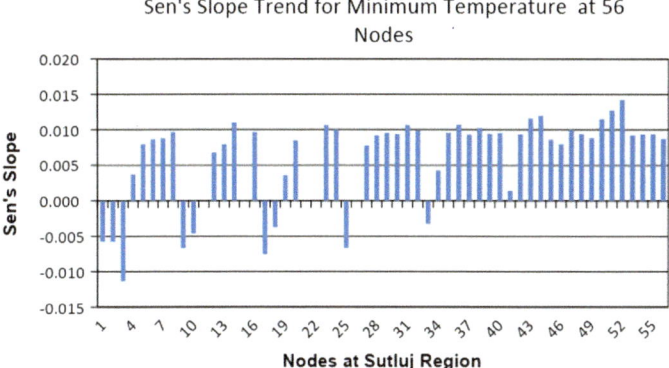

Fig. 4 Sen's slope for minimum temperature at 56 nodes

6.1.3 Precipitation Trend in the Region

P value at this node is less than 0.01% indicating significant trend at node 26. The autocorrelation is estimated using the Hamed and Rao method used to determine Sen's slope. The value of Sen's slope estimated for node 26 is −2.229; a strong negative trend indicating decrease in precipitation is reflected through Fig. 5. Sen's slope is estimated for precipitation at all 56 nodes and plotted in Fig. 6

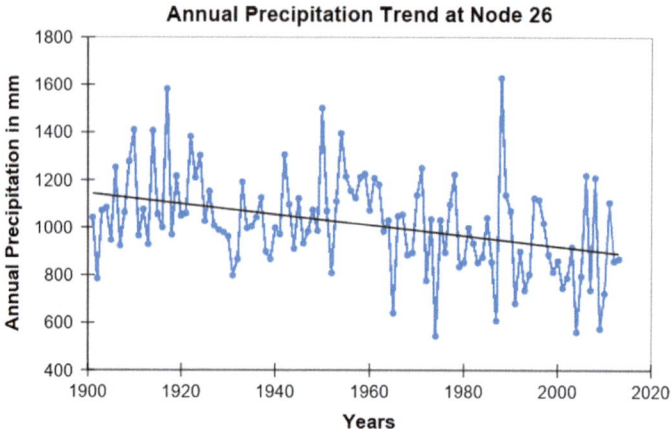

Fig. 5 Precipitation trend at node 26

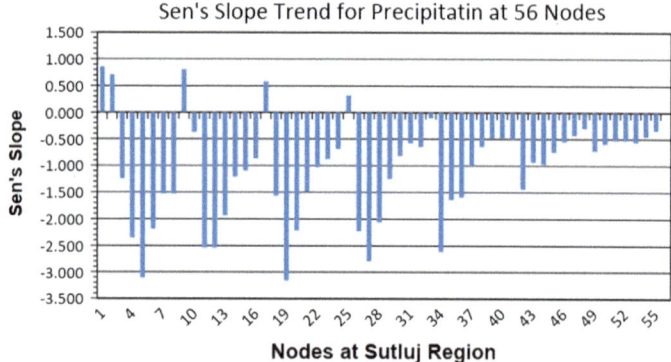

Fig. 6 Sen's slope for precipitation at 56 nodes

7 Discussion and Conclusion

Construction of smart cities is feasible only if the availability of water is predicted in advance. Availability of water can be predicted through analysis of hydrometeorological data considering climate changes. Determination of temperature and precipitation trend provided critical evidence regarding impacts of anthropogenic activities. The analysis of time series data trend reflects the changing climate at various nodes. The minimum temperature of this region reflects significant increase in trend. With the rise in minimum temperature, it is imperative that ice cap will retreat. IPCC special report (2018) indicates increase in global temperature is likely to reach 1.5° C between 2030 and 2052 compared to pre-industrial temperature due to anthropogenic activities responsible for generation of greenhouse gases. It is evident that with increase in minimum temperature and decrease in precipitation of the magnitude identified in this research will have implication on many aspects of water resource system. An immediate impact of increased temperature in this region is likely on the hydrological cycle. With increased warming, the vigor of the hydrological cycle is likely to intensify. The changes in hydrological cycle are accompanied by erratic and uncertain patterns of rainfall in the basin. As a consequence, the agricultural production could experience substantial variation. Changes in precipitation pattern would increase the magnitude as well as frequency of extreme events in the basin. The change in climate induced by enhanced warming in the region is likely to cause substantial stress on water resource systems due to urbanization and population growth. With changes in rainfall patterns, the flow regime is likely to be adversely impacted as well. For a large reservoir such as Bhakra, variations in flow patterns either in quantity or in timing can produce serious impacts on the livelihood for those engaged in the agricultural sector or allied activities of this region. The predicted increase in temperature, both in the short and the long term, would have serious implications for the operating strategies being currently followed by the reservoir managers at Bhakra. Expected warming in this region has created the need to consider the effect of climate change while formulating operating strategies for a large dam such as Bhakra that control the flow to several irrigation systems in Northern India and is a major source of water for power generation. Inferences drawn are based on observed trends for maximum temperature, minimum temperature and precipitation at 56 nodes selected in Satluj basin.

8 Appendix

```
#This code is to extract CRU data from netCDF File
cru.pre.19012013.nc <- open.ncdf("E:/cru.pre.dat.nc",
    write=FALSE, readunlim=TRUE, verbose=FALSE)
cru.pre.19012013.var <- get.var.ncdf(cru.pre.19012013.nc, "pre")
print.ncdf(cru.pre.19012013.nc)
dim(cru.pre.19012013.var)
targetpixel.ts <- matrix(data=NA, nrow=1356, ncol=1,
    dimnames=NULL)
# loop for timesteps
for(timestep in c(1:1356)) {
    cru.pre1 <- (cru.pre.19012013.var[509,203,timestep])
    targetpixel.ts[timestep,] <- c(cru.pre1)
}
write.table(targetpixel.ts, file = "E:/kb12.txt", append =
    FALSE, quote = TRUE, sep = "\t",
        col - "\n", na = "NA", dec - ".", row.names = FALSE,
        col.names = TRUE, qmethod = c("escape", "double"))
```

References

1. Hulme M (2016) 1.5 c and climate research after the Paris agreement. Nat Clim Change 6(3):222
2. Stocker T (2014) Climate change 2013: the physical science basis: working group I contribution to the fifth assessment report of the intergovernmental panel on climate change. Cambridge University Press, Cambridge
3. Yang F, Kumar A, Schlesinger ME, Wang W (2003) Intensity of hydrological cycles in warmer climates. J Clim 16(14):2419–2423. https://doi.org/10.1175/2779.1
4. Jiang T, Chen YD, Xu C, Chen X, Chen X (2007) Singh VP (2007) Comparison of hydrological impacts of climate change simulated by six hydrological models in the Dongjiang Basin. South China. J Hydrol 336(3–4):316–333. https://doi.org/10.1016/j.jhydrol.2007.01.010
5. Hamid A, Shariff M, Archer D (2014) Analysis of temperature trends in Sutluj River Basin, India. J Earth Sci Climatic Change 05(08). https://doi.org/10.4172/2157-7617.1000222, https://www.omicsonline.org/open-access/analysis-of-temperature-trends-in-sutluj-river-basin-india-2157-7617.1000222.php?aid=30960
6. Hamid AT, Sharif M, Narsimlu B (2017) Assessment of climate change impacts on streamflows in satluj river basin, India using swat model. Int J Hydrol Sci Technol 7(2):134–157
7. Radhakrishnan K, Sivaraman I, Jena SK, Sarkar S, Adhikari S (2017) A climate trend analysis of temperature and rainfall in India. Climate Change Environ Sustain 5(2):146–153
8. Kothawale D, Singh H (2017) Recent trends in tropospheric temperature over india during the period 1971–2015. Earth Space Sci 4(5):240–246
9. Sharif M (2015) Analysis of projected temperature changes over Saudi Arabia in the twenty-first century. Arab J Geosci 8(10):8795–8809
10. Rajbhandari R, Shrestha A, Kulkarni A, Patwardhan S, Bajracharya S (2014) Projected changes in climate over the Indus river basin using a high resolution regional climate model (PRECIS). Climate Dyn 19: https://doi.org/10.1007/s00382-014-2183-8
11. Gocic M, Trajkovic S (2013) Analysis of changes in meteorological variables using Mann-Kendall and Sen's slope estimator statistical tests in Serbia. Global Planet Change 100:172–182. 10.1016/J.GLOPLACHA.2012.10.014, https://www.sciencedirect.com/science/article/pii/S0921818112002032
12. Singh P, Jain SK (2002) Snow and glacier melt in the Satluj River at Bhakra Dam in the Western Himalayan region. Hydrol Sci J 47(1):93–106. https://doi.org/10.1080/02626660209492910
13. Helsel DR, Hirsch RM (1992) Statistical methods in water resources. https://pubs.usgs.gov/twri/twri4a3/html/toc.html
14. Maghsoodloo S (1975) Estimates of the quantiles of kendall's partial rank correlation coefficient. J Stat Comput Simul 4(2):155–164
15. Mann HB (1945) Nonparametric tests against trend. Econometrica J Econometric Soc, 245–259
16. Sen PK (1968) Estimates of the regression coefficient based on Kendall's Tau. J Am Stat Assoc 63(324):1379–1389
17. Karmeshu N (2012) Trend detection in annual temperature & precipitation using the Mann Kendall test–a case study to assess climate change on select states in the Northeastern United States
18. Tesemma ZK (2009) Long term hydrologic trends in the Nile basin. PhD thesis, Cornell University